Brodley

Contributions in Petroleum Geology & Engineering 4

Gas Production Engineering

Contributions in Petroleum Geology and Engineering

Series Editor: George V. Chilingar, University of Southern California

Volume 1: Geologic Analysis of Naturally Fractured Reservoirs
Volume 2: Applied Open-Hole Log Analysis
Volume 3: Underground Storage of Natural Gas
Volume 4: Gas Production Engineering
Volume 5: Properties of Oil and Natural Gases
Volume 6: Gas Reservoir Engineering
Volume 7: Hydrocarbon Phase Behavior

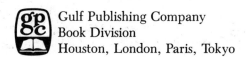

Gulf Publishing Company
Book Division
Houston, London, Paris, Tokyo

Contributions in Petroleum Geology & Engineering 4

Gas Production Engineering

Sanjay Kumar

To Professor Donald L. Katz, University of Michigan, for his innumerable contributions to natural gas engineering; and my family for their constant encouragement and support.

Contributions in Petroleum Geology and Engineering

Volume 4

Gas Production Engineering

Copyright © 1987 by Gulf Publishing Company, Houston, Texas. All rights reserved. Printed in the United States of America. This book, or parts thereof, may not be reproduced in any form without permission of the publisher.

Library of Congress Cataloging-in-Publication Data
Kumar, S. (Sanjay), 1960–
Gas production engineering.
(Contributions in petroleum geology & engineering; v. 4)
Includes index.
1. Gas drilling (Petroleum engineering) 2. Gas engineering. I. Title. II. Series.
TN880.2.K86 1987   622'.3385   87-7452
ISBN 0-87201-577-7

ISBN 0-87201-066-X (series)

Contents

Preface vii

1. **Natural Gas—Origin and Development** 1
   Introduction. What Is Natural Gas? Origin of Natural Gas. Other Sources of Gaseous Fuel. Natural Gas Production and Processing System. Questions and Problems. References.

Gas Properties

2. **Phase Behavior Fundamentals** 18
   Introduction. Qualitative Hydrocarbon Phase Behavior. Quantitative Phase Behavior and Vapor-Liquid Equilibrium. Applications. Prediction of the Phase Envelope. Questions and Problems. References.

3. **Properties of Natural Gases** 39
   Introduction. Equations of State. Critical Pressure and Temperature Determination. The Gas Compressibility Factor. Some Z-Factor Related Properties. Compressibility of Gases. Viscosity of Gases. Specific Heat for Hydrocarbon Gases. Questions and Problems. References.

Gas Processing

4. **Gas and Liquid Separation** 89
   Introduction. Separation Equipment. Types of Separators. Separation Principles. Factors Affecting Separation. Separator Design. Stage Separation. Low-Temperature Separation. Gas Cleaning. Flash Calculations. Questions and Problems. References.

5. **Gas-Water Systems and Dehydration Processing** 169
   Introduction. Water Content of Natural Gases. Gas Hydrates. Hydrate Inhibition by Additive Injection. Absorption Dehydration. Adsorption Dehydration. Dehydration by Expansion Refrigeration. Questions and Problems. References.

6. Desulfurization Processes 255
   Introduction. Removal Processes. Solid Bed Sweetening Processes. Physical Absorption Processes. Chemical Absorption—The Alkanol-Amine Processes. Chemical Absorption—The Carbonate Processes. Questions and Problems. References.

Gas Production and Flow

7. Steady-State Flow of Gas Through Pipes 275
   Introduction. Gas Flow Fundamentals. Vertical and Inclined Single-Phase Flow of Gas. Gas Flow Over Hilly Terrain. Gas Flow Through Restrictions. Temperature Profile in Flowing Gas Systems. Questions and Problems. References.

8. Multiphase Gas-Liquid Flow 365
   Introduction. Approximate Method for Two-Phase Systems. Multiphase Flow. Liquid Loading in Gas Wells. Questions and Problems. References.

9. Gas Compression 394
   Introduction. Types of Compressors. Compressor Selection. Compression Processes. Compressor Design Fundamentals. Designing Reciprocating Compressors. Designing Centrifugal Compressors. Designing Rotary Compressors. Questions and Problems. References.

10. Gas Flow Measurement 451
    Introduction. Measurement Fundamentals. Methods of Measurement. Orifice Meters. Other Types of Measurement. Questions and Problems. References.

11. Gas Gathering and Transport 529
    Introduction. Gathering Systems. Steady-State Flow in Simple Pipeline Systems. Steady-State Flow in Pipeline Networks. Unsteady-State Flow in Pipelines. Some Approximate Solutions for Transient Flow. Pipeline Economics. Questions and Problems. References.

Appendix A. General Data and Unit Conversion Factors 585

Appendix B. Computer Programs (FORTRAN Subroutines) 623

Index 651

Preface

With this book I have sought to provide a rigorous and comprehensive text that reflects a broad spectrum of natural gas engineering experience. My motivation came primarily from some gas engineering courses I taught in the United States and overseas that made me realize the limitations of the existing material and the frustration of having to use several different publications, some of which would contradict others. Notwithstanding the enormity of the task, I decided to write a book to eliminate this confusion and provide a wider and more detailed coverage of this topic. Much of the material presented here was drawn from these lectures.

This book is designed to be a moderately advanced textbook for students and a handy reference for practicing engineers. It was written with the assumption that some readers with little or no background in gas engineering will use it as their first book. To keep things interesting for the expert, I have sought to include the most current developments reported in the latest published works to the extent possible without dragging this book into an interminably long series of books on these topics. Pertinent references are included at the end of each chapter for those wanting more details. Almost every chapter includes worked examples to enhance understanding and some practice questions and problems that will aid in testing this understanding. (Answers to the questions and problems are available in a separate instructor's guide.)

Chapter 1, an introduction to the scope of natural gas engineering, can be skipped by most seasoned readers. Chapters 2 and 3 describe the properties of natural gas that are crucial in designing and operating natural gas systems. Chapters 4, 5, and 6 deal with gas processing—Chapter 4 discusses the separation of gas from oil and water from a typical reservoir; Chapter 5 provides insight into gas-water systems and dehydration processing techniques for natural gas; and Chapter 6 examines some of the important and widely used desulfurization processes for natural gas.

Chapters 7–11 present important topics in gas production and flow— Chapter 7 describes relationships for the steady state flow of gas through horizontal, inclined, and vertical pipes; Chapter 8 outlines the methods of handling multiphase flow encountered in gas production; Chapter 9 provides information on gas compression; Chapter 10 discusses gas flow measurement; and finally, Chapter 11 describes gas gathering and transmission systems, and deals with the design and modeling of the complete production and transport system. Appendix A will be helpful in handling the several frequently confusing systems of units that, unfortunately, are our legacy in engineering. Appendix B presents some general FORTRAN subroutines that will help readers write computers programs to implement many of the techniques presented in this book.

It has been my good fortune to have had close associations with many people who aided in the development of this work. I am thankful to Drs. Kern H. Guppy and George V. Chilingar for their enthusiastic interest, comments, and suggestions for this book. The figures and drawings provided by several companies and institutions are gratefully acknowledged. Many sections of the book could only be included because of the interest and curiosity of my students. They never failed to ask pertinent questions and draw the discussion to related relevant issues. Finally, the moral support from my family has been of great value in the moments of frustration during this work. I thank my father, Dr. Kundan L. Goyal, for his many discussions and suggestions for this book and his insistence that I keep at it.

I submit this work to the engineering community with the hope that it shall inspire many minds, both the young and the experienced, to greater understanding and creativity in this endless search for knowledge.

Sanjay Kumar, Ph.D.
Los Angeles, California

1
Natural Gas—Origin and Development

Introduction

The earliest records of natural gas go back to A.D. 221-263 when it was first used as a fuel in China during the Shu Han dynasty. The gas, obtained from shallow wells, was distributed through an interesting piping system: hollow bamboos. Later evidence is not found until the early 17th century, when natural gas was used on a small scale for heating and lighting in northern Italy. In the United States, natural gas was first discovered in Fredonia, New York, in 1821. It is this latter discovery that has led, for the most part, to the developments we see today. In the years following this discovery, natural gas was used merely as a fuel locally: it was difficult to store and transport and had little or no commercial value. Even in the 1920s and '30s, natural gas was only produced as an unwanted byproduct of crude oil production. Only a small amount of gas was pipelined to industrial areas for commercial use, most of it being vented to the air or flared. From these humble beginnings, the natural gas industry has burgeoned, especially in the years following World War II. Natural gas now accounts for almost one-fifth of the world's primary energy consumption, surpassing coal and second only to oil (see Tables 1-1 and 1-2).

Several factors are responsible for these developments: New processes permit the manufacture of myriad petrochemicals and fertilizers from natural gas; gas is a clean, efficient, easily combustible, low-sulfur fuel and has consequently replaced coal as a domestic, industrial, and power generation fuel in many parts of the world; difficulties in storing and transporting gas have been overcome by the development of long-distance, large-diameter steel pipelines and powerful compression equipment; liquefied natural gas (LNG), produced by liquefying natural gas by a refrigeration cycle to less

(text continued on page 4)

Table 1-1
World Energy Consumption and Fuel Shares—History, 1960, 1973 and 1979*
(Quadrillion Btu)

Region or Country	1960 Total Energy Consumed	1960 Fuel Shares (Percent) Coal	Oil	Gas	Other	1973 Total Energy Consumed	1973 Fuel Shares (Percent) Coal	Oil	Gas	Other	1979 Total Energy Consumed	1979 Fuel Shares (Percent) Coal	Oil	Gas	Other
United States**	44.2	23	45	28	4	75.1	18	47	30	5	79.3	19	48	26	7
Canada	3.9	14	49	10	27	8.1	8	46	20	26	10.0	8	41	21	30
Japan	3.9	48	36	1	15	15.3	14	79	2	5	16.1	11	74	6	9
Western Europe†	26.9	55	35	2	8	54.0	19	63	10	8	57.1	19	56	14	11
Finland/Norway/Sweden	1.8	13	49	0	38	3.9	5	58	0	37	4.9	4	49	2	45
United Kingdom/Ireland	8.0	70	29	0	1	10.4	35	51	11	3	10.3	33	44	18	5
Benelux/Denmark††	2.7	53	47	0	0	7.3	11	67	22	0	6.8	12	57	28	3
West Germany	6.2	73	25	0	2	11.5	30	58	10	2	12.4	25	53	17	5
France	3.7	51	35	3	11	8.2	15	70	7	8	8.8	16	61	10	13
Australia/Switzerland	0.9	24	34	6	36	2.0	8	58	7	27	1.7	6	41	12	41
Spain/Portugal	1.0	41	41	0	18	3.0	14	70	2	14	3.8	16	68	3	13
Italy	2.2	14	54	11	21	6.2	5	78	10	7	6.5	6	69	17	8
Greece/Turkey	0.4	41	55	0	4	1.4	21	76	0	3	2.0	35	60	0	5
Australia/New Zealand	1.4	54	43	0	3	2.9	35	50	6	9	3.8	39	42	11	8
Total OECD†	80.4	35	42	16	7	155.4	18	56	19	7	166.2	18	52	20	10
Total Non-OECD†	10.5	31	57	6	6	25.0	21	60	11	8	35.7	19	62	9	10
OPEC	1.7	3	75	21	1	5.0	1	66	30	3	7.0	1	74	22	3
Other	8.8	37	53	4	6	20.0	25	59	7	9	28.7	23	59	6	12
Total Free World†	90.9	35	43	15	7	180.3	18	56	18	8	201.9	18	54	18	10

* From D.O.E.'s 1981 Annual Report to Congress, Volume 3.
** Includes Puerto Rico, Virgin Islands, and purchases for the Strategic Petroleum Reserve.
† Numbers may not add to totals due to rounding.
†† Benelux countries are Belgium, the Netherlands, and Luxembourg.
Reprinted with permission from *API*, 1984.

Natural Gas—Origin and Development 3

Table 1-2
World Energy Consumption and Fuel Shares: Base Scenario Midprice Projections* 1985, 1990, and 1995*
(Quadrillion Btu)

Region or Country	1985 Total Energy Consumed	1985 Fuel Shares (Percent) Coal	Oil	Gas	Other	1990 Total Energy Consumed	1990 Fuel Shares (Percent) Coal	Oil	Gas	Other	1995 Total Energy Consumed	1995 Fuel Shares (Percent) Coal	Oil	Gas	Other
United States**	82.3	23	42	24	11	86.5	27	38	22	13	93.9	31	35	21	13
Canada	10.1	8	36	20	36	11.2	5	32	22	41	12.4	3	29	23	45
Japan	19.4	18	55	15	12	22.6	22	50	15	13	26.0	24	46	14	16
Western Europe†	54.6	22	46	15	17	58.8	23	42	15	20	63.3	24	39	15	22
Finland/Norway/Sweden	4.8	4	44	2	50	5.2	4	40	2	54	5.7	5	37	2	56
United Kingdom/Ireland	9.1	36	38	18	8	9.5	37	37	18	8	9.8	39	37	16	8
Benelux/Denmark††	6.5	20	46	29	5	7.0	28	43	23	6	7.6	32	41	22	5
West Germany	11.7	32	41	19	8	12.6	34	37	17	12	13.4	36	32	19	13
France	8.7	13	52	9	26	9.5	11	48	10	31	10.4	9	44	13	34
Australia/Switzerland	2.2	9	41	14	36	2.3	9	39	17	35	2.4	8	37	17	38
Spain/Portugal	3.5	14	54	3	29	3.8	10	53	5	32	4.1	10	49	5	36
Italy	6.4	20	50	20	10	7.0	23	44	21	12	7.8	24	41	22	13
Greece/Turkey	1.8	28	55	0	17	1.9	21	53	0	26	2.1	19	52	0	29
Australia/New Zealand	3.9	33	44	15	8	4.3	33	39	19	9	4.6	32	37	22	9
Total OECD†	170.3	22	45	19	14	183.3	24	40	19	17	200.2	27	37	18	18
Total Non-OECD†	45.6	20	59	11	10	55.0	19	57	11	13	66.0	19	58	11	12
OPEC	10.3	0	72	27	1	14.0	0	74	25	1	19.0	0	76	23	1
Other	35.3	25	55	6	14	41.0	25	52	6	17	47.0	26	51	7	16
Total Free World†	215.9	21	47	18	14	238.3	23	44	17	16	266.2	25	42	16	17

* From D.O.E.'s 1981 Annual Report to Congress, Volume 3.
Projection ranges are based on assumed price paths for imported oil stated in 1979 constant dollars. The low price scenario assumes a delivered world oil price of $32 per barrel, the mid-range (used above) $41 per barrel, and the high range $49 per barrel.
** Includes Puerto Rico, Virgin Islands, and purchases for the Strategic Petroleum Reserve.
† Numbers may not add to totals due to rounding.
†† Benelux countries are Belgium, the Netherlands, and Luxembourg.
Reprinted with permission from *API*, 1984.

(text continued from page 1)

than 1/600 of its original volume, can now be transported across the oceans by insulated tankers.

Gas is now a highly desirable hydrocarbon resource. It is no longer cheap; the price has gone up from a mere $0.07/Mscf (1,000 standard cubic feet) in 1950 to $2.79/Mscf currently (1987). The price of gas peaked in 1984, when gas was being sold at $4.80/Mscf, and as much as $9.00/Mscf in some areas. Production costs for natural gas have also been rising over the years. Average costs in the USA have gone up from $0.86/Mscf in 1976 to about $3.90/Mscf in 1982. But natural gas is still competitive with other fuels, as shown in Table 1-3.

Table 1-3
Cost Comparison of Various Commercial Fuels

Fuel	Price	Price, $/MM Btu
Gasoline	$1.20/gallon	9.60
Heating oil	$1.10/gallon	7.95
Low-sulfur fuel oil	$30.0/bbl	4.80
Electricity	7.5 cents/kWh	22.00
Coal	$35.0/ton	1.60
Natural gas	57.9 cents/Th*	5.79

*Th indicates the unit "Therm" used often for gas. 1 Th = 10^5 Btu.

Table 1-4 shows the production for the twenty leading natural gas producers in the world. The foremost gas producers today are the USA and USSR, with a gas production of 52 Bscf/d (billion standard cubic feet per day) and 45 Bscf/d, respectively.

What Is Natural Gas?

Natural gas is simply a naturally occurring mixture of combustible hydrocarbon (HC) gases and impurities. Typical natural gas components are shown in Table 1-5. The non-hydrocarbon components of natural gas contain two types of materials: *diluents*, such as N_2, CO_2, and water vapor, and *contaminants*, such as H_2S and other sulfur compounds. Diluents are noncombustible gases that reduce the heating value of the gas. They are not very harmful, and may actually be used sometimes as "fillers" to reduce the heat content of the supply gas. The disadvantages include greater horsepower and pipelining requirements for the same energy content of the gas, greater internal corrosion, and freezing. Contaminants are very detrimental to production and transport equipment (some are hazardous pollutants),

Natural Gas—Origin and Development 5

**Table 1-4
World Natural Gas Production—Twenty Leading Nations***
(Billion cubic feet)

Nation	1978**	Nation	1979**	Nation	1980**
1. United States	19,974.0	1. United States	20,471.3	1. United States	20,378.8
2. U.S.S.R.	13,131.6	2. U.S.S.R.	14,367.1	2. U.S.S.R.	15,355.5
3. Canada	3,133.1	3. Canada	3,646.5	3. People's Republic of China	3,469.0
4. People's Republic of China	2,352.2	4. People's Republic of China	2,832.7	4. Netherlands	2,799.6
5. Iran	1,746.9	5. Netherlands	2,717.8	5. Canada	2,668.3
6. Netherlands	1,624.5	6. United Kingdom	1,965.7	6. United Kingdom	1,500.0
7. United Kingdom	1,262.4	7. Romania	1,210.5	7. Mexico	1,190.5
8. Romania	1,211.6	8. Mexico	974.9	8. Romania	1,176.4
9. Mexico	887.5	9. Iran	913.3	9. Indonesia	1,028.4
10. West Germany	643.2	10. West Germany	826.3	10. West Germany	738.6
11. Nigeria	586.6	11. Indonesia	816.3	11. Norway	705.2
12. Indonesia	565.8	12. Italy	533.6	12. Pakistan	600.0
13. Italy	540.8	13. Norway	414.9	13. Italy	524.6
14. Venezuela	401.4	14. Venezuela	406.0	14. Venezuela	518.0
15. Libya	364.0	15. Nigeria	365.6	15. Algeria	517.0
16. Norway	351.8	16. Chile	306.4	16. Saudi Arabia	310.2
17. Algeria	328.7	17. Brunei-Malaysia	302.7	17. Argentina	296.9
18. Brunei-Malaysia	325.7	18. Algeria	301.5	18. Iran	292.0
19. Chile	320.7	19. Australia	254.1	19. Kuwait	291.3
20. Argentina	254.3	20. Argentina	241.7	20. Brunei	282.0
Total Free World	35,272.8	Total Free World	38,075.1	Total Free World	37,677.7
Total World	53,911.1	Total World	57,666.5	Total World	58,747.7

Reprinted with permission from *API*, 1984.

Table 1-4 continued

Nation	1981**	Nation	1982**	Nation	1983**
1. United States	20,177.7	1. United States	18,519.7	1. U.S.S.R.	18,903.0
2. U.S.S.R.	16,390.0	2. U.S.S.R.	17,685.3	2. United States	16,599.0
3. Netherlands	3,054.0	3. Canada	2,546.1	3. Netherlands	2,678.9
4. Canada	2,623.0	4. Netherlands	2,521.7	4. Canada	2,414.1
5. Mexico	1,486.0	5. Mexico	1,549.7	5. Mexico	1,479.5
6. Romania	1,440.0	6. United Kingdom	1,276.1	6. United Kingdom	1,395.3
7. United Kingdom	1,427.0	7. Romania	1,150.0	7. Romania	1,200.0
8. Algeria	1,149.0	8. Norway	861.9	8. Norway	861.4
9. Indonesia	1,075.0	9. Algeria	828.5	9. Algeria	707.0
10. Norway	923.0	10. Saudi Arabia	591.8	10. West Germany	605.2
11. Libya	674.0	11. West Germany	585.3	11. Venezuela	559.2
12. West Germany	636.0	12. Indonesia	568.9	12. Argentina	549.6
13. Venezuela	602.0	13. Venezuela	527.2	13. Iran	500.2
14. Pakistan	600.0	14. Italy	503.6	14. Indonesia	481.8
15. Italy	500.0	15. Australia	415.8	15. Italy	450.1
16. People's Republic of China	459.0	16. Argentina	386.2	16. Australia	423.5
17. Saudi Arabia	435.0	17. Iran	381.5	17. Saudi Arabia	381.5
18. Australia	377.0	18. China	363.7	18. China	371.0
19. Kuwait	340.0	19. France	332.6	19. Pakistan	341.4
20. Argentina	330.0	20. Pakistan	315.3	20. Brunei	316.3
Total Free World	39,187.7	Total Free World	35,424.6	Total Free World	33,701.5
Total World	58,397.7	Total World	55,682.6	Total World	55,084.5

* U.S. Energy Information Administration, United States only; other nations, *Oil and Gas Journal.*

** Comprises all gas collected and used as a fuel or as a chemical industry raw material, including gas used in oil and/or gas fields as a fuel by producers, even though it is not actually sold. United States is reported marketed production; all others may include some gross production.

Table 1-5
Typical Constituents of Natural Gas (Modified after McCain, 1974)

Category	Component	Amount, %
Paraffinic HC's	Methane (CH_4)	70 - 98%
	Ethane (C_2H_6)	1 - 10%
	Propane (C_3H_8)	trace - 5%
	Butane (C_4H_{10})	trace - 2%
	Pentane (C_5H_{12})	trace - 1%
	Hexane (C_6H_{14})	trace - 0.5%
	Heptane & higher (C_7+)	none - trace
Cyclic HC's	Cyclopropane (C_3H_6)	traces
	Cyclohexane (C_6H_{12})	traces
Aromatic HC's	Benzene (C_6H_6), others	traces
Non-hydrocarbon	Nitrogen (N_2)	trace - 15%
	Carbon dioxide (CO_2)	trace - 1%
	Hydrogen sulfide (H_2S)	trace occasionally
	Helium (He)	trace - 5%
	Other sulfur and nitrogen compounds	trace occasionally
	Water (H_2O)	trace - 5%

and the primary reason for gas conditioning and processing is to remove them as soon as possible from the gas stream. Hundreds of processes and processing plants have been developed to deal with this problem. Some of the major contaminants in natural gas are (Curry, 1981):

1. Acid gases, chiefly H_2S, and to some extent, CO_2.
2. Water vapor in excess of about 5-7 lbm/MMscf.
3. All entrained free water, or water in condensed form.
4. Any liquids in the gas, such as well inhibitors, lube oil, scrubber oil, methanol, and heavier-end hydrocarbons.
5. All solid matter, sometimes called "pipeline trash," that may be present. This includes silica (sand), pipe scale, and dirt.

Like all gases, natural gas is a homogeneous fluid of low density and viscosity. It is odorless; odor-generating additives are added to it during processing to enable detection of gas leaks. Natural gas is one of the more stable flammable gases (Curry, 1981). It is flammable within the limits of a 5-15% mixture with air, and its ignition temperature ranges from 1,100 to 1,300°F (compare this with H_2S, which is flammable within 4-46% in air at a much lower ignition temperature). Typically, natural gas has an energy content of 1,000 Btu/scf, which is an important parameter because gas these days is very often priced in terms of its energy content, rather than mass or volume.

Origin of Natural Gas

Many theories have been proposed for the origin of petroleum fluids. None is perfect, and it is almost impossible to fully explain the origins of any given reservoir. The two most widely accepted theories are *inorganic* and *organic*. According to the inorganic theory, hydrogen and carbon reacted together under the immense pressure and temperature far below the earth's surface and formed oil and gas. These hydrocarbons then migrated through porous rocks to collect in various subsurface traps. The more widely accepted organic theory states that the hydrocarbons were generated from organic matter (land and sea plants and animals) under the influence of pressure and temperature over geologic time. Layers of mud and silt settled on these dead organisms by various processes—rivers emptying huge amounts of mud and silt into the oceans, wind-blown dust and sediment, etc. These layers accumulated, continually compacted by the weight of succeeding layers to form sedimentary rock. Hydrocarbon deposits are entrapped in these sedimentary rocks (sandstones, shale, limestones), and may migrate from their place of formation (source-rock) through suitable porous and permeable strata to suitable geologic traps where they accumulate.

The type of organic matter and the temperature have an important bearing on the formation of oil or gas. Some scientists believe that terrestrial (land) plants and animals predominantly produced natural gas and some waxy crudes, whereas the aquatic (sea) organisms produced normal crude oil. Because rivers play a major role in transporting terrestrial matter to the sea, river deltas are favorable places for gas to exist. Typically, the deepest sediments, deposited in the continental rift and rich in terrestrial organic matter, are overlain by marine sediments rich in aquatic matter. Thus, a vertical sequence has been envisaged, with the gas-generating material at the bottom and the oil-generating material at the top.

It is believed that hydrocarbons generally move upward from their place of formation to their accumulation sites, displacing the sea water that originally filled the pore spaces of the sedimentary rock. This upward movement is inhibited when the oil and gas reach an impervious rock that traps or seals the reservoir.

There are many types, shapes, and sizes of geologic structures that form the reservoirs for the accumulation of oil and gas, such as anticlines and domes, fault traps, unconformities, dome and plug traps, lense traps, and combination traps. Some of these are shown in Figures 1-1 through 1-4. It has been found that methane can remain stable at depths of 40,000 ft and beyond (Barker and Kemp, 1980). The amount of methane surviving is a function of the reservoir lithology—cooler, clean sandstones are more favorable than carbonates. Thus, it can be safely assumed that considerable re-

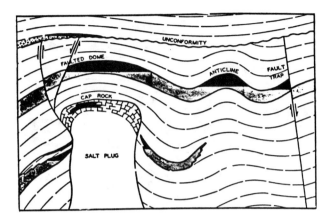

Figure 1-1. Common types of geologic features and structural traps. (From Wheeler and Whited, 1985.)

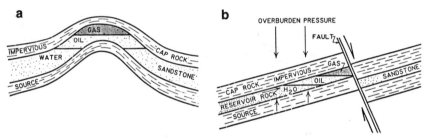

Figure 1-2. Structural traps: (a) anticline, formed by upfolding; (b) normal fault. (From McElroy, 1987.)

serves of natural gas may be found at depths of 15,000–30,000 ft that have currently not been explored.

Occurrence of Natural Gas in Conventional Reservoirs

Gas is found in sedimentary subsurface strata composed of sandstone, limestone, or dolomite. An oil reservoir always has some amount of natural gas associated with it (either free gas, or gas in solution in the oil), and some reservoirs may be completely gas reservoirs. Each well in the reservoir may produce gas with a different composition, and the composition of the gas-stream from each individual well may change as the reservoir is depleted. Thus, production equipment may need to be changed from time to time to compensate for the altered composition of the gas.

10   Gas Production Engineering

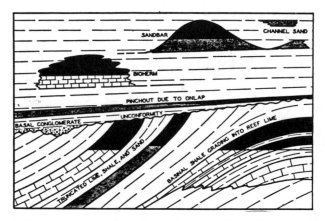

Figure 1-3. Common types of stratigraphic traps. (From Brown and Miller, 1985.)

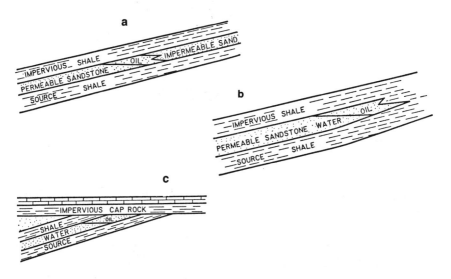

Figure 1-4. Stratigraphic traps in various geologic features: (a) "permeability pinchout"; (b) sand pinchout; (c) angular nonconformity. (From McElroy, 1987).

In addition to its composition and Btu content, natural gas is frequently characterized in terms of its nature of occurrence underground, as follows:

1. *Non-associated.* Found in reservoirs with no or minimal amounts of crude oil, non-associated gas is typically richer in CH_4, poorer in heavier components.

2. *Dissolved or associated*. Gas in solution with crude oil is termed dissolved gas, whereas the gas found in contact with the crude oil as gas cap gas is termed associated gas. Typically, associated gas is poorer in CH_4, but richer in heavier components.
3. *Gas condensates*. Gas condensates have high amounts of hydrocarbon liquids and may occur as gas in the reservoir.

The most desirable gas is the non-associated type, because it can be produced at high pressure. Associated or dissolved gas is separated from the crude oil at lower separator pressures and, therefore, entails more compression expenses. Such gas is often flared or vented. Gas condensates represent a greater amount of gas associated with the liquid than the associated or dissolved gas types, but require similar separation facilities and compression costs.

The naturally-occurring resources of gas are often classified into two categories:

1. *Proved reserves*: This figure indicates the quantities found by the drill. The estimates are proved and frequently updated by reservoir characteristics (production data, pressure transient analysis, reservoir modeling, other data) and, therefore, can be determined quite accurately (see Table 1-6).
2. *Potential reserves*: These are the additional resources of gas believed to exist in the earth, as inferred from the prevailing geologic evidence, but not actually found by the drill yet. This figure, therefore, cannot be known very precisely. It is, at best, a "guesstimate" that may vary widely from one investigator to another.

For the estimation of mineral reserves and resources, McKelvey (1972) provides an excellent guide, shown in Figure 1-5. In this figure, the *probable*

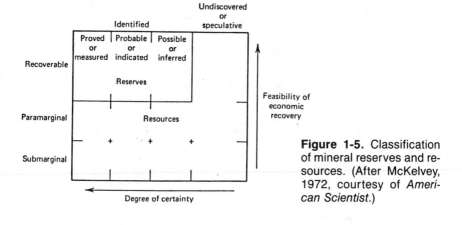

Figure 1-5. Classification of mineral reserves and resources. (After McKelvey, 1972, courtesy of *American Scientist*.)

Table 1-6
Estimated Proved World Reserves of Natural Gas*
(Billions of cubic feet)

AREA AND COUNTRY	1981*	1982*	1983*	1984*	1985
North America	375,846	376,595	378,800	375,953	372,536
Canada	90,988	91,966	92,731	101,763	99,500
Mexico	75,352	75,352	76,998	76,702	76,536
United States	209,434	209,254	209,046	197,463	196,475
Others	72	23	25	25	25
South America	107,767	109,459	109,958	112,179	113,480
Argentina	23,875	24,412	23,965	23,600	22,000
Bolivia	5,100	5,700	4,428	4,398	4,554
Brazil	2,118	2,542	2,880	2,962	3,284
Chile	4,782	4,777	5,200	5,181	5,000
Colombia	4,585	4,410	3,889	3,020	3,879
Ecuador	2,100	2,087	3,005	2,926	3,060
Peru	1,202	951	921	992	828
Trinidad	17,400	10,600	10,500	10,300	9,875
Venezuela	46,605	53,980	55,170	58,800	61,000
Others	0	0	0	0	0
Western Europe	147,711	147,103	194,742	211,474	206,099
Austria	326	385	413	381	357
Denmark	2,825	2,932	3,508	3,712	3,300
France	1,956	1,779	1,550	1,455	1,271
Germany, West	6,272	6,213	6,876	6,427	6,533
Greece	28	25	24	22	18
Italy	6,422	7,060	7,063	6,574	7,415
Netherlands	54,738	53,502	59,998	67,097	65,507
Norway	49,759	50,832	88,108	98,200	97,360
Spain	635	706	706	638	515
United Kingdom	23,400	22,354	25,143	25,603	22,883
Others	1,350	1,315	1,353	1,365	940
Eastern Europe	1,403,985	1,525,489	1,309,353	1,347,821	1,416,868
Albania	2,245	2,140	132	283	280
Bulgaria	171	167	29	177	179
Czechoslovakia	402	380	370	353	341
Germany, East	5,139	4,765	350	4,061	3,891
Hungary	3,288	3,090	2,826	4,238	3,995
Poland	4,030	3,830	1,515	3,885	3,866
Romania	9,247	13,797	1,236	7,416	6,374
USSR	1,377,000	1,495,000	1,300,000	1,324,301	1,394,872
Yugoslavia	2,463	2,320	2,895	3,108	3.072
Africa	220,560	230,995	212,944	204,891	207,441
Algeria	123,550	130,000	110,000	107,000	106,000
Angola	1,561	2,078	3,700	1,870	1,940
Cameroon	3,500	3,500	3,800	4,175	4,175
Congo	2,400	2,470	2,438	2,648	2,440
Egypt	8,820	9,740	10,305	7,060	8,300
Gabon	530	530	579	600	600
Libya	20,800	21,500	22,000	22,000	25,700
Morocco	485	508	623	620	645
Nigeria	48,346	48,911	47,568	47,025	46,400
Tunisia	6,398	6,385	6,550	3,953	3,000
Zaire	60	63	72	35	40
Others	4,110	5,310	5,309	7,905	8,201
Middle East	733,957	779,886	837,996	852,301	858,581
Abu Dhabi	85,000	84,800	85,000	84,600	84,000
Bahrain	7,956	7,825	7,400	7,206	8,000
Dubai	1,858	5,000	5,000	4,900	4,779
Iran	370,000	369,489	370,000	370,000	369,500
Iraq	25,250	25,100	25,000	24,800	24,600
Kuwait	33,000	32,900	32,745	39,800	39,600
Divided Neutral Zone	18,700	18,700	18,700	18,700	18,650
Oman	5,443	5,874	6,000	8,000	8,195
Qatar	60,000	101,500	156,844	156,000	156,000
Saudi Arabia	113,393	114,000	119,000	122,700	130,189
Sharjah	10,000	10,000	10,000	4,200	4,377
Syria	3,147	3,636	1,245	1,015	1,101
Turkey	130	1,012	1,012	10,380	9,590
Far East	166,463	162,949	171,828	192,185	207,032
Brunei	5,560	5,245	5,455	7,330	7,050
Burma	3,280	5,100	11,400	11,380	11,350
China	25,300	25,000	15,000	15,000	14,800
India	14,510	14,826	16,235	16,870	16.873
Indonesia	30,500	30,500	42,000	60,000	65,000
Japan	641	680	680	883	988
Malaysia	48,398	50,069	50,000	49,000	53,000
Pakistan	16,200	16,164	15,800	15,384	21,000
Philippines	39	38	38	34	24
Taiwan	835	785	733	820	777
Thailand	9,200	3,542	3,487	3,484	3,720
Others	12,000	11,000	11,000	12,000	12,450
Australia/Pacific	23,338	22,409	22,593	25,466	27,701
Australia	17,768	16,942	17,238	18,225	18,646
New Zealand	5,570	5,467	5,355	5,241	5,700
World Total	3,179,627	3,354,885	3,238,214	3,322,270	3,409,738

†Excludes natural gas liquids *Revised

reserves indicate the amount of gas expected to be found in close proximity to, or associated with, known producing fields in known areas with similar geological conditions. *Possible reserves* are similar to probable reserves, but in more distant regions where the uncertainty is greater. *Undiscovered, speculative,* or *potential reserves* are those that are expected to be found in areas that have not been, or only partially, explored and tested.

According to Grow (1980), the 1978 estimated US reserves (conventional reservoirs/conventional methods) are approximately 1,019 Tscf (trillion standard cubic feet). Total world reserves are 7,500 Tscf undiscovered, 2,200 Tscf proved. Current world production is 70 Tscf/year. Key areas where substantial potential exists for increasing production in the next decade are the OPEC group countries and the USSR.

Other Sources of Gaseous Fuel

Besides conventional sandstone and limestone reservoirs, sources of natural gas include: tight sands, tight shales, geopressured aquifers, and coal.

Tight Sands

Large amounts of gas are locked within very tight formations, with porosities in the range of 5-15%, irreducible water saturation of 50-70%, and ultralow permeabilities in the range of 0.001 to 1 md (millidarcy). Many geologic formations, especially in the Rocky Mountain area of the USA, contain such gas, which is not possible to produce using conventional fracturing and completion methods.

There is a need to develop special artificial fracturing techniques to produce gas economically from tight sands. Three major techniques have been proposed: nuclear explosives, chemical explosives, and massive hydraulic fracturing (MHF). Nuclear blasts are dangerous and probably not feasible at this time. Chemical explosives are also quite dangerous, and effective only in areas where natural fractures exist (which is not the case for most tight sands). MHF has been quite successful, and is presently the subject of much research. An MHF treatment typically requires the injection of a fracturing fluid (water) at high pressures for many hours to induce a fracture, followed by a fluid containing the propping agents (glass beads or sand). When pumping stops, the fluids flow back into the wellbore, leaving the proppant behind to hold the fracture open.

Tight Shales

Shales are generally rich in organic matter, finely laminated, with a permeability of the order of 1 md. Shales, predominantly composed of quartz,

14 *Gas Production Engineering*

with some kaolinite, pyrite, feldspar, and other minerals, constitute about 5% of all sedimentary rocks in the USA. The Devonian shales in eastern Kentucky and western West Virginia are well known gas producers. In these shales, production is controlled by natural fractures. The production profile exhibits a long, slow decline. The gas has a high Btu value, as high as 1,250 Btu/scf. MHF techniques are useful for tight shales also. These tight shales are an attractive source of gas and may contribute very significantly to gas production in the coming years.

Geopressured Aquifers

High-pressure brine in geopressured aquifers, which can form due to rapid subsidence, may contain up to 40 scf of natural gas per barrel of water. In the USA, such geopressured aquifers are located predominantly in a band that extends onshore and offshore from Texas to Florida along the Gulf of Mexico. An estimated 2,700 Tscf of gas reserves (unproven at this time) are associated with this region. However, no commercial means of recovering this gas have been developed to date.

Coal

Methane gas occluded in coal in minable coal beds with depths less than 3,000 ft has been estimated to be 260 Tscf in the US. This significant resource base, however, may only produce less than 40 Tscf due to practical constraints.

Another source of gas generation is coal gasification. The gas derived from coal usually has a lower heating value than natural gas. The commercial viability of coal gasification is not favorable yet.

Natural Gas Production and Processing System

Figure 1-6 outlines a typical gas production and processing system. From a design viewpoint, the total system consists of the following calculation modules:

1. The reservoir module that deals with the flow of gas (and oil) through subsurface strata.
2. Flow module for the flow of fluids from the reservoir to the wellhead at the surface.
3. Gas gathering system module. This module calculates the flow of gas through the pipeline network at the surface that is used to collect gas from several wells for separation and processing.

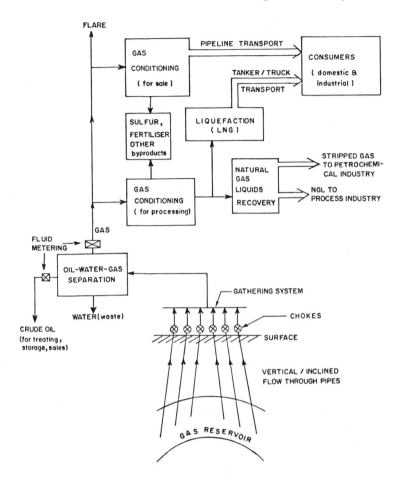

Figure 1-6. A typical gas production and processing system.

4. Separation module for calculating the amounts of gas, oil and water generated by the wellstream, and their compositions.
5. Metering devices for measuring the amount of gas and oil from the separators.
6. Gas conditioning module for the removal of contaminants from the gas.
7. Natural gas liquids recovery module.
8. Gas compression/liquefaction for economic transport by tankers, railroad, and road transport.
9. Flow module for the pipeline transport of gas to consumption sites.

This book has been designed to provide information on the design, operation, and engineering calculations for all of these modules, except the reservoir module. Chapters 2 and 3 provide the basic gas properties that are an essential input in all computations. Chapters 4 through 6 describe separation and processing for natural gas. Flow calculations are covered in Chapters 7 and 8. Chapters 9 and 10 provide information on related topics, namely, gas compression and gas metering, respectively. Finally, Chapter 11 describes gas gathering and transmission systems and the computational techniques for handling the complete production system.

Questions and Problems

1. (a) List some factors that may enhance the appeal of natural gas as a domestic and industrial fuel. (b) What are some of the factors that may reduce the appeal of natural gas as a domestic and industrial fuel?
2. Why is natural gas termed a "stable" flammable gas?
3. What impurities can one expect to find in natural gas produced from a gas field?
4. Why are gas flares a common sight in many petroleum producing areas? Can such flares be hazardous to the environment? If yes, then when and why?
5. As a petroleum engineer you are taken to tour a gas-producing field. Can you tell what type of gas (associated, dissolved, etc.) is being produced? What is the minimum information required to draw such a conclusion?
6. How are gas reserves estimated?
7. Besides conventional reservoirs, what types of gas reserves hold promise for the future? Why?

References

American Petroleum Institute, 1984. *Basic Petroleum Data Book: Petroleum Industry Statistics.* Vol. IV, No. 2 (May), API, Washington, D.C.
Barker, C. and Kemp, M. K., 1980. "Generation of Natural Gas and its Survival in the Deep Subsurface," presented at the Conf. on Natural Gas Res. Dev. in Mid-Cont. Basins: Prodn. & Expl. Techniques, Univ. of Tulsa, Tulsa, Oklahoma, March 11–12.

Brown, T. E. and Miller, S., 1985. *Layman's Guide to Oil & Gas Investments & Royalty Income*, 2nd Edition. Gulf Publishing Company, Houston, Texas.

Curry, R. N., 1981. *Fundamentals of Natural Gas Conditioning*. PennWell Publ. Co., Tulsa, Oklahoma, 118 pp.

Grow, G. C., 1980. "Future Potential Gas Supply in the United States—Current Estimates and Methods of the Potential Gas Committee," presented at the Am. Inst. of Chem. Engrs. Meeting, Philadelphia, PA, June 9.

McCain, W. D., Jr., 1974. *The Properties of Petroleum Fluids*. PennWell Publ. Co., Tulsa, Oklahoma, pp. 3–42.

McElroy, D. P., 1987. *Fundamentals of Petroleum Maps*. Gulf Publishing Company, Houston, Texas.

McKelvey, V. E., 1972. "Mineral Resource Estimates and Public Policy," *American Scientist*, 60, Jan.–Feb., 32–40.

Wheeler, R. R. and Whited, M., 1985. *Oil—From Prospect to Pipeline*, 5th Edition. Gulf Publishing Company, Houston, Texas.

2
Phase Behavior Fundamentals

Introduction

The properties exhibited by any substance depend upon its phase, namely, whether it is in the solid, liquid, or gaseous phase. Substances can be classified into two types—pure or single component, and multicomponent. Phase behavior relationships can be determined from laboratory pVT (pressure, volume, temperature) studies, or using theoretical/empirical methods such as the equations of state. These relationships are frequently shown graphically as "phase diagrams" to enhance qualitative understanding. For design purposes, however, precise quantitative phase behavior data are crucial. In this chapter, some fundamental concepts in phase equilibrium are introduced.

Qualitative Hydrocarbon Phase Behavior

A phase diagram for a single component system has three axes: p, V, and T. Usually, p,T diagrams only are used for a single component system. For multicomponent systems, the mixture composition becomes an additional variable. Generally, three types of phase diagrams are used for mixtures: (a) pressure versus temperature (p,T) diagrams, holding the composition fixed; (b) composition-composition diagrams, keeping pressure and temperature fixed; and (c) pressure-composition diagrams, keeping temperature fixed.

The Phase Rule

Gibbs' phase rule for the degrees of freedom of a system is written as:

$N = C - P + 2$

where N = number of degrees of freedom of the system
C = number of distinct chemical components (or compounds) in the system
P = number of phases in the system

The number of degrees of freedom of a system is the number of variables that must be defined to fix the physical state of the system. Usually, pressure, temperature, and composition serve as the variables.

The phase rule can easily be verified. Consider a mixture consisting of C components distributed in P phases. Then, at equilibrium, the chemical potentials (see pages xx–xx for thermodynamic criteria for equilibrium) for each component are equal in each phase:

$$\mu_{i1} = \mu_{i2} = \ldots = \mu_{iP}, \text{ for } i = 1,2,\ldots,C$$

Thus, we have (P − 1) equations for every component, or a total of C(P − 1) equations. The unknowns are the mole fractions of each component in each phase, which represent (C − 1)P unknowns, since knowing the mole fractions of (C − 1) components in any phase P fixes the mole fraction of the last component (the sum of the mole fractions equals unity). Also, the pressure p and temperature T are unknown. Thus, the total number of unknowns are (C − 1)P + 2. So, the number of degrees of freedom are:

N = number of unknowns − number of equations
= (C − 1)P + 2 − C(P − 1)
or, N = CP − P + 2 − CP + C = C − P + 2

Single-Component Systems

Figure 2-1 shows a typical p,T phase diagram for a single-component system. The equilibrium lines, AB, BC, and BD indicate the combinations of pressure and temperature at which the adjacent phases shown on the dia-

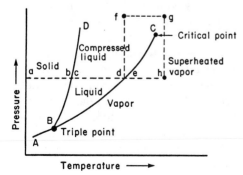

Figure 2-1. Pressure-temperature diagram for a single component system.

gram exist in equilibrium. Line AB represents the equilibrium between solid and vapor. A solid above this line will sublime directly to the vapor phase, without ever going through the liquid phase, on reduction of pressure to below line AB. An example of this is "dry ice." Line BD indicates the solid liquid equilibrium. Line BC is the equilibrium between liquid and vapor phases. At the triple point, B, all three phases exist in equilibrium. Applying the phase rule, it can be seen that for a single-component system with three phases, the degrees of freedom, $N = 1 - 3 + 2 = 0$. Hence, the triple point is a unique point that defines the whole system. The equilibrium lines, however, represent only two phases in equilibrium and the degrees of freedom $N = 1$. Therefore, specifying either the pressure or the temperature fixes the other variables for the system. In the single phase regions, $N = 2$, and both pressure and temperature must be known to determine the location (or state) of the system in the phase diagram.

The liquid-vapor equilibrium line (BC) begins at the triple point, B, and ends at the critical point, C. At the critical point, liquid and gaseous phases become indistinguishable and their *intensive* properties (that is, properties independent of the amount of fluid; for example: density, viscosity) become identical. The pressure and temperature corresponding to the critical point are termed critical pressure, p_c, and critical temperature, T_c, respectively, and are a uniquely defined numerical value for every pure substance. More formal definitions are as follows:

1. *Critical temperature*: The temperature above which a substance cannot be liquefied by the application of pressure alone.
2. *Critical pressure*: The pressure at which gas exists in equilibrium with liquid phase at the critical temperature. Also defined as the saturation pressure corresponding to the critical temperature.

Consider a constant pressure (isobaric) process represented by the line abcdeh of Figure 2-1. From a to b the system is completely solid, and the energy requirements for achieving the increase in temperature are simply proportional to the specific heat of the solid phase. From b to c, pressure and temperature remain constant, but an energy equal to the latent heat of fusion is required to convert the solid phase to liquid. The energy supplied is converted into internal energy. The temperature increase from c to d requires energy proportional to the specific heat of the liquid phase. At d the liquid is a saturated liquid—any further addition of energy will cause vaporization at constant pressure and temperature. Energy equal to the latent heat of vaporization is required to accomplish the change of phase from d to e. At e, the vapor is termed saturated. At higher temperatures, for example at h, it is called superheated. The energy spent for going from saturated vapor state at e to a superheated state at h requires an amount of energy proportional to the specific heat of the vapor.

Phase Behavior Fundamentals   21

To better understand vapor-liquid phase behavior, consider the region near the critical point in greater detail. If one proceeds from d by a compression process at constant temperature to f, liquid will at some point begin to disappear into a seemingly gaseous phase. Above the critical point at f, the system is in a "fourth" phase that exhibits properties different from gas. The properties of this fluid are in between those of gas and liquid, though not correlatable to either. For example, it is denser than regular gas, and is more compressible than a regular liquid. By going further from f to a point g beyond the critical temperature at constant pressure, from g to h at constant temperature, and finally, back to e, a phase transition from the liquid state at d to a vapor state at e can be achieved without any *abrupt* change of phase.

This shows that liquid and vapor phases are in reality quite similar: They represent separate forms of the same condition of matter, and it is possible to pass from one state to the other without an abrupt change of phase. Thus, the terms liquid and vapor have a definite meaning only in the two phase region. For conditions far removed from the two-phase region, particularly at pressure and temperature conditions above the critical point, it is impossible to define the state of the system, and is best referred to as being in the *fluid state*.

Multicomponent Systems

p,T Diagrams for a Fixed Composition

Figures 2-2a and 2-2b show typical p,T phase diagrams for a reservoir fluid. The system has three regions: (a) single phase oil, in the region above

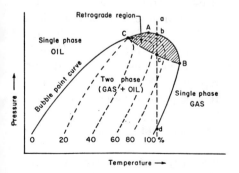 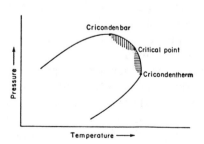

Figure 2-2a. Pressure-temperature diagram for a multicomponent reservoir fluid.

Figure 2-2b. Phase diagram showing two retrograde regions.

the bubble point (BP) curve; (b) single phase gas, in the region towards the right of the dew point (DP) curve, and (c) the two phase region, which is the portion bounded by the BP and DP curves. The BP and DP curves intersect and terminate at the critical point, C.

Point A represents the maximum pressure at which liquid and vapor may exist in equilibrium. This pressure is known as the cricondenbar. Similarly, the cricondentherm is defined as the maximum temperature at which liquid and vapor can exist in equilibrium. This is indicated as point B in Figure 2-2a. Along with the critical point, these serve as valuable tools for reconstructing approximate phase envelopes and to enhance their understanding.

The shaded region CAB, called the retrograde region, exhibits quite an interesting behavior. Usually, on lowering the pressure, a liquid phase vaporizes to gas. But in this region, on lowering the pressure, the gas at a begins to generate liquid at b (see Figure 2-2a). On further reduction in pressure, more and more liquid is condensed (refer to the quality lines), until point c is reached. Outside the retrograde region from c to d, liquid begins to vaporize until the 100% gas quality line (or the dew-point curve) is reached at d. It is evident that the retrograde region results from the shape of the phase envelope and the quality lines. The inflection points of the quality lines govern the retrograde area, which is bounded by the critical point, C, the cricondenbar at A, and the cricondentherm at B.

For naturally occurring hydrocarbon mixtures, the critical point has always been found to occur to the left of the cricondenbar (Campbell, 1984). If the critical point is to the right of the cricondenbar, two retrograde regions will occur as shown in Figure 2-2b.

p,T Diagrams for Variable Composition

The case discussed earlier concerned studying the phase diagram at a given fluid composition. When fluid is produced from a reservoir, pressure decreases over time. If we started off with an oil reservoir for example, a two-phase region will eventually be encountered at pressures below the bubble-point curve. The heavier components in the reservoir fluid will begin to vaporize, leading to altered reservoir fluid compositions. Whenever the reservoir fluid is in the two-phase region, compositions also change due to the selective removal of components in the mobile phase. Therefore, the phase equilibrium relationships change with time for a typical reservoir system, and the shifts in the phase envelopes as a function of composition must be studied carefully.

Figure 2-3 shows a p,T diagram for a binary mixture of ethane (C_2H_6) and normal-heptane (n-C_7H_{16}). It can be seen that with a shift in the mixture composition from 100% ethane to 100% n-heptane, the phase envelope shifts towards the right. This shows that the shape and location of the phase

Phase Behavior Fundamentals 23

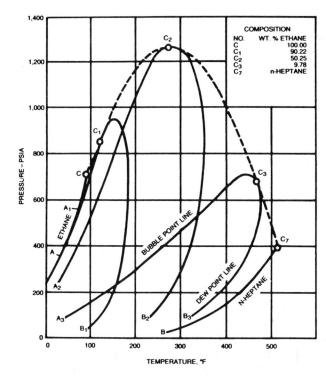

Figure 2-3. Pressure-temperature diagram for the ethane/n-heptane system. (After Stalkup, 1978, 1983; courtesy of SPE of AIME.)

envelope are composition dependent. The dashed line drawn tangent to all the phase envelopes at their critical points is called the critical locus. The critical point for a mixture is thus a function of its composition. Mixtures with more than two components have more than one critical locus, and the problem becomes quite complex. Simple mixing rules have been devised (to be discussed later) to represent this shift in terms of "pseudocritical" properties for correlation purposes. The only reliable method at this point is laboratory measurement, since none of the correlations are very exact.

Composition-Composition Diagrams (Fixed p,T)

According to the Gibbs phase rule, $N = C - P + 2$ degrees of freedom must be specified for completely defining the phase behavior of a mixture with C components distributed in P phases. Such a rigorous description, however, is neither practically feasible nor usually necessary. Ternary diagrams are used as an approximate method for representing the phase behavior of multicomponent mixtures. These diagrams represent the phase behavior of a ternary (three-component) mixture exactly; mixtures with more components are approximated by three pseudocomponents. For hydrocar-

24 Gas Production Engineering

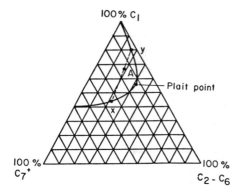

Figure 2-4. A typical ternary phase diagram for a reservoir fluid.

bon systems, the three pseudocomponents generally used are methane (C_1), ethane through hexane (C_2–C_6), and C_{7+}, which represents all hydrocarbon fractions with molecular weight greater than hexane.

Figure 2-4 shows a typical ternary phase diagram for a reservoir fluid. Each corner of the triangle represents 100% of the component indicated, whereas the side opposite to this corner represents 0% of that component. The distance in between a corner and the opposite side is divided into equal parts to represent fractions between 100% and 0% of the component. Thus, the point A in Figure 2-4 represents a mixture containing 67% C_1, 18.5% C_2–C_6, and 15% C_{7+}, lying in the two-phase region. Such a mixture, under equilibrium conditions, would result in a gas phase of composition y in equilibrium with a liquid phase of composition x. The dashed line connecting these equilibrium gas and liquid compositions, occurring on the dew-point and bubble-point curves, respectively, is called a tie-line. Clearly, there can be an infinite number of such tie-lines. The tie-lines disappear at the critical point, also called the plait point, where the liquid and gas phases become identical. Outside the phase envelope, single-phase gas occurs above the dew-point curve, whereas single-phase liquid occurs below the bubble-point curve. The size of the two-phase region depends upon the pressure and temperature. With increasing pressure, the two-phase region collapses to smaller sizes, whereas with increasing temperature it expands in size.

Lever's rule can be applied to determine the relative amounts of gas and liquid in equilibrium. For example, the mixture at A would split into a gas-liquid ratio equal to the ratio of the distance of A, along the applicable tie-line, from the bubble-point and dew-point curves, that is, Ax/Ay.

Pressure-Composition Diagrams (Fixed T)

Pressure-composition diagrams also serve as a useful method for displaying phase behavior data. These diagrams are obtained for reservoir oils by

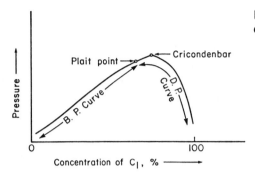

Figure 2-5. Pressure-composition diagram for a reservoir fluid.

adding the desired injection fluid to the oil in a high-pressure visual cell. The appropriate (bubble-point or dew-point) saturation pressures are measured as a function of the injection fluid mole fraction in the total mixture that results from the mixing of the injection fluid with the reservoir oil. Such diagrams are easier to obtain than the ternary diagrams, which require obtaining equilibrium samples and compositions, but their applicability is limited since they do not offer information of a general nature. Figure 2-5 shows a pressure-composition diagram for a reservoir oil as a function of the mole% methane in the mixture.

Quantitative Phase Behavior and Vapor-Liquid Equilibrium

In most petroleum production operations, it is important to know the phases present in order to design the system. In processing and other post-production operations, it is desirable to control these phases by controlling the operating conditions, or condensing or vaporizing selected components. The single-phase regions are relatively easier to define. The two-phase region is more difficult, since we need to know additional parameters such as the vapor-liquid ratio, and the compositions of both the phases, as a function of pressure and temperature.

Thermodynamic Criteria for Equilibrium

Phase equilibrium is reached when there is no net transfer of material from one phase to another. Such transfers involve a decrease in the total free energy of the system. Thus, when equilibrium is attained, the total free energy, G, also called Gibbs free energy, will be at its minimum value. This can be mathematically expressed, at a given p and T as:

$$dG = 0 \tag{2-1}$$

Note that "no net transfer" does not imply that the system is static. There is still a continuous exchange or transfer of material between the phases. But the rate of transfer of each component from one phase to another is just balanced by the same transfer occurring in the opposite direction from the latter to the former phase. Thus, there is no *net* transfer of any molecular species from one phase to another, and no net change in composition occurs in any of the phases.

An equivalent idea is expressed using the condition that $dE = 0$, where E is the total energy content of the system. Note that the derivative in Equation 2-1 is with respect to all possible variations, except those of p and T. Gibbs defined chemical potential, μ, to express these ideas. The chemical potential of any component i in a phase is the rate at which the total free energy of that phase changes as one changes the amount of the component i, keeping p, T, and amounts of all other components of the phase constant:

$$\mu_i = \left(\frac{\partial E}{\partial n_i}\right)_{V,S,n_j} = \left(\frac{\partial G}{\partial n_i}\right)_{V,S,n_j} \tag{2-2}$$

where n_i represents the moles of any component i, V is the system volume, and S is the system entropy. Chemical potentials are intensive properties and depend upon the composition of the mixture, but not on the total mass. For a single-component ideal gas, the chemical potential is given by:

$$\mu = \mu^* + RT \ln (p/p^*) \tag{2-3}$$

where R is the gas constant, and the superscript * refers to any arbitrary standard state. For a perfect gas mixture, the chemical potential for any component i, is given by:

$$\mu_i = \mu_i^* + RT \ln (p_i/p_i^*)$$

$$= \mu_i^* + RT \ln (p_i/p^*) - RT \ln y_i \tag{2-4}$$

where y_i is the mole fraction of component i in the mixture. It is not possible to ascribe absolute numeric values to quantities such as free energy and chemical potential. The ratio between two isothermal states as previously defined suffices for application purposes.

Equation 2-2 can be manipulated to yield the result that the necessary condition for equilibrium is that the chemical potentials of a given component are equal in all the phases. Thus, the chemical potential can be considered as a kind of driving force for mass transfer between phases.

Phase Behavior Fundamentals 27

The change in free energy, dG, for any substance is given by $dG = v\,dp$. For an ideal gas ($pv = RT$; see Chapter 3 for further detail), this becomes:

$$dG = RT\,d\ln p \tag{2-5}$$

To preserve this simple mathematical form of this equation for other cases where the system is not an ideal gas, Lewis defined a special function called fugacity, f, as follows:

$$dG = RT\,d\ln f \tag{2-6}$$

Integrating Equation 2-6, we can obtain:

$$G - G^* = RT\,\ln(f/f^*) = \int_{p^*}^{p} v\,dp \tag{2-7}$$

Fugacity can, therefore, be defined as a quantity whose numeric value is equal to the pressure when the substance is in the state of an ideal gas. It can be easily seen on comparing Equations 2-5 and 2-6 that fugacity has the units of pressure, and is proportional to pressure for an ideal gas. From a purely theoretical standpoint, any substance in any phase—solid, liquid, or gas—can be brought into an ideal gas state by sufficient pressure reduction at constant temperature. It can be seen also in the equations of state, which reduce to the ideal gas form, $pv = RT$, at the limit when $p = 0$ or $v = \infty$. Fugacity thus serves a very valuable purpose in equilibria studies.

Fugacity for a component i in a mixture can now be defined in terms of the chemical potential as follows:

$$\begin{aligned}\mu_i &= \mu_i^* + RT\,\ln(f_i/f_i^*) \\ &= \mu_i^* + RT\,\ln(f_i/f^*) - RT\,\ln y_i\end{aligned} \tag{2-8}$$

Using these relationships, it can be seen that the condition for equilibrium can, alternatively, be stated in terms of fugacity. A system is in equilibrium when the fugacities of a given component are equal in all the phases.

The Equilibrium Ratio

The vapor-liquid distribution coefficient, commonly known as the vapor-liquid equilibrium ratio or the equilibrium vaporization ratio, K_i, for a component i, is simply the ratio of its mole fractions in the vapor (y_i) and liquid (x_i) phases at equilibrium conditions:

$$K_i = y_i/x_i \tag{2-9}$$

Equilibrium ratio values can therefore be determined by laboratory equilibrium flash experiments by analyzing the mole fractions of the different components in the vapor and liquid phases.

The Partial Pressure Approach

K-values can be expressed in terms of partial pressures of the components. The assumption, of course, is that the gas obeys ideal gas laws and behaves as an ideal mixture. Therefore, such a concept is only applicable at low pressures, up to about 400 psi.

At equilibrium, the partial pressure of component i in the vapor phase ($=py_i$) must be equal to the vapor pressure exerted by its presence in the liquid phase ($=p_{vi}x_i$). Thus, at equilibrium:

$py_i = p_{vi}x_i$, implying that $y_i/x_i = p_{vi}/p$

So, $K_i = p_{vi}/p$ \hfill (2-10)

where p_{vi} is the vapor pressure of component i at the equilibrium pressure and temperature conditions.

This approach is very useful as a practical method for determining the equilibrium constants, but has limited applicability.

The Fugacity Approach

Non-ideal behavior can be accounted for by using the fugacity coefficients. The fugacity (or activity) coefficient, Ψ_i, for a component i is defined as the ratio of its fugacity, f_i, to its partial pressure, p_i:

$$\Psi_i = f_i/p_i \quad \text{with} \quad \lim_{p \to 0} \Psi_i = 1 \quad (2\text{-}11)$$

For a gas phase, the partial pressure, $p_i = y_ip$. To enable definition of a liquid fugacity coefficient, an equivalent hypothetical partial pressure concept is used, and liquid phase partial pressures are expressed as $p_i = x_ip$.

Fugacity coefficients are commonly written as:

$$\ln \Psi = \ln f/p = \int_0^p [(v/RT) - (1/p)]dp$$

$$= Z - 1 - \ln Z - (1/RT) \int_\infty^v [p - (RT/v)]dv \quad (2\text{-}12)$$

for a pure component, and

$$\ln \Psi_i = \ln f_i/px_i$$

$$= -\ln Z + \int_{\infty}^{V} [(1/v) - (1/RT)(\partial p/\partial n_i)]dv \qquad (2\text{-}13)$$

for a component i in a mixture, where n_i is the moles of component i in the total mixture. K can be expressed in terms of fugacity coefficients as follows:

$$K_i = \Psi_i^L/\Psi_i^V \qquad (2\text{-}14)$$

where Ψ_i^L and Ψ_i^V are the fugacity coefficients of component i in the liquid and vapor phases, respectively, coexisting under equilibrium conditions. This can be verified, by expanding the right-hand side term of Equation 2-14:

$$\Psi_i^L/\Psi_i^V = [f_i^L/(x_ip)]/[f_i^V/(y_ip)]$$

$$= (y_ip)/(x_ip) \text{ since } f_i^L = f_i^V \text{ at equilibrium}$$

$$= y_i/x_i = K_i$$

Flash Calculations

These calculations involve solving simple material balance equations for multiphase systems in order to establish the phase compositions as well as amounts upon equilibrium separation. In oil and gas systems, we generally have a two-phase oil-gas phase breakup. Water is assumed immiscible in the liquid oil (oleic) phase, and miscible only in the gaseous phase where its presence can be quantified by Dalton's law of partial pressures.

Consider F moles of a hydrocarbon mixture of composition $\{z_i\}$ enters a separation unit. At the operating conditions of the separator, it splits into L moles of liquid of composition $\{x_i\}$, and V moles of vapor of composition $\{y_i\}$. Then, by the law of conservation of mass:

$$F = L + V$$

and

$$Fz_i = Lx_i + Vy_i \text{ for each component i}$$

Changing to a unit mole basis, $F = 1$, for simplicity:

$$L + V = 1 \quad (2\text{-}15a)$$

and

$$z_i = Lx_i + Vy_i = Vx_i(L/V + K) \quad (2\text{-}15b)$$

Thus, we get

$$Vx_i = z_i/(L/V + K)$$

and

$$Vy_i = VKx_i = z_i/(L/KV + 1)$$

The following additional relationships must also be satisfied:

$$\sum_{i=1}^{n} x_i = \sum_{i=1}^{n} y_i = \sum_{i=1}^{n} z_i = 1$$

and

$$K_i = y_i/x_i$$

Thus, the amount of vapor, V, can be calculated as follows:

$$V = V \sum_{i=1}^{n} y_i = \sum_{i=1}^{n} [z_i/(L/KV + 1)] \quad (2\text{-}16)$$

If liquid, L, is desired, it can be calculated as follows:

$$L = L \sum_{i=1}^{n} x_i = \sum_{i=1}^{n} [z_i/(KV/L + 1)] \quad (2\text{-}17)$$

Equations 2-16 and 2-17 require a trial and error type of solution scheme. Both these equations are equivalent, and one has the choice of using either one; knowing either L or V is sufficient, since $L + V = 1$ (Equation 2-15a). From the standpoint of obtaining an accurate solution, these equations, however, are different. It is easy to envision that for a stream F for which V is very small, any error in the calculation of V will be far more detrimental to accuracy as compared to an error in the calculation of L. For such a system, it will therefore be necessary to solve for V using Equation 2-16 using very rigorous convergence criteria (that is, use a very low error tolerance), and then obtain L by difference. Similarly, for predominantly gas systems, Equation 2-17 would provide better accuracy.

Applications

Reservoir Behavior

Figure 2-6 shows a p,T phase diagram for a naturally occurring hydrocarbon fluid of a known composition. Consider four different reservoirs, with this same fluid, but with different initial conditions denoted by 1_i, 2_i, 3_i, and 4_i. The vertical lines represent the pressure decline in the reservoir at constant temperature. The curved lines represent the pressure and temperature changes imposed upon the reservoir fluid as it flows upward through the wellbore, and both pressure and temperature decline to the wellhead producing conditions. The subscripts s and r denote surface and reservoir conditions, respectively.

Reservoir 1, occurring above the bubble-point curve, is called an undersaturated oil reservoir. It is also called a black oil reservoir, depletion drive, dissolved gas drive, solution gas drive, or internal gas drive reservoir. Gas is

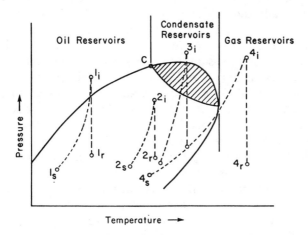

Figure 2-6. A typical pressure-temperature phase diagram for a reservoir fluid.

produced at the surface conditions (1_s is in the two-phase region), but no gas is formed in the reservoir until the pressure reaches the bubble-point pressure. Typically, higher API gravity oils release greater amounts of gas since they contain larger amounts of lighter components. Below the bubble point, the composition of the reservoir fluid starts changing. The free gas liberated in-situ eventually begins to flow toward the wellbore in increasing quantities with reduction in pressure. The producing gas-oil ratio (GOR) therefore goes up.

Reservoir 2, occurring in the two-phase region, is an oil reservoir with an initial gas cap. With declining pressure, GOR increases, and the reservoir fluid composition changes.

Reservoir 3, occurring in the retrograde region, is a gas condensate reservoir. On pressure reduction, liquid begins to condense at and beyond the dew-point curve. Since it clings to the pore spaces in the reservoir, a critical saturation has to be built up before it can flow toward the wellbore. Thus, the producing GOR (at the surface) increases. The intermediates that condense out of the gas are very valuable and their loss is a serious affair. However, as pressure is reduced further, the liquid begins to revaporize into the gaseous phase. This revaporization aids liquid recovery and, assuming that the reservoir fluid composition is constant, may lead to a decline in the producing GOR at the surface. The reservoir fluid composition, however, changes as retrograde condensation occurs: the p-T phase envelope shifts to the right, increasing retrograde liquid condensation into gas. This retrograde "loss" is greater for a greater shift of the p-T phase envelope to the right, lower reservoir temperature, and higher abandonment pressure.

Reservoir 4, occurring in the single-phase gas region, is a true gas reservoir. It is generally known as a wet gas reservoir if 4_s is inside the two-phase region, or as a dry gas reservoir if 4_s is in the single-phase gas region. In this reservoir type, the reservoir (point 4_r) is always in the single-phase gas region, regardless of the pressure. So, the reservoir fluid maintains a constant composition since there is no change of phase in the reservoir.

The producing GOR for a reservoir is an excellent indication of the reservoir type. Generally, a reservoir is classified as a gas reservoir if the GOR > 100,000 scf/stb, as an oil reservoir if the GOR < 5,000, and as a gas condensate reservoir if the GOR is between 5,000 to 100,000 scf/stb.

The relationship between the critical temperature, T_c, and the reservoir temperature, T_R, determines whether retrograde condensation occurs or not. For retrograde condensation to occur, the reservoir temperature must be in between the critical temperature and the cricondentherm. Reservoir systems in single phase with a surface GOR > 10,000 scf/stb are likely to have a high C_1 fraction, and consequently, a low T_c. Expect retrograde condensation for such reservoirs. Reservoir systems in single phase that exhibit a

GOR < 2,000 scf/stb are likely to have $T_c > T_R$, implying a dissolved gas drive reservoir.

Hydrocarbon Production and Separation

The design of the producing well and surface facilities depends upon the nature of the fluid being produced: gas, oil, or two-phase gas and oil with a given GOR, and the presence of water. Since the intermediate hydrocarbons are often the most valuable components, it is highly desirable not to lose them to the gas, and to recover as much liquid as possible from the produced hydrocarbons. For the design and operational optimization of these systems, relevant phase behavior and gas-oil equilibrium parameters must be known. Since the phase relationships are a function of the composition of the produced fluid, we can never really design equipment that is optimum under all conditions. The only recourse is to design it for the "average" condition expected over the life of the field, or to change the installation at some point where it may become economically attractive to do so.

Gas Processing and Transport

It may be necessary to remove heavier ends from the gas so that the gas does not condense liquids upon anticipated pressure-temperature changes in the transport, processing, and handling systems. To enable such calculations, at least the dew-point curve for the fluid must be known. Liquids accumulate in the piping systems, cause additional pressure drops in the lines, and are generally undesirable because of the instability caused by phase changes occurring as a function of pressure and temperature. Processing equipment must never be operated near the phase envelope boundaries where small changes in pressure, temperature, or composition can cause disproportionately large changes in vapor-liquid ratio.

Reservoir Engineering and Enhanced Oil Recovery

Reservoir engineering calculations, such as reservoir reserve estimates, predictions, and simulations using computers, require reliable phase behavior data. Nowhere is it more critical than in miscible enhanced recovery techniques, where highly detailed and accurate phase equilibrium relationships must be known over a wide range of conditions. The performance of these methods depends upon the ability of the injected fluid to be miscible, or be able to generate miscibility, with the reservoir fluid. The injection fluid must be tailored to requirements dictated solely by phase equilibrium.

Prediction of the Phase Envelope

To accurately predict the phase envelope for a multicomponent hydrocarbon system such as oil or gas is almost impossible. Experimental means must be used, since imprecise results can be quite dangerous to use in the innumerable planning, design, and operational problems. However, it becomes very necessary to use correlations and predictive methods in many instances where such studies cannot be readily performed.

As a bare minimum, an accurate compositional analysis of the reservoir fluid is essential. Estimates of the critical point, the cricondentherm, and cricondenbar can then be used in conjunction with vapor-liquid equilibrium calculations for bubble-point and dew-point curves to generate reasonable phase equilibrium curves. Sometimes, only a portion of the phase curves may be required.

Cricondentherm and Cricondenbar

Grieves and Thodos (1963) presented the following equations for the prediction of cricondentherm temperatures:

$$T_t/T_c' = (T_b'/T_b - 1)(e^{5.40x_\ell - 3.39}) + 1.01 \text{ for } 0 < x_\ell < 0.55 \tag{2-18}$$

and

$$T_t/T_c' = (T_b'/T_b - 1)(e^{6.38x_\ell - 4.38}) - 0.418x_\ell + 1.256$$
$$\text{for } 0.55 < x_\ell < 0.925 \tag{2-19}$$

For estimating cricondenbar temperatures, Grieves and Thodos (1963) suggest the following equations:

$$T_p/T_c' = (T_b'/T_b - 1)(e^{4.33x_\ell - 3.62}) + 1.008 \text{ for } 0 < x_\ell < 0.7 \tag{2-20}$$

and

$$T_p/T_c' = (T_b'/T_b - 1)(e^{6.33x_\ell - 5.14}) - 0.165x_\ell + 1.116$$
$$\text{for } 0.7 < x_\ell < 0.925 \tag{2-21}$$

where T_t = cricondentherm temperature of the mixture, °R
 T_p = cricondenbar temperature of the mixture, °R
 T_c' = pseudocritical, pseudocricondenbar, or pseudocricondentherm temperature, whichever applicable, of the mixture, °R
 T_b = normal (atmospheric) boiling point of the mixture, °R

Phase Behavior Fundamentals 35

T_b' = molar average boiling point of mixture, °R
x_ℓ = mole fraction of low-boiling component

The calculation procedure is quite tedious. The mixture composition, pure component critical temperatures, and normal boiling points are required. Values for T_b can be calculated using a mixing rule, or obtained from laboratory measurements. One proceeds by taking two successive components, calculating their T_t and T_p values, then adding one component at a time and calculating the T_t and T_p values for this new mixture until all the components have been included. Grieves and Thodos (1963) report maximum errors of less than 5% in cricondentherm, but significantly larger errors up to 13% for cricondenbar temperatures. Such a correlation is difficult to use, and the results are not very reliable.

Critical Point

Many different studies have focused on the problem of characterizing hydrocarbons for predictions of their physical properties. Watson and Nelson (1933) and Watson et al. (1935) characterized the chemical makeup of petroleum mixtures using the boiling point and specific gravity. They defined the Watson characterization factor, K, as follows:

$$K = T_b^{1/3}/\gamma \tag{2-22}$$

If SI units are used, with T_b in K (Kelvin), the right-hand side of Equation 2-6 should be multiplied by 1.21644 (= $1.8^{1/3}$). K defines the relative paraffinicity of a hydrocarbon fraction. A typical range would be from 10.0 (highly aromatic) to 13.0 (highly paraffinic). Several useful relationships have been found using the Watson characterization factor (for example, Watson et al., 1935; Smith and Watson, 1937; Simon and Yarborough, 1963; Riazi and Daubert, 1980; Whitson, 1983).

Thus, physical property correlations based upon the boiling point and specific gravity have been used for a long time, and have proven to be very reliable. Different correlation parameters have been proposed. Most of these can be expressed using the following generalized equation:

$$\theta = aT_b^b \gamma^c \tag{2-23}$$

where θ = the property
T_b = normal or cubic-average boiling point, °R
γ = specific gravity at 14.7 psia and 60°F

For critical pressures and temperatures, the constants a, b, and c, determined by Riazi and Daubert (1980) and Whitson (1983), are as follows:

θ	a	b	c	
T_c, °R	24.2787	0.58848	0.3596	
p_c, psia	3.12281×10^9	-2.3125	2.3201	for $T_b < 850°R$
p_c, psia	2.41490×10^{14}	-3.86618	4.2448	for $T_b > 850°R$

The maximum deviations were found to be 10.6%, 9.3%, and 13.2%, respectively, for these three cases (Whitson, 1983).

Bubble-Point and Dew-Point Curves

The bubble point may be defined as the p,v,T condition at which an all-liquid system releases the first (infinitesimally small) drop of gas, signaling the entry into the two-phase region. Thus, at the bubble point, the total mixture composition, $\{z_i\}$, is just equal to the liquid composition, $\{x_i\}$. Since the sum of the mole fractions equals unity, it is clear that for the gas phase released at the bubble point, $\Sigma_{i=1}^n y_i$ is equal to 1.0, and in all regions where gas is not present, this sum will be less than 1.0. Thus, at the bubble point:

$$\sum_{i=1}^n y_i = \sum_{i=1}^n K_i x_i = \sum_{i=1}^n K_i z_i = 1.0 \qquad (2\text{-}24)$$

At the dew point, the appearance of a liquid phase just begins for an all-vapor system. The system composition, $\{z_i\}$, is equal to the vapor composition, $\{y_i\}$. Also, for the just released liquid phase, the sum of the component mole fractions equals unity. Therefore:

$$\sum_{i=1}^n x_i = \sum_{i=1}^n y_i/K_i = \sum_{i=1}^n z_i/K_i = 1.0 \qquad (2\text{-}25)$$

Equations 2-24 and 2-25 can thus be used to predict the bubble-point and dew-point curves, provided K-values are known for all the components as a function of pressure and temperature. A typical procedure would involve calculating the quantities $\Sigma_{i=1}^n K_i z_i$ and $\Sigma_{i=1}^n z_i/K_i$ at different pressures and temperatures. The state of the system can be determined by applying the following rules that are true for vapor-liquid systems:

If $\Sigma_{i=1}^n K_i z_i < 1.0$, the system is single-phase all-liquid.

Phase Behavior Fundamentals   37

If $\sum_{i=1}^{p} z_i/K_i < 1.0$, the system is in the single phase all-vapor region.

If both $\sum_{i=1}^{p} K_i z_i$ and $\sum_{i=1}^{p} z_i/K_i > 1.0$, the system is in the two-phase region.

Both $\sum_{i=1}^{p} K_i z_i$ and $\sum_{i=1}^{p} z_i/K_i \not< 1.0$.

An efficient procedure to predict a bubble point is to find the pressure (or, alternatively, the temperature) at which $\sum_{i=1}^{p} K_i z_i = 1.0$ at a given temperature (or pressure). A series of such points would then constitute the bubble-point curve. A similar procedure can be used to reconstruct the dew-point curve.

Questions and Problems

1. Given the pressure and temperature, is it possible to specify the number of phases for: (a) water, and (b) a mixture of alcohol and water?
2. Name the type of phase diagram required for the following cases:
   (a) a single-component fluid being used for heat transfer in a nuclear reactor.
   (b) a two-component gas being compressed for storage as liquid.
   (c) a multi-component gas to be transported in a pipeline.
   (d) design of an oil-gas separator.
3. What is the state of a substance at: (a) its critical point, and (b) above its critical point?
4. Define fugacity. Show that the fugacity of a substance must be equal in all phases in which it may exist at equilibrium.
5. Construct a p,T phase diagram for the following mixture:

Component	mole %	T_b, °F
C_1	65	−258.7
C_2	20	−127.5
C_{3+}	15	96.9

Assume the C_{3+} fraction to be identical to n-C_5. The required hydrocarbon equilibrium ratios (K-values) can be taken from Figures 4-25 through 4-51 in Chapter 4.

6. From the phase diagram constructed in problem 5, determine the cricondenbar and cricondentherm temperatures. Compare these with the ones obtained using Grieves and Thodos' relationships.

References

Campbell, J. M., 1984. *Gas Conditioning and Processing, Vol. 1*. Campbell Petroleum Series, Norman, Oklahoma, 326pp.

Grieves, R. B. and Thodos, G., 1963. "The Cricondentherm and Cricondenbar Temperatures of Multicomponent Hydrocarbon Mixtures," *Soc. Pet. Eng. J.*, 3(4, Dec.), 287-292.

Riazi, M. R. and Daubert, T. E., 1980. "Simplify Property Predictions," *Hydr. Proc.*, 59(3, March), 115-116.

Simon, R. and Yarborough, L., 1963. "A Critical Pressure Correlation for Gas-Solvent-Reservoir Oil Systems," *J. Pet. Tech.*, 15(5, May), 556-560.

Smith, R. L. and Watson, K. M., 1937. "Boiling Points and Critical Properties of Hydrocarbon Mixtures," *Ind. and Eng. Chem.*, 29(12, Dec.), 1408-1414.

Stalkup, F. I., 1978. "Carbon Dioxide Miscible Flooding: Past, Present, and Outlook for the Future," *J. Pet. Tech.*, 30(8, Aug.), 1102-1112.

Stalkup, F. I., 1983. *Miscible Displacement*. Volume 8, SPE Monograph Series, Society of Petroleum Engineers, Richardson, Texas, 204pp.

Watson, K. M. and Nelson, E. F., 1933. "Improved Methods for Approximating Critical and Thermal Properties of Petroleum Fractions," *Ind. and Eng. Chem.*, 25(8, Aug.), 880-887.

Watson, K. M., Nelson, E. F., and Murphy, G. B., 1935. "Characterization of Petroleum Fractions," *Ind. and Eng. Chem.*, 27(12, Dec.), 1460-1464.

Whitson, C. H., 1983. "Characterizing Hydrocarbon Plus Fractions," *Soc. Pet. Eng. J.*, 23(4, Aug.), 683-694.

3
Properties of Natural Gases

Introduction

In designing gas production, processing, transport, and handling systems a complete knowledge of gas properties is crucial. For this reason, much research has been done in the measurement and prediction of hydrocarbon fluid properties. The area of property prediction continues to attract significant attention from researchers who seek to optimize design and control of gas and oil systems. The current trend is to develop mathematical equations for implementation on computers, rather than the traditional engineering charts and tables, because with computers it is far more efficient to solve equations than to interpolate in a huge domain of possible parameter values, whereas the opposite is probably true for humans.

Equations of State

All fluids follow physical laws that define their state under given physical conditions. These laws are mathematically represented as equations, which are consequently known as equations of state (EOS). These equations essentially correlate p, V, and T for any fluid and can be expressed, on a unit mole basis, as:

$\phi(p,v,T) = 0$

For practical purposes, this form is rearranged to yield a desired parameter, $v = \phi(p,T)$. Many different empirical EOS's have been developed over the years, and I will only mention the most basic and widely used among these.

40   Gas Production Engineering

Ideal Gases

The concept of an ideal gas is a hypothetical idea, but it serves as a useful tool to explain the more complex real gas behavior. An ideal gas is defined as a gas in which the molecules occupy negligible volume; there is no interaction between the molecules, that is, no attractive or repulsive forces exist between them; and collisions between the molecules are purely elastic, implying no energy loss on collision. At low pressures (≤ 400 psi) most gases exhibit an almost ideal behavior. The ideal gas law that applies to such gases can be stated as follows:

$$pV = nRT \tag{3-1}$$

where p and T are the absolute pressure and temperature of the gas, n is the number of moles, and V is the volume occupied by the gas. R, the constant of proportionality in Equation 3-1, is called the *universal gas constant*. The value of R can be easily determined from the fact that 1 lbmole (pound-mole) of any gas occupies 378.6 ft^3 at 14.73 psia and 60°F (520°R).

$$R = pV/nT = (14.73 \times 378.6)/(1 \times 520) = 10.732 \text{ (psia ft}^3\text{)/(lbmole °R)}$$

Thus, if p is in psia, T is in °R, and V is in ft^3, then the appropriate value of R is 10.732 psia-ft^3/lbmole-°R. In SI units, where p is in kPa, T is in K, and V is in m^3, the value of R to use is 8.314 kPa-m^3/kgmole-K.

Behavior of Real Gases

In general, gases do not exhibit ideal behavior. The reasons for the deviation from ideal behavior can be summarized as follows:

1. Molecules for even a sparse system, such as gas, occupy a finite volume.
2. Intermolecular forces are exerted between the molecules. Some of these forces are:
   - Electrostatic forces, also called Coulomb forces, between ions and dipoles. These are long-range forces.
   - Induced forces between a dipole and an induced dipole, for example in a mixture of a polar and a non-polar gas.
   - Attraction/repulsion forces which are generally exerted over very short distances only. These forces exist even for a perfectly non-polar gas such as argon.
3. Molecular collisions are never perfectly elastic.

The deviation from ideal behavior is greater for heavier gases because of the larger size of their molecules. Most gases compress more than an ideal gas at low pressures, whereas the opposite is true at high pressures. From an application standpoint, all equations of state for real gases must be regarded as essentially empirical in nature, since they attempt to correlate this non-ideal behavior using empirical parameters.

The Compressibility Factor Approach

To correct for non-ideality, the simplest equation of state uses a correction factor known as the gas compressibility factor, Z:

$$pV = nZRT \tag{3-2}$$

The Z-factor can therefore be considered as being the ratio of the volume occupied by a real gas to the volume occupied by it under the same pressure and temperature conditions if it were ideal. This is the most widely used real gas equation of state. The major limitation is that the gas deviation factor, Z, is not a constant. Numerous attempts have been made to define the functional dependence of Z on various other parameters that define the state of the system, and several correlations are available for the Z-factor as a result of these studies. More complex equations of state that do not represent the deviation through the Z-factor, but through other correlation constants, are often used to derive quite precise Z-factor values that can be used in Equation 3-2.

Van der Waals Equation

This equation is probably the most basic EOS and, though seldom used today, it serves as a conceptual basis for understanding and developing other equations. It corrects for the volume occupied by the molecules in a gas, by using $v - B$ as the true gas volume, instead of v; and loss in pressure exerted by the gas due to attractive forces and inelastic collisions, by using $p + A/v^2$ as the true pressure.

$$(p + A/v^2)(v - B) = RT \tag{3-3}$$

where A and B are empirical constants. This equation serves as a good approximation only for low pressures. It can also be written as:

$$v^3 - (RT + Bp)v^2/p + (A/p)v - AB/p = 0 \tag{3-4}$$

or, in terms of the Z-factor as:

$$Z^3 - Z^2(1 + Bp/RT) + Z\,Ap/(RT)^2 - ABp^2/(RT)^3 = 0 \tag{3-5}$$

At the critical point, the three roots of the cubic equation in v (Equation 3-4) are identical. Thus, if v_c represents the critical volume, then at the critical point:

$$(v - v_c)^3 = v^3 - 3v_c v^2 + 3v_c^2 v - v_c^3 = 0 \tag{3-6}$$

Comparing Equations 3-4 and 3-6, it can easily be shown that:

$$A = 3p_c v_c^2,\ B = v_c/3,\ \text{and}\ R = 8p_c v_c/3T_c \tag{3-7}$$

An alternative procedure for obtaining the values for A, B, and R is to use the fact that the critical isotherm, that is, the curve relating pressure and volume at the critical temperature, must show an inflexion point:

$$\frac{\partial p}{\partial v} = \frac{\partial^2 p}{\partial v^2} = 0 \tag{3-8}$$

These conditions are known as the Van der Waals conditions for the critical point. The first and second partial differentials of Equation 3-4 with respect to v would yield the results shown in Equation 3-7 upon substitution of parameter values relevant to the critical point and use of Equation 3-8.

Van der Waals equation may be written in the reduced form by substituting the values for A, B and R from Equation 3-7 into Equation 3-3:

$$[p + 3p_c v_c^2/v^2][v - v_c/3] = 8p_c v_c T/3T_c$$

or, $[p/p_c + 3/(v^2/v_c^2)][3v/v_c - 1]p_c v_c = 8p_c v_c T/T_c$

thus, $[p_r + 3/v_r^2][3v_r - 1] = 8T_r$ \hfill (3-9)

Benedict-Webb-Rubin Equation

This equation was developed by Benedict et al. (1940) for describing the behavior of pure, light hydrocarbons. It has found a lot of application in computing thermodynamic properties and phase equilibria for gases due to two reasons: it gives sufficient accuracy for natural gases, since natural gases are a mixture of light hydrocarbons for which this equation was originally developed; and it can be written explicitly in the reduced form, like Van der

Properties of Natural Gases   43

Waals equation, which makes it applicable in developing general correlations in terms of reduced variables.

$$p = RTd + (B_oRT - A_o - C_o/T^2)d^2 + (bRT - a)d^3 + a\alpha d^6$$
$$+ (cd^3/T^2)[(1 + \gamma d^2) \exp(-\gamma d^2)] \qquad (3\text{-}10)$$

where a, b, c, α, γ, A_o, B_o, and C_o are constants for a given gas, d is the molar density (lbmole/ft^3), and p, T are the absolute pressure and temperature, respectively.

Redlich-Kwong Equation

Redlich and Kwong (1949) proposed the following equation:

$$p = \frac{RT}{v - B} - \frac{A}{T^{0.5}v(v + B)} \qquad (3\text{-}11)$$

where A and B are constants. Other details will be discussed later.

The Redlich-Kwong (RK) equation, along with modifications by Zudkevitch and Joffe (1970) and Joffe et al. (1970) (ZJRK), and by Soave (1972) (SRK), is widely used.

Peng-Robinson Equation

Another widely used equation is the Peng and Robinson (1976) (PR) equation:

$$p = \frac{RT}{v - B} - \frac{A}{v(v + B) + B(v - B)} \qquad (3\text{-}12)$$

where A and B are functions of temperature. Note that the RK and PR equations cannot be written explicitly in the reduced form.

A General Form for Cubic Equations of State

Martin (1979) showed that all cubic equations of state can be represented in the following general form:

$$p = \frac{RT}{v} - \frac{\alpha(T)}{(v + \beta)(v + \gamma)} + \frac{\delta(T)}{v(v + \beta)(v + \gamma)} \qquad (3\text{-}13)$$

where α and δ are functions of temperature, and β and γ are constants. Coats (1985) presents the following equivalent form of Martin's equation, with $\delta = 0$:

$$Z^3 + [(m_1 + m_2 - 1)B - 1] Z^2 + [A + m_1m_2B^2 \\ - (m_1 + m_2)(B + 1)B] Z \\ - [AB + m_1m_2B^2(B + 1)] = 0 \quad (3\text{-}14)$$

where $A = \sum_{j=1}^{n} \sum_{k=1}^{n} x_j x_k A_{jk}$ (3-15a)

$B = \sum_{j=1}^{n} x_j B_j$ (3-15b)

$A_{jk} = (1 - \delta_{jk})(A_j A_k)^{0.5}$ (3-15c)

$A_j = \Omega_{aj}\, p_{rj}/T_{rj}^2$ (3-15d)

$B_j = \Omega_{bj}\, p_{rj}/T_{rj}$ (3-15e)

The δ_{jk} are the binary interaction coefficients, symmetric in j and k with $\delta_{jj} = 0$.

Theoretically, Ω_a and Ω_b are universal constants, Ω_a^o and Ω_b^o, determined by making the EOS satisfy Van der Waals' criteria at the critical point (Equation 3-8). The values of Ω_a^o and Ω_b^o for some commonly used EOS's are as follows:

RK and SRK: $\Omega_a^o = 0.4274802$, $\Omega_b^o = 0.08664035$

PR: $\Omega_a^o = 0.457235529$, $\Omega_b^o = 0.077796074$

In practice, however, Ω_a and Ω_b are treated as component- and temperature-dependent quantities, $\Omega_{ai}(T)$ and $\Omega_{bi}(T)$:

RK: $\Omega_{ai} = \Omega_a^o/T_{ri}^{0.5}$ and $\Omega_{bi} = \Omega_b^o$

SRK: $\Omega_{ai} = \Omega_a^o[1 + (0.48 + 1.574\omega_i - 0.176\omega_i^2)(1 - T_{ri}^{0.5})]^2$, and $\Omega_{bi} = \Omega_b^o$

PR: $\Omega_{ai} = \Omega_a^o[1 + (0.37464 + 1.54226\omega_i - 0.26992\omega_i^2)(1 - T_{ri}^{0.5})]^2$, and $\Omega_{bi} = \Omega_b^o$

The acentric factor (ω), first proposed by Pitzer (1955), is a measure of a fluid's deviation from the law of corresponding states (see the section on compressibility factors). It is defined as:

$$\omega = -\log(p_v^*/p_c) - 1 \quad (3\text{-}16)$$

where p_v^* is the vapor pressure of the fluid at a temperature of $0.7T_c$, and p_c is its critical pressure. As expected, ω is equal to zero for noble gases like argon, and is close to zero for gases like methane.

In these equations, putting $m_1 = 0$ and $m_2 = 1$, yields the RK or SRK equations. For the PR equation, $m_1 = 1 + 2^{0.5}$, and $m_2 = 1 - 2^{0.5}$.

Equilibrium is achieved when the fugacities for any component i are equal in both the liquid and vapor phases. Fugacities can be calculated using the following relationship for the fugacity coefficient, derived using Equation 2-13 (Coats, 1985):

$$\ln \Psi_i = \ln (f_i/px_i) = B_i(Z - 1)/B - \ln(Z - B) + [A/(m_1 - m_2)B][(2/A)$$

$$\cdot \sum_{j=1}^{n} A_{ij}x_j - B_i/B]\ln [(Z + m_2B)/(Z + m_1B)] \tag{3-17}$$

A Typical Solution Method for Cubic Equations of State

The input data include the overall composition of the mixture, $\{z_i\}$, and the critical pressure (p_{ci}), temperature (T_{ci}), acentric factor (ω_i), and critical compressibility factor (Z_{ci}) for each component i.

The objective is to determine for F moles of the feed of composition $\{z_i\}$: the moles of liquid (L) and the moles of vapor (V) generated, and the composition of these liquid ($\{x_i\}$) and vapor ($\{y_i\}$) phases that exist in equilibrium at any given pressure (p) and temperature (T). Referring to Equations 2-15 through 2-17, it is clear that if the equilibrium vaporization ratio, K_i, for each component i is known, we can achieve the objectives. Thus, the problem can alternatively be conceptualized as being that of obtaining very precise values for K_i using an equation of state.

The calculation procedure is iterative. A value is assumed for the distribution coefficient (K_i) and the equation of state is then solved with this initial guess, K_i^p. The EOS solution yields a new value, K_i^{p+1}, which is then used as the new guesstimate. This procedure is repeated till convergence is reached, within a specified tolerance, between the estimate, K_i^p, and the solution, K_i^{p+1}. A good initial guess for the distribution coefficient is very advantageous in obtaining faster convergence. A reasonable guess can be obtained from the following equation, which has been used extensively (Wilson, 1969; Peng and Robinson, 1976):

$$K_i = (1/p_{ri}) \exp [5.3727(1 + \omega_i)(1 - 1/T_{ri})] \tag{3-18}$$

This equation was obtained by assuming that the fluid obeys Raoults' law and that the logarithm of the reduced vapor pressure of each component i is

46 Gas Production Engineering

a linear function of the reciprocal reduced temperature (Peng and Robinson, 1976).

Using the K-values from Equation 3-18, flash calculations are carried out for the mixture (see Chapter 2) to calculate L, V, $\{x_i\}$, and $\{y_i\}$. Equations 3-15(a–e) are solved next. Equation 3-14, a cubic equation in Z, can now be solved to determine its roots. Since it is a cubic equation, it can have either all real roots, or one real root and two imaginary ones. The gas Z-factor is taken as the largest real root, when the equation is solved with coefficients corresponding to gas. Equation 3-14 is solved again with coefficients A, B corresponding to the liquid phase. The smallest real root is taken as the liquid Z-factor. These Z values are required in Equation 3-17.

Equation 3-17 is solved to determine fugacities in the vapor and liquid phases. If these are equal in both phases for all the components (on an individual basis), convergence has been reached. Otherwise, the new K_i values are determined using the ratio of the fugacity coefficients (see Equation 2-14).

Several modifications have been proposed by different investigators to improve the rate of convergence, or to improve the accuracy of prediction in general or for various special cases. The MVNR (minimum variable Newton-Raphson) method presented by Fussell and Yanosik (1978) and Fussell (1979) provides improved convergence and is more reliable than the method of successive substitution previously discussed. Since this is an area of active research, the reader can refer to published work in this area for specific methodologies and developments.

Critical Pressure and Temperature Determination

For pure components, physical property data are readily available. Table 3-1 shows critical temperatures (T_{ci}), pressures (p_{ci}), and other useful properties of typical components in hydrocarbon fluids. For mixtures, Kay's mixing rule can be used to find the effective critical properties:

$$p_{pc} = \sum y_i p_{ci}, \text{ and } T_{pc} = \sum y_i T_{ci} \tag{3-19}$$

where p_{pc}, T_{pc} are the pseudocritical pressure and temperature, respectively, for the mixture, and y_i is the mole fraction of component i in the mixture. These properties are termed "pseudo" because they are used as a correlation basis rather than as a very precise representation of mixture critical properties.

Equation 2-23 in Chapter 2 can be used to predict the critical properties of a hydrocarbon fluid from its normal boiling point and gravity. If the composition of the gas, $\{y_i\}$, is not known, Figure 3-1 may be used to determine

Properties of Natural Gases 47

Table 3-1
Physical Constants for Typical Natural Gas Constituents*

Compound	Molecular Weight	Critical Pressure (psia)	Critical Temp. (°R)	Crit. Comp. Factor (Z_c)	Acentric Factor (ω)	Eykman Mol Refraction** (EMR)
CH_4	16.043	667.8	343.1	0.289	0.0115	13.984
C_2H_6	30.070	707.8	549.8	0.285	0.0908	23.913
C_3H_8	44.097	616.3	665.7	0.281	0.1454	34.316
$n\text{-}C_4H_{10}$	58.124	550.7	765.4	0.274	0.1928	44.243
$i\text{-}C_4H_{10}$	58.124	529.1	734.7	0.283	0.1756	44.741
$n\text{-}C_5H_{12}$	72.151	488.6	845.4	0.262	0.2510	55.267
$i\text{-}C_5H_{12}$	72.151	490.4	828.8	0.273	0.2273	55.302
$n\text{-}C_6H_{14}$	86.178	436.9	913.4	0.264	0.2957	65.575
$n\text{-}C_7H_{16}$	100.205	396.8	972.5	0.263	0.3506	75.875
$n\text{-}C_8H_{18}$	114.232	360.6	1023.9	0.259	0.3978	86.193
$n\text{-}C_9H_{20}$	128.259	332.0	1070.4	0.251	0.4437	96.529
$n\text{-}C_{10}H_{22}$	142.286	304.0	1111.8	0.247	0.4902	106.859
N_2	28.013	493.0	227.3	0.291	0.0355	9.407
CO_2	44.010	1070.9	547.6	0.274	0.2250	15.750
H_2S	34.076	1306.0	672.4	0.266	0.0949	19.828
O_2	31.999	737.1	278.6	0.292	0.0196	8.495
H_2	2.016	188.2	59.9	0.304	$-0.2234^{\dagger}$	4.450
H_2O	18.015	3203.6	1165.1	0.230	0.3210	—

* From Edmister and Lee (1984).
** From McLeod and Campbell (1969).
† ω = 0.0 used in most correlations.

the critical pressure and temperature from the gas gravity. Corrections may be made for the presence of non-hydrocarbon components. Thomas et al. (1970) took data from Figure 3-1 and other sources to obtain the following relationship:

$$P_{pc} = 709.604 - 58.718 \gamma_g \qquad (3\text{-}20a)$$

$$T_{pc} = 170.491 + 307.344 \gamma_g \qquad (3\text{-}20b)$$

where γ_g is the gas gravity with respect to air. Thomas et al. recommend the use of this equation in allowable limits of up to 3% H_2S, 5% N_2, or a total impurity (non-hydrocarbon) content of 7%, beyond which errors in critical pressure exceed 6%. It should be noted that the gas gravity method of obtaining pseudocritical pressures and temperatures is not very accurate. If the analysis of the gas is available, it must be used in accordance with Equation 3-19.

48 Gas Production Engineering

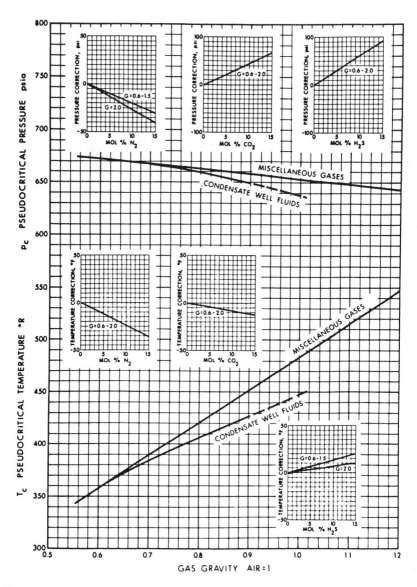

Figure 3-1. Pseudocritical properties of miscellaneous natural gases and condensate fluids. (After Brown et al., 1948; courtesy of GPSA.)

Example 3-1. The analysis of a sweet gas, in mole%, is known to be as follows: N_2 = 1.40, CH_4 = 93.0, C_2H_6 = 3.29, C_3H_8 = 1.36, $n\text{-}C_4H_{10}$ = 0.37, $i\text{-}C_4H_{10}$ = 0.23, $n\text{-}C_5H_{12}$ = 0.10, $i\text{-}C_5H_{12}$ = 0.12, C_6H_{14} = 0.08, and $C_7H_{16}^+$ = 0.05. Assume the C_7^+ fraction to exhibit the same properties as $n\text{-}C_9$.

Find the gas gravity. Also, find the critical pressure and critical temperature for this gas using (1) Kay's mixing rule, (2) Brown et al.'s method, and (3) Thomas et al.'s equations.

Solution

Comp.	y_i	M_i	p_{ci}	T_{ci}
C_1	0.9300	16.043	667.8	343.1
C_2	0.0329	30.070	707.8	549.8
C_3	0.0136	44.097	616.3	665.7
$n\text{-}C_4$	0.0037	58.124	550.7	765.4
$i\text{-}C_4$	0.0023	58.124	529.1	734.7
$n\text{-}C_5$	0.0010	72.151	488.6	845.4
$i\text{-}C_5$	0.0012	72.151	490.4	828.8
C_6	0.0008	86.178	436.9	913.4
C_{7+}	0.0005	128.259	332.0	1070.4
N_2	0.0140	28.013	493.0	227.3

$M = \Sigma\ y_i M_i = 17.54$

The gas gravity $\gamma_g = 17.54/28.97 = 0.6055$

1. $p_{pc} = \Sigma\ y_i p_{ci} = 664.47$ psia
   $T_{pc} = \Sigma\ y_i T_{ci} = 356.93\ °R$

2. From Figure 3-1,

   $p_{pc} = 670$ psia
   $T_{pc} = 360\ °R$

   Applying the correction factor for 1.4% N_2,

   $p_{pc} = 670 - 5 = 665$ psia
   $T_{pc} = 360 - 5 = 355\ °R$

3. Using Equation 3-20a:

   $p_{pc} = 709.604 - (58.718)(0.6055) = 674.05$ psia

   Using Equation 3-20b:

   $T_{pc} = 170.491 + (307.344)(0.6055) = 356.59\ °R$

The Gas Compressibility Factor

Several different correlations are available for this important parameter. The basic correlations use the corresponding states concept. According to Van der Waals' law of corresponding states, the physical characteristics of a substance are a function of its relative proximity to the critical point. This means that the deviation from ideal behavior of gases is the same if they are located at the same state relative to their critical state. Thus the relevant temperature and pressure values that express the departure of a real gas from ideal behavior are the reduced pressure, p_r, and the reduced temperature, T_r:

$$Z = f(p_r, T_r) \tag{3-21}$$

where $p_r = p/p_c$
$T_r = T/T_c$

For gas mixtures such as natural gases, the reduced parameters are denoted as pseudoreduced temperature T_{pr} ($= T/T_{pc}$), and pseudoreduced pressure, p_{pr} ($= p/p_{pc}$).

The corresponding states concept is more applicable to substances with similar molecular structure, such as neighboring hydrocarbon components of a similar paraffinic type, than to widely dissimilar substances. For more complex mixtures, it is more appropriate to consider the compressibility factor as a function of an additional parameter, ξ:

$$Z = f(p_r, T_r, \xi) \tag{3-22}$$

This additional parameter, ξ, is any term that can characterize the mixture behavior, particularly the discrepancy resulting from the assumption in Equation 3-21. Among several alternatives, the two most widely used are the critical compressibility factor, Z_c, and the acentric factor, ω. Z-factor correlations using Z_c as a basis have been used for a long time; different charts are made for different values of Z_c for manual calculations. The acentric factor, ω, has become very popular because it gives good accuracy. Most equations of state and other computer-generated solutions use the acentric factor. In addition, the Eykman molecular refraction (EMR) approach by McLeod and Campbell (1969) is of interest and is described here.

Standing-Katz Compressibility Chart

Figure 3-2 shows the Standing and Katz (1942) correlation for Z as a function of p_{pr} and T_{pr} for sweet (non-H_2S or CO_2 containing) natural gases. The

Properties of Natural Gases 51

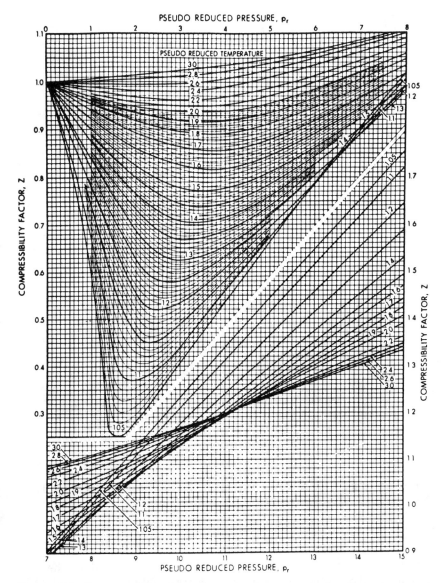

Figure 3-2. Compressibility factor for natural gases as a function of reduced pressure and temperature. (After Standing and Katz, 1942; courtesy of SPE of AIME.)

applicability of Equation 3-21 is assumed, and Kay's mixing rules are used for gas mixture properties. This chart is generally reliable for sweet natural gases with minor amounts of non-hydrocarbons such as N_2.

Wichert and Aziz (1972) proposed a correction factor, ϵ, to extend the applicability of the Standing-Katz Z-factor chart to sour gases:

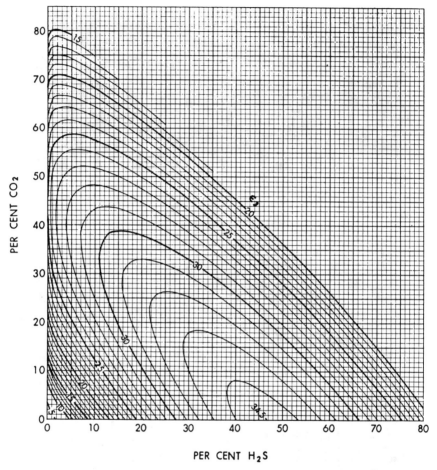

Figure 3-3. Pseudocritical correction factor, ϵ, for sour gases. (After Wichert and Aziz, 1972; courtesy of *Hydrocarbon Processing*.)

$$\epsilon = 120(A^{0.9} - A^{1.6}) + 15(B^{0.5} - B^{4.0}) \quad (3\text{-}23)$$

where A is the sum of the mole fractions of H_2S and CO_2, and B is the mole fraction of H_2S, in the gas. Figure 3-3 is a graphical representation by Wichert and Aziz (1972) of Equation 3-23. This method uses the correction factor, ϵ, to correct the pseudocritical temperature (T_{pc}) and pressure (p_{pc}) determined by Kay's rule, as follows:

$$T'_{pc} = T_{pc} - \epsilon$$

$$p'_{pc} = \frac{p_{pc} T'_{pc}}{T_{pc} + B\epsilon(1 - B)} \quad (3\text{-}24)$$

Properties of Natural Gases   53

These modified values for the critical pressure (p'_{pc}) and temperature (T'_{pc}) are used to calculate the values of reduced pressure and temperature that will give a valid Z-factor for a sour gas from the Standing-Katz correlation.

Example 3-2. For the gas composition given in Example 3-1, find the compressibility factor at 2,000 psia and 200°F using the Standing-Katz Z-factor chart.

Solution

From Example 3-1,

p_{pc} = 664.47 psia
T_{pc} = 356.93 °R

So,

p_{pr} = 2,000/664.47 = 3.010
T_{pr} = (200 + 459.67)/356.93 = 1.848

Using Figure 3-2, the compressibility factor Z = 0.905.

Example 3-3. Find the compressibility factor at 2,500 psia and 650°R for a sour gas with the following composition in mole%: CH_4 = 89.10, C_2H_6 = 2.65, C_3H_8 = 1.90, $n\text{-}C_4H_{10}$ = 0.30, $i\text{-}C_4H_{10}$ = 0.20, N_2 = 0.65, CO_2 = 1.00, H_2S = 4.20.

Solution

Comp.	y_i	p_{ci}	T_{ci}
C_1	0.8910	667.8	343.1
C_2	0.0265	707.8	549.8
C_3	0.0190	616.3	665.7
$n\text{-}C_4$	0.0030	550.7	765.4
$i\text{-}C_4$	0.0020	529.1	734.7
N_2	0.0065	493.0	227.3
CO_2	0.0100	1070.9	547.6
H_2S	0.0420	1306.0	672.4

A = 0.0420 + 0.0100 = 0.0520, and B = 0.0420

$\epsilon = 120[(0.052)^{0.9} - (0.052)^{1.6}] + 15[(0.042)^{0.5} - (0.042)^{4.0}] = 10.402$

54   Gas Production Engineering

$p_{pc} = \Sigma\ y_i p_{ci} = 696.96$ psia

$T_{pc} = \Sigma\ y_i T_{ci} = 371.77\ °R$

Applying the Wichert and Aziz correction,

$T'_{pc} = 371.77 - 10.40 = 361.37\ °R$

$p'_{pc} = (696.96)(361.37)/[371.77 + (0.042)(10.402)(1 - 0.042)]$
$= 676.69$ psia

$p_{pr} = 2,500/676.69 = 3.694$

$T_{pr} = 650/361.37 = 1.799$

Using Figure 3-2, the compressibility factor, Z = 0.90.

Curve-Fits for the Standing-Katz Correlation

Gopal (1977) found straight line fits for the Standing-Katz chart of the form:

$$Z = p_r(AT_r + B) + CT_r + D \quad (3\text{-}25)$$

where A, B, C, and D are correlation constants (see Table 3-2). Thirteen equations of this type were found to suitably represent the Standing-Katz chart, with average errors on the order of 0.6%, and maximum errors up to 2.5%. (Gopal (1977) does not report any statistical parameters for the error; it is the author's deduction from the reported results only.) Wichert and Aziz's (1972) correction factor for sour gases is applicable. Since it attempts to curve-fit the Standing-Katz chart and uses the same mixing rules, its applicability is also subject to the same limitations as the Standing-Katz chart. The major advantage of this method lies in the fact that it is not trial and error type.

Example 3-4. For the gas composition given in Example 3-1, find the compressibility factor at 2,000 psia and 200°F using Gopal's method.

Solution

From Example 3-2,

$p_{pr} = 3.010$

Table 3-2
Correlation Equations for the Standing-Katz Z-factor Chart*

Reduced Pressure p_r Range Between	Reduced Temperature, T_r Range Between	Equations	Equation Number
0.2 and 1.2	1.05 and 1.2	$p_r(\ 1.6643T_r - 2.2114) - 0.3647T_r + 1.4385$	1
	1.2+ and 1.4	$p_r(\ 0.5222T_r - 0.8511) - 0.0364T_r + 1.0490$	2
	1.4+ and 2.0	$p_r(\ 0.1391T_r - 0.2988) + 0.0007T_r^a + 0.9969$	3[b]
	2.0+ and 3.0	$p_r(\ 0.0295T_r - 0.0825) + 0.0009T_r^a + 0.9967$	4[b]
1.2+ and 2.8	1.05 and 1.2	$p_r(-1.3570T_r + 1.4942) + 4.6315T_r - 4.7009$	5[c]
	1.2+ and 1.4	$p_r(\ 0.1717T_r - 0.3232) + 0.5869T_r + 0.1229$	6
	1.4+ and 2.0	$p_r(\ 0.0984T_r - 0.2053) + 0.0621T_r + 0.8580$	7
	2.0+ and 3.0	$p_r(\ 0.0211T_r - 0.0527) + 0.0127T_r + 0.9549$	8
2.8+ and 5.4	1.05 and 1.2	$p_r(-0.3278T_r + 0.4752) + 1.8223T_r - 1.9036$	9[b]
	1.2+ and 1.4	$p_r(-0.2521T_r + 0.3871) + 1.6087T_r - 1.6635$	10[b]
	1.4+ and 2.0	$p_r(-0.0284T_r + 0.0625) + 0.4714T_r - 0.0011^a$	11
	2.0+ and 3.0	$p_r(\ 0.0041T_r + 0.0039) + 0.0607T_r + 0.7927$	12
5.4+ and 15.0	1.05 and 3.0	$p_r(\ 0.711 + 3.66T_r)^{-1.4667}$ $- 1.637/(0.319\ T_r + 0.522) + 2.071$	13

* After Gopal (1977). Courtesy of *Oil and Gas Journal*.
[a] These terms may be ignored.
[b] For a very slight loss in accuracy, Eqs. 3 and 4 and 9 and 10 can, respectively, be replaced by the following two equations:
$Z = p_r(0.0657T_r - 0.1751) + 0.0009\ T_r^a + 0.9968$
$Z = p_r(-0.2384\ T_r + 0.3695) + 1.4517\ T_r - 1.4580$
[c] Preferably use this equation for p_r up to 2.6 only. For $p_r = 2.6+$, Eq. 9 will give slightly better results. Also, preferably, use Eq. 1 for $1.08 \le T_r \le 1.19$ and $p_r \le 1.4$.

$T_{pr} = 1.848$

From Table 3-2, the applicable equation is:

$$Z = p_{pr}(-0.0284T_{pr} + 0.0625) + 0.4714T_{pr} - 0.0011$$
$$= (3.010)[(-0.0284)(1.848) + 0.0625] + (0.4714)(1.848) - 0.0011$$
$$= \underline{0.900}$$

Compressibility Factors from Equations of State

Several authors have reported results using different equations of state. An interesting comparison of the various techniques used is provided by Takacs (1976). Yarborough and Hall (1974) used the Starling-Carnahan equation of state to arrive at the following equation:

$$Z = 0.06125(p_{pr}/\rho_r T_{pr}) \exp[-1.2(1 - 1/T_{pr})^2] \tag{3-26}$$

where the reduced density, ρ_r, is calculated by trial and error from the following equation:

$$\frac{\rho_r + \rho_r^2 + \rho_r^3 - \rho_r^4}{(1 - \rho_r)^3} - (14.76/T_{pr} - 9.76/T_{pr}^2 + 4.58/T_{pr}^3)\rho_r^2$$

$$+ (90.7/T_{pr} - 242.2/T_{pr}^2 + 42.4/T_{pr}^3)\rho_r^{(2.18 + 2.82/T_{pr})}$$

$$= 0.06125(p_{pr}/T_{pr}) \exp[-1.2(1 - 1/T_{pr})^2] \qquad (3\text{-}27)$$

Dranchuk et al. (1974) presented an eight-factor trial and error type of equation for Z-factor using the Benedict-Webb-Rubin equation of state.

$$Z = 1 + (A_1 + A_2/T_r + A_3/T_r^3)\rho_r + (A_4 + A_5/T_r)\rho_r^2$$

$$+ A_5 A_6 \rho_r^5 T_r + A_7(1 + A_8\rho_r^2)(\rho_r^2/T_r^3) \exp(-A_8\rho_r^2) \qquad (3\text{-}28)$$

where the reduced density was defined as:

$$\rho_r = Z_c p_r/(ZT_r), \text{ and } Z_c \text{ was assumed to be } 0.270 \qquad (3\text{-}29)$$

and the correlation constants A_1 through A_8 are: 0.31506237, −1.04670990, −0.57832729, 0.53530771, −0.61232032, −0.10488813, 0.681570001, and 0.68446549, respectively.

Dranchuk and Abou-Kassem (1975) developed the following equation from the Starling equation of state:

$$Z = 1 + (A_1 + A_2/T_r + A_3/T_r^3 + A_4/T_r^4 + A_5/T_r^5)\rho_r$$

$$+ (A_6 + A_7/T_r + A_8/T_r^2)\rho_r^2 - A_9(A_7/T_r + A_8/T_r^2)\rho_r^5$$

$$+ A_{10}(1 + A_{11}\rho_r^2)(\rho_r^2/T_r^3) \exp(-A_{11}\rho_r^2) \qquad (3\text{-}30)$$

where the reduced density is given by Equation 3-30 as before.

By fitting Equation 3-29 from the Starling equation of state to the Standing-Katz correlation using more than 1,500 data points, Dranchuk and Abou-Kassem (1975) found the values of the eleven coefficients A_1 through A_{11} to be: 0.3265, −1.0700, −0.5339, 0.01569, −0.05165, 0.5475, −0.7361, 0.1844, 0.1056, 0.6134, and 0.7210, respectively.

The Wichert and Aziz correction for sour gases (Equations 3-23 and 3-24) is applicable to all these methods using equations of state. It is clear that all the EOS methods involve a trial and error type of solution scheme. The accuracy of these methods is within 0.5%, but for the region where $T_r = 1.0$,

$p_r > 1.0$, very large errors have been reported (Dranchuk and Abou-Kassem, 1975). Thus, these correlations should be used with caution. The accuracy of the Gopal (1977) method in this region is not apparent from his reported results (little data in this region have been reported by him), but it seemingly provides acceptable results.

The Eykman Molecular Refraction (EMR) Method

This method, developed by McLeod and Campbell (1969), uses a correlation between the EMR of a gas or liquid and its properties such as molecular weight, density, and critical properties. A different mixture combination rule is used, and the method also requires the use of a different compressibility chart.

The Eykman molecular refraction, EMR, is defined as follows:

$$\text{EMR} = [(n^2 - 1)/(n + 0.4)](M/\rho) \qquad (3\text{-}31)$$

where n = refractive index of the gas or liquid using Sodium-D yellow light
 M = molecular weight
 ρ = density in gm/cc

The Eykman molecular refraction index, EMRI, defined as

$$\text{EMRI} = \text{EMR}/M = [(n^2 - 1)/(n + 0.4)]/\rho \qquad (3\text{-}32)$$

is also used for correlation purposes. McLeod and Campbell (1969) found the following empirical relationship between EMR and M for normal paraffin hydrocarbons:

$$\text{EMR} = 2.4079 + 0.7293\,M + 0.00003268\,M^2 \qquad (3\text{-}33)$$

Alternatively, M can be expressed in terms of EMR for normal paraffin hydrocarbons as follows:

$$M = -3.2971 + 1.3714\,\text{EMR} - 0.00008156\,\text{EMR}^2 \qquad (3\text{-}34)$$

If the molecular weight, M, is not known, the EMR may be determined from the density, ρ, using the EMR versus ρ^2 plot shown in Figure 3-4.

McLeod and Campbell found that EMR correlates very well with the critical properties of hydrocarbons and also the non-hydrocarbons generally associated with natural gas. Figures 3-5 and 3-6 show the correlations determined by them. The curve fit equations indicated in Figures 3-5 and 3-6 may be used, instead of the figures, for programming purposes. In the EMR

58 *Gas Production Engineering*

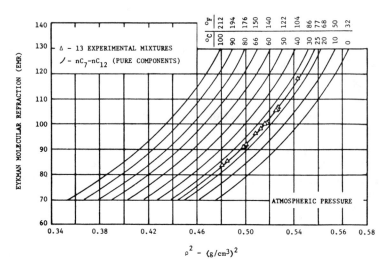

Figure 3-4. Eykman molecular refraction (EMR) versus ρ^2. (After McLeod and Campbell, 1969; courtesy of Campbell Petroleum Series.)

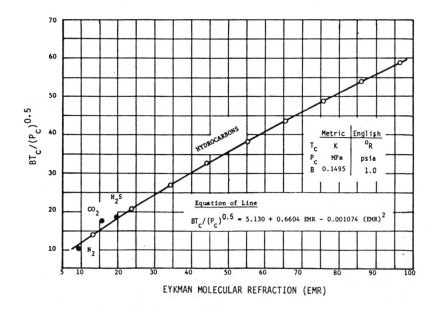

Figure 3-5. EMR versus $T_c/p_c^{0.5}$ correlation. (After McLeod and Campbell, 1969; courtesy of Campbell Petroleum Series.)

Properties of Natural Gases 59

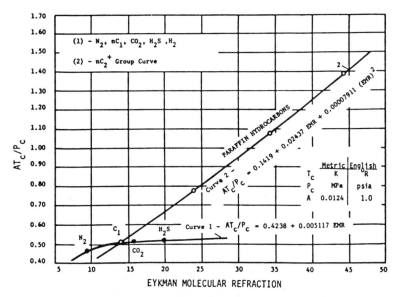

Figure 3-6. EMR versus T_c/p_c correlation. (After McLeod and Campbell, 1969; courtesy of Campbell Petroleum Series.)

versus $T_c/p_c^{0.5}$ plot (Figure 3-5), both the hydrocarbons as well as non-hydrocarbons lie on the same correlation. In the EMR versus T_c/p_c plot (Figure 3-6), two different correlations were found for two categories of components: N_2, CO_2, H_2S, and H_2 lie on curve 1; and all normal hydrocarbon components lie on the upper curve 2. The two curves intersect at a value corresponding to C_1.

So, for gases containing any non-hydrocarbon components, it is necessary to divide the mixture into two groups: C_1, N_2, CO_2, H_2S, and H_2; and hydrocarbon components above methane, that is $n\text{-}C_{2+}$. An n-component gas mixture is thus treated as a pseudobinary mixture in this method. The calculation procedure is as follows (Campbell, 1984):

1. For each of the two groups of components, find the EMR_j as follows:

$$EMR_j = \Sigma \; x_{ji}(EMR)_i \qquad (3\text{-}35)$$

where x_{ji} is the normalized mole fraction of component i in the group j it belongs to, and EMR_i is its EMR value determined from Table 3-1. For fractions such as C_{7+}, the EMR value can be directly determined using Equation 3-33 if the molecular weight is known, or using Figure 3-4 if the density is known. The normalized mole fractions, x_{ji}, are calculated as:

$$x_{ji} = x_i/X_j$$

where X_j is the sum of the mole fractions of all the components in group j.

2. Using Figure 3-6 (or the equations indicated in Figure 3-6), find the value for $(AT_c/p_c)_1$ and $(AT_c/p_c)_2$ for the two groups. The total mixture $(AT_c/p_c)_{mix}$ is calculated as:

$$(AT_c/p_c)_{mix} = X_1(AT_c/p_c)_1 + X_2(AT_c/p_c)_2$$

3. Determine $BT_c/p_c^{0.5}$ from Figure 3-5 (or the equation indicated on Figure 3-5) for the total mixture using the EMR_{mix} for the total mixture calculated as follows:

$$EMR_{mix} = X_1 EMR_1 + X_2 EMR_2$$

4. Using the mixture AT_c/p_c and $BT_c/p_c^{0.5}$ values, solve for the critical pressure, p_c, and critical temperature, T_c. In English units, A and B are 1.0, whereas in the metric or SI system of units, $A = 0.0124$ and $B = 0.1495$.
5. Calculate the reduced pressure and temperature. Finally, the compressibility factor, Z, is found from Figure 3-7.

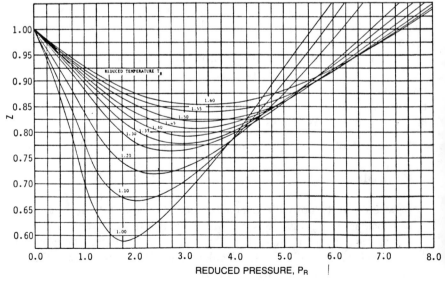

Figure 3-7. Compressibility factor chart for the EMR-method. (After McLeod and Campbell, 1969; courtesy of Campbell Petroleum Series.)

Example 3-5. For the gas composition given in Example 3-3, find the compressibility factor at 2,500 psia and 650°R using the EMR method.

Solution

Comp.	x_i	x_{ji}	EMR
N_2	0.0065	0.00685	9.407
CO_2	0.0100	0.01053	15.750
H_2S	0.0420	0.04423	19.828
C_1	0.8910	0.93839	13.984
$X_1 =$	0.9495		
C_2	0.0265	0.52475	23.913
C_3	0.0190	0.37624	34.316
$n\text{-}C_4$	0.0030	0.05941	44.243
$i\text{-}C_4$	0.0020	0.03960	44.741
$X_2 =$	0.0505		

$EMR_1 = \Sigma\, x_{1i} EMR_i = 14.2297$

$EMR_2 = \Sigma\, x_{2i} EMR_i = 29.8596$

From Figure 3-6,

$(AT_c/p_c)_1 = 0.50$

$(AT_c/p_c)_2 = 0.94$

So,

$(AT_c/p_c)_{mix} = (0.9495)(0.50) + (0.0505)(0.94) = 0.522$

$EMR_{mix} = (0.9495)(14.2297) + (0.0505)(29.8596) = 15.019$

From Figure 3-5,

$(BT_c/p_c^{0.5}) = 14.90$

In English units being used here,

$A = 1$ and $B = 1$

Thus,

$(T_c/p_c)_{mix} = 0.522$

$(T_c/p_c^{0.5})_{mix} = 14.90$

So,

$p_c^{0.5} = 14.90/0.522 = 28.54$ implying that

$p_c = 814.76$ psia

$T_c = (0.522)(814.76) = 425.30°R$

Using the values of critical pressure and temperature as previously determined,

$p_{pr} = 2,500/814.76 = 3.068$

$T_{pr} = 650/425.30 = 1.528$

Using Figure 3-7, the compressibility factor Z = 0.835.

The Stewart, Burkhardt, and Voo (SBV) Method

Stewart, Burkhardt, and Voo (1959) presented a mixing rule for pseudocritical pressure and temperature:

$$T_{pc} = K^2/J, \text{ and } p_{pc} = T_{pc}/J \qquad (3\text{-}36)$$

where $J = (1/3)\Sigma_{i=1}^n y_i(T_c/p_c)_i + (2/3)[\Sigma_{i=1}^n y_i(T_c/p_c)_i^{0.5}]^2$
$K = \Sigma_{i=1}^n y_i(T_c/p_c^{0.5})_i$

The pseudocritical pressure and temperature from this equation are then used in Pitzer's (1955) charts, extended by Satter and Campbell (1963), to determine Z^0 and Z^1 (Figures 3-8 and 3-9). Z is calculated as follows:

$$Z = Z^0 + Z^1 \Sigma_{i=1}^n y_i \omega_i \qquad (3\text{-}37)$$

This is one of the best available methods for hydrocarbon mixtures and has proven to be very accurate.

Properties of Natural Gases   63

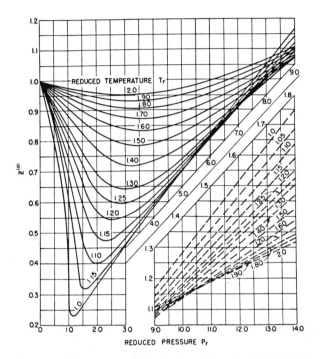

Figure 3-8. Generalized compressibility factors Z^0 as a function of reduced pressures and temperatures. (After Satter and Campbell, 1963; courtesy of SPE of AIME.)

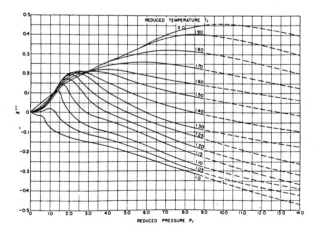

Figure 3-9. Generalized compressibility factors Z^1 as a function of reduced pressures and temperatures. (After Satter and Campbell, 1963; courtesy of SPE of AIME.)

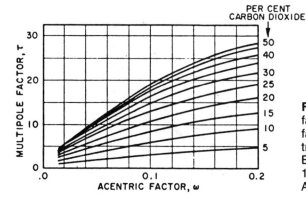

Figure 3-10a. Multipole factor, τ, versus acentric factor, ω, for CO_2 concentrations up to 50%. (After Buxton and Campbell, 1967; courtesy of SPE of AIME.)

Figure 3-10b. Multipole factor, τ, versus acentric factor, ω, for CO_2 concentrations greater than 50%. (After Buxton and Campbell, 1967; courtesy of SPE of AIME.)

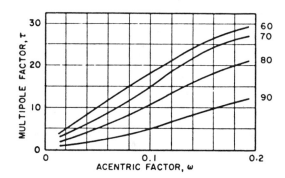

The Buxton-Campbell Approach for CO_2-Rich Gases

Sour gas correction methods described earlier have been developed for sour gases containing H_2S and CO_2. For gases that have a high CO_2 content but a low H_2S content, these methods have not been found to be satisfactory.

To extend the applicability of the SBV method to sour gases rich in CO_2, Buxton and Campbell (1967) provide a correction factor known as the multipole correction factor, τ, as shown in Figures 3-10a and b. It corrects for the deviation in the critical pressure and temperature due to the presence of CO_2. The effective acentric factor, ω_e, for use in Figures 3-10a and b is as follows:

$$\omega_e = [1/(1 - y_{CO_2})] \sum_{i=1}^{p} y_i \omega_i \qquad (3\text{-}38)$$

where $\Sigma_{i=1}^{n}$ refers to a summation over all components i *except* CO_2. The value of τ determined from Figures 3-10a and b is then used to correct the p_{pc} and T_{pc} calculated from Equation 3-36 as follows:

$$T'_{pc} = T_{pc} - \tau/A$$

$$p'_{pc} = p_{pc}(T'_{pc}/T_{pc}) \qquad (3\text{-}39)$$

where T'_{pc} and p'_{pc} are the corrected values for pseudocritical temperature and pressure, respectively, and A = 1.0 for English units, A = 1.8 if metric units are used. These corrected values are used, as in the SBV method, to determine the pseudoreduced pressure and temperature, and Z^0 and Z^1 from Figures 3-8 and 3-9. Finally, Z is determined using Equation 3-37.

Example 3-6. The analysis of a sour gas, in mole%, is known to be as follows: CH_4 = 56.1, C_2H_6 = 20.6, C_3H_8 = 5.3, CO_2 = 15.0, and H_2S = 3.0. Find the compressibility factor for this gas at 3,000 psia and 300°F. Use the Buxton-Campbell method because the CO_2 content is high.

Solution

Comp.	y_i	T_{ci}	p_{ci}	T_c/p_c	$(T_c/p_c)^{0.5}$	$T_c/p_c^{0.5}$	ω_i
C_1	0.561	343.1	667.8	0.51361	0.71667	13.2727	0.0115
C_2	0.206	549.8	707.8	0.77673	0.88132	20.6645	0.0908
C_3	0.053	665.7	616.3	1.08012	1.03929	26.8145	0.1454
CO_2	0.150	547.6	1070.9	0.51108	0.71490	16.7284	0.2250
H_2S	0.030	672.4	1306.0	0.51475	0.71746	18.6025	0.0949

$\Sigma\ y_i(T_{ci}/p_{ci}) = 0.59749$

$\Sigma\ y_i(T_{ci}/p_{ci})^{0.5} = 0.76744$

$\Sigma\ y_i(T_{ci}/p_{ci}^{0.5}) = 16.1914$

$\Sigma\ y_i\omega_i = 0.06946$, and $\Sigma\ y_i\omega_i$ for all components except CO_2 = 0.03571

So,

$K = 16.1914$

$J = (1/3)(0.59749) + (2/3)[(0.76744)^2] = 0.59181$

$$T_{pc} = K^2/J = (16.1914)^2/0.59181 = 442.98°R$$

$$p_{pc} = T_{pc}/J = 442.98/0.59181 = 748.52 \text{ psia}$$

Using Equation 3-38,

$$\omega_e = [1/(1 - 0.150)](0.03571) = 0.0420$$

From Figure 3-10b,

$$\tau = 4.3$$

So,

$$T'_{pc} = 442.98 - 4.3/1 = 438.68°R$$

$$p'_{pc} = 748.52(438.68/442.98) = 741.25 \text{ psia}$$

Thus,

$$p_{pr} = 3000/741.25 = 4.047, \text{ and } T_{pr} = 759.67/438.68 = 1.732$$

From Figure 3-8, $Z^0 = 0.880$, and from Figure 3-9, $Z^1 = 0.250$

Therefore, $Z = 0.880 + (0.250)(0.06946) = \underline{0.897}$

Some Z-Factor Related Properties

Gas properties that can be derived from the Z-factor are gas density, supercompressibility, gas formation volume factor, and expansion factor.

Gas Density

Using the gas law, the density of a gas, ρ_g, can be calculated as:

$$\rho_g = M/v = pM/ZRT \tag{3-40}$$

where M is the molecular weight of the gas. If p is in psia, T is in °R, and R is in (psia ft^3)/(lbmole °R), then ρ_g is in lbm/ft^3.

Properties of Natural Gases   67

Supercompressibility Factor

In several applications such as gas flow measurement, the factor $1/Z^{0.5}$ appears very frequently. It is called the supercompressibility factor, F_{pv}.

$$F_{pv} = 1/Z^{0.5}, \text{ or } F_{pv}^2 = 1/Z \tag{3-41}$$

Gas Formation Volume Factor

In reservoir engineering applications, one must often relate reservoir volumes to surface volumes. The formation volume factor, B_g, defined as the ratio of the volume occupied by a given mass of gas at reservoir pressure-temperature conditions to the volume occupied at standard (surface) conditions, is generally used. If V denotes the volume at reservoir pressure p and temperature T, and V_{sc} denotes the volume at standard pressure p_{sc} and temperature T_{sc}, then:

$$B_g = V/V_{sc} = (nZRT/p)/(nZ_{sc}RT_{sc}/p_{sc})$$

$$= (p_{sc}ZT)/(pZ_{sc}T_{sc}) \tag{3-42}$$

In oil-field practice, generally the standard conditions are taken to be 14.73 psia ($= p_{sc}$) and 60°F ($= T_{sc}$). At these conditions, Z_{sc} can be assumed to be unity. B_g therefore becomes:

$$B_g = 0.0283 \, ZT/p \text{ ft}^3/\text{scf} \tag{3-43}$$

The expansion factor, E, is simply the reciprocal of the formation volume factor, B_g. Thus, E is given by:

$$E = (pZ_{sc}T_{sc})/(p_{sc}ZT)$$

$$= 35.30 \, p/ZT \text{ scf/ft}^3 \tag{3-44}$$

Example 3-7. For the sweet gas given in Example 3-1, find: density in lbm/ft³, and formation volume factor, at 2,000 psia and 200°F.

Solution

From Example 3-1, the molecular weight M = 17.54

68 Gas Production Engineering

At 2,000 psia and 200°F, from Example 3-2, the compressibility factor $Z = 0.905$

From Equation 3-40,

$$\rho_g = (2,000)(17.54)/[(0.905)(10.732)(459.67 + 200)] = \underline{5.475 \text{ lbm/ft}^3}$$

From Equation 3-43,

$$B_g = (0.0283)(0.905)(459.67 + 200)/2,000 = \underline{0.008448 \text{ ft}^3/\text{scf}}$$

Compressibility of Gases

Compressibility of any substance is a measure of the change in its volume upon changes in pressure. It can also be considered as a measure of the change in density with pressure. Gases exhibit high compressibility, which makes the storage of gas under pressure a common occurrence.

The isothermal compressibility, c, of any substance is defined as:

$$c = -(1/V)(\partial V/\partial p)_T \qquad (3\text{-}45)$$

Using the ideal gas law, $(\partial V/\partial p)_T = -nRT/p^2$, and the compressibility (c_g) for an ideal gas can be expressed as:

$$c_g = -(p/nRT)(-nRT/p^2) = 1/p \qquad (3\text{-}46)$$

Thus, the compressibility of an ideal gas is simply equal to the inverse of the pressure. With increasing pressure, the compressibility decreases. For a real gas:

$$V = nZRT/p \text{ and } (\partial V/\partial p)_T = nRT[(1/p)(\partial Z/\partial p) - Z/p^2]$$

and the compressibility, c_g, is given by:

$$c_g = -(p/nZRT)nRT[(1/p)(\partial Z/\partial p) - Z/p^2]$$

$$= 1/p - (1/Z)(\partial Z/\partial p) \qquad (3\text{-}47)$$

Or, in terms of reduced variables:

$$c_g = \frac{1}{p_{pc}p_{pr}} - \frac{1}{Zp_{pc}}\left[\frac{\partial Z}{\partial p_{pr}}\right]_{T_{pr}}$$

The pseudoreduced compressibility, c_{pr}, is thus given by:

$$c_{pr} = c_g p_{pc} = \frac{1}{p_{pr}} - \frac{1}{Z}\left[\frac{\partial Z}{\partial p_{pr}}\right]_{T_{pr}} \quad (3\text{-}48)$$

Differentiating Equation 3-29 with respect to pseudoreduced pressure, we get:

$$\frac{\partial \rho_r}{\partial p_{pr}} = (0.27/T_{pr})\left[1/Z - (p_{pr}/Z^2)\frac{\partial Z}{\partial p_{pr}}\right]$$

The derivative, $\partial Z/\partial p_{pr}$, in Equation 3-48 can be expressed as follows:

$$\frac{\partial Z}{\partial p_{pr}} = \frac{\partial Z}{\partial \rho_r}\frac{\partial \rho_r}{\partial p_{pr}}$$

$$= \left(\frac{\partial Z}{\partial \rho_r}\right)(0.27/T_{pr})\left[1/Z - (p_{pr}/Z^2)\frac{\partial Z}{\partial p_{pr}}\right]$$

$$= \frac{0.27}{ZT_{pr}}\frac{\partial Z}{\partial \rho_r} - \left\{\frac{0.27\,p_{pr}\partial Z}{Z^2 T_{pr}\partial \rho_r}\right\}\frac{\partial Z}{\partial p_{pr}}$$

Rearranging this equation results in:

$$\left[1 + \frac{0.27 p_{pr}\partial Z}{Z^2 T_{pr}\partial \rho_r}\right]\frac{\partial Z}{\partial p_{pr}} = \frac{0.27}{ZT_{pr}}\frac{\partial Z}{\partial \rho_r}$$

Thus, $\dfrac{\partial Z}{\partial p_{pr}} = \dfrac{0.27}{ZT_{pr}}\left\{\dfrac{(\partial Z/\partial \rho_r)}{1 + (\rho_r/Z)(\partial Z/\partial \rho_r)}\right\}$

Substituting for the derivative, $\partial Z/\partial p_{pr}$, in Equation 3-48, the reduced compressibility can be written as follows:

$$c_{pr} = \frac{1}{p_{pr}} - \frac{0.27}{Z^2 T_{pr}}\left\{\frac{(\partial Z/\partial \rho_r)}{1 + (\rho_r/Z)(\partial Z/\partial \rho_r)}\right\} \quad (3\text{-}49)$$

Using the Z-factor equation derived from the BWR equation of state by Dranchuk et al. (Equation 3-28), Mattar et al. (1975) evaluated the derivative $\partial Z/\partial \rho_r$ as follows:

$$\partial Z/\partial \rho_r = (A_1 + A_2/T_{pr} + A_3/T_{pr}^3) + 2(A_4 + A_5/T_{pr})\rho_r + 5A_5 A_6 \rho_r^4/T_{pr}$$
$$+ (2A_7 \rho_r/T_{pr}^3)(1 + A_8\rho_r^2 - A_8^2\rho_r^4)\exp(-A_8\rho_r^2) \quad (3\text{-}50)$$

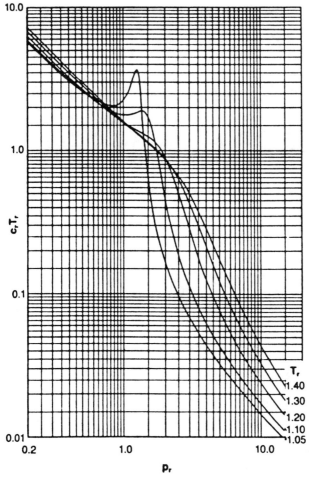

Figure 3-11. $c_r T_r$ as a function of reduced temperature and pressure in the range $1.05 \leqslant T_r \leqslant 1.4$, and $0.2 \leqslant p_r \leqslant 15.0$. (After Mattar et al., 1975; courtesy of the *Journal of Canadian Petroleum Technology*.)

For hand calculations, Mattar et al. (1975) also presented this correlation in a graphical form as shown in Figures 3-11 and 3-12.

Example 3-8. Find the compressibility for the sweet gas given in Example 3-1 at 2,000 psia and 200°F.

Solution

From Example 3-1,

$p_{pc} = 664.47$ psia

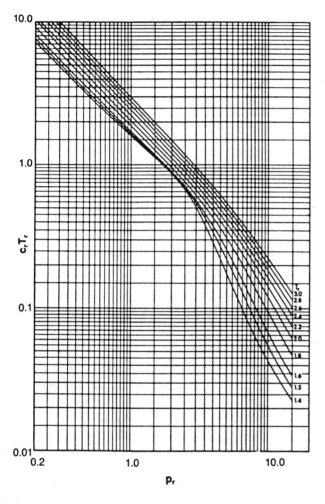

Figure 3-12. $c_r T_r$ as a function of reduced temperature and pressure in the range $1.4 \leqslant T_r \leqslant 3.0$, and $0.2 \leqslant p_r \leqslant 15.0$. (After Mattar et al., 1975; courtesy of the *Journal of Canadian Petroleum Technology*.)

From Example 3-2, at 2,000 psia and 200°F,

$p_{pr} = 3.010$

$T_{pr} = 1.848$

From Figure 3-12,

$c_{pr} T_{pr} = 0.620$

Thus,

$c_{pr} = 0.620/1.848 = 0.3355$

$c_g = c_{pr}/p_{pc} = 0.3355/664.47 = \underline{0.000505 \text{ psi}^{-1}}$

Viscosity of Gases

The viscosity of a fluid, a measure of its resistance to flow, is defined as the ratio of the shear force per unit area to the local velocity gradient:

$$\mu = (F/A)/(dv/dL)$$

The viscosity, μ, as defined is called *dynamic viscosity*. In addition, the ratio of the dynamic viscosity of a fluid to its density, known as *kinematic viscosity* (ν), is also used in many flow problems:

$$\nu = \mu/\rho$$

Viscosity is usually measured in centipoises (cp), with 1 cp equal to 0.01 g/(cm sec), or 6.72×10^{-4} lbm/(ft sec). The most accurate method of determining viscosity is obviously to measure it for a given fluid under the desired conditions. This, however, is not generally possible. Some common methods for predicting gas viscosity are described briefly here.

Carr et al. Correlation for Natural Gases

The Carr et al. (1954) correlation requires only the gas gravity (or the molecular weight) for determining viscosity. It is the most widely used method for determining the viscosity of natural gases. The correlations were presented in graphical form, as shown in Figures 3-13, 3-14, and 3-15. Figure 3-13 is used first to calculate the viscosity at one atmosphere pressure and any given temperature. Corrections for non-hydrocarbon components N_2, CO_2, and H_2S are provided. The effect of these components is to increase the viscosity. Then, Figure 3-14 or 3-15 can be used to correct for pressure. Figures 3-14 and 3-15 use the corresponding states principle to present $\mu = f(p_r, T_r)$.

Example 3-9. Find the viscosity for the gas given in Example 3-1 at 2,000 psia and 200°F.

Solution

From Example 3-1,

M = 17.54, and the gravity γ_g = 0.6055

Properties of Natural Gases 73

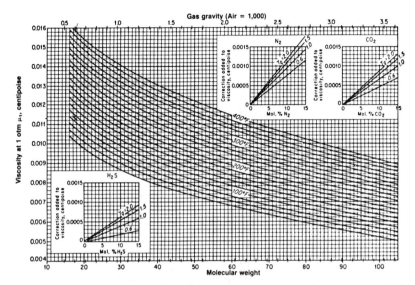

Figure 3-13. Viscosity of paraffinic hydrocarbon gases at 1.0 atmosphere. (After Carr et al., 1954; courtesy of SPE of AIME.)

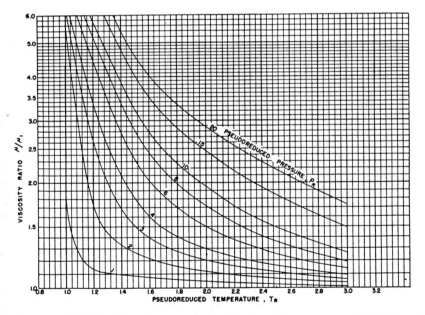

Figure 3-14. Viscosity ratio versus pseudoreduced temperature. (After Carr et al., 1954; courtesy of SPE of AIME.)

74 Gas Production Engineering

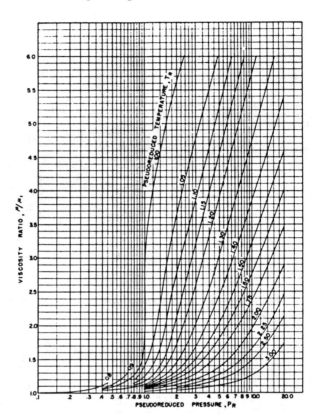

Figure 3-15. Viscosity ratio versus pseudoreduced pressure. (After Carr et al., 1954; courtesy of SPE of AIME.)

From Example 3-2, at 2,000 psia and 200°F,

$p_{pr} = 3.010$

$T_{pr} = 1.848$

From Figure 3-13,

$\mu_{1g} = 0.0128$ cp

The gas has 1.4% N_2, for which the correction factor (to be added) from Figure 3-13 is 0.00013 cp. Therefore, the corrected viscosity at 1 atmosphere,

$\mu_{1g} = 0.0128 + 0.00013 = 0.01293$ cp

Properties of Natural Gases 75

From Figure 3-14, the pressure correction,

$\mu_g/\mu_{1g} = 1.25$

So, the gas viscosity at 2,000 psia and 200°F,

$\mu_g = (1.25)(0.01293) = \underline{0.0162 \text{ cp}}$

Viscosity from Single-Component Data

If the analysis of a gas is known, it is possible to calculate the viscosity of the gas mixture from the component viscosities. First, the viscosity is determined at one atmosphere (or any "low" pressure) and the given temperature using the Herning and Zipperer (1936) mixing rule:

$$\mu_{1g} = \frac{\sum_{i=1}^{n} \mu_{1gi} y_i M_i^{0.5}}{\sum_{i=1}^{n} y_i M_i^{0.5}} \qquad (3\text{-}51)$$

where y_i = mole fraction of component i in the gas mixture
μ_{1gi} = pure component viscosity at 1 atmosphere pressure and the temperature of interest

The viscosity can be corrected to the desired pressure by using Carr et al.'s reduced pressure and temperature correlations of Figures 3-14 and 3-15, or any of the several other such correlations available. Lohrenz et al. (1964) found the chart by Baron et al. (1959) useful since it provides a good range of reduced pressure and temperature. The differences among these charts are, however very minor (Lohrenz et al., 1964).

Figure 3-16 shows the pure component viscosities as a function of temperature for some common natural gas constituents. Single component viscosities at 1 atmosphere pressure, μ_{1gi}, may also be calculated using the following relationship developed by Stiel and Thodos (1961):

$\mu_{1gi} = 34 \times 10^{-5} T_{ri}^{0.94}/\xi_i$, for $T_{ri} < 1.5$ \hfill (3-52a)

$\mu_{1gi} = 17.78 \times 10^{-5}(4.58 T_{ri} - 1.67)^{5/8}/\xi_i$, for $T_{ri} > 1.5$ \hfill (3-52b)

where $\xi_i = \dfrac{T_{ci}^{1/6}}{M_i^{1/2} p_{ci}^{2/3}}$ \hfill (3-53)

76 Gas Production Engineering

T_{ri} = reduced temperature for component i
T_{ci} = critical temperature in K
p_{ci} = critical pressure in atmospheres
M_i = molecular weight

Note that units of K for temperature and atmospheres for pressure must be used in Equation 3-53. Equations 3-52a and b are valid for pressures in the range of 0.2 to 5 atmospheres. The average error was reported by Stiel and Thodos (1961) to be 1.83% and 1.62%, respectively, for Equations 3-52a and b.

Lohrenz et al. (1964) report that this method gives an average error of 4.03% for the gas mixtures tested by them.

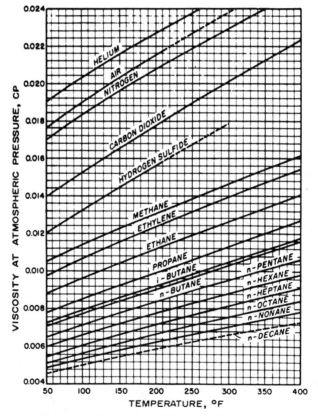

Figure 3-16. Viscosity of some natural gas constituents at low pressure. (After Carr et al., 1954; courtesy of SPE of AIME.)

Example 3-10. Find the viscosity for a gas with composition, in mole %, of $C_1 = 90.5$, $C_2 = 7.2$, and $C_3 = 2.3$, at 3,000 psia and 540°R.

Solution

Comp.	y_i	M_i	p_{ci}	T_{ci}	$M_i^{0.5}$	μ_{1gi}
C_1	0.905	16.043	667.8	343.1	4.0054	0.0110
C_2	0.072	30.070	707.8	549.8	5.4836	0.0092
C_3	0.023	44.097	616.3	665.7	6.6353	0.0082

The μ_{1gi} values are from Figure 3-16.

$M = \Sigma\ y_i M_i = 17.697$

$p_{pc} = \Sigma\ y_i p_{ci} = 669.50$ psia

$T_{pc} = \Sigma\ y_i T_{ci} = 365.31°R$

So,

$p_{pr} = 3,000/669.50 = 4.481$

$T_{pr} = 540/365.31 = 1.478$

$\Sigma\ y_i M_i^{0.5} = 4.1723$

$\Sigma\ \mu_{1gi} y_i M_i^{0.5} = 0.04476$ cp

Therefore,

$\mu_{1g} = 0.04476/4.1723 = 0.01073$

From Figure 3-15, the pressure correction,

$\mu_g/\mu_{1g} = 1.95$

Thus,

$\mu_g = (1.95)(0.01073) = \underline{0.0209\ cp}$

Lee et al. Correlation for Natural Gases

Lee et al. (1966) provide an analytic expression for viscosity that can be used for programming purposes:

$$\mu_g = K \exp(X\rho_g^y) \qquad (3\text{-}54)$$

where $K = \dfrac{10^{-4}(9.4 + 0.02M)T^{1.5}}{209 + 19M + T}$

$X = 3.5 + 986/T + 0.01M$

$y = 2.4 - 0.2X$

and μ_g is in cp, ρ_g is in g/cm³, T is in °R. Equation 3-54 reproduced experimental data with a maximum error of 8.99% (Lee et al., 1966). The problem with this method is that it does not correct for impurities such as N_2, CO_2, and H_2S.

Example 3-11. Find the viscosity for the gas given in Example 3-10 at 3,000 psia and 540°R using the Lee et al. method.

Solution

From Example 3-10,

M = 17.697, p_{pr} = 4.481

T_{pr} = 1.478

From Figure 3-2, the compressibility factor,

Z = 0.78

From Equation 3-40, the gas density,

ρ_g = (3,000)(17.697)/(0.78)(10.732)(540) = 11.745 lbm/ft³
    = (11.745)(1.601846)10⁻² = 0.18814 g/cm³

Now,

K = 10⁻⁴[9.4 + (0.02)(17.697)](540^{1.5})/
    [209 + (19)(17.697) + 540] = 0.011278

$X = 3.5 + 986/540 + (0.01)(17.697) = 5.50290$

$y = 2.4 - (0.2)(5.50290) = 1.29942$

Using Equation 3-54, the gas viscosity at 3,000 psia and 540°R is:

$\mu_g = (0.011278)\exp[(5.50290)(0.18814^{1.29942})] = \underline{0.0211 \text{ cp}}$

Specific Heat For Hydrocarbon Gases

One of the basic thermodynamic quantities is specific heat, defined as the amount of heat required to raise the temperature of a unit mass of a substance through unity. It is an intensive property of a substance. It can be measured at constant pressure (c_p), or at constant volume (c_v), resulting in two distinct specific heat values. In terms of basic thermodynamic quantities, molal enthalpy (h) and molal internal energy (u), the specific heats, c_p and c_v can be written as:

$$c_p = (\partial h/\partial T)_p \tag{3-55}$$

and

$$c_v = (\partial u/\partial T)_v \tag{3-56}$$

where h is the molal enthalpy (Btu/lbmole), u is the molal internal energy (Btu/lbmole), c_p is the molal specific heat at constant pressure (Btu/lbmole-°R), and c_v is the molal specific heat at constant volume (Btu/lbmole-°R). Using Maxwell's relationships, it can be shown that:

$$c_p - c_v = -T\frac{(\partial p/\partial T)_v^2}{(\partial p/\partial v)_T} \tag{3-57}$$

For an ideal gas, where $pv = RT$, c_p and c_v are a function of temperature only. Furthermore, the right side of Equation 3-57 becomes equal to the gas constant R. Therefore, for an ideal gas:

$$c_p - c_v = R \tag{3-58}$$

Note that the units for c_p and c_v are Btu/(lbmole °R) in Equation 3-58. So R must be expressed in the same units:

$R = 10.732 \text{ (lb}_f \text{ ft}^3\text{)}/(\text{in.}^2 \text{ lbmole °R}) \times 144 \text{ in.}^2/\text{ft}^2 \times (1/778.2) \text{ Btu}/(\text{lb}_f \text{ ft})$
$= 1.986 \text{ Btu}/(\text{lbmole °R})$

80 Gas Production Engineering

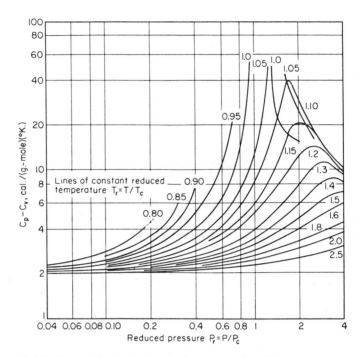

Figure 3-17. Generalized plot for the heat capacity difference, $c_p - c_v$, versus p_{pr} and T_{pr} for real gases. (After Edmister, 1948; reprinted from *Chemical Engineers' Handbook*, 1973; courtesy of McGraw-Hill Book Co.)

For a real gas, c_p and c_v are functions of both pressure as well as temperature. A real gas equation of state can be used to determine the right side of Equation 3-57, and thus the difference between c_p and c_v. Knowing either one of c_p or c_v, the other can then be determined. Figure 3-17 shows a generalized plot for the heat capacity difference, $c_p - c_v$, versus p_r and T_r.

c_p Determination

The low-pressure isobaric heat capacity, c_p^o, can be determined primarily by two methods: using the gas gravity, if the gas composition is not known; and using a weighted molal average (Kay's rule), if the gas composition is known.

Hankinson et al. (1969) give the following relationship for calculating c_p^o in Btu/(lbmole °F), from the gas gravity, γ_g, at any temperature T (°F):

$$c_p^o = A + BT + C\gamma_g + D\gamma_g^2 + E(T\gamma_g) + FT^2 \qquad (3\text{-}59)$$

where for a temperature range of 0 to 200°F:

A = 4.6435, B = -0.0079997, C = 5.8425, D = 1.1533, E = 0.020603, F = $9.849(10^{-6})$

and for a temperature range of 0 to 600°F:

A = 3.7771, B = − 0.0011050, C = 7.5281, D = 0.65621,
E = 0.014609, F = 0.0

Thomas et al. (1970) report average errors of 1.01% and 1.37% for the temperature range 0–200°F and 0–600°F, respectively, with Equation 3-59.

The effective mixture specific heat at low pressure, c_p^o, can be determined more precisely using Kay's rule:

$$c_p^o = \Sigma_{i=1}^n y_i c_{pi}^o \qquad (3\text{-}60)$$

where the c_{pi}^o for some common constituents of natural gas can be obtained from Table 3-3, or the data by Touloukian et al. (1970) shown in Figure 3-18.

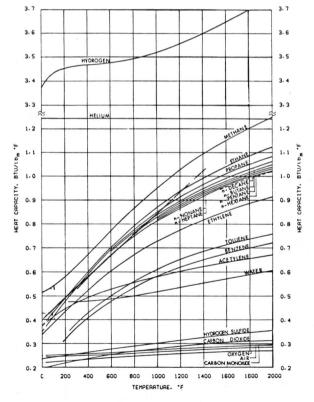

Figure 3-18. Heat capacity of gases at 1 atmosphere, c_p^o, as a function of temperature. (After Touloukian et al., 1970; courtesy of IFI/Plenum.)

Table 3-3*
Molal Heat Capacity (Ideal-Gas State), Btu/(lb mol-°R**)

Gas	Chemical formula	Mol wt	0°F	50°F	60°F	100°F	150°F	200°F	250°F	300°F
Methane	CH_4	16.043	8.23	8.42	8.46	8.65	8.95	9.28	9.64	10.01
Ethyne (Acetylene)	C_2H_2	26.038	9.68	10.22	10.33	10.71	11.15	11.55	11.90	12.22
Ethene (Ethylene)	C_2H_4	28.054	9.33	10.02	10.16	10.72	11.41	12.09	12.76	13.41
Ethane	C_2H_6	30.070	11.44	12.17	12.32	12.95	13.78	14.63	15.49	16.34
Propene (Propylene)	C_3H_6	42.081	13.63	14.69	14.90	15.75	16.80	17.85	18.88	19.89
Propane	C_3H_8	44.097	15.65	16.88	17.13	18.17	19.52	20.89	22.25	23.56
1-Butene (Butylene)	C_4H_8	56.108	17.96	19.59	19.91	21.18	22.74	24.26	25.73	27.16
cis-2-Butene	C_4H_8	56.108	16.54	18.04	18.34	19.54	21.04	22.53	24.01	25.47
trans-2-Butene	C_4H_8	56.108	18.84	20.23	20.50	21.61	23.00	24.37	25.73	27.07
iso-Butane	C_4H_{10}	58.124	20.40	22.15	22.51	23.95	25.77	27.59	29.39	31.11
n-Butane	C_4H_{10}	58.124	20.80	22.38	22.72	24.08	25.81	27.55	29.23	30.90
iso-Pentane	C_5H_{12}	72.151	24.94	27.17	27.61	29.42	31.66	33.87	36.03	38.14
n-Pentane	C_5H_{12}	72.151	25.64	27.61	28.02	29.71	31.86	33.99	36.08	38.13
Benzene	C_6H_6	78.114	16.41	18.41	18.78	20.46	22.45	24.46	26.34	28.15
n-Hexane	C_6H_{14}	86.178	30.17	32.78	33.30	35.37	37.93	40.45	42.94	45.36
n-Heptane	C_7H_{16}	100.205	34.96	38.00	38.61	41.01	44.00	46.94	49.81	52.61
Ammonia	NH_3	17.031	8.52	8.52	8.52	8.52	8.52	8.53	8.53	8.53
Air		28.964	6.94	6.95	6.95	6.96	6.97	6.99	7.01	7.03
Water	H_2O	18.015	7.98	8.00	8.01	8.03	8.07	8.12	8.17	8.23
Oxygen	O_2	31.999	6.97	6.99	7.00	7.03	7.07	7.12	7.17	7.23
Nitrogen	N_2	28.013	6.95	6.95	6.95	6.96	6.96	6.97	6.98	7.00
Hydrogen	H_2	2.016	6.78	6.86	6.87	6.91	6.94	6.95	6.97	6.98
Hydrogen sulfide	H_2S	34.076	8.00	8.09	8.11	8.18	8.27	8.36	8.46	8.55
Carbon monoxide	CO	28.010	6.95	6.96	6.96	6.96	6.97	6.99	7.01	7.03
Carbon dioxide	CO_2	44.010	8.38	8.70	8.76	9.00	9.29	9.56	9.81	10.05

* Courtesy of Gas Processors Suppliers Association.
** Data source: Selected values of properties of hydrocarbons, API Research Project 44. Exceptions: "Air," Keenan and Keyes, *Thermodynamic Properties of Air*, Wiley, 3rd Printing 1947. "Ammonia," Edw. R. Grabl, "Thermodynamic Properties of Ammonia at High Temperatures and Pressures," *Petr. Processing*, April 1953. "Hydrogen Sulfide," L. B. West, *Chem. Eng. Progress*, 44, 287, 1948.

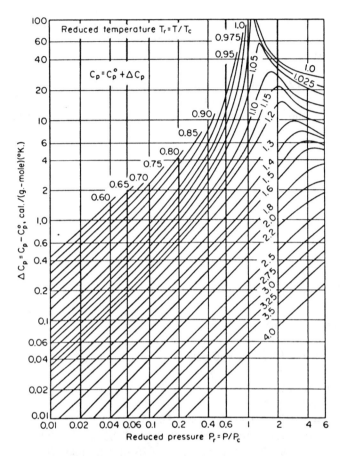

Figure 3-19. Isothermal pressure correction, $c_p - c_p^o$, to the molar heat capacity of gases. (After Edmister, 1950; reprinted from *Chemical Engineers' Handbook*, 1973; courtesy of McGraw-Hill Book Co.)

The isobaric specific heat, c_p, at the pressure and temperature of interest can subsequently be determined by correcting c_p^o for pressure. To correct for pressure, Edmister's (1950) correlation for $c_p - c_p^o$ versus T_r, p_r shown in Figure 3-19 can be used. Note that the units for c_p in Figure 3-19 are calories/(gmmole K), which is exactly equivalent to Btu/(lbmole °F).

c_v Determination

The specific heat at constant volume, c_v, is usually determined indirectly. First, c_p is calculated using any of the previous methods. Then, Figure 3-17 is used to determine $c_p - c_v$, and finally c_v is calculated. Thomas et al. (1970) suggest using the BWR equation of state to calculate c_v, but it is a

84 Gas Production Engineering

cumbersome method and is subject to limitations of non-hydrocarbon gas content of less than 7%, a reduced temperature greater than 1.1, and a reduced pressure between 0.4 and 15.

Example 3-12. Find c_p^o for the gas given in Example 3-10 at 540°R (80.33°F) using (1) gas gravity method of Hankinson et al., and (2) the gas analysis. Use the value for c_p^o from the gas analysis method to find c_p and c_v at 3,000 psia and 540°R for this gas.

Solution

From Example 3-10,

$$M = 17.697$$

$$p_{pr} = 4.481$$

$$T_{pr} = 1.478$$

1. The gas gravity,

$$\gamma_g = 17.697/28.97 = 0.6109$$

From Equation 3-59,

$$\begin{aligned}c_p^o &= 4.6435 + (-0.0079997)(80.33) + (5.8425)(0.6109) \\ &+ (1.1533)(0.6109^2) + (0.020603)(80.33)(0.6109) \\ &+ (9.849)10^{-6}(80.33^2) = 9.075 \text{ Btu/(lbmole °R)}\end{aligned}$$

2.

Comp.	y_i	M_i	c_{pi}^o Btu/(lbm °F)	c_{pi}^o Btu/(lbmole °F)
C_1	0.905	16.043	0.530	8.503
C_2	0.072	30.070	0.425	12.780
C_3	0.023	44.097	0.425	18.741

The c_{pi}^o values in Btu/(lbm °F), obtained from Figure 3-18, are converted into Btu/(lbmole °F) by multiplying by the molecular weight, M_i, in lbm/lbmole. Thus,

$$c_p^o = \Sigma\ y_i c_{pi}^o = 9.046 \text{ Btu/(lbmole °F)} = 9.046 \text{ Btu/(lbmole °R)}$$

To find c_p and c_v:

Properties of Natural Gases 85

From Figure 3-19, at p_{pr} = 4.481, and T_{pr} = 1.478, the specific heat correction,

$(c_p - c_p^o)$ = 6.5

Therefore,

c_p = 9.046 + 6.5 = 15.546 Btu/(lbmole °R)

From Figure 3-17,

$(c_p - c_v)$ = 7.0

Therefore,

c_v = 15.546 − 7.0 = 9.546 Btu/(lbmole °R).

Questions and Problems

1. A gas pipeline consists of a 10-mile long 12-in. ID pipe section in series with a 20-mile long section of 8-in. ID. How many Mscf of gas must be removed to blowdown this gas pipeline from an average pressure of 1,500 psia to an average pressure of 300 psia? Assume that at the prevailing temperature of 85°F the gas has a compressibility factor of 0.95 at 1,500 psia, and 0.85 at 300 psia.
2. We have a sweet gas of known composition: N_2 = 1%, C_1 = 89%, C_2 = 5%, and C_{3+} = 5%. Assume the C_{3+} fraction to be equivalent to n-C_5. Find: (i) the gas gravity, and (ii) the pseudo-critical pressure and temperature for the gas.
3. For the gas in Problem 2, find the gas compressibility factor at 1,200 psia and 95°F using all the applicable methods given in Chapter 3.
4. A sour gas is known to have the following composition: N_2 = 8.5%, H_2S = 5.4%, CO_2 = 0.5%, C_1 = 77.6%, C_2 = 5.8%, C_3 = 2.0%, n-C_4 = 0.1%, and i-C_4 = 0.1%. This gas is being sold at a contract pressure of 1,000 psia at 120°F. What is the error introduced in the calculation of the sales volume by assuming Z = 0.85? Use the best estimates of Z-factor from the available correlations.
5. For the gas given in Problem 2, find the viscosity and compressibility at 200 psia and 80°F.
6. For the gas given in Problem 4, find the viscosity and compressibility at 800 psia and 65°F.

References

Baron, J. D., Roof, J. G., and Wells, F. W., 1959. "Viscosity of Nitrogen, Methane, Ethane, and Propane at Elevated Temperature and Pressure," *J. Chem. Eng. Data*, 4(3, July), 283–288.

Benedict, M., Webb, G. B., and Rubin, L. C., 1940. "An Empirical Equation for Thermodynamic Properties of Light Hydrocarbons and Their Mixtures," *J. Chem. Phys.*, 8(4, Apr.), 334–345.

Brown, G. G., Katz, D. L., Oberfell, G. B., and Alden, R. C., 1948. *Natural Gasoline and Volatile Hydrocarbons*. NGAA, Tulsa, OK.

Buxton, T. S. and Campbell, J. M., 1967. "Compressibility Factors for Lean Natural Gas-Carbon Dioxide Mixtures at High Pressure," *Soc. Pet. Eng. J.*, 7(1, Mar.), 80–86.

Campbell, J. M., 1984. *Gas Conditioning and Processing, Vol. 1*. Campbell Petroleum Series, Norman, Oklahoma, 326pp.

Carr, N. L., Kobayashi, R., and Burrows, D. B., 1954. "Viscosity of Hydrocarbon Gases Under Pressure," *Trans.*, AIME, 201, 264–272.

Chemical Engineers' Handbook, 1973. R. H. Perry and C. H. Chilton (eds.), McGraw-Hill Book Co., New York, 5th edition.

Coats, K. H., 1985. "Simulation of Gas Condensate Reservoir Performance," *J. Pet. Tech.*, 37(10, Oct.), 1870–1886.

Dranchuk, P. M. and Abou-Kassem, J. H., 1975. "Calculation of Z-Factors for Natural Gases Using Equations of State," *J. Cdn. Pet. Tech.*, 14(3, July–Sept.), 34–36.

Dranchuk, P. M., Purvis, R. A., and Robinson, D. B., 1974. "Computer Calculation of Natural Gas Compressibility Factors Using the Standing and Katz Correlation," *Inst. of Petr. Tech. Series*, No. IP 74-008.

Edmister, W. C., 1948. *Petrol. Refiner* (Nov.), p. 613. Cited reference on page 3-238 in: *Chemical Engineers' Handbook*, 5th ed., 1973, edited by R. H. Perry and C. H. Chilton, McGraw-Hill Book Co., New York.

Edmister, W. C., 1950. *Petrol. Engr.* (Dec.), p. C-16. Cited reference on page 3-237 in: *Chemical Engineers' Handbook*, 5th ed., 1973, edited by R. H. Perry and C. H. Chilton, McGraw-Hill Book Co., New York.

Edmister, W. C. and Lee, B. I., 1984. *Applied Hydrocarbon Thermodynamics, Vol. 1*. (2nd ed.) Gulf Publ. Co., Houston, Texas, 233pp.

Fussell, D. D. and Yanosik, J. L., 1978. "An Iterative Sequence for Phase Equilibria Calculations Incorporating the Redlich-Kwong Equation of State," *Soc. Pet. Eng. J.*, 18(3, June), 173–182.

Fussell, L. T., 1979. "A Technique for Calculating Multiphase Equilibria," *Soc. Pet. Eng. J.*, 19(4, Aug.), 203–210.

Gas Processors Suppliers Association, 1981. *Engineering Data Book*, 9th ed. (5th revision), GPSA, Tulsa, Oklahoma.

Gopal, V. N., 1977. "Gas Z-Factor Equations Developed for Computer," *O. and Gas J.*, 75(32, Aug. 8), 58–60.

Hankinson, R. W., Thomas, L. K., and Phillips, K. A., 1969. "Predict Natural Gas Properties," *Hydr. Proc.*, 48(4, Apr.), 106–108.

Herning, F. and Zipperer, L., 1936. "Calculation of the Viscosity of Technical Gas Mixtures from the Viscosity of Individual Gases," *Gas u. Wasserfach* 79(49), 69.

Joffe, J., Schroeder, G. M., and Zudkevitch, D., 1970. "Vapor-liquid Equilibria with the Redlich-Kwong Equation of State," *AIChE J.*, 16(3, May), 496–498.

Lee, A. L., Gonzalez, M. H., and Eakin, B. E., 1966. "The Viscosity of Natural Gases," *J. Pet. Tech.*, 18(8, Aug.), 997–1000.

Lohrenz, J., Bray, B. G., and Clark, C. R., 1964. "Calculating Viscosities of Reservoir Fluids from Their Compositions," *J. Pet. Tech.*, 16(10, Oct.), 1171–1176.

Martin, J. J., 1979. "Cubic Equations of State—Which?" *Ind. and Eng. Chem. Fund.*, 18(2, May), 81–97.

Mattar, L., Brar, G. S, and Aziz, K., 1975. "Compressibility of Natural Gases," *J. Cdn. Pet. Tech.*, 14(4, Oct.–Dec.), 77–80.

McLeod, W. and Campbell, J. M., 1969. "Prediction of the Critical Temperature and Pressure, and Density of Natural Gas," *Proc. 48th Ann. Conv. NGPA* (Nov.), Nat. Gas Process. Assn., Tulsa, Oklahoma.

Peng, D.-Y. and Robinson, D. B., 1976. "A New Two-Constant Equation of State," *Ind. and Eng. Chem. Fund.* 15(1, Feb.), 59–64.

Peng, D.-Y. and Robinson, D. B., 1976. "Two- and Three-Phase Equilibria Calculations for Systems Containing Water," *Cdn. J. Chem. Eng.*, 54 (6, Dec.), 595–599.

Pitzer, K. S., Lippman, D. Z., Curl, R. F., Jr., Huggins, C. M., and Peterson, D. E., 1955. "The Volumetric and Thermodynamic Properties of Fluids. II—Compressibility Factor, Vapor Pressure, and Entropy of Vaporization," *J. Am. Chem. Soc.*, 77(13, July 5), 3433–3440.

Redlich, O. and Kwong, J. N. S., 1949. "On the Thermodynamics of Solutions. V—An Equation of State. Fugacities of Gaseous Solutions," *Chem. Reviews* 44, 233–244.

Reid, R. C., Prausnitz, J. M., and Sherwood, T. K., 1977. *The Properties of Gases and Liquids.* (3rd ed.), McGraw-Hill Book Co., Inc., New York, 688pp.

Satter, A. and Campbell, J. M., 1963. "Non-Ideal Behavior of Gases and Their Mixtures," *Soc. Pet. Eng. J.*, 3(4, Dec.), 333–347.

Soave, G., 1972. "Equilibrium Constants from a Modified Redlich-Kwong Equation of State," *Chem. Eng. Sci.* 27(6, June), 1197–1203.

Standing, M. B. and Katz, D. L., 1942. "Density of Natural Gases," *Trans. AIME*, 146, 140–149.

Stewart, W. E., Burkhardt, S. F., and Voo, D., 1959. "Prediction of Pseudocritical Parameters for Mixtures," Paper presented at the AIChE Meet., Kansas City, MO.

Stiel, L. I. and Thodos, G., 1961. "The Viscosity of Nonpolar Gases at Normal Pressures," *AIChE J.*, 7(4, Dec.), 611–615.

Takacs, G., 1976. "Comparisons Made for Computer Z-Factor Calculations," *O. and Gas J.*, 74(51, Dec. 20), 64–66.

Thomas, L. K., Hankinson, R. W., and Phillips, K. A., 1970. "Determination of Acoustic Velocities for Natural Gas," *J. Pet. Tech.*, 22(7, July), 889–895.

Touloukian, Y. S., Kirby, R. K., Taylor, R. E., and Lee, T. Y. R., 1970. "Specific Heat—Nonmetallic Liquids and Gases," *Thermophysical Properties of Matter*, IFI/Plenum, New York City, 6.

Wichert, E. and Aziz, K., 1972. "Calculation of Z's for Sour Gases," *Hydr. Proc.*, 51(5, May), 119–122.

Wilson, G. M., 1969. "A Modified Redlich-Kwong Equation of State, Application to General Physical Data Calculations," Paper no. 15C presented at the AIChE 65th National Meet., Cleveland, Ohio, May 4–7.

Yarborough, L. and Hall, K. R., 1974. "How to Solve Equation of State for Z-Factors," *O. and Gas J.*, 72(7, Feb. 18), 86–88.

Zudkevitch, D. and Joffe, J., 1970. "Correlation and Prediction of Vapor-Liquid Equilibria with the Redlich-Kwong Equation of State," *AIChE J.*, 16(1, Jan.), 112–119.

4
Gas and Liquid Separation

Introduction

Only rarely does a reservoir yield almost pure natural gas. Typically, a produced hydrocarbon stream is a complex mixture of several hydrocarbons, intimately mixed with water, in the liquid and gaseous states. Often, solids and other contaminants are also present. The produced stream may be unstable, with components undergoing rapid phase transitions as the stream is produced from a several hundred feet deep reservoir with a high temperature and pressure to surface conditions. It is important to remove any solids and contaminants, and to separate the produced stream into water, oil, and gas, which are handled and transported separately. Gas and liquid separation operations involve the separation and stabilization of these phases into saleable products. Generally, intermediate hydrocarbons in the liquid state fetch a higher price; therefore, it is desirable to maximize liquid recovery.

Field processing of natural gas includes:

1. Gas and liquid separation operations to remove the free liquids—crude oil, hydrocarbon condensate, and water, and the entrained solids.
2. Recovery of condensable hydrocarbon vapors. Stage separation, or low temperature separation techniques are used.
3. Further cleaning of the gas and oil streams after separation.
4. Gas dehydration processing to remove from the gas condensable water vapor that may lead to the formation of hydrates under certain conditions (see Chapter 5).
5. Removal of contaminants or otherwise undesirable components, such as hydrogen sulfide and other corrosive sulfur compounds, and carbon dioxide (discussed in Chapter 6).

This chapter describes gas-liquid separation and gas cleaning techniques.

Separation Equipment

Separators operate basically upon the principle of pressure reduction to achieve separation of gas and liquid from an inlet stream. Further refinement of the gas and liquid streams is induced by allowing the liquid to "stand" for a period of time, so that any dissolved gas in the liquid can escape by the formation of small gas bubbles that rise to the liquid surface; and removing the entrained liquid mist from the gas by gravity settling, impingement, centrifugal action, and other means. Turbulent flow allows gas bubbles to escape more rapidly than laminar flow, and many separators therefore have sections where turbulence is induced for this purpose. On the other hand, for the removal of liquid droplets from the gas by gravity settling, turbulence is quite detrimental to removal efficiency. Thus, the design of a separator comprises different modules assembled to achieve different functions in a single vessel. Equilibrium is attained in the piping and equipment just upstream of the separator, the separator itself serving only as a "wide" spot in the line to refine the vapor and liquid streams resulting from this basic separation.

To efficiently perform its separation functions, a well designed separator must (Campbell, 1984; Beggs, 1984):

1. Control and dissipate the energy of the wellstream as it enters the separator, and provide low enough gas and liquid velocities for proper gravity segregation and vapor-liquid equilibrium. For this purpose, a tangential inlet to impart centrifugal motion to the entering fluids is generally used.
2. Remove the bulk of the liquid from the gas in the primary separation section. It is desirable to quickly achieve good separation at this stage.
3. Have a large settling section, of sufficient volume to refine the primary separation by removing any entrained liquid from the gas, and handle any slugs of liquid (usually known as "liquid surges").
4. Minimize turbulence in the gas section of the separator to ensure proper settling.
5. Have a mist extractor (or eliminator) near the gas outlet to capture and coalesce the smaller liquid particles that could not be removed by gravity settling.
6. Control the accumulation of froths and foams in the vessel.
7. Prevent re-entrainment of the separated gas and liquid.
8. Have proper control devices for controlling the back-pressure and the liquid level in the separator.
9. Provide reliable equipment for ensuring safe and efficient operations. This includes pressure gauges, thermometers, devices for indicating

the liquid level, safety relief valves to prevent blowup in case the gas or liquid outlets are plugged, and gas and liquid discharge ("dump") valves.

Types of Separators

Separators can be categorized into three basic types: *vertical, horizontal* (single-tube or double-tube), and *spherical*. Selection of a particular type depends upon the application, and the economics. Advantages and disadvantages for these separator types are cited from Campbell (1955).

Vertical Separators

The wellstream enters the vertical separator (Figure 4-1) tangentially through an inlet diverter that causes an efficient primary separation by three simultaneous actions on the stream: *gravity settling, centrifugation,* and *impingement* of the inlet fluids against the separator shell in a thin film. The gas from the primary separation section flows upwards, while the liquid falls downward into the liquid accumulation section. A conical baffle is provided as a separation between the liquid accumulation section and the primary separation section to ensure an undisturbed liquid surface for proper liquid level control and release of any dissolved gas. The smaller liquid droplets that are carried along by the upwards rising gas stream are removed in the centrifugal baffles near the top. Finally, a mist extractor at the gas outlet removes any entrained liquid droplets from the gas in the micron size range. The liquid particles coalesce and accumulate, until they become sufficiently heavy to fall into the liquid accumulation section.

Advantages

A vertical separator can handle relatively large liquid slugs without carryover into the gas outlet. It thus provides better surge control, and is often used on low to intermediate gas-oil ratio (GOR) wells and wherever else large liquid slugs are expected. Vertical vessels can handle more sand. A false cone bottom can be easily fitted to handle sand production. Liquid level control is not as critical in a vertical separator. The tendency of the liquid to revaporize is also minimized, because less surface area is available to the liquid for evaporation. It occupies less floor space, a particularly important advantage for operations on an offshore platform where floor area is at a premium.

92 Gas Production Engineering

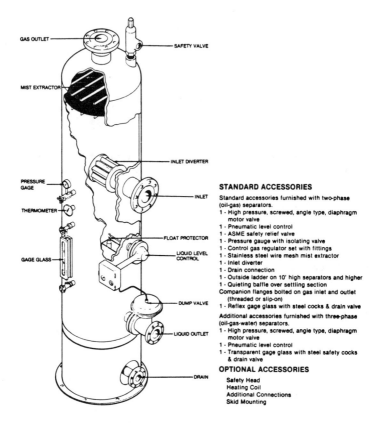

Figure 4-1. A conventional vertical separator. (Courtesy of HTI-Superior, a Berry Industries Company.)

Disadvantages

Vertical vessels are more expensive to fabricate, and also more expensive to transport to location. A vertical separator for the same capacity is usually larger than a horizontal separator, since the upwards flowing gas in the vertical separator opposes the falling droplets of liquid.

Horizontal Separators

These separators may be of a single-tube (Figure 4-2) or a double-tube (Figure 4-3) design. In the single-tube horizontal separator, the wellstream upon entering through the inlet, strikes an angle baffle and then the separator shell, resulting in an efficient primary separation similar to the vertical

HORIZONTAL LOW PRESSURE SEPARATORS

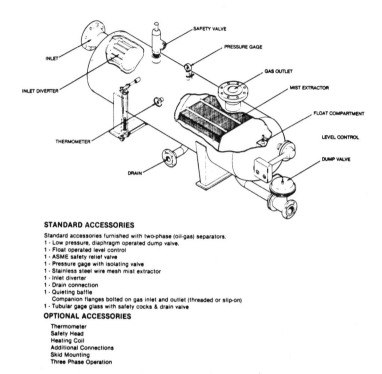

STANDARD ACCESSORIES

Standard accessories furnished with two-phase (oil-gas) separators.
1 - Low pressure, diaphragm operated dump valve.
1 - Float operated level control
1 - ASME safety relief valve
1 - Pressure gage with isolating valve
1 - Stainless steel wire mesh mist extractor
1 - Inlet diverter
1 - Drain connection
1 - Quieting baffle
   Companion flanges bolted on gas inlet and outlet (threaded or slip-on)
1 - Tubular gage glass with safety cocks & drain valve

OPTIONAL ACCESSORIES

Thermometer
Safety Head
Heating Coil
Additional Connections
Skid Mounting
Three Phase Operation

Figure 4-2. A horizontal single-tube separator. (Courtesy of HTI-Superior, a Berry Industries Company.)

separator. The liquid drains into the liquid accumulation section, via horizontal baffles as shown in Figure 4-2. These baffles act as sites for further release of any dissolved gas. Gas flows horizontally in a horizontal separator. It strikes baffles placed at an angle of 45°, thereby releasing entrained liquid by impingement. A mist extractor is usually provided near the gas outlet.

In the double-tube type, the upper tube acts as the separator section, while the lower tube merely functions as a liquid accumulation section. Thus, the double-tube separator is similar to a single-tube separator, but with a greater liquid capacity. The liquid generated in the primary separation section near the inlet is immediately drained out into the lower tube. The wet gas flows through the baffles in the upper separator tube at higher velocities. Additional liquid generated is drained into the lower section through the liquid drains provided along the length of the separator.

94   Gas Production Engineering

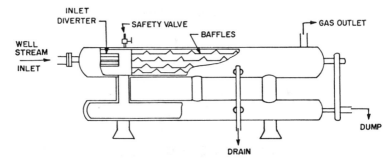

Figure 4-3. A horizontal double-tube separator.

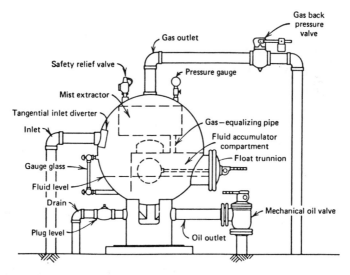

Figure 4-4. A spherical low-pressure separator. (After Sivalls, 1977; courtesy of C. R. Sivalls and the University of Oklahoma.)

Advantages

Horizontal separators have a much greater gas-liquid interface area, permitting higher gas velocities. They can, therefore, handle large gas volumes economically and efficiently. They are cheaper to fabricate and ship than vertical separators. They are also easier and cheaper to install and service. Horizontal separators minimize turbulence and foaming. For a given capacity, horizontal separators are smaller and cheaper than vertical separators. Horizontal separators are almost always used for high GOR wells, for foaming well streams, and for liquid-liquid separation (Beggs, 1984).

Disadvantages

Liquid level control is critical for horizontal separators, and the surge space is rather limited. They are much harder to clean, and are therefore not advisable to use where the well produces a lot of sand. They occupy a lot of space. The space requirements, however, can be minimized by stacking several of these on top of each other for stage separation operations.

Spherical Separators

The spherical separator is designed to make optimum use of all the known means of gas and liquid separation such as gravity, low velocity, centrifugal force, and surface contact (Craft et al., 1962). An inlet flow diverter spreads the entering wellstream tangentially against the separator wall. The liquid is split into two streams that come together after going halfway around the circular vessel wall and then fall into the liquid accumulation section. Liquid droplets from the gas are removed mostly by the velocity reduction imposed upon the gas inside the vessel. A mist extractor is used for the final removal of smaller liquid droplets in the gas (see Figure 4-4).

Advantages

Spherical separators are very inexpensive, cheaper than either the vertical or the horizontal separators. They are very compact, and offer better clean-out and bottom drain features than even the vertical type. Spherical separators are applicable to well streams with low to intermediate GOR's.

Disadvantages

Liquid level control is critical to the spherical separator performance. They have very limited surge capacity and liquid settling section. Because of the limited internal space, it is difficult to use a spherical separator for three-phase (gas-oil-water) separation.

Separation Principles

The several different techniques applied for separation processing can be broadly classified into two categories: *mechanical separation* and *chemical separation* techniques based upon thermodynamic vapor-liquid equilibrium principles. Mechanical separation includes several techniques that will be discussed later in the section on gas cleaning. In separators, the mechanical

96 Gas Production Engineering

separation methods that are applied are of three types: centrifugal action, gravity settling, and impingement.

Centrifuge Separation

Consider a centrifuge of radius R_2, height h, and inner shaft radius R_1, as shown in Figure 4-5. Feed enters at a volumetric rate q. As the centrifuge rotates at an angular speed ω, the heavier liquid droplets are thrown outward to the centrifuge walls as shown in Figure 4-5. The residence time t for the fluid in the centrifuge is given by:

t = centrifuge volume/volumetric flow rate of fluid

$$= \pi(R_2^2 - R_1^2)h/q \qquad (4\text{-}1)$$

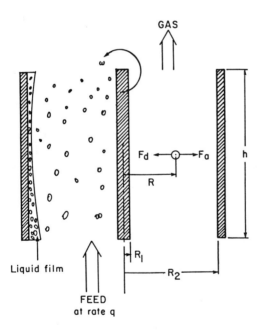

Figure 4-5. Forces acting on a particle in a gas stream in a centrifuge.

In the analysis that follows, it is assumed for simplicity that the liquid droplets are spherical, with a uniform diameter d_p. The area A projected by a droplet is therefore equal to $(\pi/4)d_p^2$. As shown in Figure 4-5, there are two forces acting on any droplet at a radius R:

1. The drag force, F_d, due to friction, given by:

$$F_d = (1/2)C_d\rho_g v^2 A = (\pi/8)C_d\rho_g v^2 d_p^2$$

where C_d = drag (or friction) coefficient
ρ_g = gas density
v = droplet velocity

2. The force F_a, due to angular acceleration, given by:

$$F_a = (\pi/6)d_p^3\rho_l\omega^2 R$$

where ρ_l = liquid density

At equilibrium conditions, the sum of forces on the droplet must be equal to zero. Therefore, the two forces F_d and F_a, acting in opposite directions must be equal in magnitude. So:

$$(\pi/8)C_d\rho_g v^2 d_p^2 = (\pi/6)d_p^3\rho_l\omega^2 R$$

or,

$$v^2 = \frac{4d_p\rho_l\omega^2 R}{3C_d\rho_g}$$

Therefore,

$$v = \frac{dR}{dt} = \frac{(4d_p\rho_l\omega^2 R)^{0.5}}{(3C_d\rho_g)^{0.5}} \tag{4-2}$$

Separating the variables,

$$\frac{dR}{R^{0.5}} = \frac{2(d_p\rho_l)^{0.5}\omega}{(3C_d\rho_g)^{0.5}}\,dt$$

and integrating, we get:

$$2(R_2^{0.5} - R_1^{0.5}) = 2(d_p\rho_l)^{0.5}\,\omega t/(3C_d\rho_g)^{0.5}$$

Solving for t:

$$t = \frac{(R_2^{0.5} - R_1^{0.5})(3C_d\rho_g)^{0.5}}{(d_p\rho_l)^{0.5}\omega} \tag{4-3}$$

Substituting for t from Equation 4-1 into 4-3:

$$\frac{\pi(R_2^2 - R_1^2)h}{q} = \frac{(R_2^{0.5} - R_1^{0.5})(3C_d\rho_g)^{0.5}}{(d_p\rho_l)^{0.5}\omega}$$

Solving for the droplet diameter d_p:

$$d_p^{0.5} = \frac{(R_2^{0.5} - R_1^{0.5})(3C_d\rho_g)^{0.5}q}{\pi h(R_2^2 - R_1^2)\rho_l^{0.5}\omega}$$

$$d_p = \frac{3(R_2^{0.5} - R_1^{0.5})^2 C_d \rho_g q^2}{\pi^2 h^2 (R_2^2 - R_1^2)^2 \rho_l \omega^2} \tag{4-4}$$

Equation 4-4 gives the size of the smallest droplet that can be removed by the centrifuge. Thus, to decrease the droplet size d_p that can be removed, we have three options: decrease q, which is not quite feasible; increase the height, h, of the centrifuge; or increase the rotational speed, ω, of the centrifuge. To achieve good separation in a centrifuge, it is therefore essential to have as large a centrifuge column as possible, and to operate the centrifuge at as high a speed as possible.

In practice, centrifuge separation involves using several relatively small, standard-size cyclones in parallel. Centrifuges (or cyclones) perform best when the gas flows at a constant rate and pressure for which the equipment has been designed. At lower rates, separation is poorer, whereas higher rates lead to an excessive pressure drop through the centrifuge or cyclone. Properly operated, centrifuge separation can usually handle liquid droplets down to a size of about 2 microns.

Gravity Settling

Consider a liquid droplet of diameter d_p suspended in a gas stream flowing at a rate q. The forces on the droplet are:

1. Force F_g due to gravity, given by:

$$F_g = (\pi/6)d_p^3 g(\rho_l - \rho_g)$$

where g = acceleration due to gravity, and
2. Force F_d due to friction, given by:

$$F_d = (\pi/8)C_d\rho_g v^2 d_p^2$$

At equilibrium, $F_d = F_g$. Therefore:

$$(\pi/8)C_d\rho_g v^2 d_p^2 = (\pi/6)d_p^3 g(\rho_l - \rho_g)$$

Solving for the droplet velocity, v:

$$v = \frac{[4gd_p(\rho_l - \rho_g)]^{0.5}}{(3C_d\rho_g)^{0.5}} \tag{4-5}$$

Equation 4-5 for the droplet velocity is often written as:

$$v = K[(\rho_l - \rho_g)/\rho_g]^{0.5} \tag{4-6}$$

This is known as the Souders-Brown equation. The constant K relates to the liquid droplet diameter and the drag coefficient, and is called the separation coefficient.

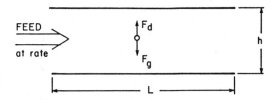

Figure 4-6. Forces acting on a particle in a gas stream in a gravity settling chamber.

The residence time for the droplet in a separation chamber of circular cross-section, with length L and diameter h, as shown in Figure 4-6, is given by:

$$t = (\pi h^2 L)/4q \tag{4-7}$$

The velocity v at which the droplet falls in the vertical direction is given by:

$$v = dh/dt$$

Upon integration,

$$v = h/t \text{ assuming a constant } v$$

Substituting for t from Equation 4-7 and rearranging, we get:

$$q = (\pi hL/4)\, v$$

Substituting for v from Equation 4-5 into Equation 4-7b:

$$q = (\pi hL/4)\, \frac{[4gd_p(\rho_l - \rho_g)]^{0.5}}{(3C_d\rho_g)^{0.5}} \tag{4-8}$$

Thus, to allow smaller droplets to settle, we must maximize the height h and length L.

Gravity settling can effectively remove liquid droplets of size greater than about 80 microns. For removing smaller particles, the required chamber size becomes too large for any practical application.

Impingement

In the impingement process, gas with entrained liquid particles strikes a surface such as a baffle plate, or wire mesh. The gas flows around this flow obstruction, but the momentum of the liquid droplet tends to move it straight ahead and impinge and collect on the surface. In practice, the flow direction is altered several times to achieve better gas-liquid separation.

The design parameters for an effective impingement surface are the distance across the flow path necessary to stop the liquid droplet, since this determines the size of the impingement surface needed to remove droplets up to a given size; and the pressure drop resulting from the flow of gas through the impingement medium. The impingement surface size can be determined using laws governing the flow of particles suspended in a fluid, such as the terms derived in Equations 4-2 and 4-5 for the particle velocity in a centrifuge and a gravity settling chamber, respectively. The pressure drop caused by the impingement media is also of concern, because a greater pressure drop is synonymous, in engineering terms, to higher compression costs for the gas.

Impingement techniques can handle liquid particles of size greater than about 5 microns.

Factors Affecting Separation

As is true for all vapor-liquid equilibrium processes, gas-liquid separation is affected by the separator operating pressure, temperature, and the composition of the fluid feed to the separator. From the general phase diagrams for naturally occurring hydrocarbon mixtures, such as those in Chapter 2, it is clear that as the pressure increases, or the temperature decreases, there will be greater liquid recovery, up to a point referred to as the optimum. Vapor-liquid equilibrium flash calculations will yield the optimum pressure and temperature quite easily. From a practical standpoint, however, it may not be possible to operate at this optimum point because of the costs involved, operational problems, or enhanced storage system vapor losses.

Generally, with increasing pressure, the gas capacity of the separator increases. The reasons are the effect pressure has on gas and liquid densities, actual flowing volume, and the allowable velocity through the separator. With increasing temperature, the separator capacity usually decreases, due to the effect of temperature on the flowing volume and gas and liquid densities.

In actual oil-field practice, economics is the foremost concern. The idea is to maximize the income from a given wellstream. The type of separator chosen, the operating conditions, and the product mix (gas versus oil), all are subject to an economic analysis. In addition, the product sales specifications must be considered. For example, it is usually desirable to produce more liquid because it sells at a higher price in most cases; however, doing so may strip the gas of its intermediate components to an extent that it may not meet pipeline specifications for the energy (or Btu) content per unit volume.

Controlling the Operating Conditions

The two parameters that can be controlled are the pressure and temperature. For pressure, back-pressure valves are used. Limitations in separator operating pressure are imposed by its obvious relationship to the wellhead pressure and the transmission line pressure. The wellhead pressure, in turn, relates to the flowing characteristics of the well, while the transmission line pressure requirements are usually fixed by its design. One of the goals, besides vapor-liquid separation considerations, is to minimize gas compression expenses.

Temperature control is usually available to a certain extent. The wellstream flowing temperature is generally higher than the usual separator operating temperatures, and the wellstream therefore, has to be cooled to the desired temperature. The most widely used cooling method is the simultaneous pressure and temperature reduction by expansion of the wellstream through a choke (Joule-Thomson effect), although other cooling methods such as heat exchangers, cooling towers, and refrigeration, are also used. In the final analysis, it is really the economics that determines whether the wellstream should be cooled, the extent to which it should be cooled, and the cooling method applicable.

Separator Design

The design aspects encountered by a petroleum engineer only involve choosing the correct separator size for a given field installation. Separator sizing is essentially quoted in terms of the gas capacity, and the liquid capacity of the separator. Other parameters, such as pressure drop through the separator, are specified for a given design by the manufacturer and are beyond the scope of the present discussion.

Separator Design Using Basic Separation Principles

Gas Capacity

The Souders-Brown relationship (Equation 4-6) has been traditionally used for calculating the gas capacity of gas-liquid separators:

$$v_g = K\,[(\rho_l - \rho_g)/\rho_g]^{0.5}$$

where v_g = allowable gas velocity at the operating conditions, ft/sec
ρ_l = liquid density at the operating conditions, lbm/ft^3
ρ_g = gas density at the operating conditions, lbm/ft^3
K = separation coefficient

The separation coefficient, K, is an empirical constant given as follows (Craft et al., 1962; Sivalls, 1977):

Separator type	Range of K	Most commonly used K value
Vertical	0.06 to 0.35	0.117 without mist extractor
		0.167 with a mist extractor
Horizontal	0.40 to 0.50	0.382 with a mist extractor
Spherical*	—	0.35 with a mist extractor

Besides calculating the diameter of the separator required for a given gas capacity, the Souders-Brown relationship can also be used for other designs such as bubble cap or trayed towers for dehydration and desulfurization units, and for sizing mist eliminators. The K values given by Sivalls (1977) for these are as follows:

Wire mesh mist eliminators	K = 0.35
Bubble cap trayed columns	K = 0.16 for 24 in. spacing.
Valve tray columns	K = 0.18 for 24 in. spacing.

Using Equation 4-6, the gas capacity at operating conditions, q_g, in ft^3/sec is given by:

$$q_g = Av_g = (\pi/4)(D^2)K\,[(\rho_l - \rho_g)/\rho_g]^{0.5} \qquad (4\text{-}9)$$

---

* The gas capacity of spherical separators is based upon the capacity of the mist extractor.

where A = cross-sectional area of the separator, ft²
       D = internal diameter of the vessel, ft

Note that the gas velocity v_g is based upon total separator area, and it is therefore more appropriate to refer to it as the superficial gas velocity. The gas capacity at standard conditions (14.7 psia and 60°F), q_{gsc}, generally reported in units of MMscfd (million standard cubic feet per day), is thus given by:

$$q_{gsc} = \frac{2.40 D^2 K p (\rho_l - \rho_g)^{0.5}}{Z(T + 460) \rho_g^{0.5}} \qquad (4\text{-}10)$$

where q_{gsc} = gas capacity at standard conditions, MMscfd
      p = operating pressure, psia
      T = operating temperature, °F
      Z = gas compressibility factor at the operating conditions

Equation 4-9 or 4-10 can be used to calculate the separator diameter required to handle a given gas rate, or to calculate the gas rate that a separator of a given size can handle. The area of the mist extractor required, A_m, can be obtained as follows:

$$A_m = q_g / v_m \qquad (4\text{-}11)$$

where v_m is the gas velocity through the mist extractor, determined using Equation 4-6 with K = 0.35 for mist extractor (wire mesh type).

Liquid Capacity

The liquid capacity of a separator depends upon the volume of the separator available to the liquid, and the retention time of the liquid within the separator (Sivalls, 1977):

$$W = 1440 V_L / t \qquad (4\text{-}12)$$

where W = liquid capacity, bbl/day
       V_L = liquid settling volume, bbl
       t = retention time, min (1440 is the conversion factor to convert bbl/day into bbl/min)

The liquid settling volume, V_L, can be calculated as follows:

$V_L = 0.1399 D^2 h$ for vertical separators

$V_L = 0.1399D^2(L/2)$ for horizontal single-tube separators

$V_L = 0.1399D^2L$ for horizontal double-tube separators

$V_L = 0.0466D^3(D/2)^{0.5}$ for spherical separators*

where h = height of liquid column above the bottom of the liquid outlet in the vertical separator, ft
L = separator length (height), ft

For good separation, a sufficient retention time, t, must be provided. From field experience, the following liquid retention times have been suggested by Sivalls (1977):

Oil-gas separation	1 min.
High pressure oil-water-gas separation	2 to 5 min.
Low pressure oil-water-gas separation	5 to 10 min. at >100°F
	10 to 15 min. at 90°F
	15 to 20 min. at 80°F
	20 to 25 min. at 70°F
	25 to 30 min. at 60°F

Vessel Design Considerations

Some of the basic factors that must be considered in designing separators are (Lockhart et al., 1986):

1. The length to diameter ratio, L/D, for a horizontal or vertical separator should be kept between 3 and 8, due to considerations of fabrication costs, foundation costs, etc.
2. For a vertical separator, the vapor-liquid interface (at which the feed enters) should be at least 2 ft from the bottom and 4 ft from the top of the vessel. This implies a minimum vertical separator height (length) of 6 ft.
3. For a horizontal separator, the feed enters just above the vapor-liquid interface that may be off-centered to adjust for a greater gas (or liquid) capacity as needed. The vapor-liquid interface, however, must be kept at least 10 in. from the bottom and 16 in. from the top of the vessel. This implies a minimum horizontal separator diameter of 26 in.

In practice, novel design techniques violate these rules of thumb by providing additional features. Therefore, standard vertical separators less than

---

* Spherical separators are generally operated at half-full of liquid conditions; the relationship mentioned assumes this case. Also, the volume is increased by a factor of $(D/2)^{0.5}$ because spherical separators have greater surge capacity due to their shape.

6 ft and standard horizontal separators of diameter less than 26 in. are available and have been used successfully.

High-pressure separators are generally used for high-pressure, high gas-liquid ratio (gas and gas condensate) wells. In this case, the gas capacity of the separator is usually the limiting factor. Low-pressure separators, used generally for low gas-liquid ratio at low pressures, are subject to the opposite constraint—they require a high liquid capacity. The separator chosen must satisfy both the gas as well as liquid capacities. Also, the liquid discharge (or dump) valve should be designed for the pressure drop available and the liquid flow rate (Sivalls, 1977).

Note that as the gas-liquid ratio (GLR) increases, the retention time t decreases. The volume of the separator occupied by gas, V_G, is given by:

$$V_G = V - V_L = V - Wt$$

because $V_L = Wt$ by Equation 4-12.

where V = total separator volume

V_G is also given by:

$$V_G = q_g t = W \cdot GLR \cdot t$$

Therefore:

$$W \cdot GLR \cdot t = V - Wt$$

On rearranging:

$$t = \frac{V}{W(GLR + 1)}$$

Thus, for a fixed separator volume V and liquid capacity W, as GLR increases, retention time t decreases.

Example 4-1. A separator, to be operated at 1,000 psia, is required to handle a wellstream with gas flow rate 7 MMscfd at a GLR = 40 bbl/MMscf. Determine the separator size required, for: (1) a vertical separator, (2) a horizontal single-tube separator, and (3) spherical separator. Assume a liquid (oil + water) density of 52 lbm/ft^3, ideal gas with gravity = 0.80, an operating temperature equal to 110°F, a retention time t = 3 min., and 1/2 full of liquid conditions.

Solution

Gas density, ρ_g = pM/ZRT = (1000)(28.97)(0.8)/(1)(10.73)(570) = 3.789 lbm/ft³

From Equation 4-10:

$$D^2 = \frac{q_{gsc}Z(T+460)\rho_g^{0.5}}{2.40Kp(\rho_l - \rho_g)^{0.5}} = \frac{(7)(1)(570)(3.789^{0.5})}{K(2.40)(1000)(52-3.789)^{0.5}}$$

$$= 0.466065/K$$

For a retention time t of 3 min., the liquid settling volume required for each separator type is (Equation 4-12):

V_L = Wt/1440 = (40)(7)(3)/1440 = 0.583 bbl

1. For a vertical separator with mist extractor, K = 0.167. Therefore, the diameter of the vertical separator required is:

   D = [0.466065/0.167]$^{0.5}$ = 1.67 ft (= 20 in.)

   The liquid capacity, $V_L = 0.1399D^2h$ = 0.583

   Therefore, h = 0.583/[(0.1399)(1.67²)] = 1.49 ft. Thus, the minimum separator length of 6 ft should be used. The L/D ratio = 6/1.67 = 3.6

   So, a vertical separator of size 20 in. × 6 ft is required.

2. For a horizontal separator with mist extractor, K = 0.382. Therefore, the diameter of the horizontal separator required is:

   D = [0.466065/0.382]$^{0.5}$ = 1.10 ft (= 13.20 in.). Thus, the minimum separator diameter of 26 in. (= 2.17 ft) should be used.

   The liquid capacity, $V_L = 0.1399D^2(L/2)$ = 0.583
   Therefore, L = (2)(0.583)/[(0.1399)(2.17²)] = 1.77 ft. The L/D ratio = 1.77/2.17 = 0.816, which does not satisfy design criteria. For a minimum L/D ratio = 3, separator length required = 3 × 2.17 = 6.5 ft.

   So, a horizontal separator of size 26 in. × 6.5 ft is required.

3. For a spherical separator with mist extractor, K = 0.35. Therefore, the diameter of the spherical separator to handle the required gas capacity is:

$D = [0.466065/0.35]^{0.5} = 1.15$ ft

The liquid capacity, $V_L = 0.0466 D^3 (D/2)^{0.5} = 0.583$

Therefore, $D = [(2^{0.5})(0.583)/0.0466]^{1/3.5} = 2.27$ ft based upon liquid capacity requirements.

So, a spherical separator of diameter 2.27 ft (= 27 in.) is required.

Separator Design Using Actual Separator Performance Charts

The Souders-Brown relationship provides only an approximate approach. A better design can usually be made using actual manufacturers' field test data that account for the dependence of gas capacity on the separator height (for vertical) or length (for horizontal). Field experience shows this dependence that is not accounted for by the Souders-Brown equation. Figures 4-7 through 4-14 show the gas capacity charts for various separator types as a function of their sizes and operating pressures. For horizontal separators, a

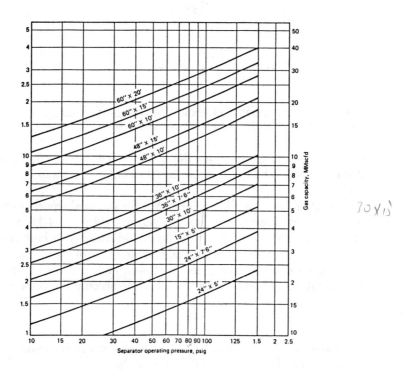

Figure 4-7. Gas capacity of vertical low-pressure separators. (After Sivalls, 1977; courtesy of C. R. Sivalls and the University of Oklahoma.)

108   Gas Production Engineering

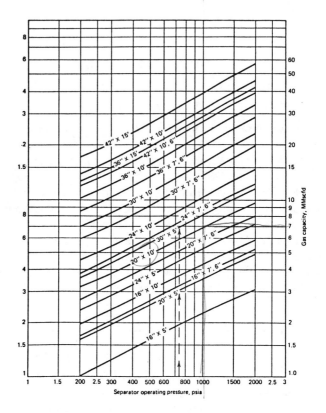

Figure 4-8. Gas capacity of vertical high-pressure separators. (After Sivalls, 1977; courtesy of C. R. Sivalls and the University of Oklahoma.)

correction has to be made for the extent of liquid (one-quarter full, one-third full, or one-half full) in the liquid section as shown in Figures 4-10 to 4-12.

The actual liquid settling volume, V_L, for standard separators can be determined from their specifications given in Tables 4-1–4-11. Figures 4-15 and 4-16 are the sizing charts for the liquid capacity of horizontal single-tube separators, based upon a liquid retention time of 1 min. Corrections for retention times other than 1 min. are linear, and are shown in these charts. Such charts are useful because they take into account the working pressure.

Example 4-2. Repeat Example 4-1, using actual manufacturers' field test data, and assuming ¼ full of liquid conditions.

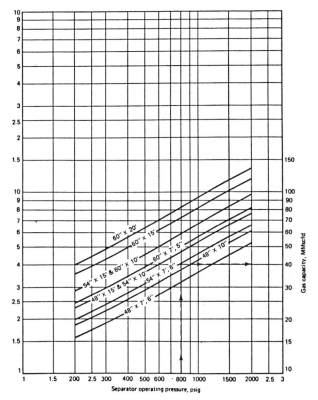

Figure 4-9. Gas capacity of vertical high-pressure separators. (After Sivalls, 1977; courtesy of C. R. Sivalls and the University of Oklahoma.)

Solution

1. From Figure 4-8, a 30-in. × 5-ft vertical separator is required. From Table 4-4, $V_L = 1.76$ bbl, which is greater than the required V_L of 0.583 bbl. Hence, use a 30 in. × 5 ft high-pressure vertical separator.

2. From Figure 4-11, a 12 in. × 10 ft horizontal separator is required, assuming ¼ full liquid conditions.

   Liquid capacity from Figure 4-15 is $(380)(0.33) = 125.4$ bbl/day, whereas $(7)(40) = 280$ bbl/day liquid capacity is required.

   Choosing a size of 24 in. × 5 ft from Figure 4-15, the liquid capacity is $(720)(0.33) = 237.6$ bbl/day, which is still insufficient.

   (text continued on page 114)

110   Gas Production Engineering

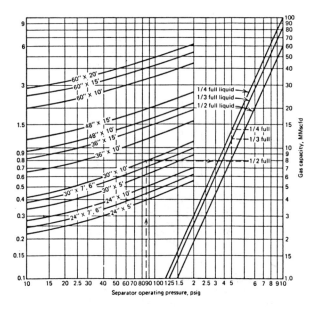

Figure 4-10. Gas capacity of horizontal low-pressure separators. (After Sivalls, 1977; courtesy of C. R. Sivalls and the University of Oklahoma.)

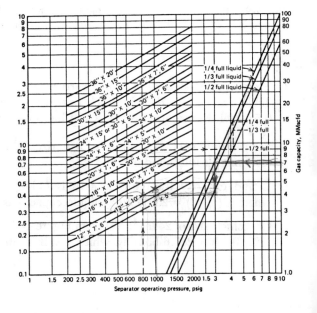

Figure 4-11. Gas capacity of horizontal high-pressure separators. (After Sivalls, 1977; courtesy of C. R. Sivalls and the University of Oklahoma.)

Gas and Liquid Separation   111

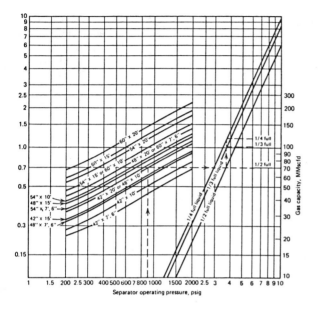

Figure 4-12. Gas capacity of horizontal high-pressure separators. (After Sivalls, 1977; courtesy of C. R. Sivalls and the University of Oklahoma.)

Figure 4-13. Gas capacity of spherical low-pressure separators. (After Sivalls, 1977; courtesy of C. R. Sivalls and the University of Oklahoma.)

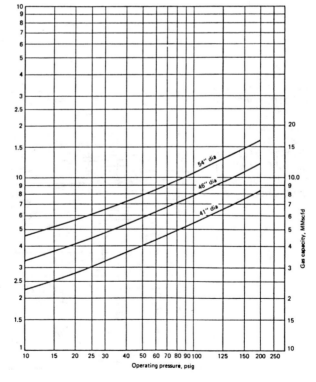

112   Gas Production Engineering

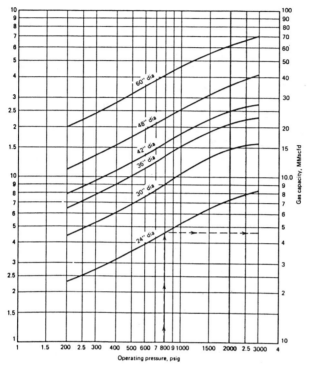

Figure 4-14. Gas capacity of spherical high-pressure separators. (After Sivalls, 1977; courtesy of C. R. Sivalls and the University of Oklahoma.)

Figure 4-15. Liquid capacity of horizontal single-tube high-pressure separators. (After Sivalls, 1977; courtesy of C. R. Sivalls and the University of Oklahoma.)

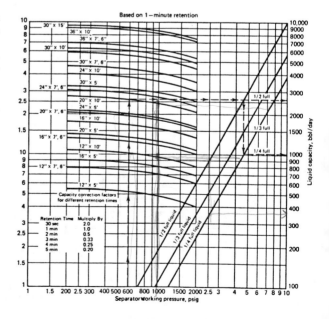

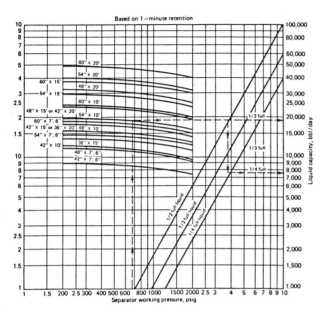

Figure 4-16. Liquid capacity of horizontal single-tube high-pressure separators. (After Sivalls, 1977; courtesy of C. R. Sivalls and the University of Oklahoma.)

Table 4-1
Specifications of Standard Vertical Low-pressure Separators*

Size, Dia × Ht	Working Pressure, psi	Inlet and Gas Outlet Conn.	Oil Outlet Conn.	Standard Valves Oil or Oil and Water	Gas	Shipping Weight, lb
24' × 5'	125	2" Thd	2" Thd	2"	2"	950
24" × 7½'	125	2" Thd	2" Thd	2"	2"	1150
30" × 10'	125	3" Thd	3" Thd	2"	2"	2000
36" × 5'	125	4" Thd	2" Thd	2"	2"	2000
36" × 7½'	125	4" Thd	3" Thd	2"	2"	2350
36" × 10'	125	4" Thd	4" Thd	2"	2"	2700
48" × 10'	125	6" Flg	4" Thd	2"	2"	3400
48" × 15'	125	6" Flg	4" Thd	2"	2"	4500
60" × 10'	125	6" Flg	4" Thd	3"	3"	5200
60" × 15'	125	6" Flg	4" Thd	3"	3"	6400
60" × 20'	125	6" Flg	4" Thd	3"	3"	7600

* After Sivalls, 1977.

Table 4-2
Settling Volumes of Standard Vertical Low-pressure Separators*
(125 psi WP)

Size, Dia × Ht	Settling Volume, bbl	
	Oil-Gas Separators	Oil-Gas-Water Separators**
24" × 5'	0.65	1.10
24" × 7½'	1.01	1.82
30" × 10'	2.06	3.75
36" × 5'	1.61	2.63
36" × 7½'	2.43	4.26
36" × 10'	3.04	5.48
48" × 10'	5.67	10.06
48" × 15'	7.86	14.44
60" × 10'	9.23	16.08
60" × 15'	12.65	12.93
60" × 20'	15.51	18.64

* After Sivalls, 1977.
** Total settling volume is usually split even between oil and water.

(text continued from page 109)

Choosing the next available size of 20 in. × 10 ft from Figure 4-15, the liquid capacity is (980)(0.33) = 323.4 bbl/day, which is sufficient. The same result could alternatively be obtained from Table 4-8, where the smallest horizontal separator size with a V_L greater than 0.583 is 20 in. × 10 ft for a ¼ full of liquid condition.

Hence, use a 20 in. × 10 ft high-pressure horizontal separator.

3. From Figure 4-14, a 30 in. diameter spherical separator is required.

Liquid settling volume, V_L, from Table 4-11 is only 0.30 bbl, whereas 0.583 bbl is required. From Table 4-11, the smallest spherical separator with a V_L greater than 0.583 is of diameter 42 in.

Hence, use a 42 in. diameter high-pressure spherical separator.

Stage Separation

When two or more equilibrium separation stages are used in series, the process is termed "stage separation." The storage tank, also called stock tank because it is generally kept at standard conditions (14.7 psia and 60°F) to

Table 4-3
Specifications of Standard Vertical High Pressure Separators*

Size, Dia × Ht	Working Pressure, psi**	Inlet and Gas Outlet Connection	Standard Liquid Valve	Shipping Weight, lb
16" × 5'	1000	2" Thd	1"	1,100
16" × 7½'		2" Thd	1"	1,200
16" × 10'		2" Thd	1"	1,500
20" × 5'	1000	3" Flg	1"	1,600
20" × 7½'		3" Flg	1"	1,900
20" × 10'		3" Flg	1"	2,200
24" × 5'	1000	3" Flg	1"	2,500
24" × 7½'		3" Flg	1"	2,850
24" × 10'		3" Flg	1"	3,300
30" × 5'	1000	4" Flg	1"	3,200
30" × 7½'		4" Flg	1"	3,650
30" × 10'		4" Flg	1"	4,200
36" × 7½'	1000	4" Flg	1"	5,400
36" × 10'		4" Flg	1"	6,400
36" × 15'		4" Flg	1"	8,700
42" × 7½'	1000	6" Flg	2"	7,700
42" × 10'		6" Flg	2"	9,100
42" × 15'		6" Flg	2"	12,000
48" × 7½'	1000	6" Flg	2"	10,400
48" × 10'		6" Flg	2"	12,400
48" × 15'		6" Flg	2"	16,400
54" × 7½'	1000	6" Flg	2"	12,300
54" × 10'		6" Flg	2"	14,900
54" × 15'		6" Flg	2"	20,400
60" × 7½'	1000	6" Flg	2"	17,500
60" × 10'		6" Flg	2"	20,500
60" × 15'		6" Flg	2"	26,500
60" × 20'		6" Flg	2"	32,500

* After Sivalls, 1977.
** Other standard working pressures available are 230, 500, 600, 1200, 1440, 1500, and 2000 psi.

enable accurate metering of the produced hydrocarbons,* constitutes the last separation stage. Therefore, as shown in Figure 4-17, a two-stage separation includes one separator and one stock tank, a three stage separation means two separators and one stock-tank, and so on.

---

* This is not always true. The pressure can be controlled, but the storage tank temperature is subject largely to the prevailing atmospheric temperature, and it is too expensive to heat or cool the contents of the tank to 60°F. Modern digital metering devices can easily correct for temperature or pressure.

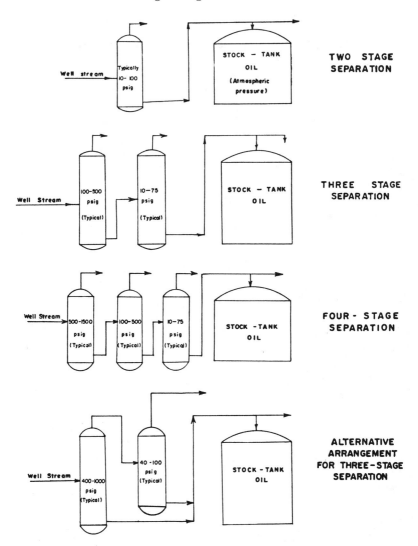

Figure 4-17. Schematic diagrams for stage separation.

Readers who have a background in phase behavior will immediately recognize that liquid recovery is enhanced on using a greater number of equilibrium stages for separation. This same principle is employed in designing columns with multiple equilibrium trays for separation purposes (Chapter 5). The ideal separation process is one in which the pressure is reduced infinitesimally at each stage, requiring an infinite number of stages. Such a process, termed "differential liberation," would maximize the liquid recovery and yield a more stable stock tank liquid. In practice, this is not feasible and differential liberation is approximated quite closely by using a few separation stages (two, three, or four). The latter process, where the pressure is reduced in very large steps, is termed "flash liberation." Flash vaporization also differs from differential vaporization in another respect: in differential vaporization, the vapor is removed as it is formed, but in flash vaporization, the liquid and vapor are kept in intimate contact until equilibrium is achieved. For this reason, flash vaporization is also termed "equilibrium vaporization" or "flash equilibrium vaporization," whereas differential vaporization is called "differential liberation."

Table 4-4
Settling Volumes of Standard Vertical High-pressure Separators*
(230 to 2,000 psi WP**)

Size, Dia × Ht	Settling Volume, bbl†	
	Oil-Gas Separators	Oil-Gas-Water Separators††
16" × 5'	0.27	0.44
16" × 7½'	0.41	0.72
16" × 10'	0.51	0.94
20" × 5'	0.44	0.71
20" × 7½'	0.65	1.15
20" × 10'	0.82	1.48
24" × 5'	0.66	1.05
24" × 7½'	0.97	1.68
24" × 10'	1.21	2.15
30" × 5'	1.13	1.76
30" × 7½'	1.64	2.78
30" × 10'	2.02	3.54
36" × 7½'	2.47	4.13
36" × 10'	3.02	5.24
36" × 15'	4.13	7.45
42" × 7½'	3.53	5.80
42" × 10'	4.29	7.32
42" × 15'	5.80	10.36
48" × 7½'	4.81	7.79
48" × 10'	5.80	9.78
48" × 15'	7.79	13.76
54" × 7½'	6.33	10.12
54" × 10'	7.60	12.65
54" × 15'	10.12	17.70
60" × 7½'	8.08	12.73
60" × 10'	9.63	15.83
60" × 15'	12.73	22.03
60" × 20'	15.31	27.20

* After Sivalls, 1977.
** Standard working pressures available are 230, 500, 1000, 1200, 1440, 1500, and 2000 psi.
† Based on 1000 psi WP separators.
†† Total settling volume is usually split even between oil and water.

Table 4-5
Specifications of Standard Horizontal Low-pressure Separators*

Size, Dia × Ht	Working Pressure, psi	Inlet and Gas Outlet Connection	Oil Outlet Connection	Standard Valves		Shipping Weight, lb
				Oil or Oil and Water	Gas	
24″ × 5′	125	2″ Thd	2″ Thd	2″	2″	1,000
24″ × 7½′	125	2″ Thd	2″ Thd	2″	2″	1,200
24″ × 10′	125	3″ Thd	2″ Thd	2″	2″	1,600
30″ × 5′	125	3″ Thd	3″ Thd	2″	2″	1,200
30″ × 7½′	125	3″ Thd	3″ Thd	2″	2″	1,600
30″ × 10′	125	4″ Thd	4″ Thd	2″	2″	2,100
36″ × 10′	125	4″ Thd	4″ Thd	2″	2″	2,900
36″ × 15′	125	4″ Thd	4″ Thd	2″	2″	3,800
48″ × 10′	125	6″ Flg	4″ Thd	2″	2″	3,500
48″ × 15′	125	6″ Flg	4″ Thd	3″	3″	4,600
60″ × 10′	125	6″ Flg	4″ Thd	3″	3″	6,200
60″ × 15′	125	6″ Flg	4″ Thd	3″	3″	8,100
60″ × 20′	125	6″ Flg	4″ Thd	4″	4″	10,000

*After Sivalls, 1977.

Table 4-6
Settling Volumes of Standard Horizontal Low-pressure Separators*
(125 psi WP)

Size, Dia × Length	Settling Volume, bbl		
	½ Full	⅓ Full	¼ Full
24" × 5'	1.55	0.89	0.59
24" × 7½'	2.22	1.28	0.86
24" × 10'	2.89	1.67	1.12
30" × 5'	2.48	1.43	0.94
30" × 7½'	3.54	2.04	1.36
30" × 10'	4.59	2.66	1.77
36" × 10'	6.71	3.88	2.59
36" × 15'	9.76	5.66	3.79
48" × 10'	12.24	7.07	4.71
48" × 15'	17.72	10.26	6.85
60" × 10'	19.50	11.24	7.47
60" × 15'	28.06	16.23	10.82
60" × 20'	36.63	21.21	14.16

* After Sivalls, 1977.

Two-stage separation is desirable for wellstreams with low gas-oil ratios such as low-API gravity oils, and low flowing pressures. Three-stage separation is most applicable to wellstreams with intermediate to high gas-oil ratios (intermediate-API gravity oils), and intermediate flowing pressures. Four-stage separation is used for high gas-oil ratio (high-API gravity) wellstreams, and high flowing pressures. Four-stage separation is also used where high-pressure gas is needed. From an economic standpoint, a three-stage separation is usually optimum for most oil-field operations. Sivalls (1977) reports that the actual increase in liquid recovery for three-stage separation over two-stage will vary from 2 to 12%, depending upon the wellstream composition, and the operating pressures and temperatures. However, additional recoveries as high as 20 to 25% have been reported for some cases.

For the complex mixture of hydrocarbons produced from an oil well, precise stage separation calculations are very difficult and require the use of a computer. For practical purposes, it is quite satisfactory to use other shortcut methods and correlations that yield acceptable results. The simplest of these methods assumes an equal pressure ratio between the stages for optimum performance (Campbell, 1984):

$$r = (p_1/p_s)^{1/n} \tag{4-13}$$

where r = pressure ratio
       n = number of stages − 1

(text continued on page 123)

Table 4-7
Specifications of Standard Horizontal High-pressure Separators*

Size, Dia × Ht	Working Pressure, psi**	Inlet and Gas Outlet Connection	Standard Liquid Valve	Shipping Weight, lb
12³/₄″ × 5′	1000	2″ Thd	1″	1,100
12³/₄″ × 7¹/₂′		2″ Thd	1″	1,200
12³/₄″ × 10′		2″ Thd	1″	1,300
16″ × 5′	1000	2″ Thd	1″	1,400
16″ × 7¹/₂′		2″ Thd	1″	1,750
16″ × 10′		2″ Thd	1″	2,100
20″ × 5′	1000	3″ Flg	1″	1,800
20″ × 7¹/₂′		3″ Flg	1″	2,300
20″ × 10′		3″ Flg	1″	2,900
24″ × 5′	1000	4″ Flg	1″	2,200
24″ × 7¹/₂′		4″ Flg	1″	3,000
24″ × 10′		4″ Flg	1″	3,800
24″ × 15′		4″ Flg	1″	5,400
30″ × 5′	1000	4″ Flg	1″	3,200
30″ × 7¹/₂′		4″ Flg	1″	4,300
30″ × 10′		4″ Flg	1″	5,500
30″ × 15′		4″ Flg	2″	7,800
36″ × 7¹/₂′	1000	6″ Flg	2″	6,100
36″ × 10′		6″ Flg	2″	7,500
36″ × 15′		6″ Flg	2″	10,200
36″ × 20′		6″ Flg	2″	12,000
42″ × 7¹/₂′	1000	6″ Flg	2″	8,200
42″ × 10′		6″ Flg	2″	9,900
42″ × 15′		6″ Flg	2″	13,400
42″ × 20′		6″ Flg	2″	16,900
48″ × 7¹/₂′	1000	8″ Flg	2″	10,900
48″ × 10′		8″ Flg	2″	12,700
48″ × 15′		8″ Flg	2″	17,500
48″ × 20′		8″ Flg	2″	22,100
54″ × 7¹/₂′	1000	8″ Flg	2″	13,400
54″ × 10′		8″ Flg	2″	16,000
54″ × 15′		8″ Flg	2″	21,200
54″ × 20′		8″ Flg	2″	26,400
60″ × 7¹/₂′	1000	8″ Flg	2″	16,700
60″ × 10′		8″ Flg	2″	19,900
60″ × 15′		8″ Flg	2″	26,400
60″ × 20′		8″ Flg	2″	32,900

* After Sivalls, 1977.
** Other standard working pressures available are 230, 500, 600, 1200, 1440, 1500, and 2000 psi.

Table 4-8
Settling Volumes of Standard Horizontal High-pressure Separators*
(230 to 2,000 psi WP**)

Size Dia × Len	Settling Volume, bbl†		
	½ Full	⅓ Full	¼ Full
12¾" × 5'	0.38	0.22	0.15
12¾" × 7½'	0.55	0.32	0.21
12¾" × 10'	0.72	0.42	0.28
16" × 5'	0.61	0.35	0.24
16" × 7½'	0.88	0.50	0.34
16" × 10'	1.14	0.66	0.44
20" × 5'	0.98	0.55	0.38
20" × 7½'	1.39	0.79	0.54
20" × 10'	1.80	1.03	0.70
24" × 5'	1.45	0.83	0.55
24" × 7½'	2.04	1.18	0.78
24" × 10'	2.63	1.52	0.01
24" × 15'	3.81	2.21	1.47
30" × 5'	2.43	1.39	0.91
30" × 7½'	3.40	1.96	1.29
30" × 10'	4.37	2.52	1.67
30" × 15'	6.30	3.65	2.42
36" × 7½'	4.99	2.87	1.90
36" × 10'	6.38	3.68	2.45
36" × 15'	9.17	5.30	3.54
36" × 20'	11.96	6.92	4.63
42" × 7½'	6.93	3.98	2.61
42" × 10'	8.83	5.09	3.35
42" × 15'	12.62	7.30	4.83
42" × 20'	16.41	9.51	6.32
48" × 7½'	9.28	5.32	3.51
48" × 10'	11.77	6.77	4.49
48" × 15'	16.74	9.67	6.43
48" × 20'	21.71	12.57	8.38
54" × 7½'	12.02	6.87	4.49
54" × 10'	15.17	8.71	5.73
54" × 15'	12.49	12.40	8.20
54" × 20'	27.81	16.08	10.68
60" × 7½'	15.05	8.60	5.66
60" × 10'	18.93	10.86	7.17
60" × 15'	26.68	15.38	10.21
60" × 20'	34.44	19.90	13.24

* After Sivalls, 1977.
** Standard working pressures available are 230, 500, 600, 1000, 1200, 1440, 1500 and 2000 psi.
† Based on 1000 psi WP separator.

122   Gas Production Engineering

Table 4-9
Specifications of Standard Spherical Separators*

Diameter	Working Pressure, psi	Inlet and Gas Outlet Connection	Standard Liquid Valve	Shipping Weight, lb
41"	125	4" Thd	2"	1000
46"		4" Thd	3"	1300
54"		4" Thd	4"	1700
42"	250	3" Flg	2"	1100
48"		4" Flg	2"	1400
60"		6" Flg	2"	3400
24"	1000**	2" Flg	1"	1300
30"		2" Flg	1"	1400
36"		3" Flg	1"	1800
42"		4" Flg	2"	2800
48"		4" Flg	2"	3700
60"		6" Flg	2"	4300

* After Sivalls, 1977.
** Other standard working pressures available are 500, 600, 1200, 1440, 2000, and 3000 psi.

Table 4-10
Settling Volumes of Standard Spherical Low-pressure Separators*
(125 psi WP)

Size, OD	Settling Volume, bbl
41"	0.77
46"	1.02
54"	1.60

* After Sivalls, 1977.

Table 4-11
Settling Volumes of Standard Spherical High-pressure Separators*
(230 to 3,000 psi WP)**

Size, OD	Settling Volume, bbl[†]
24"	0.15
30"	0.30
36"	0.54
42"	0.88
48"	1.33
60"	2.20

* After Sivalls, 1977.
** Standard working pressures available are 230, 500, 600, 1000, 1200, 1440, 1500, 2000, and 3000 psi.
† Based on 1000 psi WP separator.

(text continued from page 119)

p_1 = pressure at the first (or high-stage) separator
p_s = pressure at the stock tank

Thus, $p_2 = p_s r^{n-1}$, $p_3 = p_s r^{n-2}$, and in general, the pressure p_i for separator stage i is given by:

$$p_i = p_s r^{n-i+1} \tag{4-14}$$

Note that Equations 4-13 and 4-14 bear no relationship with the magnitude of separation (i.e, the liquid/vapor ratio).

Whinery and Campbell (1958) studied three-stage separation for several different types of wellstreams. Figure 4-18 from their work shows the relationship between the pressure at the second stage and stock-tank liquid recovery. From an empirical analysis of these results, Whinery and Campbell (1958) derived simple relationships to calculate the optimum pressure in the second stage for a three-stage separation. They found that two cases can be specified:

1. For streams with specific gravity > 1.0 (air = 1.0):

$$p_2 = A p_1^{0.686} + (A - 0.057)/0.0233 \tag{4-15}$$

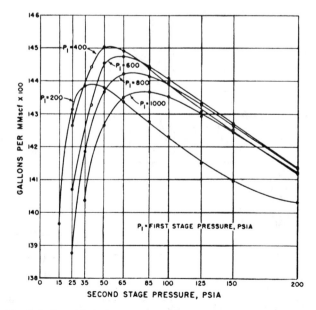

Figure 4-18. Relationship between second-stage pressure (psia) and stock-tank recovery (gal/MMscf). (After Whinery and Campbell, 1958; courtesy of SPE of AIME.)

2. For streams with specific gravity < 1.0 (air = 1.0):

$$p_2 = Ap_1^{0.765} + (A + 0.028)/0.012$$

where p_1, p_2 = pressures in the first and second stages, respectively (the third stage is assumed to be the stock tank at standard conditions of 14.7 psia and 60°F)

A = a function of the stock tank pressure and the system composition

Whinery and Campbell found that composition could be expressed in terms of the specific gravity of the feed and the mole% of $C_1 + C_2 + C_3$ components in it, as shown in Figure 4-19. Example 4-3 illustrates the calculation procedure.

Example 4-3. For a wellstream having a composition shown as follows, find the optimum second-stage pressure for a three-stage separation, if $p_1 = 800$ psia. Use the Whinery-Campbell method, and compare the result with the equal pressure ratio assumption.

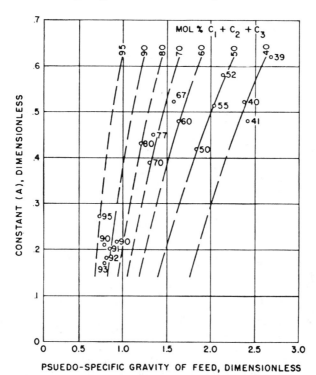

Figure 4-19. Relationship between A and pseudo-specific gravity of feed (T = 80°F). (After Whinery and Campbell, 1958; courtesy of SPE of AIME.)

Solution

Comp.	Mol. fr., y_i	Mol. wt., M_i
C_1	0.35	16.043
C_2	0.25	30.070
C_3	0.15	44.097
C_4	0.10	58.124
C_5	0.08	72.151
C_6	0.04	86.178
C_{7+} (C_9)	0.03	128.259

$\Sigma\ y_i M_i = 38.626$

Therefore, the specific gravity with respect to air = $38.626/28.97 = 1.33$

Mole% $C_1 + C_2 + C_3 = 0.35 + 0.25 + 0.15 = 0.75$

From Figure 4-19, A = 0.43

Using Equation 4-15 for wellstream gravity > 1,

$p_2 = (0.43)(800^{0.686}) + (0.43 - 0.057)/0.0233 = \underline{58.18 \text{ psia.}}$

The Equal Pressure Ratio Case. Using Equation 4-13, the pressure ratio with n = 3 − 1 = 2 is r = $(800/14.7)^{1/2} = 7.377$
From Equation 4-14, $p_2 = (14.7)(7.377^{2-2+1}) = \underline{108.44 \text{ psia.}}$

This example illustrates that the equal pressure ratio assumption is far from precise. Even the Whinery-Campbell method is only an approximation. Wherever possible, flash calculations using equations of state must be carried out for determining the amounts of vapor and liquid that will be recovered, and the composition of these vapor and liquid streams from the separators.

Low Temperature Separation

Low-temperature separation (LTX) units are based upon the principle that lowering the operating temperature of a separator increases the liquid recovery. Besides recovering more liquid from the gas than a normal-temperature separator, LTX units have another advantage: they are able to dehydrate the gas, frequently to pipeline specifications.

A low-temperature separation unit consists of a high-pressure separator, one or more pressure-reducing chokes*, and heat exchange equipment. When the high-pressure wellstream is expanded by pressure reduction through a choke, the temperature is also reduced. This irreversible adiabatic (no exchange of heat with surroundings) simultaneous pressure and temperature reduction process is called the Joule-Thomson or throttling effect. The heat content of the gas remains the same across the choke, but its temperature and pressure are reduced.

The choke is mounted directly at the inlet of the high-pressure separator, so that any hydrates formed downstream of the choke fall into the bottom settling section of the LTX unit, where they are heated and melted by the heating coils provided. The wellstream enters the separator at a low temperature and pressure, and a better separation results. A correlation by the Gas Processors Suppliers Association (GPSA) to determine the temperature drop through the choke for a given pressure drop is shown in Figure 4-20. Example 4-4 illustrates the procedure for using this correlation.

Example 4-4. A 0.7 gravity gas, at 150°F is expanded through a choke, so that its pressure is reduced by 2,000 psi. What is the temperature drop if the initial pressure is: (1) 4,000 psia, and (2) 2,500 psia; (3) What is the final temperature if the gas is initially at 4,000 psia and 150°F and the pressure is reduced to 200 psia.

Solution

1. From Figure 4-20, for an initial pressure = 4,000 psia and a pressure drop = 2,000 psia, the temperature drop = 38°F.

2. From Figure 4-20, for an initial pressure = 2,500 psia and a pressure drop = 2,000 psia, the temperature drop = 86.5°F.

3. From Figure 4-20, for an initial pressure = 4,000 psia and a pressure drop = 4,000 − 200 = 3,800 psia, the temperature drop = 132°F.

   Therefore, the final temperature = 150 − 132 = 18°F.

Example 4-4 illustrates that the cooling process is more effective for gases at high pressures, where a large pressure drop can be provided to the gas in an LTX unit. As a rule of thumb, at least a 2,500- to 3,000-psi pressure drop must be available before an LTX unit can be considered for use. Otherwise, it may not be economical.

---

* A choke is essentially an orifice plate, with a specified hole size, for controlling the rate of flow or reducing the pressure of the flowstream.

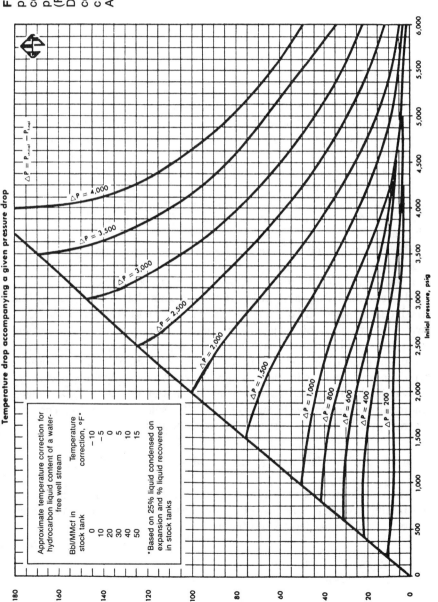

Figure 4-20. Temperature drop accompanying a given pressure drop. (From Engineering Data Book, 1981; courtesy of Gas Processors Suppliers Association.)

128 Gas Production Engineering

The lowest temperature at which an LTX unit can be used is governed by the properties of the material used in its construction. Generally, carbon steel is used, for which embrittlement occurs below a temperature of −20°F. For this reason, LTX units in an oil field are usually operated only in the temperature range of 0 to 20°F.

Gas Cleaning

Gas cleaning is important for pipeline transmission systems in order to reduce operational problems, and to maximize their operating efficiency. It is even more important in other instances, such as for gas storage, and for sales to consumers. Gas cleaning is also necessary to prevent solution/catalyst contamination in downstream processes on the gas, such as dehydration and sweetening. Some of the cleanup occurs initially at the wellhead by such means as drips, filters, and syphons. Another phase of cleanup is carried out in the gas-liquid separators. Further cleaning is generally required before the gas arrives at a processing plant, and before any processing is begun. A clean gas transmission averages about 2 lbm/MMscf particulate matter in the gas.

Gas cleaning involves the removal of two types of materials: (1) gross solids and liquids, called "pipeline trash," and (2) minute solids (particulate matter) and liquids (aerosols). Pipeline trash is also referred to as sludge, and is essentially anything in the flowing stream that is not gas. Pipeline trash consists generally of liquids such as heavier-end liquid hydrocarbons, water, chemicals (amines, glycols, methanol) carried over from processing operations, and solids such as drilling mud, pipeline scale, other dirt picked up by the gas during production and transport operations, and gas hydrates. Pipeline trash collects in the sags and low spots in the pipeline, and its removal is absolutely necessary. Particulate matter and aerosols are much more difficult to remove because of their ultra-small particle size. They are carried over by the gas as suspended solids/liquids.

General Equation for Particles Suspended in a Gas

The well known general expression for the terminal velocity, v, of a particle falling through a fluid under the influence of a force that exerts an acceleration a on the particle, is written as:

$$v = \left[\frac{4 a d_p^{n+1} (\rho_p - \rho_g)}{3 C_d \mu_g^n \rho_g^{1-n}} \right]^{1/(2-n)} \qquad (4\text{-}17)$$

where v = terminal velocity of the falling particle, ft/sec
   a = acceleration on the particle, ft/sec^2
   d_p = particle diameter, ft
   ρ_p = density of the particle, lbm/ft^3
   ρ_g, μ_g = density, lbm/ft^3, and viscosity, lbm/ft-sec, respectively, of the fluid through which the particle is falling

The drag coefficient, C_d, and the exponent n in Equation 4-17 are as follows (Lapple, 1984):

Flow regime	N_{Re}	Law	C_d	n
Laminar	<0.3	Stokes	24.0	1.0
Intermediate	0.3–10^3	Intermediate	18.5	0.6
Turbulent	10^3–2(10^5)	Newton	0.44	0

where the Reynolds number, N_{Re}, for the case of particles suspended in a gas is defined as being equal to $d_p v \rho_g / \mu_g$. The equations derived earlier for centrifugal separation (Equation 4-2) and for gravity settling (Equation 4-5) can be verified as being applicable to the turbulent flow regime.

The Reynolds number is controlled largely by the particle size, d_p, since other factors vary within a relatively small range for a given application. Thus, Newton's law would be applicable to larger particles that would possess a higher Reynolds number. This can be understood in physical terms by visualizing that larger particles fall so rapidly that they create considerable turbulence. For intermediate size particles, some turbulence is generated, and the intermediate law is used. Smaller particles settle much more slowly, a laminar flow regime is maintained, and Stokes law is used to calculate the settling rate.

As we proceed to smaller particle sizes, smaller than about 3 microns, Stokes law is no longer valid. The particles are now so small that they slip between the gas molecules at a rate greater than that predicted by Stokes law. The Cunningham correction, K_c, is used to correct the settling velocity, v_{Stokes}, from Stokes law:

$$v = K_c v_{Stokes} \tag{4-18}$$

This is called the Stokes-Cunningham equation. The correction K_c is determined as follows:

$$K_c = 1 + K_d \frac{3\mu_g(\pi M)^{0.5}}{\rho_g d_p (8 g_c RT)^{0.5}} \tag{4-19}$$

where M = molecular weight of the gas
R = gas constant, 1,546 ft-lbf/lbmole-°R (= 10.732 psia-ft^3/lbmole-°R)
T = temperature in °R
K_d = dimensionless proportionality factor
g_c = conversion factor (= 32.17 lbm-ft/lbf-sec^2)

For particles smaller than 3 microns, a random motion, known as Brownian movement, also begins to occur. Its effect is superimposed upon the particle settling velocity, and for particles under 0.1 micron, Brownian movement becomes the dominant phenomenon. Since gas cleaning is never really pursued to such levels, Brownian motion will not be discussed here.

Gas Cleaning Methods

There are several different techniques for separating liquid and solid particles from gas, such as gravity settling, centrifugal action, impingement, filtration, scrubbing, and electrostatic precipitation. Figure 4-21 shows the range of applicability of these methods in terms of the size of the particles that can be removed. Note: 1 micron = 10^{-4} cm.

Maintenance requirements are generally proportional to the removal capability. Thus, methods with greater particle removal capability usually require more elaborate maintenance schedules. In cleaning methods that use a physical separation device (cleaning element, demister, filter, etc.), once the cleaning element has accumulated to its capacity, there are three possible results, all of which are detrimental to the cleaning process (Curry, 1981):

1. The incoming particles bypass the cleaning element and enter the outflowing clean gas stream. This is the most common problem.
2. The pressure differential becomes too high, leading to rupture or dislodgement of the cleaning element.
3. The cleaning element may become so impregnated that it interrupts the flow completely.

Proper maintenance is therefore essential for these devices.

Some common gas cleaning methods, besides the gravity settling and centrifugal methods discussed earlier, are described below.

Impingement

There are several types of impingement separators as shown in Figure 4-22. The mist extraction section in an oil and gas separator uses impinge-

Gas and Liquid Separation   131

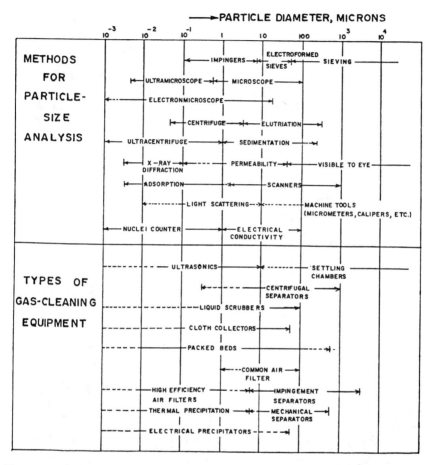

Figure 4-21. Gas cleaning methods and their range of application. (Data from C. E. Lapple, 1984.)

ment methods of two basic types: wire-mesh pad and vane-type mist extractors.

A wire-mesh separator consists of a 4–6 in. thick pad of fine wire (diameter of 0.003–0.011 in.) knitted into a mesh fabric with a high void volume (97–99%). The pad is mounted in a horizontal position, with the vapor flowing upwards. Liquid droplets impinge on the wire and collect and coalesce with other droplets, until their size becomes large enough to drop downwards through the void spaces. The droplets then cling to the bottom of the wire-mesh by surface tension forces. Here they coalesce again to a bigger size, until they become sufficiently large to overcome surface tension

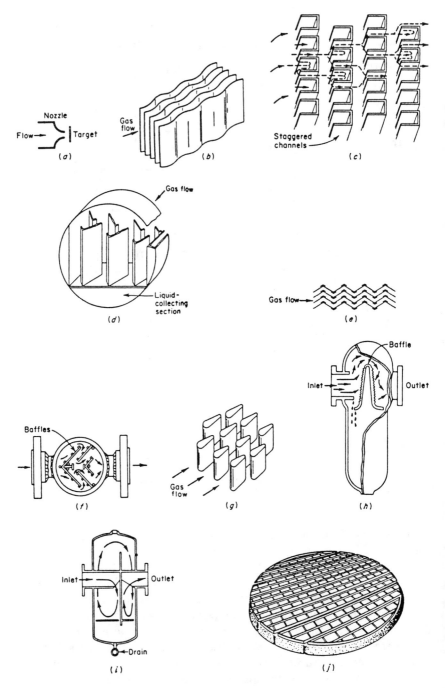

Figure 4-22. Types of impingement separators. (After Perry and Green (eds.), *Chemical Engineers' Handbook,* 1984; courtesy of Mc-Graw Hill Book Co.)

and the force resulting from the upward flow of the gas. A flow velocity of 5–10 ft/sec provides maximum operation efficiency (Curry, 1981). A higher rate would flood the pad and lead to re-entrainment of the liquid, whereas a lower rate may permit the liquid to circumvent the pad and avoid impingement. A mist extractor should not be used where the gas has a high concentration of solids, since this may block the flow of gas and result in high pressure drops across the pad. For liquid droplets, wire-mesh pads are efficient, can remove droplets down to 4 microns in size, and have a high handling capacity.

The vane-type design uses an intricate array of metal plates, called vanes, with liquid collection pockets. The vane-type mist extractor is mounted such that the gas stream flows horizontally through the vanes. During this flow, a change in direction is induced several times, resulting in a centrifugal action that aids the primary impingement separation mechanism in removing the finer liquid droplets entrained in the gas. The liquid droplets are forced into the liquid collection pockets, out of the gas flow path, and drain out by gravity. The pressure drop across a vane-type mist extractor is very small, it can handle solids in the flowing gas stream, and can remove droplets down to about 40 microns in size.

Another development of interest is the fiber mist eliminator. This uses a packed bed of fibers between two concentric screens. Liquid mist collects on the fiber surface as a film, which is moved horizontally by the gas, and downwards by gravity, into the liquid collection area. Fiber mist collectors offer extremely high collection efficiencies up to 99.98%, and can handle mists smaller than 3 microns (Brink, 1963).

Filters

These have been traditionally used to remove solid particles by using a filtration medium that allows only the gas to pass through. Bag filters, using woven fabric or compressed felt fabric as the filtration medium, are widely used. But these materials break down in the presence of liquids. Synthetic materials such as glass fiber overcome this disadvantage. Even with these synthetic materials, liquid separation with filters is always a problem. Liquid particles find their way through the filter pad and coalesce on the downstream side of the pad in the form of a liquid film. The gas passing through this film leads to the formation of bubbles, which disintegrate and are re-entrained in the gas.

Scrubbers

A scrubber is defined as any equipment that uses a liquid to aid the removal of particles from a gas. It is similar to a separator, except that a scrubber is designed to separate only small volumes of gas and liquid, and it may use some fluid such as oil to more effectively remove particles from the gas. Unlike separators, surge capacity is neither necessary nor provided. A scrubber, although similar to a separator, must never be used for gas-liquid separation in the field, where even a relatively dry gas well may produce some liquid surge and lead to severe separation problems. They may, however, be used as secondary separation devices.

The three types of scrubbers used in gas cleaning operations are: dry scrubbers, oil-bath scrubbers, and cartridge-type scrubbers. Dry scrubbers are similar to centrifuges, using centrifugal force to effect the separation of solids and liquids from gas. Oil-bath scrubbers are extensively used. They cause solid particulate matter to impinge on a surface constantly flushed with oil, and cling to it. The circulating oil washes the dirt down into a settling chamber, where the dirt settles down by gravity. Clean oil is recirculated. It is important in an oil-bath scrubber to ensure that the oil is clean, and that the proper level of oil is maintained. A less-than-clean oil or too low an oil level will reduce cleaning efficiency, whereas a high oil level will cause carryover of the oil into the gas. Dry and oil-bath type of scrubbers can be effective down to almost a 4-micron particle size.

The most effective scrubbers are the cartridge-type scrubbers (Curry, 1981). They use cleaning cartridges stacked in parallel, in different configurations, that can remove solid particulate matter down to a size of 0.3 micron. For liquid removal, mist extractors are also provided sometimes. Cartridge scrubbers, however, require constant maintenance and cartridge replacement, and are more expensive than oil-bath and dry scrubbers.

Electric Precipitators

These units induce an electrical charge that attracts particulate matter. A strong electrostatic field is provided that ionizes the gas to some extent. The particles suspended in this partially ionized gas become charged and migrate under the action of the applied electric field. The gas is retained for a long enough time for the particles to migrate to a collection surface.

Currently, the application of electric precipitators to large volumes of high-pressure gas is quite limited. In the future, though, these pollution-free devices may prove to be more useful.

Flash Calculations

Two-phase flash calculation techniques were described in Chapter 2. This section describes some methods that can be used to accelerate convergence to the correct result. The nomenclature used here is the same as in Chapter 2. F moles of feed with a composition $\{z_i\}$ split into L moles of liquid with a composition $\{x_i\}$ and V moles of vapor with a composition $\{y_i\}$. For simplicity, it is assumed in the present discussion that flash calculations are carried out on the basis of a unit mole of feed (F = 1). Thus,

$$F = 1 = L + V$$

and

$$Fz_i = z_i = Lx_i + Vy_i$$

Also, y_i and x_i are related by the equilibrium ratio, K_i:

$$K_i = y_i/x_i$$

Determination of Equilibrium Ratios

To enable flash calculations, the equilibrium ratio, K, is required. Experimental values for the particular system are the most reliable source for K-values. The next best thing is to use equations of state for consistent predictions. For manual calculations, K-value charts are also useful, though they may not be very accurate.

The calculation of K-values from equations of state has been discussed in Chapter 3. These methods are used for implementation on a computer, and provide internally consistent values for all thermodynamic properties in a convenient form.

The chart method for estimating K-values is subject to an important constraint: K-values have been found to exhibit some dependence on the mixture from which they are measured, and therefore there is no single K-value chart that is superior for all possible mixtures that may be encountered. A sharp distinction exists in K-values for the same components in a natural gas liquid versus a crude oil. These differences among K-values for different mixtures have been characterized by the concept of *convergence pressure*, which represents the composition of the vapor and liquid phases in equilibrium. Convergence pressure, at a given system temperature, is defined as the pressure at which the K-values of a fixed-composition system converge toward a common value of unity. Thus, convergence pressure for a system is the pressure (at the given system temperature) at or beyond which vapor-liquid separation is no longer possible. This is similar to the concept of critical pressure, and quite often the convergence pressure is taken to be the critical pressure of the system at the given temperature.

Figures 4-24 through 4-50 show the K-value charts for H_2S, N_2, and hydrocarbons C_1 through C_{10} for convergence pressures of 1,000 psia and 3,000 psia. It can be seen in these figures that at relatively low pressures, K-values are almost independent of convergence pressure; at higher pressures approaching the convergence pressure, K-values are very sensitive to convergence pressure.

For CO_2, the K-value can be calculated from methane and ethane K-values as follows (GPSA, 1981):

$$K_{CO_2} = (K_{C_1} K_{C_2})^{0.5}$$

The convergence pressure p_k, required for selecting the appropriate K-value chart, has been found to be a function of two parameters—temperature and the liquid phase composition. Temperature of the system is generally known, but the composition of the exit liquid stream from the vapor-liquid separation is not known in advance (we are trying to determine both the vapor as well as liquid stream compositions using K-value data). Therefore, the procedure for determining the convergence pressure is iterative, as described in the following (GPSA, 1981):

1. Assume a liquid-phase composition. The total feed composition may be used as a first guess if no better estimates can be made.
2. Identify the lightest hydrocarbon component which is present at least 0.1 mole% in the liquid phase.
3. Calculate the *weighted* average critical pressure and temperature for the remaining heavier components in the liquid to form a pseudo-binary system. A shortcut approach good for most hydrocarbon systems is to compute the weighted average critical temperature only.
4. Trace the critical locus of the binary on Figure 4-23. When the averaged pseudo-heavy component of the binary is between two real hydrocarbons, an interpolation of the two critical loci must be made.

(text continued on page 164)

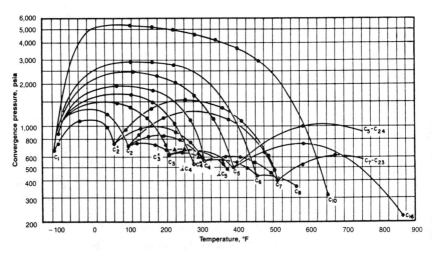

Figure 4-23. Convergence pressure p_k of binary hydrocarbon mixtures. (From *Engineering Data Book,* 1981; courtesy of GPSA.)

Gas and Liquid Separation 137

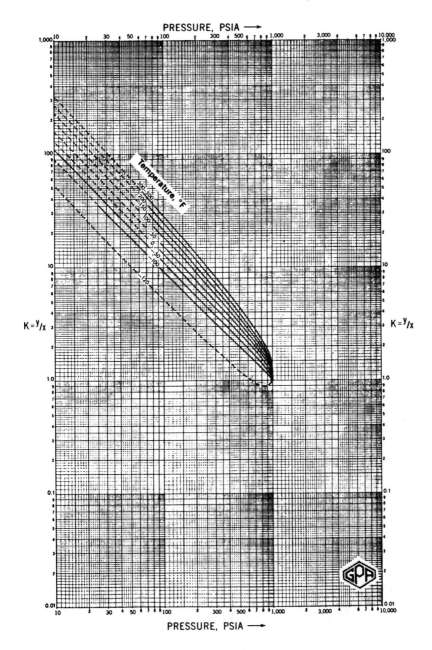

Figure 4-24. Equilibrium ratio, K, for methane for a convergence pressure of 1,000 psia. (From *Engineering Data Book*, 1981; courtesy of GPSA.)

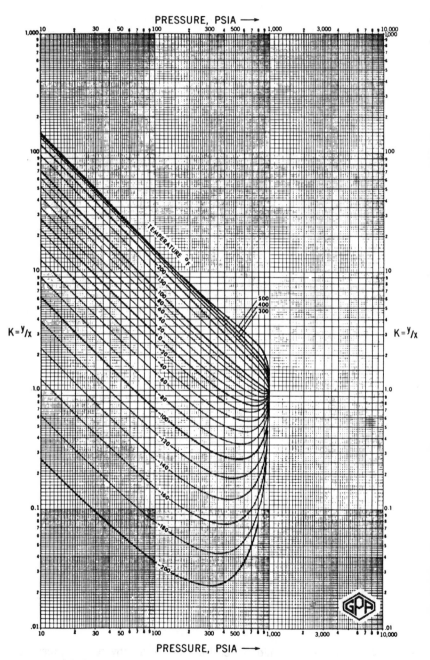

Figure 4-25. Equilibrium ratio, K, for ethylene for a convergence pressure of 1,000 psia. (From *Engineering Data Book,* 1981; courtesy of GPSA.)

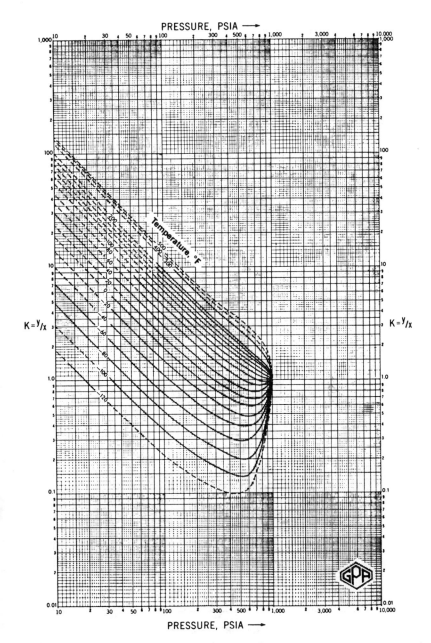

Figure 4-26. Equilibrium ratio, K, for ethane for a convergence pressure of 1,000 psia. (From *Engineering Data Book*, 1981; courtesy of GPSA.)

140 Gas Production Engineering

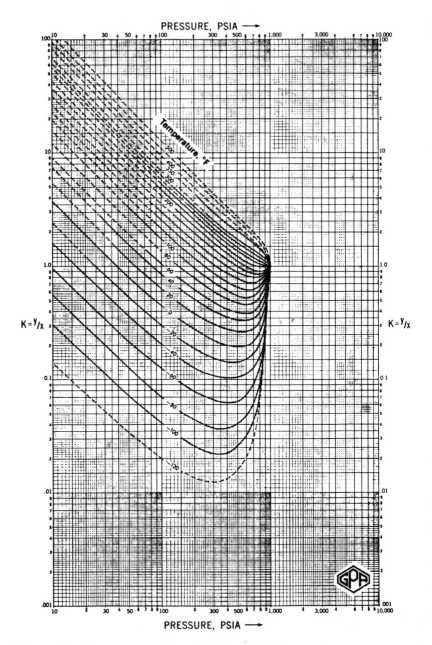

Figure 4-27. Equilibrium ratio, K, for propane for a convergence pressure of 1,000 psia. (From *Engineering Data Book*, 1981; courtesy of GPSA.)

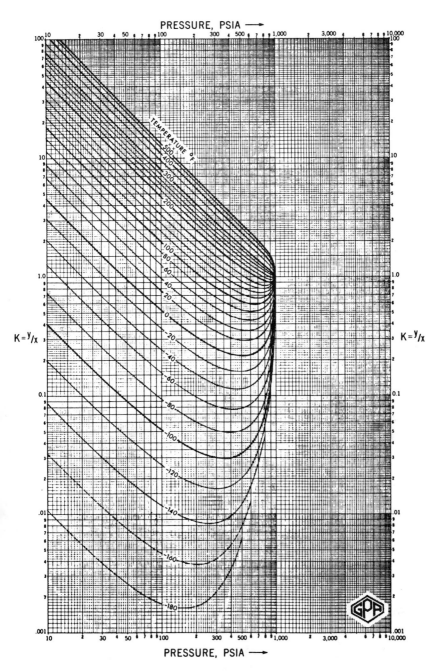

Figure 4-28. Equilibrium ratio, K, for propylene for a convergence pressure of 1,000 psia. (From *Engineering Data Book,* 1981; courtesy of GPSA.)

142 Gas Production Engineering

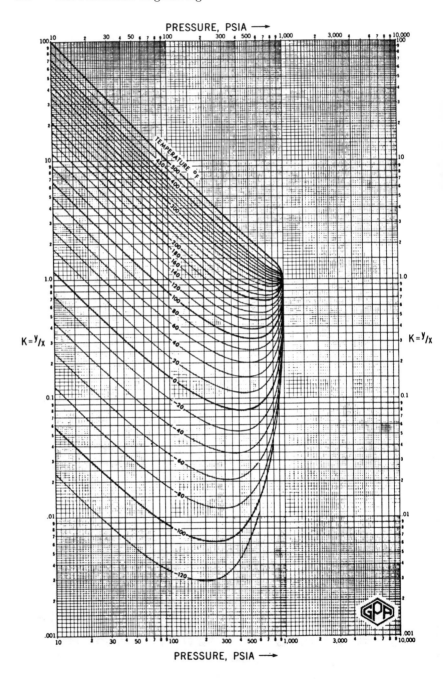

Figure 4-29. Equilibrium ratio, K, for iso-butane for a convergence pressure of 1,000 psia. (From *Engineering Data Book,* 1981; courtesy of GPSA.)

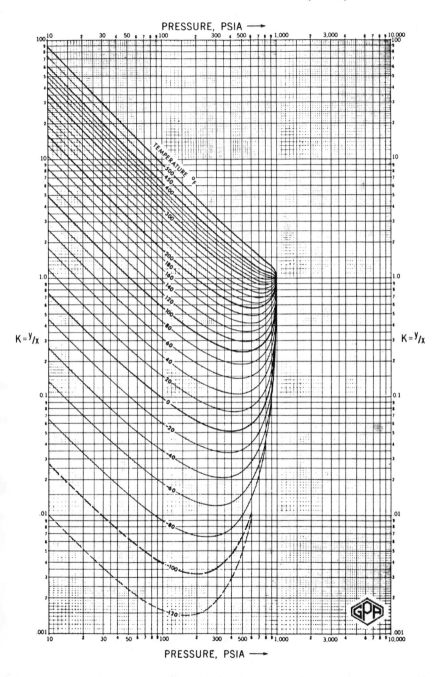

Figure 4-30. Equilibrium ratio, K, for n-butane for a convergence pressure of 1,000 psia. (From *Engineering Data Book,* 1981; courtesy of GPSA.)

144 Gas Production Engineering

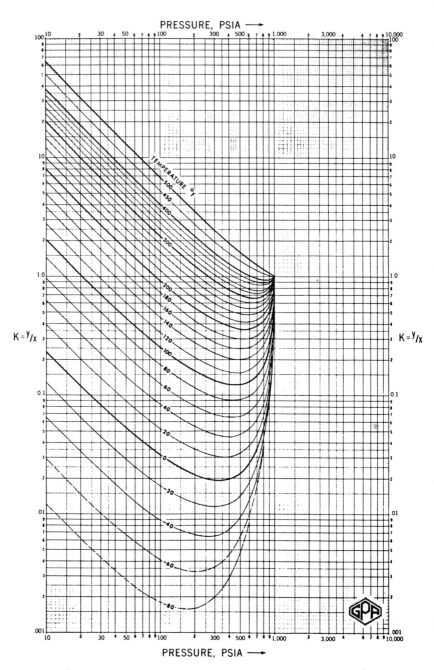

Figure 4-31. Equilibrium ratio, K, for i-pentane for a convergence pressure of 1,000 psia. (From *Engineering Data Book,* 1981; courtesy of GPSA.)

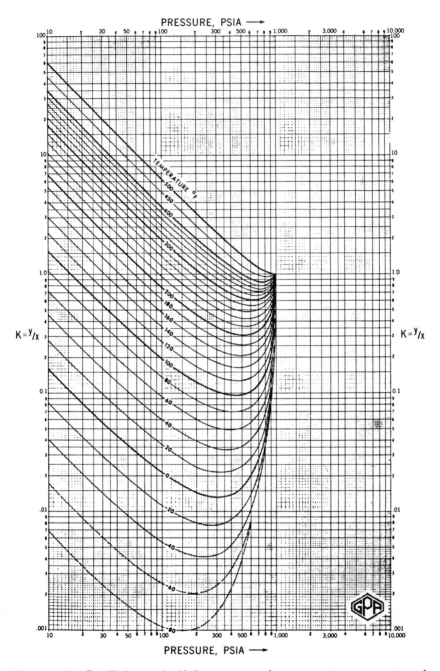

Figure 4-32. Equilibrium ratio, K, for n-pentane for a convergence pressure of 1,000 psia. (From *Engineering Data Book,* 1981; courtesy of GPSA.)

146 Gas Production Engineering

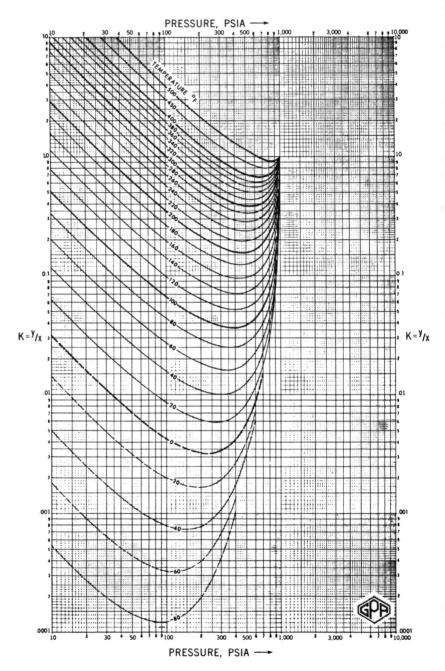

Figure 4-33. Equilibrium ratio, K, for hexane for a convergence pressure of 1,000 psia. (From *Engineering Data Book,* 1981; courtesy of GPSA.)

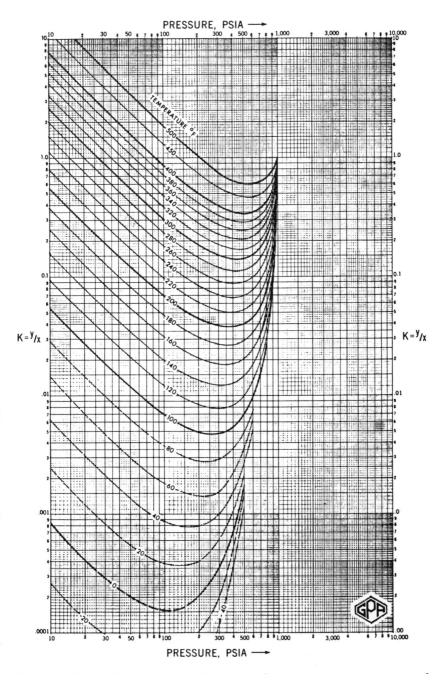

Figure 4-34. Equilibrium ratio, K, for octane for a convergence pressure of 1,000 psia. (From *Engineering Data Book,* 1981; courtesy of GPSA.)

148 *Gas Production Engineering*

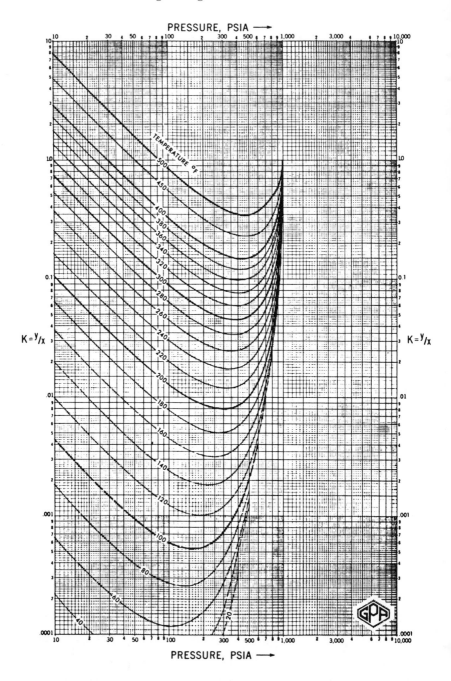

Figure 4-35. Equilibrium ratio, K, for decane for a convergence pressure of 1,000 psia. (From *Engineering Data Book,* 1981; courtesy of GPSA.)

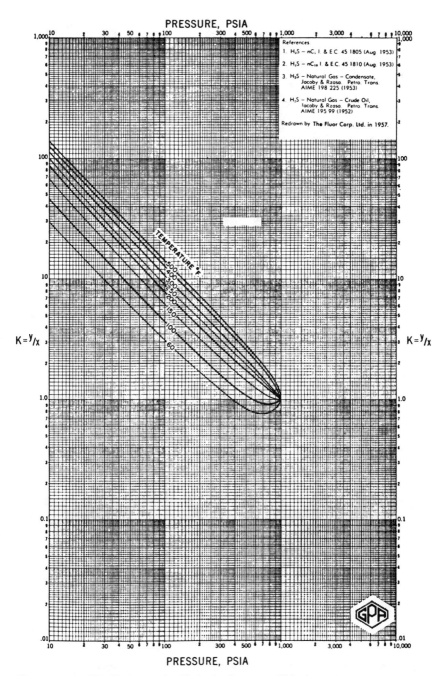

Figure 4-36. Equilibrium ratio, K, for hydrogen sulfide for a convergence pressure of 1,000 psia. (From *Engineering Data Book,* 1981; courtesy of GPSA.)

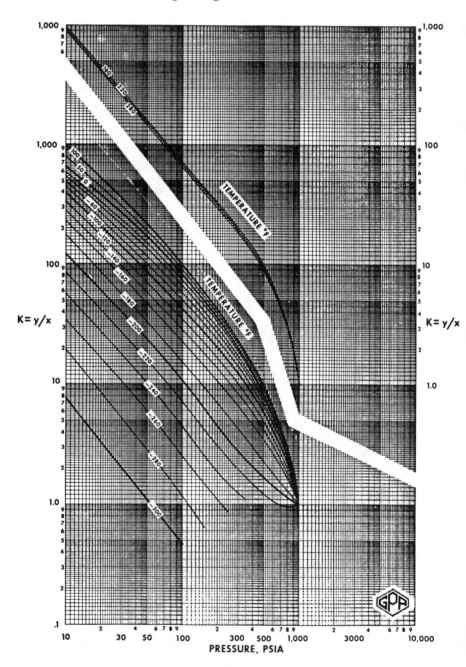

Figure 4-37. Equilibrium ratio, K, for nitrogen for a convergence pressure of 1,000 psia. (From *Engineering Data Book,* 1981; courtesy of GPSA.)

Gas and Liquid Separation 151

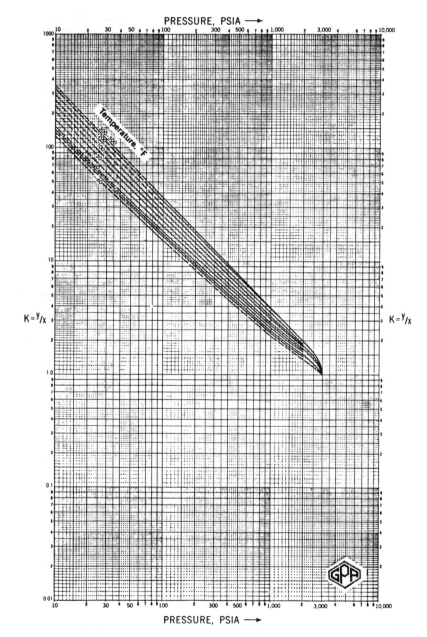

Figure 4-38. Equilibrium ratio, K, for methane for a convergence pressure of 3,000 psia. (From *Engineering Data Book,* 1981; courtesy of GPSA.)

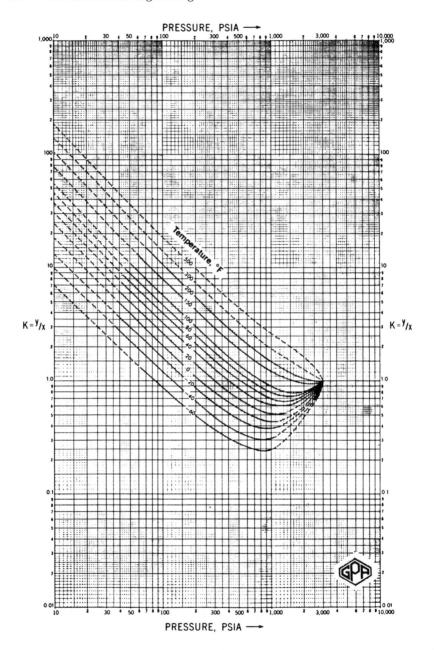

Figure 4-39. Equilibrium ratio, K, for ethane for a convergence pressure of 3,000 psia. (From *Engineering Data Book,* 1981; courtesy of GPSA.)

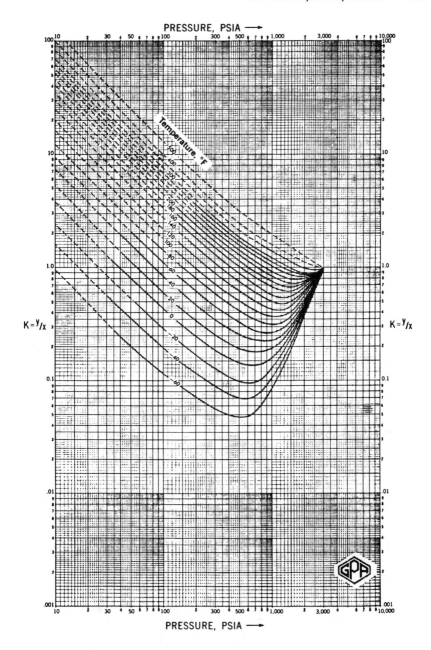

Figure 4-40. Equilibrium ratio, K, for propane for a convergence pressure of 3,000 psia. (From *Engineering Data Book,* 1981; courtesy of GPSA.)

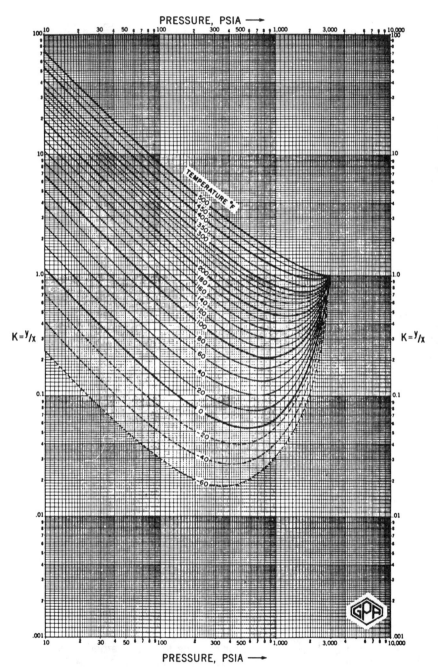

Figure 4-41. Equilibrium ratio, K, for iso-butane for a convergence pressure of 3,000 psia. (From *Engineering Data Book,* 1981; courtesy of GPSA.)

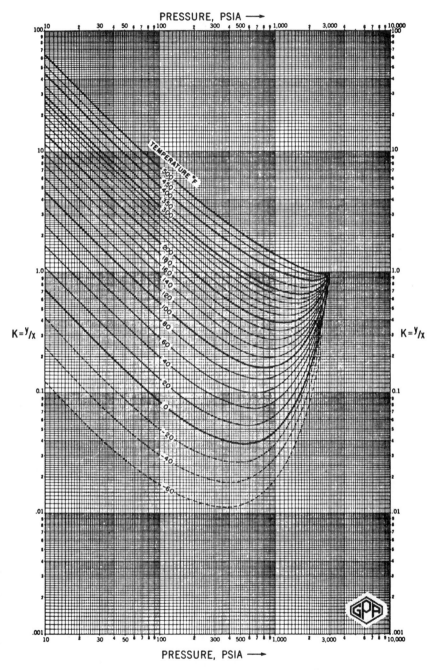

Figure 4-42. Equilibrium ratio, K, for n-butane for a convergence pressure of 3,000 psia. (From *Engineering Data Book,* 1981; courtesy of GPSA.)

156 *Gas Production Engineering*

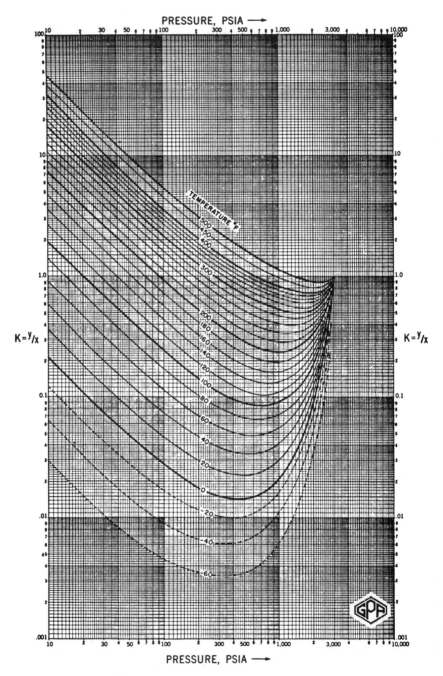

Figure 4-43. Equilibrium ratio, K, for i-pentane for a convergence pressure of 3,000 psia. (From *Engineering Data Book,* 1981; courtesy of GPSA.)

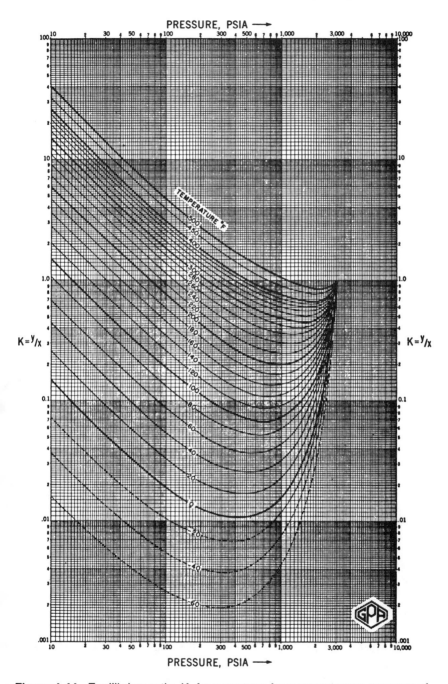

Figure 4-44. Equilibrium ratio, K, for n-pentane for a convergence pressure of 3,000 psia. (From *Engineering Data Book,* 1981; courtesy of GPSA.)

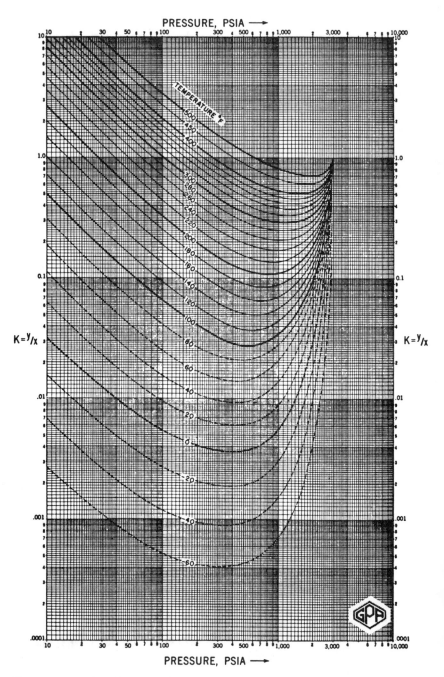

Figure 4-45. Equilibrium ratio, K, for hexane for a convergence pressure of 3,000 psia. (From *Engineering Data Book,* 1981; courtesy of GPSA.)

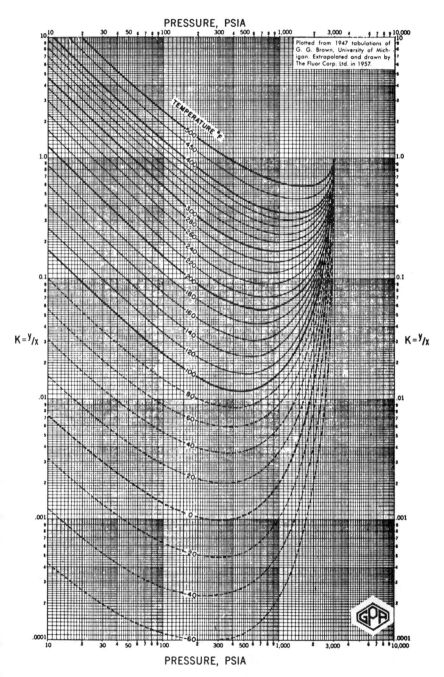

Figure 4-46. Equilibrium ratio, K, for heptane for a convergence pressure of 3,000 psia. (From *Engineering Data Book,* 1981; courtesy of GPSA.)

160 Gas Production Engineering

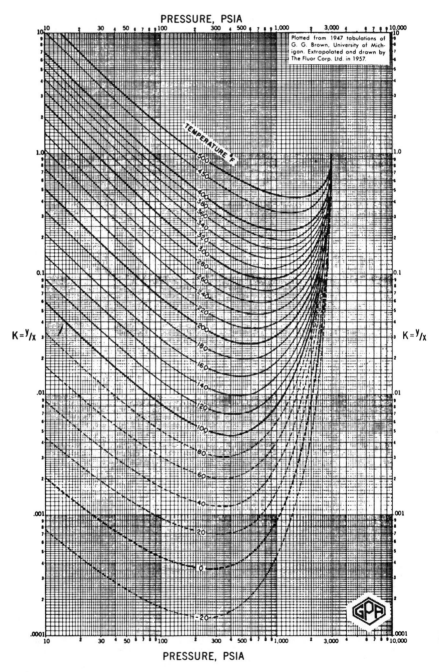

Figure 4-47. Equilibrium ratio, K, for octane for a convergence pressure of 3,000 psia. (From *Engineering Data Book,* 1981; courtesy of GPSA.)

Gas and Liquid Separation 161

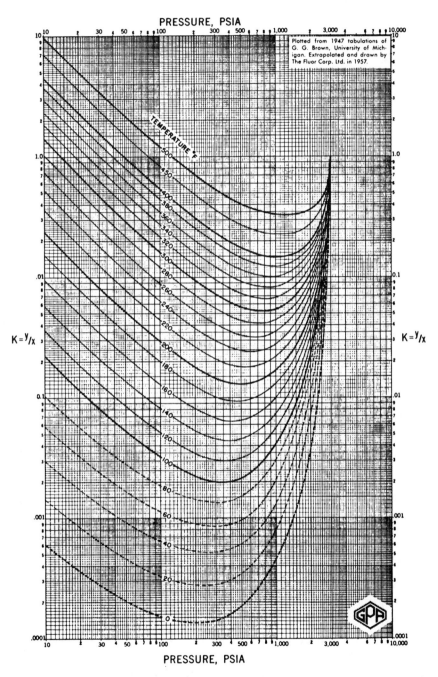

Figure 4-48. Equilibrium ratio, K, for nonane for a convergence pressure of 3,000 psia. (From *Engineering Data Book,* 1981; courtesy of GPSA.)

162 *Gas Production Engineering*

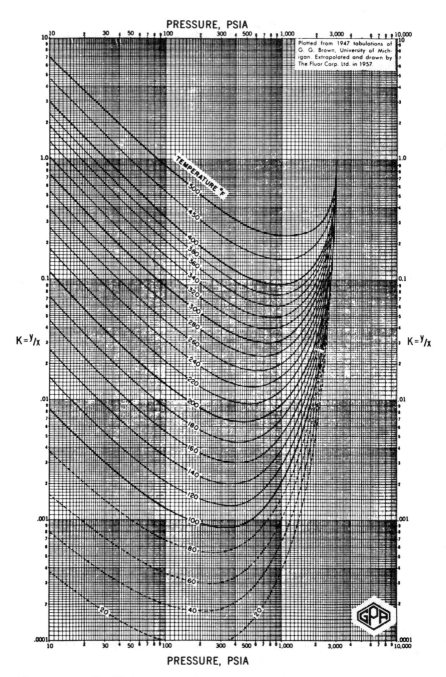

Figure 4-49. Equilibrium ratio, K, for decane for a convergence pressure of 3,000 psia. (From *Engineering Data Book,* 1981; courtesy of GPSA.)

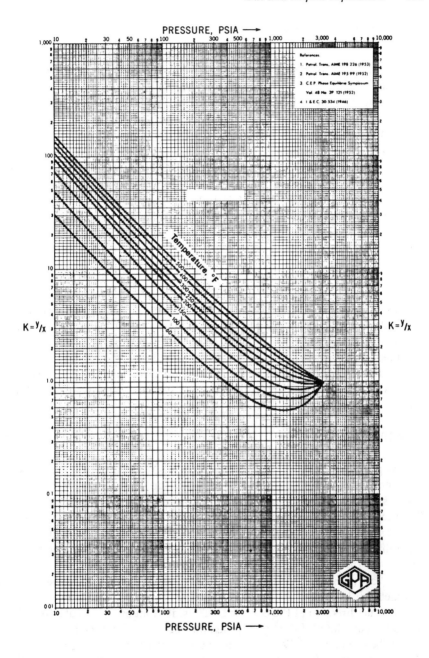

Figure 4-50. Equilibrium ratio, K, for hydrogen sulfide for a convergence pressure of 3,000 psia. (From *Engineering Data Book,* 1981; courtesy of GPSA.)

(text continued from page 136)

5. Read the convergence pressure p_k at the temperature corresponding to that of the desired flash conditions.
6. Using p_k from Step 5, together with the system temperature and pressure, obtain K-values for the components from the appropriate convergence-pressure K-value charts. Interpolate for convergence pressure between the charts, if necessary.
7. Make a flash calculation with the feed composition and the K-values from Step 6.
8. Repeat Steps 2 to 7 until the assumed and calculated p_k check within an acceptable tolerance.

Flash Calculation Methods

Simple Trial and Error Scheme

This is the simplest method. A value of the vapor-liquid ratio is assumed. Using this $(V/L)_{a1}$ in Equation 2-16 (or 2-17) along with the appropriate component K values, a new V (or L) is obtained. The value of L (or V) is obtained by difference (L = 1 − V). If this calculated $(V/L)_c$ is equal to the assumed $(V/L)_{a1}$ within a specified tolerance, the assumption was correct. Otherwise, a new value $(V/L)_{a2}$ for the liquid-vapor ratio is assumed as follows:

$$(V/L)_{a2} = (V/L)_c$$

This procedure is repeated till convergence is reached. After one or two iterations, it becomes quite obvious whether the system is predominantly gaseous (V < 0.5), or liquid (V > 0.5). For a predominantly gaseous system, Equation 2-16 is used to calculate V, and L is obtained by difference, whereas if the system is predominantly liquid, Equation 2-17 is used and V is obtained by difference. The use of either Equation 2-16 or 2-17 is entirely analogous; the only difference is in the accuracy of the results.

Lockhart (1983) has made an important observation regarding the use of this method: the value of the calculated L, as compared with the assumed L, gives the direction of the correct L. Thus, if at an assumed value of L/V of 0.5 the calculated $(L/V)_c < 0.5$, then the correct L/V will be less than $(L/V)_c$. If the calculated $(L/V)_c > 0.5$, then the correct L/V will be greater than $(L/V)_c$. Of course, if the calculated $(L/V)_c = 0.5$ exactly for an assumed L/V of 0.5, then convergence has been reached.

Method of Successive Approximations

This method is similar to the simple trial and error scheme just outlined. The only difference is in assuming a new value at the end of an unconverged trial (Lockhart et al., 1986):

$$(V/L)_{a2} = (V/L)_c + [(V/L)_c - (V/L)_{a1}]$$

Thus, the new assumed value of V/L is made as far beyond $(V/L)_c$ as $(V/L)_c$ is beyond $(V/L)_{a1}$. This scheme gives a much better convergence rate.

Lockhart-McHenry Method

Lockhart and McHenry (1958) proposed a method that reduces the number of trials required for flash-equilibrium calculations for multicomponent mixtures significantly. In this method, a multicomponent mixture is treated as a pseudo-binary mixture with a "light" component, and a "heavy" component. Components with a K value greater than 1.0 are grouped into the light component, and those with a K value less than 1.0 are grouped into the heavy component.

Consider 1 mole of a multicomponent feed with a mole-fraction z_l of the light components, and z_h of the heavy components. The total number of components in the mixture is n, with n_l number of light components and n_h heavy components. In this technique, a parameter V^* is defined as follows:

$$V^* = z_l/(1 - K_h^*) - z_h/(K_l^* - 1)$$

where K_h^* and K_l^* denote the K^* values for the heavy and light components, respectively, in the binary mixture. The K^* parameter is defined as follows:

$$K^* = \frac{\Sigma K_i z_i}{\Sigma z_i} \text{ for } (V)_{assumed} = 0$$

$$K^* = \frac{\Sigma z_i}{\Sigma(z_i/K_i)} \text{ for } (V)_{assumed} = 1$$

$$K^* = \frac{\Sigma z_i}{\Sigma[z_i/(K_i + 1)]} - 1 \text{ for } (V)_{assumed} = 0.5$$

$$K^* = (L/V)\left[\frac{\Sigma z_i}{\Sigma L x_i} - 1\right] \text{ for } (V)_{assumed} = 0.5 \text{ to } 1$$

$$K^* = \frac{L/V}{(\Sigma z_i / \Sigma V y_i) - 1} \text{ for } (V)_{assumed} = 0 \text{ to } 0.5$$

The summation Σ in these relationships represents the subtotals over all pure components i for the light component and the heavy component, as the case may be. Thus, in calculating K_l^* for the light component, the summation Σ is done for $i = 1$ to n_l. Similarly, for a K_h^* calculation, the summation Σ is carried out for $i = 1$ to n_h.

For different values of $V_{assumed}$, V^* is calculated. At convergence, $V^* = V_{assumed}$. Alternatively, a graphical method can be used, where two lines are plotted on a Cartesian graph: (1) V^* versus $V_{assumed}$ and (2) $V^* = V_{assumed}$. The intersection of these two lines gives the correct answer.

A Note on Flash Calculations for Stage Separation

For an isothermal flash calculation, the separator temperature is known, and is kept at a constant value by suitable heat exchange devices in the separator. The trial and error flash calculation procedure involves determining the separator pressure. If there is no heat exchanger, or the temperature is not known, an adiabatic flash calculation is made, involving a double trial and error: one for separator pressure, and another for the temperature. Usually, a separator pressure is chosen, and trial and error done for a corresponding separator temperature that makes the feed enthalpy equal to the product enthalpy. A reasonable separator temperature is assumed, and flash calculations are made exactly as for an isothermal case.

Questions and Problems

1. Describe the three major components of oil-water-gas separators.
2. What happens to separation quality and the separator when—
   (a) the gas flow rate exceeds the allowable rate through the separator?
   (b) the liquid flow rate exceeds the allowable rate through the separator?
   (c) mist extractor becomes plugged?
   (d) the wellstream produces slugs of liquid (oil and water) into the separator?
3. What type of separator(s) should be used for—
   (a) offshore production platform?
   (b) high GOR well, with liquid surges?

(c) low GOR well producing oil, water, and gas?
   (d) heavy, waxy crude (GOR can be assumed to be low for heavy oils)?
   (e) gas condensates with a hydrate problem?
   (f) nearly dry gas at high pressure?
4. A 20 ft × 10 ft 1,000 psi W.P. (working pressure) horizontal separator is operated at ½ full-of-liquid conditions. Can it be used on a well with a gas flow rate of 9 MMscf/day and a line pressure of 500 psig? If a gas back-pressure valve were put on the separator, what pressure should it be set at to handle the gas flow rate?
5. Design a vertical separator, with a mist extractor, that can handle 6.66 MMscf/day of a 0.80 gravity gas. The oil gravity is 45°API, operating pressure is 400 psia, and the operating temperature is 60°F. Use basic separation relationships for this design, and compare the results with Sivalls' charts. The well produces 153 bbl of liquids per day
6. A gas field delivers 13.4 MMscf/day of a 0.68 gravity gas and 300 bbl/day oil with a 50% water cut. For an operating pressure of 1,000 psig and temperature of 80°F—

   (a) Determine the ID of the spherical separator required to accommodate the liquid and gas, assuming a retention time of 5 min.
   (b) How much more gas can be flowed through the separator without worrying about the gas capacity?

7. A 1.25 gravity gas condensate stream at 1,200 psia and 110°F is to be separated into oil and gas using a three-stage separation.
   (a) Assuming that the mole% $C_1 + C_2 + C_3$ in the gas is 80%, what is the optimum second-stage pressure?
   (b) Choose and design the appropriate separators for each of these three stages.

8. Describe the applicability of the various gas cleaning methods in natural gas production and processing.

References

Beggs, H. D., 1984. *Gas Production Operations*. Oil & Gas Consultants International, Inc., Tulsa, Oklahoma, 287pp.

Brink, J. A., Jr., 1963. "Air Pollution Control with Fibre Mist Eliminators," *Cdn. J. Chem. Eng.*, 41(3, June), 134–138.

Campbell, J. M., 1955. "Know Your Separators," *O. & Gas J.*, 53(45, Mar. 14), 107–111.

Campbell, J. M., 1984. *Gas Conditioning and Processing, Vol. 2.* Campbell Petroleum Series, Norman, Oklahoma, 398pp.

Craft, B. C., Holden, W. R., and Graves, E. D., Jr., 1962. *Well Design: Drilling and Production.* Prentice-Hall Inc., Englewood Cliffs, New Jersey, 571pp.

Curry, R. N., 1981. *Fundamentals of Natural Gas Conditioning.* PennWell Publ. Co., Tulsa, Oklahoma, 118pp.

Gas Processors Suppliers Association, 1981. *Engineering Data Book*, 9th ed. (5th revision), GPSA, Tulsa, Oklahoma.

Ikoku, C. U., 1984. *Natural Gas Production Engineering.* John Wiley & Sons, Inc., New York, 517pp.

Lapple, C. E., 1984. Cited reference in: *Chemical Engineers' Handbook*, by R. H. Perry and D. W. Green (eds.), 6th ed. McGraw-Hill Book Co., Inc., New York, pp. 5–63.

Lockhart, F. J., 1983. *Personal communication.*

Lockhart, F. J., Chilingarian, G. V., and Kumar, S., 1986. "Separation of Oil and Gas," in: *Surface Operations in Petroleum Production, Vol. 1*, by G. V. Chilingarian, J. O. Robertson, and S. Kumar, Elsevier Scientific Publishing Company, Amsterdam.

Lockhart, F. J. and McHenry, R. J., 1958. "Figure Flash Equilibrium Easier, Quicker This Way," *Petr. Refiner*, 37(3), 209–212.

Petroleum Extension Service, 1972. *Field Handling of Natural Gas.* 3rd ed., Univ. of Texas Press, Austin, Texas, 143pp.

Sivalls, C. R., 1977. "Fundamentals of Oil and Gas Separation," Proc. Gas Conditioning Conf., Univ. of Oklahoma, 31pp.

Whinery, K. F. and Campbell, J. M., 1958. "A Method for Determining Optimum Second-Stage Pressure in Three-Stage Separation," *J. Pet. Tech.*, 10(4, Apr.), 53–54.

5
Gas-Water Systems and Dehydration Processing

Introduction

Water vapor is the most common undesirable impurity found in natural gas. By virtue of its source, natural gas is almost always associated with water, usually in the range of 400–500 lb water vapor/MMscf gas. The primary reason for the removal of water from gas is the problem of gas hydrate formation. Liquid water with natural gas may form solid, ice-like hydrates that plug flowlines and lead to severe operational problems. Other reasons for removing water are: (1) liquid water promotes corrosion, particularly in the presence of H_2S and CO_2; (2) slugging flow may result if liquid water condenses in the flowlines; and (3) water vapor reduces the heating value of the gas. For these reasons, pipeline specifications for natural gas restrict the water content to a value not greater than 6–8 lbm/MMscf.

Because most gas sweetening processes involve the use of an aqueous solution, dehydration is often done after desulfurization. Nevertheless, partial dehydration, or hydrate inhibition, are commonly necessary at the wellsite itself.

Water Content of Natural Gases

In order to design and operate dehydration processes, a reliable estimate of the water content of natural gas is essential. The water content of a gas depends upon:

1. Pressure—water content decreases with increasing pressure.
2. Temperature—water content increases with increasing temperature.

170   Gas Production Engineering

3. Salt content of the free water in equilibrium with the natural gas in the reservoir—water content decreases with increasing salt content of the associated reservoir water.
4. Gas composition—higher gravity gases usually have less water.

The terms *dew point* and *dew-point depression* are widely used in dehydration terminology. Dew point indirectly indicates the water content of a natural gas, and is defined as the temperature at which the gas is saturated with water vapor at a given pressure. The difference between the dew-point temperature of a gas stream before and after dehydration is called the dew-point depression. Consider a gas saturated with water at 500 psia and 100°F. Its dew point is 100°F, and its water content is approximately 100 lbm/MMscf gas. The gas is to be transported in a pipeline at 60°F. Under pipeline conditions of 500 psia and 60°F, the water vapor content of the gas is only about 30 lbm/MMscf. Thus, 70 lb water per MMscf of gas exist as free water in the pipeline. If the dew point of the inlet gas to the pipeline is reduced to 60°F, no free water will exist in the pipeline at the pipeline flow conditions. In other words, the dehydration facility should give a dew point depression of 100 − 60 = 40°F. In practice, although a dew point depression of 40°F is just sufficient, a 50°F dew point depression may be desirable for operational safety.

The methods available for calculating the water content of natural gases fall into three categories: (1) partial pressure approach, valid up to about 60 psia (gases exhibit almost ideal behavior up to 60 psia in most cases); (2) empirical plots; and (3) equations of state.

Partial Pressure Approach

Assuming ideal gas and ideal mixture behavior, the partial pressure of water in the gas is given by $p_w = py_w$, and also by $p_w = p_v x_w$. Thus,

$$py_w = p_v x_w \tag{5-1}$$

where p = absolute pressure of the gas
  y_w = mole fraction of water in the vapor (gas) phase
  p_v = vapor pressure of water at the system temperature
  x_w = mole fraction of water in the aqueous (liquid water) phase associated with the gas phase under equilibrium conditions

Since water is almost immiscible in the liquid phase with oil, x_w is usually assumed to be equal to unity. Thus, the water mole fraction in the gas, y_w, can be calculated as:

$$y_w = p_v/p \qquad (5\text{-}2)$$

This simple approach has limited applicability at the pressure and temperature of interest in most natural gas production, processing and transport systems.

Empirical Plots

For engineering calculations, empirical plots are widely used. Numerous investigations have resulted in several plots, such as by McCarthy et al. (1950), McKetta and Wehe (1958), the Gas Processors Suppliers Association (GPSA), Campbell (1984a), Robinson et al. (1978), and others. Since the real danger in design is to underestimate the water content, all correlations assume the gas to be completely saturated with water, and most correlations are designed to slightly overestimate the water content. Nitrogen holds about 6–9% less water than methane (Campbell, 1984a). Therefore, nitrogen is frequently included as a hydrocarbon, providing a small safety factor.

McKetta and Wehe Correlation for Sweet Gases   for $\gamma_g = 0.6$

The McKetta and Wehe correlation (1958), shown in Figure 5-1, includes correction factors for gas gravity and formation water salinity. It gives acceptable results for sweet gases.

Robinson et al. Correlation for Sour Gases

The Robinson et al. (1978) correlation for sour gases, shown in Figures 5-2 and 5-3, is based upon the SRK (Soave-Redlich-Kwong) equation of state. The hydrocarbon portion of the gas is assumed to be pure methane. Robinson et al. found that CO_2 carries only about 75% as much water as H_2S under the same conditions. To reduce the number of variables and thereby simplify the graphical representation of the correlation, Robinson et al. assumed that this condition was true throughout. Therefore, to use this correlation, one has to multiply the % CO_2 in the gas by 0.75, and add it to the % H_2S in the gas to get the effective H_2S content.

Campbell's Correlation for Sweet and Sour Gases

Campbell (1984a) presented a composite chart, shown in Figure 5-4, for sweet gases based upon earlier charts and other available data. This chart gives values very similar to the McKetta and Wehe correlation, but does not include corrections for gas gravity and water salinity.

(text continued on page 174)

172 Gas Production Engineering

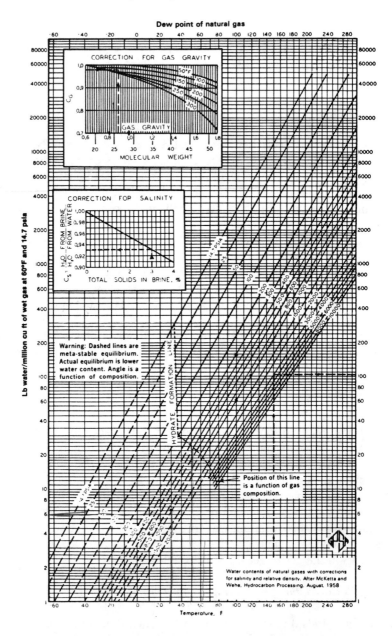

Figure 5-1. The McKetta-Wehe correlation for water content of natural gases, with corrections for water salinity and gas gravity. (After McKetta and Wehe, 1958; reprinted from *Engineering Data Book,* 1981; courtesy of GPSA.)

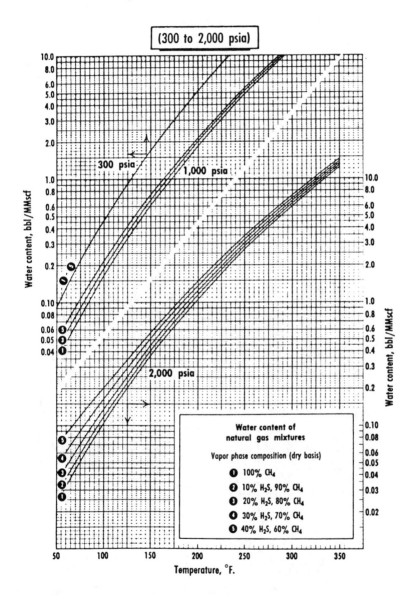

Figure 5-2. Robinson et al. correlation for water content of sour gases in the 300–2,000 psia pressure range. (After Robinson et al., 1978; courtesy of *Oil & Gas Journal*.)

(text continued from page 171)

To correct for large acid gas (H_2S and CO_2) contents, Campbell (1984a) proposes a weighted average water content, W, for the gas as follows:

$$W = y_{HC}W_{HC} + y_{CO_2}W_{CO_2} + y_{H_2S}W_{H_2S} \tag{5-3}$$

where $\quad W_{HC}$ = water content of the hydrocarbon portion of the gas from Figure 5-4

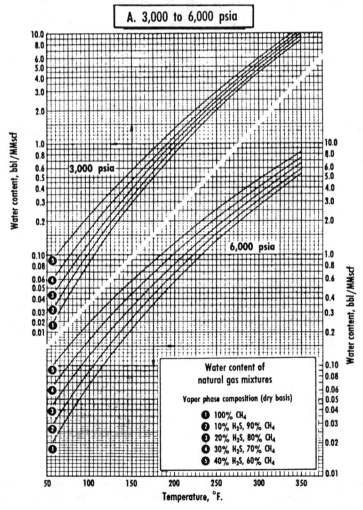

Figure 5-3. Robinson et al. correlation for water content of sour gases in the 3,000–6,000 psia pressure range, and at 10,000 psia. (After Robinson et al., 1978; courtesy of *Oil & Gas Journal*.)

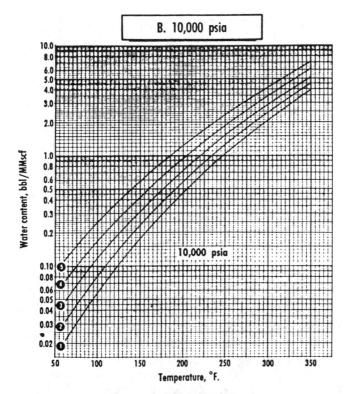

Figure 5-3. Continued.

W_{CO_2} = water content of CO_2 from Figure 5-5
W_{H_2S} = water content of H_2S from Figure 5-6
y_{HC}, y_{CO_2}, y_{H_2S} = mole fractions of hydrocarbon, CO_2, and H_2S, respectively

Example 5-1. Find the water content of a 0.75 gravity gas at 1,500 psia and 120°F.

Solution

From the McKetta and Wehe correlation (Figure 5-1), W = 78 lb H_2O/MMscf gas.
Correcting for the gas gravity, the water content, W = (0.99)(78) = 77.2 lb H_2O/MMscf gas.
Using Campbell's correlation (Figure 5-4), W = 77 lb H_2O/MMscf gas.
Thus, the two methods give very close results for most practical purposes.

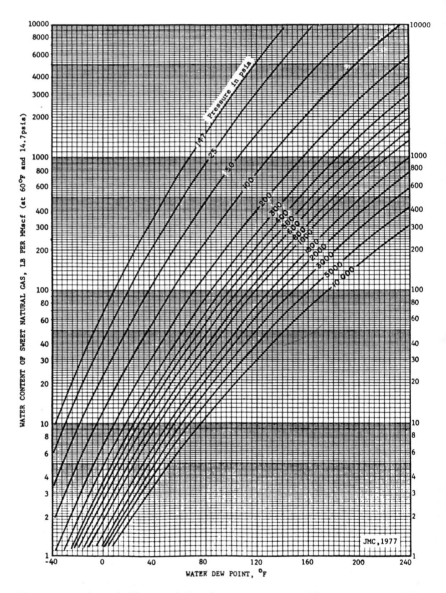

Figure 5-4. Campbell's correlation for water content of sweet gases. (After Campbell, 1984a; courtesy of Campbell Petroleum Series.)

Example 5-2. Find the water content of a gas at 1,000 psia and 100°F using (1) Campbell's method and (2) Robinson et al. method. The gas composition is as follows: $CH_4 = 80.0\%$, $C_2H_6 = 5.0\%$, $C_3H_8 = 1.5\%$, n-$C_4H_{10} = 0.5\%$, $CO_2 = 2.5\%$, $N_2 = 2.0\%$, and $H_2S = 8.5\%$.

Solution

1. $y_{HC} = 0.80 + 0.05 + 0.015 + 0.005 + 0.02 = 0.89$

   $y_{CO_2} = 0.025$

   $y_{H_2S} = 0.085$

   From Figure 5-4, $W_{HC} = 59$ lb H_2O/MMscf HC gas

   From Figure 5-5, $W_{CO_2} = 67$ lb H_2O/MMscf CO_2

   From Figure 5-6, $W_{H_2S} = 150$ lb H_2O/MMscf H_2S

   Using Equation 5-3,

   $W = (0.89)(59) + (0.025)(67) + (0.085)(150)$
   $ = \underline{66.9 \text{ lb } H_2O/\text{MMscf gas}}$

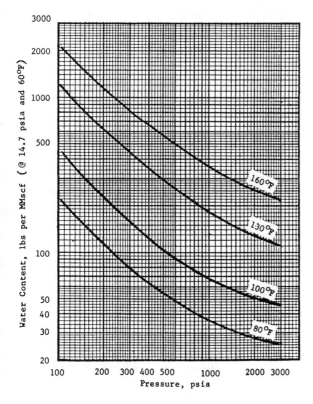

Figure 5-5. Water content of CO_2 in saturated natural gas mixtures. (After Campbell, 1984a; courtesy of Campbell Petroleum Series.)

178 Gas Production Engineering

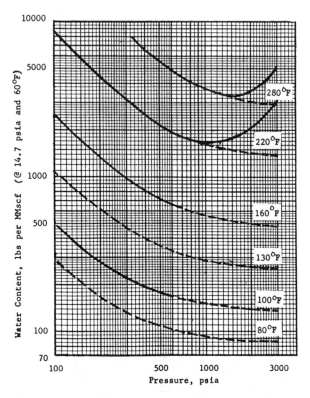

Figure 5-6. Water content of H_2S in saturated natural gas mixtures. Note: Solid curved lines are for pure H_2S only. (After Campbell, 1984a; courtesy of Campbell Petroleum Series.)

2. Effective H_2S content = $(0.75)(2.5) + 8.5 = 10.375\%$

   In Figure 5-2, the lines for 0% H_2S and 20% H_2S at 1000 psia are very close together; no line for 10% H_2S is shown. By interpolation,

   $W = 0.175$ bbl/MMscf
   $= (0.175 \text{ bbl/MMscf})(62.4 \text{ lb/ft}^3)(5.6146 \text{ ft}^3/\text{bbl})$
   $= \underline{61.3 \text{ lb } H_2O/\text{MMscf gas}}$

Equations of State

Equation of state methods are rigorous, but difficult and expensive to implement since they require the use of a computer. Referring to the vapor-liquid equilibrium principles discussed in Chapter 2, Equation 5-2 can be written for a non-ideal system as:

$$y_w = f_v/f = (f_v/f_w)(f_w/f) \tag{5-4}$$

where f_v = fugacity of water at its vapor pressure and system temperature
f_w = fugacity of water at the system pressure and temperature
f = fugacity of the total gas mixture at the system pressure and temperature

In practice, the fugacity f used is for the gas stream *without* water vapor in it. Thus, Equation 5-4 is modified in different ways to account for the discrepancy. In general, the fugacity coefficients for water at its vapor pressure (Ψ_v) and at the system pressure and temperature (Ψ_w), and for the gas at the system conditions (Ψ) are calculated using any equation of state (see Chapter 3 for further details). The mole fraction of water in the gas (in the vapor phase), y_w, at equilibrium conditions can then be calculated as:

$$y_w = (\Psi_v/\Psi_w)^a (p_v/p)^b (\Psi_w/\Psi)^c \tag{5-5}$$

where p_v = vapor pressure of water at the system temperature
p = system pressure
a, b, c = empirically determined constants

One such method has been developed by Sharma and Campbell (1984) for predicting the water content of sweet as well as sour gases based upon the EMR mixture combination rules. It has proved to be accurate and reliable (Campbell, 1984a). In this method, first the EMR critical pressure, critical temperature, and Z factor are determined using the McLeod and Campbell method discussed in Chapter 3. Thereafter, the procedure is as follows:

1. Find the value of the parameter k from Figure 5-7.

   Alternatively, k can be calculated using the following equation:

$$k = (f_v/f_w)(p/p_v)^{0.0049} \tag{5-6}$$

where the exponent 0.0049 is a semi-empirical constant. The fugacities, f_v and f_w, for water can be found from Figure 5-8 (as described in step 2).

Generally, fugacities are reported in terms of the fugacity coefficient, Ψ. Thus, Equation 5-6 can alternatively be written as:

180 Gas Production Engineering

$$k = (\Psi_v/\Psi_w)(p_v/p)^{0.9951} \tag{5-7}$$

2. Find the water fugacity coefficient, Ψ_w, from the generalized fugacity coefficient chart shown in Figure 5-8. Note that since Ψ_w represents the pure component fugacity for water, the critical pressure and temperature for water only are applicable.
3. Using Figure 5-8 and the critical pressure and temperature for the gas determined from the EMR method, find the fugacity coefficient Ψ for the gas mixture (without water as a component) at system pressure p and temperature T.

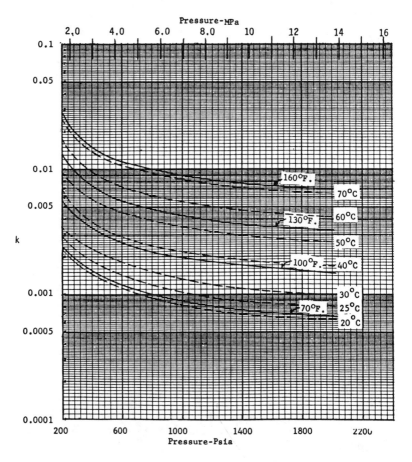

Figure 5-7. Constant k as a function of pressure and temperature. (After Campbell, 1984a; courtesy of Campbell Petroleum Series.)

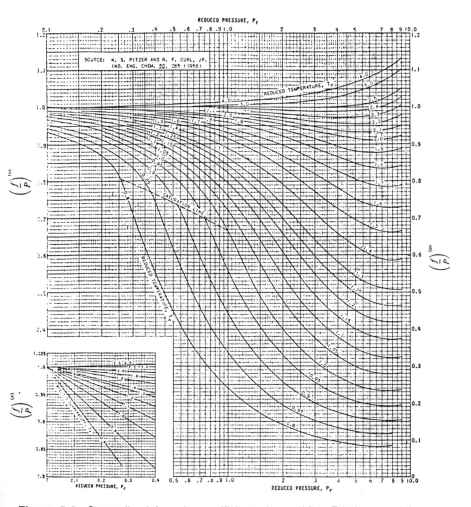

Figure 5-8. Generalized fugacity coefficient chart. (After Edmister and Lee, 1984, from data by R. F. Curl, Jr., and K. S. Pitzer, 1958, *Ind. and Engr. Chem.*, Feb. 2.)

4. Calculate the mole fraction of water in the gas (in the vapor phase), y_w, as follows:

$$y_w = k(\Psi_w/\Psi)^Z \tag{5-8}$$

where the gas compressibility factor, Z, must be that determined from the EMR method, since the EMR critical pressure and temperature were used for calculating the critical properties of the gas.

Example 5-3. Repeat Example 5-2 using the Sharma and Campbell method.

Solution

Comp.	y_i	y_{ji}	EMR
C_1	0.800	0.8602	13.984
CO_2	0.025	0.0269	15.750
H_2S	0.085	0.0914	19.828
N_2	0.020	0.0215	9.407
$Y_1 =$	0.930		
C_2	0.050	0.7143	23.913
C_3	0.015	0.2143	34.316
n-C_4	0.005	0.0714	44.243
$Y_2 =$	0.070		

$EMR_1 = \Sigma\ y_{1i}EMR_i = 14.47$

$EMR_2 = \Sigma\ y_{2i}EMR_i = 27.59$

From Figure 3-6, $(T_c/p_c)_1 = 0.51$, and $(T_c/p_c)_2 = 0.88$.

So, $(T_c/p_c)_{mix} = (0.93)(0.51) + (0.07)(0.88) = 0.5359$

And $(EMR)_{mix} = (0.93)(14.47) + (0.07)(27.59) = 15.388$

From Figure 3-5, $(T_c/p_c^{0.5}) = 15.5$

Therefore, $p_c^{0.5} = 15.5/0.5359 = 28.923$, implying that

$p_{pc} = 836.56$ psia, and $T_{pc} = (0.5359)(836.56) = 448.31°R$

Thus, $p_{pr} = 1,000/836.56 = 1.195$, and $T_{pr} = 560/448.31 = 1.249$

From Figure 3-7, the compressibility factor, $Z = 0.837$

From Figure 5-7, $k = 0.00195$

From Table 3-1, $(p_c)_{water} = 3,203.6$ psia, and $(T_c)_{water} = 1,165.1°R$.

So, $(p_r)_{water} = 1,000/3,203.6 = 0.312$, and $(T_r)_{water} = 560/1,165.1 = 0.481$

From Figure 5-8, $\Psi_w = (f/p)_w = 0.535$ on extrapolation of the T_r lines to $T_r = 0.5$.

For the gas (without water), $\Psi = 0.815$ from Figure 5-8 for a $p_{pr} = 1.195$ and $T_{pr} = 1.249$

Using Equation 5-8, the mole fraction of water is equal to:

$y_w = (0.00195)[0.535/0.815]^{0.837} = 1.371 \times 10^{-3}$

Now, the density of water vapor at standard conditions

$= (14.7)(18.015)/[(1)(10.73)(491.67)]$
$= 0.050197 \text{ lbm/scf} = 50,197 \text{ lbm/MMscf}$

Thus, the water content of the gas

$= (1.371 \times 10^{-3} \text{ MMscf water/MMscf gas})$
$\cdot (50,197 \text{ lb water/MMscf water})$
$= \underline{68.82 \text{ lb } H_2O/\text{MMscf gas}}$

Gas Hydrates

Natural gas hydrates are solid crystalline compounds, resembling ice or wet snow in appearance, but much less dense than ice. They are included in a general class of compounds known as *clathrates*, which have a structure wherein guest molecules are entrapped in a cage-like framework of the host molecules without forming a chemical bond. Natural gas hydrates are formed when natural gas components, notably methane, ethane, propane, isobutane, hydrogen sulfide, carbon dioxide, and nitrogen, enter the water lattice (which is looser than the ice lattice) and occupy the vacant lattice positions, causing the water to solidify at temperatures considerably higher than the freezing point of water. Enough gaseous molecules must enter the lattice and occupy the voids to stabilize the lattice crystal.

Hydrate formation is governed by the size of the host molecule, and its solubility in water. Size is an important parameter—the molecule should be small enough to properly orient itself within the water structure to best use the available space, and large enough to get entrapped. Smaller molecules, such as those of methane, can avoid entrapment because of their smaller size and rapid, random movement. Solubility affects the rate of formation because it governs the availability of the guest molecule to the water.

Hydrate Formulas

Two types of crystalline structures have been proposed for gas hydrates. Smaller molecules, such as methane, ethane, and hydrogen sulfide, form a body-centered cubic lattice, called Structure I. Structure II is a diamond lattice, formed by larger molecules, such as propane and isobutane. Gas mixtures form both Structure I and II type hydrates.

The number of water molecules associated with each molecule of the gaseous component included in the hydrate is known as the *hydrate number*. Since different number of guest molecules can enter the water lattice and stabilize it into a hydrate, no specific hydrate number can be given for a particular hydrate. The limiting hydrate number, a theoretical quantity determined using the size of the gas molecules and the size of the voids in the water lattice, serves as a useful parameter, although hydrate structures are stable at less than 100% occupancy of the voids. For molecules in Structure I, the limiting hydrate number is 5.75 for the smallest gases (CH_4), and 7.667 for medium sized gases (C_2H_6). For the largest molecules associated with Structure II, the limiting hydrate number is 17 (GPSA, 1981). Galloway et al. (1970) report experimental values for methane in the range of 5.8 to 6.3, and for ethane from 7.9 to 8.5. Thus, hydrate molecules can typically be represented as:

Methane:	$CH_4 \cdot 6H_2O$	Nitrogen:	$N_2 \cdot 6H_2O$
Ethane:	$C_2H_6 \cdot 8H_2O$	Carbon dioxide:	$CO_2 \cdot 6H_2O$
Propane:	$C_3H_8 \cdot 17H_2O$	Hydrogen sulfide:	$H_2S \cdot 6H_2O$
Isobutane:	$i\text{-}C_4H_{10} \cdot 17H_2O$		

Normal-butane does form a hydrate, but it is very unstable. Paraffins higher than butane do not form hydrates.

Hydrate Phase Behavior

Figure 5-9 shows the phase equilibrium diagram for a gas-water-hydrate system. Line ABCD represents the hydrate curve, HFCI is the vapor pressure curve for the hydrocarbon gas, and EBFG is the curve representing the solid-liquid equilibrium for water (or, the water freezing point curve). The hydrocarbon gas is assumed single-component to simplify the phase equilibrium representation. These lines delineate different regions in the phase equilibrium as follows:

1. Hydrates exist in the pressure-temperature region above the hydrate curve ABCD. Below the hydrate curve, and to its right, no hydrates can form.

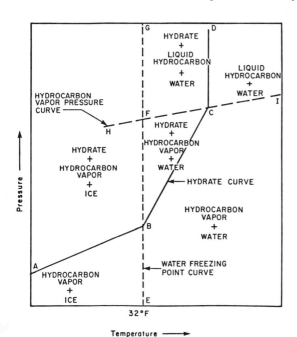

Figure 5-9. Phase equilibrium diagram for a gas-water-hydrate system.

2. Above the vapor pressure curve HFCI, the hydrocarbon exists in the liquid state.
3. Towards the left of the line EBFG, water exists in the solid form as ice. On the right of EBFG, water will be in the liquid state. In practice, the region towards the left of the line EBFG will hardly ever be encountered, except perhaps in extremely cold areas.

Note that in Figure 5-9, the hydrate curve becomes vertical at the point where it intersects the hydrocarbon vapor pressure curve. This intersection, C, therefore represents the maximum hydrate forming temperature.

Conditions Promoting Hydrate Formation

For a gas to be a hydrate former, it must satisfy two criteria: (1) it should be of the covalent bond type, with molecules smaller than 8 A° units; and (2) the gas, when in the liquid state, must be immiscible with water. The gas hydrate that is formed will be stable if the hydrate is water resistant and no Van der Waals' forces arise between the hydrate molecules. If these conditions are met, stable hydrate formation is possible under certain conditions governed by the hydrate phase equilibrium behavior. In natural gas systems, the gases mentioned earlier as being hydrate formers, satisfy these criteria. The primary conditions necessary for a natural gas to form stable hydrates can be summarized as follows:

1. Natural gas at or below the dew point with liquid water present. No hydrate formation is possible if "free" water is not present.
2. Low temperatures, at or below the hydrate formation temperature for a given pressure and gas composition.
3. High operating pressures, that may raise the hydrate formation temperature to the operating temperature.

Secondary factors that aid and accelerate hydrate formation are (GPSA, 1981):

1. High velocities, or agitation, or pressure pulsations.
2. Presence of a small "seed" crystal of hydrate.
3. Presence of H_2S and CO_2 aids hydrate formation because both these acid gases are more soluble in water than the hydrocarbons.

Prediction of Hydrate Formation

Figure 5-10 shows the hydrate forming conditions for natural gas components. Referring to Figure 5-9 it can be seen that the general procedure for hydrate prediction would involve prediction of the hydrocarbon vapor pressure curve (line HFCI), and only the portion BC of the hydrate curve (operating conditions in the region AB are never encountered, and the part CD of the hydrate curve is a vertical straight line). The hydrocarbon vapor pressure curve can be predicted using any of the several available correlations, or equation of state methods discussed in Chapters 2 and 3. Correlations for predicting the hydrate curve are discussed here.

Approximate Method for Sweet Gases

As a first approximation, the data presented in the Gas Processors Suppliers Association (GPSA) charts shown in Figures 5-11 through 5-16 can be used. These charts do not account for the presence of H_2S and CO_2.

Hydrate formation can be divided into two categories (Ikoku, 1984): (1) hydrate formation due to a decrease in temperature (or increase in pressure) with no sudden expansion (or compression), such as in flow strings and surface lines; and (2) hydrate formation due to sudden expansion, such as in flow-provers, orifices, chokes, and back-pressure regulators (sometimes bottom-hole chokes, BHC, are used to avoid sudden expansion, and consequently, avoid hydrate formation). For hydrate formation of type 1, Figure 5-11 can be used for predicting the approximate pressure-temperature con-

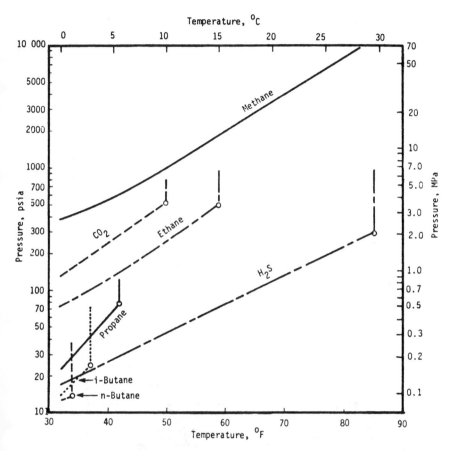

Figure 5-10. Hydrate forming conditions for natural gas components. (After Campbell, 1984a; courtesy of Campbell Petroleum Series.)

ditions for hydrate formation. Figures 5-12 to 5-16 are used for estimating hydrate formation conditions for situations of type 2. In particular, Figures 5-12 to 5-16 are used for determining the permissible expansion, without hydrate formation, of a gas with a given gravity.

Example 5-4. A 0.685 gravity gas is at 500 psia and 100°F. To what value can the temperature be reduced without hydrate formation?

Solution

From Figure 5-11, hydrates will form at or below a temperature of 54°F.

188　Gas Production Engineering

Example 5-5. How far can the pressure be lowered without expecting hydrate formation for a 0.685 gravity gas if it is initially at (a) 1,500 psia and 100°F, and (b) 1,000 psia and 100°F.

Solution

(a) From Figure 5-12 for a 0.6 gravity gas, the intersection of the 1,500 psia initial pressure line with the 100°F initial temperature curve gives a final pressure of 490 psia. Similarly, from Figure 5-13 for a 0.7 gravity gas, the final pressure is equal to 800 psia. By interpolation, the 0.685 gravity gas can be expanded without hydrate formation to a pressure = 490 + (800 − 490)(0.685 − 0.6)/(0.7 − 0.6) = 753.5 psia.

(b) Refer to Figures 5-12 and 5-13.

The 100°F initial temperature curve does not intersect the 1,000 psia initial pressure line for a 0.6 gravity as well as a 0.7 gravity gas.

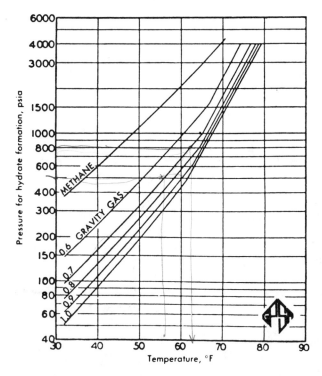

Figure 5-11. Pressure-temperature curves for approximate prediction of hydrate formation for natural gases. (From *Engineering Data Book*, 1981; courtesy of GPSA.)

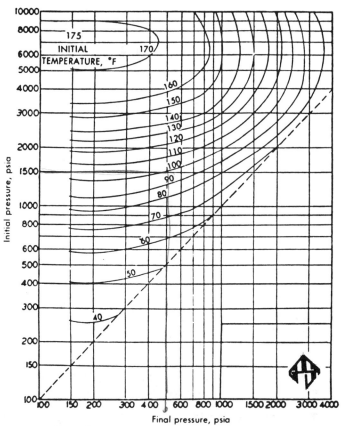

Figure 5-12. Approximate permissible expansion of a 0.6-gravity natural gas without hydrate formation. (From *Engineering Data Book*, 1981; courtesy of GPSA.)

Hence, the gas at these conditions can be expanded to any pressure without hydrate formation.

Katz et al. Method

The Katz et al. (1959) method is based upon the principle that the gases entrapped in a natural gas hydrate resemble solid solutions, because on release during the decomposition of the hydrates, they increase in density. It uses vapor-solid equilibrium ratios, K_{vs}, first proposed by Carson and Katz (1942):

$$K_{vs} = y/x_s \qquad (5\text{-}9)$$

where y = mole fraction of hydrocarbon in the gas on a water-free basis
x_s = mole fraction of hydrocarbon in the solid on a water-free basis

190 *Gas Production Engineering*

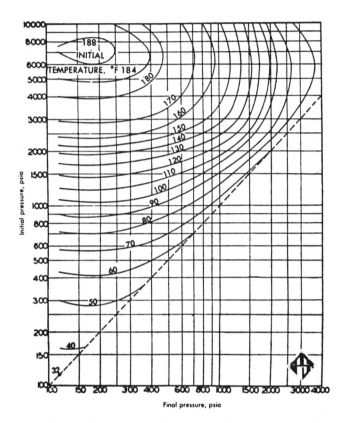

Figure 5-13. Approximate permissible expansion of a 0.7-gravity natural gas without hydrate formation. (From *Engineering Data Book*, 1981; courtesy of GPSA.)

From this definition of K_{vs}, it is clear that K_{vs} is equal to ∞ for gases that are non-hydrate formers. Thus, for hydrocarbons heavier than butane, K_{vs} is ∞. The original method assumes that nitrogen is a non-hydrate former, and that n-butane, if present in mole fractions less than 5%, has the same K_{vs} value as ethane. Theoretically, this is incorrect, but from an applications standpoint, even using a K_{vs} equal to ∞ for both nitrogen and n-butane gives acceptable results (Campbell, 1984a).

K_{vs} values for components in a gas can be found from Figures 5-17 through 5-22. Hydrates will form when the following equation is satisfied:

$$\sum_{i=1}^{n} (y_i/K_{vsi}) = 1.0 \tag{5-10}$$

Thus, calculation of the hydrate forming condition is similar to the dew point calculation for multi-component gaseous mixtures.

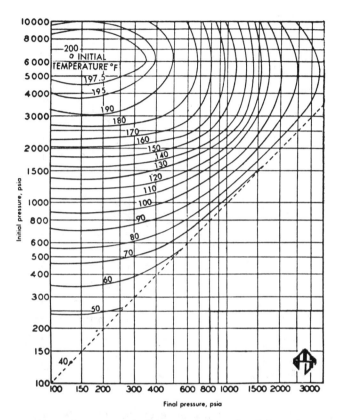

Figure 5-14. Approximate permissible expansion of a 0.8-gravity natural gas without hydrate formation. (From *Engineering Data Book*, 1981; courtesy of GPSA.)

This method accounts for H_2S and CO_2, and has proved to be very reliable up to about 1,000 psia. A gas with an H_2S content greater than 20% can be considered to behave like pure H_2S regarding hydrate forming characteristics. To account for gases that contain large amounts of nitrogen, Heinze (1971) gave the following relationship, valid up to a pressure of about 5,800 psia:

$$T_h = \frac{\sum_{i=1}^{n}(y_i K_{vsi})}{0.445} \tag{5-11}$$

where T_h = hydrate formation temperature, K

192 Gas Production Engineering

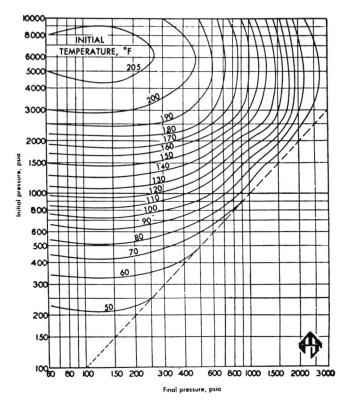

Figure 5-15. Approximate permissible expansion of a 0.9-gravity natural gas without hydrate formation. (From *Engineering Data Book*, 1981; courtesy of GPSA.)

Example 5-6. The gas given in Example 5-2 is at 500 psia. At what temperature will hydrates form?

Solution

		At 60°F		At 55°F	
Comp.	y_i	K_{vsi}	y_i/K_{vsi}	K_{vsi}	y_i/K_{vsi}
C_1	0.800	1.69	0.4734	1.61	0.4969
C_2	0.050	1.23	0.0407	0.79	0.0633
C_3	0.015	0.24	0.0625	0.106	0.1415
$n\text{-}C_4$	0.005	∞	0.0	∞	0.0
CO_2	0.025	∞	0.0	2.65	0.0094
H_2S	0.085	0.28	0.3036	0.21	0.4048
N_2	0.020	∞	0.0	∞	0.0
			0.8801		1.1159

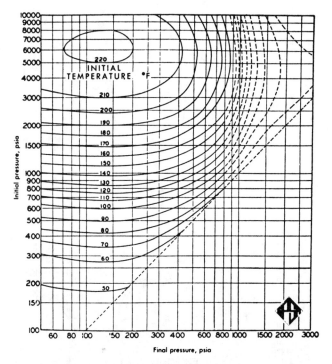

Figure 5-16. Approximate permissible expansion of a 1.0-gravity natural gas without hydrate formation. (From *Engineering Data Book*, 1981; courtesy of GPSA.)

Thus, the hydrate forming temperature corresponding to a 500 psia pressure is between 55°F and 60°F. Further guesstimates may be made to converge to the temperature value that makes $\Sigma\, y_i/K_{vsi} = 1.0$.

From a simple interpolation, the hydrate forming temperature can be calculated as:

$$T_h = 60 - (60 - 55)(1 - 0.8801)/(1.1159 - 0.8801) = \underline{57.5°F}$$

Trekell-Campbell Method for High-Pressure Gas

Trekell and Campbell (1966) provide corrections to the Katz et al. method to extend its applicability to higher pressures and account for the negative effect of the presence of non-hydrate forming molecules that are too large to fit into the voids in the water lattice. The Trekell-Campbell method uses

(text continued on page 197)

194 Gas Production Engineering

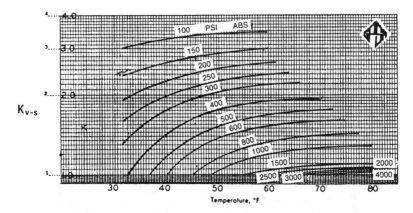

Figure 5-17. Vapor-solid equilibrium constants for methane. (After Carson and Katz, 1942; reprinted from *Engineering Data Book,* 1981; courtesy of GPSA.)

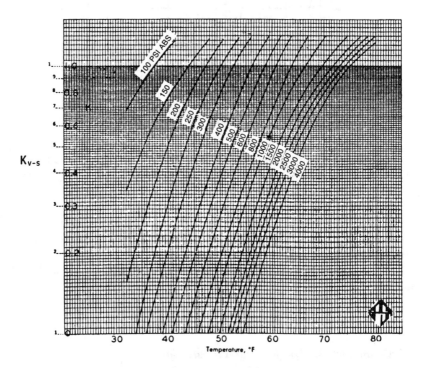

Figure 5-18. Vapor-solid equilibrium constants for ethane. (After Carson and Katz, 1942; reprinted from *Engineering Data Book,* 1981; courtesy of GPSA.)

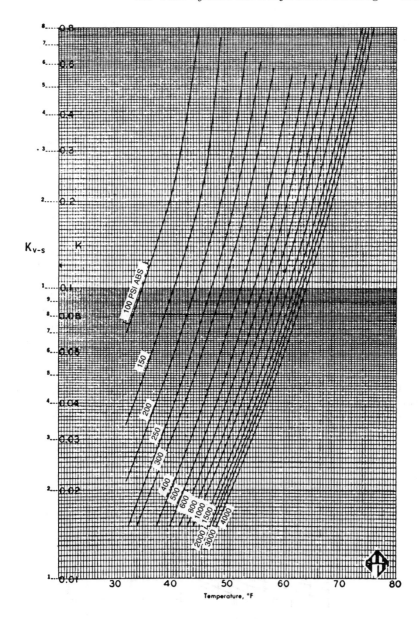

Figure 5-19. Vapor-solid equilibrium constants for propane. (After Carson and Katz, 1942; reprinted from *Engineering Data Book,* 1981; courtesy of GPSA.)

196 Gas Production Engineering

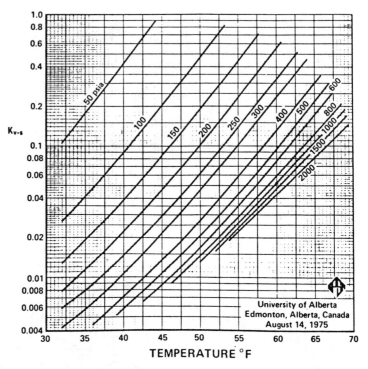

Figure 5-20. Vapor-solid equilibrium constants for iso-butane. (From *Engineering Data Book*, 1981; courtesy of GPSA.)

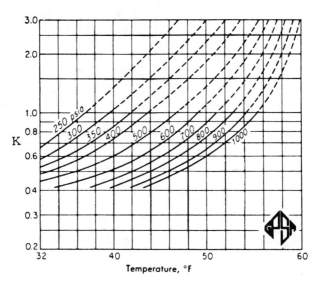

Figure 5-21. Vapor-solid equilibrium constants for CO_2. (From *Engineering Data Book*, 1981; courtesy of GPSA.)

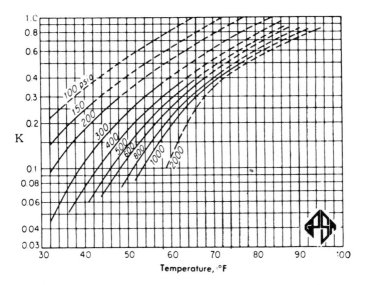

Figure 5-22. Vapor-solid equilibrium constants for H_2S. (From *Engineering Data Book*, 1981; courtesy of GPSA.)

(text continued from page 193)
methane as the reference gas. The additive effect of other hydrate forming gases, C_2, C_3, $n\text{-}C_4$, $i\text{-}C_4$, and H_2S, at different pressures is determined from Figures 5-23 to 5-28. CO_2 is ignored. Non-hydrate formers are grouped together into the pentanes plus (C_{5+}) group, and their hydrate depression effect is determined from Figures 5-29 and 5-30 as a function of the mole % pentanes plus, and the mole % pentanes plus on the basis of the gas fractions from C_2 to C_4. This latter parameter can be written as:

$$\frac{y_{C_{5+}}}{1 - y_{C_1} - y_{C_{5+}}} \, (100\%)$$

where y_{C_1}, $y_{C_{5+}}$ = mole fractions of methane and pentanes plus, respectively, in the gas.

Figures 5-23 to 5-28 indicate the hydrate forming temperature for methane at various pressures. To find the hydrate forming temperature at any pressure, the appropriate chart is used to read off the temperature displacement for the gas components as a function of their mole % in the gas. These temperature displacements are added to the methane-hydrate-forming temperature. If any pentanes plus are present, their effect (negative) is also added. The sum then gives the hydrate forming temperature at the pressure of interest. These calculations may be repeated for several pressures to obtain the hydrate curve for the gas.

(text continued on page 204)

198 Gas Production Engineering

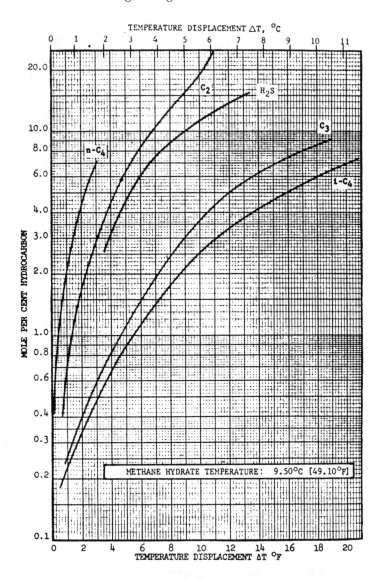

HYDRATE PREDICTION CHART FOR 6.9 MPa [1000 psia]

Figure 5-23. Trekell-Campbell temperature displacement chart for hydrate formers at 1,000 psia. (After Trekell and Campbell, 1966; reprinted from Campbell, 1984a; courtesy of Campbell Petroleum Series.)

Gas-Water Systems and Dehydration Processing 199

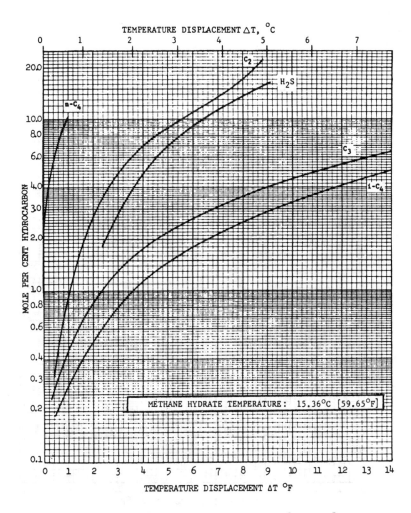

HYDRATE PREDICTION CHART FOR 13.8 MPa [2000 psia]

Figure 5-24. Trekell-Campbell temperature displacement chart for hydrate formers at 2,000 psia. (After Trekell and Campbell, 1966; reprinted from Campbell, 1984a; courtesy of Campbell Petroleum Series.)

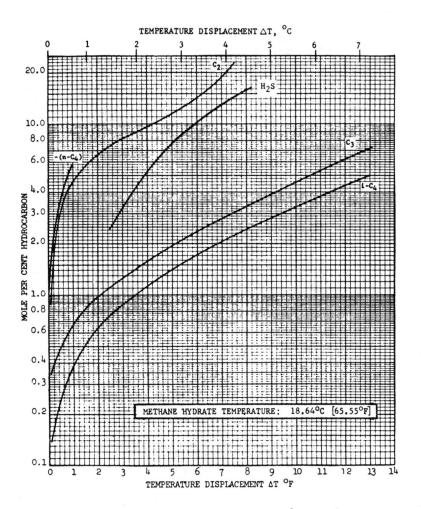

HYDRATE PREDICTION CHART FOR 20.7 MPa [3000 psia]

Figure 5-25. Trekell-Campbell temperature displacement chart for hydrate formers at 3,000 psia. (After Trekell and Campbell, 1966; reprinted from Campbell, 1984a; courtesy of Campbell Petroleum Series.)

Gas-Water Systems and Dehydration Processing 201

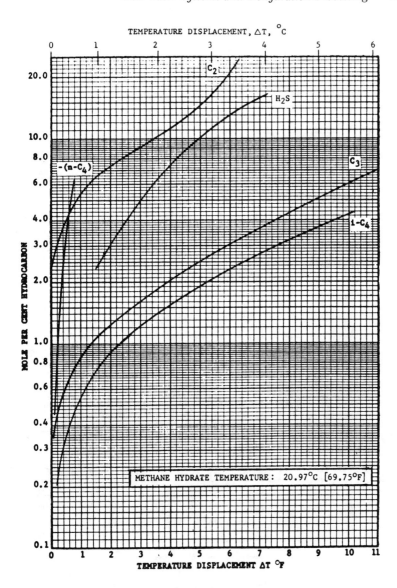

HYDRATE PREDICTION CHART FOR 27.6 MPa [4000 psia]

Figure 5-26. Trekell-Campbell temperature displacement chart for hydrate formers at 4,000 psia. (After Trekell and Campbell, 1966; reprinted from Campbell, 1984a; courtesy of Campbell Petroleum Series.)

202 *Gas Production Engineering*

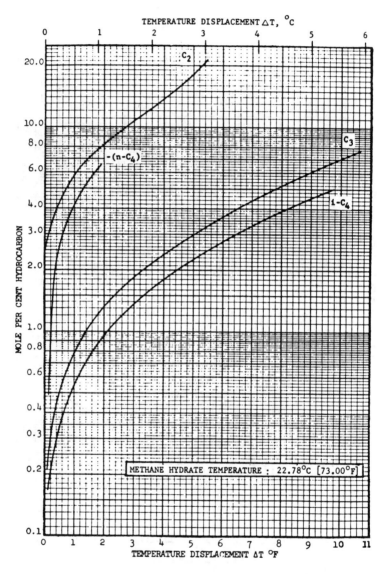

HYDRATE PREDICTION CHART FOR 34.5 MPa [5000 psia]

Figure 5-27. Trekell-Campbell temperature displacement chart for hydrate formers at 5,000 psia. (After Trekell and Campbell, 1966; reprinted from Campbell, 1984a; courtesy of Campbell Petroleum Series.)

Gas-Water Systems and Dehydration Processing 203

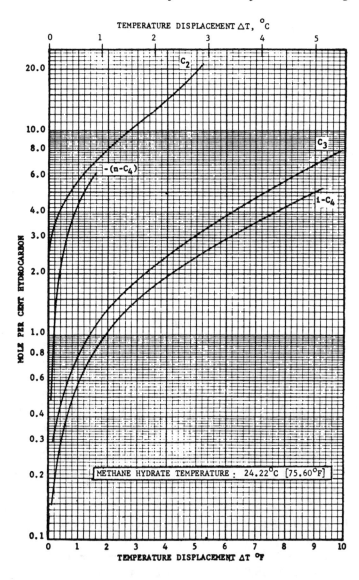

HYDRATE PREDICTION CHART FOR 41.4 MPa [6000 psia]

Figure 5-28. Trekell-Campbell temperature displacement chart for hydrate formers at 6,000 psia. (After Trekell and Campbell, 1966; reprinted from Campbell, 1984a; courtesy of Campbell Petroleum Series.)

204 Gas Production Engineering

(text continued from page 197)

The Trekell-Campbell method is applicable to gases in the pressure range of 1,000–6,000 psia.

Example 5-7. For the gas with composition given below, find the hydrate formation temperature corresponding to a pressure of 6,000 psia.

Solution

Comp.	y_i	ΔT, °F
C_1	0.810	75.60
C_2	0.050	0.75
C_3	0.025	4.1
n-C_4	0.015	0.25
i-C_4	0.010	1.9
C_{5+}	0.015	-1.17
CO_2	0.025	0.0
H_2S	0.050	0.0
	1.000	81.43

In this table, the ΔT values for C_1, C_2, C_3, n-C_4, and i-C_4 are obtained from Figure 5-28. Note that H_2S has a negligible effect on the hydrate formation temperature at high pressures and is therefore not reported in Figure 5-28.

The value of $(y_{C_{5+}})(100\%)/(1 - y_{C_1} - y_{C_{5+}}) = (0.015)(100)/(1 - 0.810 - 0.015) = 8.57$

In Figure 5-30, interpolation is made between the 8.0 and 9.5 lines to get $\Delta T = -1.17$ corresponding to $C_{5+} = 1.5\%$. Thus, the hydrate forming temperature corresponding to a 6,000 psia pressure is 81.43, or 81.4°F.

McLeod-Campbell Method for Very High Pressure Sweet Gas

For gases above 6,000 psia, the Trekell-Campbell approach is not quite valid. For the 6,000–10,000 psia pressure range, McLeod and Campbell (1961) propose the following relationship:

$$T_h = 3.89 \left(\Sigma_{i=1}^{n} y_i K_i \right)^{0.50} \tag{5-12}$$

where T_h = hydrate forming temperature, °R

The K_i values are as shown in Table 5-1. The mole fractions of the components, y_i, are normalized to the C_1 to C_4 content. All other components in the gas are ignored. The method, therefore, has limited applicability.

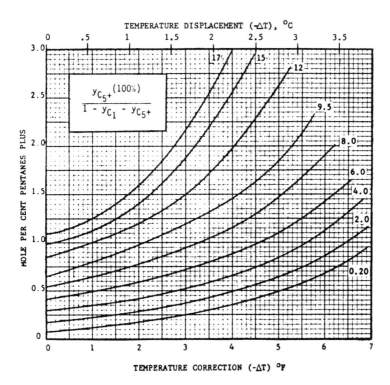

Figure 5-29. Trekell-Campbell temperature displacement chart for non-hydrate formers (pentanes plus) at 1,000 psia. (After Trekell and Campbell, 1966; reprinted from Campbell, 1984a; courtesy of Campbell Petroleum Series.)

Table 5-1
K_i Values for the McLeod-Campbell Method for Predicting Hydrate Formation*

Pressure		K_i Values				
MPa	Psia	C_1	C_2	C_3	$i\text{-}C_4$	$n\text{-}C_4$
41.4	6000	18 933	20 806	28 382	30 696	17 340
48.3	7000	18 096	20 848	28 709	30 913	17 358
55.2	8000	19 246	20 932	28 764	30 935	17 491
62.1	9000	19 367	21 094	29 182	31 109	17 868
69.0	10 000	19 489	21 105	29 200	30 935	17 868

* After McLeod and Campbell, 1961.

Example 5-8. Repeat Example 5-7, using the McLeod-Campbell method.

Solution

Comp.	y_i	y_i^*	K_i
C_1	0.810	0.8901	18,933
C_2	0.050	0.0549	20,806
C_3	0.025	0.0275	28,382
n-C_4	0.015	0.0165	17,340
i-C_4	0.010	0.0110	30,696
C_{5+}	0.015	—	0.0
CO_2	0.025	—	0.0
H_2S	0.050	—	0.0

$\Sigma \; y_i K_i = 19{,}398.78$

Using Equation 5-12,

$T_h = (3.89)(19{,}398.78^{0.5}) = 541.80°R = \underline{81.8 \; °F.}$

Note that this compares well with the value of 81.4°F determined using the Trekell-Campbell method in Example 5-7.

Equation of State Methods

Several complex computer solution methods have been developed for predicting hydrate forming conditions. Most equation of state methods are based upon the Van der Waals and Platteeuw (1959) statistical thermodynamic equation for the chemical potential of water in the hydrate lattice. The different methods essentially use different intermolecular potential functions. Van der Waals and Platteeuw (1959) used the Lennard-Jones 12-6 potential. Nagata and Kobayashi (1966) used the Kihara potential that they showed to be better than the Lennard-Jones model. Parrish and Prausnitz (1972) expanded the concept of fitting hydrate parameters to develop a generalized method for predicting hydrate dissociation pressures for gas mixtures, using the Van der Waals and Platteeuw (1959) equation and the Kihara spherical core model. Ng and Robinson (1976) have added another adjustable parameter to the Parrish and Prausnitz (1972) algorithm to improve prediction of the dissociation pressures for multicomponent gases. Holder and Hand (1982) presented a modified form of the Parrish and Prausnitz (1972) model.

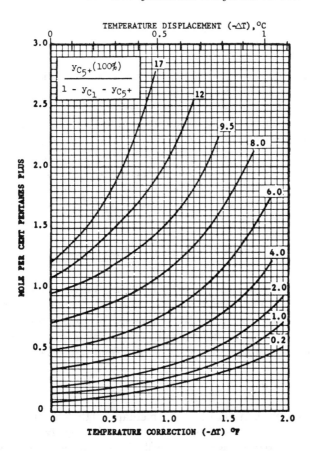

Figure 5-30. Trekell-Campbell temperature displacement chart for non-hydrate formers (pentanes plus) in the pressure range of 2,000–10,000 psia. (After Trekell and Campbell, 1966; reprinted from Campbell, 1984a; courtesy of Campbell Petroleum Series.)

The complexity of these models precludes further discussion here. These methods too can at best give only approximate answers and should be used with caution.

Preventing Hydrate Formation

The permanent solution for hydrate problems is dehydration of the gas to a sufficiently low dew point. Commonly used dehydration methods include absorption dehydration using a liquid dessicant, adsorption dehydration using a solid dessicant, and simultaneous dehydration and gas-liquid separation by expansion dehydration.

At the wellsite, two techniques are applicable: (1) heating the gas stream so that it becomes undersaturated, and maintaining flowlines and equipment at temperatures above the hydrate point; and (2) in cases where liquid water is present and the flowlines and equipment cannot be maintained above hydrate temperatures, inhibiting hydrate formation by injecting additives that depress both hydrate and freezing temperatures. This latter technique is discussed in the following section. Note that in practice, hydrates are a problem only when allowed to accumulate and grow to a size that restricts or stops the flow. Otherwise, they are of no consequence.

Hydrate Inhibition by Additive Injection

Additive injection is generally required for gas streams from producing as well as storage wells in order to prevent corrosion and hydrate formation in the gas gathering and transmission systems. In some cases, gas being injected into a reservoir for storage purposes also requires the addition of certain additives to protect the casing, tubing, and sand face (Curry, 1981).

Types of Additives

The most common additives are methanol, ethylene glycol (EG), and diethylene glycol (DEG). Methanol is used the most, because it disperses well in the gas stream, is readily available in bulk, is the least expensive, and, consequently, does not require recovery. Methanol, however, can cause contamination problems in plants. Whereas most additives are recovered and recycled, methanol recovery is often uneconomical.

Methanol injection is very useful in cases where low gas volumes prohibit dehydration processing. It is preferable in cases where hydrate problems are relatively mild, infrequent, or periodic, inhibitor injection is only a temporary phase in the field development program, or inhibition is done in conjunction with a primary dehydration system.

EG and DEG are used primarily at low-temperature processing plants for extracting natural gas liquids. The water phase of the process liquid contains the EG or DEG, which can be recovered and regenerated.

Injection Techniques

In the early years, gravity-type injection systems were being used. A gravity injection system consists of a capacity tank, a sight glass, and a valve manifold to maintain equal pressure across the chemical and to regulate its entry into the flowing gas stream. This resulted in a generally sporadic injection and poor inhibitor dispersion, leading to freeze-offs (hydrates formation) during the operations.

To counter these problems, newer systems that feed the additive at a uniform rate into the gas stream, such as the pneumatically powered chemical injection pumps, have been designed. Continuous injection is also desirable from the standpoint of field personnel who can then operate on a regular basis. Further, precise chemical requirements can be predicted and bulk purchases made at possibly lower costs.

The injection pump is usually of the positive-displacement type, so that it pumps a constant amount of chemical into the gas stream, regardless of the pressure of the gas against which it must inject the chemical. Since positive displacement pumps do not sense any buildup of discharge pressure, a rupture disc is put at the discharge end of the pump. Any brass or copper components in the pump should be avoided, since these metals are subject to severe physical breakdown in the presence of sour gases. The pump injects chemical downward into the running end of a vertically installed tee of a size about one inch. The tee functions as a sort of mixing valve. Metering devices are installed to permit chemical injection at a proportional-to-flow rate, eliminating frequent manual adjustment. As a precautionary measure in field operations, all valves and accumulator systems where free water may accumulate are precharged with methanol to prevent hydrate formation.

Prediction of Inhibitor Requirements

The weight percent inhibitor concentration in the aqueous phase, w, required to lower the gas hydrate freezing point by d°F, is given by Hammerschmidt's equation:

$$w = \frac{(dM)(100)}{K + dM} \quad (5\text{-}13)$$

where M = molecular weight of the inhibitor
K = a constant, 2,335 for methanol, 4,000 for the glycols.

Equation 5-13 can be rearranged to calculate the lowering of the gas hydrate freezing point for a given weight percent inhibitor in the aqueous phase:

$$d = \frac{wK}{100M - wM} \quad (5\text{-}14)$$

Typically, Equation 5-13 would predict injection requirements of 2–3 gallons of methanol per MMscf of gas. In practice, only about 1 to 1.25 gallons per MMscf may be required in most cases. Thus, Equation 5-13 is designed to overpredict chemical requirements so as to yield "safe" numbers. Consequently, the process of inhibition is really quite economical if the minimum required amount of chemical is injected. This is usually determined in field operations by reducing chemical amounts to a point where hydrate formation (freeze-off) begins to occur, and then keeping chemical injection just slightly above this rate. Most predictive techniques are no better than a rule of thumb; it is advisable to correctly evaluate chemical requirements by field-testing.

Absorption Dehydration

Absorption dehydration involves the use of a liquid dessicant to remove water vapor from the gas. Although many liquids possess the ability to absorb water from gas, the liquid that is most desirable to use for commercial dehydration purposes should possess the following properties:

1. High absorption efficiency.
2. Easy and economic regeneration.
3. Non-corrosive and non-toxic.
4. No operational problems when used in high concentrations.
5. No interaction with the hydrocarbon portion of the gas, and no contamination by acid gases.

The glycols, particularly ethylene glycol (EG), diethylene glycol (DEG), triethylene glycol (TEG), and tetraethylene glycol (T_4EG), come the closest to satisfying these criteria to varying degrees. Glycols are preferred because they offer a superior dew-point depression, with process reliability and lower initial capital and operating costs. For a 10-MMscf/d plant, solid dessicant will cost about 53% more than a TEG plant, and for a 50-MMscf/d plant, it will cost 33% more than TEG (Guenther, 1979). Wherever it can meet dehydration specifications, economics usually favors glycol dehydration over other processes (GPSA, 1981). The equipment is simpler to operate, maintain, and automate, and plant operation is continuous. Furthermore, glycols can be used in the presence of materials that may foul-up a solid desiccant, and can be regenerated. Glycols can be used for sour gases, but acid gas absorption necessitates certain precautions (Table 5-2A).

Of all the glycols, almost all the plants use TEG because of lower vapor losses and better dew-point depression (Table 5-2B). TEG has been success-

Gas-Water Systems and Dehydration Processing 211

Table 5-2A
Physical and Chemical Properties of Glycols*

	Method	Ethylene Glycol	Diethylene Glycol	Triethylene Glycol
Molecular weight		62.07	106.12	150.17
Specific gravity at 68°F	0.791	1.1155	1.1184	1.1255
Specific weight, lb/gal	6.59	9.292	9.316	9.375
Boiling point at 760 mm Hg, °F	149	387.7	474.4	550.4
Freezing point, °F		9.1	18.0	24.3
Surface tension at 77°F, dynes/cm		47.0	44.8	45.2
Heat of vaporization at 760 mm Hg, Btu/lb		364	232	174

* After Sivalls, 1976.

Table 5-2B
Variation of TEG Properties with Temperature*

Temp., °F	Sp Gr	Viscosity	Sp Heat, Btu/lb °F	Thermal Conductivity, Btu/(hr ft² °F/ft)
50	1.134	88	0.485	0.14
75	1.123	56	6.50	0.138
100	1.111	23	0.52	0.132
125	1.101	15.5	0.535	0.130
150	1.091	8.1	0.55	0.125
175	1.080	6.1	0.57	0.121
200	1.068	4.0	0.585	0.118
225	1.057	3.1	0.60	0.113
250	1.034	1.9	0.635	
300	1.022	1.5	0.65	

* After Sivalls, 1976.

fully used for dehydrating sweet as well as sour natural gases to effect a dew-point depression of 40 to 140°F, for operating conditions ranging from 25–2,500 psig and 40–160°F (Ikoku, 1984).

Process Flow Scheme

Figure 5-31 shows a typical flow sheet for a glycol dehydration plant described in detail by Sivalls (1976). The wet gas is first sent to a scrubber to remove any liquid water and hydrocarbons, sand, drilling mud, and other solid matter. These impurities must be thoroughly removed, because they

212   Gas Production Engineering

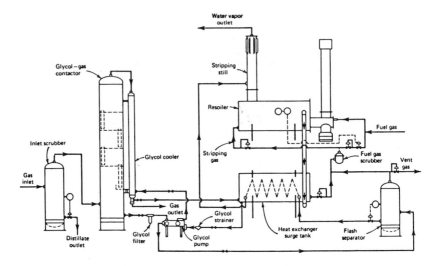

Figure 5-31. Flow diagram for a glycol dehydration plant. (After Sivalls, 1976; courtesy of C. R. Sivalls and the University of Oklahoma.)

may lead to foaming, flooding, poor efficiency, higher glycol losses, and maintenance problems in the glycol-gas contactor (also called absorber). A mist eliminator is also provided at the scrubber outlet to ensure good gas cleaning. The clean gas is then sent upward through the absorber, countercurrent to the flow of glycol. The absorber consists of several trays that act as equilibrium separation stages where water vapor from the gas is absorbed by the glycol. Glycol usually absorbs about 1 scf gas/gal at 1,000 psig absorber pressure (GPSA, 1981). Absorption is more in the presence of aromatic hydrocarbons.

Dry gas from the top of the absorber is sent through a mist eliminator (to remove any entrained glycol droplets) to a gas-glycol heat exchanger (shown as glycol cooler in Figure 5-31) where the gas cools the hot regenerated glycol before it enters the contactor. After this heat exchange, the dry gas leaves the dehydration unit. The cooled glycol enters the absorber from the top.

The wet glycol from the bottom of the glycol-gas contactor is first sent to a high-pressure filter to remove any solids that may have been acquired by it from the gas stream. The energy of this high-pressure glycol is then used to drive the glycol pump, which pumps the dry regenerated glycol to the contactor. This results in a considerable energy savings for the process. From the pump, the wet glycol flows to a low-pressure flash separator where any dissolved or entrained gas is removed. A three-phase flash separator may be required if the wet glycol also contains an appreciable amount of liquid hydrocarbons absorbed from the gas stream in the contactor. The separated

Gas-Water Systems and Dehydration Processing 213

gas is used either as fuel gas for the reboiler, or is vented to the atmosphere. Glycol from the bottom of the flash separator is sent to heat exchanger coils in the glycol surge tank to preheat it by heat exchange with the hot, dry regenerated glycol from the glycol regenerator.

The preheated wet glycol is then sent to the stripping still, which is a tower packed with ceramic saddle-type (or other types) packing, mounted on top of a reboiler. The glycol flows downwards through the packing (that acts as equilibrium separation stages for glycol-water separation) into the reboiler. Water vapor liberated from the glycol in the reboiler passes upwards through the packing, providing heat and picking up some water from the wet glycol flowing downwards. The water vapor leaves the unit from the top of the still column through an atmospheric reflux condenser that provides a partial reflux for the column (Sivalls, 1976).

In the reboiler, glycol is heated to approximately 350–400°F to reconcentrate it to 99.5% or more. In order to obtain glycol of such a high concentration, it is usually necessary to add some stripping gas to the reboiler (Sivalls, 1976). Using a valve and a pressure regulator, a small amount of gas from the fuel gas system is injected into the bottom of the reboiler through a spreader system. This gas aids the removal of water from the glycol by three mechanisms (Sivalls, 1976): (1) it "rolls" the glycol in the reboiler to allow any pockets of water vapor, entrained in the glycol due to the high viscosity of the glycol, to escape; (2) it lowers the partial pressure of the water vapor in the reboiler and still column, thereby allowing more water to vaporize from the glycol; and (3) it provides a sweeping action (a combination of absorption of water, and creation of a pressure drop for the water vapor to flow) to drive the water vapor out of the reboiler and stripping still system.

The reconcentrated glycol is sent to the shell side of the heat exchanger, which also serves as a surge tank. Here the regenerated glycol is cooled by heat exchange with the wet glycol stream, and is accumulated for feed to the glycol pump through a strainer. From the pump, it passes through the shell side of the glycol-gas heat exchanger (glycol cooler), into the top of the absorber.

Total glycol losses, excluding spillage, range from 0.05 gal/MMscf for high-pressure low-temperature gases, to as much as 0.30 gal/MMscf for low-pressure high-temperature gas streams (GPSA, 1981). These losses comprise mechanical carryover from the contactor, and vaporization losses from the contactor and regenerator.

Glycol Plant Operational Problems

Methanol, injected at the wellhead for hydrate inhibition, can lead to several operational problems in glycol dehydration plants:

1. Methanol, absorbed by the glycol along with water from the gas stream, increases the heat requirements in the regeneration system.
2. High methanol injection rates, and carryover of slugs of liquid methanol can cause flooding in the absorber and regeneration system.
3. Aqueous methanol is corrosive to carbon steel, leading to corrosion in the reboiler and still.
4. A methanol recovery unit is required at the water vapor outlet, because methanol cannot be vented directly to the atmosphere due to environmental hazards.

Any dirt or impurities can contaminate the glycol solution severely. Overheating the glycol can lead to solution decomposition, with both low as well as high boiling products. The reaction products form a sludge that collects on the heating surfaces, reducing process efficiency and in some cases, causing complete flow stoppage. This can be avoided by installing a filter at the downstream end of the glycol pump.

Glycol becomes quite corrosive with prolonged exposure to oxygen (GPSA, 1981). For this reason, a dry gas blanket is sometimes provided on the glycol surge tank. Presence of oxygen in the gas stream, however, may require special precautions. Corrosion may also be a problem in the presence of acid gases. A low pH accelerates decomposition of glycol. Generally, pH is controlled in the range 6.0-7.5, measured at a 50-50 dilution with distilled water, to avoid glycol decomposition (GPSA, 1981).

Carryover of liquid hydrocarbons can result in solution foaming. Foam inhibitors are normally used to avoid such a situation. Removal of all liquids from the gas is necessary before sending it to the contactor, because these liquids may, over long periods of time, leave crystalline deposits in the contactor. To prevent condensing of any hydrocarbons in the contactor, the inlet dry glycol must be maintained at a temperature slightly greater than that of the gas stream (Sivalls, 1976). Liquid hydrocarbons carried over into the stripping still cause severe damage. They may lead to vapor flooding in the reboiler and still due to the increased vapor load. Consequently, glycol will be carried out of the still along with the gas and water vapor. Heavier-end hydrocarbons will cause coking in the still and reboiler, and may also hinder glycol reconcentration.

Highly concentrated glycol is very viscous at low temperatures, below approximately 50°F. A heater may be required, especially in cold climatic conditions where the flowlines may solidify completely. Finally, to ensure good gas-glycol contact, all absorber trays must be completely filled with glycol at all times. Sudden liquid surges should be avoided to minimize solution carry-over losses.

Gas-Water Systems and Dehydration Processing 215

Glycol Plant Design (After Sivalls, 1976)

The primary variables required for the design include gas flow rate (MMscfd); gas gravity; operating pressure (psia); maximum working pressure in the contactor (psia); gas inlet temperature (°F); and outlet water content required in the gas (lbm/MMscf). In addition, two criteria must be selected for the design:

1. Glycol to water circulation rate. Requirements are generally in the range of 2–5 gal TEG per lb water removed (GPSA, 1981); most field installations use 2.5 to 4 gal TEG/lb H_2O removed (Ikoku, 1984).
2. Concentration of the lean TEG from the regeneration system. This ranges from 99.0% to 99.9%; most designs use a 99.5% lean glycol concentration.

The glycol to water circulation rate depends upon the lean glycol concentration, and the number (and efficiency) of trays in the absorber. Lean TEG concentration is determined by what the regenerator can deliver (which is primarily a function of the amount of stripping gas supplied to the regenerator), and the permissible lower limit of glycol viscosity the plant and equipment can handle.

Frequently, dehydration requirements are expressed in terms of dew-point depression. This can be calculated easily as the difference between the inlet and outlet gas dew-point temperatures. Sometimes, the inlet gas temperature is known, and the outlet gas water content is specified. In either case, the inlet and outlet gases are assumed to be saturated, and their water content, or dew-point temperature, is determined using any of the available correlations.

The amount of water to be removed, W_r in lbm/hr, can be calculated as follows:

$$W_r = (q/24)(W_i - W_o) \qquad (5\text{-}15)$$

where W_i, W_o = water contents, lb water/MMscf gas, for the inlet and outlet gas respectively.
A conversion factor of 24 is used in Equation 5-15 to convert the gas rate q from MMscfd to MMscf/hr.

Inlet Scrubber

The inlet scrubber is generally chosen based upon the operating pressure and gas capacity. The data shown in Tables 5-3 and 5-4, and Figure 5-32, may be used for this purpose. Generally, a two-phase scrubber with a 7.5 ft shell height is used.

Table 5-3
Specifications for Vertical Inlet Scrubbers*

Nominal WP psig	Size, OD	Nominal Gas Capacity, MMscfd**	Inlet and Gas Outlet Connection	Std Oil Valve	Shipping Weight, lb
230	16"	1.8	2"	1"	900
	20"	2.9	3"	1"	1,000
	24"	4.1	3"	1"	1,200
	30"	6.5	4"	1"	1,400
	36"	9.4	4"	1"	1,900
	42"	12.7	6"	2"	2,600
	48"	16.7	6"	2"	3,000
	54"	21.1	6"	2"	3,500
	60"	26.1	6"	2"	4,500
500	16"	2.7	2"	1"	1,000
	20"	4.3	3"	1"	1,300
	24"	6.1	3"	1"	2,100
	30"	9.3	4"	1"	2,700
	36"	13.3	4"	1"	3,800
	42"	18.4	6"	2"	4,200
	48"	24.3	6"	2"	5,000
	54"	30.6	6"	2"	5,400
	60"	38.1	6"	2"	7,500
600	16"	3.0	2"	1"	1,100
	20"	4.6	3"	1"	1,400
	24"	6.3	3"	1"	2,200
	30"	9.8	4"	1"	2,800
	36"	14.7	4"	1"	3,900
	42"	20.4	6"	2"	4,500
	48"	27.1	6"	2"	5,100
	54"	34.0	6"	2"	6,000
	60"	42.3	6"	2"	8,100
1000	16"	3.9	2"	1"	1,100
	20"	6.1	3"	1"	1,600
	24"	8.8	3"	1"	2,500
	30"	13.6	4"	1"	3,200
	36"	20.7	4"	1"	4,400
	42"	27.5	6"	2"	6,300
	48"	36.9	6"	2"	8,400
	54"	46.1	6"	2"	9,700
	60"	57.7	6"	2"	14,500
1200	16"	4.2	2"	1"	1,150
	20"	6.5	3"	1"	1,800
	24"	6.5	3"	1"	2,600
	30"	15.3	4"	1"	3,400

* After Sivalls, 1976.
** Gas capacity based on 100°F, 0.7 sp gr, and vessel working pressure.

Gas-Water Systems and Dehydration Processing 217

Table 5-3 Continued

Nominal WP psig	Size, OD	Nominal Gas Capacity, MMscfd**	Inlet and Gas Outlet Connection	Std Oil Valve	Shipping Weight, lb
	36"	23.1	4"	1"	4,700
	42"	31.0	6"	2"	6,700
	48"	40.5	6"	2"	8,500
	54"	51.4	6"	2"	11,300
	60"	62.3	6"	2"	14,500
1440	16"	4.8	2"	1"	1,500
	20"	6.7	3"	1"	2,100
	24"	11.2	3"	1"	2,800
	30"	17.7	4"	1"	3,900
	36"	25.5	4"	1"	5,400
	42"	34.7	6"	2"	7,800
	48"	45.3	6"	2"	9,200
	54"	56.1	6"	2"	12,900
	60"	69.6	6"	2"	16,000

** Gas capacity based on 100°F, 0.7 sp gr, and vessel working pressure.

Table 5-4
Settling Volumes and Liquid Capacities of Scrubbers*

	Two-phase Scrubber			Three-phase Scrubber		Liquid Capacity bbl/day††	
Size OD	Shell Height	Settling Volume, bbl**	Liquid Capacity, bbl/day†	Shell Height	Settling Volume, bbl**	Oil	Water
16"	5'	0.27	340	7½'	0.72	100	100
20"	5'	0.44	530	7½'	1.15	160	160
24"	5'	0.66	760	7½'	1.68	240	240
30"	5'	1.13	1180	7½'	2.78	400	400
36"	5'	1.73	2000	7½'	4.13	590	590
42"	5'	2.52	3000	7½'	5.80	830	830
48"	5'	3.48	4000	7½'	7.79	1120	1120
54"	5'	4.65	5000	7½'	10.12	1450	1450
60"	5'	6.01	6000	7½'	12.73	1830	1830

* After Sivalls, 1976.
** Based on nominal 1000 psig WP scrubber.
† Based on 1.0 minute retention time.
†† Based on 5.0 minute retention time.

218　　*Gas Production Engineering*

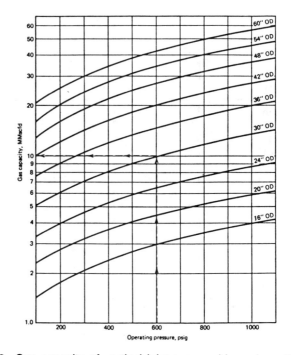

Figure 5-32. Gas capacity of vertical inlet-gas scrubbers, based upon a 0.7-gravity gas at 100°F. (After Sivalls, 1976; courtesy of C. R. Sivalls and the University of Oklahoma.)

Glycol-Gas Contactor

Contactor diameter can be selected using Figure 5-33 for a trayed-type, or Figure 5-34 for a packed-type design. Gas capacities determined using Figure 5-33 or Figure 5-34 must be corrected for operating temperature and gas specific gravity as follows:

$$q = q_b(C_t)(C_g)$$

where　　q = gas capacity at operating conditions, MMscfd
　　　　q_b = gas capacity at the base conditions of 100°F for a 0.7 gravity gas, MMscfd
　　　　C_t, C_g = correction factors for operating temperature and gas gravity, respectively, as shown in Tables 5-5 and 5-6 for both the contactor types

Additional specifications for contactors are given in Tables 5-7 through 5-10.

The approximate actual number of trays required in the contactor (for a tray-type contactor), or the depth of packing (for the packed-type contactor)

Gas-Water Systems and Dehydration Processing 219

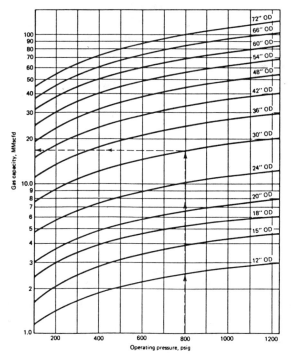

Figure 5-33. Gas capacity of trayed glycol-gas contactors, based upon a 0.7-gravity gas at 100°F. (After Sivalls, 1976; courtesy of C. R. Sivalls and the University of Oklahoma.)

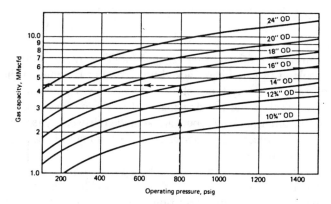

Figure 5-34. Gas capacity of packed glycol-gas contactors, based upon a 0.7-gravity gas at 100°F. (After Sivalls, 1976; courtesy of C. R. Sivalls and the University of Oklahoma.)

Table 5-5
Temperature Correction Factor, C_t, for Gas Capacity of Trayed and Packed Glycol-Gas Contactors*

Operating temperature °F	C_t Trayed	Packed
40	1.07	—
50	1.06	0.93
60	1.05	0.94
70	1.04	0.96
80	1.02	0.97
90	1.01	0.99
100	1.00	1.00
110	0.99	1.01
120	0.98	1.02

* After Sivalls, 1976.

Table 5-6
Gas Specific Gravity Correction Factor, C_g, for Gas Capacity of Trayed and Packed Glycol-Gas Contactors*

Gas specific gravity	C_g Trayed	Packed
0.55	1.14	1.13
0.60	1.08	1.08
0.65	1.04	1.04
0.70	1.00	1.00
0.75	0.97	0.97
0.80	0.93	0.94
0.85	0.90	0.91
0.90	0.88	0.88

* After Sivalls, 1976.

required, can be determined using Figure 5-35. This figure assumes that the depth of packing in feet for a packed-type column is equal to the actual number of trays for a tray-type column. This is only true when 1-in. metal pall rings are used as the packing material in the packed-type contactor. Corrections for other packing types can be made fairly easily (see, for example, *Chemical Engineers' Handbook*, 1984). A minimum height of packing equal to 4 ft should be used.

A more accurate procedure for calculating the number of trays (or the depth of packing) required is to use the McCabe-Thiele diagrams, widely used in designing equilibrium separation process equipment. The procedure for constructing this diagram, shown in Figure 5-36, is as follows:

(text continued on page 226)

Gas-Water Systems and Dehydration Processing   221

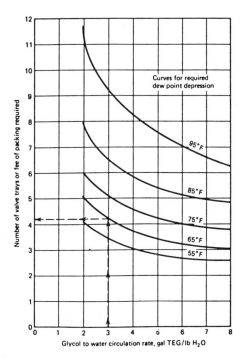

Figure 5-35. Approximate number of trays, or depth of packing, required for glycol-gas contactors. (After Sivalls, 1976; courtesy of C. R. Sivalls and the University of Oklahoma.)

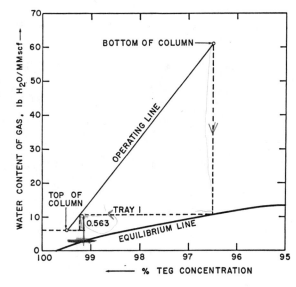

Figure 5-36. McCabe-Thiele diagram for Example 5-9.

Table 5-7
Specifications for Tray-Type Glycol-Gas Contactors*

Nominal WP psig	Nominal Size OD	Nominal Gas Capacity, MMscfd**	Gas Inlet and Outlet Size	Glycol Inlet and Outlet Size	Glycol Cooler Size	Shipping Weight, lb
250	12³/₄″	1.5	2″	¹/₂″	2″ × 4″	800
	16″	2.4	2″	³/₄″	2″ × 4″	900
	18″	3.2	3″	³/₄″	3″ × 5″	1,100
	20″	4.0	3″	1″	3″ × 5″	1,400
	24″	6.1	3″	1″	3″ × 5″	2,000
	30″	9.9	4″	1″	4″ × 6″	2,400
	36″	14.7	4″	1¹/₂″	4″ × 6″	3,200
	42″	19.7	4″	1¹/₂″	4″ × 6″	4,400
	48″	26.3	6″	2″	6″ × 8″	6,300
	54″	32.7	6″	2″	6″ × 8″	7,700
	60″	40.6	6″	2″	6″ × 8″	9,600
500	12³/₄″	2.0	2″	¹/₂″	2″ × 4″	1,000
	16″	3.2	2″	³/₄″	2″ × 4″	1,200
	18″	4.3	3″	³/₄″	3″ × 5″	1,500
	20″	5.3	3″	1″	3″ × 5″	1,700
	24″	8.3	3″	1″	3″ × 5″	2,900
	30″	13.1	3″	1″	3″ × 5″	3,900
	36″	19.2	4″	1¹/₂″	4″ × 6″	6,000
	42″	27.4	4″	1¹/₂″	4″ × 6″	7,700
	48″	35.1	6″	2″	6″ × 8″	10,000
	54″	44.5	6″	2″	6″ × 8″	12,000
	60″	55.2	6″	2″	6″ × 8″	15,300
600	12³/₄″	2.2	2″	¹/₂″	2″ × 4″	1,100
	16″	3.4	2″	³/₄″	2″ × 4″	1,300
	18″	4.5	3″	³/₄″	3″ × 5″	1,600
	20″	5.5	3″	1″	3″ × 5″	1,800
	24″	8.5	3″	1″	3″ × 5″	3,000
	30″	14.3	3″	1″	3″ × 5″	4,000
	36″	21.2	4″	1¹/₂″	4″ × 6″	6,300
	42″	29.4	4″	1¹/₂″	4″ × 6″	8,400
	48″	39.2	6″	2″	6″ × 8″	11,300
	54″	49.3	6″	2″	6″ × 8″	13,400
	60″	61.3	6″	2″	6″ × 8″	16,500

* After Sivalls, 1976.
** Gas capacity based on 100°F, 0.7 sp gr, and vessel working pressure.

Gas-Water Systems and Dehydration Processing 223

Table 5-7 Continued

Nominal WP psig	Size OD	Nominal Gas Capacity, MMscfd**	Gas Inlet and Outlet Size	Glycol Inlet and Outlet Size	Glycol Cooler Size	Shipping Weight, lb
1000	12³/₄"	2.7	2"	¹/₂"	2" × 4"	1,300
	16"	4.3	2"	³/₄"	2" × 4"	1,600
	18"	5.5	3"	³/₄"	3" × 5"	2,100
	20"	7.3	3"	1"	3" × 5"	2,600
	24"	11.3	3"	1"	3" × 5"	4,200
	30"	18.4	3"	1"	3" × 5"	5,500
	36"	27.5	4"	1¹/₂"	4" × 6"	8,500
	42"	37.1	4"	1¹/₂"	4" × 6"	11,800
	48"	49.6	6"	2"	6" × 8"	16,200
	54"	62.0	6"	2"	6" × 8"	20,200
	60"	77.5	6"	2"	6" × 8"	26,300
1200	12³/₄"	3.0	2"	¹/₂"	2" × 4"	1,500
	16"	4.7	2"	³/₄"	2" × 4"	1,900
	18"	6.0	3"	³/₄"	3" × 5"	2,300
	20"	7.8	3"	1"	3" × 5"	3,000
	24"	12.0	3"	1"	3" × 5"	4,900
	30"	20.1	3"	1"	3" × 5"	6,400
	36"	29.8	4"	1¹/₂"	4" × 6"	10,000
	42"	41.4	4"	1¹/₂"	4" × 6"	13,100
	48"	54.1	6"	2"	6" × 8"	18,400
	54"	68.4	6"	2"	6" × 8"	23,500
	60"	85.0	6"	2"	6" × 8"	29,000
1440	12³/₄"	3.1	2"	¹/₂"	2" × 4"	1,800
	16"	4.9	2"	³/₄"	2" × 4"	2,200
	18"	6.5	3"	³/₄"	3" × 5"	2,800
	20"	8.3	3"	1"	3" × 5"	3,500
	24"	13.3	3"	1"	3" × 5"	5,800
	30"	22.3	3"	1"	3" × 5"	7,500
	36"	32.8	4"	1¹/₂"	4" × 6"	11,700
	42"	44.3	4"	1¹/₂"	4" × 6"	14,400
	48"	58.3	6"	2"	6" × 8"	20,000
	54"	74.0	6"	2"	6" × 8"	25,800
	60"	91.1	6"	2"	6" × 8"	32,000

** Gas capacity based on 100°F, 0.7 sp gr and contactor working pressure.

Gas Production Engineering

Table 5-8
Specifications for Tray-Type Glycol-Gas Contactors*

Size OD	Standard Shell Height**	Standard Glycol Cooler Height**	Add to Height For Add. Tray, ea.	Glycol Charge, gal Standard**	For Each Add. Tray
123/4"	13'	9'	2'	10	1.5
16"	13'	9'	2'	13	2.2
18"	13'	9'	2'	16	2.8
20"	13'	9'	2'	19	3.6
24"	13'	9'	2'	25	5.0
30"	13'	9'	2'	38	8.2
36"	13'	9'	2'	53	11.8
42"	13'	9'	2'	73	16.8
48"	13'	9'	2'	90	20.9
54"	13'	9'	2'	112	26.6
60"	13'	9'	2'	137	32.6

* After Sivalls, 1976.
** For standard four-tray contactor.

Table 5-9
Specifications for Packed Column Glycol-Gas Contactors*

Nominal WP psig	Nominal Size OD	Nominal Gas Capacity, MMscfd**	Gas Inlet and Outlet Size	Glycol Inlet and Outlet Size	Glycol Cooler Size	Shipping Weight, lb
250	103/4"	1.1	2"	1/2"	2" × 4"	500
	123/4"	1.6	2"	1/2"	2" × 4"	600
	14"	1.9	2"	1/2"	2" × 4"	650
	16"	2.5	2"	1/2"	2" × 4"	800
	18"	3.4	3"	3/4"	3" × 5"	900
	20"	4.0	3"	3/4"	3" × 5"	1100
	24"	5.5	3"	1"	3" × 5"	1800
500	103/4"	1.5	2"	1/2"	2" × 4"	600
	123/4"	2.2	2"	1/2"	2" × 4"	700
	14"	2.6	2"	1/2"	2" × 4"	750
	16"	3.4	2"	1/2"	2" × 4"	900
	18"	4.4	3"	3/4"	3" × 5"	1000
	20"	5.5	3"	3/4"	3" × 5"	1500
	24"	7.5	3"	1"	3" × 5"	2500
600	103/4"	1.7	2"	1/2"	2" × 4"	650
	123/4"	2.4	2"	1/2"	2" × 4"	750
	14"	2.9	2"	1/2"	2" × 4"	800
	16"	3.8	2"	1/2"	2" × 4"	950
	18"	4.8	3"	3/4"	3" × 5"	1100

* After Sivalls, 1976.
** Gas capacity based on 100°F, 0.7 sp gr, and vessel working pressure.

Gas-Water Systems and Dehydration Processing 225

Table 5-9 Continued

Nominal WP psig	Nominal Size OD	Nominal Gas Capacity, MMscfd**	Gas Inlet and Outlet Size	Glycol Inlet and Outlet Size	Glycol Cooler Size	Shipping Weight, lb
	20"	6.0	3"	3/4"	3" × 5"	1700
	24"	8.1	3"	1"	3" × 5"	2700
1000	10 3/4"	2.3	2"	1/2"	2" × 4"	900
	12 3/4"	3.3	2"	1/2"	2" × 4"	1000
	14"	4.0	2"	1/2"	2" × 4"	1100
	16"	5.2	2"	1/2"	2" × 4"	1300
	18"	6.6	3"	3/4"	3" × 5"	1800
	20"	8.2	3"	3/4"	3" × 5"	2300
	24"	11.8	3"	1"	3" × 5"	3500
1200	10 3/4"	2.5	2"	1/2"	2" × 4"	1200
	12 3/4"	3.6	2"	1/2"	2" × 4"	1300
	14"	4.1	2"	1/2"	2" × 4"	1500
	16"	5.4	2"	1/2"	2" × 4"	1700
	18"	6.9	3"	3/4"	3" × 5"	2200
	20"	8.5	3"	3/4"	3" × 5"	2800
	24"	12.3	3"	3/4"	3" × 5"	4000
1440	10 3/4"	2.6	2"	1/2"	2" × 4"	1300
	12 3/4"	3.7	2"	1/2"	2" × 4"	1400
	14"	4.5	2"	1/2"	2" × 4"	1600
	16"	5.9	2"	1/2"	2" × 4"	1900
	18"	7.5	3"	3/4"	3" × 5"	2500
	20"	9.3	3"	3/4"	3" × 5"	3100
	24"	12.7	3"	1"	3" × 5"	4500

* After Sivalls, 1976.
** Gas capacity based on 100°F, 0.7 sp gr and contactor working pressure.

Table 5-10
Specifications for Packed Column Glycol-Gas Contactors*

Size, OD	Standard Shell Height	Standard Glycol Cooler Height	Standard Contacting Element**	Glycol Charge, gal
10 3/4"	9'	7'	1" × 4'	6
12 3/4"	9'	7'	1" × 4'	7
14"	9'	7'	1" × 4'	8
16"	9'	7'	1" × 4'	10
18"	9'	7'	1" × 4'	12
20"	9'	7'	1" × 4'	14
24"	9'	7'	1" × 4'	18

* After Sivalls, 1976.
** Standard contacting element is carbon steel metal pall rings of size listed in Table.

(text continued from page 220)

1. Determine the concentration of the rich TEG leaving the bottom of the contactor:

$$C_{Rich} = (C_{Lean}\rho_{Lean})/(\rho_{Lean} + 1/L_w) \qquad (5\text{-}16)$$

where C_{Rich}, C_{Lean} = concentrations of rich and lean TEG, respectively, expressed as a fraction
 ρ_{Lean} = density of lean TEG solution in lbm/gal
 L_w = glycol to water circulation rate in gal TEG/lb water

2. On the McCabe-Thiele diagram, the top of the column is indicated by the water content of the outlet gas from the contactor, and the concentration of lean TEG. Similarly, the bottom of the column is represented by the water content of the inlet gas to the contactor, and the rich TEG concentration. Draw the operating line as the straight line connecting the top and bottom of the contactor column.
3. The equilibrium line is drawn using data shown in Figure 5-37, and any of correlations for the water content of gas as a function of pressure and temperature. This line represents the water content of the gas that would exist in equilibrium with the various TEG concentrations.
4. Starting from the point representing the bottom of the column, the theoretical number of trays required are stepped off by triangulation, as shown in Figure 5-36, till the top of the column is reached.
5. Use the tray efficiency factor (or packing factors) to calculate the actual number of trays (or depth of packing) required:

Actual number of trays = Theoretical number of trays/tray efficiency

Generally, a tray efficiency equal to 25% for bubble-cap trays, and 33.3% for valve trays is used, and the figure for the actual number of trays is rounded off to the next higher integer value. For the packed-type contactor, a minimum of 4 ft height of packing must be used.

Reboiler

The glycol circulation rate in gallons per hour, L, in the plant is given by:

$$L = L_w W_i (q/24) \qquad (5\text{-}17)$$

Gas-Water Systems and Dehydration Processing 227

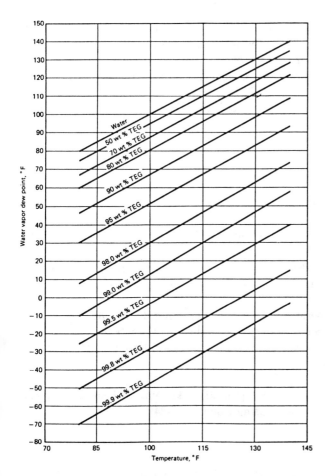

Figure 5-37. Dew points of aqueous TEG solutions versus temperature. (After Sivalls, 1976; courtesy of C. R. Sivalls and the University of Oklahoma.)

An approximate calculation of the heat required in the reboiler (also called reboiler heat load), Q in Btu/hr, can be made using the following empirical relationship:

$$Q = 2{,}000L \qquad (5\text{-}18)$$

Equation 5-18 gives good results for most high-pressure dehydration designs. A more precise estimate of the reboiler heat load may be made from the following procedure:

$$Q = Q_l + Q_w + Q_r + Q_h \qquad (5\text{-}19)$$

where Q_l = sensible heat required for glycol, Btu/hr
Q_w = heat of vaporization required for water, Btu/hr
Q_r = heat required to vaporize the reflux water in the still, Btu/hr
Q_h = heat loss from the reboiler and stripping still, Btu/hr

These heat requirements are estimated from the following relationships:

$$Q_l = L\rho c(T_2 - T_1) \tag{5-20}$$

$$Q_w = 970.3(q/24)(W_i - W_o) \tag{5-21}$$

$$Q_r = 0.25\, Q_w, \text{ assuming a 25\% reflux} \tag{5-22}$$

$$Q_h = 5{,}000 \text{ to } 20{,}000, \text{ depending upon boiler size} \tag{5-23}$$

where ρ = glycol density at the average reboiler temperature, lbm/gal
c = specific heat of the glycol at the average reboiler temperature, Btu/lbm-°F
T_1 = temperature of the glycol entering the reboiler, °F
T_2 = temperature of the glycol leaving the reboiler, °F
970.3 = heat of vaporization of water (Btu/lbm) at 212°F and 14.7 psia

If the size of the reboiler and stripping still is known, the heat loss, Q_h, can be more accurately estimated using Equation 5-24:

$$Q_h = 0.24 A_s (T_v - T_a) \tag{5-24}$$

where A_s = total exposed surface area of the reboiler and still, ft^2
T_v = average temperature (of the fluid) in the reboiler, °F
T_a = ambient air temperature (the temperature of the surroundings), °F
0.24 = an approximate heat transfer coefficient for the reboiler and stripping still, Btu/(hr ft^2 °F)

The surface area of the firebox, A (ft^2), can be determined as follows:

$$A = Q/7{,}000 \tag{5-25}$$

where 7,000 Btu/hr-ft^2 = estimated design heat flux for direct-fired reboilers

This calculation of A enables determination of the overall size of the reboiler.

Stripping Still

The diameter (or cross-sectional area) of the packed stripping still for use with the glycol reconcentrator can be estimated from Figure 5-38 as a function of the glycol to water circulation rate (gal TEG/lb water) and the glycol circulation rate (gal/hr). The size of a stripping still is governed by the vapor and liquid loading conditions at its bottom. The vapor load comprises water vapor and stripping gas flowing upwards through the still, whereas the liquid load consists of rich glycol and the reflux flowing downwards. Figure 5-38 includes stripping gas requirements, generally in the range of 2 to 10 ft^3/gal TEG circulated.

If a tray-type design is used, one theoretical tray is usually sufficient for most cases (Sivalls, 1976). For a packed-type design, a minimum of 4 ft packing height is provided, consisting generally of 1.5-in. ceramic saddle-type packing. This height should be increased with increasing glycol recon-

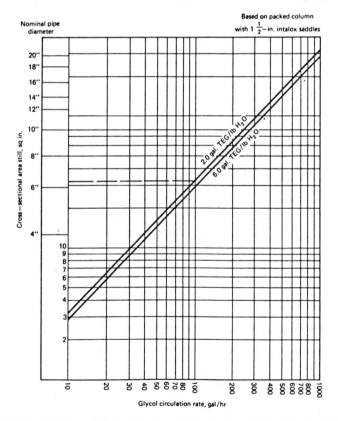

Figure 5-38. Stripping still size for glycol dehydrators. (After Sivalls, 1976; courtesy of C. R. Sivalls and the University of Oklahoma.)

230 Gas Production Engineering

centrator size, to a maximum of about 8 ft for a 1 MM Btu/hr unit (Sivalls, 1976). Additional specifications for glycol reconcentrators can be obtained using Table 5-11.

Glycol Circulating Pump

The size of the circulating pump required can be determined using the glycol circulation rate, L, and the maximum contactor operating pressure.

Table 5-11
Specifications for Glycol Reconcentrators*

Reboiler Capacity, Btu/hr	Glycol Capacity, gph**	Reboiler Size, Dia. x Len.	Heat Exchanger Surge Tank Size, Dia. x Len.	Stripping Still Size, Dia. x Ht.	Reflex Condenser Size, Dia. x Ht.
75,000	20	18" × 3', 6"	18" × 3', 6"	6⁵⁄₈" × 4', 6"	6⁵⁄₈" × 2', 0"
75,000	35	18" × 3', 6"	18" × 3', 6"	6⁵⁄₈" × 4', 6"	6⁵⁄₈" × 2', 0"
125,000	40	18" × 5'	18" × 5'	6⁵⁄₈" × 4', 6"	6⁵⁄₈" × 2', 0"
125,000	70	18" × 5'	18" × 5'	6⁵⁄₈" × 4', 6"	6⁵⁄₈" × 2', 0"
175,000	90	24" × 5'	24" × 5'	8⁵⁄₈" × 4', 6"	8⁵⁄₈" × 2', 0"
175,000	100	24" × 5'	24" × 5'	8⁵⁄₈" × 4', 6"	8⁵⁄₈" × 2', 0"
250,000	150	24" × 7'	24" × 7'	8⁵⁄₈" × 5', 0"	8⁵⁄₈" × 2', 0"
350,000	210	24" × 10'	24" × 10'	10³⁄₄" × 5', 0"	10³⁄₄" × 2', 6"
400,000	250	30" × 10'	30" × 10'	10³⁄₄" × 6', 0"	10³⁄₄" × 2', 6"
500,000	315	36" × 10'	36" × 10'	12³⁄₄" × 7', 0"	12³⁄₄" × 2', 6"
750,000	450	36" × 15'	36" × 10'	14" × 8', 0"	14" × 3', 0"
850,000	450	42" × 15'	36" × 10'	14" × 8', 0"	14" × 3', 0"
1,000,000	450	48" × 16'	36" × 10'	16" × 8', 0"	16" × 3', 0"

Flash Separator Size, Dia x Ht.	Heat Exchange Coil Size	Coil Area, sq ft	Glycol Pump Model	High-Pressure Glycol Filter Size	Glycol Charge, gal	Shipping Wt, lb
12" × 48"	¹⁄₂"	12.9	1715PV	1"	75	2,100
12" × 48"	¹⁄₂"	12.9	4015PV	1"	75	2,100
16" × 48"	¹⁄₂"	23.3	4015PV	1"	105	2,200
16" × 48"	¹⁄₂"	23.3	9015PV	1"	105	2,250
16" × 48"	¹⁄₂"	31.1	9015PV	1"	190	3,200
16" × 48"	¹⁄₂"	31.1	21015PV	1¹⁄₂"	190	3,200
16" × 48"	³⁄₄"	44.6	21015PV	1¹⁄₂"	260	3,700
20" × 48"	³⁄₄"	64.8	21015PV	1¹⁄₂"	375	4,000
20" × 48"	³⁄₄"	64.8	45015PV	2"	445	4,500
24" × 48"	1"	82.1	45015PV	2"	680	6,500
30" × 48"	1"	102.6	45015PV	2"	990	7,000
30" × 48"	1"	102.6	45015PV	2"	1175	7,500
30" × 48"	1"	102.6	45015PV	2"	1425	10,000

* After Sivalls, 1976.
** Glycol capacity is based on circulating 2.5 gal TEG/lb H₂O and is controlled by the reboiler capacity or pump capacity, whichever is smaller.

Table 5-12
Specifications for Standard High-Pressure Glycol Pumps*

Model Number	Circulation Rate—Gallons/Hour Pump Speed—Strokes/Minute Count one stroke for each discharge of pump**															
	8	10	12	14	16	18	20	22	24	26	28	30	32	34	36	38	40
1715 PV	8	10	12	14	16	18	20	22	24	26	28	30	32	34	36	38	40
4015 PV			12	14	16	18	20	22	24	26	28	30	32	34	36	38	40
9015 PV			27	31.5	36	40.5	45	49.5	54	48.5	63	67.5	72	76.5	81	85.5	90
21015 PV		66	79	92	105	118	131	144	157	171	184	197	210				
45015 PV		166	200	233	266	300	333	366	400	433	466						

* After Sivalls, 1976.
** It is not recommended to attempt to run pumps at speeds less or greater than those indicated in the above table.

Gas Consumption

Operating pressure, psig	300	400	500	600	700	800	900	1000	1100	1200	1300	1400	1500
Cu ft/gal at 14.4 and 60°F	1.7	2.3	2.8	3.4	3.7	4.5	5.0	5.6	6.1	6.7	7.2	7.9	8.3

Pump Model	Pump Conn.	Size Strainer	High-pressure Filter		Low-pressure Filter	
			Size	Elements	Size	Elements
315PV	1/4"	1/2"	1"	1-2³/₄" × 9³/₄"	1/2"	1-3" × 18"
1715PV and 815SC	1/2"	3/4"	1"	1-2³/₄" × 9³/₄"	1/2"	1-3" × 18"
4015PV and 2015SC	1/2"	3/4"	1"	1-2³/₄" × 9³/₄"	1/2"	1-3" × 18"
9015PV and 5015SC	3/4"	1"	1"	2-2³/₄" × 9³/₄"	3/4"	1-3" × 36"
21015PV and 10015SC	1"	1"	1¹/₂"	4-2³/₄" × 9³/₄"	1"	4-3" × 18"
45015PV and 20015SC	1¹/₂"	1¹/₂"	2"	8-2³/₄" × 9³/₄"	1¹/₂"	4-3" × 36"

Generally, as mentioned earlier, the glycol-powered pump that transmits energy from the rich glycol to the lean glycol, is used for glycol circulation. Table 5-12 by Sivalls (1976) shows sizing data for this type of pump. Table 5-12 also gives data on the amount of gas released by the pump as a function of the glycol circulation rate and the contactor operating pressure. For other (motor-driven) pump types, such as centrifugal pumps or positive displacement pumps, data supplied by the pump manufacturer may be used.

Glycol Flash Separator

The glycol flash separator, used downstream of the pump, should be designed for a liquid retention time of at least 5 min. The size can be calculated as follows:

$$V = Lt/60 \tag{5-26}$$

where V = settling volume required in the separator, gal
t = liquid retention time, min

Example 5-9. Size a glycol dehydration plant using Sivalls' method to meet the following requirements:

Gas flow rate = 20.0 MMscfd
Gas specific gravity = 0.7
Operating line pressure = 1,000 psig
Maximum contactor working pressure = 1,400 psig
Gas inlet temperature = 100°F
Outlet gas water content = 6.0 lb H_2O/MMscf gas

Use a tray-type contactor with bubble-cap trays, and assume the following additional criteria:

Glycol to water circulation rate = 3.5 gal TEG/lb H_2O
Lean glycol concentration = 99.5% TEG

Solution

From the McKetta and Wehe correlation (Figure 5-1):

Water content of inlet gas at 1,000 psig and 100°F = 61 lb H_2O/MMscf gas
Dew point of exit gas with 6 lb H_2O/MMscf water content = 28°F

Thus, the required dew point depression = 100 − 28 = 72°F

Using Equation 5-15, amount of water to be removed, W_r = (20/24)(61 − 6) = 45.83 lbm/hr

Inlet Scrubber Size

From Table 5-3, a 36-in. O.D. vertical scrubber is required for a 1,400-psig working pressure. The same result is obtained using Figure 5-32 (with suitable extrapolation). So, use a 36-in. × 7.5-ft vertical two-phase scrubber, with a 1,440 psig working pressure.

Contactor Size

From Figure 5-33, select a 36 in. O.D. contactor with a q_b = 27 MMscfd. From Tables 5-5 and 5-6, the correction factors for operating temperature and gas gravity are both equal to 1.0. Therefore, the contactor can handle a gas rate up to 27 MMscfd and is suitable for use.

Number of Trays Required in the Contactor

From Figure 5-35, the actual number of trays required for a 3.5 gal TEG/lb H_2O glycol to water circulation rate and 72°F dew-point depression = 4.5.

Thus, approximately 5 trays are required.

For a more precise determination, construct the McCabe-Thiele diagram.

From Table 5-2, specific gravity of lean glycol at 100°F = 1.111, implying a glycol density = (1.111)(8.34) = 9.266 lbm/gal

Using Equation 5-16, C_{Rich} = (0.995)(9.266)/(9.266 + 1/3.5) = 0.965 = 96.5%.

Thus, the operating line points are:

Top of column: 6 lb H_2O/MMscf and 99.5% TEG
Bottom of column: 61 lb H_2O/MMscf and 96.5% TEG

Data points for constructing the equilibrium line are as follows:

% TEG	Eqlbm. dew-point temp. at 100°F (Figure 5-37), °F	Water content of gas at dew-point temp. and 1,000 psig (Figure 5-1), lb H_2O/MMscf
99.6	− 10	1.1
99.0	12	3.2
98.0	30	6.3
97.0	40	9.0
96.0	47	11.7
95.0	51	13.3

The McCabe-Thiele diagram constructed using this information, as shown in Figure 5-36, gives the number of theoretical trays required = 1.563.

Therefore, the actual number of trays required = 1.563/0.25 = 6.25, or 7 trays.

This result is not in reasonable agreement with the approximate number of trays determined using Figure 5-35. It appears that Sivalls (1976) intended this figure for a valve-type contactor column. This is apparent if a tray efficiency of 0.333 is used to calculate the number of trays required from the McCabe-Thiele diagram. This gives a tray requirement of 1.563/0.333 = 4.7, or 5 trays.

234   Gas Production Engineering

On the other hand, this discrepancy could simply be a reflection on the poor quality of the correlation shown in Figure 5-35, resulting undoubtedly from the approximations involved in reducing a multivariable problem to a few select parameters.

Reboiler Heat Load

From Equation 5-17, glycol circulation rate, $L = (3.5)(61)(20/24) = 177.92$ gal/hr.

From Equation 5-18, the reboiler heat load, $Q = (2,000)(177.92) = 3.56 \times 10^5$ Btu/hr approximate.

For detailed calculations, assume $\rho c(T_2 - T_1) = 1200$. Then, using Equations 5-20 through 5-23:

$Q_l = (177.92)(1,200) = 21.35 \times 10^4$ Btu/hr

$Q_w = (970.3)(20/24)(61 - 6) = 4.45 \times 10^4$ Btu/hr

$Q_r = (0.25)(4.45)10^4 = 1.11 \times 10^4$ Btu/hr

$Q_h = 20,000 = 2.0 \times 10^4$ Btu/hr

Thus, $Q = (21.35 + 4.45 + 1.11 + 2.0)10^4 = 2.89 \times 10^5$ Btu/hr

Stripping Still

Using Figure 5-38, the minimum internal diameter required for the still = 8.2 in., based upon a 177.92 gal/hr glycol circulation rate, and 3.5 gal TEG/lb H_2O glycol to water circulation rate.

Glycol Pump

From Table 5-12, pump model number 21015 PV, with 28 strokes/min pumping speed, is required for a glycol circulation rate of 177.92 gal/hr.

Glycol Flash Separator

Using Equation 5-26, the settling volume of the flash separator required, assuming a liquid retention time of 5 min, is:

$V = (177.92)(5)/60 = 14.8$ gal $= 0.35$ bbl

From Table 5-4, a 20 in. O.D., two-phase separator is required.

Summary of Requirements

Inlet gas scrubber: 36-in. O.D. × 7.5-ft height, two-phase, 1,440-psig working pressure.
Glycol-gas contactor: 36-in. O.D., with 7 bubble-cap trays, 1,440-psig working pressure.
Glycol pump: Model number 21015 PV, with 28 strokes/min pumping speed.
Glycol flash sep.: 20-in. O.D., two-phase flash separator.
Reboiler: 289,000-Btu/hr heat load requirements.
Stripping still: 8.2-in. I.D.

Or, in terms of standard size units:

Scrubber: 36-in. O.D. × 7.5-ft height, 1,440-psig W.P.
Contactor: 36-in. O.D. × 19-ft height, 1,440-psig W.P., with 7 trays.
Reconcentrator: 24-in. × 10-ft., 350,000 Btu/hr reboiler, with 10.75-in. × 5-ft stripping still, 10.75-in. × 2.5-ft reflux condenser.
Flash sep.: 20 in. × 4 ft
Pump: Model number 21015 PV, 28 spm.

Adsorption Dehydration

Adsorption (or solid bed) dehydration is the process where a solid desiccant is used for the removal of water vapor from a gas stream. The solid desiccants commonly used for gas dehydration are those that can be regenerated and, consequently, used over several adsorption-desorption cycles. Some solid desiccants can dehydrate gas down to 1 ppm or less; these desiccants are widely used on feed streams for cryogenic processing (GPSA, 1981).

The mechanisms of adsorption on a surface are of two types: physical, and chemical. The latter process, involving a chemical reaction, is termed "chemisorption." Chemical adsorbents find very limited application in gas processing. Adsorbents that allow physical adsorption hold the adsorbate on their surface by surface forces. For physical adsorbents used in gas dehydration, the following properties are desirable (Campbell, 1984b):

1. Large surface area for high capacity. Commercial adsorbents have a surface area of 500–800 m^2/gm (= 2.4 × 10^6 to 3.9 × 10^6 ft^2/lbm).
2. Good "activity" for the components to be removed, and good activity retention with time/use. Commercial adsorbents can remove practi-

cally all the water from a natural gas to values as low as 1 ppm (parts per million).
3. High mass transfer rate, i.e., a high rate of removal.
4. Easy, economic regeneration.
5. Small resistance to gas flow, so that the pressure drop through the dehydration system is small.
6. High mechanical strength to resist crushing and dust formation. The adsorbent also must retain enough strength when "wet."
7. Cheap, non-corrosive, non-toxic, chemically inert, high bulk density, and small volume changes upon adsorption and desorption of water.

Types of Adsorbents

Some materials that satisfy these criteria, in the order of increasing cost, are: bauxite ore, consisting primarily of alumina ($Al_2O_3.xH_2O$); alumina; silica gels and silica-alumina gels; and molecular sieves. Activated carbon, a widely used adsorbent, possesses no capacity for water adsorption and is therefore not used for dehydration purposes, though it may be used for the removal of certain impurities. Bauxite also is not used much because it contains iron and is thus unsuitable for sour gases.

Alumina

A hydrated form of aluminium oxide (Al_2O_3), alumina is the least expensive adsorbent. It is activated by driving off some of the water associated with it in its hydrated form ($Al_2O_3.3H_2O$) by heating. It produces an excellent dew point depression to dew point values as low as $-100°F$, but requires much more heat for regeneration. Also, it is alkaline and cannot be used in the presence of acid gases, or acidic chemicals used sometimes for well treating. The tendency to adsorb heavy hydrocarbons is high, and it is difficult to remove these during regeneration. It has good resistance to liquids, but little resistance to disintegration due to mechanical agitation by the flowing gas.

Gels: Silica Gel and Silica-Alumina Gel

Gels are granular, amorphous solids manufactured by chemical reaction. Gels manufactured from sulfuric acid and sodium silicate reaction are called silica gels, and consist almost solely of silicon dioxide (SiO_2). Alumina gels consist primarily of some hydrated form of Al_2O_3. Silica-alumina gels are a combination of silica gel and alumina gel.

Gels can dehydrate gas to as low as 10 ppm (GPSA, 1981), and have the greatest ease of regeneration of all desiccants. They adsorb heavy hydrocarbons, but release them relatively more easily during regeneration. Since they are acidic, they can handle sour gases, but not alkaline materials such as caustic or ammonia. Although there is no reaction with H_2S, sulfur can deposit and block their surface. Therefore, gels are useful if the H_2S content is less than 5–6%.

Molecular Sieves

These are a crystalline form of alkali metal (calcium or sodium) aluminosilicates, very similar to natural clays. They are highly porous, with a very narrow range of pore sizes, and very high surface area. Manufactured by ion-exchange, molecular sieves are the most expensive adsorbents. They possess highly localized polar charges on their surface that act as extremely effective adsorption sites for polar compounds such as water and hydrogen sulfide (see also Chapter 6). Molecular sieves are alkaline and subject to attack by acids. Special acid-resistant sieves are available for very sour gases.

Since the pore size range is narrow, molecular sieves exhibit selectivity towards adsorbates on the basis of their molecular size, and tend not to adsorb bigger molecules such as the heavy hydrocarbons. Nevertheless, sieves are subject to contamination by carryover of liquids such as oil and glycol. The regeneration temperature is very high. They can produce a water content as low as 1 ppm. Molecular sieves offer a means of simultaneous dehydration and desulfurization and are therefore the best choice for sour gases.

Process Flow Scheme

The general flow scheme for solid-bed gas treating processes is shown in Figure 5-39. The adsorption process is cyclic. For continuous operation, two beds are used in parallel. Most adsorbents tend to adsorb heavy hydrocarbons, glycols, and methanol, resulting in contamination and a reduction in desiccant capacity. These components are difficult to remove during regeneration. Therefore, for efficient desiccant performance and for a longer desiccant life, the inlet gas stream is thoroughly cleaned to remove all liquids and solids.

The clean gas flows downward during dehydration through one adsorber containing a desiccant bed, while the other adsorber is being regenerated by a stream of gas from the regeneration gas heater. Gas flow is downward for dehydration so as to lessen bed disturbance due to high gas velocity, and therefore, permit higher gas rates. Regeneration gas, on the other hand, is sent upward through the adsorber to ensure thorough regeneration of the bottom of the bed, which is the last area contacted by the gas that is dehydrated. Most contamination occurs at the top of the tower, and by sending the regeneration gas upwards, these contaminants can be removed without

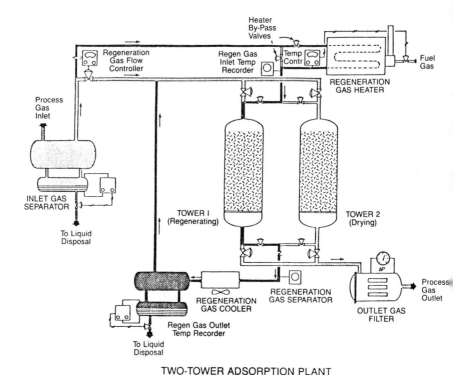

Figure 5-39. Flow diagram for a two-tower solid-bed dehydration plant. (After Campbell, 1984b; courtesy of Campbell Petroleum Series.)

flushing them through the entire bed. Sometimes, only the inlet (top) part of the tower is recharged with desiccant, in order to minimize desiccant replacement expense. Desiccant bed may also be rearranged, or the gas flow direction reversed, to obtain additional desiccant life.

The regeneration cycle consists of two parts: heating and cooling. First, regeneration gas heated to a temperature of 400–600°F, greater than the final bed regeneration temperature by about 50–100°F, is sent to the desiccant bed. Subsequently, the hot regenerated bed is cooled by letting the regeneration gas bypass, or by completely shutting off, the regeneration gas heater. This cooling gas is usually sent downward through the bed, so that only the top of the bed may adsorb any water from this gas. The hot regeneration gas and cooling gas, after flowing through the bed, are sent to the regeneration gas cooler to remove any adsorbed water. Thus, the bed acts as a sort of regenerative heat exchanger.

Oxygen, even in trace amounts, can react with hydrocarbons during the regeneration cycle, forming water and CO_2 which are adsorbed and lead to, in some cases, insufficient gas dehydration (GPSA, 1981). If oxygen is even suspected in the inlet gas stream, special design modifications are required.

The Regeneration Cycle

A typical temperature-time plot for a dry desiccant plant is shown in Figure 5-40. It shows the relationship between regeneration gas temperature at the inlet and outlet, and the ambient gas temperature for a typical 8-hr regeneration cycle.

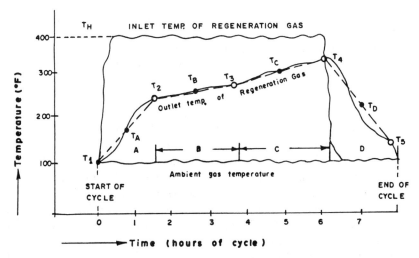

Figure 5-40. A typical temperature-time plot for regeneration gas and dessicant bed for a solid-bed dehydrator.

Refer to the regeneration-gas outlet temperature curve (curve 2) in Figure 5-40. There are four discrete intervals—A, B, C, and D—in the regeneration cycle, with average temperatures of T_A, T_B, T_C, and T_D, respectively, as shown in Figure 5-40. Initially, the hot regeneration gas heats up the tower and desiccant (from T_1 to T_2 in Figure 5-40). At about 240°F (T_2), water begins to vaporize. The bed heats up at a slower rate (the curve flattens), because a considerable portion of the heat input is used in vaporizing water from the desiccant, until point T_3 is reached where all the water in the desiccant has been desorbed. The average temperature of water desorption, T_B, is often assumed in design to be about 250°F (see Figure 5-40). Heating is continued (from T_3 to T_4 in Figure 5-40) to drive off any heavier hydrocarbons and contaminants. For a cycle time of 4 hours or greater, the bed is considered to be properly regenerated when the outlet gas temperature has reached 350–375°F (point T_4). The regeneration gas now has a high water (plus hydrocarbon and contaminant) content. This ends the heating cycle, and bed cooling is begun. The cooling cycle is usually terminated at about 125°F (point T_5), because any further cooling may cause water to condense from the wet gas stream and adsorb on the bed.

Although Figure 5-40 is for an 8-hr cycle, the relative heating and cooling times shown are typical for any cycle above 4 hours (Campbell, 1984b).

Analysis of the Adsorption Process (After Campbell, 1984b)

Figure 5-41a illustrates the movement of an adsorption zone front as a function of time. In a dry desiccant bed, the adsorbable components are adsorbed at *different* rates. A short while after the process has begun, a series of adsorption zones appear, as shown in Figure 5-41b. The distance between successive adsorption zone fronts is indicative of the length of the bed involved in the adsorption of a given component. Behind the zone, all of the entering component has been removed from the gas; ahead of the zone, the concentration of that component is zero (unless some residual adsorbed concentration exists from a previous cycle). Note the adsorption sequence: C_1 and C_2 are adsorbed almost instantaneously, followed by the heavier hydrocarbons, and finally by water that constitutes the last zone. Almost all the hydrocarbons are removed after 30–40 min and dehydration begins. Water displaces the hydrocarbons on the adsorbent surface if enough time is allowed. Thus, cycle length is critical. Very short cycle lengths are becoming increasingly popular, since they can simultaneously control water and hydrocarbon dew points at minimum cost.

Figure 5-41c shows the character of the breakthrough curve. The ordinate C/C_o is the ratio, at any time, of the mole fraction of a given component in the exit gas to its mole fraction in the entering gas. Thus, C/C_o is equal to 0 until the zone front reaches the bed outlet (breakthrough) at time t_b. Subsequently, C/C_o increases until time t_e, after which no further primary adsorption can occur. In practice, C/C_o becomes greater than 1.0 for a short period of time when one component is replaced by another. The shape of this C/C_o versus t curve is related to the efficiency of separation. Curve 1, a vertical straight line, represents an ideal case where the movement of the adsorption zone front is essentially "piston-like," whereas curves 2 and 3 are for a real case where the adsorption zone front spreads through a finite bed length. Note that the greater the spread, the more difficult it will be to delineate dry gas from partially wet gas when water breakthrough occurs.

Desiccant pellet size has a pronounced effect on the spread (or length) of the adsorption zone. As shown in Figure 5-41d, smaller the desiccant size, better and sharper is the separation obtained. Hence, the smallest size desiccant possible should be used, subject of course, to constraints of higher pres-

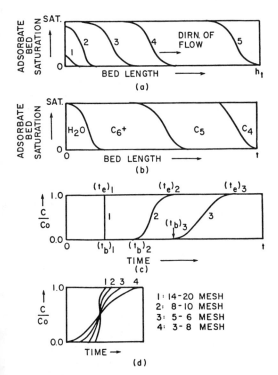

Figure 5-41. A schematic representation of the adsorption process. (After Campbell, 1984b.)

sure drops that may result from doing so. Usually a 14-mesh (Tyler screen) size is used.

Design Variables for the Adsorption Process

The three basic components of an adsorption plant are:

1. Adsorber towers.
2. Regeneration and cooling equipment—depends mainly on the adsorber towers.
3. Piping and equipment—primarily a function of the allowable pressure drop and the flow scheme desired.

In designing these components, the process variables that must be considered are (Campbell, 1984b):

1. Cycle time/length.
2. Allowable gas flow rate.
3. Desiccant capacity—design capacity as well as the effective or useful capacity.

4. Required outlet water dew point.
5. Total amount of water to be removed.
6. Regeneration requirements.
7. Allowable pressure drop.

Not all of these variables are independent; some are fixed for any calculation, and others are varied to yield an optimum design. Cycle time and dessicant capacity are the most important variables in designing an adsorption plant.

Cycle Time

Varies from less than 1 hr for a rapid cycle unit to greater than 8 hrs. The cycle time is, of course, limited by desiccant capacity so as not to exceed outlet gas specifications, and desiccant bed geometry. Also, the cycle time must be long enough to permit regeneration and cooling of the towers not currently dehydrating.

Desiccant Capacity

All desiccants degrade in service, their useful life being in the range of 1–4 years. Degradation is termed normal if there is a loss of surface area only. During its useful life, a desiccant exhibits rapid normal degradation in the initial stages, which then becomes more gradual. A more dramatic reduction in desiccant capacity occurs when the small capillary or lattice openings that contain most of the effective surface area for adsorption are blocked. This latter process, termed abnormal degradation, is caused by materials such as heavy oils, amines, glycols, and corrosion inhibitors, that cannot be removed by regeneration and effect great reductions in capacity in a very short time. No liquid water should be allowed in the inlet gas, since this generally salty water in the inlet gas will evaporate and fill the bed with salt.

Some properties of common desiccants are (Campbell, 1984b):

Material	Bulk density lbm/ft^3	Surface area m^2/gm	Design capacity lb H$_2$O/100 lb des.
Alumina	50-55	210	4-7
Alumina gel	52-55	350	7-9
Silica gel	45	750–830	7-9
Molecular sieves	43–45	650–800	9–12

The design capacity must be such that an economic desiccant life is obtained. Assuming a monolayer adsorption, the potential desiccant capacity is proportional to desiccant bulk density, and the available area for adsorption.

Adsorber Bed Design (After Campbell, 1984b)

Adsorber tower design is governed by the desiccant capacity, zone length, water loading (rate of removal of water from the gas), breakthrough time, and allowable gas flow rate and pressure drop. Relationships for these are discussed in the sequence they are calculated.

An often-used term in the context of adsorber tower design is the velocity of the gas based upon bed diameter, known as the superficial gas velocity, v_g. It is determined by converting the gas rate at standard conditions, q, to operating conditions:

$$v_g = (10^6 q/24)(ZT/p)(14.7/520)/(\pi D^2/4)$$
$$= 1499.73 qZT/(pD^2) \qquad (5\text{-}27)$$

where v_g = superficial gas velocity, ft/hr
 q = gas flow rate at standard conditions, MMscfd
 Z = gas compressibility factor at the operating temperature T (°R) and pressure p (psia)
 D = diameter of the adsorber bed, ft

Allowable Gas Flow Rate

This is generally expressed in terms of the mass flow velocity of the gas, the product of its velocity v_g and density:

$$w = 139.77 q M_g/D^2 \qquad (5\text{-}28)$$

where w = mass flow velocity, lbm/(hr ft^2)
 M_g = molecular weight of the gas

For downward flow of gas, the maximum allowable gas mass flow velocity, w, is given by:

$$w = 3,600[C\rho_g \rho_d d_p g]^{0.5}$$

where ρ_g = gas density at operating conditions, lbm/ft^3
 ρ_d = bulk desiccant density, lbm/ft^3
 d_p = average desiccant particle diameter, ft
 g = acceleration due to gravity (= 32.2), ft/sec^2
 C = empirical constant in the range of 0.025–0.033

The factor 3,600 converts the mass flow rate from lbm/(sec ft^2) to lbm/(hr ft^2). Substituting g = 32.2 ft/sec^2, w becomes:

$$w = 20{,}428.22\,[C\rho_g\rho_d d_p]^{0.5} \tag{5-29}$$

Water Loading

Also known as the rate of removal of water required per unit bed area, water loading can be calculated as follows:

$$q_w = (qW_i/24)/(\pi D^2/4) = 0.0531 q W_i/D^2 \tag{5-30}$$

where q_w = water loading, lb H$_2$O/(hr ft^2)
W_i = water content of the inlet gas, lb H$_2$O/MMscf gas

Zone Length

The zone length depends upon gas composition, flow rate, relative water saturation, and desiccant loading capability. For silica gel, the zone length, h_z, can be estimated as follows (Simpson and Cummings, 1964):

$$h_z = (297.78 q_w^{0.7895})/(v_g^{0.5506} S_r^{0.2646}) \tag{5-31}$$

where h_z = zone length, ft
S_r = relative saturation of water in the inlet gas, %

For alumina and molecular sieves, the zone length determined using Equation 5-31 is multiplied by 0.8 and 0.6, respectively.

Desiccant Capacity

The *dynamic* desiccant capacity, x_s, is shown in Figure 5-42a as a function of the relative saturation of water. The temperature correction factor is shown in Figure 5-42b. This temperature correction is required for gels and alumina, but not for molecular sieves.

The *useful* desiccant capacity, x, generally less than the dynamic capacity, x_s, is given by the following empirical equation:

$$x = x_s - C^* x_s (h_z/h_t) \tag{5-32}$$

where x = useful capacity of the desiccant, lb H$_2$O/100 lb desiccant
x_s = dynamic capacity of the desiccant, lb H$_2$O/100 lb desiccant
C^* = an empirical constant, a function of zone length

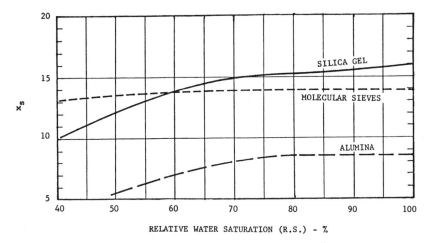
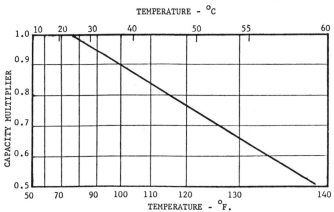

Figure 5-42. The effect of (a) relative water saturation and (b) temperature on the dynamic capacity of dessicants. (After Campbell, 1984b; courtesy of Campbell Petroleum Series.)

From extensive tests, C^* has been found to vary only in the range 0.40–0.52 for a wide range of applications, and an average $C^* = 0.45$ is generally used for design purposes.

Breakthrough Time

The breakthrough time for the water zone formed, t_b in hours, can be estimated as follows:

$$t_b = (x/100)\rho_d h_t/q_w = (0.01)x\rho_d h_t/q_w \tag{5-33}$$

where the factor 0.01 takes into account the units for x (lb H_2O/100 lb dessicant).

Minimum Bed Length

The amount of water that can be removed per cycle by the desiccant, W_c (lbm), is given by:

$$W_c = (x/100)(\rho_d)(h_t \pi D^2/4)$$

On rearranging this equation, the minimum length of desiccant bed required can be written as:

$$(h_t)_{min} = (127.32 W_c)/(x \rho_d D^2) \quad (5\text{-}34)$$

Note that $(h_t)_{min}$ is the distance from the inlet to the front of the water zone. So, if h_t is less than the total length of the bed, it means that not all of the bed is being used. From an operations perspective, it is desirable to continue until the water front reaches the end of the bed, since it prolongs desiccant life.

Example 5-10. Design an adsorber for dehydrating 20 MMscfd of a 0.7 gravity gas at 1,000 psia and 100°F. Assume a two-tower plant using an 8-hr cycle with a 15 ft long silica gel bed.

Solution

From Figure 5-1, water content of inlet gas = 61 lb H_2O/MMscf gas.
Assuming all of the inlet water is removed, the rate of removal of water = (20)(61) = 1,220 lbm/day.
Therefore, amount water to be removed per cycle, W_c = (1,220)(8/24) = 406.67 lbm/cycle.
The gas compressibility factor, $Z = 0.88$.
Assume a gas velocity of 1,800 ft/hr. Then, from Equation 5-27:
$D^2 = (1,499.73)(20)(0.88)(560)/[(1,000)(1,800)] = 8.2118$ ft², implying $D = 2.87$ ft. Choose a bed diameter of 3 ft for safety.
Then, $v_g = (1,800)(8.2118/9) = 1,642.36$ ft/hr
From Equation 5-30, water loading $q_w = (0.0531)(20)(61)/9 = 7.198$ lbm/hr-ft²
From Equation 5-31, zone length is given by:

$$h_z = (297.7)(7.198)^{0.7895}/[(1,642.36)^{0.5506}(100)^{0.2646}] = 7.094 \text{ ft}$$

Gas-Water Systems and Dehydration Processing 247

From Figures 5-42a and 5-42b, $x_s = (0.9)(16) = 14.4$ lb $H_2O/100$ lb desiccant. From Equation 5-32:

$$x = 14.4 - (0.45)(14.4)(7.094)/15 = 11.34 \text{ lb } H_2O/100 \text{ lb desiccant}$$

From Equation 5-34, $(h_t)_{min} = (127.32)(406.67)/[(11.34)(45)(9)] = 11.3$ ft

Thus, a 15 ft bed length allows a safety factor. From Equation 5-33, the breakthrough time is:

$$t_b = (0.01)(11.34)(45)(15)/7.198 = 10.63 \text{ hr}$$

Since the breakthrough time of 10.63 hr is greater than the cycle time of 8 hr, the water zone will not reach the outlet, and dry gas will be produced. The design, therefore, has a safety factor to account for future desiccant degradation.

Regeneration and Cooling Calculations (After Campbell, 1984b)

Regeneration calculations are done for establishing the regeneration gas rate, heating and cooling time intervals, heater load, and condenser load. These calculations are simply heat balances for various parts of the regeneration cycle. For each interval, the heat supplied by the gas, E_i, must be equal to the heat absorbed, Q_i:

$$Q_i = E_i \tag{5-35}$$

where i represents the interval. In addition, the total cycle time constraint must also be satisfied:

$$t_c = t_A + t_B + t_C + t_D \tag{5-36}$$

where t_c = total cycle time, hr
t_A, t_B, t_C, t_D = time spans (hr) for intervals A, B, C, and D, respectively

For these intervals, the Q_i and E_i terms are derived as follows:

Heating from T_1 to T_2 (Interval A in Figure 5-40)

The total heat load is the sum of the heat absorbed by desiccant, hydrocarbon, vessel shell, inert balls, hydrocarbon desorption, and water. Thus, the total heat load, Q_A, is given by:

$$Q_A = m_d c_d (T_2 - T_1) + m_{HC} c_{HC} (T_2 - T_1) + m_s c_s (T_2 - T_1)$$
$$+ m_{ib} c_{ib} (T_2 - T_1) + m_{HC} H_{HC} + m_w c_w (T_2 - T_1) \qquad (5\text{-}37)$$

where m = mass, lbm
c = specific heat capacity, Btu/(lbm °F), for desiccant (d), hydrocarbon (HC), vessel shell (s), inert balls (ib), and water (w)

The second last term in Equation 5-37 represents hydrocarbon desorption, with H_{HC} being its heat of desorption, usually taken to be 200 Btu/lbm. The term for heat absorbed by the vessel shell is usually multiplied by 0.7 if the vessel is insulated.

Energy available from the regeneration gas, E_A, is given by:

$$E_A = m_{rg} c_{rg} (T_H - T_A) t_A \qquad (5\text{-}38)$$

where m_{rg} = mass flow rate of the regeneration gas, lbm/hr
c_{rg} = specific heat of the regeneration gas, Btu/(lbm °F)

Heating from T_2 to T_3 (Interval B in Figure 5-40)

The total heat load is the sum of the heat absorbed by desiccant, water, water desorption, vessel shell, and inert balls. Thus, the total heat load, Q_B, is given by:

$$Q_B = m_d c_d (T_3 - T_2) + m_w c_w (T_3 - T_2) + m_w H_w$$
$$+ m_s c_s (T_3 - T_2) + m_{ib} c_{ib} (T_3 - T_2) \qquad (5\text{-}39)$$

where the heat of desorption of water, H_w, is usually taken to be 1,400 Btu/lbm for alumina and gels, and 1,800 Btu/lbm for molecular sieves. As before, the term for heat absorbed by the vessel shell is multiplied by 0.7 if the vessel is insulated. Energy available from the regeneration gas, E_B, is given by:

$$E_B = m_{rg} c_{rg} (T_H - T_B) t_B \qquad (5\text{-}40)$$

Heating from T_3 to T_4 (Interval C in Figure 5-40)

In this final cleanup step, the total heat load is the sum of the heat absorbed by desiccant, vessel shell, and inert balls, only. Thus, the total heat load, Q_C, is given by:

$$Q_C = m_d c_d (T_4 - T_3) + m_s c_s (T_4 - T_3) + m_{ib} c_{ib} (T_4 - T_3) \qquad (5\text{-}41)$$

Gas-Water Systems and Dehydration Processing 249

As before, the term for heat absorbed by the vessel shell is multiplied by 0.7 if the vessel is insulated. Energy available from the regeneration gas, E_C, is given by:

$$E_C = m_{rg}c_{rg}(T_H - T_C)t_C \qquad (5\text{-}42)$$

Cooling from T_4 to T_5 (Interval D in Figure 5-40)

The total cooling load is the sum of the cooling requirements for desiccant, vessel shell, and inert balls, only. Thus, the total cooling load, Q_D, is given by:

$$Q_D = m_d c_d (T_4 - T_5) + m_s c_s (T_4 - T_5) + m_{ib} c_{ib} (T_4 - T_5) \qquad (5\text{-}43)$$

As before, the term for the vessel shell is multiplied by 0.7 if the vessel is insulated. Assuming that the cooling gas rate is the same as the regeneration gas rate, cooling available, E_D, is given by:

$$E_D = m_{rg}c_{rg}(T_D - T_1)t_D \qquad (5\text{-}44)$$

In Equation 5-44, the unknown parameters are the regeneration gas rate, m_{rg}, and the lengths of time intervals, t_A, t_B, t_C, and t_D, a total of 5 unknowns. The number of equations are also 5: four for the heat balances, and Equation 5-36, which imposes the total cycle time constraint. For a given cycle time t_c, a solution is thus possible.

Normally, however, the maximum heat load occurs in the interval B, and the minimum regeneration gas rate m_{rg} is chosen on this basis. The minimum regeneration gas flow rate and water adsorption capacity must be enough to handle the water desorbed from the bed in time t_B. Therefore:

$$W_c = (m_{rg}t_B)(W_{T_B} - W_{T_1})(3.796 \times 10^{-4})/M_{rg} \qquad (5\text{-}45)$$

where W_c = total water removed per cycle from the gas being dehydrated, lbm
W_{T_B}, W_{T_1} = water contents of the regeneration gas at temperatures T_B and T_1, respectively, lbm water/MMscf gas
M_{rg} = molecular weight of the regeneration gas (usually the same as the gas being dehydrated)

Generally, the regeneration gas rate is about 10% of the main gas flow rate. The entering gas must be heated from a temperature T_1 to T_H, which is chosen in the range 400–600°F. Temperature T_4 must be kept as low as possible, and is generally in the range 350–500°F, less than T_H by 50–100°F (usually 100°F). T_2 and T_3 are generally equal to 230°F and 260°F. Values for specific heats needed in regeneration calculations are as follows:

Component	Specific Heat, Btu/(lbm °F)
Alumina	0.06
Silica gel	0.22
Molecular sieves	0.05
Liquid water	1.0
Vessel shell	0.12

Heater and Condenser Loads

The heater load consists of heating requirements for the regeneration gas. Thus, the heater load, Q_H in Btu/hr can be written as:

$$Q_H = m_{rg}c_{rg}(T_H - T_1) \qquad (5\text{-}46)$$

A condenser is used to cool the regeneration gas after it comes out through the adsorber bed. This cooling is done in order to separate the condensable components picked up by the regeneration gas from the desiccant bed. Cooling loads should be calculated for all the three intervals, and the condenser then designed to handle the maximum cooling load expected. Generally, condenser loads are the highest for interval B. Condenser loads can be calculated similar to heating loads, taking into account the components in the regeneration gas and their specific and latent heats.

Dehydration by Expansion Refrigeration

Dehydration can be achieved by expansion refrigeration when sufficient pressure drop is available. In this technique, the gas stream is cooled by adiabatic expansion through a choke, as discussed in Chapter 4 (in the section on "Low-Temperature Separation"), to temperatures as low as 0°F. This drop in temperature leads to condensation of the water vapor and condensable hydrocarbons associated with the gas.

Two types of expansion refrigeration techniques are used: one with a hydrate inhibitor to prevent hydrate formation that may result upon expansion of the gas, and another without an inhibitor.

Expansion refrigeration without an inhibitor is used only when the pressure drop allows the desired water dew point to be obtained without the formation of hydrates (GPSA, 1981). An inhibitor must be used if hydrate formation is expected in the precooler used upstream of the choke. Hydrates are allowed to form at this second stage (choke), and are collected in a low temperature separator (LTX). In the LTX unit, the gas entrapped in the hydrates is recovered using heating coils to heat the hydrates. These coils are often used as precoolers for the gas stream: the gas passes through the inside of the coils, providing enough heat to melt the hydrates. That is why it is important to ensure that no hydrates are expected to form at this precooling stage for the gas. Also, mist eliminators must never be used in the LTX unit, since they may be plugged by hydrates.

When hydrate inhibitors are used, the process is similar, except that the heating/cooling coils are no longer necessary, and are replaced by a gas expansion valve.

Questions and Problems

1. What are gas hydrates? List the factors that promote hydrate formation.
2. List three methods for preventing hydrate formation at wellsites.
3. Name three different types of solid desiccants used for natural gas dehydration.
4. A sour gas at 1000 psia has the following analysis: $N_2 = 8.5\%$, $H_2S = 5.4\%$, $CO_2 = 0.5\%$, $C_1 = 77.6\%$, $C_2 = 5.8\%$, $C_3 = 1.9\%$, $n\text{-}C_4 = 0.1\%$, $i\text{-}C_4 = 0.1\%$, and $i\text{-}C_5 = 0.1\%$. What is the water content of this gas at 120°F? Use all the methods applicable and compare the results.
5. For the sour gas given in Problem 4, determine the hydrate formation temperature. Use all the methods applicable and compare the results.
6. A TEG dehydration plant is to be designed to dry the following gas: $N_2 = 1.7\%$, $CO_2 = 0.3\%$, $C_1 = 65.5\%$, $C_2 = 16.6\%$, $C_3 = 8.8\%$, $n\text{-}C_4 = 3.0\%$, $i\text{-}C_4 = 1.6\%$, and $i\text{-}C_5 = 0.9\%$, $n\text{-}C_5 = 0.9\%$, $C_6 = 0.5\%$, and $C_{7+} = 0.2\%$. The C_{7+} fraction has a molecular weight of 140 and specific heat of 70 Btu/lbm-°F at 300°F. Other relevant data are as follows:
Gas flow rate = 20 MMscfd
Inlet gas pressure = 1000 psia
Inlet temperature = 100°F
Required outlet water dew point = 10°F
Glycol circulation rate = 7 gal/lb water removed from gas
Glycol reboiler temperature = 400°F
Specific gravity of glycol = 1.11
Specific heat of glycol = 0.654 Btu/lbm-°F

Gas Production Engineering

For this plant, calculate the following: *[handwritten: Design include all vessel sizes and number of trays needed. State your initial conditions and assumptions.]*

(a) Glycol circulation rate in gal/min.
(b) Glycol-glycol heat exchanger load in Btu/hr if a 25°F approach is desired. Assume that the glycol enters the absorber at 125°F.

7. A glycol (TEG) dehydrator plant is to be designed for handling 13.5 MMscfd of the sour gas of Problem 4. The glycol circulation rate is 4 gal TEG/lb water, lean glycol concentration is 99%, glycol specific gravity is 1.10, inlet gas temperature is 120°F, and the absorption tower uses bubble-cap trays. ~~For an~~ exit water content of 10 lb/MMscf gas, ~~determine the following:~~ *[handwritten: your design must address the following]*

(a) Contactor diameter.
(b) Maximum gas rate that can flow in the system.
(c) The actual number of trays needed in the contactor.
(d) Sensible heat required for the glycol, Btu/hr.
(e) Heat of vaporization required for the water, Btu/hr.
(f) Heat required to vaporize reflux water in the still, Btu/hr.
(g) Reboiler heat load, Btu/hr, assuming heat loss = 5000 Btu/hr.

8. It is desired to use glycol inhibition to prevent hydrate formation for the case given in Problem 7. Calculate the amount of TEG required in lb/day.

9. Rework Problem 6 using a two-tower desiccant plant, employing silica-gel on an 8-hour cycle. The regeneration gas leaves the heater at 550°F, and the towers are regenerated to a final temperature of 450°F. Assume internal insulation. Determine the following:

(a) Internal diameter of the towers required.
(b) Amount of silica-gel required in lbs/cycle.
(c) Regeneration gas rate, MMscfd.
(d) Regeneration heater heat load, Btu/hr.
(e) Cooling load, Btu/hr.

References

Curry, R. N., 1981. *Fundamentals of Natural Gas Conditioning*. PennWell Publ. Co., Tulsa, Oklahoma, 118pp.

Campbell, J. M., 1984a. *Gas Conditioning and Processing*, Vol. 1. Campbell Petroleum Series, Norman, Oklahoma, 326pp.

Campbell, J. M., 1984b. *Gas Conditioning and Processing*, Vol. 2. Campbell Petroleum Series, Norman, Oklahoma, 398pp.
Carson, D. B. and Katz, D. L., 1942. "Natural Gas Hydrates," *Trans., AIME*, 146, 150–158.
Chemical Engineers' Handbook, 1984. R. H. Perry and D. W. Green (eds.), McGraw-Hill Book Co., New York, 6th ed., pp. 13–27 to 13–35.
Edmister, W. C. and Lee, B. I., 1984. *Applied Hydrocarbon Thermodynamics*, Vol. 1 (2nd ed.). Gulf Publishing Company, Houston, Texas, 233pp.
Galloway, T. J., Ruska, W., Chappelear, P. S., and Kobayashi, R., 1970. "Experimental Measurement of Hydrate Numbers for Methane and Ethane and Comparison with Theoretical Values," *Ind. & Eng. Chem. Fund.*, 9(2, May), 237–243.
Gas Processors Suppliers Association, 1981. *Engineering Data Book*, 9th ed. (5th revision), GPSA, Tulsa, Oklahoma.
Guenther, J. D., 1979. "Natural Gas Dehydration," Paper presented at Seminar on Process Equipment and Systems on Treatment Platforms, Taastrup, Denmark, April 26.
Heinze, F., 1971. "Hydratbildung," *Lehrbogen 3/3 von der Bergakademie Freiberg*.
Holder, G. D. and Hand, J. H., 1982. "Multiple-Phase Equilibria in Hydrates from Methane, Ethane, Propane and Water Mixtures," *AIChE J.*, 28(3, May), 440–447.
Ikoku, C. U., 1984. *Natural Gas Production Engineering*. John Wiley & Sons, Inc., New York, 517pp.
Katz, D. L., Cornell, D., Kobayashi, R., Poettmann, F. H., Vary, J. A., Elenbaas, J. R., and Weinaug, C. F., 1959. *Handbook of Natural Gas Engineering*. McGraw-Hill Book Co., Inc., New York, 802pp.
McCarthy, E. L., Boyd, W. L., and Reid, L. S., 1950. "The Water Vapor Content of Essentially Nitrogen-Free Natural Gas Saturated at Various Conditions of Temperature and Pressure," *Trans. AIME*, 189, 241–243.
McKetta, J. J. and Wehe, A. H., 1958. "Use This Chart for Water Content of Natural Gases," *Petr. Refiner*, 37(8, Aug.), 153–154.
McLeod, H. O., Jr. and Campbell, J. M., 1961. "Natural Gas Hydrates at Pressures to 10,000 Psia," *J. Pet. Tech.*, 13(6, June), 590–594.
Nagata, I. and Kobayashi, R., 1966. "Prediction of Dissociation Pressures of Mixed Gas Hydrates From Data for Hydrates of Pure Gases with Water," *Ind. & Eng. Chem. Fund.*, 5(4, Nov.), 466–469.
Ng, H. J. and Robinson, D. B., 1976. "The Measurement and Prediction of Hydrate Formation in Liquid Hydrocarbon-Water Systems," *Ind. & Eng. Chem. Fund.*, 15(4, Nov.), 293–298.

Parrish, W. R. and Prausnitz, J. M., 1972. "Dissociation Pressures of Gas Hydrates Formed by Gas Mixtures," *Ind. & Eng. Chem., Process Design & Dev.*, 11(1), 26-35.

Petroleum Extension Service, 1972. *Field Handling of Natural Gas.* 3rd ed., Univ. of Texas Press, Austin, Texas, 143pp.

Robinson, D. B. and Ng, H. J., 1975. "Improve Hydrate Predictions," *Hydroc. Proc.*, 54(12, Dec.), 95-96.

Robinson, J. N., Wichert, E., Moore, R. G., and Heidemann, R. A., 1978. "Charts Help Estimate H_2O Content of Sour Gases," *O. & Gas J.*, 76(6, Feb. 6), 76-78.

Sharma, S. and Campbell, J. M., 1984. Unpublished. Cited reference on p. 143 in: *Gas Conditioning and Processing, Vol. 1*, by J. M. Campbell. Campbell Petroleum Series, Norman, Oklahoma, 326pp.

Simpson, E. A. and Cummings, W. P., 1964. "A Practical Way to Predict Silica Gel Performance," *Chem. Engr. Prog.*, 60 (4, Apr.), 57-60.

Sivalls, C. R., 1976. "Glycol Dehydrator Design Manual," Proc. Gas Conditioning Conf., Univ. of Oklahoma, 39pp.

Trekell, R. E. and Campbell, J. M., 1966. "Prediction of the Behavior of Hydrocarbon Clathrate Solutions," Proc. 151st Meet. of Am. Chem. Soc., Petr. Chem. Div., Pittsburg, Penn., March 23-26, p. 61.

Van der Waals, J. H. and Platteeuw, J. C., 1959. "Clathrate Solutions," *Advances in Chemical Physics, Vol. 2*, edited by I. Prigogine. Interscience Publ., a division of John Wiley & Sons, New York, 1-57.

6
Desulfurization Processes

Introduction

With increasing demands for natural gas, natural gases containing hydrogen sulfide (H_2S) are also being tapped for utilization after purification. Natural gases containing H_2S are classified as "sour," and those that are H_2S-free are called "sweet" in processing practice. Produced gases from reservoirs usually contain H_2S in concentrations ranging from barely detectable quantities to more than 0.30% (3,000 ppm). Other sulfur derivatives, besides H_2S, are usually completely insignificant or present only in trace proportions. Most contracts for the sale of natural gas require less than 4 ppm (parts per million) in the gas. Thus, a sulfur removal process must be very precise since the initial product contains only a small fractional quantity of sulfur that must be reduced several hundred times.

A characteristic feature of all H_2S-bearing natural gases is the presence of carbon dioxide (CO_2), the concentrations of which are generally in the range of 1 to 4%. In gases devoid of H_2S, however, such concentrations are rare. H_2S and CO_2 are commonly referred to as "acid gases" because they form acids or acidic solutions in the presence of water.

Reasons for Removal of H_2S and CO_2

Besides emitting a foul odor at low concentrations, H_2S is deadly poisonous and at concentrations above 600 ppm it can be fatal in just three to five minutes. Its toxicity is comparable to cyanide. Thus, it cannot be tolerated in gas that would be used as domestic fuel. Further, H_2S is corrosive to all metals normally associated with gas transporting, processing, and handling systems (although it is less corrosive to stainless steel), and may lead to premature failure of most such systems. On combustion, it forms sulfur dioxide, which is also highly toxic and corrosive. H_2S and other sulfur compounds can also cause catalyst poisoning in refinery processes.

CO_2 has no heating value and its removal may be required in some instances merely to increase the energy content of the gas per unit volume. CO_2 removal may also be required because it forms a complex, $CO_2 \cdot CO_2$, which is quite corrosive in the presence of water. For gas being sent to cryogenic plants, removal of CO_2 may be necessary to prevent solidification of the CO_2. Both the acid gases, H_2S and CO_2, promote hydrate formation, and the presence of CO_2 may not be desired for this reason also. However, if none of these situations are encountered, there is no need to remove CO_2.

Removal Processes

The processes that have been developed to accomplish gas purification vary from a simple once-through wash operation to complex multiple step recycle systems. In many cases, the process complexities arise from the need for recovery, in one form or another, of the impurity (H_2S) being removed or recovery of the materials (process chemicals) employed to remove it.

Like dehydration processes, desulfurization processes are primarily of two types: adsorption on a solid (dry process), and absorption into a liquid (wet process). There are a few processes that use other methods, such as the cellulose acetate membranes that rely upon the different diffusion rates for hydrocarbon and H_2S, and liquid fractionation techniques that exploit the relative volatility difference. Application of these latter types of processes has been quite limited.

Both the adsorption and absorption processes may be of the physical (no chemical reactions involved) or chemical type. These processes may also be classified into the following categories:

1. Non-regenerative, such as Chemsweet process (Manning, 1979), NCA process (Sun, 1980), and the Slurrisweet process (*Oil & Gas J.*, 1981). The materials used in treating the gas are not recovered in these processes.
2. Regenerative processes with recovery as H_2S. These include the physical absorption processes (water wash, Selexol, Fluor solvent, etc.), the amine processes (MEA, DEA, DGA, etc.), hot carbonate processes (Benfield, Catacarb), Alkazid processes, molecular sieves, etc.
3. Regenerative processes with recovery as elemental sulfur. The Holmes-Stretford process, and the Giammarco-Vetrocoke process fall in this category. With growing environmental concerns regarding sulfur emission, these processes have acquired a prominent role in desulfurization operations.

So many sweetening processes are in use today that it would be impossible to describe them all in detail here. Some widely used processes will be briefly mentioned, and some important ones among these will be presented in a little more detail.

Criteria for Process Selection

There are many variables in gas treating, which makes the precise definition of the area of application of a given process very difficult. Among several factors, the following are the most significant that need be considered:

1. The types and concentrations of impurities in the gas, and the degree of removal desired.
2. Selectivity of acid gas removal required, if any.
3. Temperature and pressure at which the sour gas is available, and at which the sweet gas is to be delivered.
4. Volume of the gas to be processed, and its hydrocarbon composition.
5. CO_2 to H_2S ratio in the gas.
6. Economics of the process.
7. The desirability of sulfur recovery due to environmental problems, or economics.

Besides CO_2 and H_2S, some gases may contain other sulfur derivatives such as mercaptans and carbonyl sulfide. The presence of such impurities may eliminate some of the sweetening processes. The concentrations of acid components are important to consider. Some processes effectively remove large amounts of acid gas, but not to low enough levels. Others remove to ultra-low values, but cannot handle large amounts of acid components in the incoming stream economically.

Selectivity of a process implies the degree of removal of one acid gas component relative to another. Some processes remove both H_2S and CO_2, while others are designed to selectively remove only H_2S. Generally, it is important to consider the selectivity of a process for H_2S versus CO_2 to ensure desired extent of removal of these components. For this reason, CO_2 to H_2S ratio in the gas is also an important parameter.

Operating conditions affect the performance of several sweetening processes. Also, different processes discharge the sweetened gas stream at different pressure and temperature levels relative to the conditions of the incoming stream. Thus, the inlet and outlet conditions desired are significant variables to consider. It may be economic in some cases to alter inflow conditions to suit the process, and outflow conditions to meet pipeline requirements.

Some processes are economic for large gas volumes, while others lose their economic advantage when large volumes of gas are to be treated. Gas composition should also be considered, since some processes are adversely affected by the presence of heavier fractions, while others may strip the gas of some of its hydrocarbon constituents.

The desirability of recovering sulfur as elemental sulfur reduces the choice considerably. Processes with sulfur recovery have become very important in America, and will become so worldwide in the near future due to environmental problems caused by sulfur emissions. Sulfur is a useful by-product, generally in good demand by the fertilizer and chemicals industry. Therefore, in several instances, sulfur recovery may be attractive from an economic standpoint also.

Solid Bed Sweetening Processes

These processes are based upon physical or chemical adsorption of acid gases on a solid surface. Although not as widely used as the liquid absorption processes, they offer advantages such as simplicity, high selectivity (only H_2S is generally removed), and a process efficiency almost independent of pressure. These processes are best applied to gases with moderate concentrations of H_2S, and where CO_2 is to be retained in the gas.

The Iron Sponge Process

Sour gas is passed through a bed of wood chips that have been impregnated with a special hydrated form of ferric oxide that has a high affinity for H_2S. The chemical reaction is as follows:

$$Fe_2O_3 + 3H_2S = Fe_2S_3 + 3H_2O$$

The temperature must be kept at less than 120°F. According to Kohl and Riesenfeld (1985), temperatures above 120°F and neutral or acid conditions lead to the loss of the water of crystallization of the ferric oxide. The bed then becomes very difficult to regenerate. Iron oxide is regenerated by passing oxygen/air over the bed:

$$2Fe_2S_3 + 3O_2 = 2Fe_2O_3 + 6S$$

The process operates in a batch-type reaction-regeneration cycle. Regeneration of the bed, however, is quite difficult and incurs excessive maintenance and operating costs. Also, sulfur eventually covers most of the surface of the ferric oxide particles and further regeneration becomes impossible. A continuous regeneration process has also been developed, where small amounts of oxygen or air are added along with the sour gas at the inlet. This latter process gives an improved performance, generating a higher removal efficiency as well as better regeneration.

This process offers advantages of simplicity, and absolute selectivity, which implies less gas shrinkage because CO_2 is retained in the gas. It is relatively inexpensive. There are several disadvantages, such as difficult and expensive regeneration, excessive pressure losses through the bed, inability to remove large amounts of sulfur, and sulfur disposal problems, since it does not produce sulfur in a saleable form. Details about this process are available from various sources such as Taylor (1956), Kohl and Riesenfeld (1985), and Duckworth and Geddes (1965).

Molecular Sieves

Synthetically manufactured forms of crystalline sodium-calcium alumino silicates, molecular sieves are porous in structure and have a very large surface area. The pores, formed by the removal of water of crystallization, are uniform throughout the material. Several grades of molecular sieves are available, with each grade corresponding to a very narrow range of pore sizes.

Molecular sieves remove components through a combination of a "sieving" and physical adsorption process. Because of their narrow pore sizes, they discriminate among the adsorbates on the basis of their molecular sizes. The sieves possess highly localized polar charges on their surface that act as adsorption sites for polar materials. Therefore, polar or polarizable compounds, even in low concentrations, are adsorbed on the sieve surface. They are highly selective in the removal of H_2S and other sulfur compounds from natural gas, and offer a continuously high absorption efficiency. They remove water also (a polar compound) and thus offer a means of simultaneous dehydration and desulfurization. High water content gases, however, may require upstream dehydration (Rushton and Hays, 1961).

Fails and Harris (1960) conducted several pilot studies on sieves and other adsorbents. They found that the adsorptive capacity of molecular sieves for H_2S decreases with increasing temperature, and increases with increasing H_2S/CO_2 ratio. Increasing contact time is favorable to H_2S removal up to a point, beyond which there is no effect. An optimum pressure of about 450 psia was identified by Fails and Harris, though the effect of pressure was found to be quite small.

The process flow scheme, shown in Figure 6-1, is similar to the iron sponge process. The bed is regenerated by passing a portion of the sweetened gas, preheated to about 400–600°F or more, for about 1.5 hours to heat the bed. As the temperature of the bed increases, it releases the adsorbed H_2S into the regeneration gas stream. The sour effluent regeneration gas is flared off. According to Rushton and Hays (1961), about 1–2% of the gas treated is lost in this regeneration process.

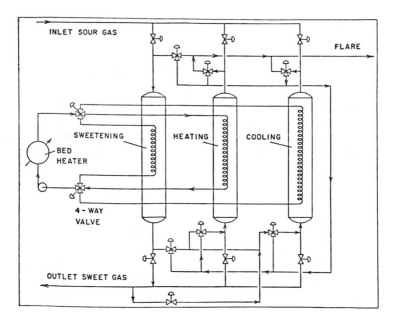

Figure 6-1. Flowsheet for a molecular sieve desulfurization process. (After Fails and Harris, 1960; courtesy of *Oil & Gas Journal*.)

Besides regeneration losses, gas is also lost by the adsorption of hydrocarbon components by the sieve. Unsaturated hydrocarbon components such as olefins and aromatics tend to be strongly adsorbed (Conviser, 1965). Molecular sieves are also prone to poisoning by several chemicals such as glycols and require thorough gas cleaning methods prior to the adsorption step. The process requires a cyclic operation, since it is batch-type, with a cycle time on the order of 2 hours. Initial capital investments are high, and regeneration requires a lot of heat. For gas streams containing CO_2, carbonyl sulfide may form as shown in the following reversible reaction:

$$H_2S + CO_2 = COS + H_2O$$

Molecular sieves tend to catalyze this reaction. The problem has been studied extensively, and new molecular sieves have been developed to retard the COS formation.

Physical Absorption Processes

These processes rely upon physical absorption of acid components as the gas sweetening mechanism. Because only physical absorption is involved, corrosion and fouling problems are generally minimal. These processes offer fair to good selectivity. The solvent used is generally recovered by flashing the acid component-rich solvent (called rich solvent) in flash tanks at successively lower pressures. Thus, little or no heat is required for regeneration or other purposes. But a sulfur recovery unit is required since these processes do not alter the acid components chemically in any manner. Some processes in this category offer simultaneous dehydration. Most solvents currently in use have a relatively high solubility for heavier hydrocarbons, particularly unsaturated and aromatic components. These components are very detrimental to the performance of most sulfur recovery processes and may yield an unsuitable, contaminated sulfur as the product. So, for sour gases containing heavy hydrocarbons (heptanes plus) or unsaturated and aromatic hydrocarbons, particular care must be taken during the regeneration step to prevent their entry into the acid gas stream that is to be sent to a sulfur recovery unit.

Water Wash (Aquasorption) Process

This process is effective for high pressure gas, with high acid gas content and high H_2S to CO_2 ratio. According to Maddox (1982), a water wash operation followed by an amine process is 12 to 15% lower in capital investment, and about 50% lower in operating expenses as compared to a single amine unit for an equivalent sweetening job. For gases with a high H_2S to CO_2 ratio, the savings can be as much as 40% in investment, and 60 to 70% in operating costs.

In the water wash sweetening process, sour gas is sent upward through a contactor, countercurrent to the water (see Figure 6-2). The partially sweetened gas from the top of the tower is sent to further treatment units (typically, an amine unit). The rich water solution from the bottom of the tower is sent to an intermediate pressure flash tank for recovery of dissolved hydrocarbons. A power recovery turbine is provided for repressurizing the water before sending it to a low pressure flash tank where all of the acid gas is removed, and the lean water obtained is recycled.

Froning et al. (1964) presented calculation procedures for estimating the performance of a water wash and provide data required for such calculations. Their results show that H_2S is about three times more soluble in water than CO_2, thereby showing that the selectivity of the process is quite good.

Selexol Process

This process uses dimethyl ether of polyethylene glycol (DMPEG) as a solvent. Solubilities in Selexol solvent of H_2S and CO_2 and other acid gas components are directly proportional to the partial pressures of these components. The solubility of H_2S is about 10 times greater than CO_2, and hydrocarbon solubility is quite small. Heavier hydrocarbons, however, have greater solubility, and intermediate flashes are generally required for hydrocarbon removal.

Different Selexol-based processes have been designed and used successfully. Hegwer and Harris (1970) describe three plants installed for a wide range of H_2S to CO_2 ratios: one plant for a high H_2S to CO_2 ratio, another for a low H_2S to CO_2 ratio, and the third for a gas containing large amounts of both H_2S and CO_2.

Figure 6-3 shows a flow scheme for a low H_2S to CO_2 ratio gas. Sour gas, pretreated to remove any solids and free liquids, is dehydrated, cooled to 40°F, and sent to the absorber for a countercurrent contact with Selexol solvent. Rich selexol from the bottom of the absorber is sent, via a surge tank to remove entrained gas that is recycled back into the absorber, and a power recovery turbine for pressurizing the solvent, to a high pressure flash. Most of the absorbed methane and some of the CO_2 is released in this flash, and is recycled to the absorber.

In the second flash stage, most of the vapor released is CO_2 and it is vented after power recovery. Finally, the Selexol is sent to the low pressure flash, which is operated at 16 psia. Here, H_2S and the remaining CO_2 are flashed off as the vapor stream, which is vented to the atmosphere.

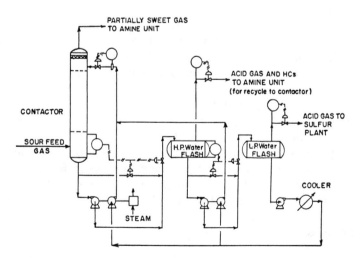

Figure 6-2. A typical water wash process. (After Froning et al., 1964; reprinted from *Hydrocarbon Processing*.)

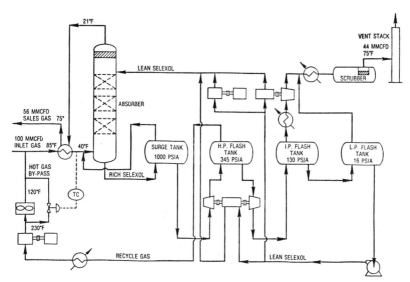

Figure 6-3. Selexol™ based plant for a low H_2S to CO_2 ratio gas. (After Raney, 1976; reprinted from *Hydrocarbon Processing*.)

Chemical Absorption—The Alkanol-Amine Processes

The alkanol-amine processes are the most prominent and widely used processes for H_2S and CO_2 removal. They offer good reactivity at low cost and good flexibility in design and operation. Some of the commonly used alkanol-amines for absorption desulfurization are: monoethanolamine (MEA), diethanolamine (DEA), triethanolamine (TEA), $\beta\beta'$ hydroxiethanolamine, usually called diglycolamine (DGA), di-isopropanolamine (DIPA), and methyldiethanolamine (MDEA). Table 6-1 shows some of the

Table 6-1
Some Characteristics of Ethanolamines*

Amine Type	Chem. Formula	Mol. Wt.	Vap. Press. at 100°F, mm Hg	Rel. Capacity %
MEA	$HOC_2H_4NH_2$	61.08	1.05	100
DEA	$(HOC_2H_4)_2NH$	105.14	0.058	58
TEA	$(HOC_2H_4)_3N$	148.19	0.0063	41
DGA	$H(OC_2H_4)_2NH_2$	105.14	0.160	58
DIPA	$(HOC_3H_6)_2NH$	133.19	0.010	46
MDEA	$(HOC_2H_4)_2NCH_3$	119.17	0.0061	51

* After Maddox, 1982.

important properties for these six alkanol-amines. Among these, MEA and DEA processes are the most widely used.

MEA has the highest reactivity, and therefore the highest acid gas removal capacity, and the lowest molecular weight among the amines. Therefore, it offers the highest removal capacity on a unit weight or unit volume basis, which implies lower solution circulation rates in a sweetening plant. MEA is chemically stable (minimal solution degradation). The reaction rate of H_2S with MEA is greater than that of CO_2. The process, however, is considered to be almost non-selective in its removal capacity for H_2S and CO_2, because even the slower reacting CO_2 is rapidly and almost completely removed. MEA is able to remove acid gases to pipeline specifications, and even beyond.

MEA, however, suffers from relatively high solution losses primarily due to two reasons: it has a relatively high vapor pressure that causes greater vaporization losses, and it reacts irreversibly with carbonyl sulfide and carbon disulfide. High vaporization losses put a limitation on the operating temperatures, which must be kept low, whereas the reactions leading to buildup of solids necessitate efficient filtration schemes.

DEA is quite similar to MEA, except it is not as reactive, so H_2S removal to pipeline specifications may cause problems in some cases; its reactions with carbonyl sulfide and carbon disulfide are slower and lead to different products, so filtration problems are less; and its vapor pressure is lower, therefore vaporization losses are less. Thus, DEA can be advantageous to use in some cases.

TEA has been almost totally replaced by MEA and DEA, primarily because it has lesser reactivity and problems are encountered in sweetening the gas to pipeline specifications. DGA has found application in recent years. It has the same reactivity as DEA, and a reasonably low vapor pressure. DIPA is also able to treat gas to pipeline specifications, and is used in the Sulfinol process by Shell, and the ADIP process. MDEA has received renewed attention because it has a fairly good selectivity for H_2S, so CO_2 can be retained in the gas; and it offers some energy saving in the regeneration step. MDEA, however, may not be commercially competitive with other amine processes (Maddox, 1982).

Solution Concentrations and Reactions

For MEA, an approximately 15% concentration (by weight) solution is generally used. DEA is used in 20–30% or more concentration. DIPA and MDEA are typically used in 30–50% concentration in aqueous solution, while DGA is used in the 40–70% by weight concentration range.

Some plants use a mixture of a glycol and an amine for simultaneous dehydration and desulfurization. Generally, a solution containing 10 to 30%

by weight MEA, 45 to 85% by weight TEG (triethylene glycol), and 5 to 25% by weight of water is used. This process removes both CO_2 and H_2S besides dehydrating the gas. At high temperatures, the mixture becomes very volatile and difficult to handle. Also, the glycol-amine combination does not dehydrate as well as an independent glycol unit.

The reactions for the ethanolamines, MEA, DEA, and TEA are as follows (Batt et al., 1980, Rahman, 1982):

1. Formation of a bisulfide of the ethanolamine on reaction with H_2S.

$$RNH_2 + H_2S \rightleftharpoons RNH_4S + \text{heat}$$

2. For MEA and DEA, with CO_2, the carbamate salt of the ethanolamine is formed. TEA does not have this reaction with CO_2.

$$2RNH_2 + CO_2 \rightleftharpoons RNHCO_2^- + RNH_3^+ + \text{heat}$$

3. In aqueous solution, with CO_2, the carbonate salt of the ethanolamine is formed. This is a slower reaction.

$$RNH_2 + H_2O + CO_2 \rightleftharpoons RNH_3HCO_3 + \text{heat}$$

General Process Flow Scheme

MEA is usually preferred over DEA because it enables a smaller circulation rate of the solution, although it has a higher vapor pressure with increased chances of greater chemical losses.

The plant is very much like an absorption dehydration system. The main differences are (Curry, 1981):

1. The contactor contains many more tray sections, up to 20 or more.
2. A side-stream reclaimer is provided in an MEA system.
3. More elaborate filtration and heat exchange equipment is required.
4. Operating temperatures are different.
5. The reflux requirements are greater.

The problems encountered are also quite similar, though in a more severe form. Some such problems are corrosion, foaming, solution losses, and poor solution filtering. There are other problems relating to the high vapor pressure of MEA. Solution loading, that is, the amount or level of liquid MEA in the contactor, is critical. The exothermic reactions in the contactor coupled with the low boiling point of MEA make solution loading a difficult problem. The presence of sulfur intensifies the corrosion problems by introducing stress corrosion.

266   Gas Production Engineering

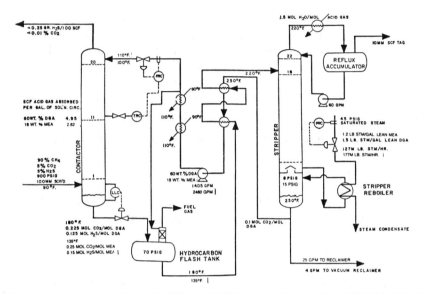

Figure 6-4. A comparison of a MEA and DGA system for treating 100 MMscfd of natural gas. (After Dingman and Moore, 1968; reprinted from *Hydrocarbon Processing*.)

A typical process flow scheme is shown in Figure 6-4. There are several variations, such as the location of the filtering system, using a packed column instead of bubble-cap or valve-type trays in the contactor and the stripper, etc. A side-stream reclaimer may or may not be used. The design of heat exchange and reflux systems vary.

Sour gas is sent upward through the contactor tower, countercurrent to the flow of MEA. The rich solution from the bottom of the contactor is sent to a flash tank where absorbed hydrocarbon gas in solution is vented. A retention time of about 2 minutes is usually sufficient. The flash tank also serves as a sediment accumulator and provisions must be made for sediment removal. After passing through a heat exchange with lean solution (which preheats this rich solution), the rich solution enters the top of the stripper where it is stripped by steam that is generated by the reboiler. Outcoming steam from the top of the stripper is sent to a condensing unit to recover MEA liquid in a reflux accumulator. The acid gases released at this stage are flared off. The liquid accumulated in the reflux is sent to the regenerative system. Lean MEA accumulated at the bottom of the stripper is continuously recirculated through the reboiler.

In the contactor (also called absorber), almost 90% of the acid gases are removed within the first three trays at the bottom. The reactions being exothermic, this results in high temperatures in the contactor in this region. Usually a 20% (by weight) MEA solution is used in the contactor, which can absorb approximately 3.5 to 4.5 scf acid gas per gallon of solution. This can be used to determine the MEA circulation rate in a preliminary design.

Besides its performance, the location of the filtering system is critical to plant operation (Curry, 1981). An efficient filtration will generally reduce foaming, corrosion, and fouling of mechanical equipment. Remember that most filtration problems result not from the failure of the filtration system, but from inefficient gas cleaning. Full-flow filtration is most desirable, although some plants use side-stream filtration where only a part of the stream is filtered. Iron sulfide (FeS) is the most common solid in MEA systems that requires filtration. Diatomaceous earth (DE) filters are commonly used to enable removal of particulate matter up to almost 1 micron. Another commonly used filter is the disposable element type, but it is quite expensive and used only where severe problems warrant its usage. Usually a secondary filtration system is installed in parallel to enable crossover and allow time for cleaning or removing a used filter.

Chemical Absorption—The Carbonate Processes

The Hot Carbonate Processes

The basic hot carbonate process uses an aqueous solution of potassium carbonate. A highly concentrated solution is used to improve process performance. Temperature is kept high to keep the potassium carbonate and the reaction products, $KHCO_3$ and KHS, in solution. Both the absorber and regenerator are operated at high temperatures of about 230–240°F, which results in considerable savings in heat exchange and heating equipment. The process requires relatively high partial pressures of CO_2, and cannot be used for gas streams that contain only H_2S. It is also difficult to treat gas to pipeline specifications with this process.

Among carbonate processes, those containing an activator to increase the activity of the hot potassium carbonate solution are more popular. Some of these are:

1. Benfield—several activators, usually DEA.
2. Catacarb—amine borates used as activators.
3. Giammarco-Vetrocoke—for CO_2 removal, arsenic trioxide most commonly used; selenous acid and tellurous acid are also used. For H_2S removal, alkaline arsenites and arsenates (usually KH_2AsO_3 and KH_2AsO_4) used.

The Giammarco-Vetrocoke (G-V) process is an important development. There are several G-V processes for different applications. There is one process for the removal of CO_2, another for a highly selective H_2S removal and subsequent sulfur recovery, and still another for the removal of both H_2S and CO_2.

The G-V CO_2 process offers substantial savings in steam consumption for solution regeneration. The arsenic trioxide inhibits corrosion, and the solution is therefore usually non-corrosive in nature. As for most conventional carbonate processes, impurities such as carbon disulfide, mercaptans, carbonyl sulfide, etc., have no detrimental effects on the solution. Jenett (1962) presented data on the performance of a G-V plant and concluded that it functioned satisfactorily with minimal problems.

The G-V process for H_2S removal produces elemental sulfur of high purity as the by-product. It is a flexible process, and by suitably selecting the pH of the solution, the arsenite and arsenate content, and the operating conditions, the process can be made to be highly selective for H_2S or remove both H_2S and CO_2. There are no problems of corrosion, solution degradation, solution foaming, or carryover. The economics of the process are also quite good (Riesenfeld and Mullowney, 1959), and it can reduce H_2S content to less than 1 ppm. The process, however, is not able to handle gas streams with H_2S concentrations greater than 1.5% (Jenett, 1962).

Another carbonate process in active use is the Alkazid process. There are three different alkazid solutions, and consequently, three different processes in this category. Alkazid DIK uses the potassium salt of diethyl glycine or dimethyl glycine. It is used for the selective removal of H_2S from gases that contain both H_2S and CO_2. Alkazid M uses sodium alanine and is effective in removing both H_2S and CO_2. Alkazid S uses a sodium phenolate mixture and is applicable to gases containing impurities such as carbon disulfide, mercaptans, HCN, etc.

The Holmes-Stretford Process

This process converts H_2S to elemental sulfur of almost 99.9% purity. The process is selective for H_2S, and can reduce the H_2S content to as low as 1 ppm. CO_2 content remains almost unaltered by this process. Operating costs are lower and the process has better flexibility in application, that is, it can be designed for larger pressure and temperature ranges.

The Holmes-Stretford process, a modification of the Stretford process developed by the North Western Gas Board in England, was developed by Peabody-Holmes. Details on the mechanism of the process, and its performance are reported in several publications (Nicklin et al., 1973; Moyes and Wilkinson, 1974; Ouwerkerk, 1978; Vasan, 1978). The process uses an aqueous solution containing sodium carbonate and bicarbonate in the ratio of approximately 1:3, resulting in a pH of about 8.5 to 9.5, and the sodium salts of 2,6 and 2,7 isomers of anthraquinone disulfonic acid (ADA). The postulated reaction mechanism involves five steps: (1) absorption of H_2S in alkali; (2) reduction of ADA by addition of hydrosulfide to a carbonyl group; (3) liberation of elemental sulfur from reduced ADA by interaction with oxygen dissolved in the solution; (4) reoxidation of the reduced ADA by air; and (5) reoxygenation of the alkaline solution, providing dissolved oxygen for step (3).

Several possible additives have been tested to increase the solution capacity for hydrogen sulfide and the rate of conversion of hydrosulfide to elemental sulfur. Alkali vanadates proved to be outstanding in reducing hydrosulfide to sulfur, with a simultaneous valence change of vanadium from five to four. In the presence of ADA, the vanadate solution can be regenerated to a five valence state. The reduced ADA is easily oxidized by air.

The circulated solution (also called Stretford solution) consists of chemicals in a dilute water solution as follows:

1. Sodium salt of 2,7 ADA. The 2,7 isomer is preferred over 2,6 because of the greater solubility of the former.
2. Sodium meta vanadate. This provides the active vanadium.
3. Sodium potassium tartrate (Rochelle salts). This is used to prevent the formation of a complex vanadium-oxygen precipitate that removes vanadium from solution.
4. A sequestering agent to prevent precipitation of metallic ions from Stretford solution, such as Chel 242 PN.
5. Sodium carbonate. Used as required to maintain total alkalinity.
6. Sodium bicarbonate added to reduce the absorption of CO_2.

The overall reaction of the Holmes-Stretford process is the atmospheric oxidation of H_2S to elemental sulfur:

$$2H_2S + O_2 = 2H_2O + 2S$$

The reaction, however, occurs in the following steps:

$$Na_2CO_3 + H_2S = NaHS + NaHCO_3 \qquad (6\text{-}1)$$

$$4NaVO_3 + 2NaHS + H_2O = Na_2V_4O_9 + 4NaOH + 2S \qquad (6\text{-}2)$$

$$Na_2V_4O_9 + 2NaOH + H_2O + 2ADA = 4NaVO_3 + 2ADA \text{ (reduced)} \quad (6\text{-}3)$$

$$2ADA \text{ (reduced)} + O_2 = 2ADA + H_2O \qquad (6\text{-}4)$$

The absorption rate of H_2S in solution (Equation 6-1) is favored by high pH, whereas the conversion to elemental sulfur (Equation 6-2) is adversely affected by pH values above 9.5. Therefore, the process is best operated within a pH range of 8.5 to 9.5.

The conversion of H_2S to sulfur is quite rapid and is essentially a function of the vanadate concentration in the solution. According to Thompson and Nicklin (1964) the reaction is second order. In practice, an excess of vanadate is used in order to avoid overloading of the solution with sulfide and subsequent formation of thiosulfate during solution regeneration.

The reduced vanadate is oxidized by ADA (Equation 6-3). Oxidation of ADA by contact with air is a fairly rapid reaction. The rate of oxidation can, however, be accelerated by the presence of small amounts of iron salts kept in solution by a chelating agent (Nicklin and Hughes, 1978). In general, a concentration of 50-100 ppm of iron combined with about 2,700 ppm of ethylenediamine tetracetic acid (EDTA) is satisfactory.

The chemicals could conceivably be used indefinitely with only minimal replenishments for losses that occur in the absorber or within the sulfur recovery unit. This, however, cannot be done indefinitely because side reactions produce undesirable dissolved solids. These solids are permitted to build up to extremely high concentrations until eventually some solution must be discarded. Primarily, sodium sulfate and sodium thiosulfate are formed in reactions such as:

$$2NaHS + 2O_2 = Na_2S_2O_3 + H_2O$$

Thiosulfate formation may be due to the fact that NaHS has not reacted with the vanadate (because of insufficient time in the absorber) and therefore is carried to the oxidizer to react with oxygen there. If there is air in the feed gas, some of the undesired material can be formed in the contactor itself. High temperature and pH also favor thiosulfate formation. The dissolved solids such as thiosulfate are allowed to build up in the solution to concentrations upwards of 250,000 ppm. It may take a year or longer after plant startup to reach this level. At these high concentrations, disposal by purging involves very little of the active chemical, and hence chemical usage is kept low.

The effluent stream from the Stretford process, containing sodium thiosulfate and in some cases sodium thiocyanate, must be treated prior to discharge. Holmes developed four alternative methods to handle effluents from this process: evaporation or spray drying, biological degradation, oxidative combustion, and reductive incineration. Of these, the reductive incineration process is the most common since it results in a zero effluent discharge, with all the products from the step being recyclable. The reductive incineration process cracks the bleed liquor containing sodium thiosulfate into a gas stream containing H_2S and CO_2 and a liquid stream containing reduced vanadium salts, all of which are recycled.

Desulfurization Processes 271

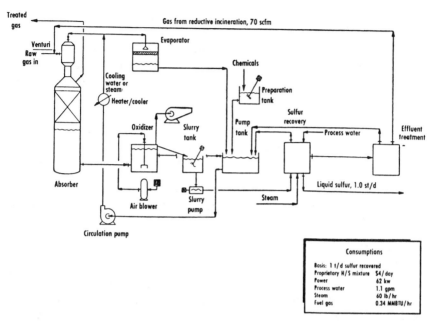

Figure 6-5. Flowsheet for Holmes-Stretford process. (After Vasan, 1978; courtesy of *Oil & Gas Journal.*)

Figure 6-5 shows the process flow scheme. In the gas absorption section, gas is first subjected to the inlet scrubber unit to remove any liquids and contaminants that may be detrimental to the process. A venturi-type absorber is used, instead of the conventional packed bed type, to reduce the size of the absorption unit and achieve a better degree of absorption since it provides a larger surface area for mass transfer. The outlet scrubber prevents carryover of any entrained Stretford solution in the gas into the sales gas stream.

The reduced solution from the bottom of the absorber is first piped through a flash drum in the oxidation section. Flash vapors may be used as incinerator fuel, or returned to compressor station and mixed with sweetened gas. Liquid from the flash drum is sent to the oxidation tank where air from the blowers contacts the liquid at the vessel bottom, and sulfur froth is isolated at the surface. The oxidized Stretford solution goes to the circulation pump and to the surge tank for recirculation to the absorber.

The sulfur processing section includes a slurry tank and an autoclave for melting sulfur. Molten sulfur from the bottom of the autoclave is cooled, flaked, and finally bagged for storage and shipment. The clear Stretford solvent is sent to the effluent treatment section.

The process is fairly easy to operate. Alkalinity, concentrations, and amounts of the active chemicals change slowly with time and replenishments need only be done quite infrequently. Corrosion problems are encountered, just as in any sweetening process. Kresse et al. (1981) provide an interesting discussion on some problems encountered, and their rectification, in two plants using this process.

Questions and Problems

1. Why must H_2S be removed from natural gas? Give at least three reasons.
2. Why is it necessary to remove CO_2? In what cases is CO_2 removal not necessary?
3. What types of desulfurization processes are available for natural gas? What factors are important in selecting the applicable process?
4. Compare the iron-sponge and molecular sieve desulfurization processes.
5. In what cases are the physical absorption processes useful?
6. Why are chemical absorption processes the most widely used desulfurization processes?
7. Compare the alkanol-amine and carbonate processes.
8. Which of the processes described in this chapter hold the most promising future? Why?

References

Batt, W. T., Vaz, R. N., Rahman, M., Mains, G. H., and Maddox, R. N., 1980. *Gas Conditioning Conference*, Univ. of Oklahoma, Norman, Oklahoma.

Conviser, S. A., 1965. "Molecular Sieves Used to Remove Mercaptans From Natural Gas," *O. & Gas J.*, 63 (49, Dec. 6), 130–133.

Curry, R. N., 1981. *Fundamentals of Natural Gas Conditioning*. PennWell Publishing Company, Tulsa, Oklahoma, 118 pp.

Dingman, J. C. and Moore, T. F., 1968. "Compare DGA and MEA Sweetening Methods," *Hydr. Proc.*, 47 (7, July), 138–140.

Duckworth, G. L. and Geddes, J. H., 1965. "Natural Gas Desulfurized by the Iron-Sponge Process," *O. & Gas J.*, 63 (37, Sept. 13), 94–96.

Fails, J. C. and Harris, W. D., 1960. "Practical Way to Sweeten Natural Gas," *O. & Gas J.*, 58 (28, July 11), 86–90.

Froning, H. R., Jacoby, R. H., and Richards, W. L., 1964. "New K-Data Show Value of Water Wash," *Hydr. Proc. & Petr. Ref.*, 43 (4, Apr.), 125–130.

Hegwer, A. M. and Harris, R. A., 1970. "Selexol Solves High H_2S/CO_2 Problem," *Hydr. Proc.*, 49 (4, Apr.), 103–104.

Jenett, E., 1962. "Six Case Studies Throw Light on the Giammarco-Vetrocoke Process," *O. & Gas J.*, 60 (18, Apr. 30), 72–77.

Kohl, A. L. and Riesenfeld, F. C., 1985. *Gas Purification*/4th ed., Gulf Publishing Co., Houston, Texas, 840 pp.

Kresse, T. J., Lindsey, E. E., and Wadleigh, T., 1981. "Stretford Plants Proving Reliable," *O. & Gas J.*, 79 (2, Jan. 12), 80–87.

Maddox, R. N., 1982. *Gas Conditioning and Processing - Vol. 4: Gas and Liquid Sweetening*/ 3rd ed., edited by J. M. Campbell, Campbell Petroleum Series, Norman, Oklahoma, 370 pp.

Manning, W. P., 1979. "Chemsweet, A New Process for Sweetening Low-Value Sour Gas," *O. & Gas J.*, 77 (42, Oct. 15), 122–124.

Moyes, A. J. and Wilkinson, J. S., 1974. "Development of the Holmes-Stretford Process," *The Chemical Engineer (Brit.)*, No. 282 (Feb.), 84–90.

Nicklin, T. and Hughes, D., 1978. "Removing H_2S From Gases and Liquids," *Chem. Abst.*, 88, 94276m.

Nicklin, T., Riesenfeld, F. C., and Vaell, R. P., 1973. "The Application of the Stretford Process to the Purification of Natural Gas," Paper presented at the 12th World Gas Conference, Nice, France, June 1973.

O. & Gas J., 1981. "New Gas-Sweetening Process Offers Economical Alternative," 79 (35, Aug. 31), 60–62.

Ouwerkerk, C., 1978. "Design for Selective H_2S Absorption," *Hydr. Proc.*, 57 (4, Apr.), 89–94.

Rahman, M., 1982. Ph.D. Thesis. Oklahoma State Univ. (May, 1982), Oklahoma.

Raney, D. R., 1976. "Remove Carbon Dioxide With Selexol," *Hydr. Proc.*, 55 (4, Apr.), 73–75.

Riesenfeld, F. C. and Mullowney, J. F., 1959. *Proc. 38th Ann. NGPA Meet.*, April 22–24.

Rushton, D. W. and Hays, W., 1961. "Selective Adsorption to Remove H_2S," *O. & Gas J.*, 59 (38, Sept. 18), 102–103.

Sun, Y-C., 1980. "Packed Beds for Best Sulfur Removal," *Hydr. Proc.*, 59 (10, Oct.), 99–102.

Taylor, D. K., 1956. "How to Desulfurize Natural Gas," *O. & Gas J.*, multiple issues of Vol. 54: Nov. 5, p.125; Nov. 19, p.260; Dec. 3, p.139; Dec. 10, p.147.

Thompson, R. J. S. and Nicklin, T., 1964. "Le Procede Stretford," paper presented at the Congress of Association Technique de l'Industrie du Gazen, France.

Vasan, S., 1978. "Holmes-Stretford Process Offers Economic H_2S Removal," *Oil & Gas J.*, 76 (1, Jan. 2), 78–80.

7
Steady-State Flow of Gas Through Pipes

Introduction

Pipes provide an economic means of producing (through tubing or casing) and transporting (via flowlines or pipelines) fluids in large volumes over great distances. They are convenient to fabricate and install, and provide an almost indefinite life span. Because flow is continuous, minimal storage facilities are required at either end (field supply end, and the consumer end). Operating costs are very low, and flow is guaranteed under all conditions of weather, with good control (an installed pipeline can usually handle a wide range of flow rates). There are no spillage or other handling losses, unless the line develops a leak, which can be easily located and fixed for surface lines.

The flow of gases through piping systems involves flow in horizontal, inclined, and vertical orientations, and through constrictions such as chokes for flow control. This chapter introduces some basic concepts for these flow types.

Gas Flow Fundamentals

All fluid flow equations are derived from a basic energy balance which, for a steady-state system (no time-dependence of flow parameters), can be expressed as:

Change in internal energy + Change in kinetic energy
+ Change in potential energy + Work done on the fluid
+ Heat energy added to the fluid
− Shaft work done by fluid on the surroundings
= 0

276 Gas Production Engineering

Thus, on a unit mass basis, the energy balance for a fluid under steady-state flow conditions can be written as:

$$dU + \frac{dv^2}{2g_c} + \frac{g}{g_c}dz + d(pV) + dQ - dw_s = 0 \tag{7-1}$$

where U = internal energy, ft-lbf/lbm
v = fluid velocity, ft/sec
z = elevation above a given datum plane, ft
p = pressure, lbf/ft^2
V = volume of a unit mass of the fluid, ft^3/lbm $\rightleftharpoons \rho$
Q = heat added to the fluid, ft-lbf/lbm
w_s = shaft work done by the fluid on the surroundings, ft-lbf/lbm
g = gravitational acceleration, ft/sec^2
g_c = conversion factor relating mass and weight*.

This basic relationship can be manipulated in several different ways. Commonly, it is converted into a mechanical energy balance using the well known thermodynamic relations for enthalpy (h):

$$dU + d(pV) = dh = Tds + Vdp$$

where h = specific fluid enthalpy, ft-lbf/lbm
T = temperature, °R
s = specific fluid entropy, ft-lbf/lbm

Equation 7-1 now becomes:

$$Tds + Vdp + \frac{dv^2}{2g_c} + (g/g_c)dz + dQ - dw_s = 0$$

For an ideal process, ds = $-$ dQ/T. Since no process is ideal (or reversible), ds $\geq$ $-$ dQ/T, or:

$$Tds = -dQ + dl_w$$

where l_w is the lost work due to irreversibilities such as friction.

---

* F = mg/g_c. In metric (or SI) units, 1 N = (1 kg)(9.8 m/sec^2)/g_c, implying that g_c = 9.8 kg m/ (N sec^2) = 1 N/N. In British units, 1 lbf = (1 lbm)(32.17 ft/sec^2)/g_c, implying that g_c = 32.17 lbm ft/(lbf sec^2).

On this further substitution, Equation 7-1 becomes:

$$Vdp + \frac{dv^2}{2g_c} + (g/g_c)dz + dl_w - dw_s = 0 \qquad (7\text{-}2)$$

Neglecting the shaft work w_s, and multiplying throughout by the fluid density, ρ:

$$dp + \frac{\rho dv^2}{2g_c} + (g/g_c)\rho dz + \rho dl_w = 0 \qquad (7\text{-}3)$$

All the terms in Equation 7-3 have units of pressure. Equation 7-3 can also be written as:

$$\Delta p + \frac{\rho \Delta v^2}{2g_c} + (g/g_c)\rho \Delta z + \rho \Delta l_w = 0$$

or,

$$\Delta p + \frac{\rho \Delta v^2}{2g_c} + (g/g_c)\rho \Delta z + \Delta p_f = 0 \qquad (7\text{-}4)$$

where Δp_f represents the pressure drop due to friction, and is dependent upon the prevailing flow conditions.

Types of Single-Phase Flow Regimes and Reynolds Number

Four types of single-phase flow are possible: *laminar, critical, transition,* and *turbulent* (see Figure 7-1). Reynolds applied dimensional analysis to flow phenomena, and concluded that the flow regime that will prevail is a function of the following dimensionless group known as the Reynolds number, N_{Re}:

$$N_{Re} = \text{inertia forces/viscous forces} = dv\rho/\mu \qquad (7\text{-}5)$$

where d = (inside) diameter of the conduit through which the fluid is flowing
v = velocity of the fluid
ρ = density of the fluid
μ = viscosity of the fluid

For cross-sections other than circular, an equivalent diameter, d_e, defined as four times the hydraulic radius, R_h, is used instead of d:

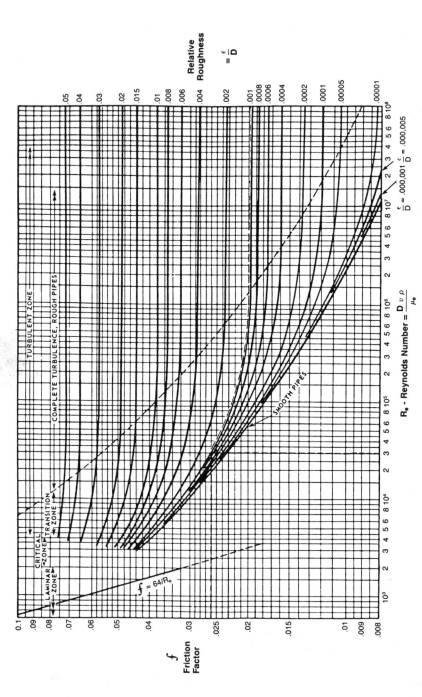

Figure 7-1. Friction factors for flow in pipes. (After Moody, 1944; courtesy of ASME.)

$$d_e = 4R_h = 4[\text{area of flow/wetted perimeter}] \qquad (7\text{-}6)$$

For example, for a flow conduit with a square cross-section (a × a):

Cross-sectional area of flow = a^2

Wetted perimeter = $4a$

Thus, $d_e = 4(a^2/4a) = a$

For flow through a casing-tubing annulus, with casing inside diameter equal to d_{ci} and tubing outside diameter equal to d_{to}:

Cross-sectional area of flow = $(\pi/4)(d_{ci}^2 - d_{to}^2)$

Wetted perimeter = $\pi(d_{ci} + d_{to})$

Thus, $d_e = \dfrac{4(\pi/4)(d_{ci}^2 - d_{to}^2)}{\pi(d_{ci} + d_{to})} = d_{ci} - d_{to}$

To verify the applicability of this approach, consider the case of a pipe with a circular cross section of diameter d:

Cross-sectional area of flow = $\pi d^2/4$

Wetted perimeter = πd

Thus, $d_e = 4(\pi d^2/4\pi d) = d$

The units for parameters in the Reynolds number should be consistent, so that a dimensionless number is obtained. One such consistent set would have d in ft, v in ft/sec, ρ in lbm/ft³, and μ in lbm/(ft sec). The dynamic fluid viscosity μ, frequently given in centipoises (cp), can be converted into lbm/(ft sec) using the conversion factor of 1 cp = 6.7197 × 10⁻⁴ lbm/(ft sec). Thus:

$$N_{Re} = \frac{d(\text{ft})v(\text{ft/sec})\rho(\text{lbm/ft}^3)}{\mu(\text{cp}) \times 6.7197 \times 10^{-4}} = 1{,}488\, \frac{d(\text{ft})v(\text{ft/sec})\rho(\text{lbm/ft}^3)}{\mu(\text{cp})} \qquad (7\text{-}7)$$

For gases, the flow rate is commonly expressed in Mscfd (thousands of standard cubic feet per day). This volumetric rate, q_{sc} (Mscfd), at standard conditions of pressure (p_{sc}, psia) and temperature (T_{sc}, °R), can be converted into mass flow rate, m in lbm/sec, as follows:

280   Gas Production Engineering

$$m = Av\rho = \frac{(1,000q_{sc})Mp_{sc}}{(24)(3,600)Z_{sc}RT_{sc}} = \frac{(3.1243 \times 10^{-2})(q_{sc})\gamma_g p_{sc}}{T_{sc}}$$

where A = cross-sectional area of flow, ft^2
M = molecular weight of the gas
R = gas constant (= 10.732 psia-ft^3/lbmole-°R)
p_{sc} = pressure at standard conditions, psia
T_{sc} = temperature at standard conditions, °R
Z_{sc} = gas compressibility at standard conditions (= 1)
γ_g = gas gravity (air = 1)

Thus, for a circular pipe of diameter d (ft), $v\rho$ is given by:

$$v\rho = (4)(3.1243 \times 10^{-2})q_{sc}\gamma_g p_{sc}/(T_{sc}\pi d^2)$$
$$= (3.9780 \times 10^{-2})q_{sc}\gamma_g p_{sc}/(T_{sc}d^2)$$

Substituting for $v\rho$ in Equation 7-9 for N_{Re}:

$$N_{Re} = \frac{d(3.9780 \times 10^{-2})q_{sc}\gamma_g p_{sc}}{T_{sc}d^2(\mu \times 6.7197 \times 10^{-4})} = \frac{59.1991 q_{sc}\gamma_g p_{sc}}{T_{sc}d\mu}$$

Generally, d is used in inches for most gas flow equations. For d in inches, N_{Re} becomes:

$$N_{Re} = \frac{710.39[p_{sc}(psia)/T_{sc}(°R)]q_{sc}(Mscfd)\gamma_g}{d(in.)\,\mu(cp)} \qquad (7\text{-}8)$$

The factor $710.39 p_{sc}/T_{sc}$ for some common standard conditions is as follows:

p_{sc}, psia	T_{sc}, °R	$710.39 p_{sc}/T_{sc}$
14.4	520 (60°F)	19.672
14.65	520 (60°F)	20.014
14.73	520 (60°F)	20.123
15.025	520 (60°F)	20.526

Thus, for most practical applications, the Reynolds number for a gas is given by:

$$N_{Re} \simeq 20 q_{sc}\gamma_g/(\mu d) \qquad (7\text{-}9)$$

where q_{sc} is in Mscfd, μ is in cp, and d is in inches.

Steady State Flow of Gas Through Pipes   281

As shown in the Moody friction factor chart (Figure 7-1), flow regime is related to Reynolds number as follows:

Flow type	N_{Re}, smooth pipes
Laminar (or viscous)	<2,000
Critical	2,000–3,000
Transition (or intermediate)	3,000–4,000 (or 10,000)
Turbulent	>4,000 (or 10,000)

Pipe Roughness

Friction to flow through a pipe is affected by pipe-wall roughness. However, pipe roughness is not easily or directly measurable, and absolute pipe roughness ϵ is, therefore, defined as the mean height of protusions in uniformly sized, tightly packed sand grains that give the same pressure gradient as the given pipe. This roughness may change with pipe use and exposure to fluids. Initially, the pipe contains mill scale that may be removed by fluids flowing inside the pipe. The fluids may also increase roughness by erosion or corrosion, or by precipitating materials that stick to the pipe wall. Thus, estimating pipe roughness is quite difficult. Usually, absolute roughness is determined by comparing the observed friction factor to that given in Moody's chart (Figure 7-1). If no roughness data are available, a value of $\epsilon = 0.0006$ in. can be used. Some typical values for roughness are shown below (*Chemical Engineers' Handbook*, 1984):

Type of pipe	ϵ, in.
Drawn tubing (brass, lead, glass)	0.00006
Aluminum pipe	0.0002
Plastic-lined or sand blasted	0.0002–0.0003
Commercial steel or wrought iron	0.0018
Asphalted cast iron	0.0048
Galvanized iron	0.006
Cast iron	0.0102
Cement-lined	0.012–0.12
Riveted steel	0.036–0.36

Commonly used well tubing and line pipe:	
New pipe	0.0005–0.0007
12-months old	0.00150
24-months old	0.00175

From dimensional analysis, it has been deduced that relative roughness, the ratio of the absolute roughness and inside pipe diameter, ϵ/d, rather than absolute roughness, affects flow through pipes.

Friction Factor

For convenience, friction factor f', defined as the ratio of the shear stress at the fluid-solid interface and the kinetic energy of the fluid per unit volume, is used in computing the magnitude of the pressure drop due to friction. For steady-state flow in a uniform circular conduit such as a pipe, this results in the well known Fanning equation:

$$\Delta p_f = 2f' L \rho v^2/(g_c d) \qquad (7\text{-}10)$$

where d is the inside pipe diameter.

Any consistent set of units can be used in the previous equations. In customary units, Δp_f is in psi, L is in ft, ρ is in lbm/ft^3, v is in ft/sec, d is in ft, and g_c is therefore equal to 32.17 $lbm\text{-}ft/lbf\text{-}sec^2$.

Friction factor f' is called the Fanning friction factor. Usually, the Moody (also called Blasius, or Darcy-Weisbach) friction factor, equal to $4f'$, is used. In terms of the Moody friction factor, f, the Fanning equation becomes:

$$\Delta p_f = f L \rho v^2/(2 g_c d) \qquad (7\text{-}11)$$

The friction factor includes, besides roughness, the flow characteristics of the flow regime. It is therefore a function of Reynolds number and relative roughness:

$$f = f(N_{Re}, \epsilon/d)$$

Laminar Single-Phase Flow

The pressure drop for laminar flow is given by the analytic Hagen-Poiseuille relationship as follows:

$$\Delta p_f = 32 \mu v L/(g_c d^2) \qquad (7\text{-}12)$$

Equating Equations 7-11 and 7-12:

$$f L \rho v^2/(2 g_c d) = 32 \mu v L/(g_c d^2)$$

or,

$$f = 64 \, \mu/dv\rho = 64/N_{Re} \qquad (7\text{-}13)$$

Thus, the friction factor is independent of pipe roughness in the laminar flow regime.

Partially-Turbulent and Fully-Turbulent Single-Phase Flow

For partially-turbulent flow, friction factor is a function of both Reynolds number and pipe roughness. For fully turbulent flow, however, the friction factor is only very slightly dependent upon Reynolds number (see Figure 7-1).

Generally, intermediate or partially turbulent flow is included in turbulent flow for purposes of developing correlations. Several correlations have been reported for the dependence of friction factors on Reynolds number and pipe wall roughness; only some of the most accurate ones are presented here.

For smooth pipes, the following relationships are applicable:

$$f = 0.5676 \, N_{Re}^{-0.3192} \text{ for intermediate flow} \tag{7-14}$$

$$f = 16 \log (N_{Re} f^{0.5}/0.7063) \text{ for partially turbulent flow} \tag{7-15}$$

$$f^{-0.5} = 2 \log (N_{Re} f^{0.5}/0.628) \text{ for fully turbulent flow (Prandtl's formula)} \tag{7-16}$$

$$f = 0.3164 \, N_{Re}^{-0.25} \text{ for } N_{Re} \text{ up to } 10^5 \text{ (Blasius formula)} \tag{7-17}$$

$$f = 0.0056 + 0.5 N_{Re}^{-0.32} \text{ for } 3{,}000 < N_{Re} < 3 \times 10^6 \text{ (Drew, Koo, and McAdams)} \tag{7-18}$$

For rough pipes, Colebrook's equation is the basis for most modern friction factor charts:

$$f^{-0.5} = -2 \log [\epsilon/(3.7d) + 0.628/(N_{Re} f^{0.5})] \tag{7-19}$$

For very rough pipes, Colebrook's equation (Equation 7-19) reduces to:

$$f^{-0.5} = -2 \log (\epsilon/3.7d) \tag{7-20}$$

whereas for smooth pipes, Colebrook's equation reduces to Prandtl's equation (Equation 7-16).

Swamee and Jain (1976) have presented an explicit correlation for friction factors as follows:

$$f^{-0.5} = 1.14 - 2 \log (\epsilon/d + 21.25/N_{Re}^{0.9}) \tag{7-21}$$

Equation 7-21 is applicable for $10^{-6} < \epsilon/d < 10^{-2}$, and $5,000 < N_{Re} < 10^8$, with errors within 1% when compared with Colebrook's equation.

In practice, solutions to the equations presented for friction factors are cumbersome, so a composite friction factor chart is used as shown in Figure 7-1.

Allowable Working Pressures for Pipes

It is desirable to operate a pipe at a high pressure in order to achieve higher throughputs. This is, however, limited by the maximum stress the pipe can handle. The maximum allowable internal working pressure can be determined using the following ANSI (1976) specification:

$$p_{max} = \frac{2(t-c)SE}{d_o - 2(t-c)Y} \qquad (7\text{-}22)$$

where p_{max} = maximum allowable internal pressure, psig
t = pipe thickness, in.
c = sum of mechanical allowances (thread and groove depth), corrosion, erosion, etc., in.
S = allowable stress (minimum yield strength) for the pipe material, psi
E = longitudinal weld joint factor (i.e., the anomaly due to weld seam), equal to 1.0 for seamless pipes, 0.8 for fusion-welded and spiral-welded, and 0.6 for butt-welded pipes
d_o = outside diameter of the pipe, in.
Y = temperature derating factor, equal to 0.4 up to 900°F, 0.5 for 950°F, and 0.7 for 1,000°F and greater

Allowable Flow Velocity in Pipes

High flow velocities in pipes can cause pipe erosion problems, especially for gases that may have a flow velocity exceeding 70 ft/sec. The velocity at which erosion begins to occur is dependent upon the presence of solid particles, their shape, etc., and is, therefore, difficult to determine precisely. The following equation can be used as a simple approach to this problem (Beggs, 1984):

$$v_e = C/\rho^{0.5} \qquad (7\text{-}23)$$

where v_e = erosional velocity, ft/sec
ρ = fluid density, lbm/ft^3
C = a constant ranging between 75 and 150

In most cases, C is taken to be 100 (Beggs, 1984). Substituting for C and the gas density (ρ = pM/ZRT), Equation 7-23 can be written as:

$$v_e = \frac{100}{(pM/ZRT)^{0.5}} = \frac{100(ZRT)^{0.5}}{(28.97p\gamma_g)^{0.5}}$$

where γ_g is the gas gravity (air = 1 basis).

The gas flow rate at standard conditions for erosion to occur, $(q_e)_{sc}$, can be obtained as follows:

$$(q_e)_{sc} = v_e\left(\frac{pT_{sc}}{ZTp_{sc}}\right)\left(\frac{\pi d^2}{4}\right)$$

$$= \left(\frac{100(pR)^{0.5}}{(28.97\gamma_g ZT)^{0.5}}\right)\left(\frac{T_{sc}}{p_{sc}}\right)\left(\frac{\pi d^2}{4}\right)$$

where $(q_e)_{sc}$ = gas flow rate for onset of erosion, scf/sec
d = diameter of the pipe, ft
p = flowing pressure, psia
T = flowing temperature, °R
R = gas constant (= 10.732 psia-ft³/lbmole-°R)
Z = gas compressibility factor at pressure p and temperature T

Substituting for p_{sc} = 14.73 psia, T_{sc} = 520 °R, and converting to field units of d in inches and $(q_e)_{sc}$ in Mscfd, we get:

$$(q_e)_{sc} = 1{,}012.435 \; d^2 \left[\frac{p}{\gamma_g ZT}\right]^{0.5} \tag{7-24}$$

where $(q_e)_{sc}$ is in Mscfd, d is in inches, p is in psia, and T is in °R.

Horizontal Flow

Many pipeline equations have been developed from the basic mechanical energy balance (Equation 7-3):

$$dp + (\rho/2g_c)dv^2 + (g/g_c)\rho dz + \rho dl_w = 0$$

Assuming horizontal, steady-state, adiabatic, isothermal flow of gas, with negligible kinetic-energy change, Equation 7-3 becomes:

$$dp + \rho dl_w = 0$$

Frictional losses for a length dL of pipe are given by (Equation 7-11):

$$\rho dl_w = (f\rho v^2/2g_c d)\, dL$$

Substituting for frictional losses:

$$dp + (f\rho v^2/2g_c d)\, dL = 0$$

Substituting for gas density ρ:

$$\rho = pM/(ZRT)$$

and gas velocity v:

$$v = q_{sc}\left(\frac{ZTp_{sc}}{pT_{sc}}\right)\left(\frac{4}{\pi d^2}\right)$$

we obtain:

$$-dp = \left(\frac{f}{2g_c d}\right)\left(\frac{pM}{ZRT}\right)\left(\frac{16 q_{sc}^2 Z^2 T^2 p_{sc}^2}{p^2 T_{sc}^2 \pi^2 d^4}\right) dL$$

or,

$$-\int \frac{p}{Z}\, dp = \frac{8fMT p_{sc}^2 q_{sc}^2}{R\pi^2 d^5 g_c T_{sc}^2} \int dL \quad (7\text{-}25)$$

Note that T is constant (or, independent of length) since isothermal flow is assumed. Otherwise, an average temperature, T_{av}, is commonly used instead of T in the previous relationship. The two types of averages used are the arithmetic average,

$$T_{av} = \frac{T_1 + T_2}{2} \quad (7\text{-}26)$$

and the log-mean temperature,

$$T_{av} = \frac{T_1 - T_2}{\ln (T_1/T_2)} \quad (7\text{-}27)$$

In practice, both these averages are quite close since temperatures T_1 and T_2 are used as absolute temperatures. Using an average temperature is practically expedient, because an analytic description of the variation of temperature along the pipeline length is rather difficult and introduces some complexity. See last section in this chapter for expressions for temperature profiles in flowing gas systems. The gas compressibility factor, Z, is made independent of temperature and pressure by using an average compressibility factor, Z_{av}, for simplicity. Integrating over the pipe length from 0 to L and pressure p_1 (at L = 0, at the upstream) to p_2 (at L = L, at the downstream end), Equation 7-25 becomes:

$$-(p_2^2 - p_1^2)/2 = \left(\frac{(8)(28.97)p_{sc}^2}{R\pi^2 g_c T_{sc}^2}\right)\left(\frac{q_{sc}^2 \gamma_g Z_{av} TfL}{d^5}\right)$$

or,

$$q_{sc}^2 = \left(\frac{Rg_c T_{sc}^2}{46.9644 p_{sc}^2}\right)\left(\frac{(p_1^2 - p_2^2)d^5}{\gamma_g Z_{av} TfL}\right) \tag{7-28}$$

Any consistent set of units can be used in Equation 7-28. In common units, with q_{sc} in Mscfd, p in psia, T in °R, d in in., L in ft, and with R = 10.732 psia ft³/lb mole-°R and g_c = 32.17 lbm ft/lbf-sec², Equation 7-28 becomes:

$$[10^3/(3,600 \times 24)]^2 q_{sc}^2 = \left(\frac{(10.732 \times 144)(32.17)(1/12)^5}{46.9644}\right)\left(\frac{T_{sc}^2}{p_{sc}^2}\right)\left(\frac{(p_1^2 - p_2^2)d^5}{\gamma_g Z_{av} TfL}\right)$$

Thus,

$$q_{sc} = 5.6353821 \left(\frac{T_{sc}}{p_{sc}}\right)\left(\frac{(p_1^2 - p_2^2)d^5}{\gamma_g Z_{av} TfL}\right)^{0.5} \tag{7-29}$$

where q_{sc} = gas flow rate measured at standard conditions, Mscfd
p_{sc} = pressure at standard conditions, psia
T_{sc} = temperature at standard conditions, °R
p_1 = upstream pressure, psia
p_2 = downstream pressure, psia
d = diameter of the pipe, ft
γ_g = gas gravity (air = 1 basis)
T = flowing temperature, °R
Z_{av} = average gas compressibility factor
f = Moody friction factor
L = length of the pipe, ft

Equation 7-29, attributed to Weymouth, is the general equation for steady-state isothermal flow of gas through a horizontal pipe. Implications of the

288   Gas Production Engineering

various assumptions made in the development of Equation 7-29 are as follows:

1. *No mechanical work:* It is assumed that no work is done on the gas between the points at which the pressures are measured. This condition can be satisfied easily by putting pressure measurement stations such that no mechanical energy is added (by compressors or pumps) between these two points.
2. *Steady-state flow:* Rarely, if ever, encountered in practice, this assumption is the major cause of discrepancies in pipeline calculations. Reasons for unsteady behavior include: pressure/flow rate pulsations or surges, liquids in the pipeline, variations in operating conditions, variations in withdrawal or supply rates, etc.
3. *Isothermal flow:* This assumption is usually met because buried pipelines are used which are not affected much by atmospheric temperature variations. Heat of compression is also dissipated rapidly, usually within a few miles downstream of the compressor station. For small temperature changes, the average temperature given by Equation 7-26 or 7-27 is generally satisfactory.
4. *Negligible kinetic energy change:* This assumption is justified because kinetic energy changes are negligible, compared to changes in pressure, for very long pipelines, such as commercial transmission lines.
5. *Constant (average) gas compressibility factor:* This is a reasonable approximation, especially if Z_{av} is computed at the average pressure given by Equation 7-32.
6. *Horizontal pipeline:* In practice, flow is never truly horizontal. Equations developed to account for elevation changes will be discussed in the section on inclined flow.

Average Pressure in a Gas Pipeline

For an incompressible fluid, the average pressure is simply the arithmetic average of the inlet (p_1) and outlet (p_2) pressures:

$$p_{av} = (p_1 + p_2)/2$$

For a gas, a compressible fluid, this is not true. Consider a pipeline AB shown in Figure 7-2. Using Equation 7-29 for gas flow, one can derive a simple formula to determine the pressure at any point C along the pipeline at a fractional distance x from the inlet end. For a flow rate q_{sc} in the pipe, the following two relationships are true (from Equation 7-29):

$$q_{sc} = 0.0775543 \left(\frac{T_{sc}}{p_{sc}}\right)\left[\frac{(p_1^2 - p_x^2)d^5}{\gamma_g Z_{av} TfLx}\right]^{0.5} \quad \longleftarrow L \text{ in miles} \quad (7\text{-}30)$$

Steady State Flow of Gas Through Pipes   289

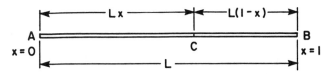

Figure 7-2. Schematic of a gas pipeline for determining the average pressure.

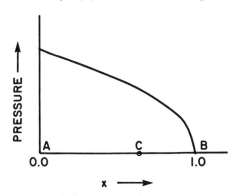

Figure 7-3. A plot of pressure versus distance in a pipeline carrying gas. (After Szilas, 1975.)

and,

$$q_{sc} = 0.0775543 \left|\frac{T_{sc}}{p_{sc}}\right|\left(\frac{(p_x^2 - p_2^2)d^5}{\gamma_g Z_{av} TfL(1 - x)}\right)^{0.5} \quad \text{L in miles} \quad (7\text{-}31)$$

Equating these Equations 7-30 and 7-31 (with the assumption that the difference in Z_{av} for the two pipe sections is negligible):

$$\frac{p_1^2 - p_x^2}{x} = \frac{p_x^2 - p_2^2}{(1 - x)}$$

Solving for p_x:

$$p_x = [p_1^2 - x(p_1^2 - p_2^2)]^{0.5}$$

This suggests a pressure profile as shown in Figure 7-3. The mean pressure is given by:

$$p_{av} = \int_0^1 p_x dx = \int_0^1 [p_1^2 - x(p_1^2 - p_2^2)]^{0.5} dx$$

$$= \frac{2}{3}\left(p_1 + \frac{p_2^2}{p_1 + p_2}\right)$$

290 Gas Production Engineering

Rearranging and multiplying both numerator and denominator by $(p_1 - p_2)$:

$$p_{av} = \frac{2}{3}\left(\frac{p_1^2 + p_1p_2 + p_2^2}{p_1 + p_2}\right)\left(\frac{p_1 - p_2}{p_1 - p_2}\right)$$

$$= \frac{2}{3}\left(\frac{p_1^3 - p_2^3}{p_1^2 - p_2^2}\right) \tag{7-32}$$

where p_{av} = average pressure, psia
p_1 = upstream pressure, psia
p_2 = downstream pressure, psia

Therefore, Z_{av} must be computed using Equation 7-32 for average pressure.

Non-Iterative Equations for Horizontal Gas Flow

Several equations for gas flow have been derived from Equation 7-29 assuming a friction factor relationship to avoid a trial and error computation. These equations differ only in the friction factor relationship assumed. As we have seen earlier, friction factors vary over a wide range with Reynolds number and pipe roughness. Thus, none of these equations is universally applicable. In gas field operations, engineers use the equation that best suits their conditions, or use their own modified versions.

Weymouth Equation

Weymouth proposed the following relationship for friction factor as a function of pipe diameter d in inches:

$$f = 0.032/d^{1/3} \tag{7-33}$$

Substituting for f from Equation 7-33 into Equation 7-29:

$$q_{sc} = 31.5027 \left(\frac{T_{sc}}{p_{sc}}\right)\left(\frac{(p_1^2 - p_2^2)d^{16/3}}{\gamma_g Z_{av} T_{av} L}\right)^{0.5} \tag{7-34}$$

This is known as the Weymouth equation for horizontal flow. It is used most often for designing gas transmission systems because it generally maximizes pipe diameter requirements for a given flow rate and pressure drop.

Steady State Flow of Gas Through Pipes   291

Panhandle (Panhandle A) Equation

This equation assumes that f is a function of Reynolds number only as follows:

$$f = 0.0768/N_{Re}^{0.1461} \tag{7-35}$$

Substituting for f from Equation 7-35 into Equation 7-29:

$$q_{sc} = \frac{5.6353821}{(0.0768)^{0.5}} \left(\frac{T_{sc}}{p_{sc}}\right) \left(\frac{p_1^2 - p_2^2}{Z_{av}T_{av}L}\right)^{0.5} \left(\frac{20 q_{sc} \gamma_g}{\mu_g d}\right)^{0.07305} \frac{d^{2.5}}{\gamma_g^{0.5}}$$

or,

$$q_{sc}^{0.92695} = 25.309468 \left(\frac{T_{sc}}{p_{sc}}\right) \left(\frac{p_1^2 - p_2^2}{Z_{av}T_{av}L}\right)^{0.5} \left(\frac{1}{\gamma_g}\right)^{0.42695} \frac{d^{2.42695}}{\mu_g^{0.07305}}$$

Thus:

$$q_{sc} = 32.6491 \left(\frac{T_{sc}}{p_{sc}}\right)^{1.07881} \left(\frac{p_1^2 - p_2^2}{Z_{av}T_{av}L}\right)^{0.53940} \left(\frac{1}{\gamma_g}\right)^{0.46060} \frac{d^{2.61821}}{\mu_g^{0.07881}} \tag{7-36}$$

The Panhandle A equation is most applicable to large diameter pipelines, at high flow rates.

Modified Panhandle (Panhandle B) Equation

One of the most widely used equations for long transmission lines, the Panhandle B equation assumes that f is a function of Reynolds number as follows:

$$f = 0.00359/N_{Re}^{0.03922} \tag{7-37}$$

The pipeline flow equation is thus given as follows:

$$q_{sc} = 109.364 \left(\frac{T_{sc}}{p_{sc}}\right)^{1.020} \left(\frac{p_1^2 - p_2^2}{Z_{av}T_{av}L}\right)^{0.510} \left(\frac{1}{\gamma_g}\right)^{0.490} \frac{d^{2.530}}{\mu_g^{0.020}} \tag{7-38}$$

The Panhandle B equation is most applicable to large diameter pipelines, at high values of Reynolds number.

A More Precise Equation for Horizontal Gas Flow: The Clinedinst Equation

The Clinedinst equation rigorously accounts for the deviation of natural gas from ideal behavior (an average gas compressibility factor, Z_{av}, is not used in this method), and the dependence of friction factor, f, on Reynolds number and pipe roughness, leading to a trial and error solution scheme. The integral $\int (p/Z)\, dp$ must be evaluated. In order to generalize this evaluation, it is converted into the following:

$$\int (p/Z)\, dp = \int [(p_{pc}p_r/Z)\, p_{pc}dp_r] = p_{pc}^2 \int (p_r/Z)\, dp_r$$

where p_{pc} = pseudocritical pressure of the flowing gas mixture, psia
p_r = reduced pressure ($p_r = p/p_{pc}$)

Values for the integral function $\int (p_r/Z)\, dp_r$ are given in Table 7-1. For this case, Equation 7-25 becomes:

$$p_{pc}^2 \left[\int_0^{p_{r1}} (p_r/Z)dp_r - \int_0^{p_{r2}} (p_r/Z)dp_r \right] = \frac{8fMTp_{sc}^2 q_{sc}^2 L}{R\pi^2 d^5 g_c T_{sc}^2}$$

Rearranging this equation, and using common units, with q_{sc} in Mscfd, p in psia, T in °R, d in in., L in ft, and substituting for R (= 10.732 psia-ft³/lb mole-°R) and g_c (= 32.17 lbm-ft/lbf-sec²):

$$q_{sc} = 7.969634 \left[\frac{p_{pc}T_{sc}}{p_{sc}} \right] \left[\frac{d^5}{\gamma_g T_{av} Lf} \right]^{0.5} \left[\int_0^{p_{r1}} (p_r/Z)dp_r - \int_0^{p_{r2}} (p_r/Z)dp_r \right]^{0.5} \quad (7\text{-}39)$$

This is known as the Clinedinst equation for horizontal flow.

Pipeline Efficiency

The pipeline equations developed here assume 100% efficient conditions. In practice, even for single-phase gas flow, some water or condensate may be present, which collects in low spots in the line over long periods of time. Some solids, such as pipe-scale and drilling muds, may also be present. To account for the reduction in pipeline capacity that will result from the presence of these materials, an efficiency factor E is generally used as a multiplying factor for the flow rate predicted by the flow equations. Only very rarely does a pipeline exhibit an E equal to 1.0; a pipeline with E greater than 0.9 is usually considered "clean." Typically, the following efficiency factors may be encountered (Ikoku, 1984):

(text continued on page 303)

Steady State Flow of Gas Through Pipes 293

Table 7-1. Values of $\int_0^{P_r} \frac{P_r}{Z} dP_r$

Multiply values in body of table by 10^{-2}; for example, read 00198 as 1.98

Pseudo-reduced pressure P_r	Pseudoreduced temperature T_r																		
	1.05	1.10	1.15	1.20	1.25	1.30	1.35	1.40	1.45	1.50	1.60	1.70	1.80	1.90	2.00	2.20	2.40	2.60	2.80	3.00
0.05	00000	00000	00000	00000	00000	00000	00000	00000	00000	00000	00000	00000	00000	00000	00000	00000	00000	00000	00000	00000
0.10	00000	00000	00000	00000	00000	00000	00000	00000	00000	00000	00000	00000	00000	00000	00000	00000	00000	00000	00000	00000
0.15	00001	00001	00001	00001	00001	00001	00001	00001	00001	00001	00001	00001	00001	00001	00001	00001	00001	00001	00001	00001
0.20	00001	00001	00001	00001	00001	00001	00001	00001	00001	00001	00001	00001	00001	00001	00001	00001	00001	00001	00001	00001
0.25	00003	00003	00003	00003	00003	00002	00002	00002	00002	00002	00002	00002	00002	00002	00002	00002	00002	00002	00002	00002
0.30	00004	00004	00004	00004	00004	00004	00004	00004	00004	00004	00004	00004	00004	00004	00004	00004	00004	00004	00004	00004
0.35	00006	00006	00006	00006	00006	00006	00006	00005	00005	00005	00005	00005	00005	00005	00005	00005	00005	00005	00005	00005
0.40	00008	00008	00008	00008	00008	00007	00007	00007	00007	00007	00007	00007	00007	00007	00007	00007	00007	00007	00007	00007
0.45	00010	00010	00010	00010	00010	00010	00010	00010	00010	00010	00009	00009	00009	00009	00009	00009	00009	00009	00009	00009
0.50	00013	00013	00013	00013	00012	00012	00012	00012	00012	00012	00012	00012	00012	00012	00012	00012	00012	00012	00012	00012
0.55	00016	00016	00016	00015	00015	00015	00015	00015	00015	00015	00015	00014	00014	00014	00014	00014	00014	00014	00014	00014
0.60	00020	00019	00019	00019	00019	00018	00018	00018	00018	00018	00018	00017	00017	00017	00017	00017	00017	00017	00017	00017
0.65	00024	00023	00023	00022	00022	00021	00021	00021	00021	00021	00021	00021	00020	00020	00020	00020	00020	00020	00020	00020
0.70	00028	00027	00027	00026	00026	00025	00025	00025	00025	00025	00024	00024	00024	00024	00024	00024	00023	00023	00023	00023
0.75	00033	00032	00031	00030	00030	00029	00029	00029	00029	00028	00028	00028	00028	00028	00027	00027	00027	00027	00027	00027
0.80	00039	00037	00036	00035	00034	00034	00033	00033	00033	00033	00032	00032	00032	00031	00031	00031	00031	00031	00031	00031
0.85	00045	00042	00041	00039	00039	00038	00038	00038	00037	00037	00037	00036	00036	00036	00035	00035	00035	00035	00035	00035
0.90	00051	00048	00047	00045	00044	00043	00043	00042	00042	00042	00041	00041	00040	00040	00040	00040	00039	00039	00039	00039
0.95	00059	00055	00053	00051	00050	00049	00048	00048	00047	00047	00046	00045	00045	00045	00045	00044	00044	00044	00044	00044
1.00	00067	00062	00059	00057	00056	00055	00054	00053	00053	00052	00051	00051	00050	00050	00050	00049	00049	00049	00049	00048
1.05	00075	00070	00066	00064	00062	00061	00060	00059	00058	00058	00057	00056	00055	00055	00055	00054	00054	00054	00054	00053
1.10	00085	00078	00074	00071	00069	00067	00066	00065	00064	00064	00063	00062	00061	00061	00060	00060	00059	00059	00059	00059
1.15	00097	00087	00082	00078	00076	00074	00073	00072	00071	00070	00069	00068	00067	00067	00066	00065	00065	00065	00065	00064
1.20	00109	00097	00090	00086	00083	00081	00080	00078	00077	00077	00075	00074	00073	00073	00072	00071	00071	00070	00070	00070
1.25	00124	00108	00099	00095	00091	00089	00087	00086	00084	00084	00082	00080	00080	00079	00078	00078	00077	00076	00076	00076

(table continued)

Table 7-1. (Continued) Values of $\int_0^{P_r} \frac{P_r}{Z} dP_r$

Multiply values in body of table by 10^{-2}; for example, read 00198 as 1.98

Pseudo-reduced pressure P_r	Pseudoreduced temperature T_r																			
	1.05	1.10	1.15	1.20	1.25	1.30	1.35	1.40	1.45	1.50	1.60	1.70	1.80	1.90	2.00	2.20	2.40	2.60	2.80	3.00
1.30	00141	00120	00109	00104	00100	00097	00095	00093	00092	00091	00089	00087	00086	00086	00085	00084	00084	00083	00083	00082
1.35	00161	00132	00120	00113	00109	00105	00103	00101	00099	00098	00096	00094	00093	00093	00092	00091	00090	00089	00089	00089
1.40	00185	00146	00131	00123	00118	00114	00112	00109	00107	00106	00104	00102	00101	00100	00099	00098	00097	00096	00096	00096
1.45	00212	00161	00143	00133	00128	00123	00121	00118	00116	00114	00112	00109	00108	00107	00106	00105	00104	00103	00103	00103
1.50	00241	00178	00155	00145	00138	00133	00130	00127	00124	00123	00120	00117	00116	00115	00114	00113	00112	00111	00111	00110
1.55	00271	00196	00169	00156	00149	00143	00139	00136	00133	00132	00128	00126	00124	00123	00122	00120	00119	00118	00118	00117
1.60	00302	00216	00183	00169	00160	00153	00149	00146	00143	00141	00137	00134	00132	00131	00130	00128	00127	00126	00126	00125
1.65	00334	00237	00198	00181	00172	00164	00160	00156	00153	00150	00146	00143	00141	00140	00138	00137	00136	00134	00134	00133
1.70	00368	00259	00214	00195	00184	00176	00171	00166	00163	00160	00156	00152	00150	00149	00147	00145	00144	00143	00142	00141
1.75	00402	00282	00231	00209	00197	00188	00182	00177	00173	00170	00166	00162	00159	00158	00156	00154	00153	00151	00151	00150
1.80	00436	00305	00249	00224	00210	00200	00193	00188	00184	00181	00176	00172	00169	00167	00165	00163	00162	00160	00160	00159
1.85	00471	00330	00267	00239	00224	00213	00206	00200	00195	00192	00186	00182	00179	00177	00175	00173	00171	00169	00169	00168
1.90	00506	00355	00286	00256	00239	00226	00218	00212	00207	00203	00197	00192	00189	00187	00185	00182	00180	00179	00178	00177
1.95	00541	00381	00307	00273	00254	00240	00231	00224	00219	00215	00208	00203	00199	00197	00195	00192	00190	00188	00187	00186
2.00	00576	00408	00327	00290	00270	00254	00244	00237	00231	00227	00219	00214	00210	00208	00205	00202	00200	00198	00197	00196
2.05	00612	00435	00349	00308	00286	00269	00258	00250	00244	00239	00231	00225	00221	00218	00216	00213	00211	00208	00207	00206
2.10	00648	00463	00372	00328	00303	00284	00272	00264	00257	00251	00243	00237	00232	00229	00227	00223	00221	00219	00218	00216
2.15	00684	00492	00395	00347	00320	00300	00287	00278	00270	00264	00256	00249	00244	00241	00238	00234	00232	00229	00228	00226
2.20	00720	00521	00419	00368	00338	00316	00302	00292	00284	00278	00268	00261	00256	00252	00250	00245	00243	00240	00239	00237
2.25	00756	00550	00444	00389	00356	00333	00318	00307	00298	00291	00281	00273	00268	00264	00261	00257	00254	00251	00250	00248
2.30	00792	00580	00469	00411	00375	00350	00334	00323	00312	00306	00295	00286	00281	00277	00273	00269	00266	00263	00261	00259
2.35	00828	00610	00495	00433	00395	00368	00351	00338	00327	00320	00308	00299	00293	00289	00286	00281	00278	00274	00272	00271
2.40	00865	00641	00521	00456	00415	00387	00368	00354	00343	00335	00322	00313	00306	00302	00298	00293	00290	00286	00284	00282
2.45	00901	00672	00548	00479	00436	00405	00385	00371	00358	00350	00337	00327	00320	00315	00311	00305	00302	00298	00296	00294
2.50	00937	00704	00576	00503	00457	00425	00403	00388	00374	00365	00351	00341	00333	00328	00324	00318	00315	00311	00308	00306
2.55	00974	00736	00603	00527	00479	00444	00421	00405	00391	00381	00366	00355	00347	00342	00337	00331	00327	00323	00321	00318
2.60	01010	00768	00632	00552	00501	00465	00440	00423	00408	00397	00382	00370	00361	00356	00351	00345	00340	00336	00334	00331
2.65	01047	00800	00660	00577	00523	00485	00459	00441	00425	00414	00397	00385	00376	00370	00365	00358	00354	00349	00347	00344
2.70	01083	00833	00689	00602	00546	00507	00479	00460	00443	00431	00413	00400	00391	00384	00379	00372	00367	00363	00360	00357

Steady State Flow of Gas Through Pipes   295

	0	1	2	3	4	5	6	7	8	9	10	11	12	13	14	15	16	17	18	19
2.80	01157	00899	00748	00655	00594	00550	00519	00498	00479	00466	00446	00432	00421	00414	00408	00401	00395	00390	00387	00384
2.85	01193	00933	00778	00682	00618	00573	00540	00518	00498	00484	00463	00448	00437	00429	00423	00415	00410	00404	00401	00398
2.90	01230	00966	00808	00709	00643	00596	00562	00538	00517	00502	00480	00464	00453	00445	00439	00430	00419	00415	00412
2.95	01267	01000	00838	00736	00668	00619	00584	00559	00536	00521	00498	00481	00469	00461	00454	00445	00439	00433	00429	00426
3.00	01304	01034	00868	00764	00694	00643	00606	00580	00556	00540	00516	00498	00486	00477	00470	00461	00454	00448	00444	00440
3.05	01340	01068	00899	00792	00720	00667	00628	00601	00577	00559	00534	00516	00502	00493	00486	00476	00470	00463	00459	00455
3.10	01377	01103	00930	00821	00746	00692	00651	00623	00597	00579	00553	00533	00519	00510	00502	00492	00485	00479	00474	00470
3.15	01414	01137	00961	00850	00773	00717	00675	00645	00618	00599	00572	00552	00537	00527	00519	00508	00501	00494	00489	00485
3.20	01451	01171	00992	00879	00800	00742	00698	00667	00640	00620	00591	00570	00555	00544	00536	00525	00517	00510	00505	00501
3.25	01488	01206	01024	00908	00827	00768	00722	00690	00661	00641	00611	00589	00573	00562	00553	00542	00534	00526	00521	00516
3.30	01525	01241	01056	00938	00855	00794	00747	00713	00683	00662	00631	00608	00591	00580	00571	00559	00550	00543	00537	00532
3.35	01562	01275	01088	00968	00883	00820	00772	00737	00706	00683	00651	00627	00609	00598	00588	00576	00567	00559	00553	00548
3.40	01599	01310	01120	00998	00911	00847	00797	00761	00729	00705	00671	00647	00628	00616	00606	00593	00584	00576	00570	00565
3.45	01636	01345	01152	01029	00940	00874	00822	00785	00752	00727	00692	00666	00647	00635	00624	00611	00602	00593	00587	00581
3.50	01674	01380	01185	01059	00969	00901	00848	00810	00775	00750	00714	00687	00667	00654	00643	00629	00619	00611	00604	00598
3.55	01711	01415	01218	01090	00998	00929	00874	00835	00799	00772	00735	00707	00687	00673	00662	00647	00637	00628	00621	00615
3.60	01748	01450	01251	01122	01027	00957	00901	00860	00823	00796	00757	00728	00707	00692	00681	00666	00655	00646	00639	00632
3.65	01785	01485	01284	01153	01057	00985	00928	00886	00848	00819	00779	00749	00727	00712	00700	00685	00674	00664	00656	00650
3.70	01822	01520	01317	01185	01087	01014	00955	00912	00872	00843	00802	00770	00747	00732	00720	00704	00692	00682	00674	00668
3.75	01859	01556	01351	01217	01117	01043	00982	00938	00897	00867	00824	00792	00768	00753	00739	00723	00711	00701	00692	00686
3.80	01897	01591	01384	01249	01148	01072	01010	00965	00923	00891	00847	00814	00789	00773	00760	00742	00730	00720	00711	00704
3.85	01934	01627	01418	01281	01178	01102	01038	00991	00949	00916	00871	00836	00811	00794	00780	00762	00750	00739	00730	00722
3.90	01971	01662	01452	01313	01209	01132	01066	01019	00975	00941	00895	00859	00833	00815	00801	00782	00769	00758	00749	00741
3.95	02009	01698	01486	01346	01240	01162	01095	01046	01001	00966	00919	00882	00855	00836	00821	00802	00789	00777	00768	00760
4.00	02046	01734	01520	01379	01272	01192	01124	01074	01027	00992	00943	00905	00877	00858	00843	00823	00809	00797	00787	00779
4.05	02083	01770	01555	01412	01304	01223	01153	01102	01054	01018	00967	00929	00899	00880	00864	00844	00829	00817	00807	00799
4.10	02121	01805	01589	01445	01335	01254	01183	01130	01082	01044	00992	00952	00922	00902	00886	00865	00850	00837	00827	00818
4.15	02158	01841	01624	01478	01368	01285	01212	01159	01109	01070	01017	00976	00945	00925	00907	00886	00870	00858	00847	00838
4.20	02196	01878	01659	01512	01400	01317	01242	01188	01137	01097	01043	01001	00968	00947	00930	00907	00891	00878	00867	00858
4.25	02233	01914	01694	01545	01432	01348	01272	01217	01165	01124	01068	01025	00992	00970	00952	00929	00913	00899	00888	00879
4.30	02271	01950	01729	01579	01465	01380	01303	01246	01193	01151	01094	01050	01016	00993	00975	00951	00934	00920	00908	00899
4.35	02308	01986	01764	01613	01498	01412	01334	01276	01222	01179	01120	01075	01040	01017	00998	00973	00956	00941	00929	00920
4.40	02346	02022	01799	01648	01531	01445	01365	01306	01250	01207	01147	01100	01064	01040	01021	00995	00978	00963	00951	00941
4.45	02383	02059	01835	01682	01564	01477	01396	01336	01280	01235	01173	01126	01089	01064	01044	01018	01000	00985	00972	00962
4.50	02421	02095	01870	01716	01598	01510	01427	01367	01309	01263	01200	01152	01113	01089	01068	01041	01022	01007	00994	00984

(table continued)

Table 7-1. (Continued) Values of $\int_0^{P_r} \frac{P_r}{Z} dP_r$

Multiply values in body of table by 10^{-2}; for example, read 00198 as 1.98

Pseudoreduced temperature T_r

Pseudo-reduced pressure P_r	1.05	1.10	1.15	1.20	1.25	1.30	1.35	1.40	1.45	1.50	1.60	1.70	1.80	1.90	2.00	2.20	2.40	2.60	2.80	3.00
4.55	02458	02132	01906	01751	01631	01543	01459	01397	01338	01291	01228	01178	01139	01113	01092	01064	01045	01029	01016	01005
4.60	02496	02168	01942	01786	01665	01576	01491	01428	01368	01320	01255	01204	01164	01138	01116	01088	01067	01051	01038	01027
4.65	02533	02205	01978	01821	01699	01609	01523	01460	01398	01349	01283	01231	01189	01163	01140	01111	01090	01074	01060	01049
4.70	02571	02242	02014	01856	01733	01643	01555	01491	01429	01379	01311	01258	01215	01188	01165	01135	01114	01097	01083	01072
4.75	02609	02279	02050	01891	01768	01677	01588	01523	01459	01408	01339	01285	01241	01213	01190	01159	01137	01120	01105	01094
4.80	02646	02316	02086	01926	01802	01710	01621	01555	01490	01438	01368	01312	01268	01239	01215	01183	01161	01143	01128	01117
4.85	02684	02352	02122	01962	01837	01745	01654	01587	01521	01468	01396	01340	01294	01265	01240	01208	01185	01167	01152	01140
4.90	02722	02389	02159	01998	01872	01779	01687	01619	01552	01498	01425	01368	01321	01291	01266	01232	01209	01191	01175	01163
4.95	02759	02427	02195	02033	01907	01813	01720	01652	01583	01529	01454	01396	01348	01317	01292	01257	01233	01215	01199	01187
5.00	02797	02464	02232	02069	01942	01848	01754	01685	01615	01559	01484	01424	01375	01344	01318	01283	01258	01239	01223	01210
5.05	02835	02501	02268	02105	01977	01883	01787	01717	01647	01590	01513	01453	01403	01371	01344	01308	01283	01263	01247	01234
5.10	02873	02538	02305	02141	02012	01918	01821	01751	01679	01621	01543	01481	01430	01398	01370	01334	01308	01288	01271	01258
5.15	02910	02575	02342	02178	02048	01953	01855	01784	01711	01653	01573	01510	01458	01425	01397	01350	01333	01313	01296	01283
5.20	02948	02613	02379	02214	02084	01988	01890	01817	01743	01684	01604	01540	01486	01453	01424	01385	01358	01338	01320	01307
5.25	02986	02650	02416	02251	02120	02023	01924	01851	01776	01716	01634	01569	01515	01480	01451	01412	01384	01363	01345	01332
5.30	03024	02688	02453	02287	02156	02059	01959	01885	01809	01748	01665	01599	01543	01508	01478	01438	01410	01388	01371	01357
5.35	03061	02725	02490	02324	02192	02095	01994	01919	01842	01780	01696	01629	01572	01536	01506	01465	01436	01414	01396	01382
5.40	03099	02763	02527	02361	02228	02130	02029	01953	01875	01813	01727	01659	01601	01565	01534	01492	01462	01440	01422	01408
5.45	03137	02800	02565	02398	02264	02166	02064	01988	01908	01845	01758	01689	01630	01593	01562	01519	01489	01466	01447	01433
5.50	03175	02838	02602	02435	02301	02203	02099	02022	01942	01878	01790	01720	01660	01622	01590	01546	01515	01492	01474	01459
5.55	03213	02876	02639	02472	02338	02239	02135	02057	01976	01911	01821	01750	01690	01651	01618	01574	01542	01519	01500	01485
5.60	03250	02913	02677	02509	02374	02275	02170	02092	02010	01944	01853	01781	01720	01680	01647	01602	01569	01546	01526	01512
5.65	03288	02951	02715	02546	02411	02312	02206	02127	02044	01978	01886	01812	01750	01710	01676	01630	01597	01573	01553	01538
5.70	03326	02989	02752	02584	02448	02348	02242	02162	02078	02011	01918	01844	01780	01740	01705	01658	01624	01600	01580	01565
5.75	03364	03027	02790	02621	02485	02385	02278	02198	02113	02045	01951	01875	01810	01770	01734	01686	01652	01627	01607	01592
5.80	03402	03065	02828	02659	02522	02422	02314	02233	02147	02079	01983	01907	01841	01800	01764	01715	01680	01655	01634	01619
5.85	03440	03103	02866	02697	02560	02459	02350	02269	02182	02113	02016	01939	01872	01830	01794	01744	01708	01683	01662	01646
5.90	03478	03141	02904	02734	02597	02496	02387	02305	02217	02147	02049	01971	01903	01860	01824	01773	01737	01711	01690	01674
5.95	03516	03179	02942	02772	02635	02533	02423	02341	02253	02182	02083	02004	01935	01891	01854	01802	01765	01739	01718	01702
														01922	01884	01831	01794	01767	01746	01730

Steady State Flow of Gas Through Pipes   297

6.05	03592	03256	03018	02848	02710	02608	02497	02413	02323	02251	02150	02069	01998	01953	01914	01861	01823	01796	01774	01758
6.10	03630	03294	03056	02886	02748	02645	02534	02450	02359	02286	02184	02102	02030	01984	01945	01891	01852	01825	01803	01786
6.15	03668	03332	03094	02924	02786	02683	02571	02486	02395	02321	02218	02135	02062	02016	01976	01921	01882	01854	01832	01815
6.20	03706	03371	03133	02963	02824	02721	02608	02523	02431	02357	02252	02168	02095	02048	02007	01951	01911	01883	01861	01844
6.25	03745	03409	03171	03001	02862	02758	02646	02560	02467	02392	02286	02202	02127	02080	02038	01982	01941	01912	01890	01873
6.30	03783	03447	03210	03039	02900	02796	02683	02597	02503	02428	02321	02235	02160	02112	02070	02012	01971	01942	01919	01902
6.35	03821	03486	03248	03078	02938	02834	02721	02634	02540	02464	02356	02269	02193	02144	02102	02043	02001	01972	01949	01931
6.40	03860	03524	03287	03116	02977	02872	02758	02671	02577	02500	02391	02303	02226	02176	02134	02074	02032	02002	01979	01961
6.45	03898	03563	03325	03155	03015	02911	02796	02708	02613	02536	02426	02338	02259	02209	02166	02106	02062	02032	02009	01991
6.50	03936	03601	03364	03193	03053	02949	02834	02746	02650	02572	02461	02372	02293	02242	02198	02137	02093	02063	02039	02021
6.55	03974	03640	03403	03232	03092	02987	02872	02783	02687	02609	02497	02407	02326	02275	02230	02169	02124	02094	02069	02051
6.60	04013	03679	03441	03271	03131	03026	02910	02821	02724	02645	02532	02441	02360	02308	02263	02201	02155	02124	02100	02082
6.65	04051	03717	03480	03309	03170	03064	02948	02859	02762	02682	02568	02476	02394	02342	02296	02233	02187	02156	02131	02113
6.70	04090	03756	03519	03348	03208	03103	02987	02897	02799	02719	02604	02511	02429	02375	02329	02265	02218	02187	02162	02144
6.75	04128	03795	03558	03387	03247	03142	03025	02935	02837	02756	02640	02547	02463	02409	02362	02297	02250	02218	02193	02175
6.80	04167	03834	03597	03426	03286	03181	03064	02973	02875	02793	02677	02582	02498	02443	02395	02330	02282	02250	02224	02206
6.85	04205	03873	03636	03465	03325	03220	03102	03012	02912	02831	02713	02618	02532	02477	02429	02363	02314	02282	02256	02237
6.90	04244	03912	03675	03504	03364	03259	03141	03050	02950	02868	02750	02654	02567	02512	02463	02396	02347	02314	02288	02269
6.95	04283	03951	03714	03543	03404	03298	03180	03089	02988	02906	02786	02690	02603	02546	02497	02429	02379	02346	02320	02301
7.00	04321	03990	03753	03582	03443	03337	03219	03127	03027	02944	02823	02726	02638	02581	02531	02462	02412	02379	02352	02333
7.05	04360	04029	03793	03622	03482	03377	03258	03166	03065	02982	02861	02762	02673	02616	02565	02496	02445	02411	02384	02365
7.10	04399	04068	03832	03661	03522	03416	03297	03205	03104	03020	02898	02799	02709	02651	02600	02530	02478	02444	02417	02398
7.15	04437	04107	03871	03701	03561	03456	03336	03244	03142	03058	02935	02835	02745	02686	02635	02564	02512	02477	02450	02431
7.20	04476	04146	03911	03740	03601	03495	03376	03283	03181	03097	02973	02872	02781	02722	02669	02598	02545	02510	02483	02464
7.25	04515	04185	03950	03780	03640	03535	03415	03322	03220	03135	03010	02909	02817	02757	02705	02632	02579	02544	02516	02497
7.30	04554	04224	03989	03819	03680	03575	03455	03362	03259	03174	03048	02946	02853	02793	02740	02667	02613	02577	02549	02530
7.35	04593	04264	04029	03859	03720	03615	03495	03401	03298	03212	03086	02983	02890	02829	02775	02701	02647	02611	02583	02563
7.40	04632	04303	04069	03899	03760	03655	03534	03441	03337	03251	03124	03021	02926	02865	02811	02736	02682	02645	02617	02597
7.45	04671	04342	04108	03938	03800	03695	03574	03481	03376	03290	03163	03058	02963	02901	02847	02771	02716	02679	02651	02631
7.50	04710	04382	04148	03978	03840	03735	03614	03520	03416	03329	03201	03096	03000	02938	02883	02807	02751	02714	02685	02665
7.55	04749	04421	04188	04018	03880	03775	03654	03560	03455	03368	03240	03134	03037	02974	02919	02842	02786	02748	02719	02699
7.60	04788	04461	04227	04058	03920	03815	03695	03600	03495	03408	03278	03172	03075	03011	02955	02878	02821	02783	02754	02733
7.65	04826	04500	04267	04098	03960	03856	03735	03640	03535	03447	03317	03210	03112	03048	02992	02913	02856	02818	02789	02768
7.70	04865	04540	04307	04138	04000	03896	03775	03681	03575	03487	03356	03248	03150	03085	03028	02949	02891	02853	02823	02803
7.75	04904	04579	04347	04178	04041	03937	03816	03721	03615	03527	03395	03287	03188	03122	03065	02985	02927	02888	02859	02838

(table continued)

298 Gas Production Engineering

Table 7-1. (Continued) Values of $\int_0^{P_r} \frac{P_r}{z} dP_r$

Multiply values in body of table by 10^{-2}; for example, read 00198 as 1.98

Pseudo-reduced pressure P_r	Pseudoreduced temperature T_r																			
	1.05	1.10	1.15	1.20	1.25	1.30	1.35	1.40	1.45	1.50	1.60	1.70	1.80	1.90	2.00	2.20	2.40	2.60	2.80	3.00
7.80	04944	04619	04387	04219	04081	03977	03856	03761	03655	03567	03435	03326	03226	03160	03102	03022	02963	02924	02894	02873
7.85	04983	04659	04427	04259	04122	04018	03897	03802	03695	03607	03474	03364	03264	03198	03139	03058	02999	02960	02929	02908
7.90	05022	04698	04467	04299	04162	04059	03937	03842	03736	03647	03513	03403	03302	03235	03176	03095	03035	02995	02965	02944
7.95	05061	04738	04507	04340	04203	04100	03978	03883	03776	03687	03553	03442	03340	03273	03214	03132	03071	03031	03001	02979
8.00	05101	04778	04548	04380	04244	04140	04019	03924	03817	03727	03593	03482	03379	03311	03251	03169	03108	03068	03037	03015
8.05	05140	04818	04588	04421	04285	04181	04060	03964	03857	03768	03633	03521	03418	03350	03289	03206	03145	03104	03073	03051
8.10	05179	04857	04628	04461	04326	04222	04101	04005	03898	03808	03673	03560	03457	03388	03327	03243	03182	03141	03110	03088
8.15	05219	04897	04668	04502	04367	04263	04142	04046	03939	03849	03713	03600	03496	03427	03365	03281	03219	03177	03146	03125
8.20	05258	04937	04708	04542	04408	04304	04184	04087	03980	03889	03754	03640	03535	03465	03404	03318	03256	03214	03183	03161
8.25	05298	04977	04749	04583	04449	04346	04225	04129	04021	03930	03794	03680	03574	03504	03442	03356	03293	03252	03220	03198
8.30	05337	05017	04789	04624	04490	04387	04266	04170	04062	03971	03835	03720	03614	03543	03481	03394	03331	03289	03257	03235
8.35	05377	05057	04830	04664	04531	04428	04308	04211	04103	04012	03875	03760	03653	03582	03520	03433	03369	03326	03294	03272
8.40	05416	05097	04870	04705	04573	04470	04349	04253	04144	04054	03916	03800	03693	03622	03559	03471	03407	03364	03332	03309
8.45	05456	05137	04911	04746	04614	04511	04391	04294	04186	04095	03957	03841	03733	03661	03598	03509	03445	03402	03369	03347
8.50	05495	05177	04951	04787	04656	04553	04433	04336	04227	04136	03998	03881	03773	03701	03637	03548	03483	03440	03407	03384
8.55	05535	05217	04992	04828	04697	04594	04474	04378	04269	04178	04039	03922	03813	03741	03677	03587	03522	03478	03445	03422
8.60	05575	05257	05032	04869	04739	04636	04516	04419	04310	04219	04081	03963	03854	03780	03716	03626	03560	03516	03483	03460
8.65	05614	05298	05073	04910	04780	04678	04558	04461	04352	04261	04122	04004	03894	03821	03756	03665	03599	03555	03522	03499
8.70	05654	05338	05114	04952	04822	04719	04600	04503	04394	04303	04163	04045	03935	03861	03796	03705	03638	03594	03560	03537
8.75	05694	05378	05155	04993	04863	04761	04642	04545	04436	04345	04205	04086	03976	03901	03836	03744	03677	03633	03599	03576
8.80	05734	05419	05195	05034	04905	04803	04684	04587	04478	04387	04247	04128	04017	03942	03876	03784	03717	03672	03638	03614
8.85	05774	05459	05236	05076	04947	04845	04726	04629	04520	04429	04289	04169	04058	03982	03917	03824	03756	03711	03677	03653
8.90	05814	05499	05277	05117	04989	04887	04768	04671	04562	04471	04331	04211	04099	04023	03957	03864	03796	03750	03716	03692
8.95	05854	05540	05318	05159	05030	04929	04810	04714	04604	04513	04373	04252	04140	04064	03998	03904	03836	03790	03756	03732
9.00	05894	05580	05359	05200	05072	04972	04852	04756	04647	04556	04415	04294	04181	04105	04039	03944	03876	03830	03795	03771
9.05	05934	05621	05400	05242	05114	05014	04895	04798	04689	04598	04457	04336	04223	04146	04079	03985	03916	03869	03835	03811
9.10	05974	05662	05441	05283	05156	05056	04937	04841	04732	04641	04500	04378	04265	04188	04121	04025	03956	03910	03875	03850
9.15	06014	05702	05483	05325	05198	05099	04980	04884	04774	04683	04542	04421	04307	04229	04162	04066	03996	03950	03915	03890
9.20	06054	05743	05524	05367	05241	05141	05023	04926	04817	04726	04585	04463	04348	04271	04203	04107	04037	03990	03955	03930
															04245	04148	04078	04031	03995	03971

Steady State Flow of Gas Through Pipes   299

9.30	06134	05825	05606	05450	05325	05226	05108	05012	04903	04812	04670	04548	04433	04355	04286	04189	04119	04071	04036	04011	
9.35	06174	05865	05648	05492	05367	05269	05151	05055	04946	04855	04713	04591	04475	04397	04328	04230	04160	04112	04077	04052	
9.40	06215	05906	05689	05534	05410	05311	05194	05098	04989	04898	04756	04633	04518	04439	04370	04272	04201	04153	04118	04093	
9.45	06255	05947	05731	05576	05452	05354	05237	05141	05032	04941	04799	04676	04560	04481	04412	04314	04242	04195	04159	04133	
9.50	06295	05988	05772	05618	05494	05397	05280	05184	05075	04984	04843	04719	04603	04524	04454	04355	04284	04236	04200	04175	
9.55	06335	06029	05814	05660	05537	05440	05323	05227	05118	05028	04886	04762	04646	04566	04497	04397	04325	04277	04241	04216	
9.60	06376	06070	05855	05702	05579	05483	05366	05270	05162	05071	04929	04806	04689	04609	04539	04439	04367	04319	04283	04257	
9.65	06416	06111	05897	05744	05622	05526	05409	05314	05205	05115	04973	04849	04732	04652	04582	04481	04409	04361	04324	04299	
9.70	06456	06152	05938	05786	05665	05569	05452	05357	05249	05158	05016	04893	04775	04695	04625	04524	04451	04403	04366	04341	
9.75	06497	06193	05980	05829	05707	05612	05495	05400	05292	05202	05060	04936	04818	04738	04668	04566	04493	04445	04408	04383	
9.80	06537	06234	06022	05871	05750	05655	05539	05444	05336	05246	05104	04980	04862	04781	04711	04609	04536	04487	04450	04425	
9.85	06578	06276	06063	05913	05793	05698	05582	05488	05380	05290	05148	05024	04905	04824	04754	04652	04578	04530	04493	04467	
9.90	06618	06317	06105	05956	05835	05741	05626	05531	05424	05334	05192	05068	04949	04868	04797	04694	04621	04572	04535	04509	
9.95	06659	06358	06147	05998	05878	05784	05669	05575	05467	05378	05236	05112	04993	04912	04841	04737	04664	04615	04578	04552	
10.00	06699	06399	06189	06040	05921	05828	05713	05619	05511	05422	05280	05156	05037	04955	04884	04780	04707	04658	04621	04595	
10.05	06740	06441	06231	06083	05964	05871	05756	05662	05556	05466	05324	05200	05081	04999	04928	04824	04750	04701	04664	04637	
10.10	06780	06482	06273	06126	06007	05915	05800	05706	05600	05510	05369	05244	05125	05043	04972	04867	04793	04744	04707	04680	
10.15	06821	06523	06315	06168	06050	05958	05844	05750	05644	05555	05413	05289	05169	05087	05016	04911	04837	04787	04750	04724	
10.20	06862	06565	06357	06211	06093	06002	05888	05794	05688	05599	05457	05333	05214	05131	05060	04954	04880	04831	04793	04767	
10.25	06902	06606	06399	06253	06136	06045	05932	05839	05733	05644	05502	05378	05258	05176	05104	04998	04924	04875	04837	04811	
10.30	06943	06648	06441	06296	06179	06089	05976	05883	05777	05688	05547	05422	05303	05220	05148	05042	04968	04918	04881	04854	
10.35	06984	06689	06483	06339	06223	06133	06020	05927	05821	05733	05592	05467	05347	05265	05193	05086	05012	04962	04925	04898	
10.40	07025	06731	06525	06382	06266	06176	06064	05971	05866	05778	05636	05512	05392	05310	05237	05131	05056	05006	04969	04942	
10.45	07065	06772	06567	06425	06309	06220	06108	06016	05911	05823	05681	05557	05437	05354	05282	05175	05100	05051	05013	04986	
10.50	07106	06814	06610	06467	06353	06264	06152	06060	05955	05868	05727	05602	05482	05399	05327	05219	05145	05095	05057	05030	
10.55	07147	06855	06652	06510	06396	06308	06196	06105	06000	05912	05772	05647	05527	05444	05372	05264	05189	05140	05102	05075	
10.60	07188	06897	06694	06553	06440	06352	06240	06149	06045	05957	05817	05693	05573	05490	05417	05309	05234	05184	05146	05119	
10.65	07229	06939	06736	06596	06483	06396	06285	06194	06090	06003	05862	05738	05618	05535	05462	05354	05279	05229	05191	05164	
10.70	07270	06980	06779	06639	06527	06440	06329	06239	06135	06048	05908	05784	05664	05580	05507	05399	05324	05274	05236	05209	
10.75	07310	07022	06821	06682	06570	06484	06374	06283	06180	06093	05953	05829	05709	05626	05553	05444	05369	05319	05281	05254	
10.80	07351	07064	06864	06726	06614	06528	06418	06328	06225	06138	05999	05875	05755	05671	05598	05489	05414	05364	05326	05299	
10.85	07392	07106	06906	06769	06658	06572	06463	06373	06270	06184	06045	05921	05801	05717	05644	05535	05459	05410	05372	05344	
10.90	07433	07148	06949	06812	06701	06616	06507	06418	06316	06229	06090	05967	05847	05763	05690	05580	05505	05455	05417	05390	
10.95	07474	07190	06991	06855	06745	06661	06552	06463	06361	06275	06136	06013	05893	05809	05735	05626	05550	05501	05463	05435	
11.00	07515	07231	07034	06898	06789	06705	06597	06508	06406	06320	06182	06059	05939	05855	05781	05672	05597	05547	05509	05481	

(table continued)

Table 7-1. (Continued) Values of $\int_0^{P_r} \frac{P_r}{Z} dP_r$

Multiply values in body of table by 10^{-2}; for example, read 00198 as 1.98

Pseudo-reduced pressure P_r	\multicolumn{17}{c	}{Pseudoreduced temperature T_r}
	1.05	1.10	1.15	1.20	1.25	1.30	1.35	1.40	1.45	1.50	1.60	1.70	1.80	1.90	2.00	2.20	2.40	2.60	2.80	3.00
11.05	07556	07273	07077	06942	06833	06749	06642	06553	06452	06366	06228	06105	05985	05901	05827	05718	05643	05593	05555	05527
11.10	07597	07315	07119	06985	06877	06794	06686	06599	06497	06412	06274	06151	06031	05947	05873	05764	05689	05639	05601	05573
11.15	07638	07357	07162	07028	06921	06838	06731	06644	06543	06458	06320	06197	06078	05993	05920	05810	05735	05685	05647	05619
11.20	07680	07399	07205	07072	06965	06882	06776	06689	06589	06503	06366	06244	06124	06040	05966	05857	05781	05731	05693	05666
11.25	07721	07441	07247	07115	07009	06927	06821	06735	06634	06549	06413	06290	06171	06086	06013	05903	05828	05778	05740	05712
11.30	07762	07483	07290	07158	07053	06972	06866	06780	06680	06595	06459	06337	06217	06133	06059	05950	05874	05825	05786	05759
11.35	07803	07525	07333	07202	07097	07016	06911	06826	06726	06642	06505	06383	06264	06180	06106	05996	05921	05871	05833	05806
11.40	07844	07568	07376	07245	07141	07061	06956	06871	06772	06688	06552	06430	06311	06227	06153	06043	05968	05918	05880	05853
11.45	07885	07610	07419	07289	07185	07106	07002	06917	06818	06734	06599	06477	06358	06274	06200	06090	06015	05965	05927	05900
11.50	07927	07652	07462	07332	07229	07150	07047	06962	06864	06780	06645	06524	06405	06321	06247	06137	06062	06012	05974	05947
11.55	07968	07694	07505	07376	07273	07195	07092	07008	06910	06827	06692	06571	06452	06368	06294	06185	06109	06060	06022	05994
11.60	08009	07736	07548	07420	07317	07240	07137	07054	06956	06873	06739	06618	06499	06415	06341	06232	06157	06107	06069	06042
11.65	08051	07779	07591	07463	07362	07285	07183	07100	07002	06919	06785	06665	06547	06463	06389	06279	06204	06154	06117	06089
11.70	08092	07821	07634	07507	07406	07330	07228	07146	07048	06966	06832	06712	06594	06510	06436	06327	06252	06202	06164	06137
11.75	08133	07863	07677	07551	07450	07375	07274	07192	07095	07013	06879	06760	06641	06558	06484	06375	06299	06250	06212	06185
11.80	08175	07905	07720	07595	07495	07420	07319	07238	07141	07059	06926	06807	06689	06605	06532	06422	06347	06298	06260	06233
11.85	08216	07948	07763	07638	07539	07465	07365	07284	07188	07106	06974	06854	06737	06653	06579	06470	06395	06346	06308	06281
11.90	08258	07990	07806	07682	07584	07510	07410	07330	07234	07153	07021	06902	06785	06701	06627	06518	06443	06394	06356	06330
11.95	08299	08033	07849	07726	07628	07555	07456	07376	07281	07200	07068	06950	06832	06749	06675	06566	06492	06442	06405	06378
12.00	08341	08075	07892	07770	07673	07600	07502	07422	07327	07246	07115	06997	06880	06797	06723	06615	06540	06491	06453	06427
12.05	08382	08117	07935	07814	07718	07646	07548	07469	07374	07293	07163	07045	06928	06845	06772	06663	06589	06540	06502	06475
12.10	08424	08160	07979	07858	07762	07691	07593	07515	07421	07340	07210	07093	06976	06893	06820	06712	06637	06558	06551	06524
12.15	08465	08202	08022	07902	07807	07736	07639	07561	07468	07387	07258	07141	07025	06942	06868	06760	06686	06637	06600	06573
12.20	08507	08245	08065	07946	07852	07781	07685	07608	07515	07435	07305	07189	07073	06990	06917	06809	06735	06686	06649	06622
12.25	08549	08287	08108	07990	07896	07827	07731	07654	07561	07482	07353	07237	07121	07038	06966	06858	06784	06735	06698	06671
12.30	08590	08330	08152	08034	07941	07872	07777	07701	07608	07529	07401	07285	07170	07087	07015	06907	06833	06785	06747	06721
12.35	08632	08373	08195	08078	07986	07918	07823	07748	07655	07576	07449	07333	07218	07136	07063	06956	06883	06834	06797	06770
12.40	08674	08415	08238	08123	08031	07963	07869	07794	07703	07624	07497	07382	07267	07185	07112	07005	06932	06883	06846	06820
12.45	08715	08458	08282	08167	08076	08009	07916	07841	07750	07671	07544	07430	07316	07233	07162	07055	06981	06933	06896	06870
12.50	08757	08500	08325	08211	08121	08054	07962	07888	07797	07719	07592	07479	07364	07282	07211	07104	07031	06983	06946	06919

Steady State Flow of Gas Through Pipes 301

12.55	08799	08543	08369	08255	08166	08100	08008	07935	07844	07766	07641	07527	07413	07332	07260	07154	07081	07033	06996	06969
12.60	08840	08586	08412	08300	08211	08145	08054	07981	07891	07814	07689	07576	07462	07381	07309	07204	07131	07083	07046	07020
12.65	08882	08628	08456	08344	08256	08191	08101	08028	07939	07862	07737	07624	07511	07430	07359	07253	07181	07133	07096	07070
12.70	08924	08671	08499	08388	08301	08237	08147	08075	07986	07909	07785	07673	07560	07479	07409	07303	07231	07183	07147	07120
12.75	08966	08714	08543	08433	08346	08282	08194	08122	08034	07957	07833	07722	07609	07529	07458	07353	07281	07234	07197	07171
12.80	09008	08757	08586	08477	08391	08328	08240	08169	08081	08005	07882	07771	07659	07579	07508	07403	07332	07284	07248	07222
12.85	09050	08800	08630	08521	08436	08374	08287	08217	08129	08053	07930	07820	07708	07628	07538	07454	07382	07335	07299	07272
12.90	09091	08842	08673	08566	08481	08420	08333	08264	08176	08101	07979	07869	07758	07678	07608	07504	07433	07386	07349	07323
12.95	09133	08885	08717	08610	08526	08466	08380	08311	08224	08149	08027	07918	07807	07728	07658	07555	07484	07437	07400	07374
13.00	09175	08928	08761	08655	08571	08512	08427	08358	08272	08197	08076	07967	07857	07778	07708	07605	07535	07488	07452	07426
13.05	09217	08971	08804	08699	08617	08557	08473	08406	08319	08245	08125	08017	07906	07828	07758	07656	07586	07539	07503	07477
13.10	09259	09014	08848	08744	08662	08603	08520	08453	08367	08293	08173	08066	07956	07878	07809	07707	07637	07590	07554	07528
13.15	09301	09057	08892	08789	08707	08649	08567	08500	08415	08341	08222	08116	08006	07928	07859	07758	07688	07642	07606	07580
13.20	09343	09100	08935	08833	08753	08696	08614	08548	08463	08390	08271	08165	08056	07978	07910	07809	07739	07693	07658	07632
13.25	09385	09143	08979	08878	08798	08742	08660	08595	08511	08438	08320	08215	08106	08029	07961	07860	07791	07745	07709	07684
13.30	09427	09186	09023	08923	08843	08788	08707	08643	08559	08486	08369	08264	08156	08079	08011	07911	07842	07797	07761	07736
13.35	09470	09229	09067	08967	08889	08834	08754	08690	08607	08535	08418	08314	08206	08130	08062	07963	07894	07849	07813	07788
13.40	09512	09272	09111	09012	08934	08880	08801	08738	08655	08583	08467	08364	08257	08180	08113	08014	07946	07901	07866	07840
13.45	09554	09315	09154	09057	08980	08926	08848	08786	08704	08632	08516	08414	08307	08231	08164	08066	07998	07953	07918	07892
13.50	09596	09358	09198	09102	09025	08973	08895	08833	08752	08681	08566	08464	08357	08282	08215	08117	08050	08006	07970	07945
13.55	09638	09401	09242	09146	09071	09019	08942	08881	08800	08729	08615	08514	08408	08333	08267	08169	08102	08058	08023	07997
13.60	09680	09444	09286	09191	09116	09065	08990	08929	08848	08778	08664	08564	08458	08384	08318	08221	08155	08111	08075	08050
13.65	09723	09487	09330	09236	09162	09112	09037	08977	08897	08827	08714	08614	08509	08435	08369	08273	08207	08163	08128	08103
13.70	09765	09530	09374	09281	09208	09158	09084	09025	08945	08876	08763	08664	08560	08486	08421	08326	08260	08216	08181	08156
13.75	09807	09573	09418	09326	09253	09204	09131	09073	08994	08925	08813	08715	08610	08537	08472	08378	08312	08269	08234	08209
13.80	09849	09616	09462	09371	09299	09251	09179	09121	09042	08974	08862	08765	08661	08588	08524	08430	08365	08322	08287	08262
13.85	09892	09659	09506	09416	09345	09297	09226	09169	09091	09023	08912	08815	08712	08640	08576	08483	08418	08375	08341	08316
13.90	09934	09703	09550	09461	09390	09344	09273	09217	09140	09072	08962	08866	08763	08691	08628	08535	08471	08429	08394	08369
13.95	09976	09746	09594	09506	09436	09391	09321	09265	09188	09121	09011	08916	08814	08743	08680	08588	08525	08482	08447	08423
14.00	10019	09789	09638	09551	09482	09437	09368	09313	09237	09170	09061	08967	08866	08795	08732	08641	08578	08536	08501	08477
14.05	10061	09832	09682	09596	09528	09484	09416	09361	09286	09219	09111	09018	08917	08846	08784	08694	08631	08589	08555	08531
14.10	10103	09875	09727	09641	09574	09530	09463	09409	09335	09268	09161	09069	08968	08898	08836	08747	08685	08643	08609	08585
14.15	10146	09919	09771	09687	09620	09577	09511	09458	09384	09318	09211	09119	09019	08950	08889	08800	08738	08697	08663	08639
14.20	10188	09962	09815	09732	09666	09624	09558	09506	09432	09367	09261	09170	09071	09002	08941	08853	08792	08751	08717	08693
14.25	10231	10005	09859	09777	09712	09671	09606	09554	09481	09417	09311	09221	09122	09054	08994	08907	08846	08805	08771	08747

(table continued)

302 Gas Production Engineering

Table 7-1. (Continued) Values of $\int_0^{P_r} \frac{P_r}{z} dP_r$

Multiply values in body of table by 10^{-2}; for example, read 00198 as 1.98

Pseudo-reduced pressure P_r	Pseudoreduced temperature T_r																		
	1.05	1.10	1.15	1.20	1.25	1.30	1.35	1.40	1.45	1.50	1.60	1.70	1.80	1.90	2.00	2.20	2.40	2.60	2.80	3.00
14.30	10273	10049	09903	09822	09758	09717	09654	09603	09530	09466	09362	09272	09174	09106	09047	08960	08900	08860	08826	08802
14.35	10316	10092	09948	09867	09804	09764	09701	09651	09580	09516	09412	09324	09226	09158	09099	09014	08954	08914	08880	08856
14.40	10358	10135	09992	09913	09850	09811	09749	09700	09629	09565	09462	09375	09277	09211	09152	09068	09008	08969	08935	08911
14.45	10401	10179	10036	09958	09896	09858	09797	09748	09678	09615	09512	09426	09329	09263	09205	09121	09063	09023	08990	08966
14.50	10443	10222	10081	10003	09942	09905	09845	09797	09727	09664	09563	09477	09381	09316	09258	09175	09117	09078	09044	09021
14.55	10486	10266	10125	10049	09988	09952	09892	09846	09776	09714	09613	09529	09433	09368	09311	09229	09172	09133	09099	09076
14.60	10528	10309	10169	10094	10034	09999	09940	09894	09825	09764	09664	09580	09485	09421	09364	09283	09226	09188	09155	09131
14.65	10571	10352	10214	10139	10080	10046	09988	09943	09875	09814	09714	09632	09537	09473	09418	09338	09281	09243	09210	09186
14.70	10614	10396	10258	10185	10126	10093	10036	09992	09924	09864	09765	09683	09589	09526	09471	09392	09336	09298	09265	09242
14.75	10656	10439	10302	10230	10173	10140	10084	10041	09973	09914	09816	09735	09641	09579	09524	09446	09391	09353	09321	09297
14.80	10699	10483	10347	10275	10219	10187	10132	10089	10023	09964	09866	09787	09694	09632	09578	09501	09446	09409	09376	09353
14.85	10742	10526	10391	10321	10265	10234	10180	10138	10072	10014	09917	09838	09746	09685	09632	09556	09501	09464	09432	09409
14.90	10784	10570	10436	10366	10311	10281	10228	10187	10122	10064	09968	09890	09798	09738	09685	09610	09557	09520	09488	09465
14.95	10827	10613	10480	10412	10358	10328	10276	10236	10171	10114	10019	09942	09851	09791	09739	09665	09612	09576	09544	09521
15.00	10870	10657	10525	10457	10404	10375	10324	10285	10221	10164	10070	09994	09904	09845	09793	09720	09668	09632	09600	09577

From Nisle and Poettmann, 1955; courtesy of *Petroleum Engineer*, a publication of Energy Publications, Dallas, Texas.

(text continued from page 293)

Type of gas in the flowline	Liquid content gal/MMscf	E (fraction)
Dry gas	0.1	0.92
Casing-head gas	7.2	0.77
Gas and condensate	800	0.60

For high liquid contents, such as the previous case shown above for gas with condensate, two-phase flow conditions exist. Pipeline efficiency can no longer represent the complex flow behavior, and different equations that account for two-phase flow must be used (see Chapter 8).

Transmission Factor

Equation 7-29 can be written as:

$$q_{sc} = 5.6353821 \left(\frac{T_{sc}}{p_{sc}}\right) \left[\frac{(p_1^2 - p_2^2)d^5}{\gamma_g Z_{av} T_{av} L}\right]^{0.5} \left(\frac{1}{f}\right)^{0.5}$$

The factor $(1/f)^{0.5}$ is known as the transmission factor. Several correlations have been developed for the transmission factor (derived from the friction factor), but it remains one of the most difficult parameters to evaluate accurately. Transmission factor, frequently used in pipeline literature, is just another way to represent friction factor and further discussion is, therefore, unnecessary.

Example 7-1. What is the maximum pressure at which a 10-in. nominal size pipe (O.D. = 10.75, I.D. = 10.02 in.) can be operated ? Calculate the maximum throughput for a 40-mile pipeline of this type, using (a) Weymouth equation, and (b) Clinedinst equation. Assume $\epsilon/d = 0.002$, c = 0.05, E = 1.0 (seamless pipe), Y = 0.40, S = 35,000 psi, $T_{av} = 100°F$, $\mu_g = 0.01$ cp, $\gamma_g = 0.70$. Assume 0.0 psia minimum pressure.

Solution

Pipe thickness, t = 10.75 − 10.02 = 0.73 in.

Using Equation 7-22,

$$p_{max} = \frac{2(0.73 - 0.05)(35,000)(1.0)}{10.75 - 2(0.73 - 0.05)(0.4)} = 4{,}663.92 \text{ psia}$$

To provide additional safety, the maximum pressure specification is chosen as 4,000 psia.

Thus, to calculate the maximum throughput, the upstream pressure, $p_1 = 4,000$ psia, and the downstream pressure, $p_2 = 0.0$ psia

Assume $T_{sc} = 520°R$, $p_{sc} = 14.73$ psia

From Figure 3-1, $p_{pc} = 663$ psia, $T_{pc} = 387°R$

Using Equation 7-32, $p_{av} = (2/3)[(4,000^3 - 0)/(4,000^2 - 0)] = 2,666.67$ psia.

$p_{pr} = 2,666.67/663 = 4.022$, $T_{pr} = (100 + 460)/387 = 1.447$

From Figure 3-2, $Z_{av} = 0.74$

(a) Weymouth equation (Equation 7-34):

$$q_{sc} = (31.5027)(520/14.73)\left[\frac{(4,000^2 - 0)(10.02)^{16/3}}{(0.7)(0.74)(560)(40 \times 5,280)}\right]^{0.5}$$

$= 265,206.15$ Mscfd $= \underline{265.21\text{ MMscfd}}$

(b) Clinedinst equation (Equation 7-39):

$p_{r1} = 4,000/663 = 6.033$, $p_{r2} = 0.0$, $T_{pr} = 1.447$

$$q_{sc} = (7.969634)(663)(520/14.73)\left[\frac{(10.02)^5}{(0.7)(560)(40 \times 5,280)f}\right]^{0.5}$$

$$\left[\int_0^{6.033} (p_r/Z)dp_r - \int_0^0 (p_r/Z)dp_r\right]^{0.5}$$

$$= \frac{6,515.2606}{f^{0.5}}\left[\int_0^{6.033} (p_r/Z)dp_r\right]^{0.5}$$

First trial: Assume $q_{sc} = 265$ MMscfd

Using Equation 7-9, $N_{Re} = (20)(265,000)(0.7)/[(0.01)(10.02)] = 3.7 \times 10^7$

Steady State Flow of Gas Through Pipes   305

From Figure 7-1, f = 0.024. From Table 7-1:

$$\int_0^{6.00} (p_r/Z)dp_r = 22.88$$

$$\int_0^{6.05} (p_r/Z)dp_r = 23.23$$

By interpolation:

$$\int_0^{6.033} (p_r/Z)dp_r = 22.88 + (23.23 - 22.88)(6.033 - 6.0)/(6.05 - 6.0)$$
$$= 23.11$$

Thus,

$$q_{sc} = \frac{(6,515.2606)(23.11)^{0.5}}{(0.024)^{0.5}} = 202,174.39 \text{ Mscfd} = \underline{202.17 \text{ MMscfd}}$$

A second trial is not required because the friction factor plot in Figure 7-1 shows an almost constant f in the turbulent region applicable here.

This example shows the differences in results that may sometimes occur from using different flow relationships. Most of the difference results from the friction factor values used in the two equations. From Equation 7-33, we can calculate that the Weymouth equation used an f = 0.01484, whereas an f = 0.024 was used in the Clinedinst equation.

Vertical and Inclined Single-Phase Flow of Gas

Neglecting changes in kinetic energy, and assuming that no mechanical (or shaft) work is done on or by the gas, the mechanical energy balance expressed by Equation 7-3 becomes:

$$dp + \frac{\rho g}{g_c} dz + \frac{f \rho v^2}{2g_c d} dL = 0 \tag{7-40}$$

Consider the flow scheme shown in Figure 7-4. θ is the angle of drift from the vertical, L is the total length of the pipe, and z is the vertical elevation difference between the inlet (point 1) and outlet (point 2). L and z are related as follows:

306 Gas Production Engineering

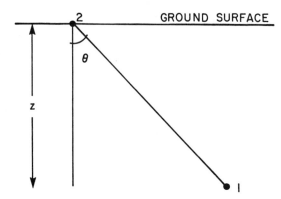

Figure 7-4. Gas flow in an inclined section.

$L = z/\cos\theta$, implying that $dL = dz/\cos\theta$

Substituting for $\cos\theta = z/L$,

$$dL = (L/z)\, dz$$

Using this relationship between L and z, Equation 7-40 can be written as:

$$-dp = \rho\left[\frac{g}{g_c} + \left(\frac{f}{2g_c d}\right)\left(\frac{L}{z}\right)v^2\right]dz$$

Substituting for gas density and velocity (as described earlier for horizontal flow):

$$-dp = \left[\frac{28.97 p \gamma_g}{ZRT}\right]\left[\frac{g}{g_c} + \left(\frac{f}{2g_c d}\right)\left(\frac{L}{z}\right)\left(\frac{16 q_{sc}^2 Z^2 T^2 p_{sc}^2}{p^2 T_{sc}^2 \pi^2 d^4}\right)\right]dz$$

or,

$$-\int_1^2 \frac{(ZT/p)\, dp}{(g/g_c) + (0.81057 f L q_{sc}^2 Z^2 T^2 p_{sc}^2)/(zp^2 T_{sc}^2 g_c d^5)} = \frac{28.97}{R}\int_1^2 \gamma_g dz \quad (7\text{-}41)$$

Almost always, γ_g is assumed constant in evaluating the right integral in Equation 7-41. Substituting for p_{sc} (= 14.73 psia), T_{sc} (= 520°R), and g_c (= 32.17 lbm ft/lbf-sec²), and converting to conventional units, Equation 7-41 can be written as:

$$\int_2^1 \frac{(ZT/p)\, dp}{1 + (6.7393 \times 10^{-4} fLq_{sc}^2 Z^2 T^2)/(zp^2 d^5)} = 0.01875 \gamma_g z \qquad (7\text{-}42)$$

where the units are: p = psia; q_{sc} = Mscfd; d = in.; T = °R; and L, z = ft.

Equation 7-42 is the general equation for vertical flow calculations. The left integral cannot be evaluated easily because Z is a complex function of p and T, and the temperature variation with depth z is not easily defined. The various simplifying assumptions made in the evaluation of this integral form the basis for the different methods that give results of varying degrees of accuracy.

Some methods for vertical flow assume an average temperature, in which case Equation 7-42 becomes:

$$\int_2^1 \frac{(Z/p)\, dp}{1 + (6.7393 \times 10^{-4} fLq_{sc}^2 Z^2 T^2)/(zp^2 d^5)} = \frac{0.01875 \gamma_g z}{T_{av}} \qquad (7\text{-}43)$$

In the left integral of Equations 7-42 and 7-43, the value of the constant is 6.7393×10^{-4}. Some authors use a standard pressure, p_{sc}, equal to 14.65 psia, in which case this constant becomes equal to 6.6663×10^{-4}. Still others use p_{sc} = 14.65 psia and diameter d in ft. In this latter case, the constant becomes equal to 2.679×10^{-9}.

Static Bottom-Hole Pressure (SBHP)

For production and reservoir engineering calculations, the shut-in (or static) bottom-hole pressure, p_{ws}, is frequently required. In many cases, it may be difficult or expensive to obtain p_{ws} by gauge measurements. Techniques have, therefore, been developed to calculate SBHP from wellhead pressure measurements.

For a shut-in well, the flow rate (q or q_{sc}) is equal to zero, and Equation 7-42 simplifies to:

$$\int_{p_{wh}}^{p_{ws}} (ZT/p)\, dp = 0.01875 \gamma_g z \qquad (7\text{-}44)$$

Equation 7-44 describes the relationship between the pressure measured at the surface (or wellhead), p_{wh}, and the pressure p_{ws} at a depth z. Note that the pipe length L has no effect on the static pressures.

Average Temperature and Z-Factor Method

This is the simplest method. An average temperature and Z-factor is used to simplify the left-hand side integral in Equation 7-44:

$$Z_{av}T_{av} \int_{p_{wh}}^{p_{ws}} (dp/p) = 0.01875\gamma_g Z$$

Upon integration:

$$\ln \frac{p_{ws}}{p_{wh}} = \frac{0.01875\gamma_g Z}{Z_{av}T_{av}}$$

or,

$$p_{ws} = p_{wh} \exp\left[\frac{0.01875\gamma_g Z}{Z_{av}T_{av}}\right] \quad (7\text{-}45)$$

Generally, this is written as:

$$p_{ws} = p_{wh}\, e^{s/2} \quad (7\text{-}46)$$

where $s = (0.0375\gamma_g Z)/(Z_{av}T_{av})$ \quad (7-47)

To determine the average temperature, the bottom-hole temperature (BHT) and the surface temperature should be known. In cases where the BHT is not known, it can be estimated from the fact that the temperature gradient is usually of the order of 0.015 °F/ft. For a static column of gas, an arithmetic average temperature is satisfactory to use. For calculating Z_{av}, the average pressure is required (in addition to average temperature). Thus a trial and error type of solution is necessary. A good initial estimate of p_{ws} can be made using Equation 7-48:

$$p_{ws} = p_{wh} + 0.25 \left(\frac{p_{wh}}{100}\right)\left(\frac{z}{100}\right) \quad (7\text{-}48)$$

Sukkar-Cornell Method

This method also uses an average temperature, but accounts for the variation of Z with pressure. Sukkar and Cornell (1955) converted the basic equation (Equation 7-43) into the pseudoreduced form:

$$\int_{p_{pr2}}^{p_{pr1}} \frac{(Z/p_{pr})dp_{pr}}{1 + BZ^2/p_{pr}^2} = \frac{0.01875\gamma_g z}{T_{av}} \quad (7\text{-}49)$$

where $B = \dfrac{(6.6663 \times 10^{-4})fLq_{sc}^2 T_{av}^2}{zd^5 p_{pc}^2} \quad (7\text{-}50)$

In Equation 7-50, p_{pc} is the pseudocritical pressure of the gas (gas mixture), and q_{sc} is the rate at 14.65 psia and 520°R (instead of 14.73 psia and 520°F). As stated earlier, a conversion constant of 6.6663×10^{-4} in Equation 7-50 implies a p_{sc} of 14.65 psia, whereas a conversion constant of 6.7393×10^{-4} (as in Equations 7-42 and 7-43) implies a p_{sc} of 14.73 psia.

For the case of SBHP, the flow rate q_{sc} is zero (or, B = 0), and Equation 7-49 becomes:

$$\int_{p_{prwh}}^{p_{prws}} (Z/p_{pr})dp_{pr} = \frac{0.01875\gamma_g z}{T_{av}} \quad (7\text{-}51)$$

Sukkar and Cornell (1955) evaluated the left integral in Equation 7-49 by numerical integration. They reported results in tabular form for reduced pressure p_{pr} in the range of 1.0 to 12, reduced temperature T_{pr} of 1.5, 1.6, and 1.7, and B in the range of 0 (for static) to 20 (flowing). Messer et al. (1974) presented an extension of the Sukkar-Cornell method, showing its applicability to deep, sour gas wells, and presented the values of the integral in Equation 7-49 for T_{pr} up to 3, p_{pr} up to 30, and B up to 70. These values are shown in Tables 7-2(a–m). The case B = 0, shown in Table 7-2a is applicable to SBHP determination.

For computer programming purposes, Messer et al. (1974) have also proposed the following equation to fit the tabular data:

$$\int_{0.2}^{p_{pr}} \frac{(Z/p_{pr})dp_{pr}}{1 + BZ^2/p_{pr}^2} = E + F\, p_{pr} + G \ln p_{pr} \quad (7\text{-}52)$$

where E, F, and G are regression constants, determined by Messer et al. (1974) as follows:
For $0 \leq B \leq 25$, $10 \leq p_{pr} \leq 30$, $1.1 \leq T_{pr} \leq 3$,

$E = (0.18011 - 0.00262B)\, T_{pr}^2 - (1.21216 - 0.01517B)\, T_{pr}$
$\quad - (0.28026 - 0.00552B)$

$F = (0.02246 - 0.00043B)\, T_{pr}^2 - (0.12792 - 0.00226B)\, T_{pr}$
$\quad + (0.21463 - 0.00274B)$

(text continued on page 336)

Table 7-2 (a)
Extended Sukkar-Cornell Integral for Bottom-hole Pressure Calculation

Reduced Temperature for $B = 0.00$

	1.1	1.2	1.3	1.4	1.5	1.6	1.7	1.8	1.9	2.0	2.2	2.4	2.6	2.8	3.0
0.20	0.0000	0.0000	0.0000	0.0000	0.0000	0.0000	0.0000	0.0000	0.0000	0.0000	0.0000	0.0000	0.0000	0.0000	0.0000
0.50	0.8387	0.8582	0.8719	0.8824	0.8897	0.8966	0.9017	0.9079	0.9082	0.9108	0.9147	0.9177	0.9194	0.9206	0.9218
1.00	1.3774	1.4440	1.4836	1.5129	1.5334	1.5514	1.5654	1.5781	1.5823	1.5889	1.5966	1.6059	1.5111	1.6148	1.6184
1.50	1.6048	1.7373	1.8078	1.8565	1.8911	1.9192	1.9422	1.9609	1.9693	1.9798	1.9951	2.0063	2.0151	2.0211	2.0274
2.00	1.7149	1.9116	2.0157	2.0642	2.1331	2.1709	2.2023	2.2273	2.2397	2.2536	2.2744	2.2893	2.3013	2.3100	2.3104
2.50	1.7995	2.0298	2.1631	2.2507	2.3138	2.3607	2.3996	2.4307	2.4469	2.4641	2.4900	2.5081	2.5234	2.5347	2.5452
3.00	1.8750	2.1255	2.2778	2.3813	2.4570	2.5125	2.5583	2.5947	2.6148	2.6354	2.6654	2.6863	2.7050	2.7189	2.7134
3.50	1.9473	2.2101	2.3746	2.4898	2.5762	2.6190	2.6909	2.7325	2.7561	2.7798	2.8138	2.8382	2.8589	2.8752	2.8896
4.00	2.0178	2.2882	2.4603	2.5945	2.6793	2.7480	2.8052	2.8515	2.8784	2.9050	2.9426	2.9699	2.9928	3.0114	3.0274
4.50	2.0689	2.3622	2.5390	2.6698	2.7715	2.8449	2.9065	2.9569	2.9867	3.0158	3.0571	3.0871	3.1119	3.1322	3.1496
5.00	2.1547	2.4330	2.6128	2.7484	2.8558	2.9330	2.9982	3.0523	3.0645	3.1158	3.1605	3.1930	3.2195	3.2413	3.2597
5.50	2.2214	2.5013	2.6833	2.8222	2.9341	3.0146	3.0828	3.1400	3.1742	3.2074	3.2557	3.2899	3.3178	3.3408	3.3600
6.00	2.2872	2.5677	2.7512	2.8926	3.0079	3.0911	3.1616	3.2215	3.2575	3.2924	3.3428	3.3795	3.4085	3.4325	3.4524
6.50	2.3522	2.6329	2.8171	2.9603	3.0781	3.1635	3.2360	3.2980	3.3355	3.3720	3.4245	3.4629	3.4931	3.5176	3.5381
7.00	2.4165	2.6971	2.8814	3.0258	3.1452	3.2324	3.3065	3.3704	3.4092	3.4470	3.5012	3.5411	3.5722	3.5973	3.6181
7.50	2.4802	2.7602	2.9442	3.0893	3.2100	3.2985	3.3740	3.4393	3.4792	3.5180	3.5738	3.6148	3.6467	3.6723	3.6934
8.00	2.5432	2.8223	3.0058	3.1612	3.2727	3.3623	3.4367	3.5052	3.5460	3.5657	3.6486	3.6847	3.7173	3.7432	3.7646
8.50	2.6057	2.8336	3.0864	3.2118	3.3338	3.4239	3.5012	3.5665	3.6101	3.6504	3.7144	3.7512	3.7844	3.8108	3.8323
9.00	2.6676	2.9441	3.1260	3.2713	3.3934	3.4838	3.5617	3.6297	3.6718	3.7126	3.7775	3.8148	3.8484	3.8750	3.8969
9.50	2.7289	3.0039	3.1847	3.3296	3.4516	3.5422	3.6204	3.6889	3.7315	3.7727	3.8382	3.8760	3.9099	3.9367	3.9588
10.00	2.7896	3.0630	3.2427	3.3870	3.5087	3.5993	3.6776	3.7465	3.7894	3.8308	3.8969	3.9350	3.9690	3.9961	4.0182
10.50	2.8499	3.1215	3.2999	3.4436	3.5647	3.6552	3.7336	3.8026	3.8456	3.8872	3.9538	3.9921	4.0262	4.0583	4.0755
11.00	2.9096	3.1794	3.3565	3.4993	3.6198	3.7100	3.7883	3.8573	3.9004	3.9421	4.0090	4.0473	4.0814	4.1086	4.1309
11.50	2.9690	3.2369	3.4126	3.5543	3.6741	3.7640	3.8420	3.9108	3.9540	3.9958	4.0627	4.1010	4.1351	4.1622	4.1845
12.00	3.0280	3.2940	3.4681	3.6086	3.7277	3.8171	3.8948	3.9634	4.0065	4.0432	4.1150	4.1532	4.1872	4.2143	4.2366
12.50	3.0867	3.3506	3.5231	3.6523	3.7806	3.8694	3.9467	4.0150	4.0579	4.0994	4.1660	4.2041	4.2380	4.2650	4.2872
13.00	3.1452	3.4068	3.5777	3.7154	3.8328	3.9211	3.9977	4.0557	4.1084	4.1495	4.2158	4.2567	4.2875	4.3144	4.3365
13.50	2.2033	3.4627	3.6319	3.7880	3.8644	3.9721	4.0480	4.1155	4.1580	4.1989	4.2645	4.3021	4.3357	4.3625	4.3846

14.00	3.2612	3.5183	3.6857	3.8200	3.9354	4.0224	4.0977	4.1547	4.2067	4.2472	4.3122	4.3494	4.3829	4.4095	4.4316
14.50	3.3189	3.5735	3.7391	3.8716	3.9859	4.0722	4.1480	4.2131	4.2546	4.2947	4.3589	4.3957	4.4289	4.4555	4.4775
15.00	3.3763	3.6285	3.7922	3.9228	4.0359	4.1215	4.1950	4.2609	4.3018	4.3414	4.4047	4.4410	4.4741	4.5005	4.6224
15.50	3.4335	3.6832	3.8450	3.9736	4.0866	4.1702	4.2428	4.8080	4.3483	4.3874	4.4497	4.4855	4.5183	4.5446	4.5663
16.00	3.4906	3.7376	3.8974	4.0240	4.1346	4.2185	4.2900	4.3546	4.3942	4.4327	4.4939	4.5291	4.5617	4.5878	4.6094
16.50	3.5474	3.7919	3.9497	4.0740	4.1833	4.2663	4.3368	4.4007	4.4395	4.4773	4.5374	4.5720	4.6042	4.6302	4.6518
17.00	3.6041	3.8459	4.0016	4.1237	4.2316	4.3138	4.3830	4.4462	4.4843	4.5213	4.5802	4.6141	4.6461	4.6719	4.6933
17.50	3.6606	3.8996	4.0533	4.1731	4.2795	4.3608	4.4289	4.4913	4.5285	4.5648	4.6223	4.6555	4.5872	4.7129	4.7341
18.00	3.7170	3.9532	4.1048	4.2221	4.3271	4.4075	4.4743	4.5359	4.5722	4.6077	4.6638	4.6963	4.7276	4.7532	4.7743
18.50	3.7732	4.0066	4.1560	4.2709	4.3744	4.4538	4.5193	4.5801	4.6154	4.6501	4.7048	4.7365	4.7675	4.7928	4.8138
19.00	3.8293	4.0599	4.2071	4.3195	4.4214	4.4998	4.5640	4.6239	4.6582	4.6921	4.7451	5.7761	4.8067	4.8319	4.8527
19.50	3.8853	4.1129	4.2579	4.3678	4.4681	4.5455	4.6053	4.6574	4.7006	4.7335	4.7850	4.8151	4.8454	4.8704	4.8911
20.00	3.9411	4.1658	4.3086	4.4158	4.5145	4.5909	4.6522	4.7104	4.7425	4.7746	4.8244	4.8536	4.8835	4.9083	4.9288
20.50	3.9969	4.2186	4.3590	4.4636	4.5606	4.6360	4.6959	4.7531	4.7841	4.8152	4.8633	4.8916	4.9211	4.9457	4.9661
21.00	4.0525	4.2712	4.4094	4.5112	4.6065	4.6808	4.7392	4.7955	4.8253	4.8554	4.9017	4.9291	4.9582	4.9827	5.0029
21.50	4.1080	4.3237	4.4595	4.5586	4.6522	4.7254	4.7822	4.8376	4.8662	4.8953	4.9397	4.9662	4.9969	5.0192	5.0392
22.00	4.1634	4.3760	4.5095	4.6058	4.6976	4.7697	4.8250	4.8794	4.9068	4.9348	4.9774	5.0027	5.0311	5.0552	5.0751
22.50	4.2187	4.4282	4.5594	4.6528	4.7428	4.8138	4.8675	4.9209	4.9470	4.9739	5.0146	5.0391	5.0670	5.0908	5.1105
23.00	4.2739	4.4803	4.6091	4.6996	4.7879	4.8577	4.9098	4.9621	4.9869	5.0128	5.0514	5.0750	5.1024	5.1260	5.1455
23.50	4.3291	4.5323	4.6587	4.7463	4.8327	4.9014	4.9518	5.0031	5.0265	5.0513	5.0879	5.1104	5.1374	5.1808	5.1802
24.00	4.3841	4.5842	4.7081	4.7928	4.8773	4.9449	4.9935	5.0438	5.0659	5.0895	5.1241	5.1455	5.1720	5.1953	5.2144
24.50	4.4391	4.6360	4.7575	4.8391	4.9217	4.9882	5.0351	5.0843	5.1050	5.1275	5.1599	5.1803	5.2083	5.2294	5.2483
25.00	4.4940	4.6877	4.8067	4.8853	4.9660	5.0312	5.0764	5.1245	5.1438	5.1651	5.1955	5.2147	5.2403	5.2631	5.2819
25.50	4.5488	4.7392	4.8558	4.9314	5.0101	5.0741	5.1176	5.1646	5.1824	5.2025	5.2307	4.2488	5.2739	5.2965	5.3151
26.00	4.6036	4.7907	4.9048	4.9772	5.0541	5.1169	5.1585	5.2044	5.2208	5.2397	5.2656	5.2826	5.3073	5.3296	5.3480
26.50	4.6583	4.8421	4.9536	5.0230	5.0979	5.1594	5.1993	5.2440	5.2589	5.2766	5.3003	5.3162	5.3403	5.3624	5.3806
27.00	4.7129	4.8934	5.0024	5.0686	5.1415	5.2019	5.2398	5.2634	5.2968	5.3132	5.3347	5.3494	5.3780	5.3950	5.4129
27.50	4.7675	4.9447	5.0511	5.1142	5.1850	5.2441	5.2802	5.3227	5.3345	5.3497	5.3688	5.3823	5.4054	5.4272	5.4460
28.00	4.8220	4.9958	5.0997	5.1595	5.2284	5.2862	5.3204	5.3617	5.3720	5.3859	5.4027	4.4150	5.4376	5.4591	5.4767
28.50	4.8764	5.0469	5.1482	5.2048	5.2716	5.3282	5.3605	5.4006	5.4096	5.4219	5.4363	5.4475	5.4695	5.4903	5.5082
29.00	4.9308	5.0979	5.1966	5.2500	5.3147	5.3700	5.4004	5.4393	5.4465	5.4577	5.4697	5.4796	5.5012	5.5223	5.5394
29.50	4.9851	5.1488	5.2450	5.2950	5.3577	5.4117	5.4401	5.4779	5.4834	5.4933	5.5029	5.5116	5.5326	5.5935	5.5704
30.00	5.0394	5.1997	5.2932	5.3400	5.4005	5.4532	5.4797	5.5163	5.5202	5.5287	5.5369	5.5433	5.5638	5.5844	5.6011

Table 7-2 (b)
Extended Sukkar-Cornell Integral for Bottom-hole Pressure Calculation

P_r	\multicolumn{20}{c}{Reduced Temperature for $B = 5.0$}
	1.1	1.2	1.3	1.4	1.5	1.6	1.7	1.8	1.9	2.0	2.2	2.4	2.6	2.8	3.0
0.20	0.0000	0.0000	0.0000	0.0000	0.0000	0.0000	0.0000	0.0000	0.0000	0.0000	0.0000	0.0000	0.0000	0.0000	0.0000
0.50	0.0226	0.0220	0.0216	0.0214	0.0212	0.0210	0.0209	0.0207	0.0207	0.0205	0.0205	0.0206	0.0204	0.0204	0.0204
1.00	0.1036	0.0983	0.0954	0.0934	0.0921	0.0909	0.0901	0.0894	0.0890	0.0886	0.0881	0.0877	0.0874	0.0871	0.0869
1.50	0.2121	0.2052	0.1995	0.1954	0.1924	0.1901	0.1882	0.1868	0.1859	0.1850	0.1938	0.1829	0.1822	0.1816	0.1811
2.00	0.3002	0.3125	0.3102	0.366	0.3034	0.3007	0.2983	0.2965	0.2954	0.2943	0.2926	0.2914	0.2904	0.2896	0.2889
2.50	0.3741	0.4046	0.4126	0.4133	0.4124	0.4107	0.4090	0.4076	0.4066	0.4056	0.4048	0.4030	0.4020	0.4012	0.4005
3.00	0.4419	0.4854	0.5032	0.5105	0.5137	0.5144	0.5143	0.5140	0.5138	0.5134	0.5125	0.5118	0.5112	0.5108	0.5103
3.50	0.5074	0.5594	0.5847	0.5983	0.6065	0.6101	0.6123	0.6138	0.6147	0.6152	0.6154	0.6155	0.6155	0.6157	0.6156
4.00	0.5715	0.6291	0.6594	0.6785	0.6915	0.6982	0.7029	0.7064	0.7087	0.7104	0.7121	0.7133	0.7140	0.7149	0.7154
4.50	0.6346	0.6957	0.7294	0.7530	0.7702	0.7797	0.7868	0.7927	0.7964	0.7994	0.8027	0.8051	0.8068	0.8084	0.8094
5.00	0.6966	0.7601	0.7960	0.8229	0.8440	0.8560	0.8653	0.8734	0.8785	0.8827	0.8879	0.8916	0.8941	0.8965	0.8980
5.50	0.7579	0.8225	0.8601	0.8895	0.9138	0.9280	0.9393	0.9493	0.9558	0.9611	0.9682	0.9732	0.9765	0.9795	0.9315
6.00	0.8185	0.8836	0.9222	0.9536	0.9803	0.9965	1.0095	1.0213	1.0289	1.0354	1.0441	1.0504	1.0547	1.0580	1.0604
6.50	0.8784	0.9437	0.9829	1.0156	1.0442	1.0620	1.0764	1.0896	1.0984	1.1060	1.1162	1.1236	1.1284	1.1324	1.1351
7.00	0.9378	1.0030	1.0423	1.0758	1.1058	1.1249	1.1406	1.1552	1.1649	1.1734	1.1848	1.1932	1.1987	1.2031	1.2060
7.50	0.9967	1.0614	1.1005	1.1346	1.1656	1.1857	1.2024	1.2182	1.2286	1.2379	1.2504	1.2597	1.2657	1.2704	1.2737
8.00	1.0551	1.1191	1.1578	1.1921	1.2237	1.2447	1.2621	1.2788	1.2900	1.2999	1.3167	1.3234	1.3299	1.3349	1.3383
8.50	1.1131	1.1761	1.2142	1.2486	1.2805	1.3020	1.3201	1.3374	1.3492	1.3596	1.3773	1.3845	1.3914	1.3967	1.4003
9.00	1.1706	1.2325	1.2698	1.3041	1.3361	1.3579	1.3764	1.3943	1.4066	1.4173	1.4357	1.4434	1.4506	1.4561	1.4599
9.50	1.2275	1.2883	1.3248	1.3687	1.3907	1.4125	1.4313	1.4497	1.4623	1.4733	1.4922	1.5003	1.5077	1.5135	1.5174
10.00	2.2841	1.3435	1.3791	1.4126	1.4443	1.4661	1.4851	1.5037	1.5165	1.5278	1.5472	1.5555	1.5630	1.5689	1.5729
10.50	1.3403	1.3983	1.4328	1.4658	1.4970	1.5187	1.5377	1.5564	1.5694	1.5808	1.6006	1.6090	1.6167	1.6226	1.6267
11.00	1.3961	1.4526	1.4860	1.5182	1.5490	1.5705	1.5894	1.6081	1.6211	1.6326	1.6526	1.6611	1.6687	1.6747	1.6789
11.50	1.4515	1.5065	1.5387	1.5701	1.6002	1.6214	1.6401	1.6587	1.6718	1.6833	1.7034	1.7118	1.7195	1.7254	1.7296
12.00	1.5067	1.5601	1.5910	1.6214	1.6509	1.6717	1.6901	1.7085	1.7215	1.7330	1.7530	1.7613	1.7689	1.7749	1.7790
12.50	1.5616	1.6133	1.6429	1.6721	1.7010	1.7213	1.7393	1.7575	1.7704	1.7817	1.8015	1.8097	1.8172	1.8231	1.8271
13.00	1.6163	1.6662	1.6944	1.7224	1.7505	1.7704	1.7879	1.8067	1.8184	1.8295	1.8489	1.8569	1.8644	1.8701	1.8742
13.50	1.6708	1.7188	1.7456	1.7722	1.7995	1.8188	1.8358	1.8532	1.8656	1.8765	1.8954	1.9032	1.9105	1.9161	1.9201

14.50	1.5753	1.5860	1.5883	1.5942	1.5476	1.5563	1.5652	1.5716	1.5794	1.5851	1.5899	1.5988	1.6016	1.6043	1.6062	1.6074
15.00	1.6261	1.6351	1.6360	1.6405	1.5942	1.6035	1.6104	1.6163	1.6237	1.6290	1.6335	1.6417	1.6443	1.6468	1.6486	1.6497
15.50	1.6767	1.6839	1.6835	1.6865	1.6405	1.6490	1.6553	1.6605	1.6575	1.6723	1.6764	1.6840	1.6862	1.6885	1.6902	1.6912
16.00	1.7271	1.7326	1.7308	1.7323	1.6865	1.6941	1.6999	1.7043	1.7108	1.7151	1.7188	1.7256	1.7274	1.7296	1.7311	1.7320
16.50	1.7775	1.7811	1.7778	1.7778	1.7323	1.7389	1.7440	1.7477	1.7537	1.7575	1.7607	1.7666	1.7679	1.7699	1.7713	1.7722
17.00	1.8277	1.8294	1.8247	1.8230	1.7778	1.7834	1.7878	1.7906	1.7961	1.7993	1.8020	1.8070	1.8078	1.8096	1.8109	1.8116
17.50	1.8778	1.8777	1.8714	1.8680	1.8230	1.8275	1.8314	1.8333	1.8382	1.8407	1.8429	1.8469	1.8472	1.8487	1.8499	1.8505
18.00	1.9278	1.9257	1.9179	1.9127	1.8680	1.8714	1.8746	1.8756	1.8799	1.8818	1.8833	1.8862	1.8859	1.8872	1.8883	1.8888
18.50	1.9777	1.9737	1.9643	1.9573	1.9127	1.9151	1.9175	1.9175	1.9212	1.9224	1.9232	1.9251	1.9242	1.9252	1.9261	1.9265
19.00	2.0276	2.0215	2.0105	2.0017	1.9573	1.9585	1.9602	1.9592	1.9622	1.9626	1.9628	1.9634	1.9619	1.9626	1.9634	1.9637
19.50	2.0773	2.0592	2.0566	2.0458	2.0017	2.0016	2.0026	2.0005	2.0029	2.0025	2.0020	2.0013	1.9992	1.9996	2.0002	2.0004
20.00	2.1269	2.1167	2.1026	2.0898	2.0458	2.0446	2.0447	2.0416	2.0433	2.0420	2.0408	2.0388	2.0359	2.0360	2.0365	2.0366
20.50	2.1765	2.1642	2.1484	2.1336	2.0898	2.0873	2.0867	2.0824	2.0833	2.0812	2.0792	2.0759	2.0723	2.0721	2.0724	2.0723
21.00	2.2260	2.2116	2.1941	2.1773	2.1336	2.1298	2.1284	2.1229	2.1232	2.1201	2.1173	2.1126	2.1082	2.1077	2.1079	2.1077
21.50	2.2754	2.2588	2.2396	2.2207	2.1773	2.1722	2.1699	2.1632	2.1627	2.1587	2.1551	2.1489	2.1438	2.1429	2.1429	2.1425
22.00	2.3248	2.3060	2.2851	2.2641	2.2207	2.2143	2.2112	2.2033	2.2020	2.1970	2.1926	2.1848	2.1789	2.1777	2.1775	2.1770
22.50	2.3741	2.3531	2.3304	2.3073	2.2641	2.2563	2.2523	2.2432	2.2411	2.2350	2.2298	2.2204	2.2137	2.2121	2.2118	2.2111
23.00	2.4233	2.4001	2.3757	2.3503	2.3073	2.2981	2.2932	2.2828	2.2799	2.2728	2.2667	2.2557	2.2461	2.2462	2.2457	2.2449
23.50	2.4725	2.4470	2.4208	2.3932	2.3503	2.3397	2.3340	2.3222	2.3185	2.3103	2.3033	2.2906	2.2822	2.2799	2.2792	2.2783
24.00	2.5216	2.4938	2.4659	2.4360	2.3932	2.3812	2.3745	2.3615	2.3569	2.3476	2.3397	2.3253	2.3160	2.3133	2.3124	2.3133
24.50	2.5706	2.5406	2.5108	2.4787	2.4360	2.4226	2.4149	2.4005	2.3951	2.3847	2.3758	2.3597	2.3494	2.3463	2.3453	2.3440
25.00	2.6196	2.5873	2.5557	2.5212	2.4787	2.4637	2.4952	2.4394	2.4331	2.4215	2.4117	2.3937	2.3826	2.3791	2.3779	2.3765
25.50	2.6685	2.6339	2.6005	2.5637	2.5212	2.5048	2.4953	2.4761	2.4709	2.4581	2.4473	2.4275	2.4155	2.4115	2.4102	2.4086
26.00	2.7174	2.6805	2.6452	2.6060	2.5637	2.5457	2.5353	2.5166	2.5085	2.4946	2.4827	2.4611	2.4481	2.4437	2.4422	2.4404
26.50	2.7663	2.7269	2.6898	2.6482	2.6060	2.5865	2.5751	2.5550	2.5459	2.5308	2.5179	2.4944	2.4804	2.4756	2.4739	2.4719
27.00	2.8151	2.7734	2.7343	2.6904	2.6482	2.6272	2.6148	2.5932	2.5832	2.5668	2.5529	2.5275	2.5124	2.5073	2.5053	2.5032
27.50	2.8638	2.8197	2.7788	2.7324	2.6904	2.6677	2.6543	2.6312	2.6203	2.6027	2.5877	2.5603	2.5443	2.5386	2.5365	2.5342
28.00	2.9125	2.8660	2.8232	2.7743	2.7324	2.7082	2.6938	2.6691	2.6573	2.6384	2.6223	2.5929	2.5758	2.5698	2.5675	2.5650
28.50	2.9612	2.9123	2.8675	2.8162	2.7743	2.7485	2.7331	2.7069	2.6941	2.6739	2.6567	3.6253	2.6072	2.6007	2.5982	2.5955
29.00	3.0098	2.9585	2.9118	2.8579	2.8162	2.7887	2.7723	2.7446	2.7307	2.7092	2.6909	2.6575	2.6383	2.6314	2.6286	2.6258
29.50	3.0584	3.0046	2.9560	2.8996	2.8579	2.8288	2.8114	2.7821	2.7673	2.7444	2.7250	2.6895	2.6692	2.6618	2.6589	2.6558
30.00	3.1069	3.0507	3.0001	2.9412	2.8996	2.8689	2.8504	2.8194	2.8036	2.7794	2.7589	2.7212	2.6999	2.6970	2.6889	2.6857
					2.9412	2.9088	2.8892	2.8567	2.8399	2.8143	2.7926	2.7528	2.7304	2.7221	2.7187	2.7153

Table 7-2 (c)
Extended Sukkar-Cornell Integral for Bottom-hole Pressure Calculation

Reduced Temperature for $B = 10.0$

P_r	1.1	1.2	1.3	1.4	1.5	1.6	1.7	1.8	1.9	2.0	2.2	2.4	2.6	2.8	3.0
0.20	0.0000	0.0000	0.0000	0.0000	0.0000	0.0000	0.0000	0.0000	0.0000	0.0000	0.0000	0.0000	0.0000	0.0000	0.0000
0.50	0.0115	0.0112	0.0110	0.0108	0.0107	0.0107	0.0106	0.0105	0.0105	0.0105	0.0104	0.0104	0.0104	0.0103	0.0103
1.00	0.0561	0.0525	0.0507	0.0494	0.0486	0.0479	0.0474	0.0470	0.0468	0.0465	0.0462	0.0460	0.0458	0.0456	0.0455
1.50	0.1292	0.1187	0.1132	0.1098	0.1074	0.1056	0.1041	0.1031	0.1024	0.1018	0.1009	0.1003	0.0997	0.0994	0.0990
2.00	0.2028	0.1968	0.1891	0.1837	0.1797	0.1767	0.1743	0.1725	0.1713	0.1703	0.1687	0.1676	0.1667	0.1660	0.1653
2.50	0.2684	0.2723	0.2677	0.2624	0.2578	0.2543	0.2513	0.2490	0.2475	0.2461	0.2440	0.2426	0.2413	0.2403	0.2394
3.00	0.3300	0.3422	0.3427	0.3399	0.3364	0.3332	0.3302	0.3278	0.3263	0.3248	0.3225	0.3210	0.3195	0.3184	0.3174
3.50	0.3897	0.4080	0.4130	0.4135	0.4123	0.4102	0.4080	0.4061	0.4047	0.4035	0.4014	0.3999	0.3985	0.3974	0.3964
4.00	0.4485	0.4708	0.4793	0.4832	0.4846	0.4841	0.4830	0.4820	0.4812	0.4803	0.4787	0.4776	0.4764	0.4755	0.4746
4.50	0.5065	0.5315	0.5423	0.5492	0.5533	0.5545	0.5547	0.5549	0.5549	0.5546	0.5538	0.5532	0.5523	0.5517	0.5511
5.00	0.5638	0.5904	0.6029	0.6122	0.6189	0.6217	0.6233	0.6248	0.6256	0.6260	0.6262	0.6263	0.6258	0.6256	0.6252
5.50	0.6204	0.6480	0.6617	0.6729	0.6818	0.6861	0.6891	0.6919	0.6934	0.6946	0.6959	0.6967	0.6967	0.6968	0.6967
6.00	0.6765	0.7045	0.7190	0.7316	0.7424	0.7481	0.7522	0.7563	0.7586	0.7605	0.7629	0.7645	0.7650	0.7654	0.7655
6.50	0.7321	0.7602	0.7752	0.7808	0.8010	0.8079	0.8131	0.8182	0.8214	0.8240	0.8273	0.8297	0.8307	0.8314	0.8317
7.00	0.7873	0.8153	0.8304	0.6447	0.8580	0.8659	0.8720	0.8781	0.8819	0.8852	0.8895	0.8925	0.8940	0.8950	0.8955
7.50	0.8421	0.8697	0.8846	0.8994	0.9134	0.9221	0.9290	0.9360	0.9404	0.9443	0.9494	0.9531	0.9550	0.9562	0.9566
8.00	0.8965	0.9236	0.9381	0.9531	0.9676	0.9770	0.9845	0.9921	0.9971	1.0015	1.0092	1.0115	1.0138	1.0152	1.0160
8.50	0.9506	0.9769	0.9909	1.0059	1.0207	1.0305	1.0385	1.0467	1.0522	1.0569	1.0653	1.0681	1.0706	1.0723	1.0732
9.00	1.0043	1.0296	1.0431	1.0580	1.0729	1.0829	1.0912	1.0999	1.1057	1.1108	1.1197	1.1228	1.1256	1.1275	1.1286
9.50	1.0575	1.0819	1.0947	1.1094	1.1242	1.1342	1.1428	1.1518	1.1579	1.1633	1.1726	1.1760	1.1790	1.1810	1.1822
10.00	1.1104	1.1338	1.1458	1.1601	1.1747	1.1847	1.1935	1.2027	1.2090	1.2145	1.2242	1.2278	1.2309	1.2331	1.2343
10.50	1.1630	1.1852	1.1964	1.2102	1.2245	1.2344	1.2432	1.2525	1.2689	1.2645	1.2746	1.2783	1.2814	1.2836	1.2850
11.00	1.2153	1.2363	1.2466	1.2598	1.2736	1.2834	1.2920	1.3013	1.3078	1.3135	1.3238	1.3275	1.3907	1.3329	1.3343
11.50	1.2674	1.2871	1.2964	1.3089	1.3222	1.3317	1.3402	1.3494	1.3559	1.3616	1.3719	1.3756	1.3788	1.3810	1.3824
12.00	1.3192	1.3376	1.3458	1.3574	1.3702	1.3794	1.3876	1.3967	1.4032	1.4088	1.4190	1.4227	1.4258	1.4280	1.4294
12.50	1.3708	1.3877	1.3949	1.4056	1.4178	1.4266	1.4345	1.4433	1.4497	1.4552	1.4653	1.4688	1.4719	1.4740	1.4753
13.00	1.4222	1.4377	1.4437	1.4533	1.4649	1.4733	1.4807	1.4893	1.4955	1.5008	1.5106	1.5140	1.5169	1.5139	1.5202

14.50	1.7791	1.8212	1.8470	1.8706		1.8960			1.9121	1.9227	1.9410	1.9485	1.9556	1.9612	1.9651
15.00	1.8330	1.8750	1.8973	1.9192		1.9436	1.9142	1.9298	1.9580	1.9681	1.9858	1.9927	1.9998	2.0053	2.0091
15.50	1.8867	1.9266	1.9472	1.9675		1.9909	1.9612	1.9760	2.0032	2.0128	2.0298	2.0364	2.0432	2.0485	2.0523
16.00	1.9402	1.9780	1.9970	2.0154		2.0377	2.0077	2.0217	2.0478	2.0570	2.0730	2.0792	2.0857	2.0910	2.0946
16.50	1.9936	2.0292	2.0465	2.0631		2.0842	2.0538	2.0669	2.0918	2.1005	2.1155	2.1212	2.1275	2.1326	2.1362
17.00	2.0469	2.0802	2.0958	2.1104		2.1303	2.0996	2.1117	2.1353	2.1434	2.1574	2.1626	2.1686	2.1736	2.1770
17.50	2.1000	2.1311	2.1449	2.1575		2.1762	2.1450	2.1561	2.1783	2.1858	2.1987	2.2032	2.2090	2.2138	2.2172
18.00	2.1530	2.1817	2.1937	2.2043		2.2217	2.1900	2.2000	2.2209	2.2276	2.2394	2.2433	2.2488	2.2535	2.2567
18.50	2.2059	2.2323	2.2424	2.2509		2.2670	2.2347	2.2437	2.2630	2.2690	2.2795	2.2828	2.2880	2.2925	2.2956
19.00	2.2587	2.2826	2.2909	2.2973		2.3120	2.2791	2.2869	2.3046	2.3100	2.3191	2.3217	2.3266	2.3309	2.3339
19.50	2.3113	2.3329	2.3393	2.3434		2.3567	2.3233	2.3299	2.3459	2.3505	2.3582	2.3600	2.3646	2.3688	2.3717
20.00	2.3639	2.3830	2.3875	2.3893		2.4012	2.3671	2.3725	2.3868	2.3906	2.3969	2.3979	2.4022	2.4062	2.4089
20.50	2.4164	2.4329	2.4355	2.4350		2.4455	2.4107	2.4148	2.4273	2.4303	2.4350	2.4353	2.4392	2.4431	2.4456
21.00	2.4688	2.4828	2.4834	2.4806		2.4895	2.4541	2.4568	2.4675	2.4696	2.4728	2.4723	2.4758	2.4795	2.4819
21.50	2.5210	2.5325	2.5311	2.5259		2.5333	2.4972	2.4986	2.5074	2.5086	2.5101	2.5088	2.5119	2.5155	2.5177
22.00	2.5733	2.5822	2.5788	2.5711		2.5770	2.5400	2.5401	2.5470	2.5472	2.5471	2.5449	2.5477	2.5510	2.5531
22.50	2.6254	2.6317	2.6263	2.6161		2.6204	2.5827	2.5814	2.5862	2.5855	2.5837	2.5806	2.5830	2.5861	2.5881
23.00	2.6774	2.6811	2.6736	2.6610		2.6637	2.6252	2.6224	2.6252	2.6235	2.6199	2.6159	2.6179	2.6209	2.6226
23.50	2.7294	2.7304	2.7209	2.7057		2.7068	2.6674	2.6632	2.6639	2.6612	2.6558	2.6508	2.6524	2.6552	2.6568
24.00	2.7813	2.7796	2.7680	2.7503		2.7497	2.7095	2.7038	2.7023	2.6986	2.6913	2.6854	2.6866	2.6892	2.6906
24.50	2.8332	2.8288	2.8151	2.7947		2.7924	2.7514	2.7441	2.7405	2.7357	2.7266	2.7197	2.7204	2.7229	2.7241
25.00	2.8849	2.8778	2.8620	2.8390		2.8351	2.7931	2.7843	2.7784	2.7726	2.7615	2.7536	2.7540	2.7562	2.7573
25.50	2.9367	2.9268	2.9088	2.8832		2.8775	2.8346	2.8243	2.8161	2.8092	2.7961	2.7872	2.7872	2.7892	2.7901
26.00	2.9883	2.9757	2.9556	2.9272		2.9198	2.8760	2.8640	2.8536	2.8456	2.8305	2.8206	2.8200	2.8192	2.8226
26.50	3.0399	3.0245	3.0022	2.9711		2.9620	2.9172	2.9037	2.8908	2.8818	2.8646	2.8536	2.8526	2.8543	2.8548
27.00	3.0915	3.0733	3.0488	3.0149		3.0040	2.9583	2.9431	2.9279	2.9177	2.8985	2.8864	2.8850	2.8864	2.8867
27.50	3.1429	3.1220	3.0953	3.0586		3.0459	2.9993	2.9824	2.9648	2.9534	2.9320	2.9189	2.9170	2.9182	2.9184
28.00	3.1944	3.1706	3.1417	3.1022		3.0877	3.0400	3.0215	3.0014	2.9889	2.9654	2.9512	2.9488	2.9498	2.9497
28.50	3.2458	3.2191	3.1880	3.1457		3.1294	3.0807	3.0604	3.0379	3.0242	2.9985	2.9832	2.9803	2.9811	2.9809
29.00	3.2971	3.2676	3.2343	3.1891		3.1711	3.1212	3.0992	3.0742	3.0593	3.0314	3.0149	3.0116	3.0122	3.0117
29.50	3.3484	3.3160	3.2804	3.2324		3.2124	3.1616	3.1379	3.1103	3.0942	3.0641	3.0465	3.0426	3.0430	3.0424
29.50	3.3484	3.3160	3.2804	3.2324		3.2124	3.1616	3.1379	3.1103	3.0942	3.0641	3.0465	3.0426	3.0430	3.0424
29.50															
29.50	3.3484														
29.50	3.3484	3.3160	3.2804	3.2324		3.2124	3.1616	3.1379	3.1103	3.0942	3.0641	3.0465	3.0426	3.0430	3.0424
29.50	3.3484	3.3160	3.2804	3.2324		3.2124	3.1616	3.1379	3.1103	3.0942	3.0641	3.0465	3.0426	3.0430	3.0424
30.00	3.3997	3.3644	3.3265	3.2756		3.2537	3.2019	3.1764	3.1463	3.1289	3.0966	3.0778	3.0735	3.0736	3.0728

Table 7-2 (d)
Extended Sukkar-Cornell Integral for Bottom-hole Pressure Calculation

P_r	\multicolumn{20}{c}{Reduced Temperature for $B = 15.0$}																			
	1.1	1.2	1.3	1.4	1.5	1.6	1.7	1.8	1.9	2.0	2.2	2.4	2.6	2.8	3.0
0.20	0.0000	0.0000	0.0000	0.0000	0.0000	0.0000	0.0000	0.0000	0.0000	0.0000	0.0000	0.0000	0.0000	0.0000	0.0000
0.50	0.0077	0.0075	0.0074	0.0073	0.0072	0.0071	0.0071	0.0071	0.0070	0.0070	0.0070	0.0070	0.0069	0.0069	0.0069
1.00	0.0385	0.0359	0.0345	0.0336	0.0330	0.0325	0.0322	0.3119	0.0317	0.0316	0.0313	0.0311	0.0310	0.0309	0.0308
1.50	0.0939	0.0838	0.0793	0.0765	0.0746	0.0732	0.0721	0.0713	0.0708	0.0703	0.0696	0.0692	0.0687	0.0685	0.0682
2.00	0.1571	0.1453	0.1371	0.1319	0.1282	0.1257	0.1236	0.1220	0.1211	0.1202	0.1189	0.1180	0.1172	0.1167	0.1161
2.50	0.2162	0.2093	0.2008	0.1943	0.1892	0.1857	0.1827	0.1804	0.1790	0.1777	0.1758	0.1745	0.1733	0.1724	0.1716
3.00	0.2725	0.2710	0.2648	0.2587	0.2533	0.2493	0.2458	0.2431	0.2413	0.2397	0.2374	0.2357	0.2342	0.2331	0.2320
3.50	0.3275	0.3302	0.3267	0.3222	0.3176	0.3138	0.3102	0.3074	0.3055	0.3038	0.3012	0.2994	0.2978	0.2964	0.2952
4.00	0.3818	0.3874	0.3862	0.3837	0.3805	0.3774	0.3743	0.3717	0.3699	0.3683	0.3657	0.3679	0.3622	0.3608	0.3596
4.50	0.4355	0.4430	0.4435	0.4431	0.4415	0.4393	0.4369	0.4349	0.4335	0.4320	0.4298	0.4281	0.4265	0.4252	0.4240
5.00	0.4887	0.4975	0.4992	0.5004	0.5006	0.4994	0.4978	0.4966	0.4956	0.4945	0.4928	0.4914	0.4900	0.4888	0.4877
5.50	0.5413	0.5508	0.5535	0.5561	0.5579	0.5577	0.5570	0.5566	0.5561	0.5554	0.5543	0.5534	0.5522	0.5512	0.5503
6.00	0.5936	0.6034	0.6066	0.6103	0.6135	0.6143	0.6144	0.6149	0.6149	0.6147	0.6143	0.6138	0.6129	0.6121	0.6113
6.50	0.6454	0.6553	0.6590	0.6634	0.6676	0.6694	0.6703	0.6715	0.6720	0.6724	0.6726	0.6727	0.6721	0.6715	0.6708
7.00	0.6969	0.7068	0.7105	0.7155	0.7205	0.7230	0.7246	0.7256	0.7276	0.7284	0.7293	0.7299	0.7296	0.7291	0.7286
7.50	0.7482	0.7577	0.7613	0.7666	0.7722	0.7754	0.7776	0.7802	0.7817	0.7829	0.7844	0.7854	0.7855	0.7852	0.7848
8.00	0.7991	0.8082	0.8114	0.8170	0.8230	0.8266	0.8293	0.8324	0.8344	0.8360	0.8391	0.8395	0.8398	0.8397	0.8394
8.50	0.8497	0.8582	0.8611	0.8666	0.8729	0.8768	0.8799	0.8835	0.8858	0.8878	0.8914	0.8920	0.8926	0.8927	0.8925
9.00	0.9000	0.9078	0.9102	0.9157	0.9220	0.9261	0.9295	0.9334	0.9360	0.9382	0.9423	0.9432	0.9440	0.9442	0.9441
9.50	0.9500	0.9570	0.9588	0.9641	0.9704	0.9746	0.9782	0.9824	0.9852	0.9876	0.9920	0.9932	0.9941	0.9944	0.9944
10.00	0.9998	1.0059	1.0071	1.0121	1.0181	1.0223	1.0260	1.0304	1.0334	1.0359	1.0407	1.0420	1.0430	1.0434	1.0435
10.50	1.0492	1.0544	1.0549	1.0595	1.0653	1.0694	1.0731	1.0776	1.0806	1.0833	1.0883	1.0897	1.0908	1.0913	1.0914
11.00	1.0985	1.1026	1.1024	1.1065	1.1119	1.1159	1.1195	1.1239	1.1271	1.1298	1.1349	1.1364	1.1375	1.1380	1.1381
11.50	1.1475	1.1506	1.1496	1.1530	1.1580	1.1618	1.1653	1.1696	1.1728	1.1755	1.1807	1.1822	1.1832	1.1837	1.1839
12.00	1.1963	1.1983	1.1964	1.1992	1.2037	1.2072	1.2105	1.2147	1.2178	1.2205	1.2256	1.2270	1.2281	1.2285	1.2287
12.50	1.2449	1.2458	1.2430	1.2449	1.2490	1.2522	1.2551	1.2592	1.2622	1.2648	1.2698	1.2711	1.2720	1.2724	1.2725
13.00	1.2934	1.2931	1.2893	1.2903	1.2939	1.2967	1.2993	1.3031	1.3060	1.3084	1.3131	1.3143	1.3152	1.3155	1.3156

14.00	1.3899	1.3870	1.3812	1.3802	1.3825	1.3845	1.3862	1.3918	1.3938	1.3977	1.3984	1.3991	1.3993	1.3992
14.50	1.4380	1.4337	1.4268	1.4247	1.4263	1.4278	1.4290	1.4339	1.4356	1.4390	1.4395	1.4400	1.4401	1.4400
15.00	1.4860	1.4803	1.4722	1.4689	1.4698	1.4708	1.4714	1.4756	1.4769	1.4797	1.4798	1.4802	1.4802	1.4800
15.50	1.5338	1.5266	1.5174	1.5129	1.5130	1.5135	1.5134	1.5168	1.5177	1.5198	1.5196	1.5197	1.5197	1.5194
16.00	1.5815	1.5728	1.5625	1.5566	1.5559	1.5558	1.5551	1.5575	1.5580	1.5594	1.5587	1.5587	1.5585	1.5582
16.50	1.6291	1.6189	1.6073	1.6001	1.5985	1.5979	1.5964	1.5978	1.5979	1.5984	1.5973	1.5971	1.5968	1.5964
17.00	1.6766	1.6649	1.6520	1.6434	1.6409	1.6397	1.6374	1.6378	1.6373	1.6370	1.6354	1.6350	1.6346	1.6341
17.50	1.7241	1.7107	1.6966	1.6865	1.6830	1.6812	1.6781	1.6773	1.6764	1.6750	1.6730	1.6723	1.6718	1.6712
18.00	1.7714	1.7564	1.7410	1.7293	1.7249	1.7225	1.7186	1.7166	1.7150	1.7127	1.7100	1.7091	1.7085	1.7078
18.50	1.8187	1.8020	1.7853	1.7720	1.7666	1.7635	1.7587	1.7554	1.7533	1.7499	1.7466	1.7455	1.7447	1.7439
19.00	1.8659	1.8475	1.8294	1.8146	1.8081	1.8043	1.7986	1.7940	1.7912	1.7866	1.7828	1.7814	1.7805	1.7796
19.50	1.9130	1.8929	1.8734	1.8569	1.8493	1.8449	1.8382	1.8322	1.8288	1.8230	1.8186	1.8169	1.8158	1.8148
20.00	1.9600	1.9382	1.9173	1.8991	1.8904	1.8853	1.8776	1.8702	1.8661	1.8590	1.8540	1.8519	1.8508	1.8496
20.50	2.0070	1.9834	1.9611	1.9412	1.9314	1.9255	1.9168	1.9079	1.9031	1.8947	1.8889	1.8866	1.8853	1.8840
21.00	2.0539	2.0285	2.0048	1.9831	1.9721	1.9655	1.9557	1.9453	1.9397	1.9300	1.9236	1.9209	1.9195	1.9180
21.50	2.1007	2.0736	2.0484	2.0248	2.0127	2.0054	1.9944	1.9824	1.9761	1.9650	1.9578	1.9549	1.9532	1.9517
22.00	2.1475	2.1185	2.0918	2.0665	2.0531	2.0450	2.0330	2.0193	2.0122	1.9997	1.9917	1.9884	1.9867	1.9850
22.50	2.1943	2.1634	2.1352	2.1080	2.0934	2.0845	2.0713	2.0560	2.0481	2.0341	2.0253	2.0217	2.0148	2.0179
23.00	2.2410	2.2082	2.1785	2.1494	2.1335	2.1239	2.1095	2.0924	2.0837	2.0681	2.0586	2.0546	2.0525	2.0506
23.50	2.2876	2.2529	2.2217	2.1906	2.1735	2.1631	2.1475	2.1286	2.1191	2.1019	2.0916	2.0872	2.0850	2.0829
24.00	2.3342	2.2976	2.2648	2.2318	2.2134	2.2021	2.1853	2.1646	2.1542	2.1355	2.1242	2.1196	2.1171	2.1149
24.50	2.3807	2.3422	2.3079	2.2728	2.2531	2.2410	2.2229	2.2005	2.1891	2.1687	2.1567	2.1516	2.1490	2.1466
25.00	2.4272	2.3867	2.3509	2.3138	2.2927	2.2798	2.2604	2.2361	2.2238	2.2017	2.1888	2.1834	2.1806	2.1780
25.50	2.4736	2.4312	2.3937	2.3546	2.3322	2.3184	2.2978	2.2715	2.2583	2.2345	2.2207	2.2149	2.2119	2.2092
26.00	2.5200	1.4756	2.4366	2.3953	2.3716	2.3569	2.3350	2.3067	2.2927	2.2671	2.2523	2.2461	2.2430	2.2401
26.50	2.5664	2.5200	2.4793	2.4360	2.4109	2.3953	2.3720	2.3418	2.3268	2.2994	2.2837	2.2771	2.2738	2.2707
27.00	2.6127	2.5643	2.5220	2.4766	2.4501	2.4336	2.4089	2.3767	2.3607	2.3315	2.3149	2.3078	2.3044	2.3011
27.50	2.6590	2.6086	2.5646	2.5170	2.4891	2.4718	2.4457	2.4115	2.3944	2.3634	2.3458	2.3384	2.3347	2.3313
28.00	2.7053	2.6520	2.6072	2.5574	2.5281	2.5098	2.4824	2.4460	2.4280	2.3951	2.3765	2.3687	2.3648	2.3612
28.50	2.7515	2.6969	2.6497	2.5977	2.5669	2.5478	2.5189	2.4805	2.4614	2.4266	2.4070	2.3987	2.3947	2.3909
29.00	2.7977	2.7410	2.6921	2.6380	2.6057	2.5856	2.5553	2.5148	2.4947	2.4579	2.4373	2.4286	2.4244	2.4205
29.50	2.8438	2.7851	2.7345	2.6781	2.6444	2.6234	2.5916	2.5489	2.5278	2.4890	2.4674	2.4583	2.4538	2.4497
30.00	2.8899	2.8291	2.7769	2.7182	2.6830	2.6610	2.6278	2.5829	2.5607	2.5200	2.4974	2.4878	2.4831	2.4788

Table 7-2 (e)
Extended Sukkar-Cornell Integral for Bottom-hole Pressure Calculation

Reduced Temperature for $B = 20.0$

P_r	1.1	1.2	1.3	1.4	1.5	1.6	1.7	1.8	1.9	2.0	2.2	2.4	2.6	2.8	3.0
0.20	0.0000	0.0000	0.0000	0.0000	0.0000	0.0000	0.0000	0.0000	0.0000	0.0000	0.0000	0.0000	0.0000	0.0000	0.0000
0.50	0.0058	0.0056	0.0055	0.0055	0.0054	0.0054	0.0053	0.0053	0.0053	0.0053	0.0052	0.0052	0.0052	0.0052	0.0052
1.00	0.0294	0.0272	0.0262	0.0255	0.0250	0.0246	0.0243	0.0241	0.0240	0.0239	0.0237	0.0236	0.0235	0.0234	0.0233
1.50	0.0740	0.0649	0.0610	0.0587	0.0572	0.0561	0.0561	0.0545	0.0541	0.0537	0.0532	0.0528	0.0525	0.0522	0.0520
2.00	0.1295	0.1156	0.1077	0.1030	0.0998	0.0976	0.0958	0.0945	0.0937	0.0930	0.0918	0.0911	0.0905	0.0900	0.0895
2.50	0.1832	0.1712	0.1614	0.1547	0.1498	0.1465	0.1438	0.1417	0.1404	0.1393	0.1376	0.1364	0.1354	0.1346	0.1339
3.00	0.2350	0.2264	0.2172	0.2099	0.2040	0.1999	0.1964	0.1937	0.1920	0.1904	0.1882	0.1867	0.1853	0.1842	0.1832
3.50	0.2860	0.2801	0.2725	0.2657	0.2597	0.2553	0.2514	0.2484	0.2463	0.2445	0.2419	0.2401	0.2384	0.2371	0.2359
4.00	0.3365	0.3326	0.3264	0.3208	0.3154	0.3111	0.3073	0.3041	0.3020	0.3000	0.2972	0.2952	0.2934	0.2919	0.2906
4.50	0.3865	0.3841	0.3790	0.3747	0.3703	0.3664	0.3629	0.3599	0.3578	0.3559	0.3531	0.3510	0.3492	0.3476	0.3462
5.00	0.4360	0.4346	0.4305	0.4273	0.4240	0.4208	0.4177	0.4151	0.4132	0.4114	0.4088	0.4068	0.4050	0.4034	0.4021
5.50	0.4852	0.4843	0.4809	0.4787	0.4765	0.4740	0.4714	0.4594	0.4678	0.4662	0.4639	0.4622	0.4604	0.4589	0.4577
6.00	0.5341	0.5335	0.5305	0.5291	0.5279	0.5261	0.5241	0.5226	0.5213	0.5201	0.5182	0.5167	0.5151	0.5137	0.5125
6.50	0.5827	0.5821	0.5794	0.5786	0.5783	0.5771	0.5756	0.5747	0.5738	0.5729	0.5714	0.5703	0.5689	0.5676	0.5665
7.00	0.6310	0.6304	0.6277	0.6274	0.6276	0.6270	0.6261	0.6257	0.6252	0.6246	0.6236	0.6228	0.6216	0.6205	0.6194
7.50	0.6791	0.6782	0.6755	0.6754	0.6761	0.6760	0.6755	0.6756	0.6754	0.6752	0.6746	0.6741	0.6732	0.6722	0.6712
8.00	0.7269	0.7257	0.7227	0.7228	0.7238	0.7241	0.7240	0.7245	0.7247	0.7247	0.7251	0.7244	0.7237	0.7227	0.7219
8.50	0.7745	0.7728	0.7695	0.7696	0.7708	0.7714	0.7716	0.7725	0.7729	0.7732	0.7740	0.7735	0.7730	0.7722	0.7714
9.00	0.8219	0.8196	0.8159	0.8160	0.8172	0.8179	0.8184	0.8195	0.8202	0.8207	0.8218	0.8216	0.8212	0.8205	0.8198
9.50	0.8690	0.8661	0.8620	0.8618	0.8631	0.8638	0.8644	0.8658	0.8666	0.8673	0.8687	0.8687	0.8684	0.8678	0.8672
10.00	0.9159	0.9123	0.9077	0.9073	0.9083	0.9091	0.9098	0.9113	0.9123	0.9131	0.9147	0.9148	0.9146	0.9141	0.9135
10.50	0.9626	0.9582	0.9530	0.9523	0.9531	0.9538	0.9545	0.9561	0.9571	0.9580	0.9599	0.9601	0.9599	0.9595	0.9589
11.00	1.0091	1.0089	0.9981	0.9969	0.9975	0.9980	0.9987	1.0002	1.0014	1.0023	1.0043	1.0045	1.0043	1.0039	1.0034
11.50	1.0554	1.0494	1.0429	1.0412	1.0414	1.0418	1.0423	1.0438	1.0450	1.0459	1.0479	1.0481	1.0479	1.0475	1.0470
12.00	1.1016	1.0946	1.0874	1.0851	1.0849	1.0851	1.0855	1.0868	1.0879	1.0688	1.0908	1.0909	1.0908	1.0903	1.0698
12.50	1.1476	1.1397	1.1317	1.1288	1.1282	1.1280	1.1282	1.1294	1.1304	1.1312	1.1331	1.1331	1.1328	1.1323	1.1318
13.00	1.1935	1.1846	1.1758	1.1721	1.1710	1.1706	1.1704	1.1714	1.1723	1.1730	1.1746	1.1745	1.1742	1.1736	1.1731

14.00		1.2833			1.2779	1.2758	1.2547	1.2542	1.2537	1.2547	1.2549	1.2559	1.2549	1.2564	1.2549	1.2542	1.2535
14.50	1.3304	1.3183	1.3068	1.3005	1.2977	1.2962	1.2948	1.2952	1.2952	1.2957	1.2949	1.2963	1.2935	1.2928
15.00	1.3759	1.3625	1.3501	1.3428	1.3394	1.3375	1.3355	1.3353	1.3352	1.3349	1.3349	1.3339	1.3331	1.3322	1.3315
15.50	1.4212	1.4067	1.3933	1.3849	1.3808	1.3784	1.3759	1.3754	1.3749	1.3743	1.3736	1.3723	1.3713	1.3704	1.3695
16.00	1.4665	1.4507	1.4363	1.4267	1.4220	1.4191	1.4150	1.4151	1.4142	1.4132	1.4118	1.4101	1.4090	1.4080	1.4071
16.50	1.5116	1.4945	1.4792	1.4684	1.4629	1.4595	1.4558	1.4544	1.4531	1.4517	1.4496	1.4475	1.4462	1.4451	1.4441
17.00	1.5567	1.5383	1.5219	1.5099	1.5036	1.4997	1.4953	1.4935	1.4916	1.4898	1.4869	1.4844	1.4829	1.4817	1.4806
17.50	1.6017	1.5820	1.5645	1.5512	1.5441	1.5397	1.5345	1.5323	1.5298	1.5275	1.5238	1.5208	1.5191	1.5178	1.5166
18.00	1.6467	1.6256	1.6069	1.5924	1.5844	1.5794	1.5735	1.5708	1.5678	1.5649	1.5603	1.5588	1.5549	1.5584	1.5522
18.50	1.6916	1.6691	1.6493	1.6334	1.6245	1.6190	1.6123	1.6090	1.6054	1.6020	1.5964	1.5924	1.5902	1.5837	1.5973
19.00	1.7364	1.7125	1.6915	1.6742	1.6644	1.6583	1.6508	1.6470	1.6427	1.6388	1.6321	1.6275	1.6252	1.6235	1.6220
19.50	1.7811	1.7558	1.7336	1.7149	1.7042	1.6975	1.6891	1.6847	1.6797	1.6752	1.6675	1.6623	1.6597	1.6579	1.6563
20.00	1.8258	1.7990	1.7757	1.7555	1.7438	1.7364	1.7271	1.7222	1.7165	1.7114	1.7025	1.6967	1.6938	1.6919	1.6902
20.50	1.8705	1.8421	1.8176	1.7959	1.7832	1.7752	1.7650	1.7595	1.7530	1.7473	1.7372	1.7308	1.7276	1.7256	1.7238
21.00	1.9150	1.8852	1.8594	1.8362	1.8225	1.8139	1.8027	1.7965	1.7893	1.7829	1.7716	1.7645	1.7611	1.7589	1.7570
21.50	1.9596	1.9282	1.9012	1.8763	1.8616	1.8523	1.8401	1.8334	1.8254	1.8183	1.8056	1.7979	1.7942	1.7918	1.7898
22.00	2.0041	1.9711	1.9429	1.9164	1.9006	1.8906	1.8774	1.8700	1.8612	1.8534	1.8394	1.8310	1.8270	1.8245	1.8223
22.50	2.0485	2.0140	1.9844	1.9563	1.9395	1.9288	1.9146	1.9065	1.8968	1.8882	1.8730	1.8638	1.8595	1.8568	1.8545
23.00	2.0929	2.0568	2.0259	1.9982	1.9782	1.9668	1.9516	1.9428	1.9322	1.9229	1.9062	1.8963	1.8916	1.8889	1.8864
23.50	2.1372	2.0995	2.0674	2.0359	2.0168	2.0047	1.9884	1.9789	1.9674	1.9573	1.9392	1.9286	1.9235	1.9206	1.9180
24.00	2.1815	2.1422	2.1087	2.0756	2.0553	2.0425	2.0250	2.0149	2.0025	1.9916	1.9719	1.9605	1.9551	1.9521	1.9493
24.50	2.2258	2.1849	2.1500	2.1151	2.0937	2.0801	2.0615	2.0507	2.0373	2.0256	2.0044	1.9922	1.9865	1.9832	1.9804
25.00	2.2700	2.2274	2.1912	2.1546	2.1319	2.1176	2.0979	2.0863	2.0719	2.0594	2.0367	2.0237	2.0176	2.0142	2.0112
25.50	2.3142	2.2700	2.2324	2.1939	2.1701	2.1550	2.1341	2.1218	2.1064	2.0980	2.0687	2.0549	2.0484	2.0449	2.0417
26.00	2.3584	2.3124	2.2735	2.2332	2.2002	2.1923	2.1702	2.1671	2.1408	2.1265	2.1005	2.0858	2.0790	2.0753	2.0720
26.50	2.4025	2.3549	2.3145	2.2724	2.2461	2.2295	2.2062	2.1923	2.1749	2.1598	2.1321	2.1166	2.1094	2.1055	2.1020
27.00	2.4466	2.3973	2.3555	2.3115	2.2840	2.2665	2.2420	2.2274	2.2089	2.1929	2.1636	2.1471	2.1395	2.1355	2.1318
27.50	2.4907	2.4396	2.3964	2.3505	2.3218	2.3035	2.2778	2.2823	2.2428	2.2258	2.1968	2.1774	2.1695	2.1652	2.1614
28.00	2.5347	2.4819	2.4373	2.3895	2.3595	2.3404	2.3134	2.2971	2.2765	2.2586	2.2258	2.2075	2.1992	2.1948	2.1908
28.50	2.5787	2.5242	2.4781	2.4284	2.3971	2.3772	2.3409	2.3118	2.3100	2.2912	2.2566	2.2375	2.2287	2.2241	2.2220
29.00	2.6227	2.5664	2.5189	2.4672	2.4146	2.4119	2.3843	2.3664	2.3435	2.3217	2.2873	2.2677	2.2540	2.2662	2.2480
29.50	2.6666	2.6085	2.5596	2.5060	2.4720	2.4504	2.4195	2.4008	2.3768	2.3560	2.3178	2.2967	2.2871	2.2822	2.2777
30.00	2.7106	2.6507	2.6003	2.5447	2.5094	2.4870	2.4547	2.4352	2.4100	2.3882	2.3481	2.3261	2.3161	2.3109	2.3063

Table 7-2 (f)
Extended Sukkar-Cornell Integral for Bottom-hole Pressure Calculation

P_r	\multicolumn{12}{c}{Reduced Temperature for $B = 25.0$}
	1.1	1.2	1.3	1.4	1.5	1.6	1.7	1.8	1.9	2.0	2.2	2.4	2.6	2.8	3.0
0.20	0.0000	0.0000	0.0000	0.0000	0.0000	0.0000	0.0000	0.0000	0.0000	0.0000	0.0000	0.0000	0.0000	0.0000	0.0000
0.50	0.0047	0.0045	0.0044	0.0044	0.0043	0.0043	0.0043	0.0042	0.0042	0.0042	0.0042	0.0042	0.0042	0.0042	0.0042
1.00	0.0237	0.0219	0.0211	0.0205	0.0201	0.0198	0.0196	0.0194	0.0193	0.0192	0.0191	0.0189	0.0189	0.0198	0.0187
1.50	0.0611	0.0529	0.0496	0.0477	0.0464	0.0454	0.0446	0.0441	0.0438	0.0435	0.0430	0.0427	0.0424	0.0422	0.0420
2.00	0.1106	0.0961	0.0888	0.0846	0.0818	0.0798	0.0783	0.0771	0.0764	0.0758	0.0749	0.0742	0.0737	0.0733	0.0729
2.50	0.1598	0.1453	0.1352	0.1287	0.1241	0.1211	0.1186	0.1188	0.1156	0.1146	0.1131	0.1171	0.1111	0.1104	0.1098
3.00	0.2079	0.1952	0.1846	0.1769	0.1711	0.1670	0.1637	0.1612	0.1596	0.1581	0.1561	0.1547	0.1534	0.1524	0.1515
3.50	0.2554	0.2444	0.2346	0.2267	0.2202	0.2156	0.2117	0.2087	0.2057	0.2049	0.2024	0.2007	0.1991	0.1978	0.1967
4.00	0.3025	0.2930	0.2840	0.2766	0.2702	0.2654	0.2613	0.2579	0.2557	0.2537	0.2508	0.2488	0.2470	0.2455	0.2442
4.50	0.3492	0.3408	0.3325	0.3260	0.3200	0.3154	0.3112	0.3078	0.3055	0.3036	0.3004	0.2982	0.2962	0.2946	0.2932
5.00	0.3957	0.3879	0.3803	0.3745	0.3693	0.3650	0.3610	0.3578	0.3555	0.3536	0.3503	0.3481	0.3461	0.3444	0.3429
5.50	0.4418	0.4345	0.4274	0.4223	0.4178	0.4139	0.4103	0.4073	0.4052	0.4031	0.4002	0.3980	0.3961	0.3943	0.3929
6.00	0.4878	0.4806	0.4739	0.4694	0.4656	0.4622	0.4589	0.4563	0.4543	0.4525	0.4498	0.4477	0.4450	0.4441	0.4428
6.50	0.5335	0.5263	0.5198	0.5158	0.5126	0.5097	0.5068	0.5045	0.5028	0.5012	0.4988	0.4969	0.4951	0.4935	0.4922
7.00	0.5790	0.5718	0.5653	0.5616	0.5589	0.5564	0.5539	0.5520	0.5506	0.5492	0.5471	0.5454	0.5437	0.5422	0.5409
7.50	0.6243	0.6169	0.6104	0.6069	0.6045	0.6024	0.6003	0.5987	0.5975	0.5966	0.5946	0.5932	0.5917	0.5902	0.5690
8.00	0.6694	0.6618	0.6550	0.6516	0.6495	0.6477	0.6459	0.6447	0.6437	0.6428	0.6415	0.6401	0.6388	0.6374	0.6362
8.50	0.7143	0.7063	0.6993	0.6960	0.6940	0.6924	0.6908	0.6899	0.6892	0.6884	0.6874	0.6882	0.6850	0.6837	0.6826
9.00	0.7591	0.7506	0.7433	0.7399	0.7380	0.7365	0.7351	0.7344	0.7338	0.7333	0.7325	0.7315	0.7304	0.7292	0.7282
9.50	0.8036	0.7946	0.7870	0.7834	0.7814	0.7800	0.7788	0.7783	0.7778	0.7774	0.7769	0.7760	0.7750	0.7739	0.7830
10.00	0.8480	0.8384	0.8303	0.8266	0.8245	0.8231	0.8219	0.8215	0.8212	0.8208	0.8205	0.8198	0.8189	0.8178	0.8169
10.50	0.8922	0.8520	0.8735	0.8695	0.8671	0.8657	0.8645	0.8641	0.8639	0.8636	0.8635	0.8628	0.8619	0.8609	0.8600
11.00	0.9362	0.9254	0.9163	0.9120	0.9094	0.9078	0.9056	0.9063	0.9061	0.9058	0.9058	0.9052	0.9043	0.9033	0.9024
11.50	0.9801	0.9686	0.9590	0.9542	0.9514	0.9496	0.9483	0.9679	0.9477	0.9475	0.9475	0.9468	0.9459	0.9449	0.9440
12.00	1.0239	1.0117	1.0014	0.9961	0.9930	0.9910	0.9896	0.9891	0.9889	0.9886	0.9885	0.9879	0.9864	0.9854	0.9850
12.50	1.0676	1.0545	1.0437	1.0378	1.0343	1.0321	1.0304	1.0298	1.0295	1.0292	1.0240	1.0283	1.0273	1.0262	1.0253
13.00	1.1111	1.0973	1.0857	1.0792	1.0753	1.0729	1.0709	1.0701	1.0698	1.0693	1.0689	1.0681	1.0670	1.0659	1.0650

14.00	1.1979	1.1823	1.1693	1.1614	1.1566	1.1535	1.1509	1.1496	1.1489	1.1481	1.1472	1.1459	1.1447	1.1435	1.1425
14.50	1.2412	1.2246	1.2109	1.2021	1.1968	1.1934	1.1904	1.1889	1.1879	1.1868	1.1855	1.1840	1.1827	1.1615	1.1604
15.00	1.2844	1.2668	1.2523	1.2427	1.2368	1.2331	1.2296	1.2278	1.2265	1.2252	1.2234	1.2217	1.2202	1.2189	1.2177
15.50	1.3275	1.3089	1.2936	1.2830	1.2766	1.2725	1.2685	1.2663	1.2647	1.2631	1.2608	1.2588	1.2572	1.2558	1.2546
16.00	1.3705	1.3509	1.3347	1.3232	1.3161	1.3116	1.3071	1.3046	1.3026	1.3007	1.2978	1.2954	1.2937	1.2922	1.2909
16.50	1.4135	1.3928	1.3757	1.3632	1.3555	1.3505	1.3455	1.3426	1.3402	1.3379	1.3343	1.3316	1.3298	1.3281	1.3268
17.00	1.4564	1.4346	1.4166	1.4031	1.3947	1.3892	1.3836	1.3803	1.3775	1.3748	1.3705	1.3674	1.3653	1.3637	1.3623
17.50	1.4992	1.4763	1.4574	1.4428	1.4336	1.4278	1.4215	1.4178	1.4145	1.4114	1.4062	1.4028	1.4005	1.3987	1.3973
18.00	1.5420	1.5180	1.4981	1.4823	1.4724	1.4661	1.4591	1.4550	1.4512	1.4476	1.4617	1.4377	1.4353	1.4334	1.4318
18.50	1.5847	1.5595	1.5387	1.5217	1.5111	1.5042	1.4965	1.4920	1.4876	1.4835	1.4767	1.4728	1.4697	1.4677	1.4660
19.00	1.6274	1.6010	1.5792	1.5610	1.5496	1.5422	1.5338	1.5287	1.5238	1.5192	1.5114	1.5065	1.5036	1.5015	1.4998
19.50	1.6700	1.6424	1.6196	1.6002	1.5879	1.5800	1.5708	1.5653	1.5597	1.5546	1.5458	1.5404	1.5373	1.5351	1.5332
20.00	1.7126	1.6837	1.6597	1.6392	1.6261	1.6176	1.6076	1.6016	1.5954	1.5997	1.5799	1.5739	1.5706	1.5692	1.5663
20.50	1.7551	1.7250	1.7001	1.6781	1.6641	1.6551	1.6443	1.6377	1.6308	1.6246	1.6137	1.6071	1.6035	1.6011	1.5990
21.00	1.7975	1.7662	1.7403	1.7169	1.7020	1.6924	1.6808	1.6736	1.6660	1.6592	1.6472	1.6400	1.6362	1.6336	1.6614
21.50	1.8400	1.8073	1.7803	1.7556	1.7398	1.7296	1.7171	1.7094	1.7011	1.6936	1.6804	1.6726	1.6685	1.6658	1.6635
22.00	1.8824	1.8484	1.8203	1.7942	1.7775	1.7667	1.7532	1.7450	1.7359	1.7278	1.7134	1.7049	1.7005	1.6977	1.6953
22.50	1.9247	1.8895	1.8603	1.8327	1.8150	1.8036	1.7892	1.7804	1.7705	1.7617	1.7460	1.7370	1.7322	1.7243	1.7267
23.00	1.9670	1.9304	1.9001	1.8711	1.8524	1.8404	1.8251	1.8156	1.8049	1.7955	1.7785	1.7687	1.7637	1.7606	1.7579
23.50	2.0093	1.9714	1.9399	1.9094	1.8898	1.8771	1.8608	1.8507	1.8392	1.8290	1.8107	1.8002	1.7949	1.7916	1.7889
24.00	2.0516	2.0122	1.9797	1.9477	1.9270	1.9136	1.8964	1.8856	1.8733	1.8623	1.8427	1.8315	1.8258	1.8224	1.8195
24.50	2.0938	2.0531	2.0193	1.9858	1.9641	1.9501	1.9318	1.9204	1.9072	1.8955	1.8744	1.8625	1.8565	1.8530	1.8499
25.00	2.1360	2.0938	2.0590	2.0239	2.0011	1.9864	1.9671	1.9550	1.9409	1.9285	1.9060	1.8933	1.8870	1.8833	1.8801
25.50	2.1761	2.1346	2.0985	2.0618	2.0380	2.0226	2.0023	1.9895	1.9745	1.9613	1.9373	1.9238	1.9172	1.9133	1.9100
26.00	2.2202	2.1753	2.1380	2.0998	2.0749	2.0588	2.0373	2.0239	2.0079	1.9939	1.9684	1.9542	1.9472	1.9431	1.9397
26.50	2.2623	2.2159	2.1775	2.1376	2.1116	2.0948	2.0723	2.0581	2.0412	2.0264	1.9994	1.9843	1.9769	1.9728	1.9692
27.00	2.3044	2.2566	2.2169	2.1754	2.1403	2.1307	2.1071	2.0923	2.0744	2.0587	2.0301	2.0142	2.0065	2.0022	1.9964
27.50	2.3464	2.2971	2.2562	2.2131	2.1848	2.1666	2.1418	2.1263	2.1074	2.0909	2.0607	2.0440	2.0359	2.0314	2.0275
28.00	2.3885	2.3377	2.2955	2.2507	2.2213	2.2024	2.1764	2.1601	2.1403	2.1229	2.0911	2.0735	2.0650	2.0603	2.0563
28.50	2.4305	2.3782	2.3348	2.2883	2.2578	2.2380	2.2110	2.1939	2.1730	2.1548	2.1213	2.1028	2.0940	2.0891	2.0849
29.00	2.4724	2.4186	2.3740	2.3258	2.2941	2.2736	2.2454	2.2276	2.2056	2.1865	2.1913	2.1320	2.1228	2.1178	2.1134
29.50	2.5144	2.4591	2.4132	2.3632	2.3304	2.3091	2.2797	2.2611	2.2331	2.2181	2.1812	2.1610	2.1514	2.1462	2.1417
30.00	2.5563	2.4995	2.4523	2.4006	2.3666	2.3446	2.3139	2.2946	2.2705	2.2496	2.2110	2.1898	2.1798	2.1744	2.1698

Table 7-2 (g)
Extended Sukkar-Cornell Integral for Bottom-hole Pressure Calculation

P_r	\multicolumn{20}{c}{Reduced Temperature for $B = 30.0$}
	1.1	1.2	1.3	1.4	1.5	1.6	1.7	1.8	1.9	2.0	2.2	2.4	2.6	2.8	3.0
0.20	0.0000	0.0000	0.0000	0.0000	0.0000	0.0000	0.0000	0.0000	0.0000	0.0000	0.0000	0.0000	0.0000	0.0000	0.0000
0.50	0.0039	0.0038	0.0037	0.0037	0.0036	0.0036	0.0036	0.0035	0.0035	0.0035	0.0035	0.0035	0.0035	0.0035	0.0035
1.00	0.0199	0.0184	0.0176	0.0172	0.0168	0.0166	0.0164	0.0162	0.0162	0.0161	0.0169	0.0159	0.0158	0.0157	0.0157
1.50	0.0521	0.0447	0.0418	0.0401	0.0390	0.0382	0.0375	0.0371	0.0368	0.0365	0.0361	0.0358	0.0356	0.0355	0.0353
2.00	0.0967	0.0823	0.0755	0.0718	0.0692	0.0676	0.0672	0.0652	0.0646	0.0640	0.0632	0.0626	0.0621	0.0618	0.0615
2.50	0.1422	0.1264	0.1164	0.1103	0.1060	0.1033	0.1010	0.0993	0.0963	0.0974	0.0960	0.0951	0.0943	0.0937	0.0931
3.00	0.1870	0.1719	0.1608	0.1531	0.1474	0.1436	0.1404	0.1381	0.1366	0.1353	0.1334	0.1321	0.1309	0.1300	0.1292
3.50	0.2314	0.2174	0.2063	0.1980	0.1914	0.1869	0.1831	0.1801	0.1782	0.1765	0.1741	0.1725	0.1710	0.1697	0.1687
4.00	0.2756	0.2625	0.2519	0.2436	0.2367	0.2318	0.2275	0.2242	0.2219	0.2199	0.2172	0.2152	0.2135	0.2120	0.2108
4.50	0.3195	0.3071	0.2970	0.2891	0.2823	0.2773	0.2729	0.2693	0.2669	0.2647	0.2617	0.2594	0.2675	0.2559	0.2545
5.00	0.3632	0.3513	0.3416	0.3343	0.3278	0.3229	0.3186	0.3149	0.3124	0.3101	0.3069	0.3046	0.3025	0.3008	0.2993
5.50	0.4067	0.3951	0.3858	0.3789	0.3729	0.3683	0.3641	0.3605	0.3580	0.3558	0.3525	0.3501	0.3480	0.3462	0.3448
6.00	0.4500	0.4386	0.4295	0.4230	0.4175	0.4132	0.4092	0.4059	0.4035	0.4013	0.3981	0.3957	0.3937	0.3919	0.3904
6.50	0.4931	0.4817	0.4728	0.4667	0.4616	0.4576	0.4539	0.4508	0.4486	0.4465	0.4435	0.4412	0.4392	0.4374	0.4359
7.00	0.5361	0.5247	0.5158	0.5099	0.5052	0.5015	0.4981	0.4952	0.4932	0.4913	0.4884	0.4863	0.4843	0.4826	0.4812
7.50	0.5789	0.5574	0.5584	0.5527	0.5483	0.5449	0.5417	0.5391	0.5372	0.5355	0.5329	0.5309	0.5291	0.5274	0.5260
8.00	0.6216	0.6098	0.6007	0.5951	0.5909	0.5877	0.5848	0.5824	0.5808	0.5792	0.5767	0.5749	0.5732	0.5716	0.5703
8.50	0.6642	0.6521	0.6428	0.6372	0.6331	0.6301	0.6273	0.6252	0.6237	0.6223	0.6200	0.6194	0.6168	0.6152	0.6139
9.00	0.7066	0.6941	0.6846	0.6789	0.6749	0.6719	0.6693	0.6674	0.6660	0.6647	0.6627	0.6612	0.6597	0.6582	0.6570
9.50	0.7488	0.7360	0.7261	0.7204	0.7163	0.7134	0.7109	0.7091	0.7078	0.7066	0.7048	0.7034	0.7020	0.7006	0.6994
10.00	0.7909	0.7775	0.7674	0.7615	0.7573	0.7544	0.7520	0.7503	0.7491	0.7480	0.7463	0.7451	0.7436	0.7423	0.7411
10.50	0.8329	0.8181	0.8085	0.8026	0.7980	0.7951	0.7926	0.7910	0.7899	0.7888	0.7873	0.7861	0.7847	0.7833	0.7822
11.00	0.8747	0.8604	0.8494	0.8430	0.8384	0.8354	0.8329	0.8313	0.8302	0.8292	0.8277	0.8265	0.8251	0.8238	0.8227
11.50	0.9165	0.9016	0.8901	0.8833	0.8785	0.8754	0.8728	0.8711	0.8700	0.8690	0.8676	0.8664	0.8650	0.8637	0.8626
12.00	0.9581	0.9426	0.9306	0.9234	0.9183	0.9150	0.9123	0.9106	0.9095	0.9084	0.9070	0.9057	0.9043	0.9030	0.9019
12.50	0.9996	0.9835	0.9710	0.9633	0.9579	0.9544	0.9515	0.9497	0.9485	0.9474	0.9459	0.9446	0.9431	0.9417	0.9406
13.00	1.0411	1.0242	1.0112	1.0030	0.9973	0.9936	0.9904	0.9884	0.9872	0.9860	0.9842	0.9828	0.9813	0.9799	0.9787

14.00	1.1237	1.0912	1.0518	1.0753	1.0710	1.0673	1.0649	1.0634	1.0618	1.0596	1.0579	1.0563	1.0547	1.0535
14.50	1.1649	1.1459	1.1310	1.1209	1.1139	1.1094	1.1054	1.1027	1.1009	1.0992	1.0966	1.0947	1.0930	1.0914	1.0901
15.00	1.2060	1.1862	1.1707	1.1598	1.1524	1.1475	1.1431	1.1402	1.1382	1.1362	1.1332	1.1311	1.1293	1.1276	1.1263
15.50	1.2471	1.2264	1.2102	1.1986	1.1907	1.1855	1.1606	1.1774	1.1751	1.1729	1.1694	1.1670	1.1651	1.1633	1.1620
16.00	1.2881	1.2666	1.2497	1.2372	1.2287	1.2232	1.2179	1.2144	1.2117	1.2092	1.2052	1.2026	1.2005	1.1987	1.1972
16.50	1.3291	1.3067	1.2890	1.2757	1.2666	1.2607	1.2549	1.2511	1.2481	1.2453	1.2407	1.2377	1.2354	1.2335	1.2320
17.00	1.3700	1.3467	1.3282	1.3140	1.3044	1.2981	1.2917	1.2876	1.2842	1.2610	1.2757	1.2724	1.2700	1.2680	1.2665
17.50	1.4109	1.3866	1.3674	1.3522	1.3419	1.3352	1.3283	1.3238	1.3200	1.3164	1.3105	1.3067	1.3042	1.3021	1.3005
18.00	1.4517	1.4264	1.4064	1.3903	1.3794	1.3722	1.3847	1.3598	1.3555	1.3515	1.3449	1.3407	1.3380	1.3358	1.3341
18.50	1.4924	1.4662	1.4454	1.4282	1.4167	1.4091	1.4009	1.3956	1.3908	1.3864	1.3789	1.3744	1.3714	1.3692	1.3674
19.00	1.5332	1.5059	1.4843	1.4661	1.4538	1.4457	1.4370	1.4312	1.4259	1.4211	1.4127	1.4077	1.4045	1.4022	1.4003
19.50	1.5738	1.5456	1.5231	1.5038	1.4908	1.4823	1.4728	1.4666	1.4608	1.4554	1.4462	1.4407	1.4373	1.4349	1.4329
20.00	1.6145	1.5852	1.5618	1.5414	1.5277	1.5187	1.5085	1.5019	1.4954	1.4896	1.4794	1.4734	1.4698	1.4672	1.4652
20.50	1.6551	1.6247	1.6005	1.5789	1.5644	1.5549	1.5440	1.5369	1.5298	1.5235	1.5123	1.5058	1.5019	1.4993	1.4971
21.00	1.6956	1.6642	1.6391	1.6163	1.6011	1.5910	1.5794	1.5718	1.5641	1.5572	1.5449	1.5379	1.5338	1.5310	1.5288
21.50	1.7361	1.7037	1.6776	1.6537	1.6376	1.6270	1.6146	1.6065	1.5981	1.5906	1.5773	1.5697	1.5654	1.5625	1.5601
22.00	1.7766	1.7431	1.7160	1.6909	1.6740	1.6629	1.6497	1.6410	1.6320	1.6239	1.6095	1.6013	1.5967	1.5937	1.5912
22.50	1.8171	1.7824	1.7544	1.7281	1.7103	1.6987	1.6846	1.6754	1.6657	1.6570	1.6414	1.6326	1.6277	1.6246	1.6220
23.00	1.8575	1.8217	1.7928	1.7651	1.7485	1.7343	1.7194	1.7096	1.6992	1.6899	1.6731	1.6636	1.6585	1.6552	1.6525
23.50	1.8979	1.8610	1.8311	1.8021	1.7826	1.7698	1.7541	1.7437	1.7325	1.7226	1.7046	1.6945	1.6890	1.6856	1.6828
24.00	1.9383	1.9002	1.8693	1.8390	1.8186	1.8053	1.7886	1.7777	1.7657	1.7551	1.7358	1.7250	1.7193	1.7158	1.7128
24.50	1.9786	1.9393	1.9075	1.8759	1.8546	1.8406	1.8230	1.8115	1.7987	1.7874	1.7669	1.7554	1.7494	1.7457	1.7426
25.00	2.0189	1.9785	1.9456	1.9127	1.8904	1.8758	1.8573	1.8452	1.8316	1.8196	1.7977	1.7855	1.7792	1.7754	1.7722
25.50	2.0592	2.0176	1.9837	1.9493	1.9262	1.9110	1.8915	1.8788	1.8644	1.8516	1.8284	1.8155	1.8088	1.8048	1.8015
26.00	2.0995	2.0566	2.0217	1.9860	1.9618	1.9460	1.9256	1.9123	1.8970	1.8835	1.8589	1.8452	1.8382	1.8341	1.8306
26.50	2.1397	2.0957	2.0597	2.0226	1.9974	1.9810	1.9596	1.9456	1.9294	1.9152	1.8891	1.8747	1.8674	1.8631	1.8595
27.00	2.1799	2.1346	2.0976	2.0591	2.0330	2.0159	1.9934	1.9788	1.9618	1.9468	1.9192	1.9040	1.8964	1.8920	1.8882
27.50	2.2201	2.1736	2.1355	2.0955	2.0684	2.0507	2.0272	2.0119	1.9940	1.9782	1.9492	1.9332	1.9252	1.9206	1.9167
28.00	2.2603	2.2125	2.1734	2.1319	2.1038	2.0854	2.0609	2.0449	2.0261	2.0095	1.9790	1.9672	1.9538	1.9491	1.9451
28.50	2.3005	2.2514	2.2112	2.1682	2.1391	2.1200	2.0945	2.0779	2.0580	2.0407	2.0086	1.9910	1.9823	1.9774	1.9732
29.00	2.3406	2.2903	2.2490	2.2045	2.1743	2.1546	2.1280	2.1107	2.0899	2.0717	2.0380	2.0196	2.0105	2.0055	2.0012
29.50	2.3807	2.3291	2.2868	2.2407	2.2095	2.1891	2.1614	2.1434	2.1216	2.1026	2.0673	2.0481	2.0386	2.0334	2.0289
30.00	2.4208	2.3679	2.3245	2.2769	2.2446	2.2235	2.1947	2.1760	2.1533	2.1334	2.0965	2.0764	2.0666	2.0612	2.0566

Table 7-2 (h)
Extended Sukkar-Cornell Integral for Bottom-hole Pressure Calculation

Reduced Temperature for $B = 35.0$

P_r	1.1	1.2	1.3	1.4	1.5	1.6	1.7	1.8	1.9	2.0	2.2	2.4	2.6	2.8	3.0
0.20	0.0000	0.0000	0.0000	0.0000	0.0000	0.0000	0.0000	0.0000	0.0000	0.0000	0.0000	0.0000	0.0000	0.0000	0.0000
0.50	0.0033	0.0032	0.0032	0.0031	0.0031	0.0031	0.0031	0.0030	0.0030	0.0030	0.0030	0.0030	0.0030	0.0030	0.0030
1.00	0.0171	0.0158	0.0152	0.0148	0.0145	0.0143	0.0141	0.0139	0.0139	0.0139	0.0137	0.0136	0.0136	0.0135	0.0135
1.50	0.0454	0.0387	0.0361	0.0346	0.0336	0.0329	0.0323	0.0320	0.0317	0.0315	0.0311	0.0309	0.0307	0.0305	0.0304
2.00	0.0861	0.0720	0.0657	0.0623	0.0601	0.0585	0.0573	0.0564	0.0559	0.0554	0.0546	0.0542	0.0537	0.0534	0.0531
2.50	0.1283	0.1119	0.1022	0.0965	0.0925	0.0900	0.0879	0.0864	0.0855	0.0847	0.0834	0.0826	0.0819	0.0813	0.0808
3.00	0.1703	0.1538	0.1425	0.1350	0.1295	0.1259	0.1230	0.1208	0.1194	0.1182	0.1165	0.1153	0.1142	0.1134	0.1127
3.50	0.2120	0.1960	0.1844	0.1759	0.1694	0.1650	0.1613	0.1585	0.1567	0.1550	0.1528	0.1513	0.1499	0.1487	0.1478
4.00	0.2536	0.2382	0.2266	0.2179	0.2108	0.2059	0.2017	0.1984	0.1962	0.1942	0.1916	0.1897	0.1880	0.1866	0.1855
4.50	0.2950	0.2800	0.2688	0.2601	0.2529	0.2477	0.2433	0.2396	0.2372	0.2350	0.2320	0.2298	0.2279	0.2263	0.2250
5.00	0.3362	0.3216	0.3106	0.3023	0.2951	0.2899	0.2854	0.2816	0.2790	0.2766	0.2734	0.2710	0.2690	0.2672	0.2658
5.50	0.3773	0.3630	0.3522	0.3442	0.3373	0.3321	0.3276	0.3238	0.3211	0.3187	0.3153	0.3128	0.3107	0.3089	0.3074
6.00	0.4183	0.4040	0.3934	0.3857	0.3791	0.3742	0.3698	0.3660	0.3634	0.3610	0.3576	0.3550	0.3529	0.3510	0.3495
6.50	0.4591	0.4449	0.4344	0.4270	0.4207	0.4159	0.4117	0.4080	0.4055	0.4032	0.3998	0.3972	0.3951	0.3932	0.3918
7.00	0.4999	0.4856	0.4752	0.4679	0.4618	0.4573	0.4532	0.4498	0.4473	0.4451	0.4418	0.4394	0.4373	0.4354	0.4339
7.50	0.5405	0.5261	0.5156	0.5085	0.5026	0.4983	0.4944	0.4912	0.4889	0.4867	0.4836	0.4812	0.4792	0.4774	0.4759
8.00	0.5810	0.5665	0.5558	0.5487	0.5431	0.5390	0.5352	0.5822	0.5300	0.5280	0.5247	0.5227	0.5208	0.5190	0.5175
8.50	0.6214	0.6066	0.5959	0.5888	0.5832	0.5792	0.5756	0.5727	0.5707	0.5688	0.5657	0.5638	0.5619	0.5602	0.5588
9.00	0.6617	0.6466	0.6357	0.6285	0.6230	0.6191	0.6156	0.6129	0.6109	0.6091	0.6062	0.6044	0.6026	0.6009	0.5996
9.50	0.7018	0.6865	0.6753	0.6681	0.6625	0.6586	0.6552	0.6526	0.6507	0.6490	0.6462	0.6445	0.6428	0.6412	0.6398
10.00	0.7419	0.7262	0.7147	0.7073	0.7017	0.6978	0.6945	0.6919	0.6901	0.6885	0.6858	0.6842	0.6825	0.6809	0.6796
10.50	0.7818	0.7657	0.7539	0.7464	0.7406	0.7367	0.7334	0.7308	0.7291	0.7275	0.7250	0.7234	0.7217	0.7201	0.7189
11.00	0.8217	0.8051	0.7930	0.7852	0.7793	0.7753	0.7719	0.7694	0.7677	0.7661	0.7637	0.7621	0.7604	0.7589	0.7576
11.50	0.8614	0.8444	0.8319	0.8239	0.8177	0.8136	0.8102	0.8076	0.8059	0.8043	0.8019	0.8004	0.7987	0.7971	0.7958
12.00	0.9011	0.8836	0.8607	0.8623	0.8559	0.8517	0.8481	0.8655	0.8438	0.3422	0.8398	0.8381	0.8364	0.8349	0.8336
12.50	0.9407	0.9227	0.9094	0.9006	0.8939	0.8895	0.8858	0.8831	0.8813	0.8797	0.8771	0.8755	0.8737	0.8721	0.8708
13.00	0.9803	0.9617	0.9479	0.9386	0.9317	0.9271	0.9232	0.9204	0.9185	0.9168	0.9141	0.9124	0.9106	0.9089	0.9076

14.00	1.0591	1.0394	1.0246	1.0143	1.0067	1.0017	0.9973	0.9941	0.9920	0.9900	0.9869	0.9848	0.9829	0.9812	0.9798
14.50	1.0985	1.0781	1.0627	1.0519	1.0439	1.0386	1.0340	1.0305	1.0282	1.0261	1.0226	1.0205	1.0184	1.0167	1.0153
15.00	1.1377	1.1167	1.1008	1.0893	1.0809	1.0754	1.0704	1.0667	1.0642	1.0618	1.0580	1.0557	1.0536	1.0517	1.0503
15.50	1.1770	1.1552	1.1388	1.1266	1.1178	1.1120	1.1066	1.1027	1.0999	1.0973	1.0931	1.0905	1.0883	1.0864	1.0849
16.00	1.2162	1.1937	1.1767	1.1638	1.1549	1.1484	1.1426	1.1384	1.1354	1.1325	1.1278	1.1247	1.1226	1.1206	1.1191
16.50	1.2553	1.2321	1.2144	1.2008	1.1911	1.1846	1.1784	1.1739	1.1705	1.1674	1.1622	1.1590	1.1566	1.1545	1.1529
17.00	1.2944	1.2705	1.2521	1.2378	1.2275	1.2207	1.2140	1.2092	1.2055	1.2020	1.1962	1.1928	1.1901	1.1880	1.1864
17.50	1.3334	1.3087	1.2898	1.2746	1.2638	1.2566	1.2494	1.2443	1.2402	1.2364	1.2300	1.2262	1.2234	1.2212	1.2195
18.00	1.3725	1.3470	1.3273	1.3113	1.2999	1.2923	1.2846	1.2792	1.2747	1.2705	1.2634	1.2592	1.2563	1.2540	1.2522
18.50	1.4114	1.3851	1.3648	1.3479	1.3359	1.3280	1.3197	1.3139	1.3089	1.3044	1.2966	1.2970	1.2889	1.2865	1.2847
19.00	1.4504	1.4232	1.4022	1.3844	1.3718	1.3634	1.3546	1.3484	1.3430	1.3380	1.3294	1.3245	1.3212	1.3187	1.3168
19.50	1.4893	1.4613	1.4395	1.4208	1.4075	1.3988	1.3893	1.3828	1.3769	1.3714	1.3620	1.3566	1.3531	1.3506	1.3485
20.00	1.5281	1.4993	1.4768	1.4571	1.4432	1.4340	1.4239	1.4170	1.4105	1.4046	1.3944	1.3885	1.3848	1.3822	1.3800
20.50	1.5670	1.5373	1.5140	1.4933	1.4788	1.4691	1.4584	1.4510	1.4440	1.4376	1.4265	1.4201	1.4162	1.4135	1.4112
21.00	1.6058	1.5752	1.5511	1.5294	1.5142	1.5041	1.4927	1.4849	1.4773	1.4704	1.4583	1.4515	1.4473	1.4445	1.4422
21.50	1.6446	1.6130	1.5882	1.5655	1.5495	1.5390	1.5269	1.5186	1.5104	1.5030	1.4900	1.4826	1.4782	1.4752	1.4720
22.00	1.6833	1.6509	1.6252	1.6014	1.5848	1.5738	1.5609	1.5522	1.5434	1.5355	1.5214	1.5134	1.5088	1.5057	1.5032
22.50	1.7220	1.6887	1.6622	1.6373	1.6199	1.6084	1.5948	1.5856	1.5762	1.5677	1.5525	1.5440	1.5391	1.5360	1.5333
23.00	1.7607	1.7264	1.6991	1.6732	1.6550	1.6430	1.6286	1.6189	1.6088	1.5998	1.5835	1.5744	1.5693	1.5660	1.5632
23.50	1.7994	1.7641	1.7360	1.7089	1.6900	1.6775	1.6623	1.6521	1.6413	1.6317	1.6143	1.6046	1.5992	1.5957	1.5929
24.00	1.8381	1.8018	1.7729	1.7446	1.7249	1.7118	1.6959	1.6851	1.6736	1.6634	1.6448	1.6345	1.6288	1.6253	1.6223
24.50	1.8767	1.8394	1.8097	1.7802	1.7597	1.7461	1.7294	1.7180	1.7058	1.6950	1.6752	1.6642	1.6583	1.6546	1.6515
25.00	1.9153	1.8771	1.8464	1.8158	1.7944	1.7803	1.7627	1.7508	1.7379	1.7264	1.7054	1.6937	1.6875	1.6837	1.6805
25.50	1.9539	1.9146	1.8831	1.8513	1.8291	1.8144	1.7960	1.7835	1.7698	1.7577	1.7354	1.7231	1.7165	1.7126	1.7093
26.00	1.9924	1.9522	1.9198	1.8867	1.8637	1.8484	1.8291	1.8161	1.8016	1.7888	1.7652	1.7522	1.7454	1.7413	1.7378
26.50	2.0310	1.9897	1.9564	1.9221	1.8982	1.8824	1.8622	1.8486	1.8333	1.8198	1.7949	1.7812	1.7740	1.7698	1.7662
27.00	2.0695	2.0272	1.9930	1.9574	1.9326	1.9163	1.8951	1.8810	1.8649	1.8506	1.8244	1.8100	1.8025	1.7981	1.7944
27.50	2.1080	2.0647	2.0295	1.9927	1.9670	1.9501	1.9280	1.9133	1.8963	1.8814	1.8537	1.8386	1.8306	1.8262	1.8224
28.00	2.1465	2.1021	2.0661	2.0279	2.0014	1.9838	1.9608	1.9454	1.9277	1.9119	1.8829	1.8670	1.8589	1.8542	1.8502
28.50	2.1850	2.1395	2.1025	2.0631	2.0356	2.0175	1.9935	1.9775	1.9589	1.9424	1.9119	1.8953	1.8868	1.8820	1.8779
29.00	2.2234	2.1769	2.1390	2.0983	2.0698	2.0511	2.0261	2.0095	1.9900	1.9726	1.9408	1.9234	1.9146	1.9096	1.9053
29.50	2.2619	2.2142	2.1754	2.1333	2.1040	2.0846	2.0587	2.0414	2.0210	2.0030	1.9696	1.9513	1.9422	1.9370	1.9327
30.00	2.3003	2.2516	2.2118	2.1684	2.1381	2.1180	2.0912	2.0732	2.0519	2.0331	1.9982	1.9791	1.9696	1.9643	1.9598

Table 7-2 (i)
Extended Sukkar-Cornell Integral for Bottom-hole Pressure Calculation

P_r	\multicolumn{20}{c	}{Reduced Temperature for $B = 40.0$}
	1.1	1.2	1.3	1.4	1.5	1.6	1.7	1.8	1.9	2.0	2.2	2.4	2.6	2.8	3.0
0.20	0.0000	0.0000	0.0000	0.0000	0.0000	0.0000	0.0000	0.0000	0.0000	0.0000	0.0000	0.0000	0.0000	0.0000	0.0000
0.50	0.0029	0.0028	0.0028	0.0027	0.0027	0.0027	0.0027	0.0027	0.0027	0.0026	0.0026	0.0026	0.0026	0.0026	0.0026
1.00	0.0150	0.0139	0.0133	0.0129	0.0127	0.0125	0.0123	0.0122	0.0122	0.0121	0.0120	0.0119	0.0119	0.0118	0.0118
1.50	0.0403	0.0341	0.0318	0.0305	0.0296	0.0290	0.0284	0.0281	0.0279	0.0276	0.0273	0.0271	0.0270	0.0268	0.0267
2.00	0.0776	0.0640	0.0582	0.0551	0.0530	0.0517	0.0505	0.0497	0.0493	0.0488	0.0482	0.0477	0.0473	0.0471	0.0468
2.50	0.1170	0.1005	0.0912	0.0858	0.0821	0.0798	0.0779	0.0765	0.0756	0.0749	0.0738	0.0730	0.0724	0.0718	0.0714
3.00	0.1565	0.1393	0.1281	0.1208	0.1156	0.1122	0.1095	0.2074	0.1061	0.1050	0.1034	0.1023	0.1013	0.1005	0.0999
3.50	0.1958	0.1787	0.1668	0.1584	0.1520	0.1477	0.1442	0.1416	0.1398	0.1383	0.1362	0.1348	0.1335	0.1324	0.1315
4.00	0.2351	0.2182	0.2062	0.1973	0.1901	0.1853	0.1812	0.1780	0.1758	0.1740	0.1714	0.1696	0.1681	0.1667	0.1656
4.50	0.2743	0.2576	0.2457	0.2367	0.2292	0.2240	0.2195	0.2159	0.2135	0.2113	0.2084	0.2063	0.2045	0.2029	0.2017
5.00	0.3133	0.2969	0.2851	0.2762	0.2686	0.2633	0.2586	0.2548	0.2521	0.2498	0.2465	0.2442	0.2422	0.2405	0.2391
5.50	0.3523	0.3360	0.3244	0.3156	0.3081	0.3028	0.2980	0.2941	0.2913	0.2889	0.2854	0.2829	0.2808	0.2790	0.2775
6.00	0.3912	0.3750	0.3634	0.3549	0.3476	0.3423	0.3376	0.3336	0.3308	0.3283	0.3247	0.3221	0.3199	0.3181	0.3166
6.50	0.4300	0.4138	0.4023	0.3939	0.3868	0.3816	0.3770	0.3731	0.3703	0.3678	0.3642	0.3616	0.3594	0.3575	0.3560
7.00	0.4687	0.4525	0.4410	0.4328	0.4258	0.4208	0.4163	0.4124	0.4097	0.4073	0.4037	0.4011	0.3989	0.3970	0.3955
7.50	0.5073	0.4910	0.4795	0.4714	0.4646	0.4597	0.4553	0.4516	0.4490	0.4466	0.4431	0.4405	0.4383	0.4365	0.4350
8.00	0.5458	0.5294	0.5179	0.5097	0.5031	0.4983	0.4941	0.4905	0.4879	0.4856	0.4819	0.4797	0.4776	0.4758	0.4743
8.50	0.5843	0.5677	0.5560	0.5479	0.5413	0.5367	0.5325	0.5290	0.5266	0.5244	0.5208	0.5187	0.5166	0.5148	0.5133
9.00	0.6227	0.6059	0.5940	0.5859	0.5793	0.5747	0.5707	0.5673	0.5650	0.5628	0.5593	0.5573	0.5553	0.5535	0.5521
9.50	0.6609	0.6439	0.6319	0.6237	0.6171	0.6125	0.6085	0.6052	0.6030	0.6009	0.5975	0.5955	0.5936	0.5918	0.5904
10.00	0.6991	0.6818	0.6696	0.6612	0.6546	0.6500	0.6461	0.6429	0.6407	0.6386	0.6353	0.6334	0.6315	0.6298	0.6284
10.50	0.7372	0.7196	0.7071	0.6987	0.6919	0.6873	0.6833	0.6802	0.6780	0.6760	0.6728	0.6710	0.6690	0.6673	0.6660
11.00	0.7753	0.7573	0.7446	0.7359	0.7290	0.7243	9.7203	0.7172	0.7150	0.7130	0.7099	0.7081	0.7052	0.7045	0.7031
11.50	0.8132	0.7949	0.7819	0.7729	0.7659	0.7611	0.7571	0.7539	0.7517	0.7498	0.7466	0.7448	0.7429	0.7412	0.7398
12.00	0.8511	0.8324	0.8190	0.8098	0.8026	0.7977	0.7936	0.7903	0.7882	0.7862	0.7830	0.7812	0.7792	0.7775	0.7762
12.50	0.8890	0.8696	0.8561	0.8466	0.8391	0.8341	0.8299	0.8265	0.8243	0.8223	0.8190	0.8171	0.8152	0.8134	0.8121
13.00	0.9268	0.9072	0.8931	0.8832	0.8755	0.8703	0.8659	0.8624	0.8602	0.8580	0.8547	0.8527	0.8507	0.8490	0.8476

14.50	0.0188	0.9921	0.9835	0.9778	0.9727	0.9588	0.9661	0.9636	0.9596	0.9572	0.9551	0.9532	0.9517
15.00	1.0398	1.0034											
15.00	0.0558	1.0282	1.0193	1.0133	1.0079	1.0037	1.0009	0.9982	0.9939	0.9914	0.9891	0.9872	0.9856
15.50	1.0774	1.0400											
15.50	1.1149	1.0928	1.0641	1.0548	1.0486	1.0429	1.0385	1.0355	1.0328	1.0279	1.0251	1.0228	1.0208	1.0192
		1.0765												
16.00	1.1525	1.1129	1.1000	1.0903	1.0837	1.0777	1.0731	1.0698	1.0667	1.0516	1.0586	1.0561	1.0541	1.0525
16.50	1.1899	1.1297												
16.50	1.1666	1.1492	1.1357	1.1255	1.1187	1.1123	1.1075	1.1039	1.1005	1.0949	1.0917	1.0891	1.0870	1.0853
17.00	1.2274	1.1855	1.1713	1.1607	1.1536	1.1468	1.1417	1.1378	1.1341	1.1280	1.1245	1.1218	1.1196	1.1179
17.50	1.2648	1.2034	1.2068	1.1958	1.1884	1.1811	1.1757	1.1714	1.1675	1.1608	1.1570	1.1541	1.1519	1.1501
17.50	1.2402	1.2217												
18.00	1.3021	1.2769	1.2242	1.2307	1.2230	1.2152	1.2095	1.2049	1.2006	1.1934	1.1892	1.1862	1.1839	1.1820
18.00		1.2579												
18.50	1.3395	1.3136	1.2776	1.2655	1.2574	1.2492	1.2432	1.2382	1.2336	1.2256	1.2211	1.2180	1.2155	1.2136
		1.2940												
19.00	1.3768	1.3502	1.3128	1.3002	1.2918	1.2831	1.2767	1.2713	1.2663	1.2577	1.2528	1.2494	1.2469	1.2450
		1.3300												
19.50	1.4140	1.3868	1.3480	1.3349	1.3261	1.3168	1.3101	1.3042	1.2988	1.2894	1.2842	1.2806	1.2780	1.2760
		1.3659												
20.00	1.4513	1.4233	1.3831	1.3694	1.3602	1.3504	1.3433	1.3369	1.3311	1.3210	1.3153	1.3116	1.3089	1.3068
		1.4019												
20.50	1.4885	1.4598	1.4181	1.4038	1.3942	1.3838	1.3763	1.3695	1.3633	1.3523	1.3462	1.3422	1.3395	1.3373
		1.4377												
21.00	1.5257	1.4963	1.4530	1.4381	1.4281	1.4171	1.4093	1.4019	1.3952	1.3834	1.3768	1.3727	1.3698	1.3675
		1.4735												
21.50	1.5629	1.5327	1.4879	1.4723	1.4620	1.4503	1.4421	1.4341	1.4270	1.4143	1.4072	1.4028	1.3999	1.3975
		1.5093												
22.00	1.6001	1.5691	1.5227	1.5065	1.4957	1.4834	1.4747	1.4662	1.4586	1.4449	1.4373	1.4328	1.4297	1.4272
		1.5450												
22.50	1.6372	1.6054	1.5574	1.5406	1.5293	1.5164	1.5072	1.4982	1.4900	1.4754	1.4673	1.4625	1.4593	1.4567
		1.5807												
23.00	1.6743	1.6417	1.5920	1.5746	1.5629	1.5492	1.5396	1.5300	1.5213	1.5057	1.4970	1.4920	1.4887	1.4860
		1.6163												
23.50	1.7114	1.6780	1.6266	1.6085	1.5963	1.5820	1.5719	1.5617	1.5525	1.5358	1.5265	1.5213	1.5178	1.5151
		1.6519												
24.00	1.7485	1.7143	1.6612	1.6423	1.6297	1.6146	1.6041	1.5932	1.5834	1.5657	1.5559	1.5503	1.5468	1.5439
		1.6874												
24.50	1.7855	1.7505	1.6957	1.6761	1.6630	1.6472	1.6362	1.6246	1.6143	1.5954	1.5850	1.5792	1.5755	1.5725
		1.7229												
25.00	1.8226	1.7867	1.7301	1.7098	1.6962	1.6797	1.6582	1.6559	1.6450	1.6249	1.6139	1.6078	1.6041	1.6010
		1.7584												
25.50	1.8596	1.8229	1.7645	1.7434	1.7293	1.7120	1.7000	1.6871	1.6755	1.6543	1.6427	1.6363	1.6324	1.6292
		1.7938												
26.00	1.8966	1.8591	1.7988	1.7770	1.7624	1.7443	1.7318	1.7181	1.7059	1.6836	1.6713	1.6646	1.6606	1.6572
		1.8292												
26.50	1.9336	1.8952	1.8331	1.8105	1.7954	1.7765	1.7634	1.7491	1.7362	1.7126	1.6997	1.6927	1.6886	1.6851
		1.8645												
27.00	1.9705	1.9313	1.8673	1.8439	1.8283	1.8086	1.7950	1.7799	1.7664	1.7415	1.7279	1.7207	1.7164	1.7128
		1.8999												
27.50	2.0075	1.9674	1.9015	1.8773	1.8612	1.8406	1.8265	1.8106	1.7965	1.7703	1.7560	1.7484	1.7440	1.7403
		1.9352												
28.00	2.0444	2.0034	1.9356	1.9107	1.8940	1.8726	1.8579	1.8412	1.8264	1.7989	1.7839	1.7760	1.7715	1.7676
		1.9704												
28.50	2.0813	2.0194	1.9697	1.9439	1.9267	1.9044	1.8892	1.8717	1.8562	1.8274	1.8116	1.8035	1.7988	1.7946
		2.0057												
29.00	2.1182	2.0755	2.0038	1.9771	1.9594	1.9362	1.9204	1.9021	1.8859	1.8557	1.8393	1.8308	1.8259	1.8216
		2.0409												
29.50	2.1551	2.1114	2.0378	2.0103	1.9920	1.9680	1.9516	1.9325	1.9155	1.8840	1.8667	1.8579	1.8529	1.8487
		2.0761												
30.00	2.1920	2.1474	2.0717	2.0434	2.0246	1.9996	1.9826	1.9627	1.9450	1.9120	1.8940	1.8849	1.8797	1.8754
		2.1112												

Table 7-2 (j)
Extended Sukkar-Cornell Integral for Bottom-hole Pressure Calculation

P_r	\multicolumn{14}{c}{Reduced Temperature for $B = 45.0$}													
	1.1	1.2	1.3	1.4	1.5	1.6	1.7	1.8	1.9	2.0	2.2	2.4	2.6	2.8	3.0
0.20	0.0000	0.0000	0.0000	0.0000	0.0000	0.0000	0.0000	0.0000	0.0000	0.0000	0.0000	0.0000	0.0000	0.0000	0.0000
0.50	0.0026	0.0025	0.0025	0.0024	0.0024	0.0024	0.0024	0.0024	0.0024	0.0024	0.0023	0.0023	0.0023	0.0023	0.0023
1.00	0.0134	0.0124	0.0119	0.0115	0.0113	0.0111	0.0110	0.0109	0.0108	0.0108	0.0107	0.0106	0.0106	0.0105	0.0105
1.50	0.0362	0.0305	0.0284	0.0272	0.0264	0.0258	0.0254	0.0250	0.0248	0.0247	0.0244	0.0242	0.0240	0.0239	0.0238
2.00	0.0707	0.0576	0.0522	0.0494	0.0475	0.0462	0.0452	0.0445	0.0440	0.0436	0.0430	0.0426	0.0423	0.0420	0.0418
2.50	0.1016	0.0912	0.0823	0.0772	0.0738	0.0716	0.0699	0.0586	0.0678	0.0671	0.0661	0.0654	0.0648	0.0644	0.0640
3.00	0.1449	0.1273	0.1163	0.1093	0.1043	0.1012	0.0986	0.0967	0.0955	0.0944	0.0930	0.0919	0.0910	0.0903	0.0897
3.50	0.1821	0.1643	0.1523	0.1441	0.1378	0.1338	0.1304	0.1279	0.1263	0.1248	0.1229	0.1215	0.1203	0.1193	0.1185
4.00	0.2193	0.2015	0.1892	0.1803	0.1732	0.1685	0.1645	0.1614	0.1594	0.1576	0.1552	0.1534	0.1520	0.1507	0.1496
4.50	0.2565	0.2388	0.2264	0.2172	0.2096	0.2045	0.2001	0.1966	0.1942	0.1921	0.1893	0.1872	0.1855	0.1840	0.1828
5.00	0.2936	0.2760	0.2637	0.2544	0.2466	0.2412	0.2366	0.2327	0.2301	0.2278	0.2246	0.2223	0.2204	0.2187	0.2174
5.50	0.3306	0.3131	0.3009	0.2917	0.2838	0.2783	0.2735	0.2695	0.2667	0.2643	0.2608	0.2583	0.2562	0.2544	0.2530
6.00	0.3676	0.3501	0.3380	0.3289	0.3211	0.3158	0.3107	0.3066	0.3038	0.3012	0.2976	0.2949	0.2928	0.2909	0.2895
6.50	0.4045	0.3871	0.3750	0.3660	0.3583	0.3528	0.3480	0.3439	0.3410	0.3384	0.3347	0.3319	0.3297	0.3278	0.3264
7.00	0.4414	0.4239	0.4118	0.4029	0.3954	0.3900	0.3852	0.3811	0.3782	0.3757	0.3719	0.3692	0.3669	0.3650	0.3635
7.50	0.4782	0.4607	0.4486	0.4397	0.4323	0.4270	0.4223	0.4182	0.4154	0.4129	0.4092	0.4064	0.4042	0.4023	0.4008
8.00	0.5150	0.4973	0.4852	0.4763	0.4690	0.4638	0.4592	0.4552	0.4525	0.4500	0.4459	0.4436	0.4414	0.4395	0.4360
8.50	0.5517	0.5339	0.5216	0.5128	0.5055	0.5004	0.4959	0.4920	0.4893	0.4869	0.4828	0.4806	0.4785	0.4766	0.4751
9.00	0.5883	0.5704	0.5580	0.5492	0.5419	0.5368	0.5323	0.5286	0.5259	0.5235	0.5196	0.5174	0.5153	0.5135	0.5120
9.50	0.6248	0.6067	0.5942	0.5853	0.5780	0.5730	0.5686	0.5649	0.5623	0.5599	0.5561	0.5540	0.5519	0.5501	0.5486
10.00	0.6613	0.6430	0.6304	0.6214	0.6140	0.6090	0.6046	0.6009	0.5984	0.5961	0.5923	0.5903	0.5882	0.5864	0.5650
10.50	0.6978	0.6792	0.6664	0.6573	0.6498	0.6447	0.6606	0.6367	0.6342	0.6320	0.6283	0.6262	0.6242	0.6224	0.6210
11.00	0.7342	0.7153	0.7023	0.6930	0.6854	0.6803	0.6759	0.6723	0.6698	0.6676	0.6639	0.6619	0.6598	0.6580	0.6566
11.50	0.7705	0.7514	0.7381	0.7286	0.7209	0.7157	0.7113	0.7076	0.7051	0.7029	0.6993	0.6977	0.6952	0.6934	0.6920
12.00	0.8068	0.7874	0.7738	0.7641	0.7562	0.7509	0.7464	0.7427	0.7402	0.7380	0.7343	0.7323	0.7302	0.7284	0.7270
12.50	0.8430	0.8233	0.8094	0.7994	0.7914	0.7860	0.7814	0.7776	0.7751	0.7728	0.7690	0.7670	0.7649	0.7630	0.7616
13.00	0.8792	0.8591	0.8449	0.8347	0.8264	0.8209	0.8161	0.8122	0.8097	0.8073	0.8035	0.8013	0.7992	0.7974	0.7959

14.00	0.9514	0.9306	0.9157	0.9048	0.8961	0.8902	0.8851	0.8809	0.8782	0.8756	0.8715	0.8691	0.8669	0.8650	0.8635
14.50	0.9875	0.9663	0.9510	0.9396	0.9307	0.9246	0.9193	0.9150	0.9121	0.9094	0.9050	0.9025	0.9002	0.8983	0.8968
15.00	1.0235	1.0019	0.9863	0.9744	0.9652	0.9589	0.9533	0.9489	0.9458	0.9429	0.9382	0.9355	0.9332	0.9312	0.9297
15.50	1.0595	1.0374	1.0214	1.0091	0.9995	0.9931	0.9872	0.9825	0.9793	0.9762	0.9712	0.9684	0.9660	0.9639	0.9623
16.00	1.0955	1.0729	1.0565	1.0437	1.0338	1.0271	1.0209	1.0160	1.0125	1.0093	1.0039	1.0009	0.9984	0.9963	0.9946
16.50	1.1315	1.1084	1.0915	1.0782	1.0679	1.0609	1.0544	1.0494	1.0456	1.0422	1.0364	1.0331	1.0305	1.0283	1.0266
17.00	1.1674	1.1438	1.1265	1.1126	1.1019	1.0947	1.0878	1.0825	1.0785	1.0748	1.0685	1.0650	1.0623	1.0600	1.0583
17.50	1.2032	1.1791	1.1614	1.1469	1.1358	1.1283	1.1211	1.1155	1.1112	1.1072	1.1005	1.0967	1.0938	1.0915	1.0697
18.00	1.2391	1.2145	1.1962	1.1811	1.1696	1.1619	1.1542	1.1484	1.1437	1.1394	1.1321	1.1281	1.1250	1.1227	1.1208
18.50	1.2749	1.2497	1.2310	1.2153	1.2033	1.1953	1.1872	1.1811	1.1761	1.1715	1.1636	1.1592	1.1560	1.1536	1.1517
19.00	1.3107	1.2850	1.2658	1.2494	1.2370	1.2286	1.2200	1.2136	1.2082	1.2033	1.1948	1.1901	1.1867	1.1842	1.1823
19.50	1.3465	1.3202	1.3005	1.2834	1.2705	1.2618	1.2528	1.2460	1.2403	1.2350	1.2258	1.2207	1.2172	1.2146	1.2126
20.00	1.3823	1.3554	1.3351	1.3173	1.3039	1.2949	1.2854	1.2783	1.2721	1.2665	1.2566	1.2511	1.2474	1.2447	1.2426
20.50	1.4180	1.3905	1.3697	1.3512	1.3373	1.3279	1.3179	1.3105	1.3038	1.2978	1.2871	1.2812	1.2774	1.2746	1.2724
21.00	1.4538	1.4256	1.4043	1.3850	1.3706	1.3608	1.3503	1.3425	1.3354	1.3290	1.3175	1.3112	1.3071	1.3043	1.3020
21.50	1.4895	1.4607	1.4388	1.4187	1.4038	1.3937	1.3825	1.3744	1.3668	1.3599	1.3477	1.3409	1.3367	1.3337	1.3314
22.00	1.5251	1.4958	1.4733	1.4524	1.4369	1.4264	1.4147	1.4062	1.3981	1.3908	1.3776	1.3704	1.3660	1.3629	1.3605
22.50	1.5608	1.5308	1.5077	1.4860	1.4699	1.4591	1.4468	1.4379	1.4292	1.4215	1.4074	1.3997	1.3951	1.3919	1.3894
23.00	1.5965	1.5658	1.5421	1.5196	1.5029	1.4916	1.4788	1.4694	1.4603	1.4520	1.4371	1.4288	1.4239	1.4207	1.4181
23.50	1.6321	1.6008	1.5765	1.5531	1.5358	1.5242	1.5106	1.5009	1.4912	1.4824	1.4665	1.4577	1.4526	1.4493	1.4466
24.00	1.6677	1.6357	1.6108	1.5866	1.5687	1.5566	1.5424	1.5323	1.5219	1.5127	1.4958	1.4865	1.4811	1.4776	1.4748
24.50	1.7033	1.6706	1.6451	1.6200	1.6015	1.5890	1.5741	1.5635	1.5526	1.5428	1.5249	1.5150	1.5094	1.5058	1.5029
25.00	1.7389	1.7055	1.6794	1.6534	1.6342	1.6212	1.6057	1.5947	1.5831	1.5728	1.5538	1.5436	1.5375	1.5338	1.5308
25.50	1.7745	1.7404	1.7136	1.6867	1.6668	1.6535	1.6373	1.6257	1.6136	1.6027	1.5826	1.5716	1.5655	1.5617	1.5585
26.00	1.8100	1.7752	1.7478	1.7200	1.6995	1.6856	1.6687	1.6567	1.6439	1.6324	1.6112	1.5996	1.5933	1.5893	1.5861
26.50	1.8456	1.8101	1.7820	1.7532	1.7320	1.7177	1.7001	1.6876	1.6741	1.6621	1.6397	1.6275	1.6209	1.6168	1.6134
27.00	1.8811	1.8449	1.8162	1.7864	1.7645	1.7498	1.7314	1.7184	1.7042	1.6916	1.6681	1.6552	1.6483	1.6441	1.6406
27.50	1.9166	1.8797	1.8503	1.8195	1.7969	1.7817	1.7626	1.7491	1.7343	1.7210	1.6963	1.6828	1.6756	1.6712	1.6677
28.00	1.9521	1.9144	1.8844	1.8526	1.8293	1.8136	1.7937	1.7798	1.7642	1.7503	1.7244	1.7102	1.7027	1.6982	1.6945
28.50	1.9876	1.9492	1.9184	1.8857	1.8617	1.8455	1.8248	1.8103	1.7940	1.7795	1.7523	1.7375	1.7297	1.7251	1.7212
29.00	2.0231	1.9839	1.9525	1.9187	1.8940	1.8773	1.8558	1.8408	1.8238	1.8086	1.7801	1.7646	1.7565	1.7518	1.7478
29.50	2.0586	2.0186	1.9865	1.9517	1.9262	1.9091	1.8868	1.8712	1.8534	1.8376	1.8078	1.7916	1.7832	1.7783	1.7742
30.00	2.0941	2.0533	2.0205	1.9847	1.9584	1.9408	1.9176	1.9016	1.8830	1.8664	1.8354	1.8184	1.8097	1.8047	1.8005

Table 7-2 (k)
Extended Sukkar-Cornell Integral for Bottom-hole Pressure Calculation

P_r	\multicolumn{20}{c}{Reduced Temperature for $B = 50.0$}
	1.1	1.2	1.3	1.4	1.5	1.6	1.7	1.8	1.9	2.0	2.2	2.4	2.6	2.8	3.0
0.20	0.0000	0.0000	0.0000	0.0000	0.0000	0.0000	0.0000	0.0000	0.0000	0.0000	0.0000	0.0000	0.0000	0.0000	0.0000
0.50	0.0023	0.0023	0.0022	0.0022	0.0022	0.0022	0.0021	0.0021	0.0021	0.0021	0.0021	0.0021	0.0021	0.0021	0.0021
1.00	0.0121	0.0111	0.0107	0.0104	0.0102	0.0100	0.0099	0.0098	0.0098	0.0097	0.0096	0.0096	0.0095	0.0095	0.0095
1.50	0.0328	0.0276	0.0257	0.0246	0.0238	0.0233	0.0229	0.0226	0.0224	0.0222	0.0220	0.0218	0.0217	0.0216	0.0215
2.00	0.0649	0.0524	0.0474	0.0447	0.0430	0.0418	0.0409	0.0402	0.0398	0.0395	0.0389	0.0385	0.0382	0.0380	0.0378
2.50	0.0997	0.0835	0.0750	0.0702	0.0670	0.0650	0.0634	0.0622	0.0615	0.0608	0.0599	0.0593	0.0587	0.0583	0.0579
3.00	0.1350	0.1173	0.1066	0.0998	0.0951	0.0921	0.0897	0.0879	0.0868	0.0858	0.0844	0.0835	0.0827	0.0820	0.0814
3.50	0.1703	0.1521	0.1402	0.1322	0.1261	0.1222	0.1191	0.1167	0.1151	0.1138	0.1119	0.1100	0.1095	0.1085	0.1078
4.00	0.2057	0.1873	0.1749	0.1660	0.1591	0.1545	0.1507	0.1477	0.1457	0.1440	0.1417	0.1401	0.1387	0.1375	0.1365
4.50	0.2410	0.2226	0.2101	0.2008	0.1933	0.1882	0.1839	0.1804	0.1781	0.1761	0.1734	0.1714	0.1697	0.1683	0.1671
5.00	0.2763	0.2579	0.2454	0.2359	0.2281	0.2227	0.2181	0.2143	0.2117	0.2094	0.2063	0.2040	0.2022	0.2006	0.1993
5.50	0.3116	0.2933	0.2807	0.2712	0.2632	0.2577	0.2529	0.2488	0.2461	0.2436	0.2402	0.2377	0.2356	0.2339	0.2326
6.00	0.3469	0.3285	0.3161	0.3066	0.2985	0.2929	0.2880	0.2838	0.2809	0.2784	0.2747	0.2721	0.2700	0.2681	0.2667
6.50	0.3821	0.3638	0.3513	0.3419	0.3339	0.3282	0.3233	0.3190	0.3161	0.3135	0.3097	0.3069	0.3048	0.3029	0.3014
7.00	0.4173	0.3990	0.3865	0.3772	0.3692	0.3636	0.3587	0.3544	0.3514	0.3488	0.3450	0.3421	0.3399	0.3380	0.3365
7.50	0.4525	0.4341	0.4216	0.4123	0.4044	0.3989	0.3940	0.3897	0.3868	0.3841	0.3803	0.3774	0.3752	0.3733	0.3718
8.00	0.4876	0.4692	0.4567	0.4474	0.4395	0.4340	0.4292	0.4250	0.4221	0.4194	0.4151	0.4128	0.4105	0.4086	0.4071
8.50	0.5227	0.5042	0.4916	0.4823	0.4745	0.4690	0.4643	0.4601	0.4573	0.4547	0.4504	0.4481	0.4458	0.4439	0.4424
9.00	0.5577	0.5391	0.5264	0.5171	0.5093	0.5039	0.4992	0.4951	0.4923	0.4897	0.4855	0.4832	0.4810	0.4791	0.4777
9.50	0.5927	0.5739	0.5612	0.5518	0.5440	0.5386	0.5340	0.5299	0.5271	0.5246	0.5204	0.5182	0.5160	0.5142	0.5127
10.00	0.6277	0.6087	0.5959	0.5864	0.5786	0.5732	0.5685	0.5645	0.5618	0.5593	0.5552	0.5530	0.5508	0.5490	0.5475
10.50	0.6626	0.6435	0.6304	0.6209	0.6130	0.6076	0.6029	0.5990	0.5962	0.5938	0.5897	0.5875	0.5854	0.5835	0.5821
11.00	0.6974	0.6781	0.6649	0.6553	0.6473	0.6418	0.6372	0.6332	0.6305	0.6280	0.6240	0.6219	0.6197	0.6179	0.6184
11.50	0.7323	0.7127	0.6994	0.6896	0.6815	0.6759	0.6712	0.6672	0.6645	0.6621	0.6581	0.6558	0.6537	0.6519	0.6505
12.00	0.7670	0.7473	0.7337	0.7237	0.7155	0.7099	0.7051	0.7011	0.6984	0.6959	0.6919	0.6897	0.6875	0.6857	0.6842
12.50	0.8018	0.7818	0.7680	0.7578	0.7494	0.7437	0.7388	0.7347	0.7320	0.7295	0.7254	0.7232	0.7210	0.7192	0.7177
13.00	0.8365	0.8163	0.8022	0.7917	0.7832	0.7774	0.7724	0.7582	0.7654	0.7629	0.7587	0.7565	0.7542	0.7523	0.7509

14.50	0.9405	0.9193	0.9044	0.8930	0.8839	0.8776	0.8722	0.8645	0.8617	0.8570	0.8545	0.8521	0.8502	0.8486
15.00	0.9751	0.9536	0.9384	0.9266	0.9172	0.9108	0.9051	0.8972	0.8942	0.8893	0.8866	0.8842	0.8822	0.8806
15.50	1.0097	0.9878	0.9722	0.9601	0.9504	0.9438	0.9379	0.9297	0.9265	0.9213	0.9185	0.9160	0.9139	0.9123
16.00	1.0442	1.0220	1.0061	0.9935	0.9836	0.9768	0.9706	0.9620	0.9586	0.9531	0.9501	0.9475	0.9454	0.9438
16.50	1.0788	1.0561	1.0399	1.0269	1.0166	1.0096	1.0031	0.9941	0.9906	0.9847	0.9814	0.9788	0.9766	0.9749
17.00	1.1133	1.0902	1.0736	1.0601	1.0495	1.0423	1.0355	1.0260	1.0223	1.0160	1.0125	1.0097	1.0075	1.0058
17.50	1.1477	1.1243	1.1073	1.0933	1.0824	1.0749	1.0678	1.0578	1.0538	1.0471	1.0434	1.0405	1.0382	1.0364
18.00	1.1822	1.1583	1.1409	1.1266	1.1151	1.1074	1.0999	1.0894	1.0852	1.0779	1.0740	1.0709	1.0686	1.0668
18.50	1.2167	1.1923	1.1745	1.1595	1.1478	1.1398	1.1320	1.1209	1.1164	1.1086	1.1043	1.1012	1.0988	1.0969
19.00	1.2511	1.2263	1.2081	1.1925	1.1804	1.1721	1.1639	1.1522	1.1474	1.1390	1.1345	1.1312	1.1287	1.1268
19.50	1.2855	1.2602	1.2416	1.2254	1.2129	1.2044	1.1957	1.1834	1.1783	1.1693	1.1644	1.1609	1.1584	1.1564
20.00	1.3199	1.2942	1.2751	1.2583	1.2453	1.2365	1.2274	1.2144	1.2090	1.1993	1.1941	1.1905	1.1878	1.1858
20.50	1.3542	1.3280	1.3085	1.2911	1.2777	1.2686	1.2590	1.2453	1.2395	1.2292	1.2236	1.2198	1.2171	1.2149
21.00	1.3886	1.3619	1.3419	1.3238	1.3100	1.3005	1.2905	1.2761	1.2699	1.2589	1.2528	1.2489	1.2461	1.2439
21.50	1.4229	1.3957	1.3753	1.3565	1.3422	1.3324	1.3219	1.3067	1.3001	1.2884	1.2819	1.2778	1.2749	1.2726
22.00	1.4573	1.4295	1.4086	1.3892	1.3743	1.3643	1.3532	1.3372	1.3302	1.3177	1.3108	1.3065	1.3035	1.3011
22.50	1.4916	1.4633	1.4419	1.4218	1.4064	1.3960	1.3844	1.3676	1.3602	1.3468	1.3395	1.3350	1.3319	1.3295
23.00	1.5259	1.4971	1.4752	1.4543	1.4385	1.4277	1.4155	1.3979	1.3900	1.3758	1.3680	1.3633	1.3601	1.3576
23.50	1.5602	1.5308	1.5084	1.4868	1.4704	1.4593	1.4456	1.4280	1.4197	1.4046	1.3964	1.3914	1.3881	1.3855
24.00	1.5944	1.5646	1.5416	1.5193	1.5024	1.4908	1.4775	1.4581	1.4493	1.4333	1.4245	1.4193	1.4160	1.4133
24.50	1.6287	1.5983	1.5748	1.5517	1.5342	1.5223	1.5084	1.4880	1.4788	1.4618	1.4525	1.4471	1.4436	1.4408
25.00	1.6629	1.6319	1.6079	1.5841	1.5660	1.5537	1.5392	1.5178	1.5081	1.4902	1.4803	1.4747	1.4711	1.4682
25.50	1.6972	1.6656	1.6410	1.6164	1.5978	1.5851	1.5700	1.5476	1.5373	1.5184	1.5080	1.5021	1.4984	1.4954
26.00	1.7314	1.6992	1.6741	1.6487	1.6295	1.6164	1.6006	1.5772	1.5664	1.5465	1.5355	1.5294	1.5256	1.5225
26.50	1.7656	1.7329	1.7072	1.6809	1.6611	1.6476	1.6312	1.6068	1.5954	1.5744	1.5629	1.5565	1.5526	1.5494
27.00	1.7998	1.7665	1.7403	1.7131	1.6927	1.6788	1.6617	1.6362	1.6243	1.6022	1.5901	1.5835	1.5794	1.5761
27.50	1.8340	1.8001	1.7733	1.7453	1.7243	1.7100	1.6922	1.6656	1.6531	1.6299	1.6172	1.6103	1.6061	1.6027
28.00	1.8882	1.8337	1.8063	1.7775	1.7558	1.7410	1.7226	1.6948	1.6818	1.6574	1.6441	1.6389	1.6328	1.6291
28.50	1.9024	1.8672	1.8393	1.8096	1.7872	1.7721	1.7529	1.7240	1.7104	1.6849	1.6709	1.6634	1.6590	1.6553
29.00	1.9366	1.9008	1.8722	1.8616	1.8187	1.8030	1.7831	1.7531	1.7389	1.7122	1.6976	1.6898	1.6853	1.6815
29.50	1.9707	1.9341	1.9052	1.8787	1.8500	1.8340	1.8133	1.7821	1.7673	1.7394	1.7241	1.7160	1.7114	1.7075
30.00	2.0049	1.9678	1.9381	1.9057	1.8814	1.8649	1.8435	1.8111	1.7956	1.7664	1.7505	1.7421	1.7373	1.7333

Table 7-2 (I)
Extended Sukkar-Cornell Integral for Bottom-hole Pressure Calculation

P_r	\multicolumn{20}{c}{Reduced Temperature for $B = 60.0$}
	1.1	1.2	1.3	1.4	1.5	1.6	1.7	1.8	1.9	2.0	2.2	2.4	2.6	2.8	3.0
0.20	0.0000	0.0000	0.0000	0.0000	0.0000	0.0000	0.0000	0.0000	0.0000	0.0000	0.0000	0.0000	0.0000	0.0000	0.0000
0.50	0.0019	0.0019	0.0019	0.0018	0.0018	0.0018	0.0018	0.0018	0.0018	0.0018	0.0018	0.0018	0.0017	0.0017	0.0017
1.00	0.0101	0.0093	0.0089	0.0087	0.0085	0.0084	0.0083	0.0082	0.0081	0.0081	0.0080	0.0080	0.0080	0.0079	0.0079
1.50	0.0277	0.0232	0.0215	0.0206	0.0200	0.0195	0.0192	0.0189	0.0188	0.0186	0.0184	0.0183	0.0181	0.0181	0.0180
2.00	0.0559	0.0443	0.0399	0.0376	0.0361	0.0351	0.0343	0.0338	0.0334	0.0331	0.0326	0.0323	0.0321	0.0319	0.0317
2.50	0.0870	0.0715	0.0637	0.0594	0.0566	0.0549	0.0535	0.0524	0.0518	0.0512	0.0504	0.0499	0.0494	0.0490	0.0487
3.00	0.1189	0.1014	0.0913	0.0851	0.0808	0.0781	0.0760	0.0745	0.0734	0.0726	0.0714	0.0705	0.0698	0.0692	0.0687
3.50	0.1509	0.1325	0.1211	0.1135	0.1079	0.1043	0.1014	0.0993	0.0979	0.0966	0.0950	0.0939	0.0928	0.0920	0.0913
4.00	0.1831	0.1642	0.1521	0.1435	0.1369	0.1326	0.1291	0.1263	0.1245	0.1229	0.1209	0.1194	0.1181	0.1170	0.1161
4.50	0.2153	0.1962	0.1837	0.1745	0.1672	0.1624	0.1583	0.1551	0.1529	0.1510	0.1485	0.1466	0.1451	0.1438	0.1428
5.00	0.2475	0.2283	0.2157	0.2062	0.1984	0.1931	0.1887	0.1850	0.1826	0.1804	0.1775	0.1753	0.1736	0.1721	0.1709
5.50	0.2798	0.2606	0.2479	0.2382	0.2301	0.2245	0.2198	0.2158	0.2132	0.2108	0.2075	0.2051	0.2032	0.2016	0.2003
6.00	0.3120	0.2928	0.2801	0.2703	0.2620	0.2563	0.2515	0.2472	0.2444	0.2419	0.2383	0.2357	0.2337	0.2320	0.2306
6.50	0.3443	0.3251	0.3124	0.3026	0.2942	0.2884	0.2834	0.2791	0.2761	0.2735	0.2697	0.2670	0.2648	0.2630	0.2616
7.00	0.3766	0.3574	0.3446	0.3348	0.3264	0.3206	0.3156	0.3111	0.3081	0.3054	0.3015	0.2986	0.2964	0.2946	0.2932
7.50	0.4088	0.3896	0.3769	0.3671	0.3587	0.3529	0.3478	0.3433	0.3403	0.3375	0.3336	0.3306	0.3284	0.3265	0.3251
8.00	0.4411	0.4219	0.4091	0.3994	0.3910	0.3851	0.3801	0.3756	0.3725	0.3697	0.3657	0.3628	0.3605	0.3586	0.3572
8.50	0.4734	0.4541	0.4413	0.4316	0.4232	0.4174	0.4123	0.4079	0.4048	0.4020	0.3976	0.3951	0.3928	0.3909	0.3894
9.00	0.5056	0.4863	0.4735	0.4637	0.4554	0.4496	0.4445	0.4401	0.4370	0.4343	0.4297	0.4273	0.4251	0.4231	0.4217
9.50	0.5378	0.5185	0.5056	0.4958	0.4875	0.4817	0.4767	0.4722	0.4692	0.4665	0.4619	0.4596	0.4573	0.4554	0.4539
10.00	0.5701	0.5507	0.5377	0.5279	0.5195	0.5137	0.5087	0.5043	0.5013	0.4985	0.4940	0.4917	0.4894	0.4875	0.4861
10.50	0.6023	0.5828	0.5698	0.5599	0.5515	0.5457	0.5407	0.5363	0.5333	0.5305	0.5260	0.5237	0.5215	0.5196	0.5181
11.00	0.6344	0.6149	0.6018	0.5718	0.5833	0.5775	0.5725	0.5681	0.5651	0.5624	0.5579	0.5556	0.5534	0.5515	0.5500
11.50	0.6666	0.6469	0.6337	0.6237	0.6151	0.6093	0.6042	0.5998	0.5988	0.5942	0.5896	0.5873	0.5851	0.5832	0.5818
12.00	0.6987	0.6790	0.6656	0.6555	0.6469	0.6409	0.6359	0.6314	0.6284	0.6257	0.6212	0.6189	0.6166	0.6148	0.6133
12.50	0.7309	0.7110	0.6975	0.6872	0.6785	0.6725	0.6674	0.6629	0.6599	0.6571	0.6526	0.6503	0.6480	0.6461	0.6446
13.00	0.7630	0.7429	0.7293	0.7189	0.7101	0.7040	0.6986	0.6943	0.6912	0.6884	0.6838	0.6815	0.6792	0.6773	0.6758

14.50	0.8592	0.8387	0.8246	0.8135	0.8043	0.7979	0.7924	0.7876	0.7843	0.7813	0.7764	0.7738	0.7714	0.7694	0.7679
15.00	0.8913	0.8705	0.8562	0.8449	0.8355	0.8291	0.8233	0.8184	0.8151	0.8120	0.8069	0.8042	0.8017	0.7997	0.7982
15.50	0.9233	0.9024	0.8879	0.8763	0.8667	0.8601	0.8542	0.8492	0.8457	0.8425	0.8371	0.8343	0.8318	0.8298	0.8282
16.00	0.9554	0.9342	0.9195	0.9076	0.8978	0.8911	0.8850	0.8798	0.8762	0.8728	0.8672	0.8643	0.8617	0.8596	0.8580
16.50	0.9874	0.9660	0.9510	0.9389	0.9288	0.9219	0.9156	0.9103	0.9065	0.9030	0.8971	0.8940	0.8914	0.8892	0.8876
17.00	1.0194	0.9977	0.9826	0.9701	0.9598	0.9527	0.9462	0.9408	0.9368	0.9331	0.9269	0.9236	0.9208	0.9186	0.9170
17.50	1.0514	1.0295	1.0141	1.0012	0.9907	0.9835	0.9767	0.9711	0.9668	0.9630	0.9564	0.9529	0.9501	0.9478	0.9461
18.00	1.0834	1.0612	1.0455	1.0323	1.0215	1.0141	1.0070	1.0013	0.9968	0.9928	0.9858	0.9820	0.9791	0.9768	0.9751
18.50	1.1153	1.0929	1.0769	1.0634	1.0523	1.0447	1.0373	1.0313	1.0267	1.0224	1.0150	1.0110	1.0080	1.0056	1.0038
19.00	1.1473	1.1246	1.1083	1.0944	1.0830	1.0752	1.0675	1.0613	1.0564	1.0519	1.0440	1.0398	1.0366	1.0342	1.0324
19.50	1.1792	1.1562	1.1397	1.1253	1.1137	1.1056	1.0976	1.0912	1.0860	1.0812	1.0728	1.0683	1.0651	1.0626	1.0607
20.00	1.2112	1.1879	1.1711	1.1562	1.1443	1.1360	1.1277	1.1210	1.1155	1.1104	1.1015	1.0967	1.0933	1.0908	1.0889
20.50	1.2431	1.2195	1.2024	1.1871	1.1748	1.1663	1.1576	1.1507	1.1449	1.1395	1.1301	1.1250	1.1214	1.1188	1.1168
21.00	1.2750	1.2511	1.2337	1.2179	1.2053	1.1965	1.1875	1.1803	1.1741	1.1685	1.1584	1.1530	1.1493	1.1466	1.1446
21.50	1.3069	1.2827	1.2650	1.2487	1.2357	1.2267	1.2173	1.2099	1.2033	1.1974	1.1867	1.1800	1.1770	1.1743	1.1721
22.00	1.3388	1.3143	1.2962	1.2795	1.2661	1.2568	1.2470	1.2393	1.2324	1.2261	1.2147	1.2086	1.2046	1.2018	1.1995
22.50	1.3707	1.3458	1.3274	1.3102	1.2964	1.2869	1.2766	1.2667	1.2614	1.2547	1.2427	1.2361	1.2319	1.2291	1.2268
23.00	1.4026	1.3774	1.3586	1.3409	1.3267	1.3169	1.3062	1.2979	1.2902	1.2832	1.2705	1.2635	1.2592	1.2562	1.2538
23.50	1.4344	1.4089	1.3898	1.3715	1.3569	1.3469	1.3357	1.3271	1.3190	1.3116	1.2981	1.2908	1.2862	1.2832	1.2807
24.00	1.4663	1.4404	1.4210	1.4021	1.3871	1.3768	1.3652	1.3563	1.3477	1.3399	1.3256	1.3179	1.3131	1.3100	1.3074
24.50	1.4982	1.4719	1.4521	1.4327	1.4173	1.4066	1.3945	1.3853	1.3763	1.3681	1.3530	1.3448	1.3399	1.3366	1.3340
25.00	1.5300	1.5034	1.4832	1.4632	1.4474	1.4364	1.4238	1.4143	1.4048	1.3962	1.3803	1.3716	1.3664	1.3631	1.3604
25.50	1.5619	1.5349	1.5143	1.4937	1.4774	1.4662	1.4531	1.4432	1.4332	1.4242	1.4074	1.3983	1.3929	1.3895	1.3867
26.00	1.5937	1.5664	1.5454	1.5242	1.5075	1.4959	1.4823	1.4721	1.4616	1.4521	1.4344	1.4248	1.4192	1.4157	1.4128
26.50	1.6255	1.5978	1.5765	1.5547	1.5374	1.5255	1.5114	1.5008	1.4898	1.4799	1.4613	1.4512	1.4454	1.4417	1.4388
27.00	1.6574	1.6292	1.6075	1.5851	1.5674	1.5552	1.5405	1.5295	1.5180	1.5076	1.4881	1.4775	1.4714	1.4677	1.4646
27.50	1.6892	1.6607	1.6385	1.6155	1.5973	1.5847	1.5695	1.5582	1.5461	1.5353	1.5148	1.5036	1.4973	1.4935	1.4903
28.00	1.7210	1.6921	1.6695	1.6459	1.6272	1.6143	1.5985	1.5868	1.5742	1.5628	1.5413	1.5296	1.5231	1.5191	1.5159
28.50	1.7528	1.7235	1.7005	1.6762	1.6570	1.6438	1.6274	1.6153	1.6021	1.5903	1.5678	1.5555	1.5486	1.5447	1.5413
29.00	1.7846	1.7549	1.7315	1.7065	1.6868	1.6732	1.6563	1.6438	1.6300	1.6176	1.5941	1.5813	1.5742	1.5701	1.5666
29.50	1.8164	1.7863	1.7625	1.7168	1.7166	1.7026	1.6851	1.6722	1.6579	1.6449	1.6204	1.6070	1.5997	1.5954	1.5918
30.00	1.8482	1.8177	1.7934	1.7671	1.7463	1.7320	1.7139	1.7005	1.6856	1.6722	1.6465	1.6325	1.6249	1.6205	1.6168

Table 7-2 (m)
Extended Sukkar-Cornell Integral for Bottom-hole Pressure Calculation

P_r							Reduced Temperature for $B = 70.0$									
	1.1	1.2	1.3	1.4	1.5	1.6	1.7	1.8	1.9	2.0	2.2	2.4	2.6	2.8	3.0
0.20	0.0000	0.0000	0.0000	0.0000	0.0000	0.0000	0.0000	0.0000	0.0000	0.0000	0.0000	0.0000	0.0000	0.0000	0.0000
0.50	0.0017	0.0016	0.0016	0.0016	0.0016	0.0015	0.0015	0.0015	0.0015	0.0015	0.0015	0.0015	0.0015	0.0015	0.0015
1.00	0.0087	0.0080	0.0077	0.0074	0.0073	0.0072	0.0071	0.0070	0.0070	0.0070	0.0069	0.0069	0.0068	0.0068	0.0068
1.50	0.0240	0.0199	0.0185	0.0177	0.0172	0.0168	0.0165	0.0163	0.0161	0.0160	0.0158	0.0157	0.0156	0.0155	0.0154
2.00	0.0491	0.0385	0.0345	0.0325	0.0312	0.0303	0.0296	0.0291	0.0288	0.0285	0.0281	0.0278	0.0276	0.0274	0.0273
2.50	0.0772	0.0625	0.0554	0.0515	0.0490	0.0475	0.0462	0.0453	0.0448	0.0443	0.0435	0.0431	0.0426	0.0423	0.0420
3.00	0.1063	0.0894	0.0799	0.0742	0.0703	0.0679	0.0660	0.0646	0.0637	0.0629	0.0618	0.0611	0.0604	0.0599	0.0595
3.50	0.1356	0.1175	0.1066	0.0994	0.0943	0.0910	0.0884	0.0864	0.0851	0.0840	0.0825	0.0815	0.0806	0.0798	0.0792
4.00	0.1651	0.1464	0.1346	0.1264	0.1202	0.1162	0.1129	0.1104	0.1087	0.1073	0.1054	0.1040	0.1029	0.1018	0.1010
4.50	0.1947	0.1756	0.1634	0.1545	0.1475	0.1429	0.1391	0.1360	0.1340	0.1322	0.1299	0.1282	0.1268	0.1256	0.1246
5.00	0.2243	0.2050	0.1926	0.1833	0.1756	0.1706	0.1664	0.1629	0.1606	0.1585	0.1558	0.1538	0.1522	0.1508	0.1497
5.50	0.2540	0.2347	0.2221	0.2125	0.2045	0.1991	0.1946	0.1907	0.1881	0.1859	0.1827	0.1805	0.1787	0.1772	0.1760
6.00	0.2838	0.2644	0.2517	0.2420	0.2337	0.2281	0.2233	0.2192	0.2164	0.2140	0.2106	0.2081	0.2061	0.2045	0.2032
6.50	0.3135	0.2941	0.2815	0.2716	0.2632	0.2574	0.2525	0.2482	0.2453	0.2427	0.2390	0.2363	0.2343	0.2326	0.2313
7.00	0.3433	0.3239	0.3113	0.3014	0.2929	0.2870	0.2820	0.2775	0.2745	0.2718	0.2680	0.2652	0.2630	0.2613	0.2599
7.50	0.3732	0.3538	0.3411	0.3312	0.3226	0.3167	0.3116	0.3071	0.3040	0.3013	0.2973	0.2944	0.2922	0.2904	0.2890
8.00	0.4030	0.3836	0.3710	0.3611	0.3525	0.3465	0.3414	0.3368	0.3337	0.3309	0.3262	0.3239	0.3217	0.3198	0.3184
8.50	0.4328	0.4135	0.4009	0.3909	0.3824	0.3764	0.3713	0.3667	0.3635	0.3607	0.3560	0.3536	0.3514	0.3495	0.3481
9.00	0.4627	0.4434	0.4307	0.4208	0.4122	0.4063	0.4011	0.3965	0.3934	0.3905	0.3858	0.3834	0.3812	0.3793	0.3779
9.50	0.4926	0.4733	0.4606	0.4507	0.4421	0.4362	0.4310	0.4264	0.4233	0.4204	0.4157	0.4133	0.4110	0.4092	0.4077
10.00	0.5225	0.5031	0.4905	0.4805	0.4720	0.4660	0.4609	0.4563	0.4531	0.4503	0.4456	0.4432	0.4409	0.4390	0.4376
10.50	0.5523	0.5330	0.5203	0.5104	0.5018	0.4958	0.4907	0.4861	0.4830	0.4801	0.4754	0.4730	0.4708	0.4689	0.4675
11.00	0.5822	0.5629	0.5502	0.5402	0.5316	0.5256	0.5204	0.5159	0.5127	0.5099	0.5052	0.5028	0.5005	0.4987	0.4972
11.50	0.6121	0.5927	0.5800	0.5700	0.5613	0.5553	0.5502	0.5456	0.5424	0.5396	0.5349	0.5325	0.5303	0.5284	0.5270
12.00	0.6420	0.6226	0.6098	0.5997	0.5910	0.5850	0.5798	0.5752	0.5721	0.5692	0.5645	0.5621	0.5599	0.5580	0.5566
12.50	0.6718	0.6524	0.6396	0.6294	0.6207	0.6146	0.6094	0.6047	0.6016	0.5987	0.5940	0.5916	0.5893	0.5875	0.5860
13.00	0.7017	0.6822	0.6693	0.6591	0.6503	0.6442	0.6389	0.6342	0.6311	0.6282	0.6234	0.6210	0.6187	0.6168	0.6154

14.00	0.7615	0.7419	0.7288	0.7183	0.7093	0.7031	0.6977	0.6929	0.6897	0.6867	0.6818	0.6793	0.6770	0.6750	0.6736
14.50	0.7913	0.7717	0.7585	0.7479	0.7388	0.7325	0.7270	0.7222	0.7189	0.7158	0.7108	0.7062	0.7059	0.7039	0.7024
15.00	0.8212	0.8014	0.7881	0.7774	0.7682	0.7619	0.7562	0.7513	0.7479	0.7448	0.7397	0.7370	0.7346	0.7326	0.7311
15.50	0.8510	0.8312	0.8178	0.8069	0.7976	0.7911	0.7854	0.7804	0.7769	0.7737	0.7684	0.7656	0.7632	0.7612	0.7597
16.00	0.8809	0.8609	0.8474	0.8363	0.8269	0.8203	0.8145	0.8094	0.8058	0.8025	0.7969	0.7941	0.7916	0.7898	0.7880
16.50	0.9107	0.8907	0.8770	0.8658	0.8662	0.8495	0.8435	0.8363	0.8345	0.8311	0.8254	0.8224	0.8198	0.8178	0.8162
17.00	0.9406	0.9204	0.9066	0.8951	0.8854	0.8786	0.8724	0.8671	0.8632	0.8597	0.8537	0.8505	0.8479	0.8458	0.8442
17.50	0.9704	0.9501	0.9362	0.9245	0.9146	0.9076	0.9013	0.8958	0.8918	0.8881	0.8818	0.8765	0.8758	0.8737	0.8721
18.00	1.0002	0.9798	0.9657	0.9538	0.9437	0.9366	0.9300	0.9245	0.9203	0.9164	0.9098	0.9064	0.9036	0.9014	0.8997
18.50	1.0300	1.0095	0.9953	0.9831	0.9728	0.9656	0.9588	0.9530	0.9486	0.9446	0.9377	0.9340	0.9311	0.9289	0.9272
19.00	1.0599	1.0392	1.0248	1.0123	1.0018	0.9945	0.9874	0.9815	0.9769	0.9727	0.9654	0.9615	0.9586	0.9563	0.9545
19.50	1.0897	1.0689	1.0543	1.0415	1.0308	1.0233	1.0160	1.0099	1.0051	1.0007	0.9930	0.9889	0.9858	0.9835	0.9817
20.00	1.1195	1.0985	1.0837	1.0707	1.0597	1.0521	1.0445	1.0383	1.0332	1.0286	1.0204	1.0181	1.0129	1.0105	1.0087
20.50	1.1493	1.1282	1.1132	1.0999	1.0886	1.0808	1.0730	1.0665	1.0612	1.0564	1.0478	1.0432	1.0398	1.0374	1.0355
21.00	1.1791	1.1578	1.1426	1.1290	1.1175	1.1095	1.1014	1.0947	1.0892	1.0841	1.0749	1.0701	1.0666	1.0641	1.0622
21.50	1.2089	1.1874	1.1721	1.1581	1.1463	1.1381	1.1297	1.1229	1.1170	1.1116	1.1020	1.0968	1.0933	1.0907	1.0887
22.00	1.2387	1.2170	1.2015	1.1871	1.1751	1.1667	1.1580	1.1509	1.1448	1.1391	1.1289	1.1235	1.1198	1.1171	1.1151
22.50	1.2685	1.2466	1.2309	1.2162	1.2039	1.1953	1.1862	1.1789	1.1724	1.1665	1.1558	1.1500	1.1461	1.1434	1.1413
23.00	1.2982	1.2762	1.2602	1.2452	1.2326	1.2238	1.2144	1.2069	1.2000	1.1938	1.1825	1.1763	1.1723	1.1695	1.1674
23.50	1.3280	1.3058	1.2896	1.2742	1.2613	1.2522	1.2425	1.2347	1.2276	1.2210	1.2090	1.2026	1.1984	1.1955	1.1933
24.00	1.3578	1.3354	1.3190	1.3031	1.2899	1.2807	1.2706	1.2625	1.2550	1.2482	1.2355	1.2287	1.2243	1.2214	1.2191
24.50	1.3876	1.3650	1.3483	1.3321	1.3185	1.3090	1.2986	1.2903	1.2824	1.2752	1.2619	1.2546	1.2501	1.2471	1.2447
25.00	1.4173	1.3946	1.3776	1.3610	1.3471	1.3374	1.3265	1.3180	1.3097	1.3022	1.2881	1.2805	1.2758	1.2727	1.2702
25.50	1.4471	1.4241	1.4069	1.3899	1.3757	1.3657	1.3544	1.3456	1.3369	1.3290	1.3142	1.3062	1.3013	1.2981	1.2956
26.00	1.4769	1.4537	1.4362	1.4187	1.4042	1.3940	1.3823	1.3732	1.3641	1.3658	1.3403	1.3318	1.3267	1.3235	1.3209
26.50	1.5066	1.4832	1.4655	1.4476	1.4327	1.4222	1.4101	1.4007	1.3912	1.3625	1.3662	1.3573	1.3520	1.3487	1.3460
27.00	1.5364	1.5127	1.4948	1.4764	1.4611	1.4504	1.4379	1.4282	1.4182	1.4092	1.3920	1.3827	1.3772	1.3738	1.3710
27.50	1.5661	1.5423	1.5240	1.5052	1.4895	1.4786	1.4656	1.4556	1.4452	1.4357	1.4178	1.4079	1.4023	1.3987	1.3759
28.00	1.5959	1.5718	1.5533	1.5340	1.5179	1.5067	1.4933	1.4829	1.4721	1.4622	1.4434	1.4331	1.4272	1.4236	1.4206
28.50	1.6256	1.6013	1.5825	1.5627	1.5463	1.5348	1.5209	1.5102	1.4989	1.4886	1.4690	1.4581	1.4520	1.4483	1.4452
29.00	1.6554	1.6308	1.6117	1.5915	1.5747	1.5629	1.5485	1.5375	1.5257	1.5150	1.4944	1.4831	1.4768	1.4729	1.4698
29.50	1.6851	1.6603	1.6410	1.6202	1.6030	1.5909	1.5761	1.5547	1.5524	1.5412	1.5198	1.5079	1.5014	1.4974	1.4942
30.00	1.7148	1.6898	1.6702	1.6489	1.6313	1.6189	1.6036	1.5919	1.5791	1.5675	1.5450	1.5327	1.5259	1.5218	1.5165

* After Messer et al. (1974); courtesy of SPE (3913).

(text continued from page 309)

$$G = (0.17584 - 0.00262B) T_{pr}^2 + (1.08235 - 0.01474B) T_{pr}$$
$$- (0.81075 - 0.00771B) \quad (7\text{-}53)$$

For $20 \leq B \leq 100$, $10 \leq p_{pr} \leq 30$, $1.1 \leq T_{pr} \leq 3$,

$$E = (0.297336 - 0.001594B) T_{pr}^2 + (0.009678B - 1.664290) T_{pr}$$
$$- (0.001963B - 0.429648)$$

$$F = (0.019112 - 0.000171B) T_{pr}^2 + (0.000941B - 0.107041) T_{pr}$$
$$- (0.001259B - 0.188162)$$

$$G = (0.001435B - 0.212139) T_{pr}^2 + (0.000941B - 0.107041) T_{pr}$$
$$+ (0.006320B - 1.003890) \quad (7\text{-}54)$$

This method is simple and quite accurate: Equations 7-53 and 7-54 give errors of less than 2% in the parameter ranges associated with them. No trial and error is involved, which simplifies the calculation procedure considerably. For these reasons, the Sukkar-Cornell method is widely used.

Example 7-2. Determine the SBHP in a gas well, given the following:
Depth, $z = 5,000$ ft; gas gravity, $\gamma_g = 0.70$; wellhead temperature, $T_{wh} = 80°F$; bottom-hole temperature, $T_{ws} = 160°F$; wellhead pressure, $p_{wh} = 400$ psia.

Use (a) average temperature and Z-factor method, and (b) Sukkar-Cornell method.

Solution

$$T_{av} = (80 + 160)/2 = 120°F = 580°R$$

From Figure 3-1, $p_{pc} = 663$ psia, $T_{pc} = 387°R$

(a) Average temperature and Z-factor method:

First trial:

Using Equation 7-48, $p_{ws} = 400 + 0.25(400/100)(5,000/100) = 450$ psia

Thus, $p_{av} = (400 + 450)/2 = 425$ psia

$p_{pr} = 425/663 = 0.641$, $T_{pr} = (580)/387 = 1.499$

From Figure 3-2, $Z_{av} = 0.967$

Using Equation 7-47, $s = (0.0375)(0.7)(5,000)/[(0.967)(580)] = 0.23402$

Using Equation 7-46, $p_{ws} = 400\ e^{0.23402/2} = \underline{449.65\ \text{psia}}$

A second trial is not needed.

(b) <u>Sukkar-Cornell method:</u>

$p_{r_{wh}} = 400/663 = 0.603$, $T_{r_{wh}} = (80 + 460)/387 = 1.395$

From Table 7-2a, by interpolation

$$\int_{0.2}^{0.603} (Z/p_r)dp_r = 1.0392$$

Using Equation 7-51,

$$\int_{0.2}^{p_{rws}} (Z/p_r)dp_r - \int_{0.2}^{0.603} (Z/p_r)dp_r = \frac{(0.01875)(0.7)(5,000)}{580}$$

Thus, $\int_{0.2}^{p_{rws}}(Z/p_r)dp_r = 1.0392 + 0.11315 = 1.15235$

$T_{r_{ws}} = (160 + 460)/387 = 1.602$

From Table 7-2a:

$$\int_{0.2}^{0.6}(Z/p_r)dp_r = 1.049$$

$$\int_{0.2}^{0.7}(Z/p_r)dp_r = 1.210$$

By interpolation,

$p_{r_{ws}} = 0.6 + (1.15235 - 1.049)(0.7 - 0.6)/(1.210 - 1.049) = 0.6642$

Thus, $p_{ws} = (0.6642)(663) = \underline{440.36\ \text{psia}}$

Cullender-Smith Method

Cullender and Smith (1956) used a more rigorous approach, accounting for the dependence of Z-factor on both temperature and pressure. Taking into account the units used by Cullender and Smith, Equation 7-42 is rearranged as follows:

$$\int_{p_{wh}}^{p_{wf}} \frac{(p/TZ)^2(TZ/p)\,dp}{(p/TZ)^2/(1,000) + (6.7393 \times 10^{-4})(4)(f/4)q_{sc}^2 L/(1,000 z d^5)}$$

$$= (1,000)(0.01875)\gamma_g z$$

or,

$$\int_{p_{wh}}^{p_{wf}} \frac{(p/TZ)\,dp}{2.6957 \times 10^{-6}(f/4)q_{sc}^2 L/(zd^5) + (p/TZ)^2/1,000} = 18.75\,\gamma_g z \quad (7\text{-}55)$$

Equation 7-55 is generally expressed as:

$$\int_{p_{wh}}^{p_{wf}} \frac{(p/TZ)\,dp}{F^2 + (p/TZ)^2/1,000} = 18.75\,\gamma_g z \quad (7\text{-}56)$$

where

$$F^2 = \frac{2.6957 \times 10^{-6}(f/4)q_{sc}^2 L}{zd^5} \quad (7\text{-}57)$$

The expression in the left integral in Equation 7-56 is represented by I:

$$I = \frac{p/(TZ)}{F^2 + (p/TZ)^2/1,000} \quad (7\text{-}58)$$

For the case of SBHP determination, I reduces to:

$$I = 1,000\,(TZ/p) \quad (7\text{-}59)$$

The left integral in Equation 7-56 can be evaluated by numerical integration techniques. This is tedious, and acceptable accuracy can generally be obtained using trapezoidal or Simpson's rule*. For the static case, the integral in Equation 7-56 can be written as:

---

* For $y = f(x)$, $y_i = f(x_i)$:
Trapezoidal rule: $\int_a^b f(x)\,dx \simeq (h/2)(y_0 + 2y_1 + 2y_2 + \ldots + 2y_{n-1} + y_n)$
Simpson's rule (more accurate than the trapezoidal rule): $\int_a^b f(x)\,dx \simeq (h/3)(y_0 + 4y_1 + 2y_2 + 4y_3 + \ldots + 2y_{n-2} + 4y_{n-1} + y_n)$, for even n.

Steady State Flow of Gas Through Pipes   339

$$\int_{p_0}^{p_n} 1{,}000 \frac{TZ}{p} \, dp \simeq \frac{1}{2} \left[(p_1 - p_0)(I_1 + I_0) + (p_2 - p_1)(I_2 + I_1) + \ldots \right.$$

$$\left. \ldots + (p_n - p_{n-1})(I_n + I_{n-1}) \right]$$

where $I_n = 1{,}000(TZ/p)_n$. Greater accuracy in SBHP calculation can be obtained using more steps, requiring the use of a computer. For manual calculations, a two-step procedure is generally used, where only the value of pressure at the midpoint, p_{ms} is considered:

$$\int_{p_{wh}}^{p_{ws}} 1{,}000 \frac{TZ}{p} \, dp \simeq \frac{(p_{ms} - p_{wh})(I_{ms} + I_{wh})}{2} + \frac{(p_{ws} - p_{ms})(I_{ws} + I_{ms})}{2} \quad (7\text{-}60)$$

Substituting for the integral from Equation 7-60 into Equation 7-56:

$$(p_{ms} - p_{wh})(I_{ms} + I_{wh}) + (p_{ws} - p_{ms})(I_{ws} + I_{ms}) = 37.5 \, \gamma_g z \quad (7\text{-}61)$$

Equation 7-61 can be separated into two parts, one for each half of the flow string. For the upper half:

$$(p_{ms} - p_{wh})(I_{ms} + I_{wh}) = 37.5 \, \gamma_g z/2 \quad (7\text{-}62)$$

and, for the lower half:

$$(p_{ws} - p_{ms})(I_{ws} + I_{ms}) = 37.5 \, \gamma_g z/2 \quad (7\text{-}63)$$

As demonstrated by Cullender and Smith (1956), an accuracy equivalent to a four-step solution scheme can be obtained by using this two-step calculation scheme and then using Simpson's rule (also called parabolic interpolation) to obtain a more accurate value of bottom-hole pressure as follows:

$$\frac{p_{ws} - p_{wh}}{3} [I_{wh} + 4I_{ms} + I_{ws}] = 37.5 \, \gamma_g z \quad (7\text{-}64)$$

The following solution procedure may be used:
1. Calculate I_{wh} (using the general relationship given by Equation 7-59) at the known wellhead conditions of pressure and temperature.
2. Calculate p_{ms} from Equation 7-62, assuming $I_{ms} = I_{wh}$.
3. Using the value of p_{ms}, and temperature T_{ms} (obtained as the arithmetic average of T_{wh} and T_{ws}), calculate I_{ms} using Equation 7-59.
4. Recalculate p_{ms} from Equation 7-62.
5. If this recalculated value is not within 1 psi (or any other small pressure tolerance) of the p_{ms} calculated earlier, repeat steps 3 and 4 until this condition is met.

340   Gas Production Engineering

6. Calculate p_{ws} from Equation 7-63, assuming $I_{ws} = I_{ms}$.
7. Using the value of p_{ws}, and the bottom-hole temperature T_{ws}, calculate I_{ws} using Equation 7-59.
8. Recalculate p_{ws} from Equation 7-63.
9. If this recalculated value is not within 1 psi (or any other small pressure tolerance) of the p_{ws} calculated earlier, repeat steps 7 and 8 until this condition is met.
10. Use Equation 7-64 (Simpson's rule) to obtain a more accurate value of the bottom-hole pressure p_{ws}.

Example 7-3. Repeat Example 7-2 using the Cullender-Smith method.

Solution

For the upper half:

p_{wh} = 400 psia, T_{wh} = 80°F = 540°R

$p_{r_{wh}}$ = 400/663 = 0.603, $T_{r_{wh}}$ = 540/387 = 1.395

From Figure 3-2, Z_{av} = 0.93

Using Equation 7-59, I_{wh} = (1,000)(540)(0.93)/(400) = 1,255.5

Using Equation 7-62,

$(p_{ms} - 400)(1,255.5 + 1,255.5) = (37.5)(0.7)(5,000)/2$

Thus, p_{ms} = 426.14 psia

$p_{r_{ms}}$ = 426.14/663 = 0.643, $T_{r_{ms}}$ = 580/387 = 1.499, and Z_{ms} = 0.96

Using Equation 7-59, I_{ms} = (1,000)(580)(0.96)/(426.14) = 1,306.61

Using Equation 7-62,

p_{ms} = 400 + (37.5)(0.7)(2,500)/(1,306.61 + 1,255.5) = 425.61 psia

Thus, p_{ms} = 425.61 psia, and I_{ms} = 1,306.61

For the lower half:

Assuming $I_{ws} = I_{ms}$ = 1,306.61, and using Equation 7-63,

Steady State Flow of Gas Through Pipes   341

$p_{ws} = 425.61 + (37.5)(0.7)(2,500)/(1,306.61 + 1,306.61) = 450.72$ psia

$p_{r_{ws}} = 450.72/663 = 0.68$, $T_{r_{ws}} = 620/387 = 1.602$, and $Z_{ws} = 0.95$

Using Equation 7-59, $I_{ws} = (1,000)(620)(0.95)/(450.72) = 1,306.798$

Thus, $p_{ws} = 450.72$ psia

Using Simpson's rule (Equation 7-64):

$[(p_{ws} - 400)/3][1,255.5 + 4(1,306.61) + 1,306.798] = (37.5)(0.7)(5,000)$

or, $p_{ws} = \underline{450.55\text{ psia}}$

Flowing Bottom-Hole Pressure (FBHP)

Equation 7-42 describes the relationship between the pressure measured at the surface (or wellhead), p_{wh}, and the pressure p_{wf} at a depth z for a flowing well:

$$\int_2^1 \frac{(ZT/p)dp}{1 + (6.7393 \times 10^{-4} fL q_{sc}^2 Z^2 T^2)/(zp^2 d^5)} = 0.01875 \gamma_g z$$

For flowing wells, the temperature-depth relationship presented by Lesem et al. (1957) or by Ramey (1962) may be useful for greater accuracy. In most cases, however, the log-mean (or even the arithmetic average) temperature gives satisfactory results.

Average Temperature and Z-Factor Method

An average Z-factor (and temperature) is used, similar to the static case, to simplify the left integral in Equation 7-42:

$$Z_{av} T_{av} \int_{p_{wh}}^{p_{ws}} \frac{dp/p}{1 + 6.7393 \times 10^{-4} fL(q_{sc} Z_{av} T_{av})^2/(zp^2 d^5)} = 0.01875 \gamma_g z$$

or,

$$\int_{p_{wh}}^{p_{ws}} \frac{p\, dp}{p^2 + 6.7393 \times 10^{-4} fL(q_{sc} Z_{av} T_{av})^2/(zd^5)} = \frac{0.01875 \gamma_g z}{Z_{av} T_{av}} \qquad (7\text{-}65)$$

342 Gas Production Engineering

The integral of a function

$$\int \frac{p\, dp}{C^2 + p^2} = \frac{1}{2} \ln(C^2 + p^2)$$

Therefore, Equation 7-65 becomes:

$$\ln \frac{C^2 + p_{wf}^2}{C^2 + p_{wh}^2} = \frac{0.03750 \gamma_g z}{Z_{av} T_{av}}$$

or,

$$C^2 + p_{wf}^2 = (C^2 + p_{wh}^2)\, e^s$$

where $s = (0.0375 \gamma_g z)/(Z_{av} T_{av})$, as given by Equation 7-47.

Solving for p_{wf}:

$$p_{wf}^2 = C^2(e^s - 1) + e^s p_{wh}^2$$

Substituting for C:

$$p_{wf}^2 = e^s p_{wh}^2 + \frac{6.7393 \times 10^{-4} fL(q_{sc} Z_{av} T_{av})^2 (e^s - 1)}{zd^5}$$

or,

$$p_{wf}^2 = e^s p_{wh}^2 + \frac{2.5272 \times 10^{-5} \gamma_g Z_{av} T_{av} fL(e^s - 1) q_{sc}^2}{sd^5} \qquad (7\text{-}66)$$

This method, applicable for wells up to 8,000 ft deep, should not be used if the change in kinetic energy is significant (Young, 1967).

Example 7-4. A 3,500-ft deep well is producing a 0.6 gravity gas at 7 MMscfd through a 3-in. I.D. tubing. The flowing wellhead temperature and pressure are 95°F and 2,000 psia, respectively. BHT = 150°F, tubing roughness, $\epsilon = 0.0006$ in., and $\mu_g = 0.017$ cp (assume constant). The well is deviated, with tubing length equal to 5,000 ft. Estimate the flowing bottom-hole pressure.

Solution

Using Equation 7-9, $N_{Re} = (20)(7,000)(0.6)/[(0.017)(3)] = 1.647 \times 10^6$

$\epsilon/d = 0.0006/3 = 0.0002$

From Figure 7-1, $f = 0.014$

From Figure 3-1, $p_{pc} = 672$ psia, $T_{pc} = 358°R$

The log-mean temperature, $T_{av} = (150 - 95)/\ln(150/95) = 120.41°F$
$= 580.41°R$, using Equation 7-27.

$T_{pr} = (580.41)/358 = 1.621$

First trial:

Assuming $p_{wf} = 2,400$ psia, and using Equation 7-32,

$p_{av} = (2/3)[(2,400^3 - 2,000^3)/(2,400^2 - 2,000^2)] = 2,206.06$ psia

$p_{pr} = 2,206.06/672 = 3.283$

From Figure 3-2, $Z_{av} = 0.828$

Using Equation 7-47, $s = (0.0375)(0.6)(3,500)/[(0.828)(580.41)] = 0.16386$

Using Equation 7-66:

$p_{wf}^2 = (2,000)^2 e^{0.16386}$

$+ \dfrac{(2.5272 \times 10^{-5})(0.6)(0.828)(580.41)(0.014)(5,000)(e^{0.16386} - 1)(7,000)^2}{(0.16386)(3)^5}$

$p_{wf} = 2,196.4$ psia

Second trial:

Assuming $p_{wf} = 2,196$ psia, $p_{av} = (2/3)[(2,196^3 - 2,000^3)/$
$(2,196^2 - 2,000^2)]$
$= 2,099.51$ psia.

344   Gas Production Engineering

$p_{pr} = 2{,}099.51/672 = 3.124$

From Figure 3-2, $Z_{av} = 0.830$

Using Equation 7-47, $s = (0.0375)(0.6)(3{,}500)/[(0.830)(580.41)] = 0.16347$

Using Equation 7-66:

$$p_{wf}^2 = (2{,}000)^2 e^{0.16347}$$
$$+ \frac{(2.5272 \times 10^{-5})(0.6)(0.830)(580.41)(0.014)(5{,}000)(e^{0.16347} - 1)(7{,}000)^2}{(0.16347)(3)^5}$$

$p_{wf} = 2{,}196.0$ psia, which is quite close to the earlier result of 2,196.4 psia.

Thus, $p_{wf} = \underline{2{,}196.0 \text{ psia}}$.

Sukkar-Cornell Method

This method uses an average temperature, but accounts for the pressure-dependence of the Z-factor. Equation 7-49 stated earlier in the section on SBHP is used:

$$\int_{p_{pr2}}^{p_{pr1}} \frac{(Z/p_{pr})dp_{pr}}{1 + BZ^2/p_{pr}^2} = \frac{0.01875\gamma_g z}{T_{av}}$$

For a given rate (corrected to 14.65 psia and 520°F), B is calculated using Equation 7-50:

$$B = \frac{6.6663 \times 10^{-4} f L q_{sc}^2 T_{av}^2}{zd^5 p_{pc}^2}$$

Tables 7-2b through 7-2m, or Equations 7-53 and 7-54, are used in evaluating the value of the integral. The Sukkar-Cornell method is preferred because in addition to being simple, consistent, and accurate, it does not require any trial and error.

Example 7-5. Repeat Example 7-4 using the Sukkar-Cornell method.

Solution

Using Equation 7-50:

$$B = \frac{(6.6663 \times 10^{-4})(0.014)(5,000)(7,000)^2(582.5)^2}{(3,500)(672)^2(3)^5} = 2.02 \simeq 2.0$$

$p_{rwh} = 2,000/672 = 2.976 \simeq 3.0$

$T_{rwh} = (95 + 460)/358 = 1.55$, $T_{rwf} = (150 + 460)/358 = 1.70$

Interpolating from Tables 7-2a and 7-2b, the integral $\int_{0.2}^{3.0} I = 1.6886$

Using Equation 7-49,

$$\int_{0.2}^{P_{rwf}} I - \int_{0.2}^{3} I = (0.01875)(0.6)(3,500)/582.5$$

$$\int_{0.2}^{P_{rwf}} I = 1.7562$$

Interpolation using Tables 7-2a and 7-2b yields $p_{rwf} = 3.208$

$p_{wf} = (3.208)(672) = \underline{2,155.8 \text{ psia.}}$

Cullender-Smith Method

The Cullender and Smith (1956) method uses Equation 7-56 as stated earlier in the section on static bottom-hole pressure:

$$\int_{p_{wh}}^{p_{wf}} \frac{(p/TZ)dp}{F^2 + (p/TZ)^2/1,000} = 18.75 \, \gamma_g z$$

with the expression in the left integral represented by I:

$$I = \frac{p/(TZ)}{F^2 + (p/TZ)^2/1,000}$$

As for the static case, a two-step procedure is generally used, where only the value of pressure at the midpoint, p_{mf} is considered:

$$37.5 \, \gamma_g z = \int_{p_{wh}}^{p_{wf}} I \, dp \simeq (p_{mf} - p_{wh})(I_{mf} + I_{wh})$$

$$+ (p_{wf} - p_{mf})(I_{wf} + I_{mf}) \quad (7\text{-}67)$$

Equation 7-67 can be separated into two parts, one for each half of the flow string:

$$(p_{mf} - p_{wh})(I_{mf} + I_{wh}) = 37.5\, \gamma_g z/2 \text{ for the upper half} \quad (7\text{-}68)$$

$$(p_{wf} - p_{mf})(I_{wf} + I_{mf}) = 37.5\, \gamma_g z/2 \text{ for the lower half} \quad (7\text{-}69)$$

And finally, Simpson's rule is used to obtain a more accurate value of bottom-hole pressure p_{wf}:

$$\frac{p_{wf} - p_{wh}}{3} [I_{wh} + 4I_{mf} + I_{wf}] = 37.5\, \gamma_g z \quad (7\text{-}70)$$

Calculation of F involves the use of the Moody friction factor f, which can be found using the friction factor chart (Figure 7-1). The calculation can, however, be simplified using Nikuradse's friction factor equation for fully turbulent flow, based upon an absolute roughness of 600 μin.*, to give (ERCB, 1975):

$$F = F_r q_{sc} = \frac{1.0797 \times 10^{-4} q_{sc}}{d^{2.612}} \text{ for } d < 4.277 \text{ in.} \quad (7\text{-}71)$$

and

$$F = F_r q_{sc} = \frac{1.0337 \times 10^{-4} q_{sc}}{d^{2.582}} \text{ for } d > 4.277 \text{ in.} \quad (7\text{-}72)$$

Frequently, the kinetic energy term, neglected in the equations developed earlier, is also included. The Cullender-Smith equation (Equation 7-56) becomes:

$$\int_{p_{wh}}^{p_{wf}} \frac{(1/0.01875\gamma_g)(p/TZ) + 1.111 \times 10^{-4} q_{sc}^2/(d^4 p)}{F^2 + (p/TZ)^2/1{,}000}\, dp = 1{,}000\, z \quad (7\text{-}73)$$

where $1.111 \times 10^{-4} q_{sc}^2/(d^4 p)$ is the kinetic energy term.

The following must be borne in mind when using the Cullender-Smith method (Young, 1967):

1. Smaller integration intervals obviously result in greater accuracy in the trapezoidal integration of Equation 7-56 (or Equation 7-73). An integration interval of 1,000 ft (or less) is recommended.

---

* 1 μin. (micro-inch) = 10^{-6} in.

Steady State Flow of Gas Through Pipes   347

2. Although Simpson's rule can be used to get a better approximation, it cannot correct for inaccuracies introduced upon using large trapezoidal integration intervals.
3. Change in kinetic energy can be ignored if the well is greater than 4,000 ft deep, or the wellhead flowing pressure is above 100 psia. For accuracy in determining the pressure traverse, the kinetic energy term, however, should be included, especially if the flowing wellhead pressure is below 500 psia.
4. For the case of gas injection, a discontinuity can develop when Equation 7-56 (or Equation 7-73) is numerically integrated. When this happens, the pressure change for that interval should be set equal to zero. Also, Simpson's rule cannot be used when a discontinuity occurs.

Gas Flow Through an Annulus

Although gas wells are generally produced through tubing, some wells may be produced through the casing-tubing annulus. For this latter case, the tubing flow equations developed earlier must be modified to reflect the proper flow conduit diameter. The equivalent diameter relationship of Equation 7-6 cannot be used because the diameter exponent is 5 in the flow equations derived.

From Equation 7-11, the pressure loss due to friction is given by:

$$\Delta p_f = f L \rho v^2 / (2 g_c d)$$

The velocity $v = K_1 q_{sc}/d^2$, where K_1 is a constant. Thus, the friction loss term in the energy balance can be written as:

$$\Delta p_f = \frac{K_1^2 f L \rho q_{sc}^2}{2 g_c d^4 d}$$

For the case of annular flow, where the outside diameter of the tubing is d_{to} and the inside diameter of the casing is d_{ci}, velocity is related to diameter as follows:

$$v = \frac{K_1 q_{sc}}{d_{ci}^2 - d_{to}^2}$$

Thus, the friction term becomes:

$$\Delta p_f = \frac{K_1^2 f L \rho q_{sc}^2}{2 g_c (d_{ci}^2 - d_{to}^2)^2 (d_{ci} - d_{to})} = \frac{K_2 q_{sc}^2}{(d_{ci}^2 - d_{to}^2)^2 (d_{ci} - d_{to})}$$

Thus, for the case of annular flow, d^5 in the vertical flow equations must be replaced by:

$$(d_{ci}^2 - d_{to}^2)^2(d_{ci} - d_{to}) = (d_{ci} - d_{to})^3(d_{ci} + d_{to})^2 \qquad (7\text{-}74)$$

Limitations in Vertical Flow Calculations

The methods described for calculating bottom-hole pressure are fairly accurate for most engineering calculation purposes. Some serious uncertainties in bottom-hole pressure calculations, however, inevitably exist because of:

1. Departure of the actual temperature-depth relationship from that assumed in the method of calculation.
2. Variations in gas gravity with depth, assumed constant in the calculation methods. For condensate systems, the change in composition with depth is even more significant.
3. Inaccuracies in determination of the Z-factor from the available correlations. For pseudoreduced pressures greater than 15, no correlation charts are even available. Equation-of-state methods can be used; laboratory measurements are strongly recommended for greater reliability and accuracy.
4. Imprecise friction factor.
5. Presence of unknown amounts of liquids in the wellbore. Two-phase flow relationships are necessary if liquid amount is significant.
6. Inaccuracies in flow rate, specific gravity, and pressure measurements.

Gas Flow Over Hilly Terrain

Transmission lines often deviate considerably from the horizontal, depending upon the terrain over which they are laid. In some cases, gas wells may also exhibit sections of different slope, such as the directionally drilled wells from offshore platforms. This section on flow over hilly terrain essentially deals with the situation where gas is flowing through pipes that are non-uniform in their slope, such as a hypothetical situation shown in Figure 7-5. Any flow terrain can be reduced to the form shown in Figure 7-5 by approximating the actual flow profile with small pipe sections of uniform slope. The three approaches applicable to such a flow are described in the following sections.

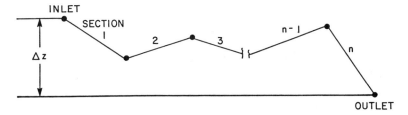

Figure 7-5. Gas flow over a hilly terrain.

Static Correction

To account for the difference in elevation between the inlet and outlet, Δz, the simplest approach is to modify the outlet pressure for the pressure exerted by a static gas column of height equal to Δz. As described earlier in the section on static bottom-hole pressure, different techniques are available for evaluating the pressure exerted by a static gas column. Since the static correction method is itself an approximation, the average temperature and Z-factor method is quite satisfactory to use. Let p_i be the inlet pressure, and p_o the outlet pressure. Then, the outlet pressure p_o must be corrected as follows (Equation 7-46):

$$p'_o = e^{s/2} p_o$$

where s is given by Equation 7-47:

$$s = (0.0375 \gamma_g \Delta z)/(Z_{av} T_{av})$$

Note that Δz is positive if the outlet is higher than the inlet (uphill flow), and negative if the inlet is higher than the outlet (downhill flow). Thus, s is positive for uphill flow, negative for downhill flow.

The flowline shown in Figure 7-5 is equivalent to a horizontal line with an upstream pressure equal to p_i, and a downstream pressure equal to $e^{s/2} p_o$. This correction can be incorporated into any horizontal flow correlation to give the flow through such a flowline. For example, the Weymouth equation for this situation can be written as:

$$q_{sc} = 31.5027 \left[\frac{T_{sc}}{p_{sc}} \right] \left[\frac{(p_i^2 - e^s p_o^2) d^{16/3}}{\gamma_g Z_{av} T_{av} L} \right]^{0.5} \tag{7-75}$$

Similar expressions can be written for the Panhandle A and B equations, and the Clinedinst equation. This approach, though imprecise, frequently gives an adequate approximation.

Flow Correction

A more rigorous correction for the flow profile accounts for inclined flow in the different sections of the pipe. Equation 7-66 gives the relationship for inclined flow, assuming an average temperature and Z-factor:

$$p_{wf}^2 = e^s p_{wh}^2 + \frac{2.5272 \times 10^{-5} \gamma_g Z_{av} T_{av} fL(e^s - 1) q_{sc}^2}{sd^5}$$

Rearranging this equation, and replacing the pressures with the inlet and outlet pressures p_i and p_o:

$$q_{sc}^2 = \frac{(p_i^2 - e^s p_o^2) sd^5}{2.5272 \times 10^{-5} \gamma_g Z_{av} T_{av} fL(e^s - 1)}$$

Thus,

$$q_{sc} = \left[\frac{(T_{sc}/520)}{(2.5272 \times 10^{-5})^{0.5}(p_{sc}/14.73)}\right]\left[\frac{(p_i^2 - e^s p_o^2)sd^5}{\gamma_g Z_{av} T_{av} fL(e^s - 1)}\right]^{0.5}$$

or,

$$q_{sc} = 5.634814 \left[\frac{T_{sc}}{p_{sc}}\right]\left[\frac{(p_i^2 - e^s p_o^2)d^5}{\gamma_g Z_{av} T_{av} fL_e}\right]^{0.5}$$

The constant 5.634814 is slightly different from the constant 5.6353821 in Equation 7-29 due to round-off errors. Thus:

$$q_{sc} = 5.63538 \left[\frac{T_{sc}}{p_{sc}}\right]\left[\frac{(p_i^2 - e^s p_o^2)d^5}{\gamma_g Z_{av} T_{av} fL_e}\right]^{0.5} \qquad (7\text{-}76)$$

where L_e is given by:

$$L_e = \frac{(e^s - 1)}{s} L \qquad (7\text{-}77)$$

Equation 7-77 gives the expression for the effective length for a single section of a flowline. For the general case of non-uniform slope where the profile is divided into a number of sections n of nearly uniform slope (as shown in Figure 7-5), the effective length is calculated as follows:

Steady State Flow of Gas Through Pipes   351

$$L_e = \frac{(e^{s_1} - 1)}{s_1} L_1 + \frac{e^{s_1}(e^{s_2} - 1)}{s_2} L_2 + \frac{e^{s_1 + s_2}(e^{s_3} - 1)}{s_3} L_3 + \ldots$$

$$+ \frac{e^{\Sigma s_n - 1}(e^{s_n} - 1)}{s_n} L_n, \quad s_i \neq 0 \tag{7-78}$$

where s_i represents the section i of the pipe.

Note that if $s_i = 0$ for any pipe section (this will happen if the section is horizontal, $z = 0$), the equivalent length term for that section is replaced by the actual length of the section.

Similar expressions can be written for the Panhandle A, Panhandle B, Clinedinst, and other equations. This average temperature and Z-factor approach is not exact, and is generally used where flow is almost isothermal and the pipe is essentially horizontal, with relatively small and few sections that are not horizontal.

General Method

If more precise calculations are necessary, the most rigorous method is to use a horizontal flow relationship such as the Clinedinst equation (Equation 7-39) for the horizontal section of the pipe, coupled with an inclined flow relationship such as the Cullender-Smith equation (Equation 7-56) for the non-horizontal portion of the flow. This results in quite complex relationships that require a computer for solution. Chapter 11 provides some insight into the modeling of gas flow in pipelines and networks. Usually, the computer program that is written for this purpose also has an economic analysis package built into it, to enable design optimization for a complete gas flow system.

Example 7-6. A section of a gas transmission system consists of three stations, A, B, and C. The 7-in. pipeline from A to B is 2 miles, and from B to C is 5 miles long. Stations A, B, and C are at elevations above sea-level of 4,000 ft, 7,000 ft, and 2,000 ft, respectively. The inlet pressure at station A is 3,000 psia, and the outlet pressure from the pipeline at station C is maintained at 2,200 psia. Assuming $\gamma_g = 0.6$, $T_{av} = 85°F$, and $f = 0.025$, find the gas flow rate through the system, and the pressure at station B.

Solution

From Figure 3-1, $p_{pc} = 672$ psia, $T_{pc} = 358°R$

$T_{av} = 85°F = 545°R$, $T_{pr} = (545)/358 = 1.522$

First trial:

Assume $p_B = 2,500$ psia

For section AB:

Using Equation 7-32, $p_{av} = (2/3)[(3,000^3 - 2,500^3)/(3,000^2 - 2,500^2)]$
$= 2,757.58$ psia

$p_{pr} = 2,757.58/672 = 4.104$

From Figure 3-2, $Z_{av} = 0.787$

Using Equation 7-47, $s_1 = (0.0375)(0.6)(7,000 - 4,000)/[(0.787)(545)]$
$= 0.15737$

For section BC:

$p_{av} = (2/3)[(2,500^3 - 2,200^3)/(2,500^2 - 2,200^2)] = 2,353.19$ psia

$p_{pr} = 2,353.19/672 = 3.502$, and, $Z_{av} = 0.780$

$s_2 = (0.0375)(0.6)(2,000 - 7,000)/[(0.780)(545)] = -0.26464$

For the complete line ABC:

$p_{av} = (2/3)[(3,000^3 - 2,200^3)/(3,000^2 - 2,200^2)] = 2,620.51$ psia

$p_{pr} = 2,620.51/672 = 3.90$, and, $Z_{av} = 0.782$

$s = (0.0375)(0.6)(2,000 - 4,000)/[(0.782)(545)] = -0.10559$

Using Equation 7-78:

$$L_e = \frac{(e^{0.15737} - 1)(2 \times 5,280)}{0.15737} + \frac{e^{0.15737}(e^{-0.26464} - 1)(5 \times 5,280)}{-0.26464}$$

$= 38,585.154$ ft

Using Equation 7-76:

$$q_{sc} = (5.63538)(520/14.73)\left[\frac{(3,000^2 - e^{-0.10559}2,200^2)(7^5)}{(0.6)(0.782)(545)(0.025)(38,585.154)}\right]^{0.5}$$

$= 111,919.16$ Mscfd $= 111.92$ MMscfd

Steady State Flow of Gas Through Pipes   353

For section AB:

Using Equation 7-77:

$$L_e = \frac{(e^{0.15737} - 1)(2 \times 5,280)}{0.15737} = 11,436.365 \text{ ft}$$

Using Equation 7-76:

$$111,919.16 = (5.63538)(520/14.73)$$

$$\left[\frac{(3,000^2 - e^{0.15737}p_B^2)(7^5)}{(0.6)(0.787)(545)(0.025)(11,436.365)}\right]^{0.5}$$

or,

$$(3,000^2 - e^{0.15737}p_B^2)^{0.5} = 1,177.093$$

$p_B = 2,550.6$ psia

Second trial:

Assume $p_B = 2,550.6$ psia

For section AB:

$p_{av} = (2/3)[(3,000^3 - 2,550.6^3)/(3,000^2 - 2,550.6^2)] = 2,781.36$ psia

$p_{pr} = 2,781.36/672 = 4.139$, and, $Z_{av} = 0.798$

$s_1 = (0.0375)(0.6)(7,000 - 4,000)/[(0.798)(545)] = 0.15520$

For section BC:

$p_{av} = (2/3)[(2,550.6^3 - 2,200^3)/(2,550.6^2 - 2,200^2)] = 2,379.62$ psia

$p_{pr} = 2,379.62/672 = 3.541$, and, $Z_{av} = 0.778$

$s_2 = (0.0375)(0.6)(2,000 - 7,000)/[(0.778)(545)] = -0.26532$

Using Equation 7-78:

$$L_e = \frac{(e^{0.15520} - 1)(2 \times 5,280)}{0.15520} + \frac{e^{0.15520}(e^{-0.26532} - 1)(5 \times 5,280)}{-0.26532}$$
$$= 38,504.546 \text{ ft}$$

Using Equation 7-76:

$$q_{sc} = (5.63538)(520/14.73)\left[\frac{(3{,}000^2 - e^{-0.10559}2{,}200^2)(7^5)}{(0.6)(0.782)(545)(0.025)(38{,}504.546)}\right]^{0.5}$$

$$= 112{,}036.24 \text{ Mscfd} = 112.04 \text{ MMscfd}$$

For section AB:

Using Equation 7-77:

$$L_e = \frac{(e^{0.15520} - 1)(2 \times 5{,}280)}{0.15520} = 11{,}423.443 \text{ ft}$$

Using Equation 7-76:

$$112{,}036.24 = (5.63538)(520/14.73)\left[\frac{(3{,}000^2 - e^{0.15520}p_B^2)(7^5)}{(0.6)(0.798)(545)(0.025)(11{,}423.443)}\right]^{0.5}$$

or,

$$[3{,}000^2 - e^{0.15520}p_B^2]^{0.5} = 1{,}185.8601$$

$p_B = 2{,}549.9$ psia $\simeq 2{,}550$ psia

Thus, the gas flow rate, q_{sc} = <u>112.04 MMscfd</u>, and the pressure at station B, p_B = <u>2,550 psia</u>.

Gas Flow Through Restrictions

In several instances in a gas production system, the gas must pass through relatively short restrictions. Chokes, consisting of a metal plate with a small hole to allow flow, are the most common restriction devices used to effect a pressure drop or reduce the rate of flow. They are capable of causing very large pressure drops: a gas can enter a choke at 5,000 psia and exit at 2,000 psia or less. Chokes have, therefore, found several applications as control devices in the oil and gas industry. Some of the purposes for which surface (and subsurface) chokes may be used are to maintain a precise wellhead flow rate, provide sand control by maintaining sufficient back pressure on the producing formation, protect surface equipment and/or prevent water coning by controlling the flow rate, and reservoir management. Subsurface (or bottom-hole) chokes are frequently used to reduce the wellhead flowing pressure, and prevent freeze-offs (hydrate formation) in the surface flow lines and controls.

The velocity of a fluid flowing through a restriction (orifice, nozzle, or choke) is expressed as follows:

$$v = \frac{K}{[1-(d_1/d_2)^4]^{0.5}} [2g(p_1 - p_2)/\rho]^{0.5} \tag{7-79}$$

where K = a constant representing the entrance/exit loss due to the change in flow diameter
d_1 = diameter at the throat of the restriction device, ft
d_2 = pipe diameter, ft
g = gravitational acceleration, ft/sec^2
p_1, p_2 = pressures at the upstream and downstream ends, respectively, of the flow restriction, lbf/ft^2
ρ = fluid density, lbm/ft^3

Equation 7-79 is generally written as:

$$v = C_d [2g(p_1 - p_2)/\rho]^{0.5} \tag{7-80}$$

where the coefficient of discharge, C_d, is given by:

$$C_d = \frac{K}{[1-(d_1/d_2)^4]^{0.5}} \tag{7-81}$$

Application of Equation 7-80 to the case of single-phase gas flow results in the well known de Saint Venant equation, assuming that the gas is perfect and that the flow is frictionless and adiabatic (see for example, Binder, 1958):

$$v_2 = C_d \left[\frac{2p_1 g_c}{\rho_1} \frac{k}{k-1} \left(1 - \left(\frac{p_2}{p_1}\right)^{(k-1)/k}\right)\right]^{0.5} \tag{7-82}$$

where k = ratio of the specific heats for the gas, c_p/c_v
subscripts 1,2 = upstream and downstream side of the choke, respectively

Substituting for gas density, $\rho_1 = (28.97 p_1 \gamma_g)/(RT_1)$,

$$v_2 = C_d \left[\frac{RT_1 g_c}{14.485 \gamma_g} \frac{k}{k-1} \left(1 - \left(\frac{p_2}{p_1}\right)^{(k-1)/k}\right)\right]^{0.5} \tag{7-83}$$

356 Gas Production Engineering

The gas velocity v_2 is related to the volumetric gas flow rate at standard conditions, q_{sc}, as before:

$$q_{sc} = v_2 A_{ch} \left(\frac{T_{sc}}{p_{sc}}\right)\left(\frac{p_2}{Z_2 T_2}\right)$$

where A_{ch} is the cross-sectional area of flow through the choke. Substituting for $A_{ch} = (\pi/4)d_{ch}^2$, and assuming perfect gas ($Z_2 = 1$):

$$q_{sc} = v_2(\pi d_{ch}^2/4) \left(\frac{T_{sc}}{p_{sc}}\right)\left(\frac{p_2}{T_2}\right) \tag{7-84}$$

For an adiabatic expansion/compression process for a perfect gas, pV^k = constant. Therefore,

$$p_1 V_1^k = p_2 V_2^k$$

implying that

$$T_2 = T_1 (p_2/p_1)^{(k-1)/k} \tag{7-85}$$

Substituting for T_2 from Equation 7-85, and for v_2 from Equation 7-83, into Equation 7-84:

$$q_{sc} = C_d \left[\frac{RT_1 g_c}{14.485\gamma_g} \frac{k}{k-1}\left(1-\left(\frac{p_2}{p_1}\right)^{(k-1)/k}\right)\right]^{0.5} (\pi d_{ch}^2/4) \left(\frac{T_{sc}}{p_{sc}}\right)\left(\frac{p_2}{T_1(p_2/p_1)^{(k-1)/k}}\right)$$

or,

$$q_{sc} = \left[\frac{\pi(Rg_c)^{0.5}T_{sc}}{(4)(14.485)^{0.5}p_{sc}}\right] C_d d_{ch}^2 \left[\frac{p_2^2 T_1}{\gamma_g T_1^2 (p_2/p_1)^{(2k-2)/k}} \frac{k}{k-1}\left(1-\left(\frac{p_2}{p_1}\right)^{(k-1)/k}\right)\right]^{0.5}$$

or,

$$q_{sc} = \left[\frac{\pi(Rg_c)^{0.5}T_{sc}}{(4)(14.485)^{0.5}p_{sc}}\right] C_d d_{ch}^2 \left[\frac{p_1^2}{\gamma_g T_1} \frac{k}{k-1} (p_2/p_1)^{2/k}\left(1-\left(\frac{p_2}{p_1}\right)^{(k-1)/k}\right)\right]^{0.5}$$

or,

$$q_{sc} = \left[\frac{\pi(Rg_c)^{0.5}T_{sc}}{(4)(14.485)^{0.5}p_{sc}}\right] C_d p_1 d_{ch}^2 \left(\frac{1}{\gamma_g T_1} \frac{k}{k-1}\left[\left(\frac{p_2}{p_1}\right)^{2/k} - \left(\frac{p_2}{p_1}\right)^{(k+1)/k}\right]\right)^{0.5} \tag{7-86}$$

Equation 7-86 is the general equation for flow through chokes. In common units, it can be written as:

$$q_{sc} = 974.61 C_d p_1 d_{ch}^2 \left(\frac{1}{\gamma_g T_1} \frac{k}{k-1} \left[\left(\frac{p_2}{p_1}\right)^{2/k} - \left(\frac{p_2}{p_1}\right)^{(k+1)/k} \right] \right)^{0.5} \quad (7\text{-}87)$$

where q_{sc} = gas flow rate, Mscfd (measured at 14.73 psia and 520°R)
d_{ch} = choke diameter, in.
p_1 = pressure at the upstream side of the choke, psia
p_2 = pressure at the downstream side of the choke, psia
T_1 = inlet (or upstream) temperature, °R

The flow through chokes (and flow restrictions in general) may be of two types: *subcritical* and *critical*.

Subcritical Flow

Flow is called subcritical when the velocity of the gas through the restriction is below the speed of sound in the gas. In the subcritical flow regime, Equations 7-86 and 7-87 apply, and the flow rate depends upon both the upstream as well as the downstream pressure. Subsurface chokes are usually designed to allow subcritical flow.

Critical Flow

Flow is called critical when the velocity of the gas through the restriction is equal to the speed of sound (about 1,100 ft/sec for air) in the gas. The maximum speed at which a pressure effect or disturbance can propagate through a gas cannot exceed the velocity of sound in the gas. Thus, once the speed of sound is attained, further increase in the pressure differential will not increase the pressure at the throat of the choke. Therefore, the flow rate cannot exceed the critical flow rate achieved when the ratio of the downstream pressure p_2 to the upstream pressure p_1 reaches a critical value, however much this ratio is decreased. Unlike subcritical flow, the flow rate in critical flow depends only upon the upstream pressure, because the pressure disturbances traveling at the speed of sound imply that a pressure disturbance at the downstream end will have no effect on the upstream pressure and/or flow rate. Surface chokes are usually designed to provide critical flow.

The choke flow relationships represented by Equations 7-86 and 7-87 are valid only in the subcritical flow regime, up to critical flow when the maximum flow velocity (equal to the speed of sound) is attained. If the pressure ratio at which critical flow occurs is represented by $(p_2/p_1)_c$, then from elementary calculus (differentiate Equation 7-87 with respect to p_2, and set the resultant expression equal to zero) it can be shown that:

$$(p_2/p_1)_c = [2/(k + 1)]^{k/(k-1)} \qquad (7\text{-}88)$$

Flow is subcritical for $(p_2/p_1) > (p_2/p_1)_c$, and critical for $(p_2/p_1) \leq (p_2/p_1)_c$. If the operating pressure ratio is less than the critical pressure ratio, the pressure ratio in Equation 7-86 or Equation 7-87 must be replaced by the critical pressure ratio, because the maximum flow rate through the choke is that which corresponds to critical flow. The critical pressure ratio $(p_2/p_1)_c$ is 0.49 for monoatomic gases, 0.53 for diatomic gases, and slightly higher for more complex gases.

Note that the analysis here assumes a perfect gas, with an adiabatic gas exponent k. No corrections have been made for deviation of gas from ideal behavior. It has been found that these relationships give very good results in field application, and a more complex analysis is generally quite unnecessary.

The value of k is also, fortunately, relatively insensitive to temperature variations as well as to the gas molecular weight for gaseous hydrocarbons. Therefore, k is usually assumed to be constant. $k \simeq 1.293$ for a 0.63 gravity gas. Generally, the value of k used is in the range of 1.25 to 1.31, implying from the relationship represented by Equation 7-88, that the flow is critical when the pressure ratio is in the range 0.5549 (for $k = 1.25$) to 0.5439 (for $k = 1.31$). As a rule of thumb, flow is assumed critical when the pressure ratio is less than or equal to 0.55, which implies a k equal to 1.275 approximately. Substituting $(p_2/p_1)_c = 0.55$, and $k = 1.275$ in Equation 7-87, we can get the well known choke design equation for critical flow:

$$q_{sc} = \frac{456.71 C_d p_1 d_{ch}^2}{(\gamma_g T_1)^{0.5}} \qquad (7\text{-}89)$$

where q_{sc} = flow rate through the choke, Mscfd
d_{ch} = choke size, in.
p_1 = upstream pressure, psia
T_1 = upstream temperature, °R
γ_g = gas gravity (air = 1)
C_d = coefficient of discharge, generally assumed to be 0.86

Example 7-7. Find the flow rate, referred to standard conditions of 14.73 psia and 520°R, of a 0.65 gravity gas through a $^{28}/_{64}$-in.-diameter choke for a downstream pressure p_2 equal to: (a) 600 psia, and (b) 300 psia. The upstream pressure $p_1 = 750$ psia, temperature $T_1 = 100°F$, and $k = 1.275$.

Solution

Using Equation 7-88:

$(p_2/p_1)_c = (2/2.275)^{1.275/0.275} = 0.55$

(a) $p_2/p_1 = 600/750 = 0.80$, which is greater than $(p_2/p_1)_c$

Using Equation 7-87,

$$q_{sc} = (974.61)(0.86)(750)(28/64)^2 \left[\left(\frac{1}{(0.65)(560)}\right)\left(\frac{1.275}{0.275}\right)(0.8^{2/1.275} - 0.8^{2.275/1.275})\right]^{0.5}$$

$= \underline{2{,}471.03 \text{ Mscfd}}$

(b) $p_2/p_1 = 300/750 = 0.40$, which is less than $(p_2/p_1)_c$

Therefore, $p_2/p_1 = (p_2/p_1)_c = 0.55$ must be used.

Using Equation 7-87,

$$q_{sc} = (974.61)(0.86)(750)(28/64)^2 \left[\left(\frac{1}{(0.65)(560)}\right)\left(\frac{1.275}{0.275}\right)(0.55^{2/1.275} - 0.55^{2.275/1.275})\right]^{0.5}$$

$= \underline{2{,}955.32 \text{ Mscfd}}$

The same result can alternatively be obtained using Equation 7-89:

$$q_{sc} = \frac{(456.71)(0.86)(750)(28/64)^2}{[(0.65)(560)]^{0.5}}$$

$= \underline{2{,}955.33 \text{ Mscfd}}$

Temperature Profile in Flowing Gas Systems

It is quite clear from the flow relationships presented so far that flow calculations require the value of the flowing temperature in order to determine the effective gas properties and pressure drop. To avoid complexity, most flow computations assume that the flowing temperature profile is linear. This assumption is not too far from reality, and generally gives quite good results. In some cases, however, more precise temperature and flow calculations may be required, such as in cases where phase changes occur during flow of the gas through the pipe.

Pressure and temperature are mutually dependent variables in flow— pressure loss depends to some extent on temperature (or heat loss), and the temperature (or heat loss) depends upon pressure that governs the changes in fluid enthalpies, overall heat transfer coefficient, and other parameters. Thus, generating a very precise temperature profile requires an enormous amount of complex, trial-and-error type of calculations for which even the amount of data available in most cases is insufficient. Thus, an approximate temperature profile, independent of pressure, is satisfactory for most engineering applications.

Flowing Temperature in (Horizontal) Pipelines

For a given inflow temperature, T_1, and surrounding soil temperature, T_s, the temperature of gas flowing in a pipeline depends upon heat exchange with the surroundings, given by the overall heat transfer coefficient; the (pressure dependent) Joule-Thomson effect* due to pressure changes caused by friction, and velocity and elevation changes; phase changes (condensation, vaporization) in the gas due to pressure and temperature changes; and energy loss (due to friction) during flow that is converted into heat.

Considering these factors, Papay (1970) has derived the following equation, assuming steady-state flow of gas, for the temperature T_{Lx} at a distance L_x from the pipeline inlet:

$$T_{Lx} = \frac{[T_s + C_4/C_2 - (C_1 C_5)/(C_2(C_2 + C_3))]C_1^{C_2/C_3}}{(C_1 + C_2 L_x)^{C_2/C_3}}$$

$$- \frac{C_4 + C_5 L_x}{C_2} + \frac{C_5(C_1 + C_3 L_x)}{C_2(C_2 + C_3)} \qquad (7\text{-}90)$$

where

$C_1 = z_{V1} c_{pL} + (1 - z_{V1}) c_{pV}$
$C_2 = k/m$
$C_3 = (z_{V2} - z_{V1})(c_{pL} - c_{pV})/L$

---

* The irreversible adiabatic simultaneous pressure and temperature reduction process that accompanies the expansion of a flowing gas by pressure reduction is called Joule-Thomson or throttling effect.

Steady State Flow of Gas Through Pipes   361

$$C_4 = \frac{p_1 - p_2}{L}[z_{V1}c_{pL}\mu_{dL} + (1 - z_{V1})c_{pV}\mu_{dV}] + \frac{z_{V2} - z_{V1}}{L}Q$$

$$+ \frac{v_2 - v_1}{L}v_1 + gh/L - \frac{k\pi d_o}{m}T_1$$

$$C_5 = \frac{(z_{V2} - z_{V1})(p_1 - p_2)}{L^2}(c_{pL}\mu_{dL} - c_{pV}\mu_{dV}) + \frac{v_2 - v_1}{L} \tag{7-91}$$

where z_V = mole fraction of vapor (gas) in the gas-liquid flowstream
p = pressure, lbf/ft^2
L = pipeline length, ft
v = fluid velocity, ft/sec
c_p = fluid specific heat at constant pressure, Btu/lbm-°F
μ_d = Joule-Thomson coefficient, ft^2-°F/lbf
m = mass flow rate, lbm/sec
Q = phase-transition heat, Btu/lbm
k = thermal conductivity, Btu/ft-sec-°F
g = gravitational acceleration, equal to 32.17 ft/sec^2
h = elevation difference between the inlet and outlet, ft
d_o = outside pipe diameter, ft
T_s = temperature of the soil or surroundings, °F

Subscripts 1 and 2 indicate the inlet and outlet ends of the pipe, respectively (except in the numbering of the constants C), and subscripts L and V represent liquid and vapor (gas), respectively.

In deriving Equation 7-90, Papay (1970) assumed that pressure, flow rate, and phase-transitions are linear functions of distance from the inlet end of the pipeline. This equation, therefore, is very accurate for short line segments. For the case where phase changes can be neglected (single-phase flow), Equation 7-90 can be simplified to:

$$T_{Lx} = T_s + (T_1 - T_s)e^{-KL_x} - \frac{\mu_{dV}(p_1 - p_2)}{KL}(1 - e^{-KL_x}) - \frac{gh}{KLc_{pV}}(1 - e^{-KL_x})$$

$$- \left(\frac{v_2 - v_1}{KLc_{pV}}\right)\left[\left(v_1 - \frac{v_2 - v_1}{KL}\right)(1 - e^{-KL_x}) + \frac{(v_2 - v_1)L_x}{L}\right] \tag{7-92}$$

where $K = \dfrac{k}{mc_{pV}}$ \hfill (7-93)

In Equation 7-92, the first two terms represent the heat exchange with the surroundings, the third term represents the Joule-Thomson effect, the fourth term accounts for the elevation changes, and the fifth term accounts for the change in velocity head. The last two terms are small and may be neglected for most practical purposes. If the pressure drop is small, then the temperature drop due to expansion is also small, and the third term may also be neglected. Neglecting these terms, Equation 7-92 simplifies to the following familiar form:

$$T_{Lx} = T_s + (T_1 - T_s) e^{-KL_x} \qquad (7\text{-}94)$$

Flowing Temperatures in Wells

The case of subsurface vertical flow is a special case of the general equation (Equation 7-90) where the temperature of the surroundings varies with distance along the flow length due to the geothermal gradient, G_T (°F/ft), of the earth.

A simplified equation presented by Ramey (1962), similar to Equation 7-94, is widely used:

$$T_{Lx} = T_1 - G_T[L_x - K^{-1}(1 - e^{-KL_x})] \qquad (7\text{-}95)$$

where L_x = distance from the bottom-hole or point of fluid entry, ft
T_{Lx} = temperature at location L_x, °F
T_1 = temperature at point of fluid entry (at L = 0), °F
G_T = geothermal gradient, °F/ft
K = k/mc_{pv}, as given by Equation 9-93

Equation 7-95 assumes that the temperature of the fluid and surroundings is equal at the point of entry, and that the heat loss is independent of time. The parameter K is quite difficult to estimate. An analysis based upon measured temperature profiles in actual gas wells is recommended, similar to an empirical equation developed by Shiu and Beggs (1980) for flowing oil wells.

Questions and Problems

1. A 12-in. ID, 80-mile gas pipeline is to deliver 100 MMscfd of gas at 250 psia. The average flowing temperature of the 0.75 gravity gas is 90°F, and ϵ/d = 0.0005 for the pipe. To what pressure must the gas be compressed at the inlet end to achieve this? Use at least two equations and compare the results.
2. What is the maximum throughput through the pipeline in Problem 1? Assume c = 0.05, E = 1.0, Y = 0.40, and S = 35,000 psi.

Steady State Flow of Gas Through Pipes   363

3. A 4,000-ft gas well requires a wellhead pressure of at least 100 psia to enable flow of the 0.85 gravity gas from the wellhead through the surface facilities. The bottom-hole temperature is 200°F and the surface temperature is 82°F. The casing is 4.5 in. ID.
   a. What is the minimum bottom-hole pressure to enable gas flow?
   b. What is the bottom-hole pressure for a gas flow rate of 10 MMscfd?
   c. Repeat part (b) assuming that the gas flows through the casing-tubing annulus, the tubing being 2.1 in. OD.
4. List some limitations of the flow equations described in this chapter for (a) horizontal flow and (b) vertical flow.
5. A gas production system consists of three sections:

   Subsurface section I: inclined, length = 5,500 ft, true vertical depth = 4000 ft
   inlet temperature = 150°F, outlet temperature = 85°F
   Surface section II: horizontal, length = 1.5 miles
   Surface section III: 2 miles long, inclined 30° from the horizontal.

   The temperature in the surface sections can be assumed to be 85°F, γ_g = 0.65, and f = 0.002. For a pressure of 4000 psia at the inlet of section I, find:

   (a) Gas flow rate through the system.
   (b) Pressure at the outlet of section III.

References

ANSI, 1976. *Code for Pressure Piping, Petroleum Refinery Piping*, ANSI B31.3, Am. Soc. of Mech. Engrs., United Engineering Center, New York, NY.
Beggs, H. D., 1984. *Gas Production Operations*. OGCI Publications, Oil & Gas Consultants International, Inc., Tulsa, Oklahoma, 287 pp.
Binder, R. C., 1958. *Advanced Fluid Mechanics*, Vol. 1. Prentice-Hall, Inc., New Jersey, 296 pp.
Chemical Engineers' Handbook, 1984. Edited by R. H. Perry and D. W. Green, McGraw-Hill Book Company, New York, 6th ed.
Cullender, M. H. and Smith, R. V., 1956. "Practical Solution of Gas-Flow Equations for Wells and Pipelines with Large Temperature Gradients," *Trans., AIME*, **207**, 281–287.

ERCB, 1975. *Theory and Practice of the Testing of Gas Wells*, 3rd ed. Energy Resources Conservation Board, Calgary, Alberta, Canada.

Ikoku, C. U., 1984. *Natural Gas Production Engineering*. John Wiley & Sons, Inc., New York, 517 pp.

Katz, D. L., Cornell, D., Kobayashi, R., Poettmann, F. H., Vary, J. A., Elenbaas, J. R., and Weinaug, C. F., 1959. *Handbook of Natural Gas Engineering*. McGraw-Hill Book Co., Inc., New York, 802 pp.

Lesem, L. B., Greytock, F., Marotta, F., and McKetta, J. J., Jr., 1957. "A Method of Calculating the Distribution of Temperature in Flowing Gas Wells," *Trans.*, *AIME*, **210**, 169–176.

Messer, P. H., Raghavan, R., and Ramey, H. J., Jr., 1974. "Calculation of Bottom-Hole Pressures for Deep, Hot, Sour Gas Wells," *J. Pet. Tech.*, **26**(1, Jan.), 85–92.

Moody, L. F., 1944. "Friction Factors for Pipe Flow," *Trans. Am. Soc. Mech. Eng.*, **66**(Nov.), 671–684.

Nisle, R. G. and Poettmann, F. H., 1955. "Calculation of the Flow and Storage of Natural Gas in Pipe," *Petrol. Engr.*, 27(1):D14; 27(2):C36; 27(3):D37.

Papay, J., 1970. "Steady Temperature Distributions in Producing Wells and Pipelines," *Koolaj es Foldgaz*, **11**. Cited reference in: *Production and Transport of Oil and Gas*, by A. P. Szilas, 1975. Developments in Petroleum Science, 3, Elsevier Scientific Publishing Co., Amsterdam, p. 563.

Ramey, H. J., Jr., 1962. "Wellbore Heat Transmission," *J. Pet. Tech.*, **14**(4, Apr.), 427–435.

Shiu, K. C. and Beggs, H. D., 1980. "Predicting Temperatures in Flowing Oil Wells," *Trans.*, *ASME: J. Energy Resour. Tech.*, **102**(1, March), 2–11.

Sukkar, Y. K. and Cornell, D., 1955. "Direct Calculation of Bottom-Hole Pressures in Natural Gas Wells," *Trans.*, *AIME*, **204**, 43–48.

Swamee, P. K. and Jain, A. K., 1976. "Explicit Equations for Pipe-Flow Problems," *J. Hydraulics Div. ASCE*, **102**(HY5, May), 657–664.

Szilas, A. P., 1975. *Production and Transport of Oil and Gas* (Developments in Petroleum Science, 3). Elsevier Scientific Publ. Co., Amsterdam, 630 pp.

Young, K. L., 1967. "Effect of Assumptions Used to Calculate Bottom-Hole Pressure in Gas Wells. *J. Pet. Tech.*, **19**(4, Apr.), 547–550.

8
Multiphase Gas-Liquid Flow

Introduction

As stated earlier, rarely do gas wells produce "dry" gas. Some liquid (oil and water) is almost always associated with it. For small amounts of liquid production, the flow may be considered single phase and the relationships for gas presented in Chapter 7 can be used with minor modifications to yield close estimates. When a large amount of liquid is associated with the gas, however, true multiphase gas-liquid flow prevails and such simplifications are no longer valid.

Although the existence of multiphase flow, defined as the simultaneous flow of free gases and liquids, has been known for a long time, its complex behavior has not yet been fully understood. The gas and liquid may exist as a homogeneous mixture, or as independent phases in other complex flow patterns such as slug, mist, emulsion, or bubble flow. The pressure drop for multiphase flow conditions is usually greater than for single-phase flow and, in some cases, the flow may be quite unsteady.

Early attempts to use modified single-phase flow relationships for multiphase flow resulted in highly overdesigned or underdesigned flowlines. An overdesigned line is not only expensive, but for two-phase flow conditions, it may result in unstable operations, with liquid slugging and pressure fluctuations. Generally, liquid and gas are transported through different flowlines at the surface, and such calculations may not always be necessary. Multiphase flow correlations are more important for wells with a significant liquid production, or for prediction of conditions for liquid loading of gas wells that are currently producing "dry" gas. This chapter outlines methods for handling multiphase flow in the context of gas engineering. For a detailed treatment of multiphase flow *per se*, readers may refer to the literature, notably Brown and Beggs (1977), and Beggs and Brill (1973).

Approximate Method for Two-Phase Systems

Flow streams with a GLR (gas-liquid ratio) greater than 10,000 scf/stb may be assumed to be single-phase gas. This is commonly the case for retrograde and wet gas reservoirs, where the oil and gas produced at the surface actually exist as gas within the reservoir. The small liquid content of the gas can be accounted for by modifying the properties that are affected by the presence of liquid. These include molecular weight, gas gravity, and gas compressibility factor (Z-factor).

Molecular weight of the flow stream consisting of gas as well as liquid can be calculated from the mixture composition, similar to calculations shown in Chapter 3 for a "dry" gas mixture. The gas specific gravity (air = 1 basis) for the total flow stream, γ_w, can be obtained by dividing the molecular weight of the flow stream by the molecular weight of air, resulting in the following expression (Craft and Hawkins, 1959):

$$\gamma_w = \frac{R_g \gamma_g + 4584 \gamma_o}{R_g + 132{,}800 \gamma_o / M_o} \qquad (8\text{-}1)$$

where R_g = gas-oil ratio, scf/stb
γ_g = specific gravity of the gas (air = 1)
γ_o = oil gravity (water = 1) at standard conditions
M_o = molecular weight of the stock-tank oil

If not known, M_o can be estimated using Equation 8-2:

$$M_o = \frac{44.29 \, \gamma_o}{1.03 - \gamma_o} = \frac{6084}{°API - 5.9} \qquad (8\text{-}2)$$

Liquid specific gravity, frequently specified in terms of °API, can be converted to specific gravity γ_o (water = 1 basis) as follows:

$$\gamma_o = \frac{141.5}{°API + 131.5} \qquad (8\text{-}3)$$

The gas compressibility factor (Z-factor) obtained from Figure 3-2 (Chapter 3) using critical pressure and temperature estimates for the total flow stream (using γ_w), can be corrected for the existence of two phases using Figure 8-1. If this two-phase Z-factor is essentially the same as the single-phase Z-factor, the mixture exists as a single gaseous phase.

Using these modified properties, calculations can be carried out with the relationships presented for gas in Chapter 7.

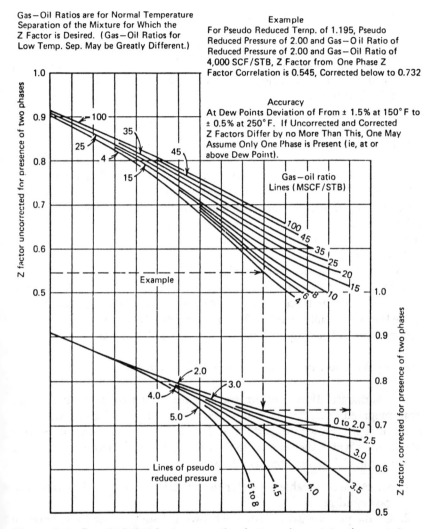

Figure 8-1. Gas deviation factor correction for two-phase natural gas systems. (After Elfrink et al., 1949; redrafted by Ikoku, 1984; courtesy of SPE.)

Multiphase Flow

For flow conditions that cannot be approximated as single-phase gas, it is necessary to use more complex procedures. Multiphase flow has been studied in detail by numerous investigators and several correlations are available. These correlations account for the flow geometry (vertical, inclined,

368   Gas Production Engineering

horizontal), flow pattern (slug, mist, bubble, plug, etc.), fluid properties (formation volume factor for oil and water, gas in solution in oil as well as water, surface tension, density and viscosity of the multiphase mixture), and other parameters related to flow (Reynolds number, friction factor, holdup, etc.). They are, however, quite complex, requiring the use of computers. Appendix B at the back of the book gives computer programs for two such methods.

For our purposes, it is satisfactory to consider a fairly good and practically expedient technique that uses pressure traverse curves. Pressure traverse curves are plots of pressure versus flow distance for a pipe of a given diameter for selected oil and gas properties at various gas/liquid ratios and flow rates. Some of the most commonly used pressure traverse curves, prepared using the correlations of Hagedorn and Brown (1965), are shown in Figure 8-2(a–v). Such curves can easily be generated for any other set of fluid properties, pipe diameters, or flow rates using the available correlations for multiphase flow.

Pressure traverse curves are simple to understand and use, fast, and fairly accurate. Since no single correlation is best over all ranges, it is better to construct a pressure traverse curve using the best applicable method for the range that we are interested in. To generate better traverse curves, a combination of several correlations and/or experimental data can be useful.

Pressure Traverse Curves for Horizontal Gas-Liquid Flow

The horizontal flow pressure traverse curves shown in Figures 8-2a–f were prepared using Eaton et al.'s (1967) correlation and give satisfactory results except for low rates and low G/L ratios. Although these curves were prepared for water, they can be used for oil, provided the free-gas/oil ratio is used for the G/L parameter, as follows:

1. Select the applicable curve for the given flowline size, flow rate, and gas/liquid ratio.
2. On the pressure axis, locate the known pressure, go vertically down to the applicable gas/liquid ratio curve, and read off the length on the length axis.
3. Correct this length for the pipeline length by: adding the pipeline length to the length determined in Step 2, if the known pressure is the outlet pressure; or subtracting the pipeline length from the length in Step 2, if the known pressure is the inlet pressure.
4. The unknown pressure is the pressure corresponding to the corrected length determined in Step 3.

Example 8-1 illustrates this procedure.

(text continued on page 380)

Multiphase Gas-Liquid Flow 369

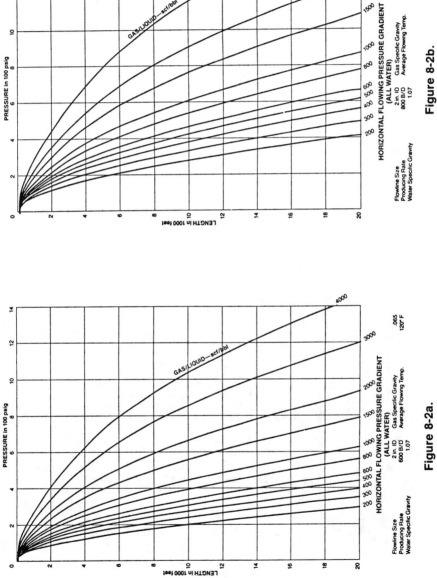

Figure 8-2b.

Figure 8-2a.

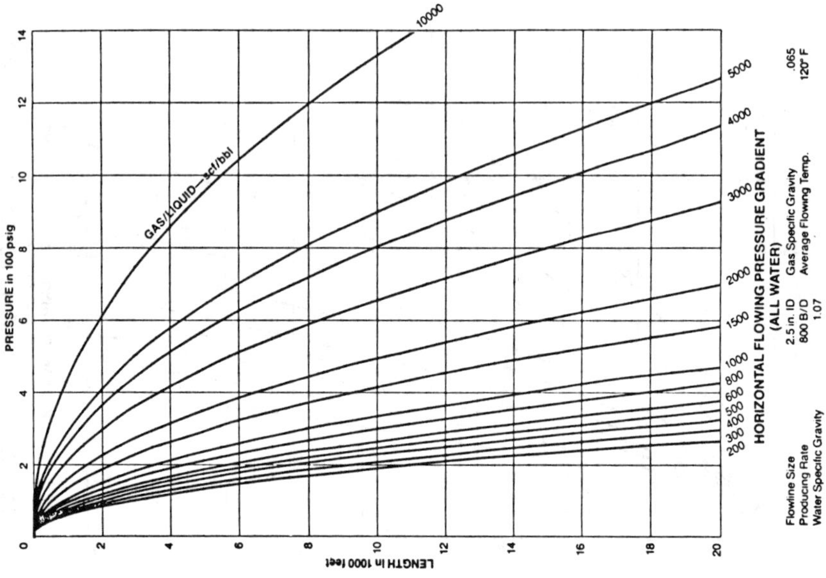

370 Gas Production Engineering

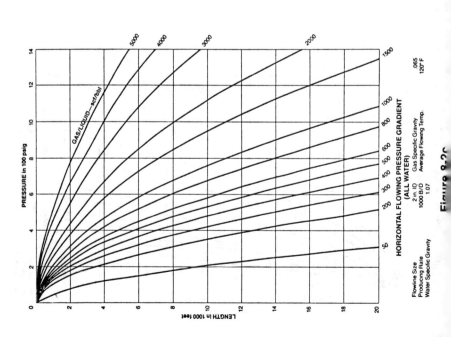

Figure 8-2d.

Figure 8-2c.

Multiphase Gas-Liquid Flow 371

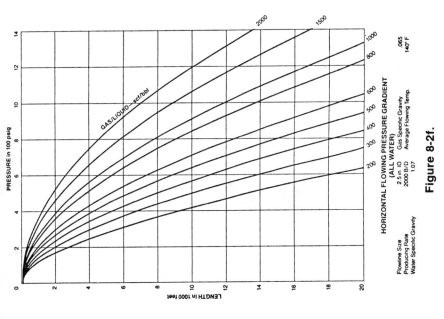

Figure 8-2f.

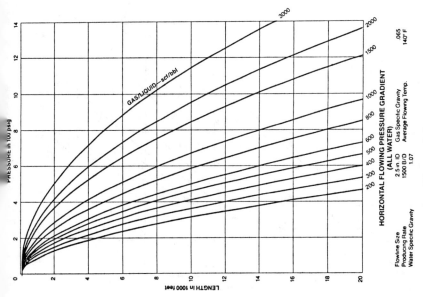

Figure 8-2e.

372 Gas Production Engineering

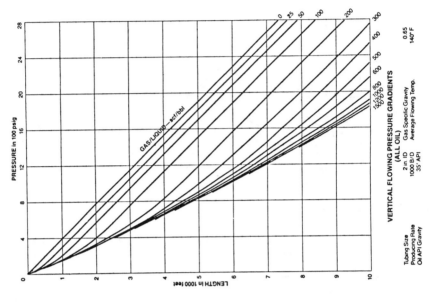

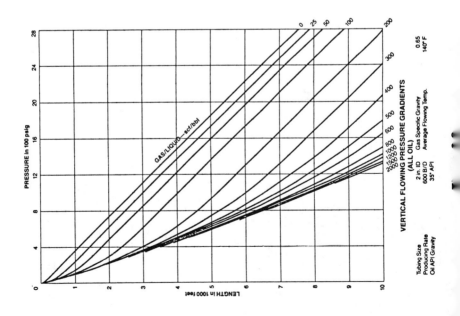

Multiphase Gas-Liquid Flow 373

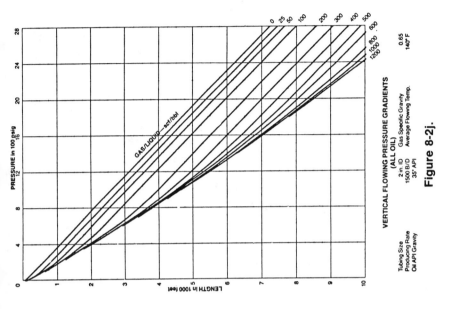

Figure 8-2j.

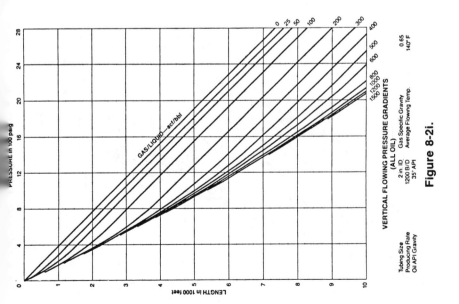

Figure 8-2i.

374 Gas Production Engineering

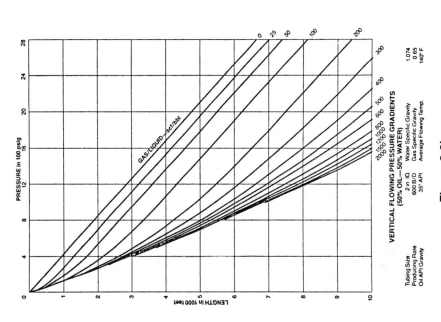

Figure 8-2I.

Figure 8-2k.

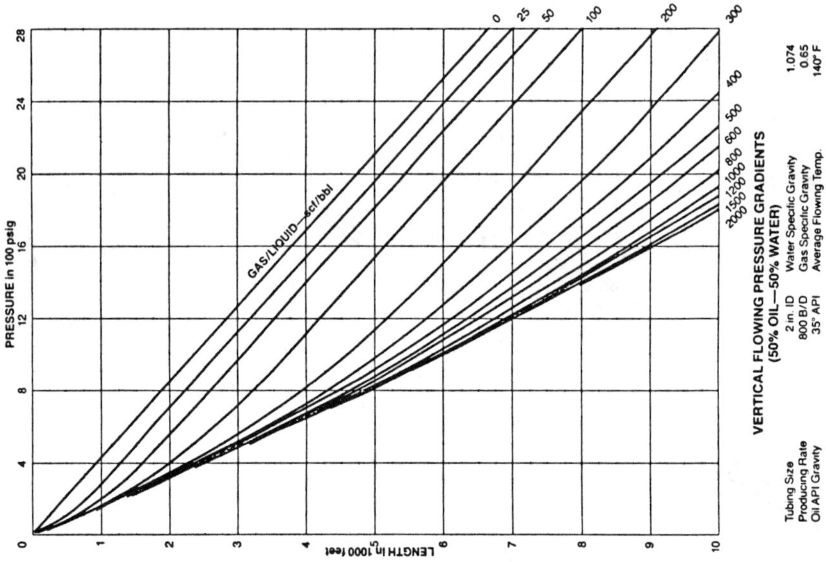

Multiphase Gas-Liquid Flow 375

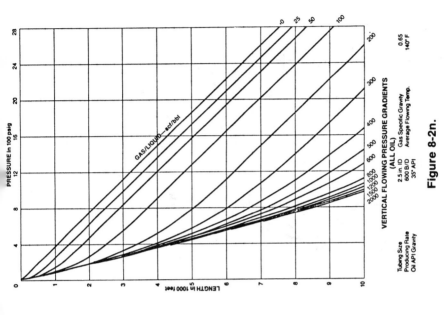

Figure 8-2n.

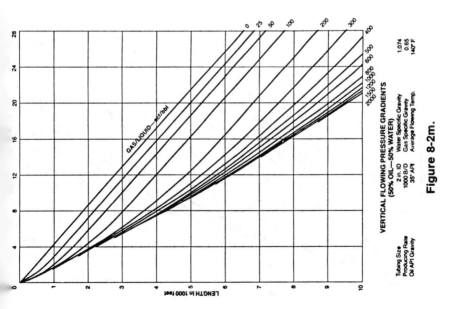

Figure 8-2m.

376 Gas Production Engineering

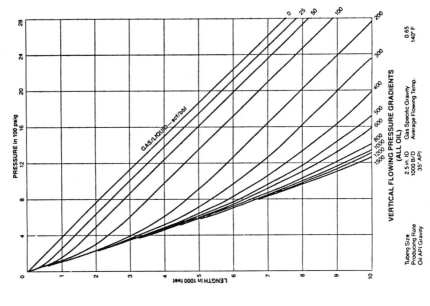

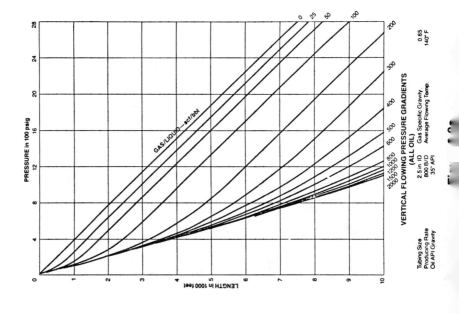

Multiphase Gas-Liquid Flow 377

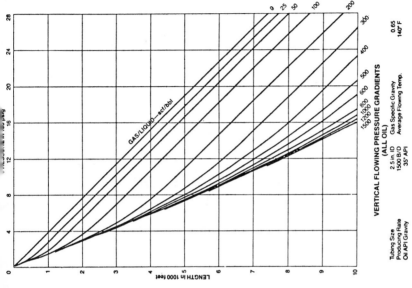

Figure 8-2r.

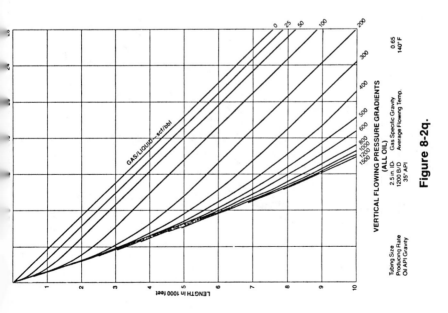

Figure 8-2q.

378 Gas Production Engineering

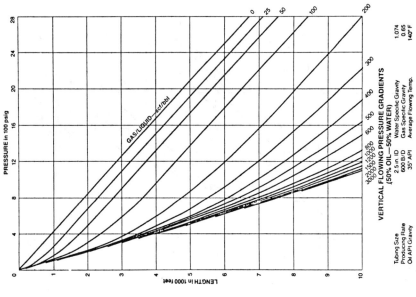

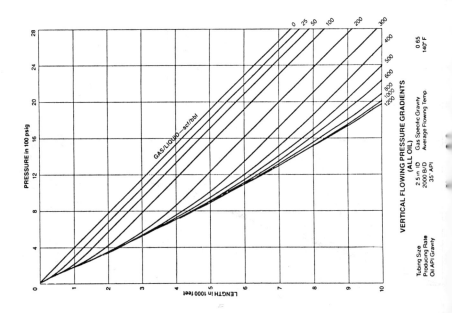

Figure 8-2t.

Multiphase Gas-Liquid Flow 379

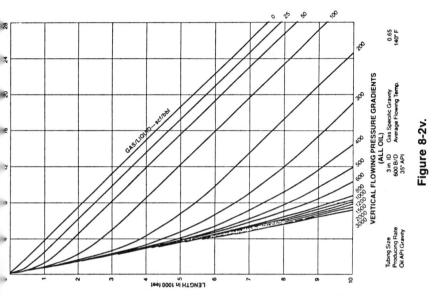

Figure 8-2v.

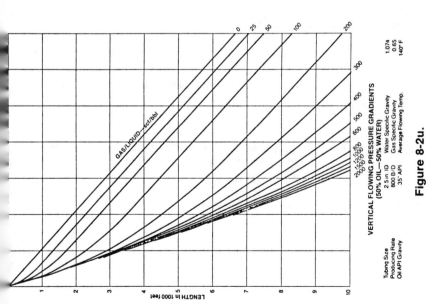

Figure 8-2u.

380 Gas Production Engineering

(text continued from page 368)

Example 8-1. A well is producing 1,500 stb/d of oil with a G/O = 800 scf/stb at a flowing wellhead pressure of 700 psig. Determine the separator pressure for a 2.5-in. ID, 9,000-ft line.

Solution

Assume that at a pressure of 700 psig there is no gas in solution. Hence free-gas/oil ratio is 800 scf/stb. Using Figure 8-1d and the procedure described earlier,

p_{sep} = 400 psig.

Pressure Traverse Curves for Vertical Gas-Liquid Flow

The vertical pressure traverse curves, shown in Figures 8-2g–v, are used in a manner similar to the curves for horizontal flow:

1. Select the applicable curve for the given tubing size, flow rate, and gas/liquid ratio.
2. On the pressure axis, locate the known pressure, go vertically down to the applicable gas/liquid ratio curve, and read off the depth on the depth axis.
3. Correct this depth by: adding the well depth to the depth determined in Step 2, if the known pressure was the surface pressure; or subtracting the well depth from the depth in Step 2, if the known pressure was the bottom-hole pressure.
4. The unknown pressure is the pressure corresponding to the corrected depth determined in Step 3.

Example 8-2. For a bottom-hole flowing pressure of 2,000 psig, a well is producing 1,000 stb/d of oil with a G/O = 300 scf/stb. Given that the tubing size is 2.5 in. and the well is 8,000 ft deep, find the flowing wellhead pressure p_{wh}.

Solution

Using Figure 8-2o for vertical flow:

A p_{wf} = 2,000 psig corresponds to 9,000 ft approximately.

Therefore, p_{wh} is equal to the pressure corresponding to a depth = 9,000 − 8,000 = 1,000 ft.

Thus, p_{wh} = 100 psig.

Handling Directional Wells

For directional wells with deviations from the vertical less than 15–20°, the true vertical depth can be used to ascertain the pressure traverse. This approximation, however, is invalid for deviations exceeding 20°, because a directional well has a greater length than a vertical well for the same depth, resulting in a greater frictional head loss. Also, holdup, defined as the volume fraction of the flow conduit occupied by liquid, differs and may be greater for inclined than for vertical flow.

For pressure traverse curves, an approximate answer can be obtained using the vertical and horizontal flow curves as follows:

1. Determine the pressure loss using only the true vertical depth and the applicable vertical flow correlation.
2. Assume a value for the inlet or outlet pressure, whichever is unknown, and calculate the average of the inlet and outlet pressures.
3. Determine the frictional pressure drop due to the extra length of the tubing (i.e., total tubing length minus true vertical depth of tubing) using the applicable horizontal flow correlation, and the average pressure estimated in step (2).
4. Calculate the total pressure loss for the deviated well. This is equal to the sum of the two pressure losses obtained in steps (1) and (3). From this new total pressure loss, find the new estimate of the inlet or outlet pressure, whichever is unknown.
5. Repeat steps 2 through 4 till the unknown pressure (inlet or outlet) from steps (2) and (4) agree within a specified tolerance.

Example 8-3. In a directionally-drilled well, the true vertical depth is equal to 5,000 ft, and the length of the 2-in. tubing is equal to 7,500 ft. Given p_{wh} = 100 psig, q = 1,000 stb/d (50% water), G/L = 800 scf/stb, find the flowing bottom-hole pressure p_{wf}.

Solution

Using Figure 8-2m for vertical flow,

p_{wf}^* = 1,100 psig for a vertical depth of 5,000 ft.

A trial and error procedure is required to determine p_{wf}. As a first guess, assume $p_{wf} = 1,150$ psig.

Then, the average pressure, $p_{av} = (p_{wh} + p_{wf})/2 = (100 + 1,150)/2 = 625$ psig.

On locating p_{av} (= 625 psig) on the horizontal flow correlation in Figure 8-2c (100% water chart is a reasonable approximation), and using the additional length of 2,500 ft (= 7,500 − 5,000), the downstream pressure is found to be 525 psig.

Therefore, the pressure loss due to friction in the extra 2,000 ft of pipe, $\Delta p_f = 625 - 525 = 100$ psig.

Thus, $p_{wf} = p_{wf}^* + \Delta p_f = 1,100 + 100 = 1,200$ psig.

Second trial: Assume $p_{wf} = 1,200$ psig.

$p_{av} = (100 + 1,200)/2 = 650$ psig. From the horizontal flow correlation, downstream pressure = 550 psig. Thus, $\Delta p_f = 650 - 550 = 100$ psig, and $p_{wf} = 1,100 + 100 = 1,200$ psig.

Thus, $p_{wf} = \underline{1,200 \text{ psig.}}$

Flow Over Inclined or Hilly Terrain

Inclined flow implies flow through pipes that deviate from the horizontal, such as flow over hills. Flanigan (1958) presented the only available method that can be applied to field problems without the use of complex computer programs. By conducting several field tests, Flanigan concluded that most of the pressure drop occurred in the uphill section of the line, and identified two main components of the pressure drop for multiphase flow in an inclined system: pressure drop due to friction, which is the predominant component in horizontal lines, and pressure drop due to the liquid head, which is the predominant component in vertical and inclined flows. The sum of these two components determines the total pressure drop.

The uphill sections are treated as equivalent vertical columns containing an equivalent amount of liquid. Because in multiphase flow the pipe is never completely filled with only liquid, Flanigan introduced the term H_F to represent the fraction of the total static pressure drop that exists as the elevation component. H_F correlates with the superficial gas velocity, v_{sg}, as follows (Flanigan, 1958):

$$H_F = \frac{1}{1 + 0.3264 \, v_{sg}^{1.006}} \qquad (8\text{-}4)$$

where H_F = elevation component of the total static pressure drop, fraction, and v_{sg} is given by:

$$v_{sg} = \frac{31194 \, q_g(ZT)_{av}}{(520) p_{av} d^2} \qquad (8\text{-}5)$$

where v_{sg} = superficial gas velocity, ft/sec
q_g = gas flow rate, MMscfd
T_{av} = average flowing temperature, °R
p_{av} = average flowing pressure, psia
Z_{av} = average gas compressibility factor (at p_{av}, T_{av})
d = inside diameter of flow conduit, in.

Baker (1960) showed that for $v_{sg} > 50$, the applicable relationship is:

$$H_F = \frac{0.00967 \, L^{0.5}}{v_{sg}^{0.7}} \qquad (8\text{-}6)$$

where L = length of the flowline, ft.

Using H_F, the pressure drop through the pipe is calculated as follows:

$$\Delta p = \frac{\rho_l H_F \Sigma H}{144} \qquad (8\text{-}7)$$

where ρ_l = liquid density, lbm/ft³

Example 8-4. A 4-in. ID, 2,000 ft long flowline passes over 5 hills having the following vertical heights: 130 ft, 40 ft, 210 ft, 80 ft, and 185 ft. Given q_l = 2,000 stb/d (95% water), G/L = 1,000 scf/stb, γ_g = 0.7 (air = 1), γ_w = 1.02, γ_o = 40°API, average pressure p_{av} = 300 psig, and average temperature T_{av} = 120°F, find the pressure loss due to the hills.

Solution

For γ_g = 0.7, at p_{av} = 300 psig, and T_{av} = 120°F, the average gas compressibility factor Z_{av} from Figure 3-2 = 0.96.

Using Equation 8-5,

$$v_{sg} = \frac{(31,194)(2,000 \times 1,000 \times 10^{-6})(0.96)(580)}{(520)(300)(16)} = 13.9173 \text{ ft/sec}$$

Using Equation 8-4,

$$H_F = \frac{1}{1 + (0.3264)(13.9173)^{1.006}} = 0.1781$$

$\Sigma H = 130 + 40 + 210 + 80 + 185 = 645$ ft.

$\gamma_o = (141.5)/(131.5 + 40) = 0.8251$

Therefore, $\gamma_l = (0.95)(1.02) + (0.05)(0.8251) = 1.01$

Using Equation 8-7,

$\Delta p_{hills} = (1.01 \times 62.4)(0.1781)(645)/144 = \underline{50.28 \text{ psi.}}$

Flow Through Chokes

The generalized equation for critical multiphase flow through a choke is written as:

$$q_l = \frac{p_u d^a}{bR^c} \tag{8-8}$$

where q_l = liquid flow rate, stb/d
  p_u = upstream pressure, psia
  d = inside diameter of the choke, 64ths of an inch
  R = producing gas-liquid ratio, scf/stb
  a, b, c = empirical constants

Various investigators have proposed different values for a, b, and c. Most commonly, however, Gilbert's (1954) correlation is used, where a = 1.89, b = 10.0, and c = 0.546:

$$q_l = \frac{p_u d^{1.89}}{(10)R^{0.546}} \tag{8-9}$$

Example 8-5. (After Kumar et al., 1986). A well is producing through a 2.5-in. ID tubing, 5,000 ft deep, at a rate q_o = 1,000 stb/d with a G/O = 600 scf/stb. This well produces a large amount of sand when the oil production rate is above 1,000 stb/d; therefore, it is required to install a choke ("choke the well back"). Because hydrate problems have made it impossible to install a surface choke, a bottom-hole choke must be designed. It is proposed that the choke be installed at a depth of 4,000 ft, i.e., 1,000 ft above the bottom of the tubing. For a 1,000 stb/d oil rate, the bottom-hole pressure may be assumed to be 1,400 psig. Determine: (a) the choke size required, and (b) the flowing wellhead pressure, assuming that the flow through the choke is critical (p_u = 2p_d).

Solution

Using Figure 8-2p, the pressure at 1,000 ft above the bottom of tubing is equal to 1,175 psig.

Thus, p_u = 1,175 psig, implying that p_d = 1,175/2 = 588 psig.

(a) Using Equation 8-9,

$$d^{1.89} = \frac{(10)(1,000)(600)^{0.546}}{1175} = 279.8$$

Thus, d = $(279.8)^{1/1.89}$ = 19.71/64 in., or 20/64 inch.

(b) Using Figure 8-2p for p_d = 588 psig, the wellhead flowing pressure, p_{wh} = 100 psig.

Liquid Loading in Gas Wells

The liquid loading of a gas well refers to the accumulation of liquids in the wellbore of a flowing gas well, a common operating problem in gas production operations. Liquid loading imposes an additional back pressure on the producing formation, restricting gas flow. In high-pressure gas wells, liquid loading will result in slugging flow and reduced well deliverability; fluctuations in gas flow rate and casing pressure will be observed, caused by the buildup of casing pressure until it becomes sufficient to blow the liquid to the surface. In low-pressure gas wells, the flow rate drops drastically and casing pressure starts to build up, but because enough casing pressure cannot be achieved, liquid loading may completely "kill" the well.

Liquid loading occurs in gas wells that have liquid production (however small the amount may be), but do not provide enough energy for the continuous removal of these liquids. As stated earlier, liquid loading can be recognized by fluctuations in gas flow rate and casing pressure for high-pressure wells, or by flow stoppage and casing pressure buildup for low-pressure wells. Pressure surveys and flow computations are very helpful in determining the presence of liquid accumulations.

There are two sources of liquids in a gas well: water and hydrocarbon condensate from condensation of the hydrocarbon gas, and water from the producing reservoir. If the gas velocity is not high enough to transport these liquids to the surface, they accumulate at the bottom of the well. The back pressure on the formation due to the hydrostatic head of this liquid column further reduces the flow rate and the velocity of the produced gas. This process continues until the well dies, or produces only intermittently.

There are essentially two approaches in solving this problem: prevent liquid accumulation, or remove the liquid as it is produced. A brief description follows of some available methods.

Preventing Liquid Loading: Minimum Flow Rate for Continuous Liquid Removal

Turner et al. (1969) conceptualized two physical models for the removal of liquid droplets from gas wells: liquid film movement along the pipe walls, and entrainment of liquid droplets in the gas stream. According to the wall film model, the well is unloaded by providing a gas flow rate sufficient to cause the movement of liquids up along the walls of the casing and/or tubing pipe. The droplet model assumes that liquid is removed as entrained droplets in the flowing gas stream. In a real system, both these models probably exist, with a continuous exchange of liquid between the gas stream and the wall film. From an analysis of both these models, Turner et al. (1969) found the droplet model to be superior because, among other reasons, it agrees with observed data. The liquid film model does not represent the controlling liquid transport mechanism; the liquid film moving down along the pipe walls eventually breaks into droplets that are re-entrained in the gas stream, and the droplet model is applicable. The more complex wall film model can, therefore, be ignored here.

According to the droplet model, the minimum gas flow rate to unload a gas well is one that enables the largest liquid droplets that can exist in the gas stream to move upwards. As discussed in Chapter 4 (Equation 4-17), the terminal velocity v_t (ft/sec) of a particle of diameter d_p (ft) and density ρ_p (lbm/ft^3) falling under gravitational acceleration g in a gas of density ρ_g (lbm/ft^3) is given by:

$$v_t = \left[\frac{4gd_p(\rho_p - \rho_g)}{3C_d\rho_g}\right]^{0.5} \tag{8-10}$$

assuming Newtonian (turbulent) flow. Thus, a particle of diameter d_p will be removed by the gas stream if the velocity of the gas stream v_g is equal to the terminal velocity v_t of the particle. Equation 8-10 can therefore be written as:

$$v_g = v_t = \left[\frac{4gd_p(\rho_p - \rho_g)}{3C_d\rho_g}\right]^{0.5}$$

Substituting $C_d = 0.44$ for Newtonian flow, and replacing the particle density ρ_p by the liquid density ρ_l, we get:

$$v_g = \left[\frac{4gd_p(\rho_l - \rho_g)}{1.32\rho_g}\right]^{0.5} \tag{8-11}$$

Thus, the larger the drop, greater is the gas flow velocity required to remove it. The diameter of the drop, however, cannot be determined directly. Turner et al. (1969) used correlations for the Weber number to determine the maximum drop diameter that can exist under given flow conditions. The Weber number, N_{We}, is a dimensionless number expressing the ratio of the velocity forces that try to disintegrate the drop, and the surface tension forces that tend to keep it together:

$$N_{We} = \frac{v_g^2 \rho_g / g_c}{\sigma / d_p} = \frac{v_g^2 \rho_g d_p}{\sigma g_c} \tag{8-12}$$

The maximum drop diameter, d_{max}, can now be specified using the fact that the drop will disintegrate if the Weber number exceeds a critical value of 20 to 30:

$$d_{max} = \frac{30\sigma g_c}{\rho_g v_g^2} \tag{8-13}$$

Substituting for d_{max} from Equation 8-13 into Equation 8-11, the required minimum gas velocity is given by:

$$v_g = \left[\frac{(4)(30)\sigma g g_c(\rho_l - \rho_g)}{1.32\rho_g^2 v_g^2}\right]^{0.5}$$

Substituting for g and g_c, and solving for v_g:

$$v_g = 17.514 \frac{\sigma^{1/4}(\rho_l - \rho_g)^{1/4}}{\rho_g^{1/2}}$$

As a safety measure, Turner et al. (1969) use a 20% (approximately) safety factor, resulting in the following equation for minimum gas flow velocity:

$$v_g = 20.4 \frac{\sigma^{1/4}(\rho_l - \rho_g)^{1/4}}{\rho_g^{1/2}} \qquad (8\text{-}14)$$

For field applications, the following equations are recommended by Turner et al. (1969) for use when the liquid is either water or condensate:

$$(v_g)_{water} = \frac{5.62 \ (67 - 0.0031p)^{1/4}}{(0.0031p)^{1/2}} \qquad (8\text{-}15)$$

$$(v_g)_{condensate} = \frac{4.02 \ (45 - 0.0031p)^{1/4}}{(0.0031p)^{1/2}} \qquad (8\text{-}16)$$

Equations 8-15 and 8-16 were derived assuming the following:

1. A 0.6 gravity gas at a temperature of 120°F, to enable expressing the gas density as a function of pressure.
2. For water, σ = 60 dynes/cm, ρ_l = 67 lbm/ft^3.
3. For condensate, σ = 20 dynes/cm, ρ_l = 45 lbm/ft^3.

When both water and condensate are present, the water equation, which gives a higher required gas velocity v_g, should be used.

The gas velocity v_g in ft/sec can be converted to field units of MMscfd as follows:

$$q_g = \frac{3.06 A p v_g}{ZT} \qquad (8\text{-}17)$$

where q_g = required minimum gas flow rate, MMscfd
 A = cross-sectional area of the gas flow conduit, ft^2
 p = flowing pressure, psia
 T = flowing temperature, °R
 Z = gas compressibility factor at the flowing pressure and temperature

Liquid Unloading: Removal of Liquid as it Accumulates

Numerous methods have been proposed for the liquid unloading, or dewatering, of gas wells (Hutlas and Granberry, 1972; Libson and Henry,

Multiphase Gas-Liquid Flow   389

1980). Some of the methods that have been successfully used in field operations are briefly reviewed here.

Beam Pumping Units

In this method, liquids are produced through the tubing, whereas gas is produced through the tubing-casing annulus. It is desirable to have the tubing set as close to the bottom perforation as possible, and preferably below it (Hutlas and Granberry, 1972), so as to provide a liquid cushion to prevent gas interference problems in the downhole pump. If a liquid diverter is provided in the subsurface tubing string to allow only liquids to enter the tubing, no liquid cushion is required and gas interference problems are minimized.

One advantage of pumping units over other types of liquid removal techniques is that they do not depend on gas velocity for lift—the energy for liquid lift is provided by the pump itself. Beam pumping units are quite inexpensive and economical for shallow wells, but become very expensive for deeper and more highly pressured wells (Hutlas and Granberry, 1972). Their applicability is best for low gas rates, with a liquid production of greater than about 10 bbl/day. Beam pumping units cannot be adjusted to handle very low rates and problems may, therefore, arise if they are used for low liquid-rate wells.

Plunger Lift

This method has proved to be very successful in unloading liquids from gas wells (Libson and Henry, 1980). It uses a steel plunger, equipped with a valve, in the tubing string (see Figure 8-3). The plunger acts as an interface between the gas stored in the annulus and the liquid accumulated above the plunger in the tubing, virtually eliminating gas slippage and liquid fallback.

Both gas and liquid are allowed to flow into the tubing through a valve-type opening at the tubing bottom. When the plunger is at the bottom of the tubing, this valve is shut off, and all production goes into the casing. The casing pressure builds up, until it is sufficient to transport the plunger and all liquid and gas above it in the tubing to the surface. A motor valve, operated either by a clock or by well flow rate monitoring, may be used on the flowline to control the cyclic rate of the plunger (Libson and Henry, 1980). If a clock is used, the optimum time cycle has to be determined by trial-and-error.

Following plunger arrival at the surface, a bumper opens the valve in the plunger, and the well is allowed to flow for a period of time. This time period may be preset by a clock, or surface flow controllers may be used to allow the well to flow until the gas velocity drops to some critical value. Flow controllers are more desirable, because they permit the well to flow

390 Gas Production Engineering

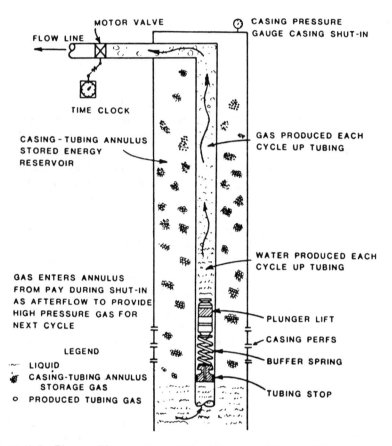

Figure 8-3. Plunger lift operations. (After Libson and Henry, 1980; courtesy of SPE.)

for an optimum time period before shut-in. Wells having sufficient deliverability to maintain rates above this critical rate for short periods are ideal candidates for flow controllers.

Gas Lift

The gas lift system also uses a liquid flow diverter in the tubing string, with liquid production through the tubing and gas production through the annulus. This system, however, uses a gas-lift valve (usually one, sometimes more than one if liquid production is quite high) mounted on the tubing at a calculated depth, and actuated (opened or closed) by any one of the several available techniques. Liquid accumulates in the tubing, until the gas-lift valve opens to lift the liquid slug to the surface.

Small Tubing String

This technique operates on the principle of increasing the gas velocity, to enable liquid carryover to the surface, by providing a reduced flow area. Small tubing strings of 1-in. diameter are quite common for such an application. This method applies best to low volume wells in which friction loss is not severe, and has proven quite useful in several field applications (Hutlas and Granberry, 1972; Libson and Henry, 1980).

Surfactant Injection

Injection of surfactants and foaming agents into the tubing-casing annulus has produced successful results in several wells (Libson and Henry, 1980). Figure 8-4 shows a schematic diagram of a surfactant injection system. Surfactant (or soap) is injected using a chemical pump and a time clock. Water is continuously unloaded in a foamed slug state through reduction in surface tension. This method has proved less successful for unloading condensate (oil), perhaps because the surface tension reduction dynamics were not suitable in the cases tested.

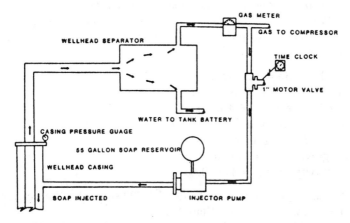

Figure 8-4. Gas well soap injection system. (After Libson and Henry, 1980; courtesy of SPE.)

Questions and Problems

1. What is multiphase flow? For what purpose are multiphase flow computations useful in the context of gas production operations and how important are they?

2. Describe the development of flow correlations for multiphase flow.
3. A well is producing 800 bbl/d oil with a gas-oil ratio of 1,500 scf/stb through a 2-in. tubing, with a safety valve installed at a depth of 2,500 ft from the surface. The total well depth is 5,000 ft. The pressures recorded at the surface and bottom-hole are 150 psig and 1,000 psig, respectively. Is the valve partially closed? If so, what is the pressure drop across the valve?
4. A 0.75 gravity gas is flowing at 0.25 MMscfd through a 2.5-in. ID tubing. It is estimated that the 35°API oil is flowing at 1,000 stb/day. A wellhead pressure of 200 psig is desired for this 6,000 ft gas well. The bottom-hole pressure is 1,100 psig. Use the approximate method for two-phase systems to answer the following:

   1. Can the gas flow under these conditions? If yes, do not answer any of the parts below.
   2. What methods are available to make this well flow?
   3. How much gas injection is required per day to keep this well flowing?
   4. What is the pipe diameter that will keep this well flowing?

5. Repeat problem 4 using multiphase flow charts.
6. Use two methods to design the required flowline to transport a two-phase gas-oil mixture from a wellhead over hills with the following conditions:

   Gas flow rate = 3.5 MMscfd
   Gas-oil ratio = 11,000 scf/stb
   Length of pipeline = 4 miles
   Total elevation of hills = 1,500 ft
   Average flowing temperature = 95°F
   Gas gravity (air = 1) = 0.68
   Oil gravity (water = 1) = 0.85

   Upstream pressure must not exceed 1,200 psig, and the downstream pressure must not be less than 250 psig.
7. Solve problem 6 assuming a horizontal pipeline (no hills).
8. Design a surface choke, assuming critical flow through it, for a gas well with the following operating conditions:

   Gas flow rate = 1.2 MMscfd
   Gas-oil ratio = 11,000 scf/stb
   Wellhead pressure = 1,200 psig

   Use two methods, and compare the results.
9. List the advantages and disadvantages of the known methods of unloading gas wells. What data would you need to consider before deciding on using any of these techniques for a specific well?

References

Beggs, H. D. and Brill, J. P., 1973. "A Study of Two-Phase Flow in Inclined Pipes," *J. Pet. Tech.*, 25(5, May), 607–617.

Brown, K. E. and Beggs, H. D., 1977. *The Technology of Artificial Lift Methods, Vol. 1.* PennWell Publ. Co., Tulsa, Oklahoma, 487pp.

Craft, B. C. and Hawkins, M. F., 1959. *Applied Petroleum Reservoir Engineering.* Prentice-Hall, Inc., Englewood Cliffs, N.J., 437pp.

Eaton, B. A., Andrews, D. E., Knowles, C. E., Silberberg, I. H., and Brown, K. E., 1967. "The Prediction of Flow Patterns, Liquid Holdup and Pressure Losses Occurring During Continuous Two-Phase Flow in Horizontal Pipelines," *Trans.*, AIME, 240, 815–828.

Elfrink, E. B., Sandberg, C. R., and Pollard, T. A., 1949. "A New Compressibility Correlation for Natural Gases and its Application to Estimates of Gas-in-Place," *Trans.*, AIME, 186, 219–223.

Flanigan, O., 1958. "Effect of Uphill Flow on Pressure Drop in Design of Two-Phase Gathering Systems," *O. & Gas J.*, 56(Mar. 10), 132–141.

Gilbert, W. E., 1954. "Flowing and Gas-Lift Well Performance," *Drilling and Production Practice*, API, 126–157.

Hagedorn, A. R. and Brown, K. E., 1965. "Experimental Study of Pressure Gradients Occurring During Continuous Two-Phase Flow in Small-Diameter Vertical Conduits," *J. Pet. Tech.*, 17(4, Apr.), 475–484.

Halliburton Energy Institute, 1976. *Tubing Size Selection.* Halliburton Services, Duncan, Oklahoma, 57pp.

Hutlas, E. J. and Granberry, W. R., 1972. "A Practical Approach to Removing Gas Well Liquids," *J. Pet. Tech.*, 24(8, Aug.), 916–922.

Ikoku, C. U., 1984. *Natural Gas Production Engineering.* John Wiley & Sons, Inc., New York, 517pp.

Kumar, S., Guppy, K. H., and Chilingar, G. V., 1986. "Design of Flowing Well Systems," In: *Surface Operations in Petroleum Production, Vol. 1*, Elsevier Scientific Publ. Co., Amsterdam, 279–326.

Libson, T. N. and Henry, J. R., 1980. "Case Histories: Identification of and Remedial Action for Liquid Loading in Gas Wells—Intermediate Shelf Gas Play," *J. Pet. Tech.*, 32(4, Apr.), 685–693.

Turner, R. G., Hubbard, M. G., and Dukler, A. E., 1969. "Analysis and Prediction of Minimum Flow Rate for the Continuous Removal of Liquids From Gas Wells," *J. Pet. Tech.*, 21(11, Nov.), 1475–1482.

9
Gas Compression

Introduction

Gas production operations often require compressors to raise the pressure of ("pressurize") the gas. One of the most important applications of compressors is for providing enough pressure to a gas for transport through transmission and distribution systems. Another transport-related application of compressors is for reducing the gas volume for shipment by tankers or for storage. In reservoir engineering operations, compressors are important in lowering the wellhead pressure below atmospheric in order to produce the well at a higher rate, for reinjection of gas for pressure maintenance or cycling, and for injection of gas into subsurface strata for underground storage. In gas processing operations, compressors are required for circulation of gas through the process or system, and for raising gas pressure to the required level for a chemical processing reaction.

This chapter presents the common types of compressors being used for these purposes in the natural gas industry. Basic elements in the design of these compressors are also discussed. Much of the material has been drawn from an Ingersoll-Rand Company (1982) publication entitled *Compressed Air and Gas Data*, an excellent reference for readers who may be interested in further details about gas compression operations.

Types of Compressors

The two major types of compressors are *positive displacement* or intermittent flow, and *continuous flow*. Positive-displacement units are those in which successive volumes of gas are confined within some type of enclosure (compression chamber) and elevated to a higher pressure. This pressure elevation or compression can be achieved in two ways—by reducing the gas volume in the enclosure, or by carrying the gas without volume change to the discharge and compressing the gas by backflow from the discharge system. Positive-displacement compressors are of two types—*reciprocating positive-displacement compressors*, and *rotary positive-displacement compressors*.

Continuous flow compressors accelerate a continually flowing gas stream and subsequently convert the velocity head into pressure. The two types of continuous flow compressors are *dynamic* and *ejector*. Dynamic compressors impart energy to a flowing gas stream by means of a rapidly rotating element, and subsequently convert the velocity head into pressure, partly in the rotating element and partly in stationary blades or diffusers. Ejectors increase the gas velocity by entraining it in a high-velocity jet of the same or another gas, and convert this kinetic energy into pressure in a long venturi-type flow chamber.

Figure 9-1 shows a breakdown of principal compressor types, and Table 9-1 describes their approximate parameters. Among these, the ones currently used in the gas industry are discussed in the following sections.

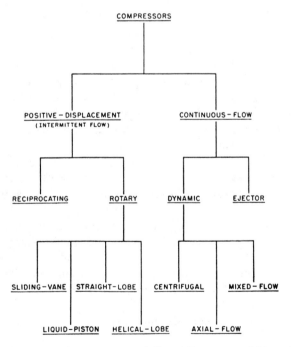

Figure 9-1. Principal compressor types. (From *Compressed Air and Gas Data*, 1982; courtesy of Ingersoll-Rand Co.)

Table 9-1
Approximate Parameters for Common Compressor Types*

Compressor Type	Approx. Maximum Pressure, psig	Approx. Maximum Power, BHP
Reciprocating	100,000	> 12,000
Vane-type Rotary	400	860
Helical-lobe Rotary	250	6,000
Centrifugal Dynamic	5,500	> 35,000
Axial-flow Dynamic	500	> 100,000

* From *Compressed Air and Gas Data*, 1982; courtesy of Ingersoll-Rand Co.

Reciprocating Positive-Displacement Compressors

Reciprocating compressors consist of a ringed piston that moves inside a cylinder. They may be of two types: *single-acting*, in which the piston compresses on only one side; and *double-acting*, in which two single-acting pistons operate in parallel inside one cylinder, thereby compressing on both sides. Besides the piston and cylinder, a suction valve and a discharge valve are provided. The suction valve opens when the pressure in the cylinder falls below the intake pressure. The discharge valve opens when the pressure in the cylinder equals or exceeds the discharge pressure.

Figure 9-2 shows the various steps in a reciprocating compressor cycle. In Figure 9-2A, the cylinder is full of the gas that is to be compressed, and is at the beginning of the compression cycle. In Figure 9-2B, the compression stroke has been completed. The discharge valve opens and the delivery stroke begins; the gas is delivered at a constant pressure as shown in Figure 9-2C. Next, the expansion stroke occurs, as shown in Figure 9-2D. Both valves remain closed. The small amount of gas trapped is expanded, leading to a reduction in pressure inside the cylinder. In the next step (Figure 9-2E), the inlet valve opens and the intake cycle begins, filling the cylinder with gas again.

Reciprocating compressors are the older type, with more moving parts and hence a lower mechanical efficiency and higher maintenance costs. They are still very widely used, and are available for all pressure and capacity ranges. Typically, they have a volumetric rate up to 30,000 cfm (cubic feet per minute) and a discharge pressure up to 10,000 psig.

Rotary Positive-Displacement Compressors

In this type, the positive action of rotating elements is used for compression and displacement. Rotary compressors usually offer low pressure differentials, but can handle large quantities of low-pressure gas at compara-

Gas Compression 397

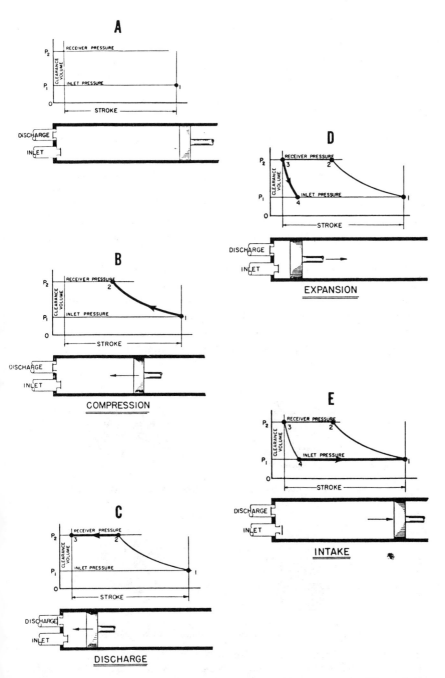

Figure 9-2. The various steps in a reciprocating compressor cycle. (From *Compressed Air and Gas Data*, 1982; courtesy of Ingersoll-Rand Co.)

tively low horsepower. They are easy to install, operate and maintain, and are used primarily in distribution systems where the pressure differential between suction and discharge is quite small. The four types of rotary compressors, indicated in Figure 9-1, are described as follows.

Sliding-Vane Compressors

The sliding-vane compressor consists of axial vanes that slide radially in a rotor mounted eccentrically in a cylindrical chamber (see Figure 9-3). It is identical in operation to the reciprocating compressor, except that it has no valves. The inlet and discharge conditions are determined by the location of the vanes that move over the inlet and discharge ports. The compression cycle begins when the leading vane of each pocket uncovers the intake port, as shown in Figure 9-3. As the rotor turns, the pocket volume decreases and gas is compressed, until the discharge port is uncovered by the leading vane of each pocket. Because this discharge point is prefixed in the design, the rotary sliding-vane compressor always compresses gas to the design pressure, regardless of the pressure in the receiver into which it is discharging.

Sliding-vane compressors can typically handle 3,000 cfm of gas, with a maximum discharge pressure of 50 psig per compression stage. They are primarily used as air compressors, boosters, and vacuum pumps.

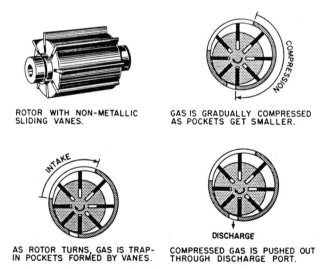

Figure 9-3. The steps involved in compression for a rotary sliding-vane compressor. (From *Compressed Air and Gas Data,* 1982; courtesy of Ingersoll-Rand Co.)

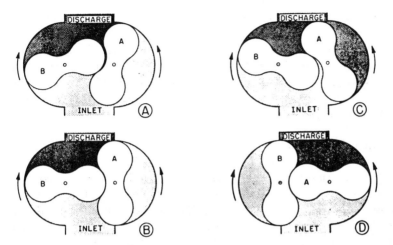

Figure 9-4. The operating cycle of a rotary two-impeller straight-lobe compressor. (From *Compressed Air and Gas Data*, 1982; courtesy of Ingersoll-Rand Co.)

Two-Impeller Straight-Lobe Compressors

Also known as *rotary blower*, the rotary two-impeller straight-lobe compressor consists of two identical rotors (or impellers) mounted symmetrically in a casing (Figure 9-4). The rotors usually have a cross section similar to the numeral eight. The two rotors intermesh, and rotate in opposite directions. One of the rotors is driven directly, while the other is driven and kept in phase by means of phasing/timing gears. The rotors do not directly compress the gas or reduce its volume and there is no internal compression; they merely transmit the gas from the inlet to the discharge. Compression occurs by backflow into the casing from the discharge line at the time the discharge port is uncovered. It is a simple device, with no contact between the rotors or between rotors and casing. The operation is shown in Figure 9-4, with the light shading indicating gas at inlet pressure, and the dark shading indicating gas at discharge pressure.

Rotary blowers can typically handle gas at 15,000 cfm, with a maximum discharge pressure of 20 psig or less per compression stage.

Liquid-Piston Compressors

Liquid-piston compressors use water or another liquid as the piston to compress and displace the gas. They are not used much in the natural gas industry.

Helical-Lobe or Spiral-Lobe Compressors

Helical-lobe compressors use two helical-type intermeshing rotors, giving a discharge pressure typically up to 150 psig per compression stage for gas flowing at 3,000 cfm. They are primarily used for vacuum and specialty applications at moderate pressures, but have not found significant application in the natural gas industry.

Dynamic Compressors

Dynamic compressors operate by transferring energy from a rotating set of blades to the gas. The rotor effects this energy transfer by changing the pressure and momentum of the gas. The momentum also is subsequently converted into pressure by reducing the gas velocity in stationary blades or diffusers. The three types of dynamic compressors are as follows.

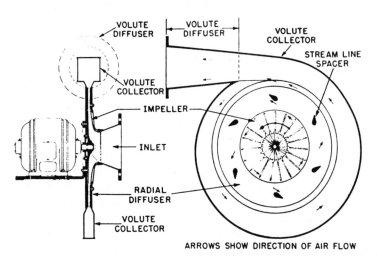

Figure 9-5. A typical overhung-impeller single-stage centrifugal compressor. (From *Compressed Air and Gas Data*, 1982; courtesy of Ingersoll-Rand Co.)

Centrifugal Compressors

In centrifugal compressors, gas flow is radial, and the energy transfer is effected predominantly by changing the centrifugal forces acting on the gas. Figure 9-5 shows a single-stage centrifugal compressor. The impeller has radial or backward leaning vanes usually between two shrouds. The mechanical action of the rapidly rotating impeller vanes force the gas through the impeller. The velocity thus generated is converted into pressure partly in the impeller itself, and partly in stationary diffusers following the impeller.

Centrifugal compressors have low maintenance costs because they have fewer moving parts—only the shaft and impeller rotate. Delivery is continuous, without cyclic variations. The discharge pressure from a single stage is about 100 psig, but the flow capacity is very high, up to about 100,000 cfm. They adapt easily to multistage operations, enabling greater compression. Multistage centrifugal compressor units with two or more impellers in series, each with its own radial diffuser, are often built as a single unit, as shown in Figure 9-6.

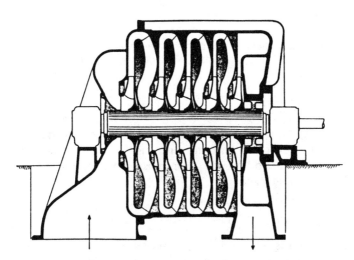

Figure 9-6. Cross-sectional view of a typical multistage uncooled centrifugal compressor. (From *Compressed Air and Gas Data,* 1982; courtesy of Ingersoll-Rand Co.)

Axial-Flow Compressors

In axial-flow compressors, gas flow is parallel to the compressor shaft (axial), because unlike centrifugal compressors, there is no vortex action. Energy is transferred by means of a number of stages of blades. Each stage consists of two rows of blades—one row (mounted on the rotor) rotating, and the next row (mounted on the casing or stator) stationary. Both the rotor and stator contribute almost equally to the pressure rise generated by the axial-flow compressor. It is a high-speed, large-capacity machine with characteristics quite different from the centrifugal type of compressor. A multistage axial-flow dynamic compressor is shown in Figure 9-7.

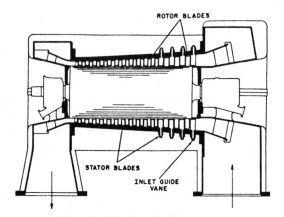

Figure 9-7. Cross-sectional view of a typical axial-flow dynamic compressor. (From *Compressed Air and Gas Data*, 1982; courtesy of Ingersoll-Rand Co.)

Mixed-Flow Compressors

These have gas flow that is in between axial and centrifugal, and combine some characteristics of both the centrifugal and axial types. Because of the long length required for each stage, mixed-flow compressors are generally not used as multistage units.

Ejectors

An ejector consists of a relatively high-pressure motive steam or gas nozzle discharging a high-velocity jet across a suction chamber into a venturi-shaped diffuser (Figure 9-8). The gas to be compressed is entrained by the jet in the suction chamber, resulting in a high velocity mixture at the pressure of the steam or gas emanating from the nozzle. The diffuser provides the compression by converting the velocity of the mixture into pressure. The drop in velocity and rise in pressure along the diffuser length is shown in Figure 9-8. Temperature changes are similar to pressure changes, and follow the pressure curve quite closely. The gas is separated from steam in condensers (coolers) that also serve to reduce the gas temperature.

Ejectors are generally used to compress from pressures below atmospheric (vacuum) to near atmospheric discharge pressures, such as for pumping waste gases and vapors out of a system. A similar design, known as a *thermal compressor*, is used for compressing gas from atmospheric to above atmospheric pressures. Thermal compressors are used where available energy in gas or steam, instead of being wasted, may be used for compression.

Ejectors generally give a relatively small compression ratio. But they have no moving parts, so that maintenance and wear is minimum. Although they can handle liquid carryover without physical damage, they should not be exposed to a steady flow of liquid.

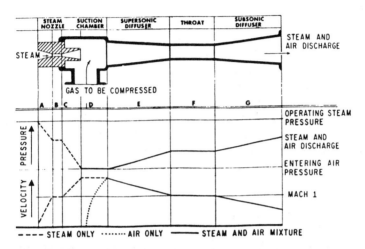

Figure 9-8. Pressure and velocity variations within a steam jet ejector handling air. (From *Compressed Air and Gas Data*, 1982; courtesy of Ingersoll-Rand Co.)

Compressor Selection

Several factors must be considered in selecting a compressor for a particular application. A multistage compression system may use different compressor types for its stages in order to optimize performance or enable more desirable operating features. Some major considerations in selecting compressors are described in this section. The discussion is limited to the two principal types of compressors used most commonly for handling natural gas: *reciprocating*, and *centrifugal dynamic* compressors. In some cases, the analysis can be quite complex and these general guidelines may not be sufficient. It is best in such cases to discuss the particular situation with the compressor manufacturers themselves.

Gas Characteristics

Gas characteristics such as the ratio of specific heats, compressibility, or moisture content, do not affect the choice of compressor type. Gas composition, however, significantly affects centrifugal compressors—many more stages are required if the inlet gas density is low. Positive-displacement compressors are not much affected by the gas molecular weight, specific gravity, or inlet density.

Flow Rate

For higher flow rates, centrifugal machines may be used for the lower compression stages, and reciprocating for the higher pressure stages. For relatively lower flow rates, the reciprocating compressor can be used for all the stages.

Flow-rate variation is another factor. Reciprocating compressors can handle enormous flow-rate variations with little loss in efficiency. Centrifugal compressors, however, do not perform efficiently below 50–90% of their rated capacity. In general, reciprocating compressors are more flexible in handling various types and sizes of gas flowstreams than centrifugal compressors.

Compression Ratios and Operating Pressures

Low compression ratios with moderate capacities favor centrifugal compressors. High compression ratios and higher pressures favor reciprocating compressors. Because too many other factors are involved, defining a precise borderline between these two types is difficult.

Variations in inlet and discharge pressures are important to consider also. In reciprocating compressors, if the inlet (suction) pressure is lowered while maintaining the same discharge pressure, the overall horsepower is lowered, the differential pressure of all but the last stage is lowered, and the pressure differential and temperature rise of the last stage are increased. If the suction pressure to the first stage is increased, horsepower of the complete system increases, pressure differential across all but the last stage is increased, while the pressure differential and temperature rise in the last stage are lowered.

In centrifugal compressors, if the suction pressure is raised, the discharge pressure will increase, frequently beyond the design point, and horsepower will also increase. If the suction pressure is lowered, the centrifugal compressor will not compress to the desired discharge pressure.

Operating Temperatures

Centrifugal compressors are less affected by high or low temperature extremes than reciprocating compressors in which temperature limitations are imposed by the lubricants. However, the maximum discharge temperature must generally be kept within some limit to avoid operating problems such as cracking of the lubricating oils. None of the compressors has any significant advantage in this respect; intercoolers are required to be used between stages to overcome high-temperature problems. Where lubricating-oil cracking and contamination problems are particularly severe, special partially-lubricated or nonlubricated reciprocating compressors may be more suitable than the centrifugal compressors.

Compressor Driver

The power source to be used for driving the compressor also influences choice. Generally, the driver is selected based upon available power source(s), the heat balance or utilization, waste gas use, and other factors. Centrifugal compressors are always preferred if the driver must be a turbine; reciprocating type are preferred if the driver must be an electric motor. This is so because motor-driven reciprocating compressors can be directly driven by the motor, whereas motor-driven centrifugal compressors require gears for increasing the speed. The opposite is true for turbines.

Foundation and Floor Space

The centrifugal compressor is usually favorable from the point of view of floor space and foundation requirements. Reciprocating compressors generally produce some unbalanced forces during their operation, requiring a foundation that will support their dead weight *plus* maintain alignment of the compressor and driver by absorbing the unbalanced forces that may be present.

Continuity of Operation

The value of lost production usually exceeds the actual cost of repairs, and continuity of operation or availability becomes an important factor. Dynamic compressors can operate uninterrupted for a longer period of time and their malfunctioning (stoppage) is more predictable than reciprocating compressors. However, the average yearly availability of both compressor types is very high and quite similar.

Capital Costs

It is difficult to compare the cost of centrifugal compressors versus reciprocating compressors. For the same volume, pressure, and other compression factors, gas specific gravity influences the cost of the centrifugal compressor but not the reciprocating compressor. Often, the cost of the driver is used as a cost criterion. As stated earlier, if the power source is conducive to turbine operations, then the centrifugal compressor will be cheaper.

Operating Costs

The operating cost, consisting mainly of the cost of power, throughout the service life of a compressor far exceeds the capital costs and is therefore a more important consideration. The reciprocating compressor is inherently more efficient than the centrifugal compressor, except at very low compression ratios. Very large volumes, low compression ratios, and low discharge pressures favor the centrifugal compressor from an operating cost viewpoint.

Maintenance Costs

It is generally accepted that for the same gas that is clean, non-corrosive, and under the same terminal pressure conditions, maintenance is less on the centrifugal compressor than the reciprocating compressor. For more difficult compression problems, maintenance requirements for both compressor types increase, and become more nearly equal, though the centrifugal compressor will frequently require a little less maintenance.

Compression Processes

Gas compression processes can be thermodynamically characterized into three types: *isothermal* compression, *isentropic* or adiabatic reversible compression, and *polytropic* compression.

Isothermal Compression

Isothermal compression occurs when the temperature is kept constant during the compression process. The pressure-volume behavior of the gas is given by:

$$p_1 V_1 = p_2 V_2 = C \qquad (9\text{-}1)$$

where p_1, p_2 = gas pressures at states 1 and 2, respectively
V_1, V_2 = gas volumes at states 1 and 2, respectively
C = constant

Isentropic Compression

The compression process is adiabatic reversible or isentropic (constant entropy) if the gas behaves as an ideal gas, no heat is added to or removed from the gas during compression, and the process is frictionless. The pressure-volume behavior of a gas under isentropic compression or expansion is given by:

$$p_1 V_1^k = p_2 V_2^k = C \tag{9-2}$$

where k is the isentropic exponent, which is equal to the ratio of the specific heat at constant pressure (c_p) and the specific heat at constant volume (c_v) for the gas (as discussed in Chapter 7 also):

$$k = c_p/c_v \tag{9-3}$$

where c_p and c_v are in Btu/lbmole.

In thermodynamic terms, c_p for ideal gases is given by:

$$c_p = \left(\frac{\partial h}{\partial T}\right)_p = f(T)$$

And c_v for ideal gases is given by:

$$c_v = \left(\frac{\partial u}{\partial T}\right)_v = \left(\frac{\partial h}{\partial T}\right)_p - R$$

$$= c_p - R = f(T)$$

implying that for ideal gases,

$$c_v = c_p - R = c_p - 1.986 \tag{9-4}$$

where h = specific molal enthalpy, Btu/lbmole
 u = specific molal internal energy, Btu/lbmole
 T = temperature, °R
 R = gas constant, equal to 1.986 Btu/lbmole

Thus, c_p and c_v are functions of temperature only for ideal gases. For real gases, however, c_p and c_v are functions of both pressure and temperature:

$$c_p, c_v = f(p, T)$$

and $c_p - c_v$ is given by Equation 3-57 (Figure 3-17):

$$c_p - c_v = -T \frac{(\partial p/\partial T)_v^2}{(\partial p/\partial V)_T}$$

The more complex the molecular structure, the higher the c_p is for the gas. Using Equation 9-4, c_v can be eliminated, and k can be expressed for ideal gases as:

$$k = \frac{c_p}{c_v} = \frac{c_p}{c_p - 1.986} \tag{9-5}$$

and

$$\frac{k}{k-1} = \frac{c_p}{1.986} \tag{9-6}$$

Polytropic Compression

For "real" gases under actual conditions (with friction and heat transfer), the compression process is polytropic, where the polytropic exponent n applies instead of the adiabatic exponent k:

$$p_1 V_1^n = p_2 V_2^n = C \tag{9-7}$$

Characteristics of Compression Processes

Isothermal compression is difficult to achieve, because it is not possible commercially to remove the heat of compression as rapidly as it is generated. Adiabatic compression is also not possible in practice because it is difficult to prevent heat exchange during the compression-expansion cycles. Most positive-displacement compressors, however, approach adiabatic behavior, and are designed using the adiabatic compression/expansion process. Other compressors are designed using the polytropic cycle, which includes the effect of non-ideal gas under non-ideal conditions, and makes no assumptions regarding the compression process. The adiabatic process is reversible, while the other two types are irreversible.

Figure 9-9 shows the theoretical zero-clearance work (or horsepower) necessary to compress a gas between any two pressure limits for various compression processes. For a given process, the area under its pressure-volume curve indicates its theoretical work requirements. The isothermal compression process gives the minimum theoretical compression work requirements (area ADEF), whereas the adiabatic compression process gives the maximum theoretical compression work requirements (area ABEF). A polytropic compression process for reciprocating compressors with water-cooled cylinders generally exhibits theoretical compression work requirements that are in between the upper and lower limits of the adiabatic and isothermal compression processes, respectively (area ACEF). On the other hand, a polytropic compression process for uncooled dynamic compressors typically has theoretical compression work requirements that exceed even the adiabatic process (area AC'EF).

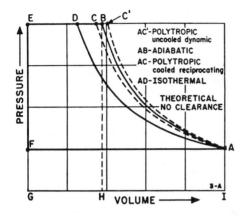

Figure 9-9. Theoretical pressure-volume relationships for various gas compression processes. (From *Compressed Air and Gas Data,* 1982; courtesy of Ingersoll-Rand Co.)

The horsepower requirements for an adiabatic or polytropic compression process can be reduced by making the process approach isothermal compression, such as by using multiple stages with intercoolers between the stages. This, in fact, is one important purpose of intercoolers between multistage compressors. Reciprocating compressors most commonly cool the gas to its original intake temperature between the compressions stages that are frequently built as a single unit driven from a single crankshaft. Dynamic compressors, however, have several stages in the same casing, normally without any intercooling. To improve compression efficiency of dynamic units, their internal diffusers or diaphragms are sometimes water-cooled, and the gas is intercooled in external heat-exchangers after every few stages of compression.

Exponents for the Isentropic and Polytropic Processes

The exponent k can be determined using Equation 9-3 if c_p and c_v are known, or from Equation 9-5 if c_p is known for the gas. As discussed in Chapter 3, the low pressure c_p^o can be determined using Equation 3-59 (by Hankinson et al.) if the gas gravity, but not the composition, is known; or by Kay's mixing rule if the gas composition is known. This low pressure c_p^o can be corrected for pressure using Figure 3-19 and the gas pseudoreduced pressure and temperature.

For natural gases in the gravity range $0.55 < \gamma_g < 1$, k at a temperature of 150°F can be determined using the following empirical relationship (Ikoku, 1984):

$$(k)_{150°F} \simeq \frac{2.738 - \log \gamma_g}{2.328} \tag{9-8}$$

Although the exponent n actually changes during compression, an effective or average value, determined experimentally for a given gas and compressor type, is generally used. The value of n may be lower or higher than k: n is usually less than k for positive-displacement and internally-cooled dynamic compressors, whereas the opposite is true for uncooled dynamic units due to internal gas friction.

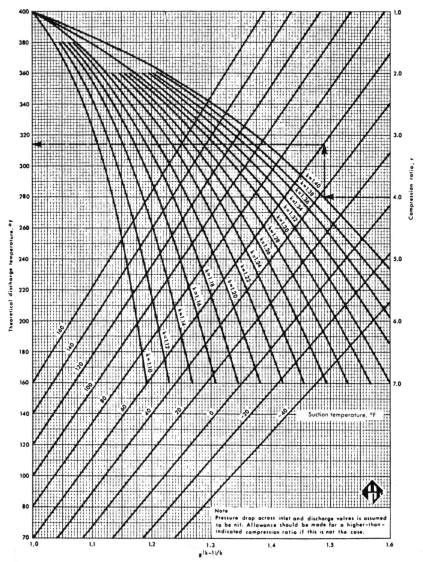

Figure 9-10. Theoretical discharge temperatures for single-stage compression. Read r to k to T_1 to T_2. (From *Engineering Data Book*, 1981; courtesy of GPSA.)

Gas Compression

The value of n is experimentally determined using the ideal gas law (pV = nRT) along with the polytropic compression/expansion relationship given by Equation 9-7, resulting in:

$$\frac{T_2}{T_1} = \left(\frac{p_2}{p_1}\right)^{(n-1)/n} = r^{(n-1)/n} \tag{9-9}$$

where p_1, p_2 = pressures at states 1 and 2, respectively
T_1, T_2 = temperatures at states 1 and 2, respectively
r = pressure ratio, equal to p_2/p_1
n = polytropic exponent

In compression terminology, p_1 and p_2 are the compressor inlet and discharge pressures, respectively, and r is called the compression ratio. Equation 9-9 relates the intake pressure and temperature and the discharge pressure and temperature for a compressor using the polytropic compression cycle, or the isentropic compression cycle if n is replaced by k:

$$\frac{T_2}{T_1} = \left(\frac{p_2}{p_1}\right)^{(k-1)/k} = r^{(k-1)/k} \tag{9-10}$$

Figure 9-10 may be used for solving Equation 9-9 or 9-10.

Another way of estimating n is by using correlations for polytropic efficiency, as shown in Figure 9-11. The polytropic efficiency, η_p, relates n and k as follows:

$$\eta_p = \frac{(k-1)/k}{(n-1)/n} \tag{9-11}$$

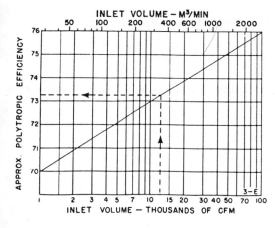

Figure 9-11. Approximate polytropic compression efficiency of a dynamic compressor versus inlet capacity. (From *Compressed Air and Gas Data*, 1982; courtesy of Ingersoll-Rand Co.)

412   Gas Production Engineering

Note that η_p relates the quantities $(k - 1)/k$ and $(n - 1)/n$, which are required more often than k or n.

Example 9-1. A gas is being compressed from 100 psia and 150°F to 2,500 psia. Determine its compression parameters (k, Z, γ_g) at the suction end. The gas has the following composition (expressed as mole fraction):

$C_1 = 0.9216$, $C_2 = 0.0488$, $C_3 = 0.0185$,
$i - C_4 = 0.0039$, $n - C_4 = 0.0055$, $i - C_5 = 0.0017$

Solution

Comp.	y_i	M_i	p_{ci}	T_{ci}	c_{pi}
C_1	0.9216	16.043	667.8	343.1	8.95
C_2	0.0488	30.070	707.8	549.8	13.78
C_3	0.0185	44.097	616.3	665.7	19.52
i-C_4	0.0039	58.124	529.1	734.7	25.77
n-C_4	0.0055	58.124	550.7	765.4	25.81
i-C_5	0.0017	72.151	490.4	828.8	31.66

$M = \Sigma y_i M_i = 17.737$ lbm/lbmole

Therefore, $\gamma_g = M/28.97 = 17.737/28.97 = \underline{0.612}$

$p_{pc} = \Sigma y_i p_{ci} = 667.313$ psia, and $T_{pc} = \Sigma y_i T_{ci} = 363.831$ °R

Therefore, $p_{pr} = p/p_{pc} = 100/667.313 = 0.150$

and $T_{pr} = T/T_{pc} = (460 + 150)/363.831 = 1.677$

From Figure 3-2, the gas Z factor at suction $\simeq \underline{1.0}$

$c_p^o = \Sigma y_i c_{pi} = 9.578$ Btu/lbmole-°R

From Figure 3-19, for $p_{pr} = 0.150$, and $T_{pr} = 1.677$, $\Delta c_p = 0.15$ Btu/lbmole-°R.

Therefore, $c_p = c_p^o + \Delta c_p = 9.578 + 0.15$
   $= 9.728$ Btu/lbmole-°R
   $= 0.548$ Btu/lbm-°R

From Equation 9-5, $k = c_p/(c_p - 1.986) = 9.728/(9.728 - 1.986)$
   $= \underline{1.2565}$

Gas Compression   413

Compressor Design Fundamentals

From the perspective of compressor users such as gas engineers, compressor design involves only the determination of compressor capacity and power requirements for a given application, in order to select the type and size of compressor required.

Multistaging

There are practical limits to the permissible amount of compression for a single compression stage. The limitations vary with the type of compressor, and include the following:

1. Discharge temperature—all types.
2. Compression efficiency (energy requirements)—all types.
3. Mechanical stress (rod loading) problems—all types.
4. Pressure rise or differential—dynamic units, and most positive-displacement units.
5. Compression ratio—dynamic units.
6. Effect of clearance—reciprocating units.

Whenever any limitation is involved, it becomes necessary to use multiple compression stages (in series). Furthermore, multistaging may be required from a purely optimization standpoint. For example, with increasing compression ratio r, compression efficiency decreases and mechanical stress and temperature problems become more severe. Therefore, if a greater compression ratio is desired, multiple compression stages must be used. Intercoolers are generally used between the stages to increase compression efficiency as well as to lower the gas temperature that may become undesirably high (as given by Equation 9-9 or 9-10), especially for high compression ratios.

Compressor Design Methods

The design of each compression stage is best considered separately because of pressure losses and temperature changes in the intercoolers and piping between the stages, condensation (if any) of water vapor from the gas, and the consequent volume changes of the gas.

The three methods used for compressor design calculations are: (1) analytic expressions derived from basic thermodynamic relationships; (2) enthalpy versus entropy charts, commonly known as Mollier diagrams, for ideal isentropic compression processes; and (3) empirical "quickie" charts, frequently provided by compressor manufacturers for quick estimates. The method to use depends upon the accuracy desired and the amount of data available. In many cases, very great accuracy is not required: compressors

414 Gas Production Engineering

are rarely chosen to satisfy a given requirement very precisely. Overdesign is quite common in order to provide an operating safety factor and to ensure that expensive compression equipment is not required to be enhanced or replaced for relatively small capacity increases in the flowing system. All three methods, therefore, find application in determining the compressor to use.

The Analytic Approach

According to the general energy equation (Equation 7-2 in Chapter 7), the theoretical work required to compress a unit mass of gas from pressure p_1 at state 1 to p_2 at state 2 is given by:

$$w = \int_{p_1}^{p_2} Vdp + \Delta v^2/2g_c + (g/g_c)\Delta z + l_w \qquad (9\text{-}12)$$

where w = work done by the compressor on the gas, ft-lbf/lbm
V = volume of a unit mass of gas, ft^3/lbm
p = pressure, lbf/ft^2
v = gas flow velocity, ft/sec
z = elevation above a datum plane, ft
l_w = lost work due to friction and irreversibilities, ft-lbf/lbm
g = gravitational acceleration (= 32.17 ft/sec^2)
g_c = conversion constant relating mass and weight (= 32.17 lbm-ft/lbf-sec^2)

Neglecting frictional losses, and the changes in kinetic and potential energies, the energy balance of Equation 9-12 can be written as:

$$w = \int_{p_1}^{p_2} Vdp$$

Substituting for V from Equation 9-2, we get:

$$w = C^{1/k} \int_{p_1}^{p_2} p^{-1/k} dp \qquad (9\text{-}13)$$

where C is a constant. Upon integration, Equation 9-13 becomes:

$$w = \frac{k}{k-1} C^{1/k}[p_2^{(k-1)/k} - p_1^{(k-1)/k}]$$

$$= \frac{k}{k-1} p_1(C/p_1)^{1/k}[(p_2/p_1)^{(k-1)/k} - 1]$$

Converting pressure p from units of lbf/ft^2 to lbf/in.^{2}, and substituting for C/p_1 from Equation 9-2, we get:

$$w = \frac{k}{k-1} (144 p_1 V_1) [(p_2/p_1)^{(k-1)/k} - 1] \tag{9-14}$$

From the ideal gas law, for a unit mass of gas:

$$p_1 V_1 = \frac{Z_1 R T_1}{M} = \frac{Z_1 R T_1}{(28.97)\gamma_g}$$

where p_1 = intake pressure, psia
V_1 = intake volume, ft³
M = molecular weight of the gas, lbm/lbmole
Z_1 = gas compressibility factor at intake conditions
R = gas constant (= 10.732 psia-ft³/lbmole-°R)
T_1 = gas intake temperature, °R
γ_g = gas specific gravity (air = 1 basis)

Substituting for R and neglecting Z_1 which is unity for intake conditions at or near atmospheric, we obtain:

$$p_1 V_1 = \frac{(10.732) T_1}{(28.97)\gamma_g} \tag{9-15}$$

Substituting for $p_1 V_1$ from Equation 9-15 into Equation 9-14:

$$w = \frac{k}{k-1} \frac{(53.345) T_1}{\gamma_g} (r^{(k-1)/k} - 1) \tag{9-16}$$

where $r = (p_2/p_1)$ is the compression ratio
w = compression work, ft-lbf/lbm

We usually need to compute the compression power per MMscfd of gas flow, rather than the work w required per pound-mass of gas. Now,

$$1 \frac{\text{ft-lbf}}{\text{lbm}} = 1 \frac{\text{ft-lbf/min}}{\text{lbm/min}}$$

The gas flow rate in lbm/min can be converted to MMscfd as follows:

$$1 \text{ lbm/min} = \frac{Z_{sc} R T_{sc}}{p_{sc} M} \text{ scf/min} = \frac{10.732 T_{sc}}{28.97 \gamma_g p_{sc}} \text{ scf/min}$$

$$= \frac{(1440)(10.732) T_{sc}}{(28.97)(10^6) \gamma_g p_{sc}} \text{ MMscf/day}$$

Using this and the fact that 1 hp = 33,000 ft-lbf/min:

$$1\frac{\text{ft-lbf}}{\text{lbm}} = \frac{(28.97)(10^6)\gamma_g p_{sc}(1/33{,}000)\text{ hp}}{(1440)(10.732)T_{sc}\text{ MMscf/day}}$$

$$= \frac{(0.0568056)\gamma_g p_{sc}}{T_{sc}}\frac{\text{hp}}{\text{MMscfd}}$$

Equation 9-16 can now be written in terms of the ideal horsepower per MMscfd, IHP/MMscfd (hp/MMscfd), required for compressing the gas through a compression ratio r as follows:

$$\frac{\text{IHP}}{\text{MMscfd}} = \frac{(0.0568056)\gamma_g p_{sc}}{T_{sc}}w$$

$$= \frac{3.0303\, p_{sc} T_1}{T_{sc}}\frac{k}{k-1}(r^{(k-1)/k} - 1) \quad (9\text{-}17)$$

where p_{sc} (psia) and T_{sc} (°R) are the standard conditions of pressure and temperature, respectively, at which the standard cubic feet is defined. Thus, the ideal (or theoretical) horsepower required for compressing q_{sc} MMscfd gas measured at pressure p_{sc} psia and temperature T_{sc} °R is given by:

$$\text{IHP} = \frac{3.0303\, q_{sc} p_{sc} T_1}{T_{sc}}\frac{k}{k-1}(r^{(k-1)/k} - 1) \quad (9\text{-}18)$$

where k is the specific heat ratio at suction conditions.

This analysis assumed an ideal gas. Where deviation from ideal-gas behavior is significant, Equation 9-18 is empirically modified in several different ways. One such modification is:

$$\text{IHP} = \frac{3.0303\, q_{sc} p_{sc} T_1 (Z_1 + Z_2)}{2 Z_1 T_{sc}}\frac{k}{k-1}(r^{(k-1)/k} - 1) \quad (9\text{-}19)$$

where Z_1, Z_2 = gas compressibility factors at the suction and discharge, respectively
T_1 = inlet temperature, °R

Temperature can be eliminated in Equation 9-19 if the gas flow rate is measured at suction temperature T_1 (°R):

$$\text{IHP} = \frac{3.0303\, q_{a1} p_a (Z_1 + Z_2)}{2 Z_1}\frac{k}{k-1}(r^{(k-1)/k} - 1) \quad (9\text{-}20)$$

where q_{a1} is the gas flow rate in MMscfd measured at any arbitrary pressure p_a psia and suction temperature T_1 °R. A similar analytic expression can be derived for the polytropic compression process also.

Gas Compression 417

The compressor discharge temperature can be calculated using Equation 9-9 for polytropic compression or Equation 9-10 for isentropic compression. The heat removed in the intercoolers or aftercooler can be calculated using the average specific heat at constant pressure, $(c_p)_{av}$, as follows:

$$\Delta H = n_g(c_p)_{av}\Delta T \qquad (9\text{-}21)$$

where ΔH = heat removed, Btu
 n_g = number of moles of gas being cooled
 $(c_p)_{av}$ = constant pressure molal specific heat of the gas at cooler pressure and average cooler temperature, Btu/lbmole-°F
 ΔT = difference in gas temperature between the inlet and outlet of the cooler, °F

Mollier Charts

Mollier charts for a gas are plots of enthalpy versus entropy as a function of pressure and temperature. Mollier charts for natural gases with specific gravities in the range of 0.6 to 1.0 are shown in Figures 9-12 through 9-17.

(text continued on page 419)

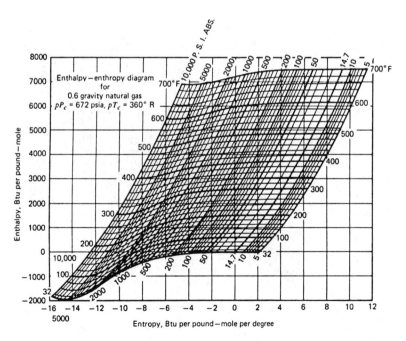

Figure 9-12. Enthalpy-entropy diagram for 0.60-gravity natural gas. (After Brown, 1945; courtesy of SPE of AIME.)

418 Gas Production Engineering

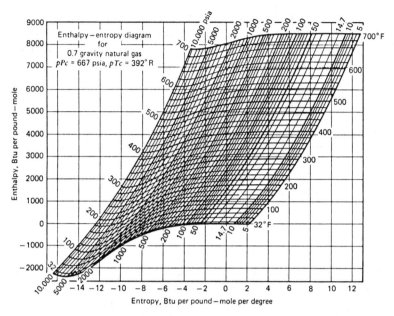

Figure 9-13. Enthalpy-entropy diagram for 0.70-gravity natural gas. (After Brown, 1945; courtesy of SPE of AIME.)

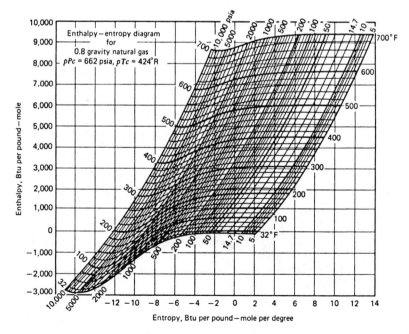

Figure 9-14. Enthalpy-entropy diagram for 0.80-gravity natural gas. (After Brown, 1945; courtesy of SPE of AIME.)

Gas Compression 419

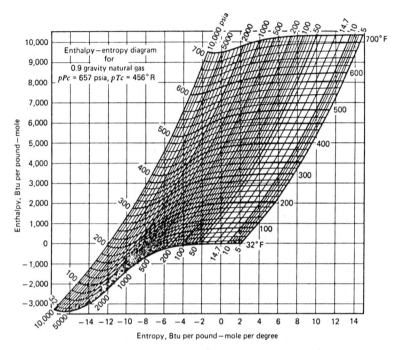

Figure 9-15. Enthalpy-entropy diagram for 0.90-gravity natural gas. (After Brown, 1945; courtesy of SPE of AIME.)

(text continued from page 417)

This method is a fairly good technique for solving compression problems for compressors that exhibit isentropic (ideal) compression, such as the reciprocating compressors, provided a Mollier chart is available for the gas being compressed.

On a Mollier chart, an isothermal compression process can be traced by following the constant temperature lines, whereas an isentropic compression process can be traced simply as a vertical line parallel to the ordinate. Intercooling, a constant-pressure (isobaric) process, is represented by following the constant pressure lines. Thus, an ideal compression process can be represented on a Mollier diagram, and the state of the gas (pressure, temperature, enthalpy, entropy) at the beginning or end of a compression process can be determined directly. Figures 9-18 and 9-19 show such a procedure for a single-stage and two-stage reciprocating compressor, respectively. In Figure 9-18, inlet gas is compressed isentropically from the inlet conditions represented by point 1 to the outlet conditions represented by point 2. Point 1 is determined by the given pressure and temperature conditions at the suction end, whereas point 2 is known from the desired compression ratio (or discharge pressure). Figure 9-19 shows a two-stage compression process with an intercooler between the two stages and an aftercooler. The dotted line 1-2

420 Gas Production Engineering

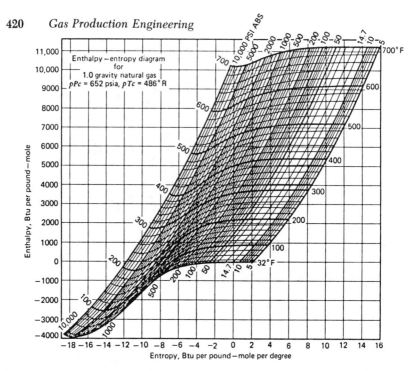

Figure 9-16. Enthalpy-entropy diagram for 1.0-gravity natural gas. (After Brown, 1945; courtesy of SPE of AIME.)

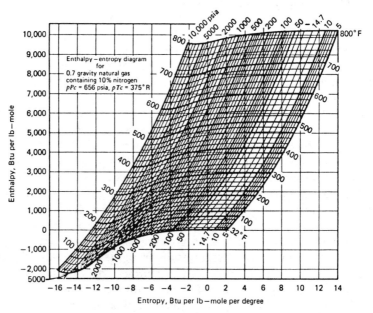

Figure 9-17. Enthalpy-entropy diagram for 0.70-gravity natural gas containing 10% nitrogen. (After Brown, 1945; courtesy of SPE of AIME.)

indicates isentropic compression in the first stage. The gas is then cooled in the intercooler at constant pressure, as shown by line 2-3. Line 3-4 shows the second isentropic compression stage, and line 4-5 shows the isobaric cooling in the aftercooler.

Neglecting heat transfer from the gas to the compression equipment and surroundings, lost work due to friction, and kinetic energy changes, the energy balance can be expressed as:

$$w = \Delta H = n_g(h_2 - h_1) \tag{9-22}$$

where w = work done by the compressor on the gas, Btu
ΔH = change in enthalpy of the gas, Btu
n_g = number of moles of gas being compressed, lbmoles
h_1, h_2 = enthalpies of the gas at the compressor inlet and discharge, respectively, Btu/lbmole

A lot of useful information regarding the compression process can be inferred from Mollier diagrams. Referring to the Mollier diagram of Figure 9-19 for example, the net change in enthalpy $(h_2 - h_1)$ is known, and the work done in the first stage of compression can be computed using Equation 9-22. Similarly, the work done in the second compression stage can be calculated using the enthalpy change $(h_4 - h_3)$. Intercooling requirements (heat removed in the intercooler) are given by the difference in enthalpy between

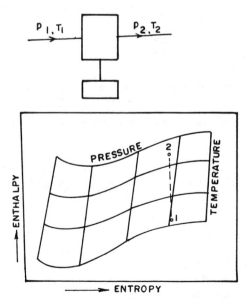

Figure 9-18. Mollier diagram of a single-stage compression process.

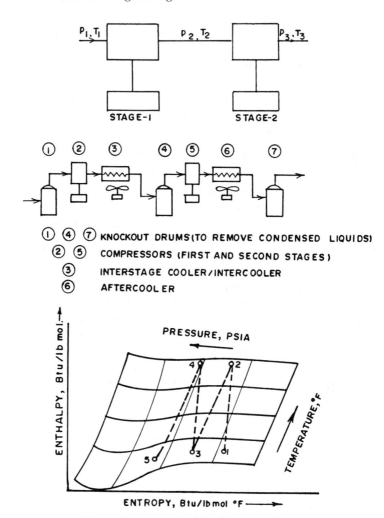

Figure 9-19. Mollier diagram of a two-stage compression process.

points 2 and 3. Similarly, the heat required to be removed in the aftercooler is equal to the difference in enthalpy between points 4 and 5.

The ideal compression power (or rate of work) required is given by:

$$P = w/t = n_g(h_2 - h_1)/t \qquad (9\text{-}23)$$

where P = compression power required, Btu/day
n_g = number of moles of gas being compressed
t = time for compression, days

h_1 = enthalpy of gas at intake, Btu/lbmole
h_2 = enthalpy of gas at discharge, Btu/lbmole

This power requirement in Btu/day can be converted to horsepower, IHP in hp, as follows:

$$\text{IHP} = \frac{1 \text{ hp}}{33,000 \text{ ft-lbf/min}} \frac{778.2 \text{ ft-lbf/min}}{\text{Btu/min}} \frac{n_g(h_2 - h_1)/t \text{ Btu/day}}{1440 \text{ min/day}}$$

$$= \frac{1.6376 \times 10^{-5} \, n_g(h_2 - h_1)}{t} \qquad (9\text{-}24)$$

where IHP = ideal compression horsepower required, hp.

Empirical "Quickie" Charts

This method uses actual compressor manufacturers' charts that relate compression horsepower requirements to the applicable compression process variables. This method has its advantages—it is simple and easy to use, and it directly gives the actual requirements that include efficiency and other factors for the actual real compressor. However, because only the most important compression variables are considered and the charts are quite specific for some assumed gas and compressor type, this method may not be very accurate or more generally applicable to all cases. Usually, it gives slightly higher results than the other two more exact methods. Further details regarding this approximate empirical method of compressor design are discussed in succeeding sections on the design of reciprocating and centrifugal compressors.

Estimating the Actual (Brake) Horsepower From Ideal Horsepower

The actual horsepower supplied by the compressor engine or driver is known as *brake horsepower* (BHP). The BHP required by a compressor is always greater than the ideal or theoretical horsepower IHP. The energy losses are represented by two types of efficiencies: *compression efficiency*, η_c, and *mechanical efficiency* of the compressor, η_m.

Compression efficiency is the ratio of the theoretical work requirement for a given process (IHP) to the actual work required within the compressor cylinder or shaft to compress and deliver the gas (GHP):

$$\eta_c = \frac{\text{IHP}}{\text{GHP}}$$

or,

$$\text{GHP} = \frac{\text{IHP}}{\eta_c} \qquad (9\text{-}25)$$

The compression efficiency η_c includes effects due to thermodynamic deviations from the theoretical compression process (such as gas turbulence and heating of the incoming gas), and fluid friction and leakage losses. For reciprocating compressors, it may also include a factor known as volumetric efficiency, to be discussed later. Compression efficiency does not include mechanical losses. Besides some factors related to the compressor itself that may affect the compression process, compression efficiency is significantly affected by the operating conditions, such as suction pressure, compression ratio, and compressor speed and loading. The degree to which these factors affect compression efficiency, however, is different for different compressor types.

Mechanical efficiency, defined as the ratio of the compression power required within the compression chamber to the actual horsepower supplied to it, includes frictional losses in compressor packings and bearings, piston rings, and other moving parts:

$$BHP = GHP + \text{mechanical losses} = \frac{GHP}{\eta_m} \qquad (9\text{-}26)$$

The mechanical efficiency, η_m, is related to the compressor type, its design details, and the mechanical condition of the unit; the operating parameters of the compression process do not affect it significantly.

The BHP can be directly computed from IHP using an overall efficiency factor, η:

$$BHP = \frac{IHP}{\eta} \qquad (9\text{-}27)$$

where the overall efficiency, η, is equal to the product of the compression and mechanical efficiencies:

$$\eta = \eta_c \eta_m$$

In most modern compressors, η_c is in the range of 83% to 93% and η_m ranges from 88% to 95% (Ikoku, 1984). Thus, the overall compression efficiency η ($= \eta_c \eta_m$) ranges from 73% to 88% for most modern compressors.

Designing Reciprocating Compressors

Number of Stages

The first parameter in designing a compression system is to determine the number of stages. As discussed earlier, it becomes necessary to use multistage compression to overcome the several limitations, such as compression ratio achievable, inherent in single-stage compression. In designing reciprocating compressors, the compression ratio r is rarely allowed to exceed a value of 4.0, and a r ≤ 6 is considered the practical limit.

Theoretically, the total power required is a minimum when perfect intercooling is provided, there is no pressure loss between the stages, and the compression ratio in each stage is the same. Thus, the optimum compression ratio for each stage is given by:

$$r_{opt} = (r_t)^{1/n_s} = (p_d/p_s)^{1/n_s} \tag{9-28}$$

where r_{opt} = optimum compression ratio per stage
r_t = total compression ratio desired
n_s = total number of stages
p_d = final discharge pressure, psia
p_s = suction pressure at the very first stage, psia

If intercoolers are provided between the stages, reduce the theoretical intake pressure of each stage by about 3% to allow for interstage pressure drop. This is equivalent to dividing the theoretical r from Equation 9-28 by $(0.97)^{1/n_s}$.

In practice, although Equation 9-28 results in minimum power, the net work or energy required varies only by a fraction of a percent for relatively large variations in the compression ratios for individual stages. This is an important fact, often used for flexibility in design for economic and technical reasons.

Example 9-2. For the problem given in Example 9-1, what is the compression ratio and how many reciprocating compressor stages are required if: (a) intercooling is not provided, and (b) intercooling is provided.

Solution

(a) Without intercooling.

From Equation 9-28, $r_{opt} = (p_d/p_s)^{1/n_s} = (2,500/100)^{1/n_s} = 25^{1/n_s}$

For $n_s = 1$, r = 25, which is too high and hence unacceptable.
For $n_s = 2$, r = 5, which is less than 6, hence acceptable.

Thus, 2 compression stages are required, giving an r = 5.

(b) With intercooling.

Allowing for interstage pressure drop, $r_{opt} = (25/0.97)^{1/n_s} = 25.773^{1/n_s}$

Repeating the procedure, for $n_s = 2$, $r = 5.077$, which is less than 6, hence acceptable.

Thus, 2 compression stages are required, giving an $r = \underline{5.077}$.

Horsepower Requirements

The ideal horsepower IHP for reciprocating compressors can be obtained using either the analytical method (Equation 9-19 or 9-20), or Mollier charts along with Equation 9-24. For the analytical method, intercooling requirements can be calculated using Equation 9-21, provided the specific heat for the gas being compressed is known. In the Mollier chart method, intercooling requirements can be determined directly from the enthalpy change indicated on the enthalpy-entropy diagram.

The IHP can be converted to brake horsepower requirements using Equation 9-27 if the overall efficiency η is known. For reciprocating compressors, volumetric (or clearance) efficiency is an important component of compression efficiency η_c and, consequently, the overall efficiency η.

Volumetric Efficiency

The volumetric or clearance efficiency, η_v, of a reciprocating compressor is defined as the ratio of the volume of gas actually delivered, corrected to suction pressure and temperature, to the piston displacement. It represents the efficiency of the compressor cylinder in compressing the gas, and accounts for gas leakage, heating of gas as it enters the compression chamber, throttling effect on valves, re-expansion of trapped gas, etc.

The theoretical volumetric efficiency, η_v, is a function of the compression ratio and clearance, as follows:

$$\eta_v = 1 - (r^{1/k} - 1)Cl \qquad (9\text{-}29)$$

where r = compression ratio
k = isentropic exponent for the compression process
Cl = clearance, fraction

Clearance is defined as the ratio of clearance volume to the piston displacement (see Figure 9-20):

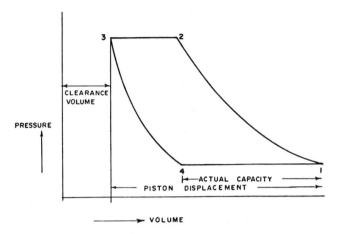

Figure 9-20. Pressure-volume diagram for the compression cycle of an actual reciprocating compressor showing the effect of clearance.

$$Cl = \frac{V_3}{V_1 - V_3} \qquad (9\text{-}30)$$

where V_1 and V_3 are the volumes on the pressure-volume diagram shown in Figure 9-20. Volume V_3, representing the volume at the cylinder end not swept by the piston, is called *clearance volume*. Clearance, usually varying from 0.04 to 0.16, limits compressor throughput, and also results in horsepower being wasted in simply compressing and re-expanding the trapped gas.

The actual volumetric efficiency includes factors for other effects such as incomplete filling of the cylinder, leakage, and friction (A), and lubrication (Lu):

$$\eta_v = 1 - A - Lu - Cl\left(\frac{Z_1 r^{1/k}}{Z_2} - 1\right) \qquad (9\text{-}31)$$

where   A = factor for incomplete filling of the cylinder, leakage, friction, etc.; it is usually between 0.03 and 0.06
Lu = compressor lubrication factor, generally 0.05 for non-lubricated compressors and zero otherwise
Z_1, Z_2 = gas compressibility factors at the suction and discharge, respectively

From Equations 9-29 and 9-31, it is obvious that the volumetric efficiency η_v increases with decreasing compression ratio, r, increasing isentropic exponent, k (for r > 1), and decreasing clearance, Cl. These parameters can be varied to alter the volumetric efficiency and, consequently, the flow capacity of a reciprocating compressor. Sometimes clearance is added to reduce the capacity for fixed operating pressure conditions, or to prevent overloading the driver under variable operating pressure conditions by reducing the capacity as compression ratio changes.

Figure 9-21 shows the effect of clearance on the volumetric efficiency of reciprocating compressors. The effect of compression ratio (all other factors

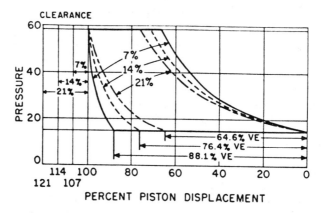

Figure 9-21. Effect of clearance on the volumetric efficiency at constant compression ratio for reciprocating compressors. (From *Compressed Air and Gas Data*, 1982; courtesy of Ingersoll-Rand Co.)

being the same) on volumetric efficiency is shown in Figure 9-22. Figure 9-23 shows the effect of the specific heat ratio or isentropic exponent, k, on the volumetric efficiency, for a fixed compression ratio and clearance.

This discussion may lead one to believe that the clearance Cl should be kept as low as possible in order to maximize the volumetric efficiency. However, lower clearance may reduce the compression efficiency η_c which depends primarily upon the valve area, just like volumetric efficiency depends primarily upon clearance. To obtain lower clearances, it is necessary to limit the valve area (size and number of valves). Thus, the designer must find the optimum clearance and valve area, depending upon the application desired for the particular compressor design.

Direct Determination of Brake Horsepower

Brake horsepower (BHP) requirements for reciprocating compressors can be estimated directly from "quickie" charts. These charts plot the brake horsepower required per million cubic feet of gas per day (BHP/MMcfd),

Gas Compression 429

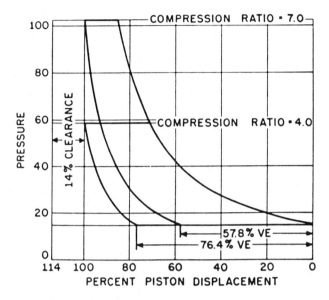

Figure 9-22. Effect of different compression ratios on the volumetric efficiency of a reciprocating compressor cylinder. (From *Compressed Air and Gas Data*, 1982; courtesy of Ingersoll-Rand Co.)

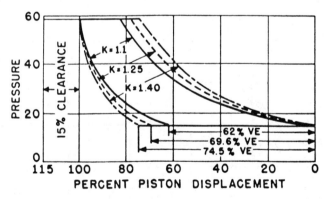

Figure 9-23. Effect of specific heat ratio k on the volumetric efficiency for a given reciprocating compressor cylinder. (From *Compressed Air and Gas Data*, 1982; courtesy of Ingersoll-Rand Co.)

referred to some defined suction pressure and actual intake temperature, versus the compression ratio r for different values of the isentropic exponent k. Figures 9-24, 9-25, and 9-26 show such charts by the Ingersoll-Rand Company (1982), based upon the analytic solution (Equation 9-20) cor-

(text continued on page 433)

430 *Gas Production Engineering*

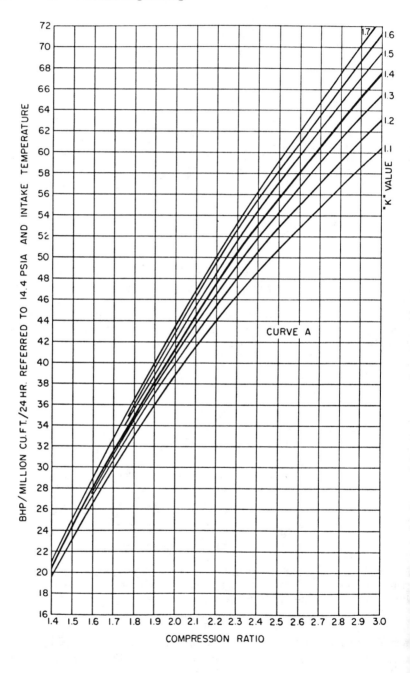

Figure 9-24. Compression power requirements for reciprocating compressors. (From *Compressed Air and Gas Data,* 1982; courtesy of Ingersoll-Rand Co.)

Gas Compression 431

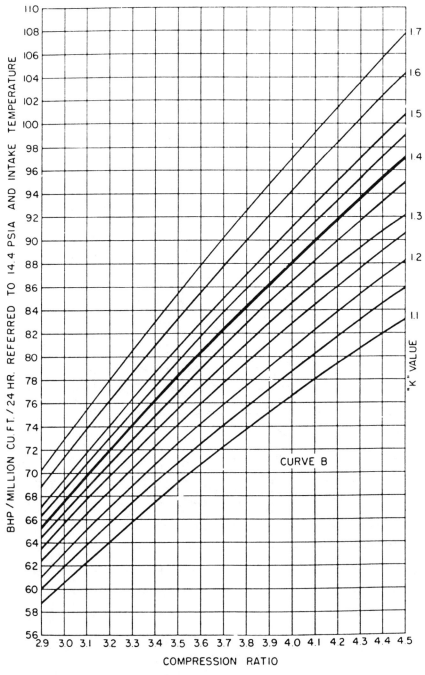

Figure 9-25. Compression power requirements for reciprocating compressors. (From *Compressed Air and Gas Data*, 1982; courtesy of Ingersoll-Rand Co.)

432 Gas Production Engineering

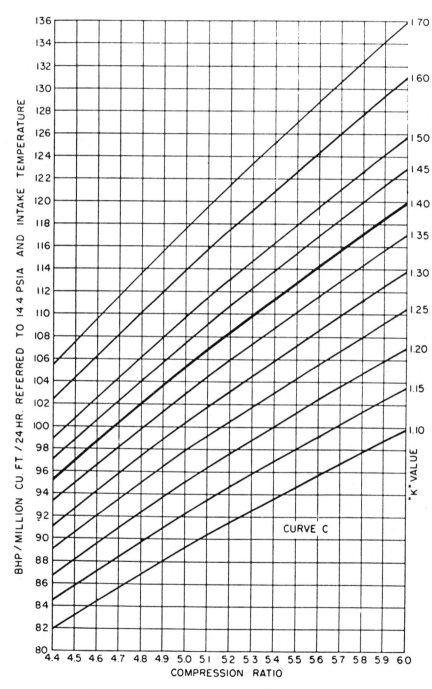

Figure 9-26. Compression power requirements for reciprocating comressors. (From *Compressed Air and Gas Data,* 1982; courtesy of Ingersoll-Rand Co.)

Gas Compression 433

(text continued from page 429)

rected for overall efficiency. These figures are applicable for designing each compression stage individually and require the following considerations:

1. Suction pressure of 14.4 psia and actual intake temperature is assumed. The correction factor for intake pressure is generally included in the calculation of BHP (as in Equation 9-32 or 9-33), or obtained from Figure 9-27.

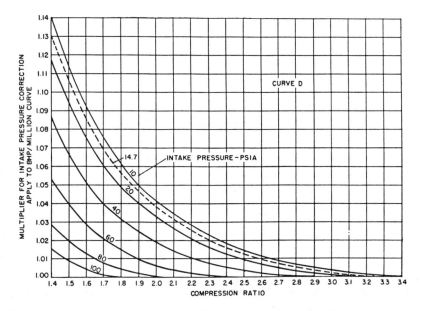

Figure 9-27. Suction pressure correction factors for compression power requirements for reciprocating compressors. (From *Compressed Air and Gas Data*, 1982; courtesy of Ingersoll-Rand Co.)

2. Gas specific gravity of 1.0 is assumed. Correction factors for gas gravities other than 1.0 are shown in Figure 9-28.
3. Gas compressibility factor of 1.0 is assumed. Corrections for gas compressibility are frequently included in the calculation of BHP (see Equation 9-33).
4. A mechanical efficiency of 95% and a compression efficiency of 83.5% is assumed. BHP values should be corrected if this is not the case.
5. The gas volume to be handled in each stage should be corrected to the actual intake temperature and moisture content at intake to that stage.
6. Allow interstage pressure drop if intercoolers are used. As discussed before, reduce the theoretical intake pressure of each stage by about 3%,

434 Gas Production Engineering

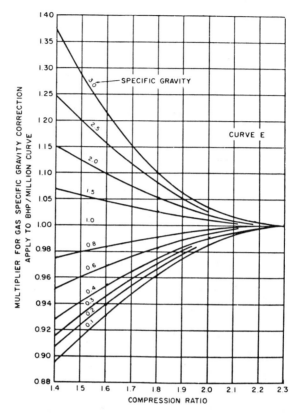

Figure 9-28. Gas specific gravity correction factors for compression power requirements for reciprocating compressors. (From *Compressed Air and Gas Data*, 1982; courtesy of Ingersoll-Rand Co.)

which is equivalent to dividing the theoretical r from Equation 9-28 by $(0.97)^{1/n_s}$.

For a given gas flow rate of q_{sc} MMscfd at standard conditions of pressure p_{sc} (psia) and temperature T_{sc} (°R), the required BHP can be computed as follows:

$$\text{BHP} = \frac{q_{sc} p_{sc} T_1}{14.4 \, T_{sc}} (\text{BHP/MMcfd})_{\text{fig}} \qquad (9\text{-}32)$$

where BHP = brake horsepower required, hp
q_{sc} = gas flow rate measured at pressure p_{sc} (psia) and temperature T_{sc} (°R), MMscfd
T_1 = gas temperature at the compressor suction (inlet), °R
$(\text{BHP/MMcfd})_{\text{fig}}$ = BHP/MMcfd from Figures 9-24 through 9-26

Gas Compression 435

Note that Equation 9-32 includes the correction factor for intake pressure and if Figure 9-27 is used for correcting the BHP for intake pressure, the factor $p_{sc}/14.4$ should be omitted. The BHP from Equation 9-32 is corrected for gas compressibility as follows:

$$BHP = \frac{q_{sc}p_{sc}T_1}{14.4\,T_{sc}} \frac{(Z_1 + Z_2)}{2Z_1} (BHP/MMcfd)_{fig} \qquad (9\text{-}33)$$

where Z_1 and Z_2 are the gas compressibility factors at the suction and discharge of the compressor, respectively. Note that this procedure should be applied for one compression stage at a time.

Compressor Speed and Stroke Length

The stroke length and/or compressor speed are selected based upon the following relationship for the flow capacity of reciprocating compressors:

$$q_{rc} = (\pi/4)d^2 L S \eta_v \qquad (9\text{-}34)$$

where q_{rc} = capacity of the reciprocating compressor, ft³/unit time
  d = piston diameter, ft
  L = stroke length, ft
  S = compressor speed, strokes/unit time
  η_v = volumetric efficiency

For double-acting reciprocating compressors, the volume occupied by the piston rod should also be accounted for. Thus, besides increasing volumetric efficiency, the actual capacity of a reciprocating compressor can be increased by increasing compressor speed and stroke length.

Example 9-3. A reciprocating compression system is to be designed to compress 5 MMcfd of the gas in Example 9-1, with intercoolers and an aftercooler that cool the gas to 150°F. Find:

(a) Brake horsepower using the analytical method.
(b) Brake horsepower using Mollier diagram method.
(c) Estimate the cooling requirements from the results of part (b).
(d) From the results of part (b), determine whether the first stage can be handled by a compressor with a speed 1,200 rpm, piston diameter = 12 in., and stroke length = 3 ft.

Assume $\eta = 0.80$, A = 0.05, Lu = 0, and Cl = 0.08. Neglect any gas compressibility factor effects.

436 Gas Production Engineering

Solution

From Examples 9-1 and 9-2:

$\gamma_g = 0.612$, k = 1.2565, r = 5.077, n_s = 2 stages.

(a) Analytical method.

Using Equation 9-18, for each of the two stages:

$$\text{IHP} = \frac{(3.0303)(5)(100)(150)(1.2565)}{(0.2565)(150)} (5.077^{0.2565/1.2565} - 1) = 2,919.1 \text{ hp}$$

Therefore, BHP for first stage = 2,919.1/0.80 = 3,648.9 hp.
and BHP for second stage = 3,648.9 hp.

(b) Mollier diagram method.

For a $\gamma_g = 0.612$, the enthalpy-entropy diagram of Figure 9-12 is approximately applicable. Use of this diagram results in:

Point	Process	p (psia), T(°F)	Enthalpy, h (Btu/lbmole)
1	Entering gas	100 psia, 150°F	880
2	After isentropic compression	500 psia, 368°F	3,310
3	After isobaric cooling	500 psia, 150°F	725
4	After isentropic compression	2,500 psia, 490°F	5,300
5	After isobaric cooling	2,500 psia, 150°F	− 200

Using the gas law, the number of moles of gas is given by

$n_g = pV/ZRT = (100)(5 \times 10^6)/(1)(10.732)(610) = 76,376.46$ lbmoles/day

From Equation 9-24,

For the first stage, IHP = $(1.6376 \times 10^{-5})(76,376.46)(3,310 - 880)/1 = 3,039.3$ hp.

Similarly, for the second stage, IHP = $(1.6376 \times 10^{-5})(76{,}376.46)(5{,}300 - 725)/1 = 5{,}722.1$ hp.

Thus, BHP for first stage = $3{,}039.3/0.80 = \underline{3{,}799.1 \text{ hp}}$.

and BHP for second stage = $5{,}722.1/0.80 = \underline{7{,}152.6 \text{ hp}}$.

(c) Cooling requirements.

Cooling load for the intercooler (between stages 1 and 2)
= $(76{,}376.46)(3{,}310 - 725) = \underline{1.97 \times 10^8 \text{ Btu/day}}$.

And cooling load for the aftercooler (after stage 2)
= $(76{,}376.46)[5{,}300 - (-200)] = \underline{4.20 \times 10^8 \text{ Btu/day}}$.

(d) Compressor speed for the first stage.

Using Equation 9-31, with $Z_1 = Z_2 = 1$,

$\eta_v = 1 - 0.05 - 0 - (0.08)(5.077^{1/1.2565} - 1) = 0.7385$

Using Equation 9-34, the flow capacity of the given compressor is

$q_{rc} = (\pi/4)(1^2)(3)(1{,}200)(0.7385) = 2{,}088.06 \text{ cf/min} = 3.0 \text{ MMcfd}$

Thus, the given compressor cannot handle 5 MMcfd gas.

Example 9-4. Assuming intercooling, repeat Example 9-3 using empirical charts.

Solution

From Example 9-3, $\gamma_g = 0.612$, r = 5.077, k = 1.2565, p_1 = 100 psia, η = 0.80.

From Figure 9-26, for the first stage, BHP/MMcfd = 98.2 hp.

From Figure 9-28, the gas gravity correction = 1.0

Applying the appropriate correction factors, the BHP required for 5 MMcfd gas measured at suction conditions of 100 psia and 150°F is given by:

438 Gas Production Engineering

$$BHP = \frac{q_{sc}p_{sc}}{14.4}(BHP/MMcfd)_{fig}\frac{(0.95)(0.835)}{\eta}$$

$$= \frac{(5)(100)(98.2)(0.95)(0.835)}{(14.4)(0.80)}$$

$$= 2{,}704.8 \text{ hp}$$

The parameters for the second stage are the same. Therefore, second stage brake horsepower required is also equal to 2,704.8 hp.

Designing Centrifugal Compressors

Analytical Method

Calculations for centrifugal compressors are very similar to those for reciprocating compressors. An analytic expression for the ideal horsepower, IHP, can be derived, similar to Equation 9-19 for reciprocating compressors, by replacing the isentropic exponent k by the polytropic exponent n:

$$IHP = \frac{3.0303 \, q_{sc} p_{sc} T_1 (Z_1 + Z_2)}{2 Z_1 T_{sc}} \frac{n}{n-1} (r^{(n-1)/n} - 1) \qquad (9\text{-}35)$$

where, as before:
IHP = theoretical horsepower required for compression, hp
q_{sc} = gas flow rate, MMscfd, measured at pressure p_{sc}, psia and temperature T_{sc}, °R
Z_1, Z_2 = gas compressibility factors at the compressor suction and discharge, respectively
T_1 = gas inlet (suction) temperature, °R
n = polytropic exponent
r = compression ratio

For centrifugal compressors, the theoretical work required for compression is represented by the polytropic head h_p, similar to the theoretical work requirement w for reciprocating compressors (see Equation 9-16):

$$h_p = \frac{(144)RT_1(Z_1 + Z_2)}{2Z_1 M} \frac{n}{n-1} (r^{(n-1)/n} - 1)$$

$$= \frac{1545 T_1(Z_1 + Z_2)}{2Z_1 M} \frac{n}{n-1} (r^{(n-1)/n} - 1) \qquad (9\text{-}36)$$

where h_p = polytropic head, ft-lbf/lbm
       R = gas constant, psia-ft^3/lbmole-°R
       M = molecular weight of the gas, lbm/lbmole
       144 = factor to convert from psia to lbf/ft^2

Thus, the ideal compression horsepower, IHP, for a gas mass flow rate of m lbm/min is given by:

$$\text{IHP} = \frac{mh_p}{33,000} \tag{9-37}$$

where IHP = ideal compression horsepower required, hp
       m = mass flow rate of the gas, lbm/min
       33,000 = factor to convert power requirement from ft-lbf/min to hp

The gas horsepower, GHP, is given by:

$$\text{GHP} = \frac{mh_p}{33,000 \, \eta_p} \tag{9-38}$$

where η_p is the polytropic compression efficiency (fraction).

Mollier Charts

Mollier charts, if available for the gas, can also be used for determining the compression parameters. The procedure outlined earlier for reciprocating compressors is used, with the assumption of isentropic compression. The polytropic head h_p is then calculated as:

$$h_p = \frac{778.2 \, (\Delta h) \eta_p}{M \eta_{ic}} \tag{9-39}$$

where h_p = polytropic head, ft-lbf/lbm
       Δh = enthalpy change for isentropic compression determined from the Mollier diagram, Btu/lbmole
       η_p = polytropic efficiency
       η_{ic} = isentropic efficiency
       M = gas molecular weight, lbm/lbmole

The factor 778.2 ft-lbf/Btu in Equation 9-39 gives the desired units conversion. Subsequently, the gas horsepower GHP can be determined using Equation 9-38.

Estimating the Actual Horsepower

The BHP for centrifugal compressors is generally estimated by adding the mechanical and hydraulic losses to the GHP:

$$BHP = GHP + HPL_m + HPL_h \qquad (9\text{-}40)$$

where HPL_m, HPL_h = mechanical and hydraulic horsepower losses, repectively, hp

Mechanical horsepower losses, consisting of losses in the compressor seals and bearings, are generally in the range of 7 to 50 hp, depending on the speed and casing size of the unit. Hydraulic horsepower losses, consisting of casing and piston leakage losses, vary between 0.3% to 2.5% of the gas horsepower GHP, depending primarily on the size of the unit. With increasing size, mechanical losses increase, whereas hydraulic losses decrease.

Number of Stages

The polytropic head, h_p, is an indication of the number of stages required for centrifugal compression. The number of stages required, n_s, is given by:

$$n_s = \frac{h_p}{9{,}500} \qquad (9\text{-}41)$$

where 9,500 ft-lbf/lbm is a common limit assigned to each centrifugal compression stage.

Equation 9-41 assumes that all impellers are running at optimum design speeds, and that each develops a polytropic head of 9,500 ft-lbf/lbm. Although the latter is true for most industrial units, machines with greater polytropic head per stage can and have been designed.

Compressor Speed

Centrifugal compressor performance is highly dependent upon speed. The capacity varies directly as the speed, S, the head developed varies as the square of the speed, and the required horsepower varies as the cube of the speed:

$q \propto S$
$h_p \propto S^2$
$BHP \propto S^3$

Gas Compression   441

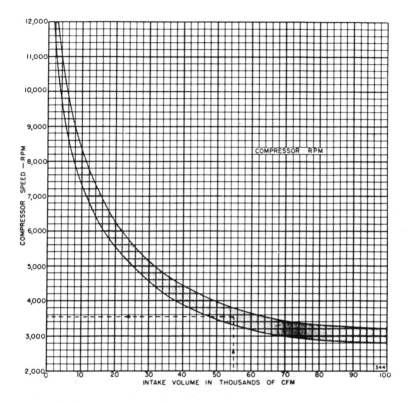

Figure 9-29. Straight-through flow centrifugal compressor speed. (From *Compressed Air and Gas Data*, 1982; courtesy of Ingersoll-Rand Co.)

Compressor speed can be estimated using Figure 9-29 as a function of the intake volume. Generally, the compressor speed is obtained from Figure 9-29 by reading directly at the center of the shaded area that indicates the range usually involved.

Empirical Charts

Centrifugal compressor parameters can also be obtained from Figures 9-30 through 9-34. Figure 9-30 shows the polytropic head, h_p, versus the compression ratio r for various intake temperatures T_1. The corrections required in the computation of the actual polytropic head for a given case are as follows:

1. For a polytropic exponent n other than 1.396, use Figure 9-31 for determining the "k" value correction factor, K_{corr}. (Surprisingly, the liter-

ature consistently refers to this "n" value correction factor as the "k" value correction factor.)
2. Appropriate adjustment for a gas molecular weight other than 28.97 (air). As shown by Equation 9-36, the polytropic head is inversely proportional to gas molecular weight M.
3. Correction for gas compressibility factors other than unity, as included in Equation 9-36.

Thus, the applicable actual head h_p is determined from the basic head $(h_p)_{fig}$ given in Figure 9-30 as follows:

$$h_p = \frac{(h_p)_{fig}(K_{corr})(28.97)(Z_1 + Z_2)}{2Z_1 M} \quad (9\text{-}42)$$

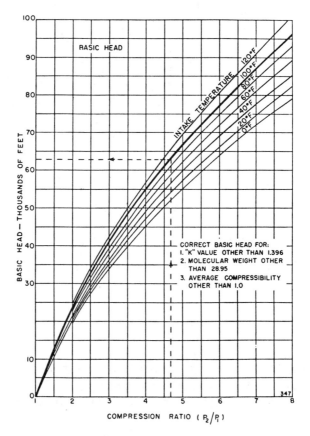

Figure 9-30. Basic head for straight-through flow centrifugal compressors. (From *Compressed Air and Gas Data,* 1982; courtesy of Ingersoll-Rand Co.)

Gas Compression 443

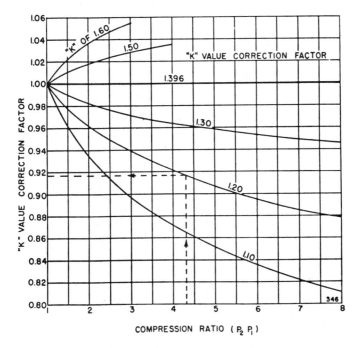

Figure 9-31. k-value correction factors for the brake horsepower requirements for straight-through flow centrifugal compressors. (From *Compressed Air and Gas Data*, 1982; courtesy of Ingersoll-Rand Co.)

Knowing the polytropic head, h_p, the number of stages required for centrifugal compression, n_s, can be calculated as before using Equation 9-41. Compressor speed can be obtained from Figure 9-29 as before.

Figure 9-32 shows the brake horsepower BHP versus the intake gas volume measured at 14.5 psia pressure and actual intake temperature. In addition to correction factors applicable to head h_p, determination of actual BHP requires a correction for intake pressure other than 14.5 psia. Thus, the actual BHP is determined from the basic brake horsepower, $(BHP)_{fig}$, given in Figure 9-32 as follows:

$$BHP = \frac{(BHP)_{fig}(p_1)(K_{corr})(28.97)(Z_1 + Z_2)}{2(14.5)Z_1 M} \qquad (9\text{-}43)$$

Basic brake horsepower requirements for intercooled centrifugal compressors are shown in Figure 9-33. Appropriate correction factors need to be applied for the polytropic exponent, intake pressure and temperature, intercooling temperature, gas gravity, and gas compressibility factor.

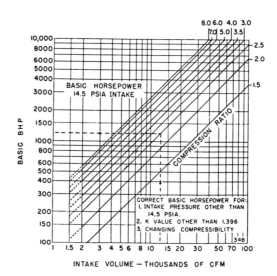

Figure 9-32. Basic brake horsepower requirements for straight-through flow centrifugal compressors. (From *Compressed Air and Gas Data,* 1982; courtesy of Ingersoll-Rand Co.)

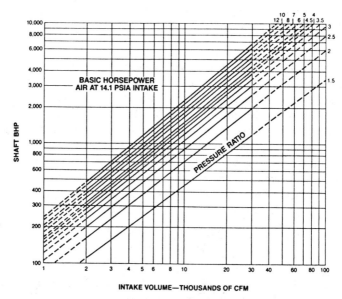

Figure 9-33. Basic brake horsepower requirements for intercooled centrifugal compressors for air at 14.1 psia and 95°F intake, with perfect intercooling (interstage inlet temperature of 95°F). (From *Compressed Air and Gas Data,* 1982; courtesy of Ingersoll-Rand Co.)

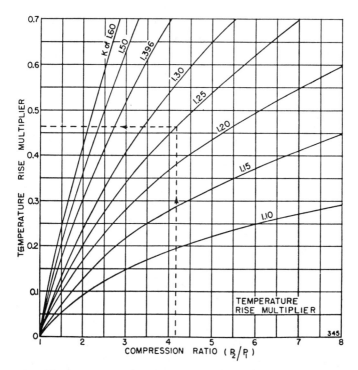

Figure 9-34. Temperature rise multiplier for determining the discharge temperature for straight-through flow centrifugal compressors. (From *Compressed Air and Gas Data*, 1982; courtesy of Ingersoll-Rand Co.)

The discharge temperature of the compressed gas from a straight-through flow centrifugal compressor can be obtained, using Figure 9-34, as follows:

$$T_2 = T_1(1 + T_{rm}) \qquad (9\text{-}44)$$

where T_1, T_2 = intake and discharge temperatures, respectively, °R
T_{rm} = temperature rise multiplier from Figure 9-34

The discharge volume, V_2, of the compressed gas can now be calculated:

$$V_2 = \frac{p_1 V_1 T_2}{p_2 T_1} \qquad (9\text{-}45)$$

where p_1, V_1, T_1 = intake pressure, volume, and temperature of the gas, respectively
p_2, V_2, T_2 = discharge pressure, volume, and temperature of the gas, respectively

Example 9-5. Find the horsepower required by a centrifugal (straight-through flow, no intercooling) compressor for compressing 10 MMcfd gas at suction conditions of 150 psia and 80°F to 500 psia. Assume $\gamma_g = 0.60$, $k = 1.296$, and $Z_1, Z_2 \simeq 1.0$.

Solution

$r = 500/150 = 3.333$, $M = (28.97)\gamma_g = (28.97)(0.60) = 17.382$ lbm/lb-mole

Flow capacity = 10 MMcfd = $(10 \times 10^6)/(24 \times 60) = 6{,}944.4$ cfm

From Figure 9-11, for $q = 6{,}944.4$ cfm, $\eta_p = 0.725$

From Equation 9-11,

$$\frac{n-1}{n} = \frac{k-1}{k\eta_p} = \frac{0.296}{(1.296)(0.725)} = 0.315$$

From Equation 9-36,

$$h_p = \frac{(1{,}545)(460+80)}{(17.382)(0.315)}(3.333^{0.315} - 1) = 70{,}292.1 \text{ ft-lbf/lbm}$$

Gas density at suction conditions is equal to

$$\rho_1 = \frac{p_1 M}{Z_1 R T_1} = \frac{(150)(17.382)}{(1)(10.732)(540)} = 0.45 \text{ lbm/ft}^3$$

Therefore, the mass flow rate, m is equal to

$$m = q_1 \rho_1 = \frac{10 \times 10^6 \text{ft}^3/\text{day}}{1{,}440 \text{ min/day}} \; 0.45 \; \frac{\text{lbm}}{\text{ft}^3} = 3{,}125 \text{ lbm/min}$$

From Equation 9-38,

$$\text{GHP} = \frac{(3{,}125)(70{,}292.1)}{(33{,}000)(0.725)} = 9{,}181.3 \text{ hp}$$

Using Equation 9-38, with the assumption that $\text{HPL}_m = 30$ hp, and $\text{HPL}_h = 1\%$ of GHP,

BHP = (1.01)(9,181.3) + 30 = 9,303 hp.

Example 9-6. For the data given in Example 9-5, design a centrifugal compressor using the empirical chart method.

Solution

1. Horsepower.

From Example 9-5, r = 3.333, M = 17.382 lbm/lbmole, k = 1.296, and intake volume = 6,944.4 cfm at 150 psia and 80°F.

From Figure 9-32, for q_1 = 6,944.4 cfm and r = 3.333, the basic BHP = 900 hp.

From Figure 9-31, for r = 3.333 and k = 1.296, K_{corr} = 0.967.

From Equation 9-43,

$$\text{BHP} = \frac{(900)(150)(0.967)(28.97)}{(14.5)(17.382)} = \underline{15,005.2 \text{ hp}}$$

2. Number of stages.

From Figure 9-30, basic head = 47,000 ft-lbf/lbm.

From Equation 9-42,

$$h_p = \frac{(47,000)(0.967)(28.97)}{(17.382)} = 75,748.33 \text{ ft-lbf/lbm.}$$

From Equation 9-41, the number of stages is equal to:

n_s = 75,748.33/9,500 = 7.97 ≈ <u>8 stages</u>

3. Compressor speed.

From Figure 9-29, for an intake volume of 6,944.4 cfm, the compressor speed S = <u>8,800 rpm</u>.

Designing Rotary Compressors

Rotary positive-displacement compressors follow the same basic theory as reciprocating positive-displacement compressors and are designed using the same theoretical relationships. There are, however, some differences and empirical corrections introduced by the specific design.

Well-designed rotary sliding-vane compressors do not have the clearance typical of reciprocating compressors. There are, however, leakage or slip losses between adjacent cells across the vanes at their edges and ends—some amount of the gas escapes past to the suction side. These losses are difficult to theorize, and are generally estimated from actual tests.

The capacity of rotary two-impeller compressors also is subject to leakage or slip losses rather than losses related to clearance. The losses are minimized by operating at or near the built-in, fixed compression ratio for which the unit is designed to give optimum performance. Low-clearance sealing is the usual method for controlling slip losses. Factors that increase slip losses are high pressure differential or rise across the compressor, and low gas density. Slip is usually independent of speed, but for more complex designs, it may be speed-dependent.

Slip is generally reported as capacity lost, cfm (cubic feet per minute) at suction conditions. It may also be accounted for as the additional rpm (rotations per minute) required to deliver a given amount of gas:

$$S_t = \frac{q_d}{V_d} + S_s \qquad (9\text{-}46)$$

where S_t = total rpm
q_d = desired capacity, cfm
V_d = displacement, cfr (cubic feet per revolution)
S_s = slip, rpm

The total rpm determined from Equation 9-46 is used in horsepower or capacity calculations, as follows:

$$BHP = 0.005 \, V_d S_t (\Delta p) \qquad (9\text{-}47)$$

where Δp is the pressure differential (or rise), and 0.005 is a constant.

Capacity control is provided by varying the speed, or by installing multiple units. The discharge temperature of a rotary compressor can theoretically be calculated using the relationship for reciprocating compressors (Equation 9-10). In practice, the discharge temperature of a rotary compressor is substantially higher than that for an equivalent water-cooled reciprocating compressor, because of the speed and compactness of the rotary that provides relatively little time or surface area for gas cooling.

Questions and Problems

1. What are the desirable traits for a compressor?
2. List the applications of the various compressor types, giving appropriate reasons for these.
3. Using basic thermodynamic principles, show that:
   (a) $c_p - c_v = R$
   (b) $c_p = 5R/2$, and $c_v = 3R/2$ for monoatomic ideal gases
4. For a reciprocating compressor, is volumetric efficiency the same as compression efficiency? Why? Also, why is polytropic efficiency primarily a function of the intake volume?
5. Discuss the basis and applicability of Equation 9-39 for determining the polytropic head of a centrifugal compressor from Mollier charts. Express the relationship of Figure 9-11 as a correlation equation.
6. Calculate the isentropic exponent at 100°F and 500 psia for a gas with the following composition (in weight %): $C_1 = 89.1$, $C_2 = 6.05$, $C_3 = 1.11$, $n\text{-}C_4 = 0.35$, $i\text{-}C_4 = 0.18$, $H_2S = 1.55$, and $CO_2 = 1.66$.

    Determine the polytropic exponent from: (1) thermodynamic charts (Chapter 3), and (2) polytropic efficiency, given an intake volume of 10,000 cfm.
7. You have been assigned to a gas field where you notice an old compressor that can be repaired and used in the future. Searching through the company records you find that the ratio of the inlet temperature (°R) to outlet temperature (°R) was 0.769 when the flow rate was 3,000 Mscfd. The inlet pressure was 1,000 psia and the horsepower was 200 hp.
    (a) What is the compression ratio (assume $k = 1.28$ and $\gamma_g = 0.6$), and the outlet pressure?
    (b) Assume that the compressor is used as the first stage in a two-stage compression system. The outlet gas from the first stage is cooled to its initial temperature. If the gas in the second stage is compressed to 7,000 psia, what is the outlet temperature? Use Mollier charts.
    (c) Determine the theoretical horsepower for the second stage in part (b) using Mollier charts.
8. A compressor is to be installed for handling gas supply into a transmission line. The compression requirements are as follows:
    Gas flow rate = 17.5 MMcfd at intake conditions of 450 psia and 90°F, discharge pressure = 6,000 psia, and gas gravity $\gamma_g = 0.83$. Intercooling to 100°F or any reasonably low temperature is desired.

450 Gas Production Engineering

Give the design specifications (number of stages, compression ratio, GHP, BHP, speed, intercooling requirements, etc.), using at least two methods, for a reciprocating compressor, and a centrifugal compressor. Compare the requirements and select the one, giving reasons, that is most suitable for this application.

References

Brown, G. G., 1945. "A Series of Enthalpy-Entropy Charts for Natural Gases," *Trans.*, AIME, 160, 65–76.

GPSA, 1981. *Engineering Data Book*, 9th ed. (5th revision). Gas Processors Suppliers Association, Tulsa, Oklahoma.

Ikoku, C. U., 1984. *Natural Gas Production Engineering*. John Wiley & Sons, New York, 517pp.

Ingersoll-Rand Company, 1982. *Compressed Air and Gas Data*. Ingersoll-Rand Co., Woodcliff Lake, NJ.

10
Gas Flow Measurement

Introduction

Gas flow measurement constitutes one of the more important auxiliary operations related to gas production and transport. It is required to enable the determination of the amount of gas being produced or sold, and also as a basic parameter for almost all of the design procedures. The produced gas stream is in a continuous state of flow from the instant it leaves the reservoir until it is consumed at the delivery end. Therefore, except for underground storage or other storages such as LNG (liquified natural gas) storage facilities, gas measurements must be done on a flowing stream of gas. Accuracy in measurement is obviously of prime importance: an error of only 1% for a typical pipeline delivering 300 MMscfd (109.5 Bscf/year) can result in an error of approximately 1.1 Bscf/year of gas which, at a typical gas price of $3.00/Mscf, would amount to a loss of $3.3 million to the buyer or seller.

Gas is most commonly measured in terms of volume because of the simplicity of such a procedure. However, to make this volumetric measurement more meaningful, base or standard pressure and temperature conditions are defined that yield measurements in standard cubic feet. This volumetric rate can be converted to mass flow rate by multiplying with the gas density at the standard pressure and temperature conditions. Since the gas density at the specified standard conditions is a constant for the particular gas under consideration, measurements in standard cubic feet are synonymous to mass flow rate measurement. However, the number of standard cubic feet measured depends upon the standard pressure and temperature conditions chosen. Table 10-1 shows some such measurement bases. The most common basis is the AGA and API recommended pressure of 14.73 psia and temperature of 60°F.

Table 10-1
Common Gas Pressure and Temperature Bases for Measurement

State	Base pressure psia	Base temperature °F
Texas, Oklahoma, Kansas, Arkansas, and Alberta (Canada)	14.65	60
California	14.7	60
AGA, API	14.73	60
U.S. Bureau of Standards, and Federal Price Commission	14.735	60
Louisiana	15.025	60

Measurement Fundamentals

Flow is one of the most difficult variables to measure (Campbell, 1984), because it cannot be measured directly like pressure and temperature. It must be inferred by indirect means, such as the pressure differential over a specified distance, speed of rotation of a rotating element, displacement rate in a measurement chamber, etc. For this and other reasons, many flow measurement techniques and devices have been developed for a wide range of applications. This discussion is limited to those devices that have found use in the oil and gas industry, primarily for natural gas measurement.

Attributes of Flow Devices

A flowmeter or measurement device is characterized using the following parameters.

Accuracy

This is a measure of a flowmeter's ability to indicate the actual flow rate within a specified flow-rate range. It is defined as the ratio of the difference between the actual and measured rates to the actual rate.

$$\text{Accuracy} = \frac{\text{Abs [Actual rate} - \text{Measured rate]}}{\text{Actual rate}} \times 100\% \qquad (10\text{-}1)$$

where Abs(x) represents the absolute value of the argument x. Accuracy is reported in either of two ways: percent of full scale, or percent of reading. For example, for a 100-MMscfd flowmeter, a $\pm 1\%$ of full scale accuracy means that the measured flow rate is within ± 1 MMscfd of the actual flow rate, regardless of the value of the flow rate. Thus, for a measured flow rate of 10 MMscfd, the actual flow rate is between 9 and 11 MMscfd, and for a measured rate of 100 MMscfd, it is between 99 and 101 MMscfd. An accuracy of ±1% of reading, however, implies that the measured flow rate is within 9.9 to 10.1 MMscfd for a measured rate of 10 MMscfd, 49.5 to 50.5 for a measured rate of 50 MMscfd, 99 to 101 MMscfd for a measured rate of 100 MMscfd, etc. Thus, the percent of reading results in a better overall performance because the error is proportional to the magnitude of the rate. Positive displacement meters and turbine meters usually have a percent of reading accuracy, whereas orifice meters and rotameters have a percent of full scale accuracy in their specifications.

Rangeability

A flowmeter's rangeability is the ratio of the maximum flow rate to the minimum flow rate at the specified accuracy.

$$\text{Rangeability} = \frac{\text{Maximum rate that can be measured}}{\text{Minimum rate that can be measured}} \quad (10\text{-}2)$$

Rangeability is usually reported as a ratio x:1. For example, a meter with maximum and minimum rates of 50 MMscfd and 10 MMscfd, respectively, for a specified accuracy of ±1%, has a rangeability of 5:1. This rangeability can be increased to 10:1 by decreasing the minimum rate by a mere 5 MMscfd to 5 MMscfd resulting in a 5- to 50-MMscfd meter, or by increasing the maximum rate by 50 MMscfd to 100 MMscfd resulting in a 10- to 100-MMscfd meter. Thus, it is important to know the flow rate range over which a quoted rangeability applies.

Repeatability

Also known as reproducibility or precision, repeatability is the ability of a meter to reproduce the same measured readings for identical flow conditions over a period of time. It is computed as the maximum difference between measured readings, sometimes expressed as a percent of full scale. Note that repeatability does not imply accuracy; a flowmeter may have very good repeatability, but a lower overall accuracy.

Linearity

This is a measure of the deviation of the calibration curve of a meter from a straight line. It can be specified over a given flow-rate range, or at a given flow rate. A linear calibration curve is desirable because it leads to a constant metering accuracy, with no portion of the scale being relatively more or less sensitive than the other. Note that a flowmeter could have a good linearity, but poor accuracy if its calibration curve is offset (shifted).

Selection of Measurement Devices

The selection of a measurement device depends upon:
1. Accuracy and reliability of the device.
2. Range of flow rate—maximum and minimum.
3. Range of flow temperature and pressure.
4. Fluid to be measured—gas or liquid, their constituents and specific gravity.
5. Maintenance requirements.
6. Expected life of the device, and its initial and operating costs.
7. Other considerations, such as simplicity, availability of power or other inputs required by the device, its susceptibility to theft or vandalism, etc.

Methods of Measurement

A brief introduction to the different fluid measurement methods, well described in the American Society of Mechanical Engineers' report on this subject (ASME, 1971), and by Corcoran and Honeywell (1975), is presented here.

Differential Pressure Method

In this method, the flow rate is computed using the pressure difference over a flow interval or restriction and other data. There are basically two types of differential pressure devices: one in which the pressure difference is measured across a flow restriction, such as the orifice meter, venturi meter, etc., and the second type where the difference in pressure is measured upon impact, such as the pitot tube. The dynamics of the relationships involved have been studied in great detail for these types of meters, and very precise and accurate results can be obtained from them. Some of the commonly used differential pressure devices are described.

Orifice Meter

This is by far the most commonly used device for metering natural gas. It consists of a flat metal plate with a circular hole, centered in a pair of flanges in a straight pipe section. The pressure differential is measured across this plate to yield the flow rate. This is a rugged, accurate, simple, and economical device, and can handle a wide range of flow rates. Orifice meters have a rangeability of about 3.5:1, with an accuracy on the order of $\pm 0.5\%$ (Corcoran and Honeywell, 1975). Details of this important meter type are discussed later in this chapter.

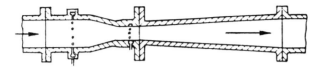

Figure 10-1. A venturi meter. (After Corcoran and Honeywell, 1975; courtesy of *Chemical Engineering*.)

Venturi Meter

This type of meter, shown in Figure 10-1, consists of a short pipe section tapering into a throat, coupled with a relatively longer diverging pipe section for pressure recovery. It is similar to an orifice meter, with the advantage of low pressure loss, and is a preferred choice where less pressure drop is available. Venturi meters have a rangeability of 3.5:1, with an accuracy of $\pm 1\%$ (Corcoran and Honeywell, 1975).

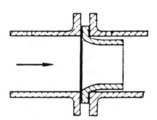

Figure 10-2. Flow nozzle. (After Corcoran and Honeywell, 1975; courtesy of *Chemical Engineering*.)

Flow Nozzles

Flow nozzles have a rounded edge that aids the handling of solids in the flow stream (see Figure 10-2). The analysis is similar to orifice meters. Flow nozzles are used for high flow rate streams, because they permit, for the same line size and pressure differential, a 60% greater rate of flow than an orifice plate (Corcoran and Honeywell, 1975). Flow nozzles have a rangeability of 3.5:1, with an accuracy of ± 1.5–2% (Corcoran and Honeywell, 1975).

Pitot (Impact) Tube

Figure 10-3 shows a pitot tube installed in a pipe section. The pitot tube measures the difference between the static pressure at the wall of the flow conduit and the flowing pressure at its impact tip where the kinetic energy of the flowing stream is converted into pressure. It gives the flow velocity

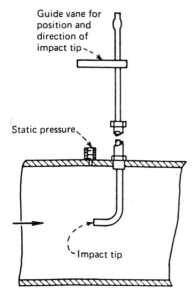

Figure 10-3. Pitot tube metering. (After Corcoran and Honeywell, 1975; courtesy of *Chemical Engineering.*)

only at a point (at the tip). To compute the mean flow velocity, the calibration must account for the velocity profile in the flow conduit. Another factor that makes the calibration of a pitot tube difficult is the low pressure differential produced by it. The tip can be easily clogged by liquids or solids. Because of the relatively poor accuracy of this device (most of the error is in measuring the static pressure), it is not used very often, except on a temporary basis.

Orifice Well Tester

This device consists of a nipple, equipped with a flange to facilitate the attachment of different sharp-edged orifice plates at its end. The device discharges the gas to the atmosphere, and only the static pressure just upstream of this plate needs to be measured. It has limited accuracy, but finds application where gas is at relatively low pressures and is being produced to the atmosphere.

Critical-Flow Provers

Similar to the orifice well tester, a critical-flow prover consists of a special nipple equipped to facilitate the attachment of orifice plates at its end, and it discharges the gas to the atmosphere. The critical-flow prover, however, is based upon the principle of critical flow of gases through flow restrictions (see critical flow through chokes in Chapter 7). In this device, critical flow is maintained, and it is only necessary to determine the upstream pressure, gas gravity, and the flowing temperature (see Equation 7-83) in order to calculate the gas flow rate.

It is important to note that the critical-flow prover uses a rounded-edge orifice, because sharp-edged orifices do not conform to critical flow theories and do not give a good repeatability.

Displacement Meters

These meters measure the volumetric displacement of the fluid at flowing conditions. The number of such known volumes through the meter per unit time, corrected to the base pressure and temperature, are counted to give the flow rate, instantaneous and/or cumulative, through the meter. Displacement meters are also called positive-displacement meters because they afford a positive volume at flowing conditions: the flow is divided into isolated measured volumes, and the number of these volumes are counted in some manner. This is in contrast with the other meter types, sometimes referred to as rate meters, in which the fluid passes without being divided into isolated quantities.

Figure 10-4 shows the two types of displacement meters commonly used: *rotary* or impeller type, and *slide-valve diaphragm* type. The rotary type

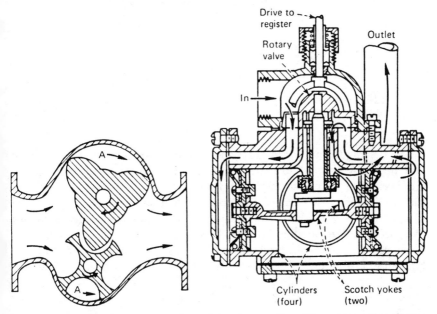

Figure 10-4. Displacement meters: (a) rotary meter, (b) reciprocating-piston meter. (After Corcoran and Honeywell, 1975; courtesy of *Chemical Engineering*.)

consists of a rotating element, whereas the diaphragm type has a piston-cylinder arrangement. Both are quite similar in operation. They contain measuring elements (or chambers) of known volume, with valves that channel the gas into and out of these measuring elements, and counters to count the number of times the measuring element is filled per unit time.

Turbine Meter

These meters are sometimes classified as positive-displacement meters. They consist of a turbine or propeller that turns at a speed proportional to the velocity of the gas flowing past it, converting linear velocity to rotational speed (see Figure 10-5). The speed of the turbine is measured as pulses that give the rate. These pulses are counted to give the instantaneous rate, or accumulated to give the cumulative rate.

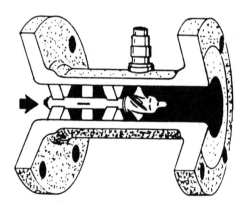

Figure 10-5. A turbine meter. (After Corcoran and Honeywell, 1975; courtesy of *Chemical Engineering*.)

The driving torque for the propeller is proportional to the fluid density and the square of the fluid velocity. Turbine meters have therefore traditionally been used for measuring liquid flow rates rather than gas flow rates. Fluctuations in velocity, caused by pressure fluctuations, turbulence, or unsteady-state flow conditions, will cause the turbine meter to give a higher than actual value. To allow sustained accuracy and trouble-free operation, filters are almost always used ahead of the turbine meter. Turbine meters typically have a rangeability up to 100:1 for gases, with an accuracy of $\pm 0.25\%$ and a repeatability of $\pm 0.05\%$ (Evans, 1973). Further details about turbine meters can be obtained from November (1972) and Evans (1973).

Elbow (Centrifugal) Meter

This device, shown in Figure 10-6, is based upon the principle of centrifugal force that is generated when the fluid changes its direction of flow along a circular path. The magnitude of this centrifugal force is governed by the pipe diameter, the radius of the circular bend in the pipe, the flow velocity, and other fluid properties. An elbow meter creates relatively little pressure loss or differential, and is therefore used primarily for control or other purposes. It finds some application in large pipes where a substantial centrifugal force is generated. Elbow meters have a rangeability of 3:1, with an accuracy of ±1% (Corcoran and Honeywell, 1975).

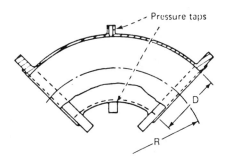

Figure 10-6. An elbow meter. (After Corcoran and Honeywell, 1975; courtesy of *Chemical Engineering*.)

Rotameter (Variable Area Meter)

A rotameter is essentially a variable orifice meter. The fluid stream is throttled by a constriction, but instead of measuring the pressure differential across a fixed sized orifice, rotameters vary the size of the flow constriction to accommodate the flow rate, keeping the differential pressure constant. Rotameters consist of a float that is free to move up and down in a vertical tube that has a gradual taper down to its base. The fluid entering at the base of the tube causes the float to rise, until the annular area between the float and the tube wall is such that the pressure drop across this constriction is just sufficient to support the float. The flow rate is directly read-off from the graduations etched upon the glass tube (see Figure 10-7).

Rotameters have a rangeability of 10:1, with an accuracy of ±1% (Corcoran and Honeywell, 1975).

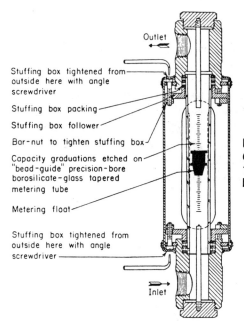

Figure 10-7. A rotameter. (From *Chemical Engineers' Handbook*, 1984; courtesy of McGraw-Hill Publishing Company.)

Vortex-Shedding Flowmeters

These devices are based upon the principle of vortex shedding, which occurs when a fluid flows past a non-streamlined (blunt) object. The flow is unable to follow the shape of the blunt object, also referred to as flow element, and separates from it. This separation of the flow leads to the formation of eddies or turbulent vortices on the surfaces along the sides of the object that grow in size as they move downstream, and are eventually shed or detached from the object (see Figures 10-8a and b). Shedding takes place alternately at either side of the object (*Chemical Engineers' Handbook*,

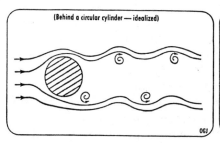 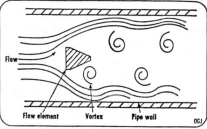

Figure 10-8. (a) Idealized Karman vortex trail behind a circular cylinder; (b) vortex flow pattern. (After Powers, 1975; courtesy of *Oil & Gas Journal*.)

1984). The rate of vortex formation and shedding is directly proportional to the volumetric flow rate, and inversely proportional to the diameter of the object. Thus, the flow rate can be inferred from vortex shedding measurements.

The formed vortices can be detected in a number of ways. Four methods are mainly used:

1. Thermistors that detect the change in temperature caused by the changing flow of the fluid as a vortex passes.
2. A magnetic pickup coil to count the oscillations imposed on a sphere or disk by the vortices as they alternate on the two sides of the object (a transverse passage is provided in the flow element to allow movement of the sphere or disk).
3. Counting the motion or the induced mechanical stresses on a vane that extends behind the blunt body and moves from side to side with the passing of the alternating vortices.
4. Ultrasonic transmitters and receivers that detect the vortices using a sonic beam.

Vortex-shedding flowmeters have a rangeability of about 15:1 or greater, an accuracy of ± 0.25 to 1% of reading, and a repeatability of ± 1% of reading (DeVries, 1982). The flow regime, however, must remain turbulent over the entire range of flow rate. Additional details on these meters can be obtained from Powers (1975) and DeVries (1982).

Ultrasonic Meters

Although several ultrasonic measurement principles are known, only two types are commercially used (Munk, 1982): the Doppler, and the contrapropagating (or transmitted energy) methods. Of these two, only the contrapropagating method is applicable to natural gas measurement. The Doppler method, using the reflections of ultrasonic energy off particles, is not applicable to natural gas, which is generally free from particulates.

The contrapropagating ultrasonic flowmeter computes the flow velocity by measuring the time difference between two ultrasonic waves traveling over the same path, but with one with the flow and the other against the flow. For this purpose, two transducers are used, as shown in Figure 10-9, that alternately transmit and receive ultrasonic pulses. This flowmeter can measure flow in either direction and indicate the direction of flow. It has a rangeability of 50:1 and an accuracy better than ± 2% (Munk, 1982).

There are some difficulties with ultrasonic meters caused by the effect of solids, gas bubbles, etc. However, because these meters can be mounted outside the pipe, they do not disturb the flow, cause no pressure loss, are portable, and offer applicability to large pipes.

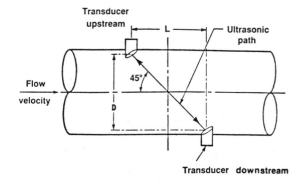

Figure 10-9. Installation geometry of a contrapropagating ultrasonic flowmeter. (After Munk, 1982; courtesy of *Oil & Gas Journal.*)

Orifice Meters

Because orifice meters are simple, accurate, relatively inexpensive, rugged, and reliable, they are the most important and widely used of the flowmeters for gases. Other devices such as turbine meters have also been used because orifice meters have a limited rangeability (typically 3.5:1), and are difficult to adapt to automation.

An orifice meter consists of a thin plate, 0.115–0.398 in. thick depending upon the pipe size and pressure, held perpendicular to the direction of flow by a pair of flanges, with a circular sharp square-edged orifice (hole) accurately machined to the required size in the center of the plate. Pressure taps are provided on the upstream as well as downstream end in the fitting that holds the orifice plate. A pressure measuring and recording device is connected to the pressure taps. The orifice fittings are designed to permit easy changing and inspection of orifice plates.

The Orifice Metering System

Orifice Types

In addition to the concentric (centered) orifice, there are two more types of orifices as shown in Figure 10-10: *eccentric* (off-center), and *segmental* (part of a circle).

The concentric type is the most common, because of its low cost, ease of fabrication, and ease of calibration. It has a rangeability of 3.5:1, with an accuracy of ±0.5%. The sharp-edged plate, however, is subject to wear and a consequent loss in accuracy.

Gas Flow Measurement 463

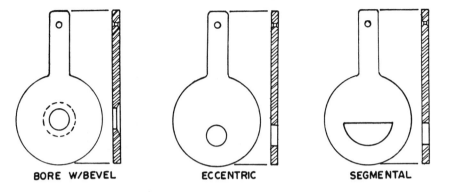

Figure 10-10. Types of orifice arrangements.

The eccentric and segmental types are very useful for two-phase flow streams, and for flow streams with suspended solids, such as dirty gases or slurries. They have a rangeability of 3:1, but with an accuracy of ±1.5–2%, they are less accurate than the concentric type. The segmental type has the additional advantage that it does not retain solids on the upstream side of the plate. Centering the hole, however, is critical to its performance, and it is recommended for use only on large pipe sizes (Corcoran and Honeywell, 1975).

Location of Pressure Taps

The magnitude of the measured pressure differential is obviously affected by the location of the points across the orifice between which it is measured. The four types of pressure tap locations (see Figure 10-11) that have been used are as follows:

1. Flange type: In this type, the pressure is measured 1 in. from the upstream face of the plate and 1 in. from the downstream face of the orifice plate. This is the most common type of pressure tap.
2. Pipe taps: In this type, the pressure is measured 2.5 pipe IDs from the upstream, and 8 pipe IDs from the downstream (where the pressure recovery is maximum) face of the orifice plate. This type requires location tolerances 10 times higher than the flange type.
3. Vena contracta: The point at which the velocity is the highest, and pressure is the lowest, is called the vena contracta. In the vena contracta type of pressure measurement, pressure is measured 1 pipe ID upstream of the plate, and at the vena contracta downstream. It is used where flow rates are fairly constant, because the location of the

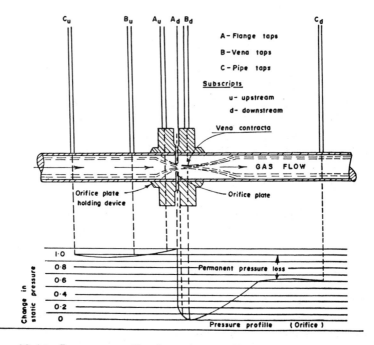

Figure 10-11. Pressure profile through an orifice meter and the relative locations of taps.

vena contracta depends upon the orifice size, and the orifice size chosen depends upon the rate. This type of pressure tap provides greater accuracy because it gives a greater pressure drop.

4. Corner type: In this type, the pressure taps are located immediately adjacent to the upstream and downstream faces of the orifice plate. The use of this design is limited, for the most part, to some European countries only.

Straightening Vanes

Straightening vanes consist of a symmetrical bundle of small diameter tubing, welded together in a concentric pattern as shown in Figure 10-12. These vanes are placed in the upstream section of the orifice meter in order to eliminate any flow irregularities, such as eddies, swirls, or cross currents caused by the pipe fittings and valves preceding the orifice meter, that may affect meter accuracy. The diameter of each of the tubes should be less than $1/4$ of the inside pipe diameter, and the length to diameter ratio for the tubes must be greater than 10 (GPSA, 1981).

Installation of vanes reduces the length of straight pipe required upstream of the orifice considerably. Vanes, however, should not be used unless abso-

Gas Flow Measurement   465

lutely necessary, because they introduce additional pressure loss, clog easily, and are subject to erosion.

Size and Location of Orifice

For commercial measurement of gases, the ratio of the orifice to pipe diameter, β, should be between 0.15 and 0.70 for meters using flange taps, and between 0.20 and 0.67 for meters using pipe taps (GPSA, 1981). The thickness of the orifice plate at the orifice edge should not exceed $1/50$ of the

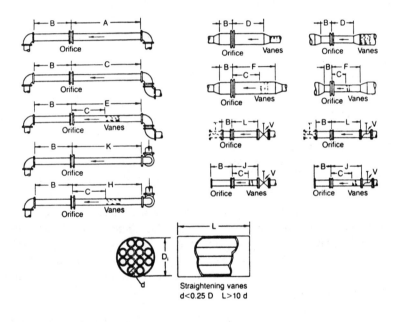

MINIMUM LENGTHS OF STRAIGHT PIPE

β	A	B	C	D	E	F	G	H	J	K	L
0.10	6.0	2.4	5.0	6.0	8.5	8.5	8.5	9.5	9.5	14.0	16.0
0.20	6.0	2.7	5.0	6.0	8.5	8.5	8.5	9.5	9.5	14.5	17.5
0.30	6.0	3.0	5.0	6.0	8.5	8.5	8.5	9.6	9.5	16.0	19.5
0.40	6.0	3.3	5.0	6.0	8.5	8.5	8.5	9.6	9.5	17.8	21.6
0.50	6.8	3.6	5.3	7.4	9.2	9.3	10.0	10.0	10.3	20.2	25.0
0.60	9.0	3.9	5.7	9.4	10.4	10.4	13.4	11.3	12.0	24.7	30.0
0.70	14.0	4.2	6.6	12.0	12.2	12.0	18.5	13.0	15.0	30.5	38.0
0.80	20.0	4.5	8.0	15.0	15.5	15.0	25.0	16.5	19.5	40.0	50.0

β = ratio diameter of orifice to inside diameter of pipe.

Figure 10-12. Proper installation of an orifice meter—required meter run and straightening vanes. (From *Engineering Data Book,* 1981; courtesy of GPSA.)

pipe diameter and ⅛ of the orifice diameter. The orifice location should be such as to have a stabilized flow to ensure proper metering. Figure 10-12 shows the minimum requirements of straight pipe section and/or vanes recommended by GPSA (GPSA, 1981) for this purpose.

Pressure Measuring and Recording

Gas is a highly compressible fluid and its density varies considerably with pressure. Because pressure variations are quite substantial through the flow path in an orifice meter, it is necessary to measure both the differential pressure as well as the flowing pressure. This is also illustrated by the orifice flow equations derived a bit later (Equation 10-12 or 10-14). The flowing pressure is often referred to as static gas pressure in gas measurement terminology.

The orifice meter is generally equipped with a two-pen recorder for recording both the static as well as the differential pressure on a circular recording chart. The chart itself has a pressure scale on it to enable reading the measured pressures that it records continuously. Static pressure is generally measured with a bourdon tube type of device that actuates the pen on the chart. Differential pressure is measured using either a mercury manometer, or a bellows meter. The bellows meter is preferable because it avoids the problems of mercury contamination, and mercury loss and the consequent change in calibration. It is important to choose pressure measuring devices that can handle about twice the maximum anticipated pressures.

Two types of circular pressure recording charts are in common use—direct-reading charts, and square-root charts. The direct-reading chart has a pressure scale with lines spaced equally apart, whereas the square-root chart records the square root of the percent of the full-scale range of the meter. Therefore, pressure from a square-root chart is determined as follows:

$$\text{Actual pressure} = \frac{\text{meter range}}{100} (\text{chart reading})^2$$

For example, for a square-root chart with a range of 50 in. × 100 psi, a differential reading of 7 is equivalent to a differential pressure of

$$\text{Differential pressure} = \frac{50}{100} (7)^2 = 24.5 \text{ in. of water}$$

and a static reading of 9 is equivalent to a static pressure of

$$\text{Static pressure} = \frac{100}{100} (9)^2 = 81 \text{ psia}$$

Gas Flow Measurement

The common chart ranges for differential pressure are from 0 to: 10, 20, 50, 100, or 200 in. of water, whereas for static pressure, the common chart ranges are from 0 to: 100, 250, 500, 1,000, or 2,500 psia.

Measurement Calculations

The relationship for orifice meters can be derived from the general energy equation (see Equation 7-2), written between two points in the flowing stream—point 1 being some point upstream of the orifice plate, and point 2 representing the orifice throat:

$$\int_1^2 V \, dp + \frac{1}{g_c} \int_1^2 v \, dv + \frac{g}{g_c} \int_1^2 dz = w_s - l_w \qquad (10\text{-}3)$$

where V = specific volume of the fluid, ft^3/lbm
 v = fluid velocity, ft/sec
 p = pressure, lbf/ft^2
 z = elevation above a given datum plane, ft
 w_s = shaft work done by the fluid on the surroundings, ft-lbf/lbm
 l_w = work energy lost due to friction, ft-lbf/lbm
 g = gravitational acceleration, ft/sec^2
 g_c = conversion factor relating mass and weight, lbm-ft/lbf-sec^2

For most meters, the elevation change between points 1 and 2 is zero, and no work is done by the flowing fluid stream. Therefore, Equation 10-3 can be written as:

$$\int_1^2 V \, dp + \frac{1}{g_c} \int_1^2 v \, dv + l_w = 0$$

Incorporating the friction loss term l_w in the compression-expansion term to avoid the complexity of referring to the friction factor, and multiplying both sides by fluid density ρ (lbm/ft^3), we get:

$$C^2 \int_1^2 dp + \frac{1}{g_c} \int_1^2 \rho \, v \, dv = 0 \qquad (10\text{-}4)$$

where C = an empirical constant

Assuming a constant, average density ρ_{av} for simplicity, and integrating Equation 10-4, we get:

$$C^2(p_2 - p_1) + \frac{\rho_{av}}{2g_c}(v_2^2 - v_1^2) = 0 \qquad (10\text{-}5)$$

Equation 10-5 uses pressure p in lbf/ft². Converting to commonly used pressure units of psia (lbf/in.²) and rearranging, we get:

$$v_2^2 - v_1^2 = \frac{2(144)g_c C^2(p_1 - p_2)}{\rho_{av}} \tag{10-6}$$

The mass flow rate, m (lbm/sec), is given by:

$$m = \rho v A \tag{10-7}$$

where A = cross-sectional area of flow, ft²

This analysis assumes steady-state flow conditions, for which the mass flow rate is constant. Equation 10-6 can now be expressed as:

$$\frac{m^2}{\rho_{av}^2}\left[\frac{1}{A_2^2} - \frac{1}{A_1^2}\right] = \frac{2(144)g_c C^2(p_1 - p_2)}{\rho_{av}}$$

or,

$$\frac{m^2}{A_2^2}[1 - (A_2/A_1)^2] = 2(144)g_c \rho_{av} C^2(p_1 - p_2) \tag{10-8}$$

Let d_1 and d_2 be the diameters of the pipe and the orifice, respectively, in inches. Defining $\beta = d_2/d_1$, and solving Equation 10-8 for m:

$$m = CA_2 \left[\frac{2(144)g_c \rho_{av}(p_1 - p_2)}{(1 - \beta^4)}\right]^{0.5}$$

or,

$$m = C\pi d_2^2 \left[\frac{g_c \rho_{av}(p_1 - p_2)}{1,152\,(1 - \beta^4)}\right]^{0.5} \tag{10-9}$$

Using the gas law, the gas density ρ_{av} can be expressed as:

$$\rho_{av} = \frac{28.97\gamma_g p_{av}}{Z_{av} R T_{av}} \tag{10-10}$$

where γ_g = gas gravity (air = 1)
 p_{av} = average pressure, psia
 T_{av} = average temperature, °R
 Z_{av} = average gas compressibility factor
 R = gas constant

The pressure differential, $p_1 - p_2$, is generally expressed in terms of inches of water. This conversion can be achieved, using the relation $\Delta p = \rho g(\Delta h)/g_c$, as follows:

$$(p_1 - p_2) \text{ psia} = \frac{(62.43 \text{ lbm/ft}^3)(g \text{ ft/sec}^2)(\Delta h \text{ in. water})}{(144 \text{ in.}^2/\text{ft}^2)(12 \text{ in.}/\text{ft})(g_c \text{ lbm-ft/lbf-sec}^2)}$$

or,

$$(p_1 - p_2) = \frac{62.43 \, \Delta h}{(144)(12)} \tag{10-11}$$

where Δh = pressure differential in inches of water.

Using Equations 10-10 and 10-11, Equation 10-9 becomes:

$$m = C\pi d_2^2 \left[\frac{(28.97)(62.43) g_c \gamma_g p_{av} \, \Delta h}{(1,152)(144)(12)(1 - \beta^4) Z_{av} RT_{av}} \right]^{0.5} \tag{10-12}$$

Gas flow is generally reported in terms of the flow rate q_{sc} in scf/hr at standard conditions, which is related to the mass flow rate m in lbm/sec as follows ($m = q\rho$):

$$m = \frac{q_{sc}}{(3,600)} \frac{(28.97)\gamma_g p_{sc}}{Z_{sc} RT_{sc}}$$

Using standard conditions of $p_{sc} = 14.73$ psia, $T_{sc} = 520°R$, and the fact that $Z_{sc} = 1$, we obtain:

$$m = \frac{(28.97)(14.73)}{(3,600)(520)R} \gamma_g q_{sc} \tag{10-13}$$

Using Equation 10-13 in Equation 10-12, substituting R = 10.73 psia-ft³/lbmole-°R, and solving for q_{sc}, we obtain:

$$q_{sc} = \frac{7,717.96 \, Cd_2^2}{[(1 - \beta^4)\gamma_g Z_{av} T_{av}]^{0.5}} [\Delta h \, p_{av}]^{0.5} \tag{10-14}$$

Note that in Equation 10-14, q_{sc} is in scf/hr, d_2 is in inches, Δh is in inches of water, p_{av} is in psia, T_{av} is in °R, and C, β, γ_g, and Z_{av} are dimensionless. Equation 10-14 is commonly expressed as:

$$q_{sc} = K_o [\Delta h \, p_{av}]^{0.5} \tag{10-15}$$

where the constant K_o is given by:

$$K_o = \frac{7{,}717.96 \, Cd_2^2}{[(1-\beta^4)\gamma_g Z_{av} T_{av}]^{0.5}}$$

In metering practice, the average pressure p_{av} is replaced by a measurable gauge pressure p_f. Factors are provided to account for this p_f being measured at the upstream or downstream, or being measured as the mean of upstream and downstream static pressures, and for the type of pipe tap. Equation 10-15 is then written in the following form:

$$q_{sc} = K \, [h_w \, p_f]^{0.5} \qquad (10\text{-}16)$$

where h_w = differential pressure at 60°F, inches of water
p_f = absolute static pressure of the flowing fluid, psia

and the constant K is expressed as a product of several different factors as follows (AGA, 1969):

$$K = F_b \, F_r \, Y \, F_{pb} \, F_{tb} \, F_{tf} \, F_g \, F_{pv} \, F_m \, F_l \, F_a \qquad (10\text{-}17)$$

where F_b = basic orifice factor, scf/hr
F_r = Reynold's number factor
Y = expansion factor
F_{pb} = pressure-base factor
F_{tb} = temperature-base factor
F_{tf} = flowing temperature factor
F_g = specific gravity factor
F_{pv} = supercompressibility factor
F_m = manometer factor (for mercury manometer)
F_l = gauge location factor
F_a = orifice thermal expansion factor

These factors, determined through extensive tests and reported for use as the industry standard by AGA, are given in Tables 10-2 through 10-14 for two pressure tap types—flange taps and pipe taps.

Basic Orifice Factor, F_b

The factor F_b is simply the constant K_o in Equation 10-15. Its value depends upon the type of pressure taps, and the pipe and orifice diameters. F_b can be obtained from Tables 10-2 and 10-7 for flange taps and pipe taps, respectively. For sizes that cannot be found in these tables, interpolation is
(text continued on page 475)

Gas Flow Measurement 471

Table 10-2
Flange Taps—Basic Orifice Factors—F_b

Base temperature = 60°F Flowing temperature = 60°F $\sqrt{h_w p_f} = \infty$
Base pressure = 14.73 psia Specific gravity = 1.0 $h_w/p_f = 0$

Pipe Sizes—Nominal and Published Inside Diameters, Inches

Orifice Diameter, in.	2			3			4	
	1.689	1.939	2.067	2.300	2.626	2.900	3.068	3.152	3.438
0.250	12.695	12.707	12.711	12.714	12.712	12.708	12.705	12.703	12.697
0.375	28.474	28.439	28.428	28.411	28.393	28.382	28.376	28.373	28.364
0.500	50.777	50.587	50.521	50.435	50.356	50.313	50.292	50.284	50.258
0.625	80.090	79.509	79.311	79.052	78.818	78.686	78.625	78.598	78.523
0.750	117.09	115.62	115.14	114.52	113.99	113.70	113.56	113.50	113.33
0.875	162.95	159.56	158.47	157.12	156.00	155.41	155.14	155.03	154.71
1.000	219.77	212.47	210.22	207.44	205.18	204.04	203.54	203.33	202.75
1.125	290.99	276.20	271.70	266.35	262.06	259.95	259.04	258.65	257.63
1.250	385.78	353.58	345.13	335.12	327.39	323.63	322.03	321.37	319.61
1.375		448.57	433.50	415.75	402.18	395.80	393.09	391.97	389.03
1.500			542.26	510.86	487.98	477.36	472.96	471.14	466.39
1.625				623.91	586.82	569.65	562.58	559.72	552.31
1.750					701.27	674.44	663.42	658.96	647.54
1.875					834.88	793.88	777.18	770.44	753.17
2.000						930.65	906.01	896.06	870.59
2.125						1091.2	1052.5	1038.1	1001.4
2.250							1223.2	1199.9	1147.7
2.375									1311.7
2.500									1498.4

Orifice Diameter, in.	4		6			8		
	3.826	4.026	4.897	5.182	5.761	6.065	7.625	7.981	8.071
0.250	12.687	12.683							
0.375	28.353	28.348							
0.500	50.234	50.224	50.197	50.191	50.182	50.178			
0.625	78.450	78.421	78.338	78.321	78.296	78.287			
0.750	113.15	113.08	112.87	112.82	112.75	112.72			
0.875	154.40	154.27	153.88	153.78	153.63	153.56	153.34	153.31	153.31
1.000	202.20	201.99	201.34	201.19	200.96	200.85	200.46	200.39	200.38
1.125	256.69	256.33	255.31	255.08	254.72	254.56	253.99	253.69	253.87
1.250	318.03	317.45	315.83	315.48	314.95	314.72	313.91	313.78	313.74
1.375	386.45	385.51	382.99	382.47	381.70	381.37	380.25	380.06	380.02
1.500	462.27	460.79	456.93	456.16	455.03	454.57	453.02	452.78	452.72
1.625	545.89	543.61	537.77	536.64	535.03	534.38	532.27	531.95	531.87
1.750	637.84	634.39	625.73	624.09	621.79	620.88	618.02	617.60	617.50
1.875	738.75	733.68	721.03	718.69	715.44	714.19	710.32	709.77	709.64
2.000	849.41	842.12	823.99	820.68	816.13	814.41	809.22	808.50	808.34
2.125	970.95	960.48	934.97	930.35	924.07	921.71	914.79	913.86	913.64
2.250	1104.7	1089.9	1054.4	1048.1	1039.5	1036.3	1027.1	1025.9	1025.6
2.375	1252.1	1231.7	1182.9	1174.2	1162.6	1158.3	1146.2	1144.7	1144.3
2.500	1415.0	1387.2	1320.9	1309.3	1293.8	1288.2	1272.3	1270.3	1269.8
2.625	1595.6	1558.2	1469.2	1453.9	1433.5	1426.0	1405.4	1402.9	1402.3
2.750	1797.1	1746.7	1628.9	1608.7	1582.1	1572.3	1545.7	1542.5	1541.8
2.875		1955.5	1801.0	1774.5	1740.0	1727.5	1693.4	1689.3	1688.4
3.000		2194.9	1986.6	1952.4	1907.8	1891.9	1848.6	1843.5	1842.3
3.125			2187.2	2143.4	2086.4	2066.1	2011.6	2005.2	2003.8
3.250			2404.2	2348.8	2276.5	2250.8	2182.6	2174.6	2172.9
3.375			2639.5	2569.8	2479.1	2446.8	2361.8	2352.0	2349.9
3.500			2895.5	2808.1	2695.1	2654.9	2654.9	2537.7	2535.0

From *Orifice Metering of Natural Gas*, 1969; courtesy of AGA. (*table continued*)

472 Gas Production Engineering

Table 10-2 Continued
Flange Taps—Basic Orifice Factors—F_b

Orifice Diameter, in.	4		6			8		
	3.826	4.026	4.897	5.182	5.761	6.065	7.625	7.981	8.071
3.625			3180.8	3065.3	2925.7	2876.0	2746.5	2731.8	2728.6
3.750				3345.5	3172.1	3111.2	2952.6	2934.8	2930.8
3.875				3657.7	3435.7	3361.5	3168.3	3146.9	3142.1
4.000					3718.2	3628.2	3394.3	3368.5	3362.9
4.250					4354.8	4216.6	3879.4	3842.3	3834.2
4.500						4900.9	4412.8	4360.5	4349.0
4.750							5000.7	4928.1	4912.2
5.000							5650.0	5551.1	5529.5
5.250							6369.3	6236.4	6207.3
5.500							7170.9	6992.0	6953.6
5.750								7830.0	7777.8
6.000									8706.9

Orifice Diameter, in.	10			12			16		
	9.564	10.020	10.136	11.376	11.938	12.090	14.688	15.000	15.250
1.000	200.20								
1.125	253.55	253.48	253.47						
1.250	313.31	313.20	313.18	312.94	312.85	312.83			
1.375	379.44	379.29	379.26	378.94	378.82	378.79			
1.500	451.95	451.76	451.72	451.30	451.14	451.10	450.53	450.48	
1.625	530.87	530.63	530.57	530.04	529.83	529.78	529.06	528.99	528.94
1.750	616.21	615.90	615.83	615.16	614.90	614.84	613.94	613.85	613.78
1.875	707.99	707.61	707.51	706.68	706.36	706.28	705.18	705.07	704.99
2.000	806.23	805.76	805.65	804.61	804.23	804.13	802.78	802.65	802.55
2.125	910.97	910.38	910.24	908.98	908.51	908.39	906.77	906.61	906.49
2.250	1,022.2	1,021.5	1,021.3	1,019.8	1,019.2	1,019.1	1,017.1	1,017.0	1,016.8
2.375	1,140.1	1,139.2	1,139.0	1,137.1	1,136.4	1,136.2	1,133.9	1,133.7	1,133.5
2.500	1,264.5	1,263.4	1,263.1	1,260.8	1,260.0	1,259.8	1,256.8	1,256.7	1,256.6
2.625	1,395.6	1,394.2	1,393.9	1,391.1	1,390.1	1,389.9	1,386.6	1,386.4	1,386.1
2.750	1,533.4	1,531.7	1,531.3	1,528.0	1,526.8	1,526.5	1,522.7	1,522.4	1,522.1
2.875	1,678.0	1,675.9	1,675.4	1,671.4	1,670.0	1,669.6	1,665.2	1,664.8	1,664.5
3.000	1,829.4	1,826.9	1,826.3	1,821.4	1,819.7	1,819.3	1,814.1	1,813.7	1,813.3
3.125	1,987.8	1,984.7	1,984.0	1,978.1	1,976.1	1,975.6	1,969.6	1,969.0	1,966.6
3.250	2,153.2	2,149.5	2,148.6	2,141.5	2,139.2	2,138.6	2,131.5	2,130.9	2,130.4
3.375	2,325.7	2,321.2	2,320.2	2,311.7	2,308.9	2,308.2	2,299.9	2,299.2	2,293.7
3.500	2,505.3	2,500.1	2,498.9	2,488.7	2,485.4	2,484.6	2,474.9	2,474.1	2,473.5
3.625	2,692.8	2,686.2	2,684.7	2,672.6	2,668.7	2,667.7	2,656.4	2,655.5	2,654.8
3.750	2,887.6	2,879.7	2,877.9	2,863.5	2,858.8	2,857.7	2,844.6	2,843.5	2,842.7
3.875	3,090.1	3,080.7	3,078.5	3,061.4	3,055.9	3,054.6	3,039.4	3,038.1	3,037.2
4.000	3,300.6	3,289.3	3,286.8	3,266.4	3,260.0	3,258.5	3,240.8	3,239.4	3,238.3
4.250	3,746.1	3,730.2	3,726.7	3,698.4	3,689.6	3,687.5	3,663.8	3,661.9	3,660.5
4.500	4,226.0	4,204.1	4,199.2	4,160.4	4,148.4	4,145.5	4,113.9	4,111.5	4,109.7
4.750	4,742.7	4,712.8	4,706.2	4,653.4	4,637.2	4,633.4	4,591.5	4,508.4	4,586.0
5.000	5,298.6	5,258.5	5,249.6	5,179.0	5,157.4	5,152.3	5,097.2	5,093.1	5,090.1
5.250	5,897.4	5,843.6	5,831.8	5,738.5	5,710.0	5,703.3	5,631.4	5,626.1	5,622.3
5.500	6,543.1	6,471.9	6,456.3	6,333.8	6,296.6	6,287.9	6,194.8	6,180.1	6,183.3
5.750	7,240.0	7,146.9	7,126.5	6,966.9	6,919.0	6,907.8	6,788.1	6,779.5	6,773.3
6.000	7,993.3	7,873.0	7,846.6	7,640.4	7,579.0	7,564.7	7,412.3	7,401.5	7,393.6
6.250	8,808.9	8,654.8	8,621.1	8,357.3	8,278.9	8,260.7	8,060.4	8,054.8	8,044.8
6.500	9,693.3	9,498.1	9,455.3	9,121.0	9,021.7	8,998.7	8,757.3	8,740.3	8,727.9

Gas Flow Measurement 473

Table 10-2 Continued
Flange Taps—Basic Orifice Factors—F_b

Orifice Diameter, in.	10			12			16		
	9.564	10.020	10.136	11.376	11.938	12.090	14.688	15.000	15.250
6.750	10,654	10,409	10,355	9,935.2	9,810.5	9,781.6	9,480.4	9,459.4	9,444.0
7.000	11,711	11,394	11,327	10,804	10,649	10,613	10,239	10,213	10,194
7.250		12,467	12,381	11,732	11,540	11,496	11,035	11,003	10,980
7.500		13,656	13,541	12,725	12,489	12,434	11,869	11,831	11,803
7.750				13,787	13,500	13,433	12,745	12,698	12,664
8.000				14,927	14,578	14,498	13,664	13,607	13,566
8.250				16,156	15,730	15,633	14,628	14,560	14,511
8.500				17,505	16,962	16,845	15,642	15,560	15,501
8.750					18,296	18,148	16,706	16,609	16,539
9.000						19,565	17,826	17,711	17,628
9.250							19,004	18,868	18,770
9.500							20,245	20,085	19,969
9.750							21,552	21,365	21,230
10.000							22,930	22,712	22,555
10.250							24,385	24,132	23,948
10.500							25,924	25,628	25,416
10.750							27,567	27,210	26,962
11.000							29,331	28,899	28,600
11.250								30,710	30,348

Orifice Diameter inch.	20			24			30		
	18.814	19.000	19.250	22.626	23.000	23.250	28.628	29.000	29.250
2.000	801.40	801.35	801.29						
2.125	905.11	905.06	904.98						
2.250	1,015.2	1,015.1	1,015.0						
2.375	1,131.6	1,131.5	1,131.4	1,130.2	1,130.1	1,130.0			
2.500	1,254.4	1,254.3	1,254.2	1,252.8	1,252.6	1,252.6			
2.625	1,383.6	1,383.5	1,383.3	1,381.7	1,381.5	1,381.4			
2.750	1,519.1	1,519.0	1,518.8	1,517.0	1,516.8	1,516.7			
2.875	1,661.0	1,660.9	1,660.7	1,658.6	1,658.4	1,658.3	1,656.0		
3.000	1,809.4	1,809.2	1,809.0	1,806.6	1,806.4	1,806.2	1,803.7	1,803.5	1,803.4
3.125	1,964.1	1,963.9	1,963.7	1,961.0	1,960.7	1,960.6	1,957.7	1,957.5	1,957.4
3.250	2,125.3	2,125.1	2,124.8	2,121.7	2,121.5	2,121.3	2,118.0	2,117.9	2,117.7
3.375	2,292.9	2,292.6	2,292.3	2,280.9	2,288.6	2,288.4	2,284.5	2,284.5	2,284.4
3.500	2,466.9	2,466.6	2,466.3	2,462.4	2,462.1	2,461.8	2,457.8	2,457.6	2,457.5
3.625	2,647.3	2,647.0	2,646.6	2,642.4	2,642.0	2,641.7	2,637.3	2,637.0	2,636.8
3.750	2,834.2	2,833.9	2,833.5	2,828.7	2,828.3	2,828.0	2,823.1	2,822.8	2,822.6
3.875	3,027.5	3,027.3	3,026.8	3,021.5	3,021.0	3,020.7	3,015.2	3,014.9	3,014.7
4.000	3,227.5	3,227.1	3,226.5	3,220.6	3,220.1	3,219.8	3,213.8	3,213.5	3,213.2
4.250	3,646.7	3,646.2	3,645.6	3,638.3	3,637.7	3,637.2	3,630.1	3,629.7	3,629.4
4.500	4,092.1	4,091.5	4,090.6	4,081.8	4,081.0	4,080.5	4,071.9	4,071.4	4,071.1
4.750	4,563.7	4,562.9	4,561.9	4,551.4	4,550.1	4,549.5	4,539.4	4,538.8	4,538.4
5.000	5,061.8	5,060.8	5,050.6	5,046.4	5,045.2	5,044.5	5,032.5	5,031.8	5,031.4
5.250	5,586.6	5,585.4	5,583.8	5,567.7	5,566.4	5,565.5	5,551.3	5,550.5	5,550.0
5.500	6,138.2	6,136.7	6,134.8	6,115.3	6,113.6	6,112.6	6,095.8	6,094.9	6,094.4
5.750	6,717.0	6,715.2	6,712.3	6,689.1	6,687.2	6,685.9	6,666.2	6,665.2	6,664.5
6.000	7,323.4	7,321.1	7,318.2	7,289.4	7,287.1	7,285.6	7,262.5	7,261.3	7,260.5

(table continued)

474 Gas Production Engineering

Table 10-2 Continued
Flange Taps—Basic Orifice Factors—F_b

Orifice Diameter inch.	20			24			30		
	18.814	19.000	19.250	22.626	23.000	23.250	28.628	29.000	29.250
6.250	7,957.5	7,954.7	7,951.2	7,916.4	7,913.6	7,911.9	7,864.7	7,883.4	7,882.5
6.500	8,620.0	8,616.5	8,612.2	8,570.2	8,566.9	8,564.8	8,533.0	8,531.4	8,530.4
6.750	9,311.1	9,306.9	9,301.6	9,251.1	9,247.2	9,244.7	9,207.4	9,205.6	9,204.4
7.000	10,031	10,026	10,020	9,959.3	9,954.6	9,951.7	9,908.0	9,905.9	9,904.6
7.250	10,782	10,776	10,768	10,695	10,669	10,686	10,635	10,633	10,631
7.500	11,562	11,555	11,546	11,459	11,452	11,448	11,388	11,386	11,384
7.750	12,374	12,365	12,354	12,250	12,243	12,238	12,168	12,165	12,163
8.000	13,218	13,207	13,194	13,071	13,062	13,056	12,975	12,971	12,969
8.250	14,095	14,082	14,066	13,920	13,910	13,903	13,809	13,805	13,802
8.500	15,005	14,990	14,971	14,799	14,787	14,779	14,669	14,665	14,661
8.750	15,950	15,933	15,911	15,708	15,693	15,684	15,557	15,552	15,548
9.000	16,932	16,911	16,885	16,648	16,630	16,620	16,473	16,466	16,462
9.250	17,950	17,926	17,895	17,618	17,596	17,585	17,416	17,409	17,404
9.500	19,007	18,979	18,943	18,620	18,597	18,582	18,387	18,379	18,373
9.750	20,104	20,071	20,030	19,655	19,628	19,611	19,386	19,377	19,371
10.000	21,243	21,205	21,157	20,723	20,692	20,672	20,414	20,403	20,396
10.250	22,426	22,332	22,326	21,825	21,789	21,767	21,471	21,458	21,450
10.500	23,654	23,603	23,538	22,926	22,921	22,895	22,556	22,542	22,533
10.750	24,931	24,672	24,797	24,134	24,087	24,058	23,672	23,656	23,646
11.000	26,257	26,190	26,104	25,344	25,290	25,257	24,817	24,799	24,787
11.250	27,636	27,559	27,460	26,592	26,531	26,492	25,992	25,972	25,959
11.500	29,070	28,982	28,870	27,878	27,809	27,766	27,199	27,176	27,161
11.750	30,562	30,462	30,334	29,205	29,126	29,077	28,437	28,411	28,394
12.000	32,116	32,001	31,856	30,574	30,485	30,429	29,706	29,677	29,659
12.500	35,417	35,270	35,084	33,444	33,330	33,259	32,343	32,306	32,283
13.000	39,003	38,817	38,581	36,502	36,357	36,267	35,114	35,068	35,039
13.500	42,913	42,673	42,375	39,762	39,581	39,467	38,025	37,968	37,932
14.000	47,244	46,921	46,523	43,241	43,015	42,874	41,082	41,012	40,968
14.500				46,958	46,679	46,505	44,291	44,206	44,151
15.000				50,934	50,591	50,378	47,622	47,557	47,490
15.500				55,192	54,774	54,513	51,202	51,075	50,993
16.000				59,759	59,251	58,935	54,923	54,769	54,671
16.500				64,701	64,060	63,670	58,835	58,649	58,531
17.000					69,288	68,792	62,950	62,728	62,586
17.500							67,282	67,017	66,848
18.000							71,844	71,530	71,330
18.500							76,653	76,282	76,046
19.000							81,725	81,289	81,012
19.500							87,079	86,568	86,244
20.000							92,734	92,140	91,761
20.500							98,728	98,025	97,564
21.000							105,130	104,280	103,750
21.500								110,980	110,340

(text continued from page 470)

not recommended; the exact equations or charts given by AGA (AGA, 1969) must be used.

Reynold's Number Factor, F_r

This factor accounts for the variation of the orifice discharge coefficient with Reynold's number. Tables 10-3 and 10-8 show the value of F_r for flange taps and pipe taps, respectively. The pressure extension, $[h_w p_f]^{0.5}$, should be some sort of average pressure extension. These F_r values have been calculated assuming average values of viscosity equal to 6.9×10^{-6} lbm/ft-sec, temperature equal to 60°F, and gas gravity equal to 0.65. These tables do not, therefore, have a general applicability to all systems. In any event, the variation in F_r in gas measurement is quite small, and is often neglected.

(text continued on page 479)

Table 10-3
"b" Values for Reynolds Number Factor F_r Determination—Flange Taps

$$F_r = 1 + \frac{b}{\sqrt{h_w p_f}}$$

Pipe Sizes—Nominal and Published Inside Diameters, Inches

Orifice Diameter, In.	2			3			4		
	1.689	1.939	2.067	2.300	2.626	2.900	3.068	3.152	3.438
0.250	0.0879	0.0911	0.0926	0.0950	0.0979	0.0999	0.1010	0.1014	0.1030
0.375	0.0677	0.0709	0.0726	0.0755	0.0792	0.0820	0.0836	0.0844	0.0867
0.500	0.0562	0.0576	0.0588	0.0612	0.0648	0.0677	0.0695	0.0703	0.0728
0.625	0.0520	0.0505	0.0506	0.0516	0.0541	0.0566	0.0583	0.0591	0.0618
0.750	0.0536	0.0485	0.0471	0.0462	0.0470	0.0486	0.0498	0.0504	0.0528
0.875	0.0595	0.0506	0.0478	0.0445	0.0429	0.0433	0.0438	0.0442	0.0460
1.000	0.0677	0.0559	0.0515	0.0458	0.0416	0.0403	0.0402	0.0403	0.0411
1.125	0.0762	0.0630	0.0574	0.0495	0.0427	0.0396	0.0386	0.0383	0.0380
1.250	0.0824	0.0707	0.0646	0.0550	0.0456	0.0408	0.0388	0.0381	0.0365
1.375		0.0772	0.0715	0.0614	0.0501	0.0435	0.0406	0.0394	0.0365
1.500			0.0773	0.0679	0.0554	0.0474	0.0436	0.0420	0.0378
1.625				0.0735	0.0613	0.0522	0.0477	0.0457	0.0402
1.750					0.0669	0.0575	0.0524	0.0500	0.0434
1.875					0.0717	0.0628	0.0574	0.0549	0.0473
2.000						0.0676	0.0624	0.0598	0.0517
2.125						0.0715	0.0669	0.0642	0.0563
2.250							0.0706	0.0685	0.0607
2.375									0.0648
2.500									0.0683

From *Orifice Metering of Natural Gas*, 1969; courtesy of AGA. *(table continued)*

Gas Production Engineering

Table 10-3 Continued
"b" Values for Reynolds Number Factor F_r Determination—Flange Taps

Orifice Diameter in.	4			6			8		
	3.826	4.026	4.897	5.189	5.761	6.065	7.625	7.981	8.071
0.250	0.1047	0.1054							
0.375	0.0894	0.0907							
0.500	0.0763	0.0779	0.0836	0.0852	0.0880	0.0892			
0.625	0.0653	0.0670	0.0734	0.0753	0.0785	0.0801			
0.750	0.0561	0.0578	0.0645	0.0665	0.0701	0.0718			
0.875	0.0487	0.0502	0.0567	0.0587	0.0625	0.0643	0.0723	0.0738	0.0742
1.000	0.0430	0.0442	0.0500	0.0520	0.0557	0.0576	0.0660	0.0676	0.0680
1.125	0.0388	0.0396	0.0444	0.0462	0.0498	0.0517	0.0602	0.0619	0.0623
1.250	0.0361	0.0364	0.0399	0.0414	0.0447	0.0464	0.0549	0.0566	0.0571
1.375	0.0347	0.0344	0.0363	0.0375	0.0403	0.0419	0.0501	0.0518	0.0523
1.500	0.0345	0.0336	0.0336	0.0344	0.0367	0.0381	0.0457	0.0474	0.0479
1.625	0.0354	0.0338	0.0318	0.0322	0.0337	0.0348	0.0418	0.0435	0.0439
1.750	0.0372	0.0350	0.0307	0.0306	0.0314	0.0322	0.0383	0.0399	0.0403
1.875	0.0398	0.0370	0.0305	0.0298	0.0298	0.0303	0.0353	0.0366	0.0371
2.000	0.0430	0.0395	0.0308	0.0296	0.0287	0.0288	0.0327	0.0340	0.0343
2.125	0.0467	0.0427	0.0318	0.0300	0.0281	0.0278	0.0304	0.0315	0.0318
2.250	0.0507	0.0462	0.0334	0.0310	0.0281	0.0274	0.0286	0.0295	0.0297
2.375	0.0548	0.0501	0.0354	0.0324	0.0286	0.0274	0.0271	0.0278	0.0280
2.500	0.0589	0.0540	0.0378	0.0342	0.0295	0.0279	0.0259	0.0264	0.0265
2.625	0.0626	0.0579	0.0406	0.0365	0.0308	0.0287	0.0251	0.0253	0.0254
2.750	0.0659	0.0615	0.0436	0.0391	0.0324	0.0300	0.0246	0.0245	0.0245
2.875		0.0647	0.0468	0.0418	0.0343	0.0314	0.0244	0.0240	0.0240
3.000		0.0673	0.0500	0.0448	0.0366	0.0332	0.0245	0.0238	0.0237
3.125			0.0533	0.0479	0.0389	0.0353	0.0248	0.0239	0.0237
3.250			0.0564	0.0510	0.0416	0.0375	0.0254	0.0242	0.0240
3.375			0.0594	0.0541	0.0443	0.0400	0.0263	0.0248	0.0244
3.500			0.0620	0.0569	0.0472	0.0426	0.0273	0.0255	0.0251
3.625			0.0643	0.0597	0.0500	0.0452	0.0286	0.0265	0.0260
3.750				0.0621	0.0527	0.0479	0.0300	0.0274	0.0271
3.875				0.0640	0.0553	0.0505	0.0316	0.0289	0.0283
4.000					0.0578	0.0531	0.0334	0.0304	0.0297
4.250					0.0620	0.0579	0.0372	0.0338	0.0330
4.500						0.0618	0.0414	0.0386	0.0366
4.750							0.0457	0.0416	0.0405
5.000							0.0500	0.0457	0.0446
5.250							0.0539	0.0497	0.0487
5.500							0.0574	0.0535	0.0524
5.750								0.0569	0.0559
6.000									0.0588

Orifice Diameter, in.	10			12			16		
	9.564	10.020	10.136	11.376	11.938	12.090	14.688	15.000	15.250
1.000	0.0738								
1.125	0.0685	0.0701	0.0705						
1.250	0.0635	0.0652	0.0656	0.0698	0.0714	0.0718			
1.375	0.0588	0.0606	0.0610	0.0654	0.0671	0.0676			
1.500	0.0545	0.0563	0.0568	0.0612	0.0631	0.0635	0.0706	0.0713	
1.625	0.0504	0.0523	0.0527	0.0573	0.0592	0.0597	0.0670	0.0678	0.0684
1.750	0.0467	0.0485	0.0490	0.0536	0.0555	0.0560	0.0636	0.0644	0.0650

Gas Flow Measurement 477

Table 10-3 Continued
"b" Values for Reynolds Number Factor F_r Determination—Flange Taps

Orifice Diameter, in.	10			12			16		
	9.564	10.020	10.136	11.376	11.938	12.090	14.688	15.000	15.250
1.875	0.0433	0.0451	0.0455	0.0501	0.0521	0.0526	0.0604	0.0612	0.0618
2.000	0.0401	0.0419	0.0414	0.0469	0.0488	0.0492	0.0572	0.0581	0.0587
2.125	0.0372	0.0389	0.0383	0.0438	0.0458	0.0463	0.0542	0.0551	0.0558
2.250	0.0346	0.0362	0.0356	0.0410	0.0429	0.0434	0.0514	0.0523	0.0529
2.375	0.0322	0.0337	0.0330	0.0383	0.0402	0.0407	0.0467	0.0496	0.0502
2.500	0.0302	0.0315	0.0308	0.0359	0.0377	0.0382	0.0461	0.0470	0.0476
2.625	0.0283	0.0296	0.0287	0.0336	0.0354	0.0358	0.0436	0.0445	0.0452
2.750	0.0267	0.0278	0.0269	0.0316	0.0332	0.0336	0.0413	0.0422	0.0428
2.875	0.0254	0.0263	0.0253	0.0297	0.0312	0.0317	0.0391	0.0399	0.0406
3.000	0.0243	0.0250	0.0252	0.0278	0.0294	0.0298	0.0370	0.0378	0.0385
3.125	0.0234	0.0239	0.0241	0.0264	0.0278	0.0282	0.0350	0.0358	0.0365
3.250	0.0226	0.0230	0.0231	0.0251	0.0263	0.0266	0.0331	0.0339	0.0346
3.375	0.0221	0.0223	0.0224	0.0239	0.0250	0.0253	0.0314	0.0321	0.0328
3.500	0.0219	0.0218	0.0218	0.0229	0.0238	0.0241	0.0298	0.0305	0.0311
3.625	0.0218	0.0214	0.0214	0.0221	0.0226	0.0230	0.0282	0.0290	0.0295
3.750	0.0218	0.0213	0.0212	0.0214	0.0219	0.0221	0.0268	0.0275	0.0281
3.875	0.0221	0.0213	0.0211	0.0208	0.0212	0.0213	0.0255	0.0262	0.0267
4.000	0.0225	0.0214	0.0212	0.0204	0.0206	0.0207	0.0243	0.0249	0.0254
4.250	0.0238	0.0222	0.0219	0.0200	0.0198	0.0198	0.0223	0.0228	0.0232
4.500	0.0256	0.0236	0.0231	0.0201	0.0195	0.0194	0.0206	0.0210	0.0213
4.750	0.0279	0.0254	0.0249	0.0207	0.0196	0.0194	0.0193	0.0196	0.0198
5.000	0.0307	0.0277	0.0270	0.0217	0.0202	0.0199	0.0184	0.0185	0.0187
5.250	0.0337	0.0303	0.0295	0.0231	0.0212	0.0208	0.0178	0.0178	0.0179
5.500	0.0370	0.0332	0.0323	0.0249	0.0226	0.0221	0.0176	0.0174	0.0174
5.750	0.0404	0.0363	0.0354	0.0270	0.0243	0.0237	0.0176	0.0174	0.0172
6.000	0.0438	0.0396	0.0386	0.0294	0.0263	0.0255	0.0180	0.0176	0.0173
6.250	0.0473	0.0437	0.0418	0.0320	0.0285	0.0277	0.0186	0.0160	0.0177
6.500	0.0505	0.0462	0.0451	0.0347	0.0309	0.0300	0.0195	0.0188	0.0183
6.750	0.0536	0.0493	0.0483	0.0376	0.0335	0.0325	0.0206	0.0198	0.0192
7.000	0.0562	0.0523	0.0513	0.0406	0.0362	0.0351	0.0220	0.0210	0.0202
7.250		0.0550	0.0540	0.0435	0.0390	0.0379	0.0235	0.0224	0.0216
7.500		0.0572	0.0564	0.0463	0.0418	0.0407	0.0252	0.0240	0.0230
7.750				0.0491	0.0446	0.0434	0.0271	0.0257	0.0246
8.000				0.0517	0.0473	0.0461	0.0291	0.0276	0.0264
8.250				0.0540	0.0498	0.0487	0.0312	0.0296	0.0283
8.500				0.0560	0.0522	0.0511	0.0334	0.0317	0.0303
8.750					0.0543	0.0534	0.0357	0.0338	0.0324
9.000					0.0553	0.0380	0.0361	0.0346	
9.250							0.0402	0.0383	0.0368
9.500							0.0425	0.0406	0.0390
9.750							0.0447	0.0427	0.0412
10.000							0.0469	0.0449	0.0434
10.250							0.0489	0.0470	0.0455
10.500							0.0508	0.0490	0.0475
10.750							0.0526	0.0509	0.0495
11.000							0.0541	0.0526	0.0513
11.250								0.0541	0.0528

(table continued)

Table 10-3 Continued
"b" Values for Reynolds Number Factor F_r Determination—Flange Taps

Orifice Diameter, in.	20			24			30		
	18.814	19.000	19.250	22.626	23.000	23.250	28.628	29.000	29.250
2.000	0.0667	0.0671	0.0676						
2.125	0.0640	0.0644	0.0649						
2.250	0.0614	0.0618	0.0622						
2.375	0.0588	0.0592	0.0597	0.0659	0.0665	0.0669			
2.500	0.0563	0.0568	0.0573	0.0636	0.0642	0.0646			
2.625	0.0540	0.0544	0.0549	0.0614	0.0620	0.0624			
2.750	0.0517	0.0521	0.0526	0.0592	0.0599	0.0603			
2.875	0.0494	0.0499	0.0504	0.0571	0.0578	0.0582	0.0662		
3.000	0.0473	0.0477	0.0483	0.0551	0.0557	0.0562	0.0644	0.0649	0.0652
3.125	0.0452	0.0457	0.0462	0.0531	0.0538	0.0542	0.0626	0.0631	0.0634
3.250	0.0433	0.0437	0.0442	0.0511	0.0520	0.0523	0.0608	0.0613	0.0616
3.375	0.0414	0.0418	0.0423	0.0493	0.0500	0.0504	0.0590	0.0596	0.0599
3.500	0.0395	0.0399	0.0405	0.0474	0.0481	0.0486	0.0574	0.0579	0.0582
3.625	0.0378	0.0382	0.0387	0.0457	0.0464	0.0468	0.0557	0.0562	0.0566
3.750	0.0361	0.0365	0.0370	0.0440	0.0447	0.0451	0.0541	0.0546	0.0550
3.875	0.0345	0.0349	0.0354	0.0423	0.0430	0.0435	0.0525	0.0530	0.0534
4.000	0.0329	0.0333	0.0339	0.0407	0.0414	0.0419	0.0509	0.0515	0.0518
4.250	0.0301	0.0304	0.0310	0.0376	0.0384	0.0388	0.0479	0.0485	0.0488
4.500	0.0275	0.0279	0.0283	0.0348	0.0355	0.0360	0.0450	0.0456	0.0460
4.750	0.0252	0.0256	0.0260	0.0322	0.0328	0.0333	0.0423	0.0429	0.0433
5.000	0.0232	0.0235	0.0239	0.0297	0.0304	0.0308	0.0397	0.0403	0.0407
5.250	0.0214	0.0217	0.0220	0.0275	0.0281	0.0285	0.0373	0.0378	0.0382
5.500	0.0199	0.0201	0.0204	0.0254	0.0260	0.0264	0.0349	0.0355	0.0359
5.750	0.0186	0.0188	0.0191	0.0236	0.0241	0.0245	0.0327	0.0333	0.0337
6.000	0.0176	0.0177	0.0179	0.0219	0.0224	0.0228	0.0306	0.0312	0.0316
6.250	0.0167	0.0168	0.0170	0.0204	0.0208	0.0212	0.0287	0.0292	0.0296
6.500	0.0161	0.0162	0.0163	0.0191	0.0195	0.0198	0.0269	0.0274	0.0277
6.750	0.0157	0.0157	0.0157	0.0179	0.0183	0.0185	0.0252	0.0257	0.0260
7.000	0.0155	0.0155	0.0154	0.0169	0.0172	0.0174	0.0236	0.0240	0.0244
7.250	0.0155	0.0154	0.0153	0.0161	0.0163	0.0165	0.0221	0.0226	0.0229
7.500	0.0157	0.0155	0.0154	0.0154	0.0156	0.0157	0.0208	0.0212	0.0215
7.750	0.0160	0.0158	0.0156	0.0148	0.0150	0.0151	0.0195	0.0199	0.0202
8.000	0.0166	0.0163	0.0160	0.0144	0.0145	0.0146	0.0184	0.0187	0.0190
8.250	0.0172	0.0169	0.0165	0.0142	0.0142	0.0142	0.0174	0.0177	0.0179
8.500	0.0180	0.0177	0.0172	0.0141	0.0140	0.0140	0.0164	0.0168	0.0170
8.750	0.0190	0.0186	0.0180	0.0141	0.0140	0.0139	0.0156	0.0159	0.0161
9.000	0.0201	0.0196	0.0190	0.0143	0.0141	0.0140	0.0149	0.0152	0.0153
9.250	0.0213	0.0208	0.0201	0.0146	0.0143	0.0141	0.0143	0.0145	0.0146
9.500	0.0226	0.0220	0.0213	0.0150	0.0146	0.0144	0.0138	0.0139	0.0141
9.750	0.0240	0.0234	0.0226	0.0155	0.0150	0.0147	0.0133	0.0135	0.0136
10.000	0.0256	0.0249	0.0240	0.0161	0.0155	0.0152	0.0130	0.0131	0.0132
10.250	0.0271	0.0264	0.0255	0.0168	0.0162	0.0158	0.0128	0.0128	0.0128
10.500	0.0288	0.0280	0.0270	0.0176	0.0169	0.0164	0.0126	0.0126	0.0126
10.750	0.0305	0.0297	0.0286	0.0185	0.0176	0.0172	0.0125	0.0125	0.0125
11.000	0.0322	0.0314	0.0303	0.0194	0.0186	0.0181	0.0125	0.0124	0.0124
11.250	0.0340	0.0332	0.0320	0.0205	0.0196	0.0190	0.0126	0.0125	0.0124
11.500	0.0358	0.0349	0.0338	0.0216	0.0207	0.0200	0.0128	0.0126	0.0125

Gas Flow Measurement

Table 10-3 Continued
"b" Values for Reynolds Number Factor F_r Determination—Flange Taps

Orifice Diameter, in.	20			24			30		
	18.814	19.000	19.250	22.626	23.000	23.250	28.628	29.000	29.250
11.750	0.0376	0.0367	0.0355	0.0228	0.0218	0.0211	0.0130	0.0128	0.0127
12.000	0.0394	0.0385	0.0373	0.0241	0.0230	0.0223	0.0134	0.0131	0.0129
12.500	0.0429	0.0420	0.0408	0.0267	0.0255	0.0248	0.0142	0.0138	0.0136
13.000	0.0463	0.0454	0.0442	0.0296	0.0282	0.0274	0.0153	0.0148	0.0145
13.500	0.0494	0.0485	0.0474	0.0326	0.0311	0.0302	0.0166	0.0160	0.0157
14.000	0.0520	0.0512	0.0502	0.0356	0.0341	0.0331	0.0182	0.0175	0.0171
14.500				0.0386	0.0370	0.0360	0.0199	0.0192	0.0187
15.000				0.0415	0.0400	0.0390	0.0218	0.0209	0.0204
15.500				0.0443	0.0426	0.0418	0.0239	0.0230	0.0224
16.000				0.0470	0.0455	0.0446	0.0260	0.0250	0.0244
16.500				0.0494	0.0480	0.0471	0.0283	0.0273	0.0266
17.000					0.0503	0.0494	0.0307	0.0296	0.0288
17.500							0.0331	0.0319	0.0312
18.000							0.0355	0.0343	0.0335
18.500							0.0379	0.0366	0.0358
19.000							0.0402	0.0390	0.0382
19.500							0.0424	0.0412	0.0404
20.000							0.0446	0.0434	0.0426
20.500							0.0466	0.0455	0.0448
21.000							0.0485	0.0475	0.0467
21.500								0.0492	0.0485

(text continued from page 475)

Expansion Factor, Y

The expansion factor, Y, accounts for the change in gas density with the pressure changes across the orifice. It is computed assuming a reversible adiabatic expansion of the gas through the orifice, and is a function of β, the ratio of the differential pressure to the absolute pressure, the type of taps, and the specific heat ratio (c_p/c_v) for the gas (generally ignored). The expansion factor can be obtained from Tables 10-4, 10-5, and 10-6 for flange taps, and Tables 10-9 and 10-10 for pipe taps. These tables indicate the pressure tap from which the absolute static pressure p_f is measured—Y_1 for upstream, Y_2 for downstream, and Y_m for static pressure recorded as the mean of the upstream and downstream static pressures. As with F_r, an average h_w/p_f should be used.

(text continued on page 489)

Gas Production Engineering

Table 10-4
Y_1 Expansion Factors—Flange Taps
Static Pressure Taken from Upstream Taps

$\dfrac{h_w}{p_n}$ Ratio $\beta = \dfrac{d}{D}$ Ratio

Ratio	0.1	0.2	0.3	0.4	0.45	0.50	0.52	0.54	0.56	0.58	0.60	0.61	0.62
0.0	1.0000	1.0000	1.0000	1.0000	1.0000	1.0000	1.0000	1.0000	1.0000	1.0000	1.0000	1.0000	1.0000
0.1	0.9989	0.9989	0.9989	0.9988	0.9988	0.9988	0.9988	0.9988	0.9988	0.9988	0.9987	0.9987	0.9987
0.2	0.9977	0.9977	0.9977	0.9977	0.9976	0.9976	0.9976	0.9976	0.9975	0.9975	0.9975	0.9975	0.9974
0.3	0.9966	0.9966	0.9966	0.9965	0.9965	0.9964	0.9964	0.9963	0.9963	0.9963	0.9962	0.9962	0.9962
0.4	0.9954	0.9954	0.9954	0.9953	0.9953	0.9952	0.9952	0.9951	0.9951	0.9950	0.9949	0.9949	0.9949
0.5	0.9943	0.9943	0.9943	0.9942	0.9941	0.9940	0.9940	0.9939	0.9938	0.9938	0.9937	0.9936	0.9936
0.6	0.9932	0.9932	0.9931	0.9930	0.9929	0.9928	0.9927	0.9927	0.9926	0.9925	0.9924	0.9924	0.9923
0.7	0.9920	0.9920	0.9920	0.9919	0.9918	0.9916	0.9915	0.9915	0.9914	0.9913	0.9912	0.9911	0.9910
0.8	0.9909	0.9909	0.9908	0.9907	0.9906	0.9904	0.9903	0.9902	0.9901	0.9900	0.9899	0.9898	0.9897
0.9	0.9898	0.9897	0.9897	0.9895	0.9894	0.9892	0.9891	0.9890	0.9889	0.9888	0.9886	0.9885	0.9885
1.0	0.9886	0.9886	0.9885	0.9884	0.9882	0.9880	0.9879	0.9878	0.9877	0.9875	0.9874	0.9873	0.9872
1.1	0.9875	0.9875	0.9874	0.9872	0.9870	0.9868	0.9867	0.9866	0.9864	0.9863	0.9861	0.9860	0.9859
1.2	0.9863	0.9863	0.9862	0.9860	0.9859	0.9856	0.9855	0.9853	0.9852	0.9850	0.9848	0.9847	0.9846
1.3	0.9852	0.9852	0.9851	0.9849	0.9847	0.9844	0.9843	0.9841	0.9840	0.9838	0.9836	0.9835	0.9833
1.4	0.9841	0.9840	0.9840	0.9837	0.9835	0.9832	0.9831	0.9829	0.9827	0.9825	0.9823	0.9822	0.9821
1.5	0.9829	0.9829	0.9828	0.9826	0.9823	0.9820	0.9819	0.9817	0.9815	0.9813	0.9810	0.9809	0.9808
1.6	0.9818	0.9818	0.9817	0.9814	0.9811	0.9808	0.9806	0.9805	0.9803	0.9800	0.9798	0.9796	0.9795
1.7	0.9806	0.9806	0.9805	0.9802	0.9800	0.9796	0.9794	0.9792	0.9790	0.9788	0.9785	0.9784	0.9782
1.8	0.9795	0.9795	0.9794	0.9791	0.9788	0.9784	0.9782	0.9780	0.9778	0.9775	0.9772	0.9771	0.9769
1.9	0.9784	0.9783	0.9782	0.9779	0.9776	0.9772	0.9770	0.9768	0.9766	0.9763	0.9760	0.9758	0.9756
2.0	0.9772	0.9772	0.9771	0.9767	0.9764	0.9760	0.9758	0.9756	0.9753	0.9750	0.9747	0.9745	0.9744
2.1	0.9761	0.9761	0.9759	0.9756	0.9753	0.9748	0.9746	0.9744	0.9741	0.9738	0.9734	0.9733	0.9731
2.2	0.9750	0.9749	0.9748	0.9744	0.9741	0.9736	0.9734	0.9731	0.9729	0.9725	0.9722	0.9720	0.9718
2.3	0.9738	0.9738	0.9736	0.9732	0.9729	0.9724	0.9722	0.9719	0.9716	0.9713	0.9709	0.9707	0.9705
2.4	0.9727	0.9726	0.9725	0.9721	0.9717	0.9712	0.9710	0.9707	0.9704	0.9700	0.9697	0.9694	0.9692
2.5	0.9715	0.9715	0.9713	0.9709	0.9705	0.9700	0.9698	0.9695	0.9692	0.9688	0.9684	0.9682	0.9680
2.6	0.9704	0.9704	0.9702	0.9698	0.9694	0.9688	0.9686	0.9683	0.9679	0.9675	0.9671	0.9669	0.9667
2.7	0.9693	0.9692	0.9691	0.9686	0.9682	0.9676	0.9673	0.9670	0.9667	0.9663	0.9659	0.9656	0.9654
2.8	0.9681	0.9681	0.9679	0.9674	0.9670	0.9664	0.9661	0.9658	0.9654	0.9650	0.9646	0.9644	0.9641
2.9	0.9670	0.9669	0.9668	0.9663	0.9658	0.9652	0.9649	0.9646	0.9642	0.9638	0.9633	0.9631	0.9628
3.0	0.9658	0.9658	0.9656	0.9651	0.9647	0.9640	0.9637	0.9634	0.9630	0.9626	0.9621	0.9618	0.9615
3.1	0.9647	0.9647	0.9645	0.9639	0.9635	0.9628	0.9625	0.9622	0.9617	0.9613	0.9608	0.9605	0.9603
3.2	0.9636	0.9635	0.9633	0.9628	0.9623	0.9616	0.9613	0.9609	0.9605	0.9601	0.9595	0.9593	0.9590
3.3	0.9624	0.9624	0.9622	0.9616	0.9611	0.9604	0.9601	0.9597	0.9593	0.9588	0.9583	0.9580	0.9577
3.4	0.9613	0.9612	0.9610	0.9604	0.9599	0.9592	0.9589	0.9585	0.9580	0.9576	0.9570	0.9567	0.9564
3.5	0.9602	0.9601	0.9599	0.9593	0.9588	0.9580	0.9577	0.9573	0.9568	0.9563	0.9558	0.9554	0.9551
3.6	0.9590	0.9590	0.9587	0.9581	0.9576	0.9568	0.9565	0.9560	0.9556	0.9551	0.9545	0.9542	0.9538
3.7	0.9579	0.9578	0.9576	0.9570	0.9561	0.9556	0.9553	0.9548	0.9543	0.9538	0.9532	0.9529	0.9526
3.8	0.9567	0.9567	0.9564	0.9558	0.9552	0.9544	0.9540	0.9536	0.9531	0.9526	0.9520	0.9516	0.9513
3.9	0.9556	0.9555	0.9553	0.9546	0.9540	0.9532	0.9528	0.9524	0.9519	0.9513	0.9507	0.9504	0.9500
4.0	0.9545	0.9544	0.9542	0.9535	0.9529	0.9520	0.9516	0.9512	0.9506	0.9501	0.9494	0.9491	0.9487

From *Orifice Metering of Natural Gas*, 1969; courtesy of AGA.

Gas Flow Measurement 481

Table 10-4 Continued
Y₁ Expansion Factors—Flange Taps
Static Pressure Taken from Upstream Taps

$\frac{h_w}{p_{f1}}$ Ratio					$\beta = \frac{d}{D}$ Ratio						
	0.63	0.64	0.65	0.66	0.67	0.68	0.69	0.70	0.71	0.72	0.73	0.74	0.75
0.0	1.0000	1.0000	1.0000	1.0000	1.0000	1.0000	1.0000	1.0000	1.0000	1.0000	1.0000	1.0000	1.0000
0.1	0.9987	0.9987	0.9987	0.9987	0.9987	0.9987	0.9986	0.9986	0.9986	0.9986	0.9986	0.9986	0.9986
0.2	0.9974	0.9974	0.9974	0.9974	0.9973	0.9973	0.9973	0.9973	0.9972	0.9972	0.9972	0.9971	0.9971
0.3	0.9961	0.9961	0.9961	0.9960	0.9960	0.9960	0.9959	0.9959	0.9958	0.9958	0.9958	0.9957	0.9957
0.4	0.9948	0.9948	0.9948	0.9947	0.9947	0.9946	0.9946	0.9945	0.9945	0.9944	0.9943	0.9943	0.9942
0.5	0.9935	0.9935	0.9934	0.9934	0.9933	0.9933	0.9932	0.9931	0.9931	0.9930	0.9929	0.9929	0.9928
0.6	0.9923	0.9922	0.9921	0.9921	0.9920	0.9919	0.9918	0.9918	0.9917	0.9916	0.9915	0.9914	0.9913
0.7	0.9910	0.9909	0.9908	0.9907	0.9907	0.9906	0.9905	0.9904	0.9903	0.9902	0.9901	0.9900	0.9899
0.8	0.9897	0.9896	0.9895	0.9894	0.9893	0.9892	0.9891	0.9890	0.9889	0.9888	0.9887	0.9886	0.9884
0.9	0.9884	0.9883	0.9882	0.9881	0.9880	0.9879	0.9878	0.9877	0.9875	0.9874	0.9873	0.9871	0.9870
1.0	0.9871	0.9870	0.9869	0.9868	0.9867	0.9865	0.9864	0.9863	0.9861	0.9860	0.9859	0.9857	0.9855
1.1	0.9858	0.9857	0.9856	0.9854	0.9853	0.9852	0.9851	0.9849	0.9848	0.9846	0.9844	0.9843	0.9841
1.2	0.9845	0.9844	0.9843	0.9841	0.9840	0.9838	0.9837	0.9835	0.9834	0.9832	0.9830	0.9828	0.9826
1.3	0.9832	0.9831	0.9829	0.9828	0.9827	0.9825	0.9823	0.9822	0.9820	0.9818	0.9816	0.9814	0.9812
1.4	0.9819	0.9818	0.9816	0.9815	0.9813	0.9812	0.9810	0.9808	0.9806	0.9804	0.9802	0.9800	0.9798
1.5	0.9806	0.9805	0.9803	0.9802	0.9800	0.9798	0.9796	0.9794	0.9792	0.9790	0.9788	0.9786	0.9783
1.6	0.9793	0.9792	0.9790	0.9788	0.9787	0.9785	0.9783	0.9781	0.9778	0.9776	0.9774	0.9771	0.9769
1.7	0.9780	0.9779	0.9777	0.9775	0.9773	0.9771	0.9769	0.9767	0.9764	0.9762	0.9760	0.9757	0.9754
1.8	0.9768	0.9766	0.9764	0.9762	0.9760	0.9758	0.9755	0.9753	0.9751	0.9748	0.9745	0.9743	0.9740
1.9	0.9755	0.9753	0.9751	0.9749	0.9747	0.9744	0.9742	0.9739	0.9737	0.9734	0.9731	0.9728	0.9725
2.0	0.9742	0.9740	0.9738	0.9735	0.9733	0.9731	0.9728	0.9726	0.9723	0.9720	0.9717	0.9714	0.9711
2.1	0.9729	0.9727	0.9725	0.9722	0.9720	0.9717	0.9715	0.9712	0.9709	0.9706	0.9703	0.9700	0.9696
2.2	0.9716	0.9714	0.9711	0.9709	0.9706	0.9704	0.9701	0.9698	0.9695	0.9692	0.9689	0.9685	0.9682
2.3	0.9703	0.9701	0.9698	0.9696	0.9693	0.9690	0.9688	0.9685	0.9681	0.9678	0.9675	0.9671	0.9667
2.4	0.9690	0.9688	0.9685	0.9683	0.9680	0.9677	0.9674	0.9671	0.9668	0.9664	0.9661	0.9657	0.9653
2.5	0.9677	0.9675	0.9672	0.9669	0.9666	0.9663	0.9660	0.9657	0.9654	0.9650	0.9646	0.9643	0.9639
2.6	0.9664	0.9662	0.9659	0.9656	0.9653	0.9650	0.9647	0.9643	0.9640	0.9636	0.9632	0.9628	0.9624
2.7	0.9651	0.9649	0.9646	0.9643	0.9640	0.9637	0.9633	0.9630	0.9626	0.9622	0.9618	0.9614	0.9610
2.8	0.9638	0.9636	0.9633	0.9630	0.9626	0.9623	0.9620	0.9616	0.9612	0.9608	0.9604	0.9600	0.9595
2.9	0.9625	0.9623	0.9620	0.9616	0.9613	0.9610	0.9606	0.9602	0.9598	0.9594	0.9590	0.9585	0.9581
3.0	0.9613	0.9610	0.9606	0.9603	0.9600	0.9596	0.9592	0.9588	0.9584	0.9580	0.9576	0.9571	0.9566
3.1	0.9600	0.9597	0.9593	0.9590	0.9586	0.9583	0.9579	0.9575	0.9571	0.9566	0.9562	0.9557	0.9552
3.2	0.9587	0.9584	0.9580	0.9577	0.9573	0.9569	0.9565	0.9561	0.9557	0.9552	0.9547	0.9542	0.9537
3.3	0.9574	0.9571	0.9567	0.9564	0.9560	0.9556	0.9552	0.9547	0.9543	0.9538	0.9533	0.9528	0.9523
3.4	0.9561	0.9558	0.9554	0.9550	0.9546	0.9542	0.9538	0.9534	0.9529	0.9524	0.9519	0.9514	0.9508
3.5	0.9548	0.9545	0.9541	0.9537	0.9533	0.9529	0.9524	0.9520	0.9515	0.9510	0.9505	0.9500	0.9494
3.6	0.9535	0.9532	0.9528	0.9524	0.9520	0.9515	0.9511	0.9506	0.9501	0.9496	0.9491	0.9485	0.9480
3.7	0.9522	0.9518	0.9515	0.9511	0.9506	0.9502	0.9497	0.9492	0.9487	0.9482	0.9477	0.9471	0.9465
3.8	0.9509	0.9505	0.9502	0.9497	0.9493	0.9488	0.9484	0.9479	0.9474	0.9468	0.9463	0.9457	0.9451
3.9	0.9496	0.9492	0.9488	0.9484	0.9480	0.9475	0.9470	0.9465	0.9460	0.9454	0.9448	0.9442	0.9436
4.0	0.9483	0.9479	0.9475	0.9471	0.9465	0.9462	0.9457	0.9451	0.9446	0.9440	0.9434	0.9428	0.9422

482 Gas Production Engineering

Table 10-5
Y_2 Expansion Factors—Flange Taps
Static Pressure Taken from Downstream Taps

$\frac{h_w}{p_{f_2}}$ Ratio	\multicolumn{11}{c}{$\beta = \frac{d}{D}$ Ratio}
	0.1	0.2	0.3	0.4	0.45	0.50	0.52	0.54	0.56	0.58	0.60	0.61	0.62
0.0	1.0000	1.0000	1.0000	1.0000	1.0000	1.0000	1.0000	1.0000	1.0000	1.0000	1.0000	1.0000	1.0000
0.1	1.0007	1.0007	1.0006	1.0006	1.0006	1.0006	1.0006	1.0006	1.0006	1.0006	1.0005	1.0005	1.0005
0.2	1.0013	1.0013	1.0013	1.0013	1.0012	1.0012	1.0012	1.0012	1.0011	1.0011	1.0011	1.0011	1.0010
0.3	1.0020	1.0020	1.0020	1.0019	1.0019	1.0018	1.0018	1.0018	1.0017	1.0017	1.0016	1.0016	1.0016
0.4	1.0027	1.0027	1.0026	1.0026	1.0025	1.0024	1.0024	1.0023	1.0023	1.0022	1.0022	1.0021	1.0021
0.5	1.0033	1.0033	1.0033	1.0032	1.0031	1.0030	1.0030	1.0029	1.0029	1.0028	1.0027	1.0027	1.0026
0.6	1.0040	1.0040	1.0040	1.0039	1.0038	1.0036	1.0036	1.0035	1.0034	1.0034	1.0033	1.0032	1.0032
0.7	1.0047	1.0047	1.0046	1.0045	1.0044	1.0043	1.0042	1.0041	1.0040	1.0039	1.0038	1.0038	1.0037
0.8	1.0054	1.0053	1.0053	1.0052	1.0050	1.0049	1.0048	1.0047	1.0046	1.0045	1.0044	1.0043	1.0042
0.9	1.0060	1.0060	1.0060	1.0058	1.0057	1.0055	1.0054	1.0053	1.0052	1.0050	1.0049	1.0048	1.0048
1.0	1.0067	1.0067	1.0066	1.0065	1.0063	1.0061	1.0060	1.0059	1.0058	1.0056	1.0055	1.0054	1.0053
1.1	1.0074	1.0074	1.0073	1.0071	1.0069	1.0067	1.0066	1.0065	1.0063	1.0062	1.0060	1.0059	1.0058
1.2	1.0080	1.0080	1.0080	1.0078	1.0076	1.0073	1.0072	1.0071	1.0069	1.0068	1.0066	1.0065	1.0064
1.3	1.0087	1.0087	1.0086	1.0084	1.0082	1.0080	1.0078	1.0077	1.0075	1.0073	1.0071	1.0070	1.0069
1.4	1.0094	1.0094	1.0093	1.0091	1.0089	1.0086	1.0084	1.0083	1.0081	1.0079	1.0077	1.0076	1.0074
1.5	1.0101	1.0101	1.0100	1.0097	1.0095	1.0092	1.0090	1.0089	1.0087	1.0085	1.0082	1.0081	1.0080
1.6	1.0108	1.0107	1.0106	1.0104	1.0101	1.0098	1.0096	1.0095	1.0093	1.0090	1.0088	1.0087	1.0085
1.7	1.0114	1.0114	1.0113	1.0110	1.0108	1.0104	1.0103	1.0101	1.0099	1.0096	1.0094	1.0092	1.0091
1.8	1.0121	1.0121	1.0120	1.0117	1.0114	1.0111	1.0109	1.0107	1.0104	1.0102	1.0099	1.0098	1.0096
1.9	1.0128	1.0128	1.0126	1.0123	1.0121	1.0117	1.0115	1.0113	1.0110	1.0108	1.0105	1.0103	1.0102
2.0	1.0135	1.0134	1.0133	1.0130	1.0127	1.0123	1.0121	1.0119	1.0116	1.0114	1.0110	1.0109	1.0107
2.1	1.0142	1.0141	1.0140	1.0136	1.0134	1.0129	1.0127	1.0125	1.0122	1.0119	1.0116	1.0114	1.0112
2.2	1.0148	1.0148	1.0147	1.0143	1.0140	1.0136	1.0133	1.0131	1.0128	1.0125	1.0122	1.0120	1.0118
2.3	1.0155	1.0155	1.0154	1.0150	1.0146	1.0142	1.0140	1.0137	1.0134	1.0131	1.0127	1.0126	1.0124
2.4	1.0162	1.0162	1.0160	1.0156	1.0153	1.0148	1.0146	1.0143	1.0140	1.0137	1.0133	1.0131	1.0129
2.5	1.0169	1.0168	1.0167	1.0163	1.0159	1.0154	1.0152	1.0149	1.0146	1.0142	1.0139	1.0137	1.0134
2.6	1.0176	1.0175	1.0174	1.0170	1.0166	1.0161	1.0158	1.0155	1.0152	1.0148	1.0144	1.0142	1.0140
2.7	1.0182	1.0182	1.0180	1.0176	1.0172	1.0167	1.0164	1.0161	1.0158	1.0154	1.0150	1.0148	1.0146
2.8	1.0189	1.0189	1.0187	1.0183	1.0179	1.0173	1.0170	1.0167	1.0164	1.0160	1.0156	1.0154	1.0151
2.9	1.0196	1.0196	1.0194	1.0189	1.0185	1.0180	1.0177	1.0173	1.0170	1.0166	1.0162	1.0159	1.0157
3.0	1.0203	1.0203	1.0201	1.0196	1.0192	1.0186	1.0183	1.0180	1.0176	1.0172	1.0167	1.0165	1.0162
3.1	1.0210	1.0210	1.0208	1.0203	1.0198	1.0192	1.0189	1.0186	1.0182	1.0178	1.0173	1.0170	1.0168
3.2	1.0217	1.0216	1.0214	1.0209	1.0205	1.0198	1.0195	1.0192	1.0188	1.0184	1.0179	1.0176	1.0173
3.3	1.0224	1.0223	1.0221	1.0216	1.0211	1.0205	1.0202	1.0198	1.0194	1.0189	1.0184	1.0182	1.0179
3.4	1.0230	1.0230	1.0228	1.0223	1.0218	1.0211	1.0208	1.0204	1.0200	1.0195	1.0190	1.0187	1.0184
3.5	1.0237	1.0237	1.0235	1.0229	1.0224	1.0217	1.0214	1.0210	1.0206	1.0201	1.0196	1.0193	1.0190
3.6	1.0244	1.0244	1.0242	1.0236	1.0231	1.0224	1.0220	1.0216	1.0212	1.0207	1.0202	1.0199	1.0196
3.7	1.0251	1.0251	1.0248	1.0243	1.0237	1.0230	1.0226	1.0222	1.0218	1.0213	1.0207	1.0204	1.0201
3.8	1.0258	1.0258	1.0255	1.0249	1.0244	1.0236	1.0233	1.0229	1.0224	1.0219	1.0213	1.0210	1.0207
3.9	1.0265	1.0264	1.0262	1.0256	1.0250	1.0243	1.0239	1.0235	1.0230	1.0225	1.0219	1.0216	1.0213
4.0	1.0272	1.0271	1.0269	1.0263	1.0257	1.0249	1.0245	1.0241	1.0236	1.0231	1.0225	1.0222	1.0218

From *Orifice Metering of Natural Gas*, 1969; courtesy of AGA.

Gas Flow Measurement 483

Table 10-5 Continued
Y_2 Expansion Factors—Flange Taps
Static Pressure Taken from Downstream Taps

$\dfrac{h_w}{p_{f2}}$ Ratio	\multicolumn{11}{c}{$\beta = \dfrac{d}{D}$ Ratio}										
	0.63	0.64	0.65	0.66	0.67	0.68	0.69	0.70	0.71	0.72	0.73	0.74	0.75
0.0	1.0000	1.0000	1.0000	1.0000	1.0000	1.0000	1.0000	1.0000	1.0000	1.0000	1.0000	1.0000	1.0000
0.1	1.0005	1.0005	1.0005	1.0005	1.0005	1.0004	1.0004	1.0004	1.0004	1.0004	1.0004	1.0004	1.0004
0.2	1.0010	1.0010	1.0010	1.0010	1.0009	1.0009	1.0009	1.0009	1.0008	1.0008	1.0008	1.0008	1.0007
0.3	1.0015	1.0015	1.0015	1.0014	1.0014	1.0014	1.0013	1.0013	1.0013	1.0012	1.0012	1.0011	1.0011
0.4	1.0021	1.0020	1.0020	1.0019	1.0019	1.0018	1,0018	1.0017	1.0017	1.0016	1.0016	1.0015	1.0014
0.5	1.0026	1.0025	1.0025	1.0024	1.0024	1.0023	1.0022	1.0022	1.0021	1.0020	1.0020	1.0019	1.0018
0.6	1.0031	1.0030	1.0030	1.0029	1.0028	1.0028	1.0027	1.0026	1.0025	1.0025	1.0024	1.0023	1.0022
0.7	1.0036	1.0036	1.0035	1.0034	1.0033	1.0032	1.0032	1.0031	1.0030	1.0029	1.0028	1.0027	1.0026
0.8	1.0042	1.0041	1.0040	1.0039	1.0038	1.0037	1.0036	1.0035	1.0034	1.0033	1.0032	1.0030	1.0029
0.9	1.0047	1.0046	1.0045	1.0044	1.0043	1.0042	1.0041	1.0040	1.0038	1.0037	1.0036	1.0034	1.0033
1.0	1.0052	1.0051	1.0050	1.0049	1.0048	1.0047	1.0045	1.0044	1.0043	1.0041	1.0040	1.0038	1.0037
1.1	1.0057	1.0056	1.0055	1.0054	1.0053	1.0051	1.0050	1.0049	1.0047	1.0046	1.0044	1.0042	1.0041
1.2	1.0062	1.0061	1.0060	1.0059	1.0058	1.0056	1.0055	1.0053	1.0052	1.0050	1.0048	1.0046	1.0044
1.3	1.0068	1.0066	1.0065	1.0064	1.0062	1.0061	1.0059	1.0058	1.0056	1.0054	1.0052	1.0050	1.0048
1.4	1.0073	1.0072	1.0070	1.0069	1.0067	1.0066	1.0064	1.0062	1.0060	1.0058	1.0056	1.0054	1.0052
1.5	1.0078	1.0077	1.0076	1.0074	1.0072	1.0070	1.0069	1.0067	1.0065	1.0063	1.0060	1.0058	1.0056
1.6	1.0084	1.0082	1.0081	1.0079	1.0077	1.0075	1.0073	1.0071	1.0069	1.0067	1.0065	1.0062	1.0060
1.7	1.0089	1.0088	1.0086	1.0084	1.0082	1.0080	1.0078	1.0076	1.0074	1.0071	1.0069	1.0066	1.0064
1.8	1.0094	1.0093	1.0091	1.0089	1.0087	1.0085	1.0083	1.0080	1.0078	1.0076	1.0073	1.0070	1.0068
1.9	1.0100	1.0098	1.0096	1.0094	1.0092	1.0090	1.0088	1.0085	1.0083	1.0080	1.0077	1.0074	1.0071
2.0	1.0105	1.0103	1.0101	1.0099	1.0097	1.0095	1.0092	1.0090	1.0087	1.0084	1.0081	1.0078	1.0075
2.1	1.0111	1.0109	1.0106	1.0104	1.0102	1.0100	1.0097	1.0094	1.0092	1.0089	1.0086	1.0083	1.0079
2.2	1.0116	1.0114	1.0112	1.0109	1.0107	1.0104	1.0102	1.0099	1.0096	1.0093	1.0090	1.0087	1.0083
2.3	1.0121	1.0119	1.0117	1.0114	1.0112	1.0109	1.0106	1.0104	1.0101	1.0098	1.0094	1.0091	1.0087
2.4	1.0127	1.0124	1.0122	1.0120	1.0117	1.0114	1.0111	1.0108	1.0105	1.0102	1.0098	1.0095	1.0091
2.5	1.0132	1.0130	1.0127	1.0125	1.0122	1.0119	1.0116	1.0113	1.0110	1.0106	1.0103	1.0099	1.0095
2.6	1.0138	1.0135	1.0133	1.0130	1.0127	1.0124	1.0121	1.0118	1.0114	1.0111	1.0107	1.0103	1.0099
2.7	1.0143	1.0140	1.0138	1.0135	1.0132	1.0129	1.0126	1.0122	1.0119	1.0115	1.0111	1.0107	1.0103
2.8	1.0148	1.0146	1.0143	1.0140	1.0137	1.0134	1.0131	1.0127	1.0124	1.0120	1.0116	1.0112	1.0107
2.9	1.0154	1.0151	1.0148	1.0145	1.0142	1.0139	1.0136	1.0132	1.0128	1.0124	1.0120	1.0116	1.0111
3.0	1.0160	1.0157	1.0154	1.0150	1.0147	1.0144	1.0140	1.0137	1.0133	1.0129	1.0124	1.0120	1.0116
3.1	1.0165	1.0162	1.0159	1.0156	1.0152	1.0149	1.0145	1.0141	1.0137	1.0133	1.0129	1.0124	1.0120
3.2	1.0170	1.0167	1.0164	1.0161	1.0158	1.0154	1.0150	1.0146	1.0142	1.0138	1.0133	1.0128	1.0124
3.3	1.0176	1.0173	1.0170	1.0166	1.0163	1.0159	1.0155	1.0151	1.0147	1.0142	1.0138	1.0133	1.0128
3.4	1.0181	1.0178	1.0175	1.0171	1.0168	1.0164	1.0160	1.0156	1.0151	1.0147	1.0142	1.0137	1.0132
3.5	1.0187	1.0184	1.0180	1.0177	1.0173	1.0169	1.0165	1.0160	1.0156	1.0151	1.0146	1.0141	1.0136
3.6	1.0192	1.0189	1.0186	1.0182	1.0178	1.0174	1.0170	1.0165	1.0161	1.0156	1.0151	1.0146	1.0140
3.7	1.0198	1.0195	1.0191	1.0187	1.0183	1.0179	1.0175	1.0170	1.0165	1.0160	1.0155	1.0150	1.0144
3.8	1.0204	1.0200	1.0196	1.0192	1.0188	1.0184	1.0180	1.0175	1.0170	1.0165	1.0160	1.0154	1.0148
3.9	1.0209	1.0206	1.0202	1.0198	1.0194	1.0189	1.0185	1.0180	1.0175	1.0170	1.0164	1.0159	1.0153
4.0	1.0215	1.0211	1.0207	1.0203	1.0199	1.0194	1.0190	1.0185	1.0180	1.0174	1.0169	1.0163	1.0157

Table 10-6
Y_m Expansion Factors—Flange Taps
Static Pressure Mean of Upstream and Downstream

$\dfrac{h_w}{p_{fm}}$ Ratio							$\beta = \dfrac{d}{D}$ Ratio					
	0.1	0.2	0.3	0.4	0.45	0.50	0.52	0.54	0.56	0.58	0.60	0.61	0.62
0.0	1.000	1.0000	1.0000	1.0000	1.0000	1.0000	1.0000	1.0000	1.0000	1.0000	1.0000	1.0000	1.0000
0.1	0.9998	0.9998	0.9998	0.9997	0.9997	0.9997	0.9997	0.9997	0.9997	0.9996	0.9996	0.9996	0.9996
0.2	0.9995	0.9995	0.9995	0.9995	0.9994	0.9994	0.9994	0.9994	0.9993	0.9993	0.9993	0.9993	0.9992
0.3	0.9993	0.9993	0.9993	0.9992	0.9992	0.9991	0.9991	0.9990	0.9990	0.9990	0.9989	0.9989	0.9989
0.4	0.9991	0.9990	0.9990	0.9990	0.9989	0.9988	0.9988	0.9987	0.9987	0.9986	0.9986	0.9985	0.9985
0.5	0.9988	0.9988	0.9988	0.9987	0.9986	0.9985	0.9985	0.9984	0.9984	0.9983	0.9982	0.9982	0.9981
0.6	0.9986	0.9986	0.9986	0.9984	0.9984	0.9982	0.9982	0.9981	0.9980	0.9979	0.9978	0.9978	0.9977
0.7	0.9984	0.9984	0.9983	0.9982	0.9981	0.9980	0.9979	0.9978	0.9977	0.9976	0.9975	0.9974	0.9974
0.8	0.9982	0.9981	0.9981	0.9980	0.9978	0.9977	0.9976	0.9975	0.9974	0.9973	0.9971	0.9971	0.9970
0.9	0.9979	0.9979	0.9978	0.9977	0.9976	0.9974	0.9973	0.9972	0.9971	0.9969	0.9968	0.9967	0.9966
1.0	0.9977	0.9977	0.9976	0.9974	0.9973	0.9971	0.9970	0.9969	0.9968	0.9966	0.9964	0.9964	0.9963
1.1	0.9975	0.9975	0.9974	0.9972	0.9970	0.9968	0.9967	0.9966	0.9964	0.9963	0.9961	0.9960	0.9959
1.2	0.9972	0.9972	0.9972	0.9970	0.9968	0.9965	0.9964	0.9963	0.9961	0.9959	0.9958	0.9956	0.9955
1.3	0.9970	0.9970	0.9969	0.9967	0.9965	0.9962	0.9961	0.9960	0.9958	0.9956	0.9954	0.9953	0.9952
1.4	0.9968	0.9968	0.9967	0.9965	0.9963	0.9960	0.9958	0.9957	0.9955	0.9953	0.9951	0.9950	0.9948
1.5	0.9966	0.9966	0.9965	0.9962	0.9960	0.9957	0.9955	0.9954	0.9952	0.9950	0.9947	0.9946	0.9945
1.6	0.9964	0.9964	0.9962	0.9960	0.9957	0.9954	0.9952	0.9951	0.9949	0.9946	0.9944	0.9942	0.9941
1.7	0.9962	0.9961	0.9960	0.9957	0.9955	0.9951	0.9950	0.9948	0.9946	0.9943	0.9940	0.9939	0.9938
1.8	0.9959	0.9959	0.9958	0.9955	0.9952	0.9949	0.9947	0.9945	0.9942	0.9940	0.9937	0.9936	0.9934
1.9	0.9957	0.9957	0.9956	0.9953	0.9950	0.9946	0.9944	0.9942	0.9939	0.9937	0.9934	0.9932	0.9930
2.0	0.9955	0.9955	0.9954	0.9950	0.9947	0.9943	0.9941	0.9939	0.9936	0.9934	0.9930	0.9929	0.9927
2.1	0.9953	0.9953	0.9951	0.9948	0.9945	0.9940	0.9938	0.9936	0.9933	0.9930	0.9927	0.9925	0.9923
2.2	0.9951	0.9951	0.9949	0.9946	0.9942	0.9938	0.9936	0.9933	0.9930	0.9927	0.9924	0.9922	0.9920
2.3	0.9949	0.9948	0.9947	0.9943	0.9940	0.9935	0.9933	0.9930	0.9927	0.9924	0.9920	0.9918	0.9916
2.4	0.9947	0.9946	0.9945	0.9941	0.9937	0.9932	0.9930	0.9927	0.9924	0.9921	0.9917	0.9915	0.9913
2.5	0.9945	0.9944	0.9943	0.9939	0.9935	0.9930	0.9927	0.9924	0.9921	0.9918	0.9914	0.9912	0.9910
2.6	0.9943	0.9942	0.9941	0.9936	0.9932	0.9927	0.9924	0.9922	0.9918	0.9915	0.9911	0.9908	0.9906
2.7	0.9940	0.9940	0.9938	0.9934	0.9930	0.9924	0.9922	0.9919	0.9915	0.9912	0.9907	0.9905	0.9903
2.8	0.9938	0.9938	0.9936	0.9932	0.9928	0.9922	0.9919	0.9916	0.9912	0.9908	0.9904	0.9902	0.9899
2.9	0.9936	0.9936	0.9934	0.9929	0.9925	0.9919	0.9916	0.9913	0.9910	0.9905	0.9901	0.9898	0.9896
3.0	0.9934	0.9934	0.9932	0.9927	0.9923	0.9917	0.9914	0.9910	0.9906	0.9902	0.9898	0.9895	0.9892
3.1	0.9932	0.9932	0.9930	0.9925	0.9920	0.9914	0.9911	0.9908	0.9904	0.9899	0.9894	0.9892	0.9889
3.2	0.9930	0.9930	0.9928	0.9923	0.9918	0.9912	0.9908	0.9905	0.9901	0.9896	0.9891	0.9889	0.9886
3.3	0.9928	0.9928	0.9926	0.9920	0.9916	0.9909	0.9906	0.9902	0.9898	0.9893	0.9888	0.9885	0.9882
3.4	0.9926	0.9926	0.9924	0.9918	0.9913	0.9906	0.9903	0.9899	0.9895	0.9890	0.9885	0.9882	0.9879
3.5	0.9924	0.9924	0.9922	0.9916	0.9911	0.9904	0.9900	0.9896	0.9892	0.9887	0.9882	0.9879	0.9876
3.6	0.9922	0.9922	0.9920	0.9914	0.9909	0.9901	0.9898	0.9894	0.9889	0.9884	0.9879	0.9876	0.9872
3.7	0.9921	0.9920	0.9918	0.9912	0.9906	0.9899	0.9895	0.9891	0.9886	0.9881	0.9876	0.9872	0.9869
3.8	0.9919	0.9918	0.9916	0.9910	0.9904	0.9896	0.9893	0.9888	0.9884	0.9878	0.9872	0.9869	0.9866
3.9	0.9917	0.9916	0.9914	0.9907	0.9902	0.9894	0.9890	0.9886	0.9881	0.9875	0.9869	0.9866	0.9863
4.0	0.9915	0.9914	0.9912	0.9905	0.9899	0.9891	0.9887	0.9883	0.9878	0.9872	0.9866	0.9863	0.9859

From *Orifice Metering of Natural Gas*, 1969; courtesy of AGA.

Gas Flow Measurement

Table 10-6 Continued
Y_m Expansion Factors—Flange Taps
Static Pressure Mean of Upstream and Downstream

$\frac{h_w}{p_{fm}}$ Ratio	0.63	0.64	0.65	0.66	0.67	0.68	0.69	0.70	0.71	0.72	0.73	0.74	0.75
0.0	1.0000	1.0000	1.0000	1.0000	1.0000	1.0000	1.0000	1.0000	1.000	1.0000	1.0000	1.0000	1.0000
0.1	0.9996	0.9996	0.9996	0.9996	0.9996	0.9996	0.9995	0.9995	0.9995	0.9995	0.9995	0.9995	0.9994
0.2	0.9992	0.9992	0.9992	0.9992	0.9991	0.9991	0.9991	0.9991	0.9990	0.9990	0.9990	0.9989	0.9989
0.3	0.9988	0.9988	0.9988	0.9987	0.9987	0.9987	0.9986	0.9986	0.9986	0.9985	0.9985	0.9984	0.9984
0.4	0.9984	0.9984	0.9984	0.9983	0.9983	0.9982	0.9982	0.9981	0.9981	0.9980	0.9980	0.9979	0.9978
0.5	0.9981	0.9980	0.9980	0.9979	0.9979	0.9978	0.9977	0.9977	0.9976	0.9975	0.9975	0.9974	0.9973
0.6	0.9977	0.9976	0.9976	0.9975	0.9974	0.9974	0.9973	0.9972	0.9971	0.9970	0.9970	0.9969	0.9968
0.7	0.9973	0.9972	0.9972	0.9971	0.9970	0.9969	0.9968	0.9968	0.9966	0.9966	0.9964	0.9964	0.9962
0.8	0.9969	0.9968	0.9968	0.9967	0.9966	0.9965	0.9964	0.9963	0.9962	0.9961	0.9960	0.9958	0.9957
0.9	0.9966	0.9965	0.9964	0.9963	0.9962	0.9961	0.9960	0.9958	0.9957	0.9956	0.9955	0.9953	0.9952
1.0	0.9962	0.9961	0.9960	0.9959	0.9958	0.9956	0.9955	0.9954	0.9952	0.9951	0.9950	0.9948	0.9946
1.1	0.9958	0.9957	0.9956	0.9955	0.9953	0.9952	0.9951	0.9949	0.9948	0.9946	0.9945	0.9943	0.9941
1.2	0.9954	0.9953	0.9952	0.9951	0.9949	0.9948	0.9946	0.9945	0.9943	0.9942	0.9940	0.9938	0.9936
1.3	0.9951	0.9949	0.9948	0.9947	0.9945	0.9944	0.9942	0.9940	0.9939	0.9937	0.9935	0.9933	0.9931
1.4	0.9947	0.9946	0.9944	0.9943	0.9941	0.9939	0.9938	0.9936	0.9934	0.9932	0.9930	0.9928	0.9926
1.5	0.9943	0.9942	0.9940	0.9939	0.9937	0.9935	0.9933	0.9931	0.9929	0.9927	0.9925	0.9923	0.9920
1.6	0.9940	0.9938	0.9936	0.9935	0.9933	0.9931	0.9929	0.9927	0.9925	0.9922	0.9920	0.9918	0.9915
1.7	0.9936	0.9934	0.9932	0.9931	0.9929	0.9927	0.9925	0.9922	0.9920	0.9918	0.9915	0.9913	0.9910
1.8	0.9932	0.9930	0.9929	0.9927	0.9925	0.9923	0.9920	0.9918	0.9916	0.9913	0.9910	0.9908	0.9905
1.9	0.9929	0.9927	0.9925	0.9923	0.9921	0.9918	0.9916	0.9914	0.9911	0.9908	0.9906	0.9903	0.9900
2.0	0.9925	0.9923	0.9921	0.9919	0.9917	0.9914	0.9912	0.9909	0.9907	0.9904	0.9901	0.9898	0.9895
2.1	0.9922	0.9919	0.9917	0.9915	0.9913	0.9910	0.9908	0.9905	0.9902	0.9899	0.9896	0.9893	0.9890
2.2	0.9918	0.9916	0.9914	0.9911	0.9909	0.9906	0.9903	0.9901	0.9898	0.9895	0.9891	0.9888	0.9885
2.3	0.9914	0.9912	0.9910	0.9907	0.9905	0.9902	0.9899	0.9896	0.9893	0.9890	0.9887	0.9883	0.9880
2.4	0.9911	0.9908	0.9906	0.9903	0.9901	0.9898	0.9895	0.9892	0.9889	0.9885	0.9882	0.9878	0.9874
2.5	0.9907	0.9905	0.9902	0.9900	0.9897	0.9894	0.9891	0.9888	0.9884	0.9881	0.9877	0.9873	0.9870
2.6	0.9904	0.9901	0.9898	0.9896	0.9893	0.9890	0.9887	0.9883	0.9880	0.9876	0.9872	0.9868	0.9864
2.7	0.9900	0.9898	0.9895	0.9892	0.9889	0.9886	0.9882	0.9879	0.9875	0.9872	0.9868	0.9864	0.9860
2.8	0.9897	0.9894	0.9891	0.9888	0.9885	0.9882	0.9878	0.9875	0.9871	0.9867	0.9863	0.9859	0.9854
2.9	0.9893	0.9890	0.9887	0.9884	0.9881	0.9878	0.9874	0.9870	0.9867	0.9863	0.9858	0.9854	0.9850
3.0	0.9890	0.9887	0.9884	0.9881	0.9877	0.9874	0.9870	0.9866	0.9862	0.9858	0.9854	0.9849	0.9845
3.1	0.9886	0.9883	0.9880	0.9877	0.9873	0.9870	0.9866	0.9862	0.9858	0.9854	0.9849	0.9844	0.9840
3.2	0.9883	0.9880	0.9876	0.9873	0.9870	0.9866	0.9862	0.9858	0.9854	0.9849	0.9845	0.9840	0.9835
3.3	0.9879	0.9876	0.9873	0.9869	0.9866	0.9862	0.9858	0.9854	0.9849	0.9845	0.9840	0.9835	0.9830
3.4	0.9876	0.9873	0.9869	0.9866	0.9862	0.9858	0.9854	0.9850	0.9845	0.9840	0.9835	0.9830	0.9825
3.5	0.9872	0.9869	0.9866	0.9862	0.9858	0.9854	0.9850	0.9845	0.9841	0.9836	0.9831	0.9826	0.9820
3.6	0.9869	0.9866	0.9862	0.9858	0.9854	0.9850	0.9846	0.9841	0.9836	0.9831	0.9826	0.9821	0.9815
3.7	0.9866	0.9862	0.9858	0.9855	0.9850	0.9846	0.9842	0.9837	0.9832	0.9827	0.9822	0.9816	0.9810
3.8	0.9862	0.9859	0.9855	0.9851	0.9847	0.9842	0.9838	0.9833	0.9828	0.9823	0.9817	0.9812	0.9806
3.9	0.9859	0.9855	0.9851	0.9847	0.9843	0.9838	0.9834	0.9829	0.9824	0.9818	0.9813	0.9807	0.9801
4.0	0.9856	0.9852	0.9848	0.9844	0.9839	0.9835	0.9830	0.9825	0.9819	0.9814	0.9808	0.9802	0.9796

Table 10-7
F_b Basic Orifice Factors—Pipe Taps

Basic temperature = 60°F Flowing temperature = 60°F $\sqrt{h_w p_f} = \infty$
Base pressure = 14.73 psia Specific gravity = 1.0 $h_w/p_f = 0$
Pipe Sizes—Nominal and Published Inside Diameters, Inches

Orifice Diameter, in.	2		3				4		
	1.689	1.939	2.067	2.300	2.626	2.900	3.068	3.152	3.430
0.250	12.850	12.813	12.800	12.782	12.765	12.753	12.748	12.745	12.737
0.375	29.359	29.097	29.005	28.882	28.771	28.710	28.682	28.669	28.634
0.500	53.703	52.816	52.401	52.019	51.591	51.353	51.243	51.196	51.064
0.625	87.212	84.919	84.083	82.922	81.795	81.142	80.835	80.703	80.332
0.750	132.23	126.86	124.99	122.45	120.06	118.67	118.00	117.70	116.86
0.875	192.74	181.02	177.08	171.92	167.23	164.58	163.31	162.76	161.17
1.000	275.45	251.10	243.27	233.30	224.56	219.76	217.52	216.55	213.79
1.125	391.93	342.98	327.98	309.43	293.79	285.48	281.66	280.02	275.42
1.250		465.99	437.99	404.52	377.36	363.41	357.12	354.45	347.03
1.375			583.96	524.68	478.68	455.82	445.74	441.48	429.83
1.500				679.10	602.45	565.79	549.94	543.31	525.40
1.625					755.34	697.43	672.95	662.81	635.76
1.750					946.99	856.37	819.05	803.77	763.51
1.875						1050.4	993.98	971.19	911.98
2.000						1290.7	1205.6	1171.8	1085.5
2.125							1465.1	1415.0	1289.7
2.250									1532.0
2.375									1822.8

Orifice Diameter, in.	4		6				8		
	3.826	4.026	4.897	5.189	5.761	6.065	7.625	7.981	8.071
0.250	12.727	12.722							
0.375	26.598	28.584							
0.500	50.936	50.886	50.739	50.705	50.652	50.628			
0.625	79.974	79.835	79.436	79.349	79.217	79.162			
0.750	116.05	115.73	114.81	114.61	114.32	114.20			
0.875	159.57	158.94	157.11	156.71	156.13	155.89	155.10	154.99	154.96
1.000	211.03	209.91	206.62	205.91	204.84	204.41	203.00	202.80	202.75
1.125	270.90	269.10	263.71	262.51	260.71	259.98	257.62	257.28	257.20
1.250	339.87	337.05	328.73	326.85	324.02	322.86	319.10	318.56	318.44
1.375	418.79	414.51	402.06	399.30	395.08	393.33	387.62	386.81	386.62
1.500	508.76	502.38	484.20	480.23	474.20	471.69	463.39	462.19	461.92
1.625	611.11	601.80	575.73	570.14	561.73	558.24	546.61	544.92	544.53
1.750	727.54	714.16	677.38	669.63	658.08	653.33	637.51	635.19	634.65
1.875	860.17	841.19	789.99	779.40	763.77	757.39	736.34	733.23	732.52
2.000	1011.7	985.04	914.57	900.28	879.38	870.93	843.34	839.29	838.35
2.125	1185.3	1148.4	1052.3	1033.2	1005.6	994.52	958.78	953.58	952.38
2.250	1385.4	1334.3	1204.7	1179.4	1143.2	1128.8	1083.0	1076.4	1074.9
2.375	1617.2	1547.3	1373.4	1340.2	1293.1	1274.6	1216.3	1208.0	1206.1
2.500	1887.6	1792.3	1560.5	1517.2	1456.4	1432.7	1359.2	1348.8	1346.5
2.625	2206.0	2075.9	1768.3	1712.3	1634.3	1604.3	1512.0	1499.2	1496.3
2.750		2407.0	1999.8	1927.6	1828.3	1790.3	1675.4	1659.7	1656.1
2.875			2258.5	2165.9	2039.9	1992.2	1849.9	1830.6	1826.3
3.000			2548.6	2430.2	2271.2	2211.6	2036.0	2012.7	2007.3
3.125			2875.2	2724.4	2524.3	2450.1	2234.7	2206.4	2199.9
3.250			3244.8	3052.8	2801.8	2709.9	2446.5	2412.4	2404.7
3.375			3665.6	3420.9	3106.9	2993.3	2672.5	2631.6	2622.3
3.500				3835.7	3443.0	3303.0	2913.7	2864.7	2853.7
3.625				4305.7	3914.4	3642.3	3171.1	3112.7	3099.6
3.750					4226.3	4014.8	3446.0	3376.6	3361.0
3.875					4684.9	4425.1	3739.9	3657.6	3639.2
4.000					5197.7	4878.4	4054.2	3957.0	3935.4

From *Orifice Metering of Natural Gas*, 1969; courtesy of AGA.

Gas Flow Measurement 487

Table 10-7 Continued
F_b Basic Orifice Factors—Pipe Taps

Orifice Diameter, in.	4			6			8		
	3.826	4.026	4.897	5.189	5.761	6.065	7.625	7.981	8.071
4.250							4751.4	4616.6	4586.6
4.500							5554.7	5369.0	5327.9
4.750							6485.3	6231.1	6175.2
5.000							7571.4	7224.3	7148.7
5.250							8850.3	8376.3	8274.0
5.500								9723.8	9585.1

Orifice Diameter, in.	10			12			18		
	9.564	10.020	10.136	11.376	11.938	12.090	14.688	15.000	15.250
1.000	202.16								
1.125	256.22	256.01	255.96						
1.250	316.90	316.56	316.49	315.84	315.57	315.51			
1.375	384.29	383.79	383.68	382.66	382.30	382.22			
1.500	458.52	457.79	457.63	456.16	455.64	455.52	453.92	453.78	
1.625	539.72	538.69	538.45	536.38	535.66	535.48	533.27	533.07	532.93
1.750	628.03	626.61	626.29	623.44	622.45	622.20	619.18	618.92	618.73
1.875	723.61	721.70	721.27	717.43	716.10	715.78	711.73	711.39	711.13
2.000	826.63	824.12	823.54	818.48	816.73	816.30	810.99	810.53	810.19
2.125	937.28	934.02	933.27	926.72	924.44	923.88	917.01	916.43	915.99
2.250	1,055.7	1,051.6	1,050.6	1,042.3	1,039.4	1,038.7	1,029.9	1,092.6	1,028.6
2.375	1,182.2	1,177.0	1,175.8	1,165.3	1,161.6	1,160.7	1,149.7	1,148.8	1,148.1
2.500	1,316.9	1,310.5	1,309.0	1,295.9	1,291.4	1,290.2	1,276.5	1,275.4	1,274.5
2.625	1,460.0	1,452.1	1,450.3	1,434.3	1,428.7	1,427.4	1,410.5	1,409.1	1,408.0
2.750	1,611.8	1,602.3	1,600.1	1,580.7	1,573.9	1,572.2	1,551.7	1,549.9	1,548.6
2.875	1,772.5	1,761.0	1,758.4	1,735.1	1,726.9	1,724.9	1,700.1	1,698.1	1,696.5
3.000	1,942.5	1,928.8	1,925.6	1,897.8	1,888.1	1,885.7	1,856.1	1,853.6	1,851.7
3.125	2,122.1	2,105.7	2,102.0	2,069.0	2,057.5	2,054.7	2,019.5	2,016.6	2,014.3
3.250	2,311.6	2,292.2	2,287.8	2,248.9	2,235.4	2,232.1	2,190.7	2,187.2	2,184.5
3.375	2,511.5	2,488.6	2,483.4	2,437.7	2,421.8	2,418.0	2,369.6	2,365.5	2,362.4
3.500	2,722.3	2,695.3	2,689.1	2,635.6	2,617.2	2,612.6	2,556.5	2,551.7	2,548.1
3.625	2,944.3	2,912.7	2,905.5	2,843.0	2,821.6	2,816.3	2,751.4	2,745.9	2,741.7
3.750	3,178.1	3,141.2	3,132.7	3,060.2	3,035.3	3,029.3	2,954.5	2,948.1	2,943.3
3.875	3,424.3	3,381.3	3,371.5	3,287.4	3,258.7	3,251.7	3,165.9	3,158.6	3,153.1
4.000	3,683.5	3,633.5	3,622.1	3,524.9	3,492.0	3,483.9	3,385.8	3,377.5	3,371.2
4.250	4,243.8	4,176.8	4,161.6	4,032.8	3,989.5	3,979.0	3,851.6	3,840.9	3,832.8
4.500	4,865.1	4,776.2	4,756.1	4,587.1	4,530.8	4,517.2	4,353.4	4,339.8	4,329.6
4.750	5,554.9	5,437.9	5,411.5	5,191.5	5,119.0	5,101.5	4,892.9	4,875.8	4,862.9
5.000	6,322.2	6,169.2	6,134.9	5,850.6	5,757.8	5,735.4	5,471.9	5,450.5	5,434.3
5.250	7,177.7	6,978.9	6,934.4	6,569.4	6,451.5	6,423.2	6,092.5	6,065.9	6,045.9
5.500	8,134.1	7,877.2	7,820.0	7,354.1	7,205.1	7,169.5	6,757.0	6,724.1	6,699.4
5.750	9,207.0	8,876.3	8,803.1	8,211.4	8,024.2	7,979.6	7,468.0	7,427.6	7,397.4
6.000	10,415	9,991.2	9,897.8	9,149.5	8,915.4	8,859.8	8,228.5	8,179.2	8,142.3
6.250	11,783	11,240	11,121	10,178	9,886.1	9,817.2	9,041.6	8,981.7	8,937.3
6.500	13,340	12,644	12,492	11,307	10,945	10,860	9,911.2	9,838.7	9,764.7
6.750		14,230	14,038	12,550	12,103	11,998	10,841	10,754	10,689
7.000		16,035	15,790	13,923	13,371	13,242	11,837	11,732	11,654
7.250				15,442	14,762	14,604	12,902	12,777	12,684
7.500				17,131	16,294	16,101	14,044	13,894	13,783
7.750				19,017	17,986	17,750	15,268	15,090	14,959
8.000					19,861	19,572	16,583	16,371	16,216

(table continued)

Table 10-7 Continued
F_b Basic Orifice Factors—Pipe Taps

Orifice Diameter, in.	10			12			18		
	9.564	10.020	10.136	11.376	11.938	12.090	14.688	15.000	15.250
8.250					21,947	21,593	17,996	17,746	17,561
8.500							19,517	19,221	19,003
8.750							21,156	20,807	20,551
9.000							22,926	22,515	22,214
9.250							24,841	24,356	24,003
9.500							26,917	26,346	25,932
9.750							29,172	28,501	28,014
10.000							31,629	30,839	30,268
10.250							34,315	33,383	32,713
10.500								36,160	35,372

Orifice Diameter, in.	20			24			30		
	18.814	19.000	19.250	22.626	23.000	23.250	28.628	29.000	29.250
2.000	806.71	806.57	806.40						
2.125	911.51	911.35	911.13						
2.250	1,022.9	1,022.7	1,022.4						
2.375	1,141.0	1,140.7	1,140.4	1,136.8	1,136.5	1,136.3			
2.500	1,265.7	1,265.4	1,265.0	1,260.6	1,260.2	1,259.9			
2.625	1,397.2	1,396.8	1,396.3	1,390.9	1,390.5	1,390.2			
2.750	1,535.5	1,535.0	1,534.4	1,527.9	1,527.3	1,527.0			
2.875	1,680.7	1,680.1	1,679.3	1,671.5	1,670.9	1,670.4	1,663.8		
3.000	1,832.7	1,832.1	1,831.2	1,821.9	1,821.1	1,820.6	1,812.7	1,812.3	1,812.0
3.125	1,991.8	1,991.0	1,990.0	1,978.9	1,978.0	1,977.4	1,968.1	1,967.7	1,967.4
3.250	2,158.0	2,157.0	2,155.8	2,142.8	2,141.7	2,141.0	2,130.2	2,129.6	2,129.3
3.375	2,331.3	2,330.2	2,328.7	2,313.5	2,312.3	2,311.5	2,298.8	2,298.2	2,297.7
3.500	2,511.9	2,510.6	2,508.8	2,491.2	2,489.7	2,488.8	2,474.1	2,473.3	2,472.9
3.625	2,699.7	2,698.2	2,696.2	2,675.8	2,674.0	2,673.0	2,656.0	2,655.2	2.654.6
3.750	2,895.0	2,893.2	2,890.9	2,867.4	2,865.4	2,864.1	2,844.6	2,843.7	2,843.0
3,875	3,097.7	3,095.7	3,093.0	3,066.0	3,063.8	3,062.3	3,040.0	3,038.9	3,038.2
4.000	3,308.0	3,305.7	3,302.7	3,271.8	3,269.2	3,267.6	3,242.2	3,240.9	3,240.1
4.250	3,751.6	3,748.7	3,744.8	3,705.0	3,701.7	3,699.6	3,666.9	3.665.3	3,664.3
4.500	4,226.8	4,223.0	4,218.1	4,167.6	4,163.4	4,160.7	4,119.3	4,117.3	4,116.0
4.750	4,734.1	4,729.4	4,723.3	4,660.0	4,654.8	4,651.4	4,599.6	4,597.1	4,595.4
5.000	5,274.6	5,268.7	5,261.1	5,183.0	5,176.4	5,172.3	5,108.2	5,105.0	5,103.0
5.250	5,849.0	5,841.9	5,832.6	5,737.1	5,729.1	5,723.9	5,645.4	5,641.4	5,639.1
5.500	6,458.6	6,449.9	6,438.7	6,322.9	6,313.2	6,307.0	6,211.8	6,207.2	6,204.2
5.750	7,104.4	7,094.0	7,080.4	6,941.3	6,929.7	6,922.2	6,807.7	6,802.1	6,798.5
6.000	7,787.9	7,775.4	7,759.1	7,592.8	7,579.0	7,570.1	7,433.6	7,426.9	7,422.6
6.250	8,510.4	8,495.4	8,476.0	8,278.3	8,262.0	8,251.5	8,089.9	8,082.0	8,076.9
6.500	9,273.4	9,255.6	9,232.5	8,998.8	8,979.5	8,967.1	8,777.2	8,768.0	8,761.9
6.750	10,079	10,058	10,030	9,755.0	9,732.4	9,717.9	9,496.0	9,485.2	9,478.1
7.000	10,928	10,903	10,871	10,548	10,522	10,505	10,247	10,234	10,226
7.250	11,823	11,794	11,756	11,379	11,348	11,329	11,030	11,016	11,006
7.500	12,767	12,733	12,689	12,249	12,214	12,191	11,847	11,830	11,819
7.750	13,762	13,722	13,670	13,160	13,119	13,093	12,697	12,678	12,665
8.000	14,810	14,763	14,703	14,113	14,065	14,035	13,582	13,560	13,546
8.250	15,914	15,860	15,791	15,109	15,054	15,020	14,501	14,477	14,461
8.500	17,078	17,015	16,935	16,150	16,087	16,048	15,457	15,429	15,411
8.750	18,305	18,232	18,129	17,237	17,166	17,121	16,450	16,418	16,397
9.000	19,598	19,515	19,408	18,373	18,292	18,241	17,480	17,444	17,421

Gas Flow Measurement   489

Table 10-7 Continued
F_b Basic Orifice Factors—Pipe Taps

Orifice Diameter, in.	20			24			30		
	18.814	19.000	19.250	22.626	23.000	23.250	28.628	29.000	29.250
9.250	20,963	20,866	20,743	19,560	19,468	19,409	18,548	18,508	18,482
9.500	22,402	22,292	22,151	20,800	20,695	20,628	19,656	19,611	19,582
9.750	23,923	23,796	23,634	22,094	21,976	21,900	20,805	20,754	20,721
10.000	25,529	25,384	25,198	23,447	23,312	23,227	21,995	21,938	21,901
10.250	27,227	27,061	26,849	24,859	24,708	24,612	23,228	23,165	23,124
10.500	29,023	28,834	28,592	26,335	26,164	26,056	24,505	24,434	24,358
10.750	30,925	30,709	30,434	27,878	27,685	27,563	25,827	25,749	25,698
11.000	32,940	32,694	32,381	29,490	29,273	29,136	27,196	27,109	27,052
11.250	35,078	34,798	34,443	31,175	30,932	30,779	28,613	28,156	28,453
11.500	37,348	37,030	36,626	32,938	32,666	32,494	30,080	29,972	29,903
11.750	39,761	39,400	38,941	34,783	34,478	34,285	31,598	31,479	31,402
12.000	42,330	41,920	41,399	36,714	36,373	36,158	33,169	33,038	32,953
12.500	47,991	47,461	46,790	40,855	40,429	40,161	36,478	36,318	36,215
13.000	54,463	53,778	52,914	45,406	44,877	44,544	40,024	39,829	39,704
13.500				50,420	49,763	49,352	43,823	43,589	43,437
14.000				55,959	55.147	54,638	47,898	47,615	47,433
14.500				62,099	61,094	60,468	52,271	51,932	51,714
15.000				68,929	67,687	66,915	56,967	56,562	56,301
15.500				76,562	75,025	74,074	62,017	61,533	61,223
16.000					83,231	82,055	67,453	66,878	66,509
16.500							73,314	72,630	72,193
17.000							79,641	78,831	78,313
17.500							86,485	85,525	84,913
18.000							93,900	92,765	92,042
18.500							101,950	100,610	99,758
19.000							110,720	109,130	108,130
19.500							120,300	118,420	117,230
20.000							130,780	128,560	127,150

(text continued from page 479)

Pressure-Base Factor, F_{pb}

This factor corrects for cases where the base (standard) pressure, p_b in psia, at which flow is to be measured is other than 14.73 psia:

$$F_{pb} = \frac{14.73}{p_b} \qquad (10\text{-}18)$$

$F_{pb} = 1$ for a desired base pressure of 14.73 psia, because the flow relationship assumes a standard pressure of 14.73 psia.

(text continued on page 498)

Table 10-8
"b" Values for Reynolds Number Factor F_r Determination—Pipe Taps

$$F_r = 1 + \frac{b}{\sqrt{h_w p_f}}$$

Pipe Sizes—Nominal and Published Inside Diameters, Inches

Orifice Diameter, in.	2			3			4		
	1.689	1.939	2.067	2.300	2.626	2.900	3.068	3.152	3.438
0.250	0.1105	0.1091	0.1087	0.1081	0.1078	0.1078	0.1080	0.1081	0.1084
0.375	0.0890	0.0878	0.0877	0.0879	0.0888	0.0898	0.0905	0.0908	0.0918
0.050	0.0758	0.0734	0.0729	0.0728	0.0737	0.0750	0.0758	0.0763	0.0778
0.625	0.0693	0.0647	0.0635	0.0624	0.0624	0.0634	0.0642	0.0646	0.0662
0.750	0.0675	0.0608	0.0586	0.0559	0.0546	0.0548	0.0552	0.0555	0.0568
0.875	0.0684	0.0602	0.0570	0.0528	0.0497	0.0488	0.0488	0.0489	0.0496
1.000	0.0702	0.0614	0.0576	0.0522	0.0473	0.0452	0.0445	0.0443	0.0443
1.125	0.0708	0.0635	0.0595	0.0532	0.0469	0.0435	0.0422	0.0417	0.0407
1.250		0.0650	0.0616	0.0552	0.0478	0.0434	0.0414	0.0406	0.0387
1.375			0.0629	0.0574	0.0496	0.0443	0.0418	0.0408	0.0379
1.500				0.0590	0.0518	0.0460	0.0431	0.0418	0.0382
1.625					0.0539	0.0482	0.0450	0.0435	0.0392
1.750					0.0553	0.0504	0.0471	0.0456	0.0408
1.875						0.0521	0.0492	0.0477	0.0427
2.000						0.0532	0.0508	0.0495	0.0448
2.125							0.0519	0.0509	0.0467
2.250									0.0483
2.375									0.0494

Orifice Diameter, in.	4		6				8		
	3.826	4.026	4.897	5.189	5.761	6.065	7.625	7.981	8.071
0.250	0.1087	0.1091							
0.375	0.0932	0.0939							
0.500	0.0799	0.0810	0.0850	0.0862	0.0883	0.0895			
0.625	0.0685	0.0697	0.0747	0.0762	0.0789	0.0802			
0.750	0.0590	0.0602	0.0655	0.0672	0.0703	0.0718			
0.875	0.0513	0.0524	0.0575	0.0592	0.0625	0.0642	0.0716	0.0730	0.0733
1.000	0.0453	0.0461	0.0506	0.0523	0.0556	0.0573	0.0652	0.0668	0.0662
1.125	0.0408	0.0412	0.0448	0.0464	0.0495	0.0512	0.0592	0.0609	0.0613
1.250	0.0376	0.0377	0.0401	0.0413	0.0442	0.0458	0.0538	0.0555	0.0560
1.375	0.0358	0.0353	0.0363	0.0373	0.0397	0.0412	0.0489	0.0506	0.0510
1.500	0.0350	0.0340	0.0334	0.0340	0.0360	0.0372	0.0445	0.0462	0.0466
1.625	0.0351	0.0336	0.0313	0.0315	0.0329	0.0339	0.0404	0.0421	0.0425
1.750	0.0358	0.0340	0.0300	0.0298	0.0304	0.0311	0.0369	0.0384	0.0388
1.875	0.0371	0.0349	0.0293	0.0287	0.0285	0.0290	0.0338	0.0352	0.0355
2.000	0.0388	0.0363	0.0292	0.0281	0.0273	0.0273	0.0311	0.0323	0.0327
2.125	0.0407	0.0360	0.0297	0.0281	0.0265	0.0262	0.0288	0.0298	0.0301
2.250	0.0427	0.0398	0.0305	0.0285	0.0261	0.0258	0.0268	0.0277	0.0280
2.375	0.0445	0.0417	0.0316	0.0293	0.0262	0.0253	0.0252	0.0259	0.0261
2.500	0.0460	0.0435	0.0330	0.0304	0.0267	0.0254	0.0239	0.0244	0.0246
2.625	0.0472	0.0450	0.0345	0.0317	0.0274	0.0258	0.0230	0.0232	0.0233
2.750		0.0462	0.0362	0.0331	0.0264	0.0265	0.0224	0.0224	0.0224
2.875			0.0379	0.0347	0.0295	0.0274	0.0220	0.0218	0.0218
3.000			0.0395	0.0364	0.0308	0.0285	0.0219	0.0214	0.0213

From *Orifice Metering of Natural Gas*, 1969; courtesy of AGA.

Gas Flow Measurement 491

Table 10-8 Continued
"b" Values for Reynolds Number Factor F_r Determination—Pipe Taps

Orifice Diameter, in.	4		6				8		
	3.826	4.026	4.897	5.189	5.761	6.065	7.625	7.981	8.071
3.125			0.0410	0.0380	0.0323	0.0297	0.0220	0.0213	0.0211
3.250			0.0422	0.0394	0.0338	0.0311	0.0223	0.0214	0.0212
3.375			0.0432	0.0408	0.0353	0.0325	0.0228	0.0216	0.0214
3.500				0.0419	0.0367	0.0339	0.0235	0.0221	0.0218
3.625				0.0428	0.0381	0.0354	0.0243	0.0227	0.0224
3.750					0.0393	0.0367	0.0252	0.0234	0.0230
3.875					0.0404	0.0380	0.0262	0.0243	0.0238
4.000					0.0413	0.0391	0.0273	0.0252	0.0246
4.250							0.0296	0.0273	0.0268
4.500							0.0321	0.0296	0.0290
4.750							0.0344	0.0320	0.0314
5.000							0.0364	0.0342	0.0336
5.250							0.0381	0.0361	0.0356
5.500								0.0377	0.0372

Orifice Diameter, in.	10			12			16		
	9.564	10.020	10.136	11.376	11.938	12.090	14.688	15.000	15.250
1.000	0.0728								
1.125	0.0674	0.0690	0.0694						
1.250	0.0624	0.0641	0.0646	0.0687	0.0704	0.0708			
1.375	0.0576	0.0594	0.0599	0.0643	0.0661	0.0666			
1.500	0.0532	0.0550	0.0555	0.0601	0.0620	0.0625	0.0697	0.0705	
1.625	0.0490	0.0509	0.0514	0.0561	0.0580	0.0585	0.0662	0.0670	0.0676
1.750	0.0452	0.0471	0.0476	0.0523	0.0543	0.0548	0.0628	0.0636	0.0642
1.875	0.0417	0.0436	0.0440	0.0488	0.0508	0.0513	0.0594	0.0603	0.0610
2.000	0.0385	0.0403	0.0407	0.0454	0.0475	0.0480	0.0563	0.0572	0.0578
2.125	0.0355	0.0372	0.0377	0.0423	0.0443	0.0449	0.0532	0.0541	0.0548
2.250	0.0329	0.0345	0.0349	0.0394	0.0414	0.0419	0.0503	0.0512	0.0519
2.375	0.0305	0.0320	0.0324	0.0367	0.0387	0.0392	0.0475	0.0484	0.0492
2.500	0.0283	0.0298	0.0301	0.0342	0.0361	0.0366	0.0449	0.0458	0.0466
2.625	0.0265	0.0277	0.0281	0.0319	0.0337	0.0342	0.0424	0.0433	0.0440
2.750	0.0248	0.0260	0.0262	0.0298	0.0316	0.0320	0.0400	0.0409	0.0417
2.875	0.0234	0.0244	0.0246	0.0279	0.0295	0.0300	0.0378	0.0387	0.0394
3.000	0.0222	0.0230	0.0232	0.0262	0.0277	0.0281	0.0356	0.0365	0.0372
3.125	0.0212	0.0218	0.0220	0.0244	0.0260	0.0264	0.0336	0.0345	0.0352
3.250	0.0204	0.0209	0.0210	0.0232	0.0245	0.0249	0.0317	0.0326	0.0332
3.375	0.0199	0.0201	0.0202	0.0220	0.0232	0.0235	0.0300	0.0308	0.0314
3.500	0.0195	0.0195	0.0196	0.0210	0.0220	0.0222	0.0263	0.0291	0.0297
3.625	0.0193	0.0191	0.0191	0.0200	0.0209	0.0212	0.0268	0.0275	0.0281
3.750	0.0192	0.0188	0.0188	0.0193	0.0200	0.0202	0.0254	0.0261	0.0267
3.875	0.0193	0.0187	0.0186	0.0187	0.0192	0.0194	0.0240	0.0247	0.0253
4.000	0.0195	0.0187	0.0186	0.0182	0.0185	0.0187	0.0228	0.0235	0.0240
4.250	0.0203	0.0192	0.0189	0.0176	0.0176	0.0177	0.0207	0.0213	0.0217
4.500	0.0215	0.0200	0.0197	0.0175	0.0172	0.0171	0.0190	0.0194	0.0198
4.750	0.0230	0.0212	0.0208	0.0178	0.0171	0.0170	0.0176	0.0180	0.0182
5.000	0.0248	0.0228	0.0223	0.0185	0.0174	0.0173	0.0166	0.0168	0.0170

(table continued)

Table 10-8 Continued
"b" Values for Reynolds Number Factor F_r Determination—Pipe Taps

Orifice Diameter, in.	10			12			16		
	9.564	10.020	10.136	11.376	11.938	12.090	14.688	15.000	15.250
5.250	0.0267	0.0244	0.0239	0.0194	0.0181	0.0178	0.0160	0.0161	0.0162
5.500	0.0287	0.0263	0.0257	0.0207	0.0190	0.0186	0.0156	0.0156	0.0156
5.750	0.0307	0.0282	0.0276	0.0221	0.0202	0.0197	0.0155	0.0154	0.0153
6.000	0.0326	0.0302	0.0295	0.0231	0.0215	0.0210	0.0157	0.0154	0.0153
6.250	0.0343	0.0320	0.0316	0.0253	0.0230	0.0224	0.0161	0.0157	0.0154
6.500	0.0358	0.0336	0.0331	0.0270	0.0246	0.0239	0.0167	0.0162	0.0159
6.750		0.0351	0.0346	0.0288	0.0262	0.0256	0.0174	0.0169	0.0164
7.000		0.0363	0.0359	0.0304	0.0279	0.0272	0.0184	0.0177	0.0172
7.250				0.0320	0.0295	0.0288	0.0195	0.0187	0.0181
7.500				0.0334	0.0310	0.0304	0.0206	0.0198	0.0191
7.750				0.0347	0.0325	0.0318	0.0219	0.0209	0.0202
8.000					0.0338	0.0332	0.0232	0.0222	0.0214
8.250					0.0349	0.0344	0.0246	0.0235	0.0227
8.500							0.0259	0.0248	0.0240
8.750							0.0273	0.0262	0.0253
9.000							0.0286	0.0276	0.0267
9.250							0.0299	0.0288	0.0280
9.500							0.0311	0.0300	0.0292
9.750							0.0322	0.0312	0.0304
10.000							0.0332	0.0323	0.0315
10.250							0.0341	0.0333	0.0326
10.500								0.0341	0.0335

Orifice Diameter, in.	20			24			30		
	18.814	19.000	19.250	22.626	23.000	23.250	28.628	29.000	29.250
2.000	0.0663	0.0667	0.0672						
2.125	0.0635	0.0639	0.0644						
2.250	0.0609	0.0613	0.0618						
2.375	0.0583	0.0588	0.0593	0.0658	0.0665	0.0669			
2.500	0.0558	0.0562	0.0568	0.0635	0.0642	0.0646			
2.625	0.0534	0.0539	0.0544	0.0613	0.0620	0.0624			
2.750	0.0510	0.0515	0.0520	0.0591	0.0598	0.0603			
2.875	0.0488	0.0492	0.0498	0.0570	0.0577	0.0582	0.0667		
3.000	0.0466	0.0470	0.0476	0.0549	0.0556	0.0561	0.0649	0.0654	0.0657
3.125	0.0445	0.0449	0.0455	0.0529	0.0536	0.0541	0.0630	0.0636	0.0639
3.250	0.0425	0.0429	0.0435	0.0509	0.0516	0.0521	0.0613	0.0616	0.0622
3.375	0.0406	0.0410	0.0416	0.0490	0.0497	0.0502	0.0595	0.0601	0.0604
3.500	0.0387	0.0391	0.0397	0.0471	0.0479	0.0484	0.0578	0.0584	0.0587
3.625	0.0369	0.0373	0.0379	0.0454	0.0461	0.0466	0.0561	0.0567	0.0571
3.750	0.0352	0.0356	0.0362	0.0436	0.0444	0.0449	0.0545	0.0550	0.0554
3.875	0.0336	0.0340	0.0346	0.0419	0.0427	0.0432	0.0528	0.0534	0.0538
4.000	0.0320	0.0324	0.0330	0.0403	0.0411	0.0416	0.0513	0.0518	0.0522
4.250	0.0291	0.0295	0.0301	0.0372	0.0380	0.0385	0.0482	0.0488	0.0492
4.500	0.0265	0.0269	0.0274	0.0343	0.0351	0.0356	0.0453	0.0459	0.0463
4.750	0.0242	0.0246	0.0250	0.0316	0.0324	0.0328	0.0425	0.0431	0.0435
5.000	0.0221	0.0225	0.0229	0.0292	0.0299	0.0303	0.0399	0.0405	0.0409
5.250	0.0203	0.0206	0.0210	0.0269	0.0276	0.0280	0.0374	0.0380	0.0384

Gas Flow Measurement 493

Table 10-8 Continued
"b" Values for Reynolds Number Factor F_r Determination—Pipe Taps

Orifice Diameter, in.	20			24			30		
	18.814	19.000	19.250	22.626	23.000	23.250	28.628	29.000	29.250
5.500	0.0188	0.0190	0.0194	0.0248	0.0255	0.0259	0.0350	0.0356	0.0360
5.750	0.0175	0.0177	0.0180	0.0230	0.0236	0.0240	0.0328	0.0334	0.0338
6.000	0.0164	0.0165	0.0168	0.0212	0.0218	0.0222	0.0307	0.0313	0.0317
6.250	0.0155	0.0156	0.0158	0.0197	0.0202	0.0206	0.0287	0.0293	0.0297
6.500	0.0148	0.0149	0.0150	0.0184	0.0189	0.0192	0.0269	0.0274	0.0278
6.750	0.0143	0.0144	0.0145	0.0172	0.0176	0.0179	0.0252	0.0257	0.0260
7.000	0.0141	0.0141	0.0141	0.0162	0.0166	0.0168	0.0236	0.0241	0.0244
7.250	0.0140	0.0140	0.0139	0.0153	0.0156	0.0158	0.0221	0.0226	0.0229
7.500	0.0140	0.0140	0.0139	0.0146	0.0148	0.0150	0.0207	0.0212	0.0215
7.750	0.0142	0.0141	0.0140	0.0140	0.0142	0.0144	0.0195	0.0199	0.0202
8.000	0.0146	0.0144	0.0142	0.0136	0.0138	0.0138	0.0183	0.0187	0.0190
8.250	0.0151	0.0148	0.0146	0.0133	0.0134	0.0132	0.0173	0.0177	0.0179
8.500	0.0156	0.0154	0.0151	0.0132	0.0132	0.0130	0.0164	0.0167	0.0169
8.750	0.0163	0.0160	0.0157	0.0131	0.0130	0.0130	0.0155	0.0158	0.0161
9.000	0.0171	0.0168	0.0163	0.0131	0.0130	0.0130	0.0148	0.0151	0.0153
9.250	0.0180	0.0176	0.0171	0.0133	0.0131	0.0130	0.0142	0.0144	0.0146
9.500	0.0189	0.0185	0.0180	0.0136	0.0133	0.0132	0.0136	0.0138	0.0140
9.750	0.0198	0.0194	0.0189	0.0139	0.0136	0.0134	0.0132	0.0133	0.0134
10.000	0.0209	0.0204	0.0198	0.0143	0.0140	0.0135	0.0128	0.0129	0.0130
10.250	0.0219	0.0214	0.0208	0.0148	0.0144	0.0142	0.0125	0.0126	0.0127
10.500	0.0230	0.0225	0.0219	0.0154	0.0150	0.0147	0.0123	0.0124	0.0124
10.750	0.0241	0.0236	0.0229	0.0160	0.0155	0.0152	0.0122	0.0122	0.0122
11.000	0.0252	0.0247	0.0240	0.0168	0.0162	0.0158	0.0121	0.0121	0.0121
11.250	0.0263	0.0261	0.0251	0.0175	0.0169	0.0165	0.0122	0.0121	0.0121
11.500	0.0273	0.0268	0.0262	0.0183	0.0176	0.0172	0.0122	0.0121	0.0122
11.750	0.0284	0.0278	0.0272	0.0191	0.0184	0.0180	0.0124	0.0123	0.0122
12.000	0.0293	0.0288	0.0282	0.0200	0.0192	0.0190	0.0126	0.0124	0.0123
12.500	0.0312	0.0307	0.0301	0.0218	0.0210	0.0204	0.0132	0.0130	0.0128
13.000	0.0327	0.0323	0.0318	0.0236	0.0228	0.0222	0.0140	0.0137	0.0135
13.500				0.0254	0.0246	0.0240	0.0150	0.0146	0.0143
14.000				0.0272	0.0264	0.0258	0.0161	0.0156	0.0153
14.500				0.0289	0.0280	0.0275	0.0173	0.0168	0.0165
15.000				0.0304	0.0296	0.0291	0.0166	0.0181	0.0177
15.500				0.0310	0.0311	0.0306	0.0200	0.0194	0.0190
16.000					0.0323	0.0318	0.0215	0.0209	0.0204
16.500							0.0230	0.0223	0.0219
17.000							0.0244	0.0238	0.0233
17.500							0.0259	0.0252	0.0248
18.000							0.0272	0.0266	0.0261
18.500							0.0286	0.0279	0.0275
19.000							0.0298	0.0292	0.0288
19.500							0.0309	0.0303	0.0299
20.000							0.0318	0.0313	0.0310

Table 10-9*
Y_1 Expansion Factors—Pipe Taps
Static Pressure Taken from Upstream Taps

$\dfrac{h_w}{p_{f_1}}$ Ratio										$\beta = \dfrac{d}{D}$ Ratio		
	0.1	0.2	0.3	0.4	0.45	0.50	0.52	0.54	0.56	0.58	0.60	0.61
0.0	1.0000	1.0000	1.0000	1.0000	1.0000	1.0000	1.0000	1.0000	1.0000	1.0000	1.0000	1.0000
0.1	0.9990	0.9989	0.9988	0.9985	0.9984	0.9982	0.9981	0.9980	0.9979	0.9978	0.9977	0.9976
0.2	0.9981	0.9979	0.9976	0.9971	0.9968	0.9964	0.9962	0.9961	0.9959	0.9957	0.9954	0.9953
0.3	0.9971	0.9968	0.9964	0.9956	0.9952	0.9946	0.9944	0.9941	0.9938	0.9935	0.9931	0.9929
0.4	0.9962	0.9958	0.9951	0.9942	0.9936	0.9928	0.9925	0.9921	0.9917	0.9913	0.9908	0.9906
0.5	0.9952	0.9947	0.9939	0.9927	0.9919	0.9910	0.9906	0.9902	0.9897	0.9891	0.9885	0.9882
0.6	0.9943	0.9937	0.9927	0.9913	0.9903	0.9892	0.9887	0.9882	0.9876	0.9870	0.9862	0.9859
0.7	0.9933	0.9926	0.9915	0.9898	0.9887	0.9874	0.9869	0.9862	0.9856	0.9848	0.9840	0.9835
0.8	0.9923	0.9916	0.9903	0.9883	0.9871	0.9857	0.9850	0.9843	0.9835	0.9826	0.9817	0.9811
0.9	0.9914	0.9905	0.9891	0.9869	0.9855	0.9839	0.9831	0.9823	0.9814	0.9805	0.9794	0.9788
1.0	0.9904	0.9895	0.9878	0.9854	0.9839	0.9821	0.9812	0.9803	0.9794	0.9783	0.9771	0.9764
1.1	0.9895	0.9884	0.9866	0.9840	0.9823	0.9803	0.9794	0.9784	0.9773	0.9761	0.9748	0.9741
1.2	0.9885	0.9874	0.9854	0.9825	0.9807	0.9785	0.9775	0.9764	0.9752	0.9739	0.9725	0.9717
1.3	0.9876	0.9863	0.9842	0.9811	0.9791	0.9767	0.9756	0.9744	0.9732	0.9718	0.9702	0.9694
1.4	0.9866	0.9853	0.9830	0.9796	0.9775	0.9749	0.9737	0.9725	0.9711	0.9696	0.9679	0.9670
1.5	0.9857	0.9842	0.9818	0.9782	0.9758	0.9731	0.9719	0.9705	0.9690	0.9674	0.9656	0.9646
1.6	0.9847	0.9832	0.9805	0.9767	0.9742	0.9713	0.9700	0.9685	0.9670	0.9652	0.9633	0.9623
1.7	0.9837	0.9821	0.9793	0.9752	0.9726	0.9695	0.9681	0.9666	0.9649	0.9631	0.9610	0.9599
1.8	0.9828	0.9811	0.9781	0.9738	0.9710	0.9677	0.9662	0.9646	0.9628	0.9609	0.9587	0.9576
1.9	0.9818	0.9800	0.9769	0.9723	0.9694	0.9659	0.9643	0.9626	0.9608	0.9587	0.9565	0.9552
2.0	0.9809	0.9790	0.9757	0.9709	0.9678	0.9641	0.9625	0.9607	0.9587	0.9566	0.9542	0.9529
2.1	0.9799	0.9779	0.9745	0.9694	0.9662	0.9623	0.9606	0.9587	0.9566	0.9544	0.9519	0.9505
2.2	0.9790	0.9768	0.9732	0.9680	0.9646	0.9605	0.9587	0.9567	0.9546	0.9522	0.9496	0.9481
2.3	0.9780	0.9758	0.9720	0.9665	0.9630	0.9587	0.9568	0.9548	0.9525	0.9500	0.9473	0.9458
2.4	0.9770	0.9747	0.9708	0.9650	0.9613	0.9570	0.9550	0.9528	0.9505	0.9479	0.9450	0.9434
2.5	0.9761	0.9737	0.9696	0.9636	0.9597	0.9552	0.9531	0.9508	0.9484	0.9457	0.9427	0.9411
2.6	0.9751	0.9726	0.9684	0.9621	0.9581	0.9534	0.9512	0.9489	0.9463	0.9435	0.9404	0.9387
2.7	0.9742	0.9716	0.9672	0.9607	0.9565	0.9516	0.9493	0.9469	0.9443	0.9414	0.9381	0.9364
2.8	0.9732	0.9705	0.9659	0.9592	0.9549	0.9498	0.9475	0.9449	0.9422	0.9392	0.9358	0.9340
2.9	0.9723	0.9695	0.9647	0.9578	0.9533	0.9480	0.9456	0.9430	0.9401	0.9370	0.9335	0.9316
3.0	0.9713	0.9684	0.9635	0.9563	0.9517	0.9462	0.9437	0.9410	0.9381	0.9348	0.9312	0.9293
3.1	0.9704	0.9674	0.9623	0.9549	0.9501	0.9444	0.9418	0.9390	0.9360	0.9327	0.9290	0.9269
3.2	0.9694	0.9663	0.9611	0.9534	0.9485	0.9426	0.9400	0.9371	0.9339	0.9305	0.9267	0.9246
3.3	0.9684	0.9653	0.9599	0.9519	0.9469	0.9408	0.9381	0.9351	0.9319	0.9283	0.9244	0.9222
3.4	0.9675	0.9642	0.9587	0.9505	0.9452	0.9390	0.9362	0.9331	0.9298	0.9261	0.9221	0.9199
3.5	0.9665	0.9632	0.9574	0.9490	0.9436	0.9372	0.9343	0.9312	0.9277	0.9240	0.9198	0.9175
3.6	0.9656	0.9621	0.9562	0.9476	0.9420	0.9354	0.9324	0.9292	0.9257	0.9218	0.9175	0.9151
3.7	0.9646	0.9611	0.9550	0.9461	0.9404	0.9336	0.9306	0.9272	0.9236	0.9196	0.9152	0.9128
3.8	0.9637	0.9600	0.9538	0.9447	0.9388	0.9318	0.9287	0.9253	0.9216	0.9175	0.9129	0.9104
3.9	0.9627	0.9590	0.9526	0.9432	0.9372	0.9301	0.9268	0.9233	0.9195	0.9153	0.9106	0.9081
4.0	0.9617	0.9579	0.9514	0.9417	0.9356	0.9283	0.9249	0.9213	0.9174	0.9131	0.9083	0.9057

From *Orifice Metering of Natural Gas*, 1969; courtesy of AGA.

Gas Flow Measurement 495

Table 10-9 Continued
Y_1 Expansion Factors—Pipe Taps
Static Pressure Taken from Upstream Taps

$\dfrac{h_w}{p_{f1}}$ Ratio						$\beta = \dfrac{d}{D}$ Ratio		
	0.62	0.63	0.64	0.65	0.66	0.67	0.68	0.69	0.70
0.0	1.0000	1.0000	1.0000	1.0000	1.0000	1.0000	1.0000	1.0000	1.0000
0.1	0.9976	0.9975	0.9974	0.9973	0.9972	0.9971	0.9970	0.9969	0.9968
0.2	0.9951	0.9950	0.9948	0.9947	0.9945	0.9943	0.9941	0.9938	0.9935
0.3	0.9927	0.9925	0.9923	0.9920	0.9917	0.9914	0.9911	0.9907	0.9903
0.4	0.9903	0.9900	0.9897	0.9893	0.9890	0.9886	0.9881	0.9876	0.9871
0.5	0.9879	0.9875	0.9871	0.9867	0.9862	0.9857	0.9851	0.9845	0.9839
0.6	0.9854	0.9850	0.9845	0.9840	0.9834	0.9828	0.9822	0.9814	0.9806
0.7	0.9830	0.9825	0.9819	0.9813	0.9807	0.9800	0.9792	0.9784	0.9774
0.8	0.9806	0.9800	0.9794	0.9787	0.9779	0.9771	0.9762	0.9753	0.9742
0.9	0.9782	0.9775	0.9768	0.9760	0.9752	0.9742	0.9733	0.9722	0.9710
1.0	0.9757	0.9750	0.9742	0.9733	0.9724	0.9714	0.9703	0.9691	0.9677
1.1	0.9733	0.9725	0.9716	0.9707	0.9696	0.9685	0.9673	0.9660	0.9645
1.2	0.9709	0.9700	0.9690	0.9680	0.9669	0.9757	0.9643	0.9629	0.9613
1.3	0.9685	0.9675	0.9664	0.9653	0.9641	0.9628	0.9614	0.9598	0.9581
1.4	0.9660	0.9650	0.9639	0.9627	0.9614	0.9599	0.9584	0.9567	0.9548
1.5	0.9636	0.9625	0.9613	0.9600	0.9586	0.9571	0.9554	0.9536	0.9516
1.6	0.9612	0.9600	0.9587	0.9573	0.9558	0.9542	0.9525	0.9505	0.9484
1.7	0.9587	0.9575	0.9561	0.9547	0.9531	0.9514	0.9495	0.9474	0.9452
1.8	0.9563	0.9550	0.9535	0.9520	0.9503	0.9485	0.9465	0.9443	0.9419
1.9	0.9539	0.9525	0.9510	0.9493	0.9476	0.9456	0.9435	0.9412	0.9387
2.0	0.9515	0.9500	0.9484	0.9467	0.9448	0.9428	0.9406	0.9381	0.9355
2.1	0.9490	0.9475	0.9458	0.9440	0.9420	0.9399	0.9376	0.9351	0.9323
2.2	0.9466	0.9450	0.9432	0.9413	0.9393	0.9371	0.9346	0.9320	0.9290
2.3	0.9442	0.9425	0.9406	0.9387	0.9365	0.9342	0.9317	0.9289	0.9258
2.4	0.9418	0.9400	0.9381	0.9360	0.9338	0.9313	0.9287	0.9258	0.9226
2.5	0.9393	0.9375	0.9355	0.9333	0.9310	0.9285	0.9257	0.9227	0.9194
2.6	0.9369	0.9350	0.9329	0.9307	0.9282	0.9256	0.9227	0.9196	0.9161
2.7	0.9345	0.9325	0.9303	0.9280	0.9255	0.9227	0.9198	0.9165	0.9129
2.8	0.9321	0.9300	0.9277	0.9253	0.9227	0.9199	0.9168	0.9134	0.9097
2.9	0.9296	0.9275	0.9252	0.9227	0.9200	0.9170	0.9138	0.9103	0.9064
3.0	0.9272	0.9250	0.9226	0.9200	0.9172	0.9142	0.9108	0.9072	0.9032
3.1	0.9248	0.9225	0.9200	0.9173	0.9144	0.9113	0.9079	0.9041	0.9000
3.2	0.9223	0.9200	0.9174	0.9147	0.9117	0.9084	0.9049	0.9010	0.8968
3.3	0.9199	0.9175	0.9148	0.9120	0.9089	0.9056	0.9019	0.8979	0.8935
3.4	0.9175	0.9150	0.9122	0.9093	0.9062	0.9027	0.8990	0.8948	0.8903
3.5	0.9151	0.9125	0.9097	0.9067	0.9034	0.8999	0.8960	0.8918	0.8871
3.6	0.9126	0.9100	0.9071	0.9040	0.9006	0.8970	0.8930	0.8887	0.8839
3.7	0.9102	0.9075	0.9045	0.9013	0.8979	0.8941	0.8900	0.8856	0.8806
3.8	0.9078	0.9050	0.9019	0.8987	0.8951	0.8913	0.8871	0.8825	0.8774
3.9	0.9054	0.9025	0.8993	0.8960	0.8924	0.8884	0.8841	0.8794	0.8742
4.0	0.9029	0.9000	0.8968	0.8933	0.8896	0.8856	0.8811	0.8763	0.8710

Gas Production Engineering

Table 10-10
Y_2 Expansion Factors—Pipe Taps
Static Pressure, Taken from Downstream Taps

$\dfrac{h_w}{p_{f2}}$ Ratio	\multicolumn{9}{c}{$\beta = \dfrac{d}{D}$ Ratio}
	0.1	0.2	0.3	0.4	0.45	0.50	0.52	0.54	0.56	0.58
0.0	1.0000	1.0000	1.0000	1.0000	1.0000	1.0000	1.0000	1.0000	1.0000	1.0000
0.1	1.0008	1.0008	1.0006	1.0003	1.0002	1.0000	0.9999	0.9998	0.9997	0.9996
0.2	1.0017	1.0015	1.0012	1.0007	1.0004	1.0000	0.9999	0.9997	0.9995	0.9993
0.3	1.0025	1.0023	1.0018	1.0010	1.0006	1.0000	0.9998	0.9995	0.9992	0.9989
0.4	1.0034	1.0030	1.0024	1.0014	1.0008	1.0001	0.9997	0.9994	0.9990	0.9986
0.5	1.0042	1.0038	1.0030	1.0018	1.0010	1.0001	0.9997	0.9992	0.9988	0.9982
0.6	1.0051	1.0045	1.0036	1.0021	1.0012	1.0001	0.9996	0.9991	0.9985	0.9979
0.7	1.0059	1.0053	1.0041	1.0025	1.0014	1.0002	0.9996	0.9990	0.9983	0.9975
0.8	1.0068	1.0060	1.0047	1.0028	1.0016	1.0002	0.9995	0.9988	0.9980	0.9972
0.9	1.0076	1.0068	1.0053	1.0032	1.0018	1.0002	0.9995	0.9987	0.9978	0.9969
1.0	1.0085	1.0075	1.0059	1.0036	1.0021	1.0003	0.9994	0.9986	0.9976	0.9965
1.1	1.0093	1.0083	1.0065	1.0039	1.0023	1.0003	0.9994	0.9984	0.9974	0.9962
1.2	1.0102	1.0091	1.0071	1.0043	1.0025	1.0004	0.9994	0.9983	0.9972	0.9959
1.3	1.0110	1.0098	1.0077	1.0047	1.0027	1.0004	0.9994	0.9982	0.9970	0.9956
1.4	1.0119	1.0106	1.0083	1.0051	1.0030	1.0004	0.9993	0.9981	0.9968	0.9953
1.5	1.0127	1.0113	1.0089	1.0054	1.0032	1.0005	0.9993	0.9980	0.9966	0.9950
1.6	1.0136	1.0121	1.0096	1.0058	1.0034	1.0006	0.9993	0.9979	0.9964	0.9947
1.7	1.0144	1.0128	1.0102	1.0062	1.0036	1.0006	0.9992	0.9978	0.9962	0.9944
1.8	1.0153	1.0136	1.0108	1.0066	1.0039	1.0007	0.9992	0.9977	0.9960	0.9941
1.9	1.0161	1.0144	1.0114	1.0070	1.0041	1.0008	0.9992	0.9976	0.9958	0.9938
2.0	1.0170	1.0151	1.0120	1.0073	1.0044	1.0008	0.9992	0.9975	0.9956	0.9935
2.1	1.0178	1.0159	1.0126	1.0077	1.0046	1.0009	0.9992	0.9974	0.9954	0.9932
2.2	1.0187	1.0167	1.0132	1.0081	1.0048	1.0010	0.9992	0.9973	0.9952	0.9929
2.3	1.0195	1.0174	1.0138	1.0085	1.0051	1.0010	0.9992	0.9972	0.9950	0.9927
2.4	1.0204	1.0182	1.0144	1.0089	1.0053	1.0011	0.9992	0.9971	0.9949	0.9924
2.5	1.0212	1.0189	1.0150	1.0093	1.0056	1.0012	0.9992	0.9971	0.9947	0.9921
2.6	1.0221	1.0197	1.0156	1.0097	1.0058	1.0013	0.9992	0.9970	0.9945	0.9919
2.7	1.0229	1.0205	1.0162	1.0101	1.0061	1.0014	0.9992	0.9969	0.9944	0.9916
2.8	1.0238	1.0212	1.0169	1.0104	1.0063	1.0014	0.9992	0.9968	0.9942	0.9914
2.9	1.0246	1.0220	1.0175	1.0108	1.0066	1.0015	0.9992	0.9968	0.9941	0.9911
3.0	1.0255	1.0228	1.0181	1.0112	1.0068	1.0016	0.9993	0.9967	0.9939	0.9908
3.1	1.0264	1.0235	1.0187	1.0116	1.0071	1.0017	0.9993	0.9966	0.9938	0.9906
3.2	1.0272	1.0243	1.0193	1.0120	1.0074	1.0018	0.9993	0.9966	0.9936	0.9904
3.3	1.0280	1.0250	1.0199	1.0124	1.0076	1.0019	0.9993	0.9965	0.9935	0.9901
3.4	1.0289	1.0258	1.0206	1.0128	1.0079	1.0020	0.9994	0.9965	0.9933	0.9899
3.5	1.0298	1.0266	1.0212	1.0133	1.0082	1.0021	0.9994	0.9964	0.9932	0.9896
3.6	1.0306	1.0273	1.0218	1.0137	1.0084	1.0022	0.9994	0.9964	0.9931	0.9894
3.7	1.0314	1.0281	1.0224	1.0141	1.0087	1.0024	0.9994	0.9963	0.9929	0.9892
3.8	1.0323	1.0289	1.0230	1.0145	1.0090	1.0025	0.9995	0.9963	0.9928	0.9890
3.9	1.0332	1.0296	1.0237	1.0149	1.0093	1.0026	0.9995	0.9963	0.9927	0.9888
4.0	1.0340	1.0304	1.0243	1.0153	1.0095	1.0027	0.9996	0.9962	0.9926	0.9885

From *Orifice Metering of Natural Gas*, 1969; courtesy of AGA.

Gas Flow Measurement 497

Table 10-10 Continued
Y_2 Expansion Factors—Pipe Taps
Static Pressure, Taken from Downstream Taps

$\beta = \dfrac{d}{D}$ Ratio

0.60	0.61	0.62	0.63	0.64	0.65	0.66	0.67	0.68	0.69	0.70
1.0000	1.0000	1.0000	1.0000	1.0000	1.0000	1.0000	1.0000	1.0000	1.0000	1.0000
0.9995	0.9994	0.9994	0.9993	0.9992	0.9991	0.9990	0.9989	0.9988	0.9987	0.9986
0.9990	0.9989	0.9988	0.9986	0.9985	0.9983	0.9981	0.9979	0.9977	0.9974	0.9972
0.9986	0.9984	0.9982	0.9979	0.9977	0.9974	0.9972	0.9969	0.9965	0.9962	0.9958
0.9981	0.9978	0.9976	0.9972	0.9969	0.9966	0.9962	0.9958	0.9954	0.9949	0.9944
0.9976	0.9973	0.9970	0.9966	0.9962	0.9958	0.9953	0.9948	0.9942	0.9936	0.9930
0.9972	0.9968	0.9964	0.9959	0.9954	0.9949	0.9944	0.9938	0.9931	0.9924	0.9916
0.9967	0.9962	0.9958	0.9953	0.9947	0.9941	0.9935	0.9928	0.9920	0.9912	0.9902
0.9962	0.9957	0.9952	0.9946	0.9940	0.9933	0.9926	0.9918	0.9909	0.9899	0.9889
0.9958	0.9952	0.9946	0.9940	0.9932	0.9925	0.9917	0.9908	0.9898	0.9887	0.9875
0.9954	0.9947	0.9940	0.9933	0.9925	0.9917	0.9908	0.9898	0.9887	0.9875	0.9862
0.9949	0.9942	0.9935	0.9927	0.9918	0.9909	0.9899	0.9888	0.9876	0.9863	0.9848
0.9945	0.9937	0.9929	0.9920	0.9911	0.9901	0.9890	0.9878	0.9865	0.9851	0.9835
0.9941	0.9932	0.9924	0.9914	0.9904	0.9893	0.9881	0.9868	0.9854	0.9839	0.9822
0.9936	0.9928	0.9918	0.9908	0.9897	0.9885	0.9872	0.9859	0.9844	0.9827	0.9809
0.9932	0.9923	0.9912	0.9902	0.9890	0.9877	0.9864	0.9849	0.9833	0.9815	0.9796
0.9928	0.9918	0.9907	0.9896	0.9883	0.9870	0.9855	9.9840	0.9822	0.9804	0.9783
0.9924	0.9913	0.9902	0.9889	0.9876	0.9862	0.9847	0.9830	0.9812	0.9792	0.9770
0.9920	0.9908	0.9896	0.9883	0.9870	0.9854	0.9838	0.9821	0.9801	0.9780	0.9757
0.9916	0.9904	0.9891	0.9877	0.9863	0.9847	0.9830	0.9811	0.9791	0.9769	0.9744
0.9912	0.9899	0.9886	0.9872	0.9856	0.9840	0.9822	0.9802	0.9781	0.9757	0.9732
0.9908	0.9895	0.9881	0.9866	0.9849	0.9832	0.9813	0.9793	0.9770	0.9746	0.9719
0.9904	0.9890	0.9876	0.9860	0.9843	0.9825	0.9805	0.9784	0.9760	0.9734	0.9706
0.9900	0.9886	0.9870	0.9854	0.9836	0.9817	0.9797	0.9774	0.9750	0.9723	0.9694
0.9896	0.9881	0.9865	0.9848	0.9830	0.9810	0.9789	0.9765	0.9740	0.9712	0.9681
0.9893	0.9877	0.9860	0.9842	0.9823	0.9803	0.9780	0.9756	0.9730	0.9701	0.9669
0.9889	0.9873	0.9855	0.9837	0.9817	0.9796	0.9772	0.9747	0.9720	0.9690	0.9657
0.9885	0.9868	0.9850	0.9831	0.9811	0.9788	0.9764	0.9738	0.9710	0.9679	0.9644
0.9882	0.9864	0.9846	0.9826	0.9804	0.9781	0.9757	0.9730	0.9700	0.9668	0.9632
0.9878	0.9860	0.9841	0.9820	0.9798	0.9774	0.9749	0.9721	0.9690	0.9657	0.9620
0.9874	0.9856	0.9836	0.9815	0.9792	0.9767	0.9741	0.9712	0.9681	0.9646	0.9608
0.9871	0.9852	0.9831	0.9809	0.9786	0.9760	0.9733	0.9703	0.9671	0.9635	0.9596
0.9867	0.9848	0.9826	0.9804	0.9780	0.9754	0.9725	0.9695	0.9661	0.9625	0.9584
0.9864	0.9843	0.9822	0.9798	0.9774	0.9747	0.9718	0.9686	0.9652	0.9614	0.9572
0.9860	0.9839	0.9817	0.9793	0.9768	0.9740	0.9710	0.9678	0.9642	0.9603	0.9561
0.9857	0.9835	0.9812	0.9788	0.9762	0.9733	0.9702	0.9669	0.9633	0.9593	0.9549
0.9854	0.9832	0.9808	0.9783	0.9756	0.9727	0.9695	0.9661	0.9623	0.9582	0.9537
0.9850	0.9828	0.9803	0.9778	0.9750	0.9720	0.9688	0.9652	0.9614	0.9572	0.9526
0.9847	0.9824	0.9799	0.9772	0.9744	0.9713	0.9680	0.9644	0.9605	0.9562	0.9514
0.9844	0.9820	0.9794	0.9767	0.9738	0.9707	0.9673	0.9636	0.9596	0.9551	0.9503
0.9840	0.9816	0.9790	0.9762	0.9732	0.9700	0.9665	0.9628	0.9586	0.9541	0.9491

(text continued from page 489)

Temperature-Base Factor, F_{tb}

This factor corrects for cases where the base (standard) temperature, T_b in °R, at which flow is to be measured is other than 520°R:

$$F_{tb} = \frac{T_b}{520} \qquad (10\text{-}19)$$

As expected, $F_{tb} = 1$ for a desired base temperature of 520°R.

Flowing Temperature Factor, F_{tf}

The flowing temperature factor corrects for cases where the flowing temperature, T_f (°R), is not 520°R, using the fact that the gas flow rate varies inversely as the square root of the absolute flow temperature:

$$F_{tf} = \left[\frac{520}{T_f}\right]^{0.5} \qquad (10\text{-}20)$$

Specific Gravity Factor, F_g

The basic orifice factor, F_b, is determined assuming a gas gravity of 1.0. So, a correction for gas gravity is required, as follows:

$$F_g = \frac{1}{\gamma_g^{0.5}} \qquad (10\text{-}21)$$

Supercompressibility Factor, F_{pv}

This factor corrects for the deviation of an actual gas from ideal-gas behavior. It is calculated as follows:

$$F_{pv} = \frac{Z_b}{Z^{0.5}} \qquad (10\text{-}22)$$

where Z_b, Z = gas compressibility factors at the base (generally assumed to be equal to 1.0) and operating conditions, respectively.

Because of the tremendous variations in gas compressibility factors with gas composition, pressure, and temperature, it is advisable to determine F_{pv} experimentally or through proven empirical techniques. AGA (AGA, 1969) provides two approximate empirical methods for determining the supercompressibility of natural gas mixtures: the specific gravity method, and the heating value method.

The specific gravity method uses the specific gravity, and the carbon dioxide (CO_2) and nitrogen (N_2) contents of the gas to calculate the pressure and temperature adjustment indices, f_{pg} and f_{tg}, respectively, as follows:

$$f_{pg} = \gamma_g - 13.84y_C + 5.420y_N$$

$$f_{tg} = \gamma_g - 0.472y_C - 0.793y_N \qquad (10\text{-}23)$$

where γ_g = specific gravity of the gas (air = 1)
 y_C = mole fraction of CO_2 in the gas
 y_N = mole fraction of N_2 in the gas

These f_{pg} and f_{tg} values are used to determine the pressure and temperature correction factors from Tables 10-11a and 10-11b, that are added to the actual flowing pressure and temperature of the gas, respectively. These corrected pressure and temperature values are used in Table 10-11e to estimate the supercompressibility factor, F_{pv}.

The heating value method uses the specific gravity (γ_g), total heating value (H_t in Btu/scf), and the CO_2 content of the gas to calculate the pressure and temperature adjustment indices, f_{ph} and f_{th}, respectively, as follows:

$$f_{ph} = \gamma_g - 0.0005688H_t + 3.690y_C$$

$$f_{th} = \gamma_g - 0.001814H_t + 2.641y_C \qquad (10\text{-}24)$$

These f_{ph} and f_{th} values are used to determine the pressure and temperature correction factors from Tables 10-11c and 10-11d, that are added to the actual flowing pressure and temperature of the gas, respectively. These corrected pressure and temperature values are used in Table 10-11e to estimate the supercompressibility factor, F_{pv}.

This method gives reasonable results for a gas with a gravity within 0.75, and non-hydrocarbon content of less than 12 mole% N_2 and/or 5 mole% CO_2.

Table 10-11a
Supercompressibility Pressure Adjustments, Δp
(Based on Specific Gravity Method)

Pressure Adjustment Index, f_{pa}	0	200	400	600	800	1000	1200	1400	1600	1800	2000	2200	2400	2600	2800	3000
									Pressure, psig							
−0.7	0	−11.32	−22.65	−33.97	−45.30	−56.62	−67.94	−79.27	−90.59	−101.92	−113.24	−124.56	−135.89	−147.21	−158.54	−169.86
−0.6	0	−10.50	−21.00	−31.49	−41.99	−52.49	−62.99	−73.49	−83.98	−94.48	−104.98	−115.48	−125.98	−136.47	−146.97	−157.47
−0.5	0	−9.67	−19.33	−29.00	−38.66	−48.33	−58.00	−67.66	−77.33	−86.99	−96.66	−106.33	−115.99	−125.66	−135.32	−144.99
−0.4	0	−8.83	−17.65	−26.48	−35.30	−44.13	−52.96	−61.78	−70.61	−79.43	−88.26	−97.09	−105.91	−114.74	−123.56	−132.39
−0.3	0	−7.98	−15.96	−23.93	−31.91	−39.89	−47.87	−55.85	−63.82	−71.80	−79.78	−87.76	−95.74	−103.71	−111.69	−119.67
−0.2	0	−7.12	−14.25	−21.37	−28.50	−35.62	−42.74	−49.87	−56.99	−64.12	−71.24	−78.36	−85.49	−92.61	−99.74	−106.86
−0.1	0	−6.26	−12.52	−18.78	−25.04	−31.30	−37.56	−43.82	−50.08	−56.34	−62.60	−68.86	−75.12	−81.38	−87.64	−93.90
0	0	−5.39	−10.78	−16.17	−21.56	−26.95	−32.34	−37.73	−43.12	−48.51	−53.90	−59.29	−64.68	−70.07	−75.46	−80.85
+0.1	0	−4.51	−9.02	−13.54	−18.05	−22.56	−27.07	−31.58	−36.10	−40.61	−45.12	−49.63	−54.14	−58.66	−63.17	−67.68
+0.2	0	−3.63	−7.25	−10.88	−14.50	−18.13	−21.76	−25.38	−29.01	−32.63	−36.26	−39.89	−43.51	−47.14	−50.76	−54.39
+0.3	0	−2.73	−5.46	−8.20	−10.93	−13.66	−16.39	−19.12	−21.86	−24.59	−27.32	−30.05	−32.78	−35.52	−38.25	−40.98
+0.4	0	−1.83	−3.66	−5.49	−7.32	−9.15	−10.98	−12.81	−14.64	−16.47	−18.30	−20.13	−21.96	−23.79	−25.62	−27.45
+0.5	0	−0.92	−1.84	−2.76	−3.68	−4.60	−5.52	−6.43	−7.35	−8.27	−9.19	−10.11	−11.03	−11.95	−12.87	−13.79
+0.6	0	0	0	0	0	0	0	0	0	0	0	0	0	0	0	0
+0.7	0	0.93	1.86	2.78	3.71	4.64	5.57	6.49	7.42	8.35	9.28	10.20	11.13	12.06	12.99	13.91
+0.8	0	1.86	3.73	5.59	7.46	9.32	11.18	13.05	14.91	16.78	18.64	20.50	22.37	24.23	26.10	27.96
+0.9	0	2.81	5.62	8.42	11.23	14.04	16.85	19.66	22.46	25.27	28.08	30.89	33.70	36.50	39.31	42.12
+1.0	0	3.76	7.52	11.29	15.05	18.81	22.57	26.33	30.10	33.86	37.62	41.38	45.14	48.91	52.67	56.43
+1.1	0	4.73	9.45	14.18	18.90	23.63	28.36	33.08	37.81	42.53	47.26	51.99	56.71	61.44	66.16	70.89
+1.2	0	5.70	11.40	17.09	22.79	28.49	34.19	39.89	45.58	51.28	56.98	62.68	68.38	74.07	79.77	85.47
+1.3	0	6.68	13.36	20.04	26.72	33.40	40.08	46.76	53.44	60.12	66.80	73.48	80.16	86.84	93.52	100.20
+1.4	0	7.67	15.34	23.01	30.68	38.35	46.02	53.69	61.36	69.03	76.70	84.37	92.04	99.71	107.38	115.05
+1.5	0	8.67	17.34	26.01	34.68	43.35	52.02	60.69	69.36	78.03	86.70	95.37	104.04	112.71	121.38	130.05
+1.6	0	9.68	19.36	29.04	38.72	48.40	58.08	67.76	77.44	87.12	96.80	106.48	116.16	125.84	135.52	145.20
+1.7	0	10.70	21.40	32.10	42.80	53.50	64.20	74.90	85.60	96.30	107.00	117.70	128.40	139.10	149.80	160.50
+1.8	0	11.73	23.46	35.19	46.92	58.65	70.38	82.11	93.84	105.57	117.30	129.03	140.76	152.49	164.22	175.95
+1.9	0	12.77	25.54	38.31	51.08	63.85	76.62	89.39	102.16	114.93	127.70	140.47	153.24	166.01	178.78	191.55
+2.0	0	13.82	27.64	41.46	55.28	69.10	82.92	96.74	110.56	124.38	138.20	152.02	165.84	179.66	193.48	207.30

Note: Factors for intermediate values of pressure adjustment index and pressure should be interpolated.
From *Orifice Metering of Natural Gas*, 1969; courtesy of AGA.

Gas Flow Measurement 501

Table 10-11b
Supercompressibility Temperature Adjustments, ΔT
(Based on Specific Gravity Method)

Temperature Adjustment Index, f_{tg}	Temperature, °F									
	0	20	40	60	80	100	120	140	160	180	200
0.45	75.16	78.43	81.70	84.97	88.24	91.50	94.77	98.04	101.31	104.58	107.84
0.46	69.41	72.43	75.45	78.47	81.49	84.50	87.52	90.54	93.56	96.58	99.59
0.47	63.76	66.53	69.30	72.07	74.84	77.62	80.39	83.16	85.93	88.70	91.48
0.48	58.24	60.77	63.30	65.83	68.36	70.90	73.43	75.96	78.49	81.02	83.56
0.49	52.81	55.10	57.40	59.70	61.99	64.29	66.58	68.88	71.18	73.47	75.77
0.50	47.52	49.58	51.65	53.72	55.78	57.85	59.91	61.98	64.05	66.11	68.18
0.51	42.33	44.17	46.01	47.85	49.69	51.53	53.37	55.21	57.05	58.89	60.73
0.52	37.25	38.87	40.48	42.10	43.72	45.34	46.96	48.58	50.20	51.82	53.44
0.53	32.26	33.67	35.07	36.47	37.88	39.28	40.68	42.08	43.49	44.89	46.29
0.54	27.38	28.57	29.76	30.95	32.14	33.33	34.52	35.71	36.90	38.09	39.28
0.55	22.60	23.58	24.56	25.54	26.52	27.51	28.49	29.47	30.45	31.44	32.42
0.56	17.90	18.68	19.46	20.23	21.01	21.79	22.57	23.35	24.12	24.90	25.68
0.57	13.29	13.87	14.45	15.03	15.61	16.18	16.76	17.34	17.92	18.50	19.07
0.58	8.78	9.16	9.54	9.92	10.30	10.68	11.07	11.45	11.83	12.21	12.59
0.59	4.35	4.54	4.73	4.92	5.10	5.29	5.48	5.67	5.86	6.05	6.24
0.60	0	0	0	0	0	0	0	0	0	0	0
0.61	−4.27	−4.45	−4.64	−4.82	−5.01	−5.19	−5.38	−5.57	−5.75	−5.94	−6.12
0.62	−8.45	−8.82	−9.19	−9.56	−9.93	−10.29	−10.66	−11.03	−11.40	−11.76	−12.13
0.63	−12.57	−13.11	−13.66	−14.21	−14.75	−15.30	−15.85	−16.39	−16.94	−17.48	−18.03
0.64	−16.61	−17.33	−18.05	−18.77	−19.49	−20.22	−20.94	−21.66	−22.38	−23.10	−23.83
0.65	−20.57	−21.47	−22.36	−23.25	−24.15	−25.04	−25.94	−26.83	−27.73	−28.62	−29.52
0.66	−24.47	−25.53	−26.60	−27.66	−28.72	−29.79	−30.85	−31.91	−32.98	−34.04	−35.11
0.67	−28.29	−29.52	−30.76	−31.99	−33.22	−34.45	−35.68	−36.91	−38.14	−39.37	−40.60
0.68	−32.06	−33.45	−34.84	−36.24	−37.63	−39.03	−40.42	−41.81	−43.21	−44.60	−46.00
0.69	−35.75	−37.30	−38.86	−40.41	−41.97	−43.52	−45.08	−46.63	−48.19	−49.74	−51.30
0.70	−39.38	−41.10	−42.81	−44.52	−46.23	−47.95	−49.66	−51.37	−53.08	−54.80	−56.51
0.71	−42.95	−44.82	−46.69	−48.56	−50.42	−52.29	−54.16	−56.03	−57.90	−59.76	−61.63
0.72	−46.46	−48.48	−50.50	−52.52	−54.54	−56.56	−58.58	−60.60	−62.62	−64.64	−66.66
0.73	−49.91	−52.08	−54.25	−56.42	−58.59	−60.76	−62.93	−65.10	−67.27	−69.44	−71.61
0.74	−53.31	−55.63	−57.95	−60.27	−62.59	−64.90	−67.22	−69.54	−71.86	−74.18	−76.49
0.75	−56.67	−59.14	−61.60	−64.06	−66.53	−68.99	−71.46	−73.92	−76.38	−78.85	−81.31

Note: Factors for intermediate values of temperature adjustment index and temperture should be interpolated.
From *Orifice Metering of Natural Gas*, 1969; courtesy of AGA.

Manometer Factor, F_m

This factor is required only where a mercury manometer is used for measuring the differential pressure. It compensates for the different heads of gas above the two mercury columns of the manometer. It is generally negligible, and is totally ignored for pressures below 500 psia. Table 10-12 gives this correction factor as a function of gas gravity, flowing pressure, and ambient temperature.

(text continued on page 517)

Table 10-11c
Supercompressibility Pressure Adjustments, Δp
(Based on Heating Value Method)

TABLE A.35(d)
Supercompressibility Pressure Adjustments, ΔP

Based on carbon dioxide content, heating value and specific gravity pressure adjustment index = $f_{pb} = G - 0.5688\,H_v/1000 - 3.690X_c$

Pressure Adjustment Index, f_{pb}	0	200	400	600	800	1000	1200	1400	1600	1800	2000	2200	2400	2600	2800	3000
									Pressure, psig							
−0.22	0	−11.25	−22.51	−33.76	−45.02	−56.27	−67.52	−78.78	−90.03	−101.29	−112.54	−123.79	−135.05	−146.30	−157.56	−168.81
−0.20	0	−10.27	−20.54	−30.80	−41.07	−51.34	−61.61	−71.88	−82.14	−92.41	−102.68	−112.95	−123.22	−133.48	−143.75	−154.02
−0.18	0	−9.27	−18.54	−27.82	−37.09	−46.36	−55.63	−64.90	−74.18	−83.45	−92.72	−101.99	−111.26	−120.54	−129.81	−139.08
−0.16	0	−8.27	−16.53	−24.80	−33.06	−41.33	−49.60	−57.86	−66.13	−74.39	−82.66	−90.93	−99.19	−107.46	−115.72	−123.99
−0.14	0	−7.25	−14.50	−21.74	−28.99	−36.24	−43.49	−50.74	−57.98	−65.23	−72.48	−79.73	−86.98	−94.22	−101.47	−108.72
−0.12	0	−6.22	−12.44	−18.66	−24.88	−31.10	−37.32	−43.54	−49.76	−55.98	−62.20	−68.42	−74.64	−80.86	−87.08	−93.30
−0.10	0	−5.18	−10.36	−15.55	−20.73	−25.91	−31.09	−36.27	−41.46	−46.64	−51.82	−57.00	−62.18	−67.37	−72.55	−77.73
−0.08	0	−4.13	−8.26	−12.40	−16.53	−20.66	−24.79	−28.92	−33.06	−37.19	−41.32	−45.45	−49.58	−53.72	−57.85	−61.98
−0.06	0	−3.07	−6.14	−9.21	−12.28	−15.35	−18.42	−21.49	−24.56	−27.63	−30.70	−33.77	−36.84	−39.91	−42.98	−46.05
−0.04	0	−2.00	−3.99	−5.99	−7.99	−9.98	−11.98	−13.98	−15.97	−17.97	−19.97	−21.96	−23.96	−25.96	−27.96	−29.95
−0.02	0	−0.91	−1.82	−2.74	−3.65	−4.56	−5.47	−6.38	−7.29	−8.21	−9.12	−10.03	−10.94	−11.85	−12.76	−13.68
0.00	0	0.18	0.37	0.56	0.74	0.92	1.11	1.30	1.48	1.67	1.85	2.04	2.22	2.41	2.59	2.78
+0.02	0	1.29	2.59	3.88	5.18	6.47	7.77	9.06	10.35	11.65	12.95	14.24	15.53	16.82	18.12	19.41
+0.04	0	2.42	4.83	7.25	9.66	12.08	14.50	16.91	19.33	21.74	24.16	26.58	28.99	31.41	33.82	36.24
+0.06	0	3.55	7.10	10.65	14.20	17.75	21.30	24.85	28.40	31.95	35.50	39.05	42.60	46.15	49.70	53.25
+0.08	0	4.70	9.39	14.09	18.78	23.48	28.18	32.87	37.57	42.26	46.96	51.66	56.35	61.05	65.74	70.44
+0.10	0	5.86	11.71	17.57	23.42	29.28	35.14	40.99	46.85	52.70	58.56	64.42	70.27	76.13	81.98	87.84
+0.12	0	7.03	14.06	21.08	28.11	35.14	42.17	49.20	56.22	63.25	70.28	77.31	84.34	91.36	98.39	105.52
+0.14	0	8.22	16.43	24.65	32.86	41.08	49.30	57.51	65.73	73.94	82.16	90.38	98.59	106.81	115.02	123.24
+0.16	0	9.42	18.83	28.25	37.66	47.08	56.50	65.91	75.33	84.74	94.16	103.58	112.99	122.41	131.82	141.24
+0.18	0	10.63	21.26	31.89	42.52	53.15	63.78	74.41	85.04	95.67	106.30	116.93	127.56	138.19	148.82	159.45
+0.20	0	11.86	23.72	35.57	47.43	59.29	71.15	83.01	94.86	106.72	118.58	130.44	142.30	154.15	166.01	177.87
+0.22	0	13.10	26.20	39.30	52.40	65.50	78.60	91.70	104.80	117.90	131.00	144.10	157.20	170.30	183.40	196.50

Note: Factors for intermediate values of pressure adjustment index and pressure should be interpolated.

From *Orifice Metering of Natural Gas*, 1969; courtesy of AGA.

Gas Flow Measurement 503

Table 10-11d
Supercompressibility Temperature Adjustments, ΔT
(Based on Heating Value Method)

Temperature Adjustment Index, f_{tb}	Temperature, °F									
	0	20	40	60	80	100	120	140	160	180	200
2.10	56.03	58.46	60.90	63.34	65.77	68.21	70.64	73.08	75.52	77.95	80.39
2.12	53.13	55.44	57.75	60.06	62.37	64.68	66.99	69.30	71.61	73.92	76.23
2.14	50.19	52.37	54.55	56.73	58.91	61.10	63.28	65.46	67.64	69.82	72.01
2.16	47.29	49.34	51.40	53.46	55.51	57.57	59.62	61.68	63.74	65.79	67.85
2.18	44.46	46.40	48.33	50.26	52.20	54.13	56.06	58.00	59.93	61.86	63.80
2.20	41.64	43.45	45.26	47.08	48.89	50.70	52.51	54.32	56.13	57.94	59.75
2.22	38.86	40.54	42.24	43.92	45.61	47.30	48.99	50.68	52.37	54.06	55.75
2.24	36.10	37.66	39.24	40.80	42.37	43.94	45.51	47.08	48.65	50.22	51.79
2.26	33.37	34.82	36.28	37.73	39.18	40.63	42.08	43.53	44.98	46.43	47.88
2.28	30.67	32.01	33.34	34.67	36.01	37.34	38.67	40.01	41.34	42.68	44.01
2.30	28.01	29.23	30.44	31.66	32.88	34.10	35.32	36.53	37.75	38.97	40.19
2.32	25.37	26.47	27.58	28.68	29.78	30.88	31.99	33.09	34.19	35.30	36.40
2.34	22.76	23.75	24.74	25.73	26.72	27.71	28.70	29.69	30.68	31.67	31.66
2.36	20.18	21.05	21.93	22.81	23.68	24.56	25.44	26.32	27.19	28.07	28.95
2.38	17.62	18.39	19.16	19.92	20.69	21.45	22.22	22.99	23.75	24.52	25.28
2.40	15.09	15.75	16.40	17.06	17.72	18.37	19.03	19.69	20.34	21.00	21.65
2.42	12.59	13.14	13.69	14.24	14.78	15.33	15.88	16.43	16.98	17.52	18.07
2.44	10.12	10.56	11.00	11.44	11.88	12.32	12.76	13.20	13.64	14.08	14.52
2.46	7.67	8.00	8.34	8.67	9.00	9.34	9.67	10.00	10.34	10.67	11.00
2.48	5.24	5.47	5.70	5.93	6.16	6.38	6.61	6.84	7.07	7.30	7.52
2.50	2.85	2.97	3.10	3.22	3.34	3.46	3.59	3.71	3.84	3.96	4.08
2.52	0.47	0.49	0.51	0.53	0.55	0.57	0.60	0.62	0.64	0.66	0.68
2.54	−1.88	−1.96	−2.04	−2.12	−2.20	−2.29	−2.37	−2.45	−2.53	−2.61	−2.69
2.56	−4.20	−4.39	−4.57	−4.75	−4.94	−4.12	−5.30	−5.48	−5.67	−5.85	−6.03
2.58	−6.50	−6.79	−7.07	−7.35	−7.64	−7.92	−8.20	−8.48	−8.77	−9.05	−9.33
2.60	−8.79	−9.17	−9.55	−9.93	−10.31	−10.70	−11.08	−11.46	−11.84	−12.22	−12.61
2.62	−11.04	−11.52	−12.00	−12.48	−12.96	−13.44	−13.92	−14.41	−14.89	−15.37	−15.85
2.64	−13.28	−13.85	−14.43	−15.01	−15.58	−16.16	−16.74	−17.32	−17.89	−18.47	−19.05
2.66	−15.49	−16.16	−16.84	−17.51	−18.18	−18.86	−19.53	−20.20	−20.88	−21.55	−22.22
2.68	−17.68	−18.45	−19.22	−19.98	−20.75	−21.52	−22.29	−23.06	−23.83	−24.60	−25.36
2.70	−19.85	−20.71	−21.58	−22.44	−23.30	−24.16	−25.03	−25.89	−26.75	−27.62	−28.48
2.72	−22.00	−22.95	−23.91	−24.87	−25.82	−26.78	−27.74	−28.69	−29.65	−30.60	−31.56
2.74	−24.12	−25.17	−26.22	−27.27	−28.32	−29.37	−30.42	−31.46	−32.51	−33.56	−34.61
2.76	−26.23	−27.37	−28.51	−29.65	−30.79	−31.93	−33.07	−34.21	−35.35	−36.49	−37.63
2.78	−28.31	−29.54	−30.78	−32.01	−33.24	−34.47	−35.70	−36.93	−38.16	−39.39	−40.62
2.80	−30.38	−31.70	−33.02	−34.34	−35.66	−36.98	−38.30	−39.62	−40.94	−42.26	−43.59
2.82	−32.42	−33.84	−35.24	−36.65	−38.06	−39.47	−40.88	−42.29	−43.70	−45.11	−46.52
2.84	−34.45	−35.95	−37.45	−38.95	−40.45	−41.94	−43.44	−44.94	−46.44	−47.94	−49.43
2.86	−36.46	−38.04	−39.63	−41.22	−42.80	−44.38	−45.97	−47.56	−49.14	−50.73	−52.31
2.88	−38.45	−40.12	−41.80	−43.47	−45.14	−46.81	−48.48	−40.15	−51.82	−53.50	−55.17
2.90	−40.42	−42.18	−43.94	−45.69	−47.45	−49.21	−50.96	−52.72	−54.48	−56.24	−57.99
2.92	−42.37	−44.21	−46.06	−47.90	−49.74	−51.58	−53.42	−55.27	−57.11	−58.95	−60.79
2.94	−44.31	−46.23	−48.16	−50.09	−52.01	−53.94	−55.86	−57.79	−59.72	−61.64	−63.57
2.96	−46.23	−48.24	−50.25	−52.26	−54.27	−56.28	−58.29	−60.30	−62.31	−64.32	−66.33
2.98	−48.12	−50.21	−52.30	−54.39	−56.48	−58.58	−60.67	−62.76	−64.85	−66.94	−69.04
3.00	−50.00	−52.18	−54.35	−56.52	−58.70	−60.87	−63.05	−65.22	−67.39	−69.57	−71.74
3.02	−51.89	−54.14	−56.40	−58.66	−60.91	−63.17	−65.42	−67.68	−69.94	−72.19	−74.45
3.04	−53.73	−56.06	−58.40	−60.74	−63.07	−65.40	−67.74	−70.08	−72.42	−74.75	−77.09
306	−55.57	−57.98	−60.40	−62.82	−65.23	−67.65	−70.06	−72.48	−74.90	−77.31	−79.73
3.08	−57.36	−59.86	−62.35	−64.84	−67.34	−69.83	−72.33	−74.82	−77.31	−79.81	−82.30
3.10	−59.16	−61.73	−64.30	−66.87	−69.44	−72.02	−74.59	−77.16	−79.73	−82.30	−84.88
3.12	−60.95	−63.60	−66.25	−68.90	−71.55	−74.20	−76.85	−79.50	−82.15	−84.80	−87.45
3.14	−62.70	−65.42	−68.15	−70.88	−73.60	−76.33	−79.05	−81.78	−84.51	−87.23	−89.96
3.16	−64.45	−67.25	−70.05	−72.85	−75.65	−78.46	−81.26	−84.06	−86.86	−89.66	−92.47
3.18	−66.19	−69.07	−71.95	−74.83	−77.71	−80.58	−83.46	−86.34	−89.22	−92.10	−94.97
3.20	−67.90	−70.85	−73.80	−76.75	−79.70	−82.66	−85.61	−88.56	−91.51	−94.46	−97.42

Note: Factors for intermediate values of temperature adjustment index and temperature should be interpolated.

From *Orifice Metering of Natural Gas*, 1969; courtesy of AGA.

Table 10-11e*
F_{pv} Supercompressibility Factors
(Base Data—0.6 Specific Gravity Hydrocarbon Gas)

p_f psig	Temperature, °F							
	-40	-35	-30	-25	-20	-15	-10	-5
0	1.0000	1.0000	1.0000	1.0000	1.0000	1.0000	1.0000	1.0000
20	1.0031	1.0030	1.0029	1.0028	1.0027	1.0026	1.0025	1.0024
40	1.0062	1.0060	1.0059	1.0057	1.0055	1.0053	1.0051	1.0049
60	1.0093	1.0091	1.0089	1.0086	1.0083	1.0080	1.0077	1.0074
80	1.0125	1.0122	1.0119	1.0115	1.0111	1.0107	1.0103	1.0099
100	1.0158	1.0154	1.0148	1.0145	1.0139	1.0134	1.0129	1.0125
120	1.0192	1.0186	1.0178	1.0175	1.0168	1.0162	1.0156	1.0151
140	1.0227	1.0218	1.0209	1.0205	1.0198	1.0190	1.0183	1.0177
160	1.0262	1.0251	1.0241	1.0235	1.0228	1.0218	1.0210	1.0202
180	1.0297	1.0285	1.0274	1.0265	1.0258	1.0246	1.0237	1.0228
200	1.0333	1.0319	1.0307	1.0296	1.0288	1.0275	1.0265	1.0255
220	1.0369	1.0335	1.0340	1.0328	1.0317	1.0304	1.0291	1.0282
240	1.0406	1.0388	1.0373	1.0360	1.0347	1.0334	1.0321	1.0309
260	1.0444	1.0424	1.0407	1.0392	1.0377	1.0364	1.0350	1.0337
280	1.0482	1.0461	1.0442	1.0425	1.0408	1.0394	1.0379	1.0365
300	1.0522	1.0499	1.0478	1.0459	1.0441	1.0425	1.0409	1.0393
320	1.0562	1.0537	1.0514	1.0494	1.0474	1.0456	1.0439	1.0422
340	1.0602	1.0575	1.0551	1.0529	1.0507	1.0488	1.0469	1.0451
360	1.0642	1.0614	1.0589	1.0564	1.0541	1.0520	1.0500	1.0480
380	1.0684	1.0654	1.0627	1.0601	1.0576	1.0553	1.0531	1.0510
400	1.0727	1.0695	1.0666	1.0638	1.0611	1.0586	1.0563	1.0540
420	1.0771	1.0737	1.0706	1.0675	1.0646	1.0620	1.0595	1.0571
440	1.0816	1.0779	1.0746	1.0713	1.0682	1.0654	1.0627	1.0601
460	1.0862	1.0822	1.0787	1.0752	1.0719	1.0688	1.0660	1.0632
480	1.0909	1.0866	1.0828	1.0791	1.0756	1.0723	1.0693	1.0664
500	1.0956	1.0910	1.0869	1.0830	1.0793	1.0759	1.0727	1.0696
520	1.1004	1.0956	1.0911	1.0869	1.0830	1.0794	1.0761	1.0728
540	1.1055	1.1002	1.0955	1.0910	1.0868	1.0830	1.0795	1.0760
560	1.1106	1.1051	1.1000	1.0952	1.0908	1.0868	1.0830	1.0793
580	1.1159	1.1100	1.1045	1.0995	1.0948	1.0906	1.0865	1.0826
600	1.1213	1.1149	1.1091	1.1038	1.0989	1.0944	1.0901	1.0860
620	1.1267	1.1200	1.1138	1.1082	1.1030	1.0982	1.0937	1.0894
640	1.1323	1.1252	1.1186	1.1127	1.1072	1.1021	1.0973	1.0928
660	1.1379	1.1305	1.1236	1.1172	1.1114	1.1060	1.1010	1.0963
680	1.1439	1.1359	1.1286	1.1218	1.1156	1.1099	1.1047	1.0998
700	1.1499	1.1413	1.1336	1.1265	1.1199	1.1138	1.1083	1.1033
720	1.1562	1.1469	1.1388	1.1313	1.1245	1.1181	1.1123	1.1069
740	1.1626	1.1528	1.1442	1.1363	1.1291	1.1225	1.1162	1.1106
760	1.1692	1.1587	1.1496	1.1413	1.1337	1.1267	1.1202	1.1143
780	1.1759	1.1647	1.1551	1.1464	1.1384	1.1311	1.1242	1.1180
800	1.1826	1.1708	1.1607	1.1516	1.1432	1.1355	1.1283	1.1217
820	1.1894	1.1769	1.1663	1.1568	1.1480	1.1399	1.1324	1.1255
840	1.1967	1.1835	1.1723	1.1622	1.1528	1.1443	1.1365	1.1293
860	1.2041	1.1901	1.1783	1.1676	1.1577	1.1488	1.1407	1.1332
880	1.2116	1.1968	1.1843	1.1731	1.1627	1.1533	1.1449	1.1373

From *Orifice Metering of Natural Gas*, 1969; courtesy of AGA.

Gas Flow Measurement 505

Table 10-11e Continued
F_{pv} Supercompressibility Factors
(Base Data—0.6 Specific Gravity Hydrocarbon Gas)

p_f psig	Temperature, °F							
	-40	-35	-30	-25	-20	-15	-10	-5
900	1.2191	1.2035	1.1903	1.1786	1.1677	1.1579	1.1491	1.1410
920	1.2269	1.2103	1.1965	1.1842	1.1728	1.1625	1.1534	1.1450
940	1.2347	1.2173	1.2028	1.1899	1.1780	1.1674	1.1577	1.1490
960	1.2427	1.2245	1.2093	1.1956	1.1832	1.1721	1.1620	1.1530
980	1.2509	1.2318	1.2157	1.2014	1.1884	1.1768	1.1663	1.1570
1000	1.2591	1.2391	1.2221	1.2072	1.1936	1.1815	1.1706	1.1610
1020	1.2673	1.2464	1.2286	1.2131	1.1990	1.1864	1.1751	1.1650
1040	1.2756	1.2537	1.2351	1.2190	1.2044	1.1913	1.1796	1.1690
1060	1.2839	1.2611	1.2418	1.2250	1.2098	1.1962	1.1841	1.1731
1080	1.2922	1.2685	1.2485	1.2310	1.2152	1.2011	1.1886	1.1772
1100	1.3008	1.2759	1.2552	1.2370	1.2206	1.2060	1.1933	1.1813
1120	1.3091	1.2834	1.2619	1.2431	1.2260	1.2109	1.1978	1.1854
1140	1.3176	1.2909	1.2686	1.2492	1.2315	1.2159	1.2023	1.1896
1160	1.3259	1.2985	1.2753	1.2552	1.2370	1.2209	1.2068	1.1939
1180	1.3337	1.3056	1.2820	1.2612	1.2425	1.2258	1.2111	1.1979
1200	1.3412	1.3127	1.2883	1.2669	1.2477	1.2305	1.2154	1.2018
1220	1.3486	1.3196	1.2946	1.2726	1.2529	1.2352	1.2197	1.2058
1240	1.3559	1.3264	1.3009	1.2783	1.2580	1.2399	1.2240	1.2098
1260	1.3628	1.3329	1.3071	1.2839	1.2631	1.2446	1.2283	1.2138
1280	1.3692	1.3390	1.3128	1.2894	1.2682	1.2493	1.2326	1.2176
1300	1.3754	1.3448	1.3184	1.2947	1.2732	1.2540	1.2369	1.2214
1320	1.3812	1.3505	1.3240	1.3000	1.2782	1.2586	1.2411	1.2252
1340	1.3867	1.3561	1.3294	1.3053	1.2832	1.2631	1.2451	1.2289
1360	1.3917	1.3611	1.3344	1.3101	1.2878	1.2675	1.2491	1.2326
1380	1.3961	1.3655	1.3388	1.3145	1.2920	1.2715	1.2530	1.2362
1400	1.4002	1.3699	1.3432	1.3186	1.2960	1.2754	1.2568	1.2398
1420	1.4037	1.3738	1.3473	1.3228	1.3000	1.2792	1.2604	1.2432
1440	1.4069	1.3774	1.3508	1.3264	1.3038	1.2830	1.2640	1.2466
1460	1.4096	1.3805	1.3540	1.3298	1.3072	1.2864	1.2673	1.2498
1480	1.4118	1.3833	1.3571	1.3331	1.3105	1.2894	1.2703	1.2530
1500	1.4137	1.3857	1.3597	1.3357	1.3132	1.2924	1.2735	1.2558
1520	1.4152	1.3878	1.3621	1.3384	1.3161	1.2954	1.2763	1.2586
1540	1.4164	1.3896	1.3643	1.3408	1.3186	1.2979	1.2788	1.2612
1560	1.4172	1.3910	1.3661	1.3428	1.3207	1.3004	1.2813	1.2638
1580	1.4177	1.3922	1.3677	1.3445	1.3228	1.3027	1.2838	1.2661
1600	1.4179	1.3930	1.3690	1.3462	1.3247	1.3047	1.2860	1.2683
1620	1.4179	1.3936	1.3700	1.3476	1.3263	1.3064	1.2878	1.2702
1640	1.4176	1.3938	1.3708	1.3488	1.3278	1.3079	1.2895	1.2720
1660	1.4170	1.3939	1.3713	1.3497	1.3289	1.3094	1.2912	1.2738
1680	1.4162	1.3936	1.3716	1.3504	1.3300	1.3108	1.2928	1.2755
1700	1.4151	1.3932	1.3718	1.3510	1.3309	1.3119	1.2940	1.2769
1720	1.4139	1.3926	1.3715	1.3513	1.3317	1.3130	1.2951	1.2782
1740	1.4126	1.3919	1.3712	1.3514	1.3321	1.3137	1.2961	1.2793
1760	1.4111	1.3909	1.3707	1.3513	1.3321	1.3143	1.2970	1.2804
1780	1.4094	1.3897	1.3701	1.3511	1.3322	1.3148	1.2977	1.2812

(table continued)

506 Gas Production Engineering

Table 10-11e Continued
F_{pv} Supercompressibility Factors
(Base Data—0.6 Specific Gravity Hydrocarbon Gas)

p_f psig	Temperature, °F							
	−40	−35	−30	−25	−20	−15	−10	−5
1800	1.4075	1.3884	1.3693	1.3507	1.3323	1.3151	1.2983	1.2819
1820	1.4056	1.3870	1.3684	1.3502	1.3324	1.3153	1.2988	1.2826
1840	1.4035	1.3855	1.3673	1.3496	1.3321	1.3153	1.2990	1.2831
1860	1.4012	1.3837	1.3661	1.3488	1.3317	1.3152	1.2991	1.2835
1880	1.3989	1.3818	1.3647	1.3478	1.3312	1.3150	1.2992	1.2838
1900	1.3965	1.3799	1.3632	1.3468	1.3305	1.3146	1.2990	1.2839
1920	1.3940	1.3779	1.3617	1.3457	1.3298	1.3142	1.2989	1.2840
1940	1.3914	1.3758	1.3601	1.3444	1.3289	1.3136	1.2986	1.2841
1960	1.3888	1.3737	1.3584	1.3431	1.3279	1.3129	1.2982	1.2839
1980	1.3861	1.3714	1.3566	1.3416	1.3267	1.3120	1.2977	1.2836
2000	1.3834	1.3691	1.3547	1.3400	1.3254	1.3110	1.2971	1.2833
2020	1.3806	1.3667	1.3527	1.3384	1.3241	1.3100	1.2963	1.2828
2040	1.3778	1.3642	1.3506	1.3368	1.3228	1.3089	1.2955	1.2823
2060	1.3749	1.3617	1.3484	1.3351	1.3212	1.3078	1.2947	1.2817
2080	1.3720	1.3591	1.3462	1.3332	1.3196	1.3065	1.2937	1.2809
2100	1.3690	1.3565	1.3439	1.3312	1.3180	1.3052	1.2926	1.2801
2120	1.3660	1.3539	1.3416	1.3292	1.3164	1.3039	1.2915	1.2793
2140	1.3630	1.3513	1.3392	1.3271	1.3147	1.3025	1.2903	1.2784
2160	1.3600	1.3486	1.3367	1.3250	1.3129	1.3010	1.2891	1.2774
2180	1.3569	1.3459	1.3343	1.3228	1.3110	1.2994	1.2878	1.2764
2200	1.3538	1.3431	1.3318	1.3206	1.3091	1.2978	1.2864	1.2753
2220	1.3507	1.3402	1.3295	1.3184	1.3071	1.2961	1.2850	1.2741
2240	1.3476	1.3373	1.3268	1.3162	1.3051	1.2943	1.2835	1.2729
2260	1.3444	1.3344	1.3243	1.3139	1.3031	1.2925	1.2820	1.2716
2280	1.3412	1.3315	1.3217	1.3116	1.3011	1.2907	1.2804	1.2702
2300	1.3380	1.3286	1.3191	1.3092	1.2990	1.2889	1.2788	1.2688
2320	1.3349	1.3257	1.3164	1.3068	1.2969	1.2870	1.2772	1.2674
2340	1.3317	1.3228	1.3137	1.3044	1.2947	1.2851	1.2755	1.2659
2360	1.3285	1.3199	1.3110	1.3019	1.2925	1.2831	1.2737	1.2643
2380	1.3254	1.3170	1.3083	1.2994	1.2903	1.2811	1.2719	1.2627
2400	1.3223	1.3141	1.3056	1.2969	1.2880	1.2790	1.2700	1.2611
2420	1.3191	1.3112	1.3029	1.2944	1.2857	1.2769	1.2682	1.2594
2440	1.3159	1.3082	1.3002	1.2919	1.2734	1.2748	1.2663	1.2577
2460	1.3128	1.3052	1.2975	1.2894	1.2811	1.2727	1.2644	1.2560
2480	1.3096	1.3022	1.2948	1.2869	1.2788	1.2706	1.2624	1.2542
2500	1.3064	1.2992	1.2921	1.2843	1.2764	1.2684	1.2604	1.2524
2520	1.3033	1.2963	1.2893	1.2817	1.2741	1.2663	1.2585	1.2506
2540	1.3001	1.2934	1.2864	1.2792	1.2717	1.2642	1.2566	1.2488
2560	1.2970	1.2904	1.2835	1.2766	1.2693	1.2620	1.2546	1.2470
2580	1.2939	1.2875	1.2807	1.2740	1.2669	1.2597	1.2525	1.2451
2600	1.2909	1.2846	1.2780	1.2714	1.2645	1.2575	1.2505	1.2433
2620	1.2878	1.2817	1.2753	1.2687	1.2620	1.2553	1.2484	1.2414
2640	1.2847	1.2787	1.2725	1.2661	1.2596	1.2530	1.2462	1.2394
2660	1.2816	1.2758	1.2697	1.2635	1.2572	1.2507	1.2441	1.2375
2680	1.2785	1.2729	1.2670	1.2609	1.2547	1.2484	1.2420	1.2356

Gas Flow Measurement 507

Table 10-11e Continued
F_{pv} Supercompressibility Factors
(Base Data—0.6 Specific Gravity Hydrocarbon Gas)

p_f psig	Temperature, °F							
	-40	-35	-30	-25	-20	-15	-10	-5
2700	1.2754	1.2700	1.2643	1.2584	1.2523	1.2461	1.2399	1.2336
2720	1.2723	1.2670	1.2614	1.2557	1.2498	1.2438	1.2377	1.2315
2740	1.2693	1.2641	1.2587	1.2531	1.2473	1.2414	1.2355	1.2295
2760	1.2663	1.2612	1.2559	1.2505	1.2448	1.2391	1.2334	1.2275
2780	1.2633	1.2584	1.2532	1.2479	1.2424	1.2368	1.2312	1.2255
2800	1.2603	1.2555	1.2504	1.2454	1.2400	1.2345	1.2290	1.2234
2820	1.2573	1.2526	1.2476	1.2427	1.2374	1.2322	1.2268	1.2213
2840	1.2543	1.2497	1.2448	1.2401	1.2349	1.2298	1.2246	1.2193
2860	1.2513	1.2469	1.2421	1.2375	1.2324	1.2274	1.2224	1.2172
2880	1.2483	1.2441	1.2394	1.2349	1.2300	1.2251	1.2202	1.2152
2900	1.2454	1.2413	1.2368	1.2324	1.2276	1.2228	1.2180	1.2131
2920	1.2424	1.2384	1.2341	1.2298	1.2252	1.2205	1.2158	1.2110
2940	1.2395	1.2356	1.2314	1.2272	1.2227	1.2181	1.2135	1.2089
2960	1.2366	1.2328	1.2287	1.2246	1.2202	1.2157	1.2112	1.2067
2980	1.2338	1.2301	1.2261	1.2221	1.2178	1.2134	1.2091	1.2047
3000	1.2309	1.2273	1.2234	1.2195	1.2153	1.2111	1.2069	1.2027

p_f psig	Temperature, °F										
	0	5	10	15	20	25	30	35	40	45	50	55
0	1.0000	1.0000	1.0000	1.0000	1.0000	1.0000	1.0000	1.0000	1.0000	1.0000	1.0000	1.0000
20	1.0023	1.0022	1.0022	1.0021	1.0020	1.0020	1.0019	1.0018	1.0018	1.0017	1.0016	1.0016
40	1.0048	1.0047	1.0045	1.0044	1.0042	1.0041	1.0040	1.0038	1.0037	1.0036	1.0034	1.0033
60	1.0071	1.0069	1.0067	1.0065	1.0063	1.0061	1.0059	1.0057	1.0054	1.0053	1.0051	1.0049
80	1.0096	1.0093	1.0090	1.0087	1.0084	1.0081	1.0078	1.0076	1.0073	1.0070	1.0068	1.0066
100	1.0121	1.0117	1.0113	1.0109	1.0105	1.0102	1.0098	1.0095	1.0091	1.0088	1.0085	1.0083
120	1.0146	1.0141	1.0136	1.0131	1.0127	1.0122	1.0118	1.0114	1.0110	1.0106	1.0103	1.0100
140	1.0170	1.0164	1.0158	1.0152	1.0148	1.0142	1.0138	1.0132	1.0128	1.0124	1.0120	1.0116
160	1.0195	1.0188	1.0182	1.0176	1.0169	1.0163	1.0158	1.0152	1.0147	1.0142	1.0138	1.0133
180	1.0220	1.0213	1.0206	1.0198	1.0191	1.0184	1.0178	1.0171	1.0166	1.0160	1.0155	1.0150
200	1.0245	1.0237	1.0229	1.0220	1.0213	1.0206	1.0198	1.0192	1.0185	1.0179	1.0173	1.0167
220	1.0272	1.0263	1.0254	1.0244	1.0235	1.0227	1.0219	1.0211	1.0204	1.0197	1.0191	1.0184
240	1.0298	1.0288	1.0277	1.0267	1.0257	1.0248	1.0239	1.0231	1.0223	1.0215	1.0208	1.0201
260	1.0324	1.0313	1.0302	1.0291	1.0280	1.0270	1.0260	1.0250	1.0242	1.0234	1.0226	1.0219
280	1.0351	1.0339	1.0327	1.0315	1.0303	1.0292	1.0281	1.0271	1.0261	1.0252	1.0244	1.0236
300	1.0379	1.0365	1.0352	1.0339	1.0326	1.0314	1.0303	1.0291	1.0281	1.0271	1.0262	1.0253
320	1.0406	1.0391	1.0377	1.0363	1.0349	1.0336	1.0324	1.0312	1.0300	1.0290	1.0280	1.0270
340	1.0434	1.0417	1.0401	1.0386	1.0372	1.0358	1.0344	1.0332	1.0320	1.0308	1.0298	1.0287
360	1.0462	1.0444	1.0427	1.0411	1.0395	1.0380	1.0366	1.0353	1.0340	1.0328	1.0316	1.0305
380	1.0491	1.0471	1.0453	1.0436	1.0420	1.0404	1.0388	1.0374	1.0361	1.0347	1.0334	1.0322
400	1.0519	1.0498	1.0479	1.0461	1.0444	1.0427	1.0410	1.0395	1.0381	1.0366	1.0352	1.0340
420	1.0548	1.0526	1.0506	1.0486	1.0468	1.0450	1.0433	1.0417	1.0401	1.0386	1.0371	1.0358
440	1.0577	1.0553	1.0531	1.0511	1.0492	1.0472	1.0453	1.0437	1.0421	1.0405	1.0389	1.0375
460	1.0606	1.0581	1.0558	1.0536	1.0516	1.0496	1.0476	1.0458	1.0441	1.0425	1.0408	1.0393
480	1.0636	1.0609	1.0585	1.0562	1.0540	1.0519	1.0498	1.0479	1.0461	1.0444	1.0427	1.0411

(table continued)

Table 10-11e Continued
F_{pv} Supercompressibility Factors
(Base Data—0.6 Specific Gravity Hydrocarbon Gas)

p_f psig	\multicolumn{11}{c}{Temperature, °F}										
	0	5	10	15	20	25	30	35	40	45	50	55
500	1.0667	1.0639	1.0613	1.0588	1.0565	1.0543	1.0521	1.0501	1.0482	1.0464	1.0446	1.0429
520	1.0697	1.0667	1.0639	1.0613	1.0588	1.0565	1.0543	1.0522	1.0503	1.0484	1.0465	1.0447
540	1.0727	1.0696	1.0667	1.0640	1.0613	1.0588	1.0564	1.0543	1.0523	1.0503	1.0483	1.0465
560	1.0759	1.0726	1.0695	1.0666	1.0639	1.0612	1.0587	1.0565	1.0544	1.0523	1.0502	1.0483
580	1.0790	1.0757	1.0724	1.0693	1.0665	1.0637	1.0611	1.0587	1.0565	1.0543	1.0521	1.0501
600	1.0822	1.0787	1.0753	1.0721	1.0691	1.0661	1.0634	1.0609	1.0586	1.0562	1.0540	1.0519
620	1.0853	1.0816	1.0781	1.0747	1.0716	1.0685	1.0656	1.0631	1.0607	1.0582	1.0559	1.0538
640	1.0886	1.0848	1.0811	1.0775	1.0742	1.0710	1.0680	1.0653	1.0628	1.0602	1.0578	1.0556
660	1.0919	1.0879	1.0840	1.0802	1.0767	1.0735	1.0704	1.0675	1.0649	1.0623	1.0598	1.0574
680	1.0953	1.0910	1.0869	1.0830	1.0793	1.0760	1.0728	1.0698	1.0670	1.0643	1.0617	1.0593
700	1.0986	1.0941	1.0898	1.0857	1.0819	1.0784	1.0751	1.0720	1.0691	1.0663	1.0636	1.0611
720	1.1020	1.0973	1.0928	1.0885	1.0847	1.0810	1.0775	1.0742	1.0712	1.0684	1.0656	1.0630
740	1.1054	1.1005	1.0958	1.0914	1.0873	1.0835	1.0799	1.0766	1.0734	1.0704	1.0675	1.0648
760	1.1089	1.1038	1.0989	1.0943	1.0900	1.0860	1.0822	1.0788	1.0756	1.0725	1.0694	1.0667
780	1.1124	1.1070	1.1019	1.0972	1.0927	1.0885	1.0846	1.0810	1.0777	1.0745	1.0714	1.0685
800	1.1159	1.1103	1.1050	1.1000	1.0954	1.0911	1.0870	1.0833	1.0798	1.0765	1.0733	1.0704
820	1.1193	1.1135	1.1080	1.1029	1.0981	1.0936	1.0894	1.0856	1.0819	1.0785	1.0752	1.0722
840	1.1229	1.1169	1.1112	1.1057	1.1008	1.0962	1.0919	1.0879	1.0841	1.0805	1.0771	1.0740
860	1.1265	1.1202	1.1143	1.1087	1.1037	1.0989	1.0943	1.0902	1.0863	1.0826	1.0792	1.0759
880	1.1301	1.1236	1.1175	1.1117	1.1064	1.1015	1.0968	1.0925	1.0885	1.0847	1.0811	1.0778
900	1.1337	1.1270	1.1206	1.1146	1.1091	1.1040	1.0991	1.0947	1.0906	1.0867	1.0830	1.0795
920	1.1373	1.1303	1.1237	1.1175	1.1118	1.1066	1.1016	1.0970	1.0928	1.0887	1.0849	1.0813
940	1.1410	1.1338	1.1269	1.1205	1.1146	1.1092	1.1041	1.0994	1.0950	1.0908	1.0868	1.0832
960	1.1448	1.1372	1.1301	1.1234	1.1175	1.1119	1.1065	1.1016	1.0971	1.0928	1.0887	1.0850
980	1.1485	1.1407	1.1334	1.1265	1.1203	1.1145	1.1090	1.1039	1.0992	1.0948	1.0906	1.0868
1000	1.1520	1.1440	1.1365	1.1294	1.1230	1.1170	1.1114	1.1062	1.1013	1.0968	1.0925	1.0885
1020	1.1558	1.1475	1.1397	1.1324	1.1258	1.1196	1.1138	1.1084	1.1035	1.0988	1.0945	1.0904
1040	1.1595	1.1509	1.1428	1.1353	1.1285	1.1222	1.1163	1.1107	1.1057	1.1008	1.0964	1.0922
1060	1.1633	1.1544	1.1461	1.1383	1.1313	1.1249	1.1188	1.1131	1.1078	1.1028	1.0983	1.0940
1080	1.1669	1.1578	1.1492	1.1411	1.1340	1.1273	1.1211	1.1153	1.1099	1.1048	1.1001	1.0957
1100	1.1707	1.1612	1.1524	1.1441	1.1368	1.1299	1.1235	1.1175	1.1120	1.1069	1.1020	1.0976
1120	1.1744	1.1647	1.1555	1.1471	1.1395	1.1325	1.1259	1.1198	1.1141	1.1088	1.1038	1.0993
1140	1.1781	1.1681	1.1587	1.1501	1.1423	1.1350	1.1282	1.1220	1.1163	1.1109	1.1057	1.1011
1160	1.1819	1.1716	1.1619	1.1531	1.1451	1.1377	1.1307	1.1243	1.1184	1.1128	1.1075	1.1028
1180	1.1858	1.1751	1.1651	1.1559	1.1478	1.1402	1.1331	1.1265	1.1205	1.1148	1.1094	1.1046
1200	1.1895	1.1784	1.1682	1.1588	1.1505	1.1427	1.1354	1.1287	1.1225	1.1167	1.1113	1.1063
1220	1.1932	1.1819	1.1714	1.1617	1.1532	1.1453	1.1377	1.1308	1.1245	1.1186	1.1131	1.1080
1240	1.1968	1.1852	1.1745	1.1646	1.1558	1.1477	1.1401	1.1331	1.1266	1.1206	1.1149	1.1097
1260	1.2005	1.1886	1.1776	1.1675	1.1585	1.1502	1.1425	1.1353	1.1287	1.1225	1.1167	1.1114
1280	1.2040	1.1918	1.1805	1.1703	1.1611	1.1526	1.1446	1.1374	1.1307	1.1244	1.1184	1.1130
1300	1.2075	1.1951	1.1836	1.1730	1.1637	1.1550	1.1469	1.1395	1.1327	1.1263	1.1202	1.1147
1320	1.2109	1.1983	1.1867	1.1758	1.1663	1.1574	1.1492	1.1417	1.1347	1.1281	1.1219	1.1163
1340	1.2144	1.2016	1.1897	1.1786	1.1689	1.1599	1.1514	1.1437	1.1366	1.1299	1.1237	1.1180
1360	1.2178	1.2048	1.1926	1.1814	1.1714	1.1622	1.1536	1.1458	1.1386	1.1317	1.1253	1.1195
1380	1.2210	1.2078	1.1954	1.1840	1.1739	1.1645	1.1557	1.1476	1.1404	1.1334	1.1270	1.1211
1400	1.2244	1.2108	1.1983	1.1866	1.1763	1.1667	1.1577	1.1496	1.1422	1.1352	1.1287	1.1226
1420	1.2276	1.2137	1.2010	1.1892	1.1786	1.1689	1.1598	1.1516	1.1441	1.1369	1.1303	1.1241
1440	1.2307	1.2166	1.2037	1.1918	1.1810	1.1712	1.1602	1.1536	1.1459	1.1386	1.1318	1.1256
1460	1.2336	1.2193	1.2062	1.1942	1.1833	1.1732	1.1639	1.1554	1.1476	1.1402	1.1333	1.1270
1480	1.2365	1.2220	1.2088	1.1966	1.1856	1.1754	1.1658	1.1572	1.1493	1.1418	1.1349	1.1285

Gas Flow Measurement 509

Table 10-11e Continued
F_{pv} Supercompressibility Factors
(Base Data—0.6 Specific Gravity Hydrocarbon Gas)

p_f psig	\multicolumn{10}{c}{Temperature, °F}									
	0	5	10	15	20	25	30	35	40	45	50	55
1500	1.2394	1.2247	1.2112	1.1989	1.1877	1.1774	1.1678	1.1591	1.1510	1.1434	1.1364	1.1299
1520	1.2421	1.2273	1.2137	1.2012	1.1900	1.1795	1.1697	1.1608	1.1526	1.1450	1.1378	1.1313
1540	1.2447	1.2298	1.2160	1.2034	1.1921	1.1815	1.1716	1.1626	1.1543	1.1466	1.1393	1.1327
1560	1.2469	1.2320	1.2182	1.2054	1.1940	1.1834	1.1733	1.1642	1.1559	1.1480	1.1407	1.1340
1580	1.2492	1.2343	1.2204	1.2075	1.1960	1.1853	1.1752	1.1660	1.1575	1.1495	1.1420	1.1352
1600	1.2514	1.2365	1.2225	1.2095	1.1979	1.1871	1.1769	1.1676	1.1590	1.1510	1.1435	1.1366
1620	1.2535	1.2386	1.2245	1.2114	1.1998	1.1889	1.1786	1.1692	1.1606	1.1524	1.1448	1.1378
1640	1.2555	1.2406	1.2265	1.2132	1.2015	1.1905	1.1802	1.1707	1.1620	1.1537	1.1461	1.1390
1660	1.2573	1.2423	1.2282	1.2149	1.2032	1.1921	1.1817	1.1722	1.1633	1.1550	1.1473	1.1401
1680	1.2591	1.2441	1.2299	1.2166	1.2049	1.1938	1.1832	1.1736	1.1647	1.1563	1.1485	1.1413
1700	1.2606	1.2457	1.2315	1.2182	1.2064	1.1953	1.1847	1.1751	1.1661	1.1575	1.1496	1.1424
1720	1.2620	1.2471	1.2331	1.2198	1.2079	1.1967	1.1861	1.1764	1.1674	1.1587	1.1508	1.1435
1740	1.2633	1.2485	1.2345	1.2213	1.2093	1.1980	1.1874	1.1776	1.1686	1.1600	1.1519	1.1445
1760	1.2645	1.2497	1.2357	1.2227	1.2106	1.1993	1.1887	1.1787	1.1697	1.1610	1.1529	1.1455
1780	1.2656	1.2509	1.2370	1.2239	1.2118	1.2005	1.1899	1.1799	1.1708	1.1621	1.1539	1.1464
1800	1.2665	1.2519	1.2381	1.2251	1.2130	1.2017	1.1910	1.1810	1.1718	1.1631	1.1549	1.1473
1820	1.2674	1.2529	1.2392	1.2262	1.2141	1.2028	1.1921	1.1821	1.1728	1.1640	1.1558	1.1482
1840	1.2680	1.2538	1.2401	1.2272	1.2151	1.2038	1.1930	1.1830	1.1738	1.1649	1.1566	1.1490
1860	1.2685	1.2545	1.2410	1.2281	1.2160	1.2047	1.1939	1.1839	1.1747	1.1658	1.1575	1.1498
1880	1.2690	1.2551	1.2417	1.2289	1.2169	1.2056	1.1948	1.1848	1.1755	1.1667	1.1583	1.1506
1900	1.2694	1.2556	1.2424	1.2296	1.2177	1.2064	1.1956	1.1856	1.1763	1.1675	1.1591	1.1514
1920	1.2697	1.2561	1.2429	1.2303	1.2184	1.2072	1.1964	1.1864	1.1770	1.1682	1.1598	1.1521
1940	1.2699	1.2564	1.2434	1.2309	1.2191	1.2078	1.1971	1.1871	1.1777	1.1689	1.1605	1.1528
1960	1.2700	1.2566	1.2438	1.2314	1.2197	1.2085	1.1978	1.1877	1.1784	1.1696	1.1612	1.1534
1980	1.2700	1.2568	1.2441	1.2318	1.2203	1.2092	1.1985	1.1884	1.1790	1.1701	1.1617	1.1540
2000	1.2699	1.2569	1.2443	1.2321	1.2207	1.2097	1.1990	1.1890	1.1796	1.1707	1.1623	1.1545
2020	1.2698	1.2570	1.2445	1.2324	1.2210	1.2101	1.1994	1.1894	1.1801	1.1712	1.1628	1.1550
2040	1.2695	1.2569	1.2446	1.2326	1.2213	1.2104	1.1998	1.1898	1.1805	1.1716	1.1632	1.1554
2060	1.2691	1.2568	1.2446	1.2327	1.2215	1.2107	1.2002	1.1902	1.1809	1.1721	1.1637	1.1559
2080	1.2686	1.2565	1.2445	1.2328	1.2216	1.2109	1.2005	1.1906	1.1813	1.1725	1.1640	1.1563
2100	1.2680	1.2561	1.2443	1.2328	1.2217	1.2110	1.2008	1.1909	1.1816	1.1728	1.1643	1.1566
2120	1.2674	1.2556	1.2440	1.2327	1.2217	1.2111	1.2009	1.1912	1.1819	1.1730	1.1646	1.1569
2140	1.2666	1.2551	1.2437	1.2325	1.2217	1.2112	1.2010	1.1914	1.1821	1.1733	1.1649	1.1572
2160	1.2658	1.2545	1.2433	1.2322	1.2216	1.2112	1.2011	1.1915	1.1823	1.1735	1.1651	1.1574
2180	1.2650	1.2538	1.2428	1.2319	1.2214	1.2111	1.2011	1.1916	1.1824	1.1736	1.1653	1.1576
2200	1.2640	1.2531	1.2423	1.2315	1.2212	1.2110	1.2011	1.1916	1.1825	1.1737	1.1654	1.1577
2220	1.2631	1.2523	1.2417	1.2311	1.2209	1.2108	1.2010	1.1916	1.1825	1.1738	1.1655	1.1578
2240	1.2621	1.2514	1.2410	1.2307	1.2206	1.2106	1.2009	1.1916	1.1825	1.1738	1.1655	1.1579
2260	1.2610	1.2505	1.2402	1.2302	1.2202	1.2103	1.2007	1.1915	1.1825	1.1738	1.1656	1.1579
2280	1.2600	1.2495	1.2394	1.2296	1.2197	1.2100	1.2004	1.1913	1.1824	1.1738	1.1656	1.1579
2300	1.2588	1.2485	1.2386	1.2289	1.2192	1.2096	1.2001	1.1911	1.1823	1.1737	1.1656	1.1579
2320	1.2576	1.2475	1.2378	1.2282	1.2186	1.2092	1.1998	1.1909	1.1821	1.1736	1.1655	1.1579
2340	1.2563	1.2465	1.2369	1.2275	1.2180	1.2087	1.1995	1.1906	1.1819	1.1734	1.1654	1.1578
2360	1.2549	1.2454	1.2360	1.2267	1.2173	1.2081	1.1991	1.1903	1.1816	1.1732	1.1652	1.1576
2380	1.2535	1.2442	1.2350	1.2258	1.2166	1.2076	1.1987	1.1899	1.1813	1.1730	1.1650	1.1574
2400	1.2521	1.2430	1.2339	1.2249	1.2158	1.2070	1.1983	1.1895	1.1810	1.1727	1.1648	1.1572
2420	1.2507	1.2418	1.2329	1.2240	1.2150	1.2063	1.1977	1.1891	1.1806	1.1724	1.1646	1.1570
2440	1.2491	1.2405	1.2318	1.2231	1.2142	1.2056	1.1971	1.1886	1.1802	1.1720	1.1643	1.1567
2460	1.2475	1.2391	1.2306	1.2221	1.2134	1.2049	1.1965	1.1881	1.1798	1.1716	1.1639	1.1565
2480	1.2459	1.2377	1.2294	1.2210	1.2125	1.2041	1.1958	1.1875	1.1793	1.1712	1.1636	1.1562

(table continued)

510 Gas Production Engineering

Table 10-11e Continued
F_{pv} Supercompressibility Factors
(Base Data—0.6 Specific Gravity Hydrocarbon Gas)

p_f psig	\multicolumn{12}{c}{Temperature, °F}
	0	5	10	15	20	25	30	35	40	45	50	55
2500	1.2443	1.2363	1.2282	1.2199	1.2115	1.2033	1.1951	1.1869	1.1787	1.1708	1.1631	1.1558
2520	1.2427	1.2349	1.2269	1.2188	1.2106	1.2025	1.1944	1.1863	1.1782	1.1703	1.1626	1.1555
2540	1.2411	1.2335	1.2256	1.2176	1.2096	1.2016	1.1936	1.1856	1.1776	1.1698	1.1621	1.1551
2560	1.2395	1.2320	1.2242	1.2164	1.2086	1.2007	1.1928	1.1849	1.1770	1.1693	1.1617	1.1546
2580	1.2378	1.2304	1.2228	1.2152	1.2075	1.1997	1.1919	1.1842	1.1764	1.1687	1.1612	1.1542
2600	1.2361	1.2287	1.2214	1.2140	1.2064	1.1987	1.1910	1.1834	1.1757	1.1681	1.1606	1.1537
2620	1.2343	1.2271	1.2200	1.2127	1.2052	1.1977	1.1901	1.1825	1.1749	1.1675	1.1601	1.1532
2640	1.2325	1.2256	1.2185	1.2113	1.2040	1.1966	1.1892	1.1817	1.1742	1.1668	1.1596	1.1527
2660	1.2308	1.2240	1.2170	1.2100	1.2028	1.1955	1.1882	1.1808	1.1734	1.1661	1.1590	1.1522
2680	1.2290	1.2223	1.2155	1.2086	1.2015	1.1944	1.1872	1.1798	1.1725	1.1653	1.1584	1.1516
2700	1.2272	1.2206	1.2139	1.2072	1.2003	1.1933	1.1862	1.1789	1.1717	1.1646	1.1577	1.1510
2720	1.2253	1.2189	1.2124	1.2058	1.1990	1.1922	1.1852	1.1780	1.1709	1.1638	1.1569	1.1503
2740	1.2234	1.2172	1.2108	1.2044	1.1977	1.1910	1.1841	1.1771	1.1700	1.1630	1.1562	1.1497
2760	1.2216	1.2155	1.2092	1.2029	1.1964	1.1898	1.1830	1.1761	1.1691	1.1622	1.1554	1.1490
2780	1.2197	1.2138	1.2077	1.2014	1.1950	1.1885	1.1818	1.1750	1.1682	1.1613	1.1547	1.1483
2800	1.2178	1.2120	1.2060	1.1999	1.1936	1.1872	1.1806	1.1740	1.1672	1.1605	1.1539	1.1476
2820	1.2159	1.2102	1.2044	1.1983	1.1922	1.1859	1.1794	1.1729	1.1662	1.1596	1.1531	1.1468
2840	1.2140	1.2084	1.2027	1.1968	1.1908	1.1846	1.1782	1.1718	1.1652	1.1586	1.1522	1.1461
2860	1.2120	1.2066	1.2010	1.1952	1.1893	1.1832	1.1770	1.1707	1.1642	1.1577	1.1513	1.1453
2880	1.2100	1.2048	1.1993	1.1937	1.1878	1.1818	1,1757	1.1696	1.1632	1.1568	1.1504	1.1445
2900	1.2081	1.2029	1.1976	1.1921	1.1863	1.1804	1.1744	1.1685	1.1621	1.1558	1.1495	1.1437
2920	1.2062	1.2011	1.1959	1.1905	1.1848	1.1790	1.1731	1.1673	1.1610	1.1547	1.1486	1.1428
2940	1.2042	1.1992	1.1942	1.1888	1.1832	1.1776	1.1718	1.1660	1.1599	1.1537	1.1477	1.1420
2960	1.2023	1.1974	1.1924	1.1872	1.1817	1.1761	1.1705	1.1648	1.1587	1.1527	1.1468	1.1411
2980	1.2004	1.1956	1.1907	1.1856	1.1802	1.1747	1.1692	1.1635	1.1576	1.1516	1.1458	1.1402
3000	1.1984	1.1937	1.1889	1.1839	1.1786	1.1733	1.1678	1.1622	1.1564	1.1505	1.1448	1.1393

p_f psig	\multicolumn{13}{c}{Temperature °F}
	60	65	70	75	80	85	90	95	100	105	110	115	120
0	1.0000	1.0000	1.0000	1.0000	1.0000	1.0000	1.0000	1.0000	1.000	1.000	1.0000	1.000	1.0000
20	1.0016	1.0015	1.0014	1.0014	1.0014	1.0013	1.0013	1.0012	1.0012	1.0012	1.0011	1.0011	1.0010
40	1.0032	1.0031	1.0030	1.0029	1.0028	1.0027	1.0027	1.0026	1.0025	1.0024	1.0023	1.0022	1.0022
60	1.0047	1.0046	1.0045	1.0043	1.0042	1.0040	1.0039	1.0038	1.0037	1.0036	1.0035	1.0033	1.0032
80	1.0064	1.0062	1.0061	1.0058	1.0056	1.0054	1.0052	1.0051	1.0049	1.0047	1.0046	1.0044	1.0043
100	1.0080	1.0078	1.0075	1.0073	1.0071	1.0068	1.0066	1.0064	1.0061	1.0059	1.0058	1.0056	1.0055
120	1.0097	1.0094	1.0091	1.0088	1.0085	1.0082	1.0079	1.0076	1.0073	1.0071	1.0069	1.0067	1.0065
140	1.0112	1.0109	1.0105	1.0102	1.0099	1.0095	1.0092	1.0088	1.0085	1.0083	1.0080	1.0078	1.0076
160	1.0129	1.0125	1.0121	1.0117	1.0112	1.0108	1.0105	1.0101	1.0098	1.0095	1.0092	1.0089	1.0087
180	1.0145	1.0140	1.0136	1.0131	1.0126	1.0122	1.0118	1.0114	1.0111	1.0107	1.0103	1.0100	1.0098
200	1.0162	1.0156	1.0151	1.0146	1.0140	1.0135	1.0131	1.0127	1.0123	1.0119	1.0115	1.0111	1.0108
220	1.0178	1.0172	1.0166	1.0160	1.0154	1.0149	1.0145	1.0140	1.0136	1.0131	1.0126	1.0122	1.0119
240	1.0194	1.0188	1.0181	1.0175	1.0168	1.0163	1.0158	1.0153	1.0148	1.0143	1.0138	1.0133	1.0129
260	1.0211	1.0204	1.0197	1.0190	1.0183	1.0177	1.0171	1.0165	1.0160	1.0155	1.0150	1.0144	1.0139
280	1.0228	1.0220	1.0212	1.0205	1.0197	1.0191	1.0185	1.0178	1.0173	1.0167	1.0162	1.0155	1.0150
300	1.0244	1.0236	1.0228	1.0220	1.0212	1.0205	1.0199	1.0192	1.0185	1.0179	1.0173	1.0167	1.0162
320	1.0261	1.0252	1.0243	1.0235	1.0227	1.0219	1.0212	1.0205	1.0198	1.0191	1.0185	1.0178	1.0173
340	1.0277	1.0267	1.0258	1.0249	1.0241	1.0233	1.0225	1.0217	1.0209	1.0203	1.0196	1.0189	1.0183
360	1.0294	1.0284	1.0273	1.0264	1.0256	1.0247	1.0238	1.0230	1.0222	1.0215	1.0207	1.0200	1.0194
380	1.0311	1.0300	1.0289	1.0279	1.0270	1.0261	1.0252	1.0243	1.0234	1.0227	1.0219	1.0211	1.0204

Gas Flow Measurement 511

Table 10-11e Continued
F_{pv} Supercompressibility Factors
(Base Data—0.6 Specific Gravity Hydrocarbon Gas)

p_f					Temperature °F					
psig	60	65	70	75	80	85	90	95	100	105	110	115	120
400	1.0328	1.0317	1.0305	1.0294	1.0285	1.0275	1.0265	1.0256	1.0246	1.0238	1.0230	1.0223	1.0215
420	1.0345	1.0333	1.0321	1.0309	1.0299	1.0289	1.0279	1.0269	1.0259	1.0250	1.0242	1.0234	1.0226
440	1.0361	1.0349	1.0336	1.0324	1.0313	1.0302	1.0292	1.0281	1.0272	1.0262	1.0253	1.0244	1.0236
460	1.0378	1.0365	1.0351	1.0339	1.0327	1.0315	1.0305	1.0294	1.0285	1.0275	1.0265	1.0255	1.0247
480	1.0395	1.0381	1.0367	1.0254	1.0341	1.0329	1.0318	1.0307	1.0297	1.0287	1.0276	1.0267	1.0258
500	1.0413	1.0398	1.0384	1.0370	1.0356	1.0344	1.0332	1.0320	1.0309	1.0298	1.0288	1.0278	1.0269
520	1.0430	1.0414	1.0399	1.0385	1.0371	1.0357	1.0345	1.0333	1.0321	1.0310	1.0299	1.0289	1.0279
540	1.0447	1.0431	1.0415	1.0400	1.0385	1.0371	1.0358	1.0346	1.0334	1.0322	1.0310	1.0300	1.0289
560	1.0465	1.0448	1.0432	1.0416	1.0400	1.0385	1.0372	1.0359	1.0346	1.0334	1.0322	1.0311	1.0300
580	1.0482	1.0464	1.0447	1.0431	1.0415	1.0399	1.0385	1.0372	1.0358	1.0346	1.0333	1.0322	1.0310
600	1.0499	1.0481	1.0463	1.0446	1.0430	1.0414	1.0399	1.0384	1.0370	1.0358	1.0345	1.0333	1.0321
620	1.0517	1.0497	1.0479	1.0461	1.0445	1.0428	1.0412	1.0397	1.0383	1.0369	1.0356	1.0344	1.0331
640	1.0534	1.0514	1.0495	1.0476	1.0460	1.0442	1.0426	1.0410	1.0396	1.0381	1.0368	1.0355	1.0341
660	1.0552	1.0530	1.0511	1.0492	1.0474	1.0456	1.0439	1.0423	1.0408	1.0393	1.0379	1.0366	1.0352
680	1.0570	1.0547	1.0527	1.0507	1.0488	1.0470	1.0453	1.0436	1.0420	1.0405	1.0390	1.0377	1.0363
700	1.0587	1.0563	1.0543	1.0522	1.0502	1.0483	1.0466	1.0449	1.0432	1.0416	1.0401	1.0387	1.0373
720	1.0605	1.0580	1.0559	1.0537	1.0517	1.0497	1.0479	1.0461	1.0444	1.0428	1.0412	1.0398	1.0383
740	1.0622	1.0597	1.0575	1.0553	1.0531	1.0510	1.0492	1.0474	1.0456	1.0440	1.0424	1.0409	1.0393
760	1.0640	1.0614	1.0591	1.0568	1.0546	1.0524	1.0505	1.0487	1.0468	1.0451	1.0435	1.0419	1.0403
780	1.0658	1.0631	1.0607	1.0583	1.0560	1.0538	1.0519	1.0500	1.0480	1.0463	1.0446	1.0430	1.0414
800	1.0676	1.0648	1.0623	1.0598	1.0575	1.0552	1.0532	1.0513	1.0492	1.0474	1.0456	1.0440	1.0424
820	1.0693	1.0665	1.0639	1.0613	1.0589	1.0566	1.0545	1.0524	1.0504	1.0485	1.0467	'1.0450	1.0434
840	1.0711	1.0681	1.0654	1.0628	1.0603	1.0580	1.0558	1.0536	1.0517	1.0497	1.0478	1.0460	1.0443
860	1.0728	1.0697	1.0670	1.0643	1.0617	1.0593	1.0571	1.0549	1.0529	1.0500	1.0489	1.0471	1.0453
880	1.0745	1.0714	1.0686	1.0658	1.0631	1.0607	1.0584	1.0562	1.0540	1.0519	1.0500	1.0481	1.0463
900	1.0762	1.0730	1.0701	1.0673	1.0646	1.0620	1.0597	1.0574	1.0552	1.0530	1.0510	1.0491	1.0473
920	1.0779	1.0746	1.0716	1.0688	1.0660	1.0634	1.0610	1.0586	1.0563	1.0541	1.0520	1.0501	1.0482
940	1.0797	1.0763	1.0733	1.0703	1.0675	1.0649	1.0623	1.0599	1.0575	1.0553	1.0531	1.0511	1.0492
960	1.0814	1.0779	1.0748	1.0718	1.0689	1.0662	1.0636	1.0610	1.0586	1.0563	1.0541	1.0521	1.0501
980	1.0831	1.0795	1.0763	1.0732	1.0703	1.0675	1.0648	1.0622	1.0597	1.0574	1.0552	1.0530	1.0510
1000	1.0847	1.0811	1.0778	1.0746	1.0717	1.0687	1.0660	1.0634	1.0608	1.0585	1.0562	1.0539	1.0519
1020	1.0865	1.0827	1.0794	1.0761	1.0730	1.0701	1.0673	1.0646	1.0619	1.0595	1.0572	1.0549	1.0529
1040	1.0882	1.0843	1.0809	1.0775	1.0744	1.0714	1.0685	1.0658	1.0631	1.0606	1.0582	1.0559	1.0538
1060	1.0900	1.0860	1.0825	1.0790	1.0758	1.0727	1.0697	1.0670	1.0642	1.0617	1.0592	1.0569	1.0547
1080	1.0916	1.0875	1.0839	1.0804	1.0771	1.0740	1.0709	1.0681	1.0654	1.0628	1.0602	1.0578	1.0556
1100	1.0933	1.0891	1.0854	1.0819	1.0785	1.0753	1.0722	1.0692	1.0665	1.0638	1.0612	1.0588	1.0565
1120	1.0950	1.0908	1.0870	1.0834	1.0800	1.0766	1.0734	1.0703	1.0676	1.0649	1.0623	1.0598	1.0574
1140	1.0966	1.0924	1.0885	1.0848	1.0814	1.0779	1.0746	1.0716	1.0687	1.0659	1.0633	1.0607	1.0583
1160	1.0983	1.0939	1.0899	1.0862	1.0826	1.0791	1.0758	1.0727	1.0698	1.0669	1.0643	1.0616	1.0592
1180	1.1000	1.0955	1.0914	1.0875	1.0839	1.0804	1.0771	1.0738	1.0708	1.0679	1.0652	1.0625	1.0601
1200	1.1016	1.0970	1.0928	1.0889	1.0851	1.0816	1.0782	1.0750	1.0718	1.0689	1.0661	1.0634	1.0610
1220	1.1032	1.0985	1.0942	1.0902	1.0864	1.0828	1.0794	1.0760	1.0729	1.0699	1.0671	1.0643	1.0618
1240	1.1048	1.1001	1.0957	1.0916	1.0876	1.0840	1.0805	1.0771	1.0739	1.0709	1.0681	1.0652	1.0626
1260	1.1064	1.1015	1.0971	1.0929	1.0889	1.0852	1.0816	1.0781	1.0748	1.0719	1.0690	1.0661	1.0635
1280	1.1079	1.1030	1.0985	1.0942	1.0901	1.0863	1.0827	1.0791	1.0758	1.0728	1.0699	1.0670	1.0643
1300	1.1094	1.1044	1.0999	1.0955	1.0913	1.0875	1.0838	1.0802	1.0768	1.0737	1.0707	1.0678	1.0651
1320	1.1110	1.1059	1.1012	1.0968	1.0925	1.0886	1.0849	1.0812	1.0778	1.0746	1.0716	1.0686	1.0659

(table continued)

Table 10-11e Continued
F_{pv} Supercompressibility Factors
(Base Data—0.6 Specific Gravity Hydrocarbon Gas)

p_f psig	Temperature °F											
	60	65	70	75	80	85	90	95	100	105	110	115	120
1340	1.1125	1.1073	1.1025	1.0980	1.0937	1.0897	1.0859	1.0822	1.0788	1.0755	1.0725	1.0695	1.0667
1360	1.1140	1.1087	1.1039	1.0993	1.0949	1.0909	1.0870	1.0833	1.0797	1.0764	1.0733	1.0703	1.0675
1380	1.1154	1.1100	1.1052	1.1005	1.0961	1.0920	1.0881	1.0843	1.0806	1.0773	1.0741	1.0711	1.0682
1400	1.1168	1.1114	1.1065	1.1017	1.0973	1.0931	1.0891	1.0853	1.0816	1.0782	1.0750	1.0719	1.0690
1420	1.1183	1.1128	1.1078	1.1030	1.0985	1.0941	1.0902	1.0863	1.0825	1.0791	1.0759	1.0727	1.0697
1440	1.1197	1.1141	1.1090	1.1042	1.0995	1.0952	1.0912	1.0873	1.0834	1.0800	1.0767	1.0735	1.0705
1460	1.1210	1.1154	1.1103	1.1053	1.1006	1.0962	1.0921	1.0882	1.0843	1.0808	1.0775	1.0742	1.0712
1480	1.1225	1.1167	1.1115	1.1064	1.1016	1.0973	1.0931	1.0891	1.0852	1.0816	1.0783	1.0750	1.0719
1500	1.1238	1.1179	1.1126	1.1075	1.1027	1.0983	1.0941	1.0900	1.0861	1.0825	1.0791	1.0758	1.0727
1520	1.1251	1.1191	1.1138	1.1087	1.1038	1.0993	1.0950	1.0909	1.0870	1.0833	1.0799	1.0766	1.0734
1540	1.1263	1.1204	1.1150	1.1098	1.1049	1.1003	1.0960	1.0918	1.0879	1.0842	1.0807	1.0773	1.0741
1560	1.1276	1.1215	1.1161	1.1108	1.1059	1.1012	1.0969	1.0927	1.0887	1.0850	1.0815	1.0780	1.0748
1580	1.1288	1.1227	1.1172	1.1119	1.1068	1.1022	1.0978	1.0935	1.0896	1.0858	1.0823	1.0788	1.0755
1600	1.1301	1.1238	1.1183	1.1129	1.1078	1.1031	1.0987	1.0944	1.0904	1.0866	1.0830	1.0795	1.0762
1620	1.1312	1.1249	1.1193	1.1139	1.1088	1.1041	1.0995	1.0952	1.0912	1.0873	1.0837	1.0802	1.0768
1640	1.1323	1.1260	1.1203	1.1149	1.1097	1.1049	1.1004	1.0960	1.0920	1.0881	1.0844	1.0809	1.0775
1660	1.1334	1.1270	1.1213	1.1158	1.1106	1.1058	1.1012	1.0968	1.0927	1.0888	1.0851	1.0815	1.0781
1680	1.1345	1.1281	1.1223	1.1167	1.1115	1.1066	1.1020	1.0976	1.0934	1.0895	1.0858	1.0822	1.0787
1700	1.1335	1.1290	1.1232	1.1176	1.1124	1.1074	1.1028	1.0984	1.0942	1.0903	1.0865	1.0828	1.0793
1720	1.1366	1.1300	1.1241	1.1185	1.1132	1.1082	1.1036	1.0992	1.0950	1.0910	1.0872	1.0835	1.0799
1740	1.1376	1.1309	1.1250	1.1193	1.1139	1.1089	1.1044	1.0999	1.0957	1.0917	1.0878	1.0841	1.0805
1760	1.1385	1.1318	1.1258	1.1201	1.1147	1.1097	1.1051	1.1006	1.0964	1.0923	1.0884	1.0847	1.0811
1780	1.1393	1.1326	1.1266	1.1209	1.1154	1.1104	1.1058	1.1012	1.0970	1.0929	1.0890	1.0853	1.0816
1800	1.1402	1.1334	1.1273	1.1216	1.1161	1.1111	1.1064	1.1019	1.0976	1.0935	1.0896	1.0858	1.0821
1820	1.1410	1.1342	1.1281	1.1223	1.1168	1.1118	1.1071	1.1025	1.0982	1.0941	1.0902	1.0863	1.0826
1840	1.1418	1.1349	1.1288	1.1230	1.1175	1.1124	1.1077	1.1031	1.0988	1.0947	1.0907	1.0868	1.0831
1860	1.1426	1.1357	1.1295	1.1237	1.1181	1.1130	1.1083	1.1037	1.0994	1.0952	1.0911	1.0873	1.0836
1880	1.1433	1.1364	1.1302	1.1243	1.1187	1.1137	1.1089	1.1043	1.0999	1.0957	1.0916	1.0877	1.0840
1900	1.1440	1.1371	1.1309	1.1249	1.1193	1.1142	1.1094	1.1048	1.1004	1.0962	1.0920	1.0881	1.0844
1920	1.1447	1.1378	1.1315	1.1255	1.1199	1.1148	1.1099	1.1053	1.1009	1.0967	1.0925	1.0886	1.0848
1940	1.1454	1.1384	1.1321	1.1261	1.1204	1.1153	1.1104	1.1058	1.1014	1.0971	1.0929	1.0890	1.0852
1960	1.1460	1.1389	1.1326	1.1266	1.1209	1.1158	1.1109	1.1063	1.1019	1.0976	1.0934	1.0894	1.0856
1980	1.1465	1.1394	1.1331	1.1271	1.1214	1.1163	1.1114	1.1068	1.1023	1.0980	1.0938	1.0898	1.0860
2000	1.1470	1.1399	1.1336	1.1276	1.1219	1.1168	1.1119	1.1073	1.1027	1.0984	1.0942	1.0902	1.0864
2020	1.1475	1.1403	1.1340	1.1280	1.1223	1.1172	1.1123	1.1077	1.1031	1.0988	1.0946	1.0906	1.0867
2040	1.1480	1.1408	1.1344	1.1284	1.1227	1.1176	1.1127	1.1081	1.1034	1.0992	1.0950	1.0909	1.0870
2060	1.1484	1.1412	1.1349	1.1288	1.1231	1.1180	1.1131	1.1085	1.1038	1.0995	1.0953	1.0912	1.0873
2080	1.1488	1.1416	1.1353	1.1291	1.1234	1.1184	1.1134	1.1088	1.1042	1.0999	1.0956	1.0915	1.0876
2100	1.1491	1.1419	1.1355	1.1294	1.1237	1.1186	1.1137	1.1091	1.1045	1.1002	1.0959	1.0917	1.0878
2120	1.1494	1.1422	1.1358	1.1297	1.1240	1.1189	1.1140	1.1094	1.1048	1.1004	1.0961	1.0919	1.0880
2140	1.1497	1.1425	1.1361	1.1300	1.1243	1.1192	1.1143	1.1096	1.1050	1.1006	1.0963	1.0921	1.0882
2160	1.1499	1.1427	1.1363	1.1302	1.1245	1.1194	1.1145	1.1098	1.1052	1.1008	1.0965	1.0923	1.0884
2180	1.1501	1.1429	1.1365	1.1304	1.1248	1.1196	1.1147	1.1100	1.1054	1.1010	1.0967	1.0925	1.0886
2200	1.1503	1.1431	1.1367	1.1306	1.1250	1.1198	1.1149	1.1102	1.1056	1.1012	1.0969	1.0927	1.0888
2220	1.1504	1.1432	1.1368	1.1308	1.1252	1.1200	1.1151	1.1104	1.1058	1.1014	1.0971	1.0928	1.0890
2240	1.1505	1.1433	1.1369	1.1310	1.1254	1.1201	1.1152	1.1105	1.1059	1.1015	1.0972	1.0930	1.0891
2260	1.1505	1.1434	1.1370	1.1311	1.1256	1.1203	1.1153	1.1107	1.1060	1.1016	1.0973	1.0931	1.0892
2280	1.1505	1.1434	1.1371	1.1312	1.1257	1.1204	1.1154	1.1108	1.1061	1.1017	1.0974	1.0932	1.0893
2300	1.1505	1.1434	1.1371	1.1312	1.1258	1.1205	1.1155	1.1109	1.1062	1.1018	1.0975	1.0933	1.0894
2320	1.1504	1.1434	1.1371	1.1312	1.1258	1.1205	1.1156	1.1110	1.1063	1.1019	1.0976	1.0934	1.0895

Gas Flow Measurement 513

Table 10-11e Continued
F_{pv} Supercompressibility Factors
(Base Data—0.6 Specific Gravity Hydrocarbon Gas)

p_f						Temperature °F						
psig	60	65	70	75	80	85	90	95	100	105	110	115	120
2340	1.1503	1.1433	1.1371	1.1312	1.1258	1.1205	1.1156	1.1110	1.1063	1.1020	1.0977	1.0935	1.0896
2360	1.1502	1.1432	1.1370	1.1312	1.1258	1.1205	1.1156	1.1110	1.1063	1.1020	1.0978	1.0936	1.0897
2380	1.1501	1.1431	1.1369	1.1312	1.1257	1.1205	1.1156	1.1110	1.1063	1.1020	1.0978	1.0937	1.0897
2400	1.1499	1.1429	1.1368	1.1311	1.1256	1.1205	1.1156	1.1110	1.1063	1.1020	1.0978	1.0937	1:0897
2420	1.1497	1.1428	1.1367	1.1310	1.1256	1.1205	1.1156	1.1110	1.1063	1.1020	1.0978	1.0937	1.0897
2440	1.1495	1.1426	1.1366	1.1309	1.1255	1.1204	1.1155	1.1109	1.1063	1.1020	1.0978	1.0937	1.0897
2460	1.1493	1.1424	1.1365	1.1308	1.1254	1.1203	1.1154	1.1108	1.1062	1.1019	1.0977	1.0936	1.0896
2480	1.1491	1.1422	1.1363	1.1306	1.1253	1.1201	1.1153	1.1107	1.1061	1.1018	1.0976	1.0935	1.0895
2500	1.1488	1.1420	1.1361	1.1304	1.1251	1.1200	1.1152	1.1106	1.1060	1.1017	1.0975	1.0934	1.0894
2520	1.1485	1.1417	1.1358	1.1302	1.1249	1.1198	1.1151	1.1105	1.1059	1.1016	1.0974	1.0933	1.0893
2540	1.1482	1.1414	1.1356	1.1300	1.1247	1.1196	1.1149	1.1103	1.1057	1.1014	1.0973	1.0932	1.0892
2560	1.1478	1.1411	1.1352	1.1297	1.1244	1.1194	1.1147	1.1101	1.1055	1.1013	1.0972	1.0931	1.0891
2580	1.1474	1.1408	1.1349	1.1294	1.1242	1.1191	1.1145	1.1099	1.1053	1.1011	1.0970	1.0930	1.0890
2600	1.1470	1.1404	1.1345	1.1290	1.1239	1.1189	1.1142	1.1097	1.1051	1.1010	1.0968	1.0929	1.0889
2620	1.1466	1.1400	1.1341	1.1287	1.1236	1.1186	1.1139	1.1094	1.1049	1.1008	1.0967	1.0927	1.0887
2640	1.1461	1.1396	1.1337	1.1283	1.1232	1.1183	1.1136	1.1091	1.1047	1.1006	1.0965	1.0925	1.0885
2660	1.1456	1.1392	1.1333	1.1279	1.1229	1.1180	1.1133	1.1088	1.1045	1.1004	1.0963	1.0923	1.0883
2680	1.1450	1.1387	1.1329	1.1275	1.1225	1.1177	1.1130	1.1085	1.1042	1.1001	1.0960	1.0921	1.0881
2700	1.1445	1.1382	1.1325	1.1270	1.1221	1.1173	1.1127	1.1082	1.1039	1.0998	1.0958	1.0918	1.0879
2720	1.1440	1.1377	1.1320	1.1266	1.1217	1.1170	1.1123	1.1079	1.1036	1.0995	1.0955	1.0917	1.0876
2740	1.1434	1.1371	1.1315	1.1262	1.1213	1.1166	1.1120	1.1076	1.1033	1.0992	1.0952	1.0914	1.0874
2760	1.1428	1.1366	1.1310	1.1257	1.1208	1.1162	1.1116	1.1072	1.1030	1.0989	1.0949	1.0911	1.0872
2780	1.1421	1.1360	1.1305	1.1252	1.1204	1.1157	1.1112	1.1069	1.1027	1.0986	1.0946	1.0908	1.0869
2800	1.1414	1.1354	1.1299	1.1247	1.1199	1.1153	1.1108	1.1065	1.1024	1.0983	1.0943	1.0904	1.0866
2820	1.1408	1.1349	1.1294	1.1242	1.1194	1.1148	1.1104	1.1061	1.1020	1.0979	1.0940	1.0901	1.0863
2840	1.1401	1.1343	1.1288	1.1237	1.1188	1.1144	1.1099	1.1057	1.1016	1.0975	1.0936	1.0898	1.0860
2860	1.1394	1.1336	1.1282	1.1231	1.1183	1.1139	1.1094	1.1052	1.1012	1.0972	1.0933	1.0895	1.0857
2880	1.1387	1.1330	1.1276	1.1225	1.1177	1.1134	1.1090	1.1047	1.1008	1.0968	1.0929	1.0892	1.0854
2900	1.1379	1.1324	1.1270	1.1219	1.1172	1.1128	1.1085	1.1042	1.1003	1.0964	1.0925	1.0888	1.0851
2920	1.1371	1.1316	1.1263	1.1213	1.1166	1.1123	1.1079	1.1037	1.0998	1.0959	1.0921	1.0884	1.0847
2940	1.1364	1.1309	1.1256	1.1207	1.1160	1.1117	1.1074	1.1032	1.0993	1.0955	1.0917	1.0880	1.0844
2960	1.1355	1.1302	1.1249	1.1201	1.1155	1.1111	1.1069	1.1027	1.0988	1.0950	1.0913	1.0876	1.0840
2980	1.1347	1.1294	1.1242	1.1194	1.1149	1.1105	1.1063	1.1022	1.0983	1.0945	1.0908	1.0871	1.0835
3000	1.1339	1.1288	1.1235	1.1187	1.1142	1.1099	1.1058	1.1017	1.0978	1.0941	1.0904	1.0867	1.0831

p_f						Temperature, °F						
psig	125	130	135	140	145	150	155	160	165	170	175	180	185
0	1.0000	1.0000	1.0000	1.0000	1.0000	1.0000	1.0000	1.0000	1.0000	1.0000	1.0000	1.0000	1.0000
20	1.0010	1.0010	1.0010	1.0010	1.0010	1.0009	1.0009	1.0009	1.0008	1.0008	1.0008	1.0008	1.0007
40	1.0022	1.0020	1.0020	1.0020	1.0019	1.0018	1.0018	1.0017	1.0016	1.0016	1.0016	1.0015	1.0014
60	1.0032	1.0030	1.0030	1.0029	1.0028	1.0027	1.0027	1.0026	1.0024	1.0023	1.0023	1.0022	1.0021
80	1.0042	1.0040	1.0039	1.0039	1.0038	1.0036	1.0035	1.0034	1.0032	1.0031	1.0030	1.0029	1.0028
100	1.0053	1.0051	1.0049	1.0048	1.0047	1.0045	1.0044	1.0042	1.0040	1.0039	1.0038	1.0037	1.0035
120	1.0063	1.0061	1.0059	1.0057	1.0056	1.0054	1.0052	1.0050	1.0048	1.0047	1.0045	1.0044	1.0042
140	1.0074	1.0071	1.0068	1.0066	1.0065	1.0063	1.0060	1.0058	1.0056	1.0055	1.0053	1.0051	1.0049
160	1.0084	1.0081	1.0078	1.0076	1.0074	1.0072	1.0069	1.0067	1.0064	1.0063	1.0061	1.0058	1.0056
180	1.0094	1.0091	1.0088	1.0085	1.0083	1.0081	1.0078	1.0075	1.0072	1.0070	1.0068	1.0065	1.0063
200	1.0104	1.0101	1.0097	1.0094	1.0092	1.0089	1.0086	1.0083	1.0080	1.0078	1.0075	1.0073	1.0070
220	1.0115	1.0111	1.0107	1.0104	1.0101	1.0098	1.0095	1.0092	1.0088	1.0086	1.0083	1.0080	1.0077

(table continued)

514 Gas Production Engineering

Table 10-11e Continued
F_{pv} Supercompressibility Factors
(Base Data—0.6 Specific Gravity Hydrocarbon Gas)

p_f psig	\multicolumn{11}{c}{Temperature, °F}										
	125	130	135	140	145	150	155	160	165	170	175	180	185
240	1.0125	1.0121	1.0117	1.0114	1.0110	1.0107	1.0103	1.0100	1.0096	1.0094	1.0090	1.0087	1.0084
260	1.0135	1.0132	1.0128	1.0123	1.0119	1.0116	1.0112	1.0109	1.0104	1.0102	1.0098	1.0095	1.0091
280	1.0146	1.0142	1.0137	1.0132	1.0128	1.0125	1.0121	1.0117	1.0112	1.0109	1.0105	1.0102	1.0098
300	1.0157	1.0152	1.0146	1.0141	1.0137	1.0134	1.0130	1.0125	1.0121	1.0116	1.0112	1.0109	1.0105
320	1.0167	1.0161	1.0156	1.0151	1.0146	1.0142	1.0138	1.0133	1.0129	1.0124	1.0119	1.0116	1.0112
340	1.0177	1.0171	1.0165	1.0160	1.0155	1.0151	1.0146	1.0141	1.0137	1.0132	1.0127	1.0122	1.0118
360	1.0187	1.0181	1.0175	1.0169	1.0164	1.0159	1.0154	1.0149	1.0144	1.0139	1.0134	1.0129	1.0125
380	1.0197	1.0191	1.0185	1.0179	1.0173	1.0168	1.0163	1.0157	1.0152	1.0146	1.0141	1.0136	1.0131
400	1.0208	1.0201	1.0195	1.0189	1.0182	1.0177	1.0171	1.0165	1.0160	1.0154	1.0149	1.0143	1.0138
420	1.0218	1.0211	1.0204	1.0198	1.0191	1.0185	1.0179	1.0173	1.0167	1.0161	1.0156	1.0150	1.0144
440	1.0228	1.0220	1.0213	1.0207	1.0200	1.0193	1.0187	1.0181	1.0175	1.0168	1.0162	1.0156	1.0151
460	1.0238	1.0229	1.0222	1.0216	1.0219	1.0202	1.0196	1.0189	1.0182	1.0175	1.0169	1.0163	1.0157
480	1.0248	1.0239	1.0232	1.0225	1.0218	1.0211	1.0204	1.0197	1.0190	1.0183	1.0176	1.0169	1.0163
500	1.0259	1.0249	1.0242	1.0234	1.0227	1.0220	1.0212	1.0205	1.0197	1.0190	1.0183	1.0176	1.0170
520	1.0269	1.0258	1.0251	1.0243	1.0235	1.0228	1.0220	1.0212	1.0204	1.0197	1.0190	1.0183	1.0177
540	1.0279	1.0268	1.0260	1.0252	1.0243	1.0236	1.0228	1.0220	1.0212	1.0204	1.0197	1.0190	1.0183
560	1.0289	1.0278	1.0269	1.0261	1.0252	1.0245	1.0236	1.0227	1.0219	1.0211	1.0204	1.0196	1.0189
580	1.0299	1.0288	1.0279	1.0270	1.0261	1.0253	1.0244	1.0235	1.0226	1.0218	1.0210	1.0202	1.0195
600	1.0309	1.0298	1.0288	1.0279	1.0270	1.0261	1.0251	1.0242	1.0233	1.0225	1.0217	1.0209	1.0201
620	1.0319	1.0308	1.0298	1.0288	1.0278	1.0269	1.0259	1.0250	1.0241	1.0232	1.0223	1.0215	1.0207
640	1.0329	1.0317	1.0307	1.0296	1.0287	1.0277	1.0267	1.0257	1.0248	1.0239	1.0230	1.0222	1.0214
660	1.0340	1.0327	1.0316	1.0305	1.0295	1.0285	1.0275	1.0265	1.0255	1.0246	1.0237	1.0228	1.0220
680	1.0350	1.0337	1.0325	1.0314	1.0304	1.0293	1.0282	1.0272	1.0262	1.0253	1.0244	1.0235	1.0226
700	1.0359	1.0346	1.0334	1.0323	1.0312	1.0301	1.0290	1.0279	1.0269	1.0259	1.0250	1.0241	1.0231
720	1.0369	1.0355	1.0343	1.0331	1.0320	1.0309	1.0298	1.0287	1.0276	1.0266	1.0257	1.0247	1.0237
740	1.0379	1.0365	1.0352	1.0340	1.0328	1.0316	1.0305	1.0294	1.0283	1.0273	1.0263	1.0253	1.0243
760	1.0388	1.0374	1.0361	1.0349	1.0336	1.0324	1.0313	1.0301	1.0290	1.0280	1.0269	1.0259	1.0249
780	1.0398	1.0384	1.0371	1.0358	1.0344	1.0332	1.0320	1.0308	1.0297	1.0286	1.0275	1.0265	1.0255
800	1.0408	1.0393	1.0380	1.0366	1.0353	1.0340	1.0327	1.0315	1.0303	1.0292	1.0281	1.0271	1.0260
820	1.0418	1.0402	1.0388	1.0374	1.0360	1.0347	1.0334	1.0322	1.0310	1.0299	1.0287	1.0277	1.0266
840	1.0427	1.0412	1.0396	1.0382	1.0368	1.0355	1.0342	1.0329	1.0317	1.0306	1.0294	1.0283	1.0272
860	1.0437	1.0421	1.0405	1.0391	1.0376	1.0362	1.0349	1.0336	1.0324	1.0312	1.0300	1.0288	1.0277
880	1.0446	1.0430	1.0414	1.0399	1.0384	1.0370	1.0356	1.0343	1.0330	1.0318	1.0306	1.0294	1.0282
900	1.0455	1.0439	1.0423	1.0407	1.0392	1.0377	1.0363	1.0350	1.0336	1.0324	1.0311	1.0299	1.0287
920	1.0464	1.0448	1.0431	1.0415	1.0400	1.0385	1.0371	1.0357	1.0343	1.0330	1.0317	1.0305	1.0293
940	1.0474	1.0457	1.0440	1.0423	1.0408	1.0393	1.0378	1.0363	1.0350	1.0336	1.0323	1.0310	1.0298
960	1.0483	1.0465	1.0448	1.0431	1.0415	1.0400	1.0385	1.0370	1.0356	1.0342	1.0329	1.0315	1.0303
980	1.0492	1.0473	1.0456	1.0439	1.0422	1.0407	1.0391	1.0376	1.0363	1.0348	1.0334	1.0321	1.0308
1000	1.0501	1.0481	1.0463	1.0446	1.0429	1.0413	1.0398	1.0383	1.0369	1.0354	1.0340	1.0326	1.0313
1020	1.0509	1.0490	1.0471	1.0454	1.0437	1.0420	1.0404	1.0389	1.0375	1.0360	1.0345	1.0331	1.0318
1040	1.0518	1.0498	1.0480	1.0461	1.0444	1.0428	1.0412	1.0396	1.0381	1.0365	1.0350	1.0336	1.0323
1060	1.0527	1.0506	1.0487	1.0469	1.0452	1.0435	1.0419	1.0403	1.0387	1.0371	1.0356	1.0341	1.0328
1080	1.0535	1.0514	1.0495	1.0476	1.0459	1.0442	1.0425	1.0409	1.0393	1.0377	1.0361	1.0346	1.0333
1100	1.0544	1.0522	1.0503	1.0484	1.0465	1.0448	1.0431	1.0415	1.0399	1.0383	1.0367	1.0352	1.0338
1120	1.0552	1.0530	1.0510	1.0491	1.0472	1.0454	1.0437	1.0420	1.0404	1.0388	1.0372	1.0357	1.0343
1140	1.0561	1.0538	1.0518	1.0498	1.0478	1.0460	1.0443	1.0426	1.0410	1.0394	1.0378	1.0362	1.0348
1160	1.0569	1.0546	1.0525	1.0505	1.0485	1.0467	1.0450	1.0432	1.0415	1.0399	1.0382	1.0367	1.0352
1180	1.0577	1.0554	1.0533	1.0512	1.0492	1.0473	1.0456	1.0438	1.0421	1.0404	1.0388	1.0372	1.0357
1200	1.0585	1.0562	1.0540	1.0519	1.0499	1.0479	1.0461	1.0443	1.0426	1.0410	1.0393	1.0377	1.0361
1220	1.0593	1.0569	1.0547	1.0526	1.0506	1.0486	1.0467	1.0449	1.0432	1.0415	1.0398	1.0381	1.0365

Gas Flow Measurement 515

Table 10-11e Continued
F_{pv} Supercompressibility Factors
(Base Data—0.6 Specific Gravity Hydrocarbon Gas)

p_f					Temperature, °F						
psig	125	130	135	140	145	150	155	160	165	170	175	180	185
1240	1.0601	1.0577	1.0554	1.0532	1.0512	1.0492	1.0473	1.0455	1.0437	1.0420	1.0403	1.0386	1.0370
1260	1.0610	1.0585	1.0561	1.0539	1.0518	1.0498	1.0479	1.0461	1.0443	1.0425	1.0407	1.0391	1.0375
1280	1.0618	1.0592	1.0568	1.0546	1.0524	1.0504	1.0485	1.0466	1.0448	1.0430	1.0412	1.0395	1.0379
1300	1.0626	1.0600	1.0575	1.0552	1.0530	1.0510	1.0490	1.0471	1.0453	1.0435	1.0417	1.0399	1.0383
1320	1.0633	1.0607	1.0583	1.0559	1.0536	1.0515	1.0496	1.0477	1.0457	1.0439	1.0421	1.0404	1.0386
1340	1.0640	1.0614	1.0590	1.0565	1.0542	1.0521	1.0501	1.0482	1.0462	1.0444	1.0426	1.0408	1.0390
1360	1.0647	1.0621	1.0597	1.0572	1.0548	1.0527	1.0506	1.0487	1.0467	1.0449	1.0431	1.0412	1.0394
1380	1.0654	1.0628	1.0603	1.0578	1.0554	1.0532	1.0511	1.0492	1.0472	1.0453	1.0435	1.0416	1.0398
1400	1.0661	1.0635	1.0610	1.0585	1.0560	1.0537	1.0516	1.0497	1.0477	1.0458	1.0439	1.0420	1.0402
1420	1.0669	1.0641	1.0616	1.0591	1.0566	1.0543	1.0522	1.0501	1.0481	1.0462	1.0443	1.0424	1.0406
1440	1.0676	1.0648	1.0622	1.0596	1.0571	1.0548	1.0527	1.0506	1.0486	1.0466	1.0447	1.0428	1.0410
1460	1.0683	1.0655	1.0629	1.0603	1.0577	1.0554	1.0532	1.0510	1.0490	1.0470	1.0451	1.0432	1.0414
1480	1.0690	1.0662	1.0635	1.0609	1.0583	1.0560	1.0538	1.0515	1.0494	1.0474	1.0454	1.0435	1.0417
1500	1.0697	1.0668	1.0641	1.0614	1.0588	1.0565	1.0542	1.0520	1.0499	1.0478	1.0458	1.0439	1.0421
1520	1.0704	1.0675	1.0647	1.0620	1.0594	1.0570	1.0547	1.0525	1.0503	1.0482	1.0462	1.0443	1.0425
1540	1.0711	1.0681	1.0653	1.0626	1.0600	1.0575	1.0552	1.0530	1.0507	1.0486	1.0466	1.0446	1.0428
1560	1.0717	1.0687	1.0658	1.0631	1.0605	1.0579	1.0556	1.0534	1.0511	1.0490	1.0470	1.0450	1.0432
1580	1.0724	1.0693	1.0664	1.0637	1.0610	1.0584	1.0561	1.0538	1.0516	1.0494	1.0473	1.0453	1.0435
1600	1.0730	1.0699	1.0670	1.0642	1.0615	1.0589	1.0566	1.0543	1.0520	1.0498	1.0477	1.0457	1.0439
1620	1.0736	1.0705	1.0675	1.0647	1.0620	1.0593	1.0570	1.0547	1.0524	1.0502	1.0481	1.0460	1.0442
1640	1.0743	1.0711	1.0681	1.0652	1.0625	1.0597	1.0574	1.0551	1.0528	1.0505	1.0484	1.0464	1.0445
1660	1.0748	1.0716	1.0686	1.0657	1.0630	1.0602	1.0578	1.0554	1.0531	1.0509	1.0488	1.0467	1.0447
1680	1.0754	1.0721	1.0691	1.0662	1.0634	1.0606	1.0582	1.0558	1.0535	1.0512	1.0491	1.0470	1.0450
1700	1.0759	1.0726	1.0696	1.0667	1.0639	1.0611	1.0586	1.0562	1.0539	1.0516	1.0494	1.0473	1.0453
1720	1.0765	1.0732	1.0701	1.0672	1.0643	1.0615	1.0591	1.0567	1.0542	1.0519	1.0497	1.0476	1.0456
1740	1.0770	1.0737	1.0707	1.0677	1.0648	1.0620	1.0595	1.0571	1.0546	1.0522	1.0500	1.0479	1.0459
1760	1.0776	1.0742	1.0711	1.0681	1.0652	1.0624	1.0598	1.0574	1.0549	1.0525	1.0503	1.0482	1.0462
1780	1.0781	1.0747	1.0716	1.0686	1.0656	1.0628	1.0602	1.0577	1.0553	1.0528	1.0505	1.0484	1.0464
1800	1.0786	1.0752	1.0720	1.0690	1.0659	1.0631	1.0605	1.0580	1.0556	1.0531	1.0508	1.0487	1.0466
1820	1.0791	1.0757	1.0725	1.0694	1.0663	1.0635	1.0608	1.0583	1.0559	1.0534	1.0511	1.0490	1.0469
1840	1.0796	1.0761	1.0729	1.0698	1.0667	1.0639	1.0612	1.0586	1.0562	1.0537	1.0514	1.0492	1.0471
1860	1.0800	1.0765	1.0733	1.0702	1.0671	1.0643	1.0615	1.0589	1.0564	1.0539	1.0516	1.0495	1.0473
1880	1.0805	1.0769	1.0737	1.0706	1.0675	1.0647	1.0618	1.0592	1.0567	1.0542	1.0519	1.0497	1.0475
1900	1.0809	1.0773	1.0741	1.0709	1.0678	1.0650	1.0621	1.0595	1.0569	1.0544	1.0521	1.0499	1.0477
1920	1.0813	1.0777	1.0745	1.0713	1.0682	1.0653	1.0625	1.0598	1.0572	1.0546	1.0523	1.0501	1.0479
1940	1.0817	1.0781	1.0749	1.0716	1.0685	1.0656	1.0628	1.0600	1.0574	1.0548	1.0524	1.0503	1.0480
1960	1.0820	1.0784	1.0752	1.0719	1.0688	1.0659	1.0630	1.0602	1.0576	1.0550	1.0526	1.0505	1.0482
1980	1.0823	1.0787	1.0754	1.0721	1.0690	1.0661	1.0632	1.0605	1.0578	1.0552	1.0528	1.0506	1.0483
2000	1.0826	1.0790	1.0757	1.0724	1.0693	1.0664	1.0635	1.0607	1.0580	1.0554	1.0530	1.0508	1.0485
2020	1.0830	1.0793	1.0760	1.0727	1.0696	1.0667	1.0638	1.0609	1.0582	1.0556	1.0532	1.0509	1.0487
2040	1.0833	1.0796	1.0762	1.0729	1.0698	1.0669	1.0640	1.0611	1.0584	1.0558	1.0534	1.0511	1.0488
2060	1.0836	1.0799	1.0765	1.0732	1.0701	1.0671	1.0642	1.0613	1.0586	1.0560	1.0536	1.0512	1.0489
2080	1.0839	1.0801	1.0767	1.0734	1.0703	1.0673	1.0644	1.0615	1.0588	1.0562	1.0537	1.0514	1.0490
2100	1.0841	1.0804	1.0770	1.0736	1.0705	1.0674	1.0645	1.0617	1.0589	1.0563	1.0538	1.0515	1.0491
2120	1.0843	1.0807	1.0772	1.0738	1.0707	1.0676	1.0647	1.0619	1.0590	1.0565	1.0540	1.0516	1.0492
2140	1.0845	1.0809	1.0774	1.0740	1.0709	1.0677	1.0648	1.0620	1.0591	1.0566	1.0541	1.0517	1.0493
2160	1.0847	1.0811	1.0776	1.0742	1.0711	1.0679	1.0650	1.0621	1.0592	1.0567	1.0542	1.0518	1.0493
2180	1.0849	1.0813	1.0778	1.0744	1.0712	1.0680	1.0651	1.0622	1.0593	1.0568	1.0543	1.0519	1.0494

(table continued)

Table 10-11e Continued
F_{pv} Supercompressibility Factors
(Base Data—0.6 Specific Gravity Hydrocarbon Gas)

p_f psig	Temperature, °F											
	125	130	135	140	145	150	155	160	165	170	175	180	185
2200	1.0851	1.0814	1.0780	1.0746	1.0713	1.0681	1.0652	1.0623	1.0594	1.0569	1.0544	1.0520	1.0495
2220	1.0853	1.0816	1.0781	1.0747	1.0714	1.0682	1.0653	1.0624	1.0595	1.0569	1.0544	1.0521	1.0495
2240	1.0854	1.0817	1.0782	1.0748	1.0715	1.0683	1.0654	1.0625	1.0596	1.0570	1.0545	1.0521	1.0495
2260	1.0855	1.0818	1.0783	1.0749	1.0716	1.0684	1.0654	1.0625	1.0596	1.0570	1.0545	1.0521	1.0496
2280	1.0856	1.0819	1.0784	1.0750	1.0717	1.0685	1.0655	1.0626	1.0597	1.0571	1.0546	1.0521	1.0496
2300	1.0856	1.0819	1.0784	1.0750	1.0718	1.0685	1.0655	1.0626	1.0597	1.0571	1.0546	1.0521	1.0496
2320	1.0857	1.0820	1.0785	1.0751	1.0719	1.0686	1.0655	1.0626	1.0597	1.0571	1.0546	1.0522	1.0496
2340	1.0857	1.0821	1.0785	1.0752	1.0719	1.0687	1.0656	1.0627	1.0598	1.0572	1.0547	1.0522	1.0497
2360	1.0858	1.0821	1.0786	1.0752	1.0719	1.0687	1.0656	1.0627	1.0598	1.0572	1.0547	1.0522	1.0497
2380	1.0858	1.0821	1.0786	1.0752	1.0719	1.0687	1.0656	1.0627	1.0598	1.0572	1.0547	1.0522	1.0497
2400	1.0859	1.0822	1.0787	1.0752	1.0719	1.0687	1.0657	1.0628	1.0599	1.0572	1.0546	1.0521	1.0496
2420	1.0859	1.0822	1.0787	1.0752	1.0719	1.0687	1.0657	1.0628	1.0599	1.0572	1.0546	1.0521	1.0496
2440	1.0859	1.0822	1.0787	1.0752	1.0719	1.0687	1.0657	1.0628	1.0599	1.0572	1.0546	1.0521	1.0496
2460	1.0858	1.0822	1.0786	1.0751	1.0718	1.0687	1.0657	1.0627	1.0598	1.0571	1.0545	1.0521	1.0495
2480	1.0858	1.0822	1.0786	1.0751	1.0718	1.0687	1.0657	1.0627	1.0598	1.0571	1.0545	1.0520	1.0495
2500	1.0857	1.0821	1.0785	1.0750	1.0717	1.0686	1.0657	1.0627	1.0598	1.0571	1.0545	1.0510	1.0494
2520	1.0856	1.0820	10.784	1.0749	1.0716	1.0685	1.0656	1.0627	1.0598	1.0571	1.0544	1.0519	1.0493
2540	1.0855	1.0819	1.0783	1.0748	1.0716	1.0684	1.0655	1.0626	1.0597	1.0570	1.0543	1.0518	1.0493
2560	1.0854	1.0818	1.0782	1.0747	1.0715	1.0683	1.0654	1.0625	1.0596	1.0569	1.0542	1.0517	1.0492
2580	1.0853	1.0816	1.0781	1.0746	1.0714	1.0682	1.0653	1.0624	1.0595	1.0567	1.0541	1.0516	1.0491
2600	1.0852	1.0814	1.0779	1.0745	1.0713	1.0681	1.0651	1.0622	1.0594	1.0566	1.0540	1.0515	1.0490
2620	1.0850	1.0813	1.0778	1.0744	1.0712	1.0680	1.0650	1.0621	1.0593	1.0565	1.0539	1.0514	1.0489
2640	1.0848	1.0811	1.0776	1.0742	1.0710	1.0679	1.0648	1.0619	1.0591	1.0563	1.0537	1.0513	1.0488
2660	1.0846	1.0809	1.0774	1.0741	1.0709	1.0678	1.0647	1.0618	1.0590	1.0562	1.0536	1.0512	1.0487
2680	1.0844	1.0807	1.0772	1.0739	1.0707	1.0676	1.0645	1.0616	1.0588	1.0560	1.0534	1.0511	1.0485
2700	1.0842	1.0805	1.0770	1.0737	1.0705	1.0675	1.0644	1.0615	1.0587	1.0559	1.0533	1.0509	1.0484
2720	1.0840	1.0803	1.0768	1.0734	1.0703	1.0673	1.0643	1.0614	1.0586	1.0558	1.0532	1.0507	1.0482
2740	1.0838	1.0801	1.0766	1.0732	1.0701	1.0671	1.0641	1.0612	1.0584	1.0556	1.0530	1.0505	1.0480
2760	1.0835	1.0799	1.0764	1.0730	1.0699	1.0669	1.0639	1.0610	1.0582	1.0555	1.0528	1.0503	1.0478
2780	1.0833	1.0796	1.0762	1.0728	1.0697	1.0667	1.0638	1.0608	1.0580	1.0553	1.0527	1.0501	1.0476
2800	1.0830	1.0793	1.0759	1.0726	1.0695	1.0664	1.0635	1.0606	1.0577	1.0551	1.0525	1.0499	1.0474
2820	1.0827	1.0791	1.0757	1.0724	1.0693	1.0662	1.0633	1.0604	1.0575	1.0549	1.0523	1.0497	1.0472
2840	1.0824	1.0788	1.0754	1.0721	1.0690	1.0659	1.0631	1.0602	1.0573	1.0547	1.0521	1.0495	1.0470
2860	1.0822	1.0786	1.0752	1.0719	1.0688	1.0656	1.0629	1.0600	1.0571	1.0544	1.0518	1.0493	1.0468
2880	1.0819	1.0783	1.0749	1.0716	1.0685	1.0653	1.0626	1.0598	1.0569	1.0542	1.0516	1.0491	1.0466
2900	1.0816	1.0780	1.0746	1.0713	1.0682	1.0650	1.0623	1.0595	1.0566	1.0539	1.0513	1.0488	1.0463
2920	1.0812	1.0777	1.0743	1.0710	1.0679	1.0648	1.0620	1.0592	1.0564	1.0537	1.0511	1.0486	1.0461
2940	1.0808	1.0773	1.0740	1.0707	1.0676	1.0646	1.0617	1.0589	1.0561	1.0534	1.0509	1.0484	1.0458
2960	1.0805	1.0770	1.0737	1.0704	1.0673	1.0643	1.0614	1.0586	1.0558	1.0531	1.0506	1.0481	1.0456
2980	1.0801	1.0767	1.0734	1.0701	1.0670	1.0640	1.0611	1.0583	1.0556	1.0529	1.0504	1.0479	1.0454
3000	1.0797	1.0764	1.0731	1.0698	1.0667	1.0637	1.0608	1.0580	1.0554	1.0527	1.0501	1.0476	1.0451

Note: Factors for intermediate values of pressure and temperature should be interpolated.

Gas Flow Measurement 517

Table 10-12
F_m Mercury Manometer Factors

Specific Gravity, γ_g	Flowing Pressure, psig						
	0	500	1000	1500	2000	2500	3000
Ambient Temperature = 0°F							
0.55	1.0000	0.9989	0.9976	0.9960	0.9943	0.9930	0.9921
0.60	1.0000	0.9988	0.9972	0.9952	0.9932	0.9919	0.9910
0.65	1.0000	0.9987	0.9967	0.9941	0.9920	0.9908	0.9900
0.70	1.0000	0.9985	0.9961	0.9927	0.9907	0.9896	0.9890
0.75	1.0000						
Ambient Temperature = 40°F							
0.55	1.0000	0.9990	0.9979	0.9967	0.9954	0.9942	0.9932
0.60	1.0000	0.9989	0.9976	0.9962	0.9946	0.9933	0.9923
0.65	1.0000	0.9988	0.9973	0.9955	0.9937	0.9923	0.9913
0.70	1.0000	0.9987	0.9970	0.9947	0.9926	0.9912	0.9903
0.75	1.0000	0.9986	0.9965	0.9937	0.9915	0.9902	0.9893
Ambient Temperature = 80°F							
0.55	1.0000	0.9991	0.9981	0.9971	0.9960	0.9950	0.9941
0.60	1.0000	0.9990	0.9979	0.9967	0.9955	0.9943	0.9933
0.65	1.0000	0.9989	0.9977	0.9963	0.9948	0.9935	0.9925
0.70	1.0000	0.9988	0.9974	0.9958	0.9940	0.9926	0.9915
0.75	1.0000	0.9987	0.9971	0.9951	0.9931	0.9916	0.9906
Ambient Temperature = 120°F							
0.55	1.0000	0.9992	0.9983	0.9974	0.9965	0.9956	0.9948
0.60	1.0000	0.9991	0.9981	0.9971	0.9960	0.9950	0.9941
0.65	1.0000	0.9990	0.9979	0.9967	0.9955	0.9944	0.9934
0.70	1.0000	0.9989	0.9977	0.9963	0.9950	0.9937	0.9926
0.75	1.0000	0.9988	0.9975	0.9959	0.9943	0.9929	0.9918

Notes: Factors for intermediate values of pressure, temperature, and specific gravity should be interpolated. This table is for use with mercury manometer type recording gauges that have gas in contact with the mercury surface. The real gas density is equal to γ_g ($Z_{sc,air}/Z_{sc,gas}$); thus for all practical purposes, the real gas relative density is approximately equal to the gas specific gravity, γ_g.
From *Orifice Metering of Natural Gas*, 1969; courtesy of AGA.

(text continued from page 501)

Gauge Location Factor, F_l

The gauge location factor, F_l, given in Table 10-13, is used where orifice meters are installed at locations other than sea-level elevation and 45° latitude. This is also a very small correction.

Table 10-13
F₁—Gauge Location Factors (Gravitation Correction Factors for Manometer Factor Adjustment)
(Based on Elevation and Latitude, Applicable Unadjusted Factors in Preceding Table)

Degrees latitude	Sea level	Gauge elevation above sea level—feet				
		2,000'	4,000'	6,000'	8,000'	10,000'
0 (Equator)	0.9987	0.9986	0.9985	0.9984	0.9983	0.9982
5	0.9987	0.9986	0.9985	0.9984	0.9983	0.9982
10	0.9988	0.9987	0.9986	0.9985	0.9984	0.9983
15	0.9989	0.9988	0.9987	0.9986	0.9985	0.9984
20	0.9990	0.9989	0.9988	0.9987	0.9986	0.9985
25	0.9991	0.9990	0.9989	0.9988	0.9987	0.9986
30	0.9993	0.9992	0.9991	0.9990	0.9989	0.9988
35	0.9995	0.9994	0.9993	0.9992	0.9991	0.9990
40	0.9998	0.9997	0.9996	0.9995	0.9994	0.9993
45	1.0000	0.9999	0.9998	0.9997	0.9996	0.9995
50	1.0002	1.0001	1.0000	0.9999	0.9998	0.9997
55	1.0004	1.0003	1.0002	1.0001	1.0000	0.9999
60	1.0007	1.0006	1.0005	1.0004	1.0003	1.0002
65	1.0008	1.0007	1.0006	1.0005	1.0004	1.0003
70	1.0010	1.0009	1.0008	1.0007	1.0006	1.0005
75	1.0011	1.0010	1.0009	1.0008	1.0007	1.0006
80	1.0012	1.0011	1.0010	1.0009	1.0008	1.0007
85	1.0013	1.0012	1.0011	1.0010	1.0009	1.0008
90 (Pole)	1.0013	1.0012	1.0011	1.0010	1.0009	1.0008

Note: While F_1 values are strictly manometer factors, to account for guages being operated under gravitational forces that depart from standard location; it is suggested that it be combined with other flow constants. In which instance, F_1 becomes a location factor constant and F_m, the manometer factor agreeable with standard gravity remains a variable factor, subject to change with specific gravity, ambient temperature, and static pressure.

From *Orifice Metering of Natural Gas*, 1969; courtesy of AGA.

Orifice Thermal Expansion Factor, F_a

This factor accounts for the expansion or contraction of the orifice hole with flowing temperature, calculated as follows:

$F_a = 1 + [0.0000185 \, (T_f - 528)]$ for stainless steel

$F_a = 1 + [0.0000159 \, (T_f - 528)]$ for monel (10-25)

where T_f = gas flowing temperature at the orifice, °R

Orifice Meter Selection

Several factors need to be considered in choosing an orifice metering system:

1. Flow rate: flow rate uniformity, maximum and minimum flow rates expected.
2. Pressure: expected static and differential pressures, and their range; permissible pressure variations.

The size of the orifice affects the range of flow rates that can be measured, and the pressure differential that will be obtained. A well designed metering system can only be achieved if all these factors are carefully considered in choosing the size and type of orifice, and the pressure measuring devices.

Example 10-1. An orifice meter with a 2-in. orifice, equipped with pipe taps using upstream static pressure connections in a 6-in. nominal (6.065-in. internal diameter) pipeline, shows an average differential head = 60-in. water and an average upstream static pressure = 90 psia. The flowing temperature is 50°F, and the gas gravity is 0.65. Using a base pressure of 14.9 psia and base temperature of 50°F, calculate the gas flow rate indicated by the meter.

Solution

$\beta = 2/6.065 = 0.3298$.

Average $(h_w p_f)^{0.5} = [(60)(90)]^{0.5} = 73.485$.

Average $h_w/p_f = 60/90 = 0.6667$.

From Table 10-7, $F_b = 870.93$.

From Table 10-8, $b = 0.0273$. Therefore, $F_r = 1 + 0.0273/73.485 = 1.00037$.

From Table 10-9, by suitable interpolation, $Y_1 = 0.9914$.

From Equation 10-18, for $p_b = 14.9$ psia, $F_{pb} = 14.73/14.9 = 0.9886$.

From Equation 10-19, for $T_b = 50°F$, $F_{tb} = (460 + 50)/520 = 0.9808$.

From Equation 10-20, for $T_f = 50°F$, $F_{tf} = (520/510)^{0.5} = 1.0098$.

From Equation 10-21, for $\gamma_g = 0.65$, $F_g = 1/(0.65)^{0.5} = 1.2403$.

From Figure 3-1, for $\gamma_g = 0.65$, $p_{pc} = 670$ psia, and $T_{pc} = 375°R$.

Thus, $p_{pr} = 90/670 = 0.134$, $T_{pr} = 510/375 = 1.36$, and Z from Figure 3-2 is equal to 0.98.

Therefore, $F_{pv} = 1/(0.98)^{0.5} = 1.010$.

For $\gamma_g = 0.65$, $p_f = 90 - 14.7 = 75.3$ psig, and $T_f = 50°F$, $F_m = 0.9998$ from Table 10-12.

Neglecting F_l and F_a, and using Equation 9-17,

$K = (870.93)(1.00037)(0.9914)(0.9886)(0.9808)(1.0098)(1.2403)(1.010)$
$(0.9998) = 1,059.23$

Using Equation 9-16,

$q = (1,059.23)(73.485) = \underline{77,837 \text{ ft}^3/\text{hr}}$

Factors Affecting Orifice Meter Accuracy

According to the Petroleum Extension Service (1972), the following are the sources of constant errors (errors that are constant over time for an installed meter):

1. Incorrect estimate of orifice size.
2. Convex or concave contouring of the orifice plate.
3. Thick or dull orifice edge.
4. Eccentricity of orifice with respect to the pipe.
5. Incorrect estimate of pipe diameter.
6. Excessive recess between the end of pipe and the face of the orifice plate.
7. Excessive pipe roughness.

The Petroleum Extension Service (1972) also lists the following as the most common sources of variable errors:

1. Flow disturbances, caused by insufficient provisions for flow stabilization, or by irregularities in the pipe, welding, etc.
2. Imprecise location of the pressure taps.
3. Pulsating flow.
4. Buildup of solids or sediment on the upstream face of the orifice plate.
5. Liquid accumulation in the bottom of a horizontal pipe run, or in pipe sags, or in meter body.

6. Differences or changes in prevailing operating conditions from those used for calculation purposes.
7. Incorrect zero adjustment of the meter.
8. Non-uniform calibration characteristic of the meter.
9. Corrosion or deposits in the meter internals, or contaminated mercury.
10. Emulsification of liquids with mercury.
11. Leakage around the orifice plate.
12. Formation of hydrates in meter piping or body.
13. Incorrect pen movement on chart, such as incorrect arc for the pens, or excessive friction between pen and chart.
14. Chart malfunctions—incorrect range, incorrect rotation time.
15. Overdampening of the meter response.

Common Measurement Problems

Some of the common measurement problems encountered in gas metering are (Petroleum Extension Service, 1972): hydrate formation (freezing); pulsating flow; slugging; and sour gas.

Hydrate Formation

Hydrates may be formed at the orifice, or in the meter piping or internals, whenever the gas temperature falls below the hydrate-forming temperature for the gas. Such an instance should be recorded on the meter charts, and the estimated static and differential pressure lines should be drawn in.

Hydrate formation can be prevented using any of the following (see Chapter 5):

1. Gas dehydration.
2. Use of hydrate inhibitors.
3. Installation of heaters along the line or near the meter.
4. Other methods—elimination of pipe leaks, enlarging meter piping and valves, and replacing needle valves with plug or gate valves (Petroleum Extension Service, 1972).

Pulsating Flow

Pulsating flow is flow comprising sudden changes in pressure and flow rate of the flowing fluid. Common sources of such flow in gas measurement are (Petroleum Extension Service, 1972):

1. Reciprocating systems—compressors, or engines.
2. Improperly sized, loose, or worn valves and regulators.
3. Two-phase flow conditions.
4. Intermitters on wells and automatic drips.

Pulsating flow can be a source of considerable metering errors. There is no known method to correct for such a flow. The Petroleum Extension Service (1972) outlines the following methods to reduce pulsating flow and/or diminish its effect on orifice flow measurement:

1. Locate the meter along the flowline in a position where pulsations are minimized.
2. Reduce the amplitude of the pulsations by placing a volume capacity, flow restriction, or specially designed filter between the pulsation source and the meter.
3. Operate at pressure differentials as high as possible, by using a smaller diameter orifice, or by allowing flow only through a limited number of tubes in a multiple tube installation. The same can also be achieved by using smaller sized tubes, keeping the same orifice size and maintaining as high a pressure differential as possible.

Slugging

Slugging refers to the accumulation of liquids in the gas flowline. In low-pressure lines, the liquid accumulates at low spots in the line, restricting gas flow until enough pressure is built up for the gas to blow through the liquid. In high-pressure lines, liquid is swept through to the orifice and beyond. Both these situations result in flow disturbances that cause erratic and inaccurate measurements. A common method of preventing slugging flow is the installation of liquid accumulators in the flowline.

Sour Gas

As discussed in Chapters 5 and 6, sour gas is detrimental to all flow equipment for two reasons: corrosion and accelerated hydrate formation. Sour gas in a closed line causes little or no corrosion; it is the hydrogen sulfide in the surrounding atmosphere that affects the measurement equipment (Petroleum Extension Service, 1972). Common preventive measures to ensure proper gas metering include using hydrogen sulfide resistant components in the meters, and sealing the meters against the atmosphere.

Gas Flow Measurement   523

Other Types of Measurements

Mass Flow Rate Measurement

In recent years, fluids are being handled near their critical region, where the fluid density changes very rapidly with small changes in the flowing conditions. Consequently, several instruments and techniques have been developed to measure mass flow rate. There are essentially two types of mass flowmeters: *true mass flowmeters* that respond directly to mass flow rate, and *inferential mass flowmeters* that infer the mass flow rate from separate volumetric flow rate and density measurements.

True Mass Flowmeters

Several types of true mass flowmeters have been developed, including the following (*Chemical Engineers' Handbook*, 1984):

1. Axial-flow, transverse-momentum mass flowmeter.
2. Radial-flow, transverse-momentum mass flowmeter.
3. Gyroscopic transverse-momentum mass flowmeter.
4. Magnus-effect mass flowmeter.
5. Thermal mass flowmeter—commonly uses vibrating tubes and heat transfer.

Of these, the axial-flow transverse-momentum mass flowmeter is the most commonly used. Also known as an angular-momentum mass flowmeter, it uses axial flow through an impeller and a turbine in series. The impeller imparts angular momentum to the fluid, which in turn supplies a torque to the turbine. The mass flow rate of the fluid through the meter is obtained by measuring this torque which is proportional to the impeller's rotational speed and the mass flow rate.

Inferential Mass Flowmeters

The several types of inferential mass flowmeters can be classified into three major categories (*Chemical Engineers' Handbook*, 1984):

1. Head-type meters with density compensation. In this type of mass flow measurement, a head meter, such as an orifice or venturi, is used along with a densitometer. The signal from the head meter, proportional to ρv^2, is multiplied with the density ρ given by the densitometer. The square root of the product thus obtained is proportional to the mass flow rate.

524 Gas Production Engineering

2. Head-type meters with velocity compensation. This type of mass measurement uses a head meter and a velocity meter (pitot tube, or a turbine meter). The signal from the head meter is divided by the velocity signal from a velocity meter to obtain a signal proportional to the mass flow rate.
3. Volume or velocity-type meters with density compensation. In this method, the signal from a velocity meter (such as turbine meter or sonic velocity meter) or a volume (displacement) meter (such as rotary meter or reciprocating piston meter), is multiplied by the signal from a densitometer to generate a signal proportional to the mass flow rate.

Various types of densitometers are available for determining the density of a flowing gas stream, based upon different principles, such as buoyant force on a fluid-supported float, radiation attenuation, and piezoelectric crystals that respond to pressure.

For the popular orifice-meter/densitometer combination for mass flow rate measurement, AGA (AGA, 1969) gives the following equation:

$$m = 1.0618 \, F_b \, F_r \, Y \, F_m \, F_l \, F_a \, (h_w \omega_g)^{0.5} \qquad (10\text{-}26)$$

where m = mass flow rate of the gas, lbm/hr
h_w = differential pressure across the orifice, in. of water
ω_g = gas specific weight, lbf/ft^3, equal to (gas density) $\times$ (g/g_c)

F_b, F_r, Y, F_m, F_l, and F_a are the orifice factors as described earlier.

Natural Gas Liquids Measurement Using Orifice Meters

Natural gas liquids are generally measured under static conditions using conventional tank-gauging methods. For flowing liquid streams, orifice meters are used quite often. For volumetric flow rate measurement in gallons per hour, the American Meter Company (1973) provides the following equation:

$$q = F_b \, F_{gt} \, F_{sl} \, F_r \, (h_w)^{0.5} \qquad (10\text{-}27)$$

where q = liquid flow rate, gal/hr
h_w = differential pressure across the orifice, in. of water
F_b = basic orifice factor
F_{gt} = specific gravity factor (for temperature correction)
F_{sl} = factor for seals (if required)
F_r = Reynolds number factor

For mass flow rate measurement in pounds per day, the following equation by the Foxboro Company (1961) may be used:

$$m_L = 68{,}045\, Sd^2\, F_a\, F_m\, F_c\, F_p\, (\gamma_L h_w)^{0.5} \qquad (10\text{-}28)$$

where m_L = liquid flow rate, lbm/day
h_w = differential pressure across the orifice, in. of water
S = a constant determined by the orifice and pipe diameters
d = inside diameter of the meter tube, in.
F_a = orifice thermal expansion factor (= 1.0 for temperatures between 23 and 99°F)
F_m = manometer factor
F_c = viscosity factor, generally assumed to be unity
F_p = correction factor for liquid compressibility
γ_L = specific gravity of the flowing liquid stream

Equation 10-28 is often used in the following simplified form:

$$m_L = 68{,}045\, Sd^2\, [\gamma_L h_w]^{0.5} \qquad (10\text{-}29)$$

Two-Phase Systems

Flowmeter accuracy is generally quite poor for measuring two-phase flowstreams. No good method is currently known—most techniques give only an approximate value. Murdock (1962) gives the following equation for two-phase flow through orifice meters:

$$m_{TP} = \frac{359 K_g Y_g F_a d^2 [\rho_{g1}(h_w)_{TP}]^{0.5}}{(1-X) + 1.26 X (K_g Y_g / K_L)(\rho_{g1}/\rho_{L1})^{0.5}} \qquad (10\text{-}30)$$

where m_{TP} = mass flow rate of two-phase flow, lbm/hour
K_g, K_L = orifice flow coefficients for gas and liquid, respectively
Y_g = expansion factor for the orifice
F_a = orifice thermal expansion factor
d = orifice diameter, in.
ρ_g, ρ_L = gas and liquid densities, respectively, lbm/ft^3
$(h_w)_{TP}$ = effective differential head for the two-phase flow, in. of water
X = liquid weight fraction in the flowstream

and subscript 1 represents the value at the orifice inlet.

For metering two-phase flow, the Petroleum Extension Service (1972) recommends the following precautions:

1. Keep pressure and temperature as high as possible at the meter.
2. Use a free-water knockout upstream of the meter.
3. Use a vertical meter run that may improve the differential pressure and flow volume relationship in some, if not all, cases.
4. Determine a meter factor to correct the metering results, using test data from (periodic) separator tests.
5. Connect manifold lead lines to bottom of bellows-type meter with self-draining pots installed above orifice fitting.

Questions and Problems

1. What would be the features of an "ideal" gas measurement device?
2. Why is gas measured in terms of volume? Compare the requirements for volumetric versus mass flow rate measurement.
3. Discuss the problems in two-phase measurement. Address the issue of flashing of liquids into vapor (a common occurrence for undersaturated steam and all except dead oils), and what strategies can be adopted in such a situation.
4. Is a linear meter necessarily better?
5. List the advantages and applicability of the various flow measurement devices.
6. Compare the orifice types, including their effect on gas flow measurement.
7. Discuss, giving reasons, the important factors that affect orifice meter accuracy. What operating conditions need be assured to enable accurate orifice metering of natural gas?
8. Given static and differential pressure recordings on an orifice meter chart, show the calculations that are required for computing flow volumes for the case of: (a) a linear orifice-meter chart, and (b) L-10 square-root chart.
9. Calculate the gas production from a well with the following orifice-meter information:
   Readings from a square-root chart with a chart range of 50 in. × 100 psi: differential = 7.1, static = 8.5 taken at downstream flange-type tap.
   Pipe diameter = 3 in., orifice diameter = 0.5 in., γ_g = 0.72, flowing temperature = 95°F.

Neglect F_m, F_l, and F_a. Assume base conditions of 14.73 psia and 60°F, and that the gas has (in mole%): CO_2 = 1.2, N_2 = 0.58, and H_2S = 0.96.

10. It is suspected that the chart recording device is malfunctioning in a field orifice measurement system. A test with a mass flow meter indicates a gas rate of 18,000 lbm/hr. For the following conditions, determine if the orifice meter chart is in error:

Pipe diameter = 8-in. nominal (8.071 in. ID)
Orifice diameter = 3.0 in.
γ_g = 0.63
Flowing temperature = 85°F
Static pressure reading = 110 psia
Differential pressure reading = 175.5 in. water
Pipe taps downstream

If there is an error, find its dollar value, given that gas sells for $1.0 per Mscf (measured at 14.73 psia, 60°F).

References

American Meter Company, 1973. *Orifice Meter Constants: Handbook E-2*, revised.
AGA, 1969. *Orifice Metering of Natural Gas*. Gas Measurement Committee Report No. 3 (Revised), American Gas Association, New York.
ASME, 1971. *Fluid Meters—Their Theory and Application*, 6th edition. Report of ASME Research Committee on Fluid Meters, The American Society of Mechanical Engineers, New York.
Campbell, J. M., 1984. *Gas Conditioning and Processing*, Vol. 1. Campbell Petroleum Series, Norman, Oklahoma, 326pp.
Chemical Engineers' Handbook, 1984. R. H. Perry and D. W. Green (eds.). McGraw-Hill Book Co., New York, 6th ed.
Corcoran, W. S. and Honeywell, J., 1975. "Practical Methods for Measuring Flows," *Chem. Eng.*, 82(14, July 7), 86–92.
DeVries, E. A., 1982. "Facts and Fallacies of Vortex Flowmeters," *Hydr. Proc.*, 61(8, Aug.), 75–76.
Evans, H. J., 1973. "Turbine Meters Gain in Gas Measurement," *Oil & Gas J.*, 71(34, Aug. 20), 67–69.
Foxboro Company, 1961. *Principles and Practices of Flowmeter Engineering*, 8th edition. Foxboro Company, Foxboro, Massachussets.
GPSA, 1981. *Engineering Data Book*, 9th ed. (5th revision). Gas Processors Suppliers Association, Tulsa, Oklahoma.

Munk, W. D., 1982. "Ultrasonic Flowmeter Offers New Approach to Large-Volume Gas Measurement," *Oil & Gas J.*, 80(36, Sept. 6), 111–117.

Murdock, J. W., 1962. "Two-Phase Flow Measurement With Orifices," *Trans., ASME: J. Basic Eng.*, 84(4, Dec.), 419–433.

November, M. H., 1972. "How to Use High-Capacity Axial-Flow Turbine Meters for Gas Measurement," *Oil & Gas J.*, 70(14, Apr. 3), 69–77.

Petroleum Extension Service, 1972. *Field Handling of Natural Gas*, 3rd edition. University of Texas Press, Austin, Texas, 143pp.

Powers, L., 1975. "Vortex Shedding Provides Accurate Flow," *Oil & Gas J.*, 73(31, Aug. 4), 84-88.

11
Gas Gathering and Transport

Introduction

Natural gas produced from several wells in a given area is collected and brought to field separation and processing facilities via a system of pipes known as a gathering system. Processed or partially processed gas is then sent to the trunk lines that transport the gas to consumers. Gas is often distributed via pipeline grids that introduce a lot of complexity into the flow computations. This chapter briefly describes gathering systems and the transport of gas through pipeline networks, building upon the concepts for steady state flow through a single pipe described in Chapter 7. Some basic elements of unsteady state gas flow, encountered quite often in pipeline practice, are also introduced.

Gathering Systems

The surface flow gathering system consists of the section of pipe and fittings that serve to transmit the produced fluid from the wellhead to the field treatment facilities (generally, the oil-water-gas separators). Production systems with extremely high capacity wells may provide individual separation, metering, and possibly treatment, facilities to each of the wells. Because these single well systems are seldom economical, it is quite common to design gathering and separation facilities that enable combined handling of several wellstreams.

The two basic types of gathering systems are *radial*, and *axial*. In the radial system (Figure 11-1a), flowlines emanating from several different wellheads converge to a central point where facilities are located. Flowlines are usually terminated at a header, which is essentially a pipe large enough to

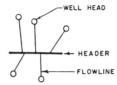
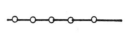

Figure 11-1b. An axial gathering system.

Figure 11-1a. A radial gathering system.

handle the flow of all the flowlines. In the axial gathering system, several wells produce into a common flowline (Figure 11-1b).

For larger leases, these two basic systems are modified a little. The well-center gathering system (Figure 11-2a) uses a radial gathering philosophy at the local level for individual wells, as well as at the global level for groups of wells. The common-line or trunk-line gathering system uses an axial gathering scheme for the groups of wells that, in turn, use a radial gathering scheme (Figure 11-2b). The trunk-line gathering system is more applicable to relatively larger leases, and to cases where it is undesirable or impractical to build the field processing facilities at a central point.

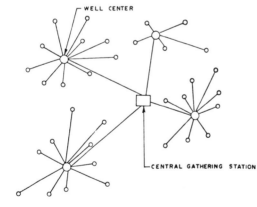

Figure 11-2a. Well-center gathering system.

Figure 11-2b. Trunk-line gathering system.

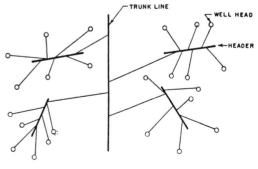

It is obvious that very complex metering facilities are required to measure the production of individual wells simultaneously. Generally, a test header is used to route fluids from a single well through the metering system. This well test header also provides the means to control the production from individual wells and to conduct well tests on individual wells.

The choice between the gathering systems is usually economic. The cost of the several small sections of pipe used in the well-center system is compared to the cost of a single large pipe for the trunk-line system. Technical feasibility may be another criterion. The gathering system may have to be buried a few feet beneath the surface, favoring one system over another in terms of cost and ease of maintenance. The production characteristics of the field are also important to consider. These include current and estimated future production distribution over the wells in the field, wellhead flowing pressures, future development of the field, and the possibility of the development of underground storage operations.

Steady-State Flow in Simple Pipeline Systems

The term "simple" is used here to indicate the gas pipeline systems that can be handled with minor modifications to the flow relationships presented in Chapter 7. The one feature for simplicity is that gas flows in at one end, and flows out at the other end; no flow occurs at any other point in the piping system. Such a scheme is often used for increasing the throughput of a pipeline while maintaining the same pressure and pressure drop (such as when new gas wells have been developed that must use the existing pipeline) or for operating a pipeline at a lower pressure (pressure-deration) while maintaining the same throughput. The latter may be required when the pipeline has "aged" or corroded.

The three possible ways of handling these requirements are to replace a portion of the pipeline with a larger one (pipelines in series), place one or more pipelines in parallel along the complete length of the existing line (pipelines in parallel), or place one or more pipelines in parallel only partially along the length of the existing line (series-parallel or looped lines). For each of these systems, relationships will be derived here, based upon the basic equation by Weymouth (Equation 7-29) for steady-state flow of gas through pipes, that reduce the set of pipelines to a single pipeline that is equivalent to the set in terms of the pressure drop and flow capacity.

The Weymouth equation (Equation 7-29) can be written as:

$$q = K_1 \, [d^5/fL]^{0.5} \qquad (11\text{-}1)$$

where K_1 is a constant, given by

$$K_1 = 5.635382 \left(\frac{T_{sc}}{p_{sc}}\right) \left[\frac{p_1^2 - p_2^2}{\gamma_g (TZ)_{av}}\right]^{0.5} \quad (11\text{-}2)$$

Note that q denotes q_{sc} throughout this chapter for simplicity. Equation 11-1 can be written as:

$$L = Kd^5/(fq^2) \quad (11\text{-}3)$$

Consider two pipelines A and B, of lengths L_A and L_B, and diameters d_A and d_B, respectively. We can equate these two lines A and B using Equation 11-3. For example, the length L_{eBA} of a line of diameter d_A that will have the same pressure drop as line B of length L_B and diameter d_B (i.e., the equivalent length of line B in terms of the diameter of line A) is given by

$$L_{eBA} = L_B(f_B/f_A)(d_A/d_B)^5 \quad (11\text{-}4)$$

Alternatively, the equivalent diameter d_{eBA} of line B may be used:

$$d_{eBA} = d_B[(f_A L_A)/(f_B L_B)]^{1/5} \quad (11\text{-}5)$$

where d_{eBA} is the diameter of a pipe of length L_A and friction factor f_A equivalent to the line B of length L_B, diameter d_B, and friction factor f_B.

Series Pipelines

Consider three pipelines A (length L_A and diameter d_A), B (length L_B and diameter d_B), and C (length L_C and diameter d_C), connected in series as shown in Figure 11-3. The inlet and outlet pressures for the system are p_1

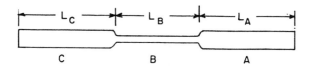

Figure 11-3. Pipelines in series.

and p_2, respectively. For this system, the flow rates through each of the pipe legs are equal:

$$q_A = q_B = q_C = q_t \quad (11\text{-}6)$$

where the subscript t indicates the total for the system. The pressure drops through the pipe legs, however, are not equal. The total pressure drop is equal to the sum of the pressure drops in each of the pipe legs. Thus,

$$\Delta p_A \neq \Delta p_B \neq \Delta p_C \qquad (11\text{-}7)$$

and

$$\Delta p_t = \Delta p_A + \Delta p_B + \Delta p_C \qquad (11\text{-}8)$$

From Equation 7-29 (or, Equations 11-1 and 11-2), we know that the pressure drop Δp in a pipe section is proportional to the length L, all other factors being the same. Substituting in Equation 11-8,

$$L_e = L_A + L_{eBA} + L_{eCA} \qquad (11\text{-}9)$$

where L_e = equivalent length of the total system
 L_A = length of segment A
 L_{eBA}, L_{eCA} = equivalent lengths of segments B and C, respectively

Therefore, the three lines A, B, and C in series are equivalent to a single line of diameter d_A and length L_e given by Equation 11-9.

Let q_{old} be the old flow rate for the pipeline of length $L_A + L_B + L_C$ and diameter d_A, and q_{new} be the new flow rate obtained by altering the sections B and C of the pipeline to two pipe sections of length L_B and diameter d_B and length L_C and diameter d_C. Using the fact that q is proportional to $(1/L)^{0.5}$, we get

$$\frac{q_{new}}{q_{old}} = \frac{1/(L_e)^{0.5}}{1/(L_A + L_B + L_C)^{0.5}} = \left(\frac{L_A + L_B + L_C}{L_e}\right)^{0.5} \qquad (11\text{-}10)$$

The fractional increase in flow capacity, Δq, is given by:

$$\Delta q = \frac{q_{new} - q_{old}}{q_{old}} = \frac{q_{new}}{q_{old}} - 1$$

$$= \left(\frac{L_A + L_B + L_C}{L_e}\right)^{0.5} - 1 \qquad (11\text{-}11)$$

Parallel Pipelines

Consider pipelines A (length L_A, diameter d_A), B (length L_B, diameter d_B), and C (length L_C, diameter d_C) in parallel, as shown in Figure 11-4.

534 Gas Production Engineering

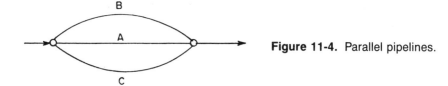

Figure 11-4. Parallel pipelines.

The inlet and outlet pressures for this system are p_1 and p_2, respectively. Because the pipelines are in parallel with a common inlet and outlet, the pressure drop through each of them is the same, but the flow rates are not. The total flow rate, however, is the sum of the flow rates through each of the pipe legs. Thus,

$$\Delta p_A = \Delta p_B = \Delta p_C = \Delta p_t \qquad (11\text{-}12)$$

$$q_A \ne q_B \ne q_C \qquad (11\text{-}13)$$

and

$$q_t = q_A + q_B + q_C \qquad (11\text{-}14)$$

Let the length and diameter of the pipeline equivalent to the three lines A, B, and C in parallel be L_e and d_e, respectively. Then, substituting for q in Equation 11-14 from Equation 11-1:

$$\left(\frac{d_e^5}{f_e L_e}\right)^{0.5} = \left(\frac{d_A^5}{f_A L_A}\right)^{0.5} + \left(\frac{d_B^5}{f_B L_B}\right)^{0.5} + \left(\frac{d_C^5}{f_C L_C}\right)^{0.5} \qquad (11\text{-}15)$$

In Equation 11-15, any value may be assumed for two of the three unknowns, d_e, f_e, and L_e, and the third calculated. Equation 11-15 can also be expressed as:

$$L_e^{0.5} = \frac{1}{(d_A^5 f_e/d_e^5 f_A L_A)^{0.5} + (d_B^5 f_e/d_e^5 f_B L_B)^{0.5} + (d_C^5 f_e/d_e^5 f_C L_C)^{0.5}}$$

Choosing $d_e = d_A$, we obtain:

$$L_e = \left[\frac{1}{(f_e/f_A L_A)^{0.5} + (d_B^5 f_e/d_A^5 f_B L_B)^{0.5} + (d_C^5 f_e/d_A^5 f_C L_C)^{0.5}}\right]^2 \qquad (11\text{-}16)$$

Assuming $L_A = L_B = L_C$, Equation 11-16 reduces to:

$$L_e = \left[\frac{L_A^{0.5}}{(f_e/f_A)^{0.5} + (d_B^5 f_e/d_A^5 f_B)^{0.5} + (d_C^5 f_e/d_A^5 f_C)^{0.5}}\right]^2 \quad (11\text{-}17)$$

The ratio of the new flow rate for lines in parallel to the old flow rate for the single line A is given, as before for the series pipeline, by

$$\frac{q_{new}}{q_{old}} = \frac{1/(L_e)^{0.5}}{1/(L_A)^{0.5}} = \left(\frac{L_A}{L_e}\right)^{0.5} \quad (11\text{-}18)$$

where L_e is obtained using Equation 11-16 or 11-17, as the case may be. The fractional increase in flow capacity, Δq, can then be calculated using the following relationship used earlier for series pipelines:

$$\Delta q = \frac{q_{new}}{q_{old}} - 1$$

A special case of two lines in parallel, also called a fully-looped line, with both lines of equal length, deserves mention. Substituting for L_e from Equation 11-17 into Equation 11-18 for such a system:

$$\frac{q_{new}}{q_{old}} = \frac{L_A^{0.5}}{L_A^{0.5}/[(f_e/f_A)^{0.5} + (d_B^5 f_e/d_A^5 f_B)^{0.5}]}$$

$$= \left(\frac{f_e}{f_A}\right)^{0.5} + \left(\frac{d_B^5 f_e}{d_A^5 f_B}\right)^{0.5}$$

$$= \left[\left(\frac{f_B}{f_A}\right)^{0.5} + \left(\frac{d_B}{d_A}\right)^{2.5}\right]\left(\frac{f_e}{f_B}\right)^{0.5} \quad (11\text{-}19)$$

Looped Pipelines

A looped pipeline is one in which only a part of the line has a parallel segment. The original pipeline is looped to some distance with another line to increase the flow capacity. In the looped line shown in Figure 11-5, the original line having two segments A and C of the same diameter is looped with a segment B. A looped system may be considered to be a combination of series and parallel sections. For the two lines A and B in parallel, Equation 11-15, 11-16, or 11-17 can be used. The resultant equivalent pipe for the parallel segments A and B is then combined in series with the segment C.

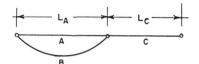

Figure 11-5. Looped pipelines.

Relatively simplified expressions are often found in the literature for the case when the loop-line B has the same length as A. For this case, the equivalent length for the parallel section A-B, $(L_e)_{AB}$, can be determined using Equation 11-17:

$$(L_e)_{AB} = \left[\frac{L_A^{0.5}}{(f_e/f_A)^{0.5} + (d_B^5 f_e/d_A^5 f_B)^{0.5}}\right]^2$$

Then, the equivalent length for the total system, L_e, becomes:

$$L_e = L_C + (L_e)_{AB}$$

$$= L_C + \left[\frac{L_A^{0.5}}{(f_e/f_A)^{0.5} + (d_B^5 f_e/d_A^5 f_B)^{0.5}}\right]^2$$

The old length (without the loop-pipe B) of the system was $L_A + L_C$. From Equation 11-18, the ratio of the flow rates is

$$\frac{q_{new}}{q_{old}} = \left(\frac{L_A + L_C}{L_e}\right)^{0.5}$$

or

$$\left[\frac{q_{old}}{q_{new}}\right]^2 = \frac{L_e}{L_A + L_C}$$

$$= \frac{L_C}{L_A + L_C} + \left[\frac{L_A}{L_A + L_C}\right]\left[\frac{1}{[(f_e/f_A)^{0.5} + (d_B^5 f_e/d_A^5 f_B)^{0.5}]^2}\right] \quad (11\text{-}20)$$

Let x_f be the fractional length of a line that must be looped in order to achieve a desired flow rate. Then,

$$L_A/(L_A + L_C) = x_f$$

$$L_C/(L_A + L_C) = 1 - x_f$$

and Equation 11-20 becomes:

$$\left[\frac{q_{old}}{q_{new}}\right]^2 = (1 - x_f) + x_f \left[\frac{1}{[(f_e/f_A)^{0.5} + (d_B^5 f_e/d_A^5 f_B)^{0.5}]^2}\right]$$

or

$$x_f \left[1 - \frac{1}{[(f_e/f_A)^{0.5} + (d_B^5 f_e/d_A^5 f_B)^{0.5}]^2}\right] = 1 - (q_{old}/q_{new})^2$$

Solving for x_f:

$$x_f = \frac{1 - (q_{old}/q_{new})^2}{1 - 1/[(f_e/f_A)^{0.5} + (d_B^5 f_e/d_A^5 f_B)^{0.5}]^2} \quad (11\text{-}21)$$

Note that A indicates the original line, and B indicates the added line.

Extensions to Commonly Used Pipeline Equations

The expressions for series, parallel, and looped pipelines described contain the friction factors for the individual legs. Most non-iterative equations for gas flow, such as the Weymouth, Panhandle-A, and Panhandle-B equations, assume a friction factor correlation that simplifies the flow calculations (see Chapter 7). The resulting expressions are given below for these equations for series, parallel, and looped lines. The values of the exponents a, b, c, and d are given in Table 11-1.

Equation 11-4 for the equivalent length of a line becomes:

$$L_{e21} = L_2(d_1/d_2)^a \quad (11\text{-}22)$$

where L_{e21} is the equivalent length of line 2 in terms of the diameter of line 1.

Table 11-1
Coefficients for Series, Parallel, Looped Line Equations

Equation	a	b	c	d
Weymouth	16/3 = 5.333	0.50	8/3 = 2.667	2.0
Panhandle-A	4.854	0.5394	2.618	1.86
Panhandle-B	4.961	0.510	2.530	—

Equation 11-5 for the equivalent diameter of a line becomes:

$$d_{e21} = d_2(L_1/L_2)^{1/a} \qquad (11\text{-}23)$$

where d_{e21} is the equivalent diameter of line 2 in terms of the length of line 1.

For lines in series, the equivalent length is the sum of the individual equivalent lengths of all the pipe sections (Equation 11-9). For lines in parallel, the equivalent length can be calculated as follows (compare with Equation 11-16):

$$L_e = \left[\frac{1}{(1/L_1)^b + (1/L_{e21})^b + (1/L_{e31})^b + \ldots + (1/L_{en1})^b} \right]^{1/b} \qquad (11\text{-}24)$$

where L_{ei1} is the equivalent length of pipe leg i, determined using Equation 11-22. If all the parallel pipe legs are of equal length, Equation 11-24 simplifies to (compare with Equation 11-17):

$$L_e = d^a L \left[\frac{1}{d_1^c + d_2^c + d_3^c + \ldots + d_n^c} \right]^{1/b} \qquad (11\text{-}25)$$

If all sections in the parallel lines are of equal length, the ratio of the new to the old flow rate is (compare with Equation 11-19):

$$\frac{q_{new}}{q_{old}} = 1 + \left[\frac{d_{lp}}{d_{old}} \right]^c \qquad (11\text{-}26)$$

where d_{old} = diameter of the original single line
d_{lp} = diameter of the pipe installed in parallel with this line

For looped lines, the looping requirements, analogous to Equation 11-21, are (Campbell, 1984):

$$x_f \simeq \frac{1 - (q_{old}/q_{new})^d}{1 - 1/[1 + (d_{lp}/d_{old})^c]^d} \qquad (11\text{-}27)$$

If the diameter of original and parallel lines is the same, Equation 11-27 simplifies to:

$$x_f \simeq \frac{1 - (q_{old}/q_{new})^d}{1 - 1/[1 + 1]^d}$$

$$= \frac{1 - (q_{old}/q_{new})^d}{1 - 2^{-d}} \qquad (11\text{-}28)$$

Gas Gathering and Transport 539

Example 11-1. Gas of specific gravity 0.65 is being transported from station A to stations C and D. A single pipeline of diameter 8 in., length 3 miles, runs from station A to a pipeline junction at B. A 6-in., 2-mile pipeline connects junction B to station C, while a 4-in., 3-mile pipeline connects B to D. Given that the pressure at station A (p_A) = 400 psia, and that stations C and D are at the same pressure (p_C, p_D) = 30 psia, determine the capacity of the system. Assume that the flowing temperature is 80°F, and use the Weymouth equation.

Solution

Because $p_C = p_D$, we can consider sections BC and BD to be in parallel. One calculation strategy is to reduce the parallel lines BC and BD to a single 8 in. diameter line, and add the equivalent length thus determined to the length of the section AB which is in series.

Using Equation 11-22, the equivalent length of the 4-in. line in terms of an 8-in. diameter line is

$$L_{e4} = 3 \, (8/4)^{16/3} = 120.9525$$

Similarly, $L_{e6} = 2 \, (4/3)^{16/3} = 9.2762$

Using Equation 11-24, the 4-in. and 8-in. lines in parallel are thus equivalent to a single 8-in. diameter line of length:

$$L_e = \left[\frac{1}{(1/120.9525)^{0.5} + (1/9.2762)^{0.5}} \right]^{1/0.5} = 5.689 \text{ miles}$$

Thus, the complete system is equivalent to a single 8-in. diameter line of length = 3 + 5.689 = 8.689 miles

For $\gamma_g = 0.65$, $p_{pc} = 670$ psia, and $T_{pc} = 373°R$. Using Equation 7-32,

$$p_{av} = \left[\frac{(400)^3 - (30)^3}{(400)^2 - (30)^2} \right] = 268 \text{ psia}$$

So, $p_{pr} = 0.4$, and $T_{pr} = 1.44$, and therefore, $Z_{av} = 0.955$.
Using the Weymouth Equation (Equation 7-34),

$$q_{sc} = 31.5027 \, \frac{520}{14.73} \left[\frac{[(400)^2 - (30)^2](8)^{16/3}}{(0.65)(8.689 \times 5,280)(540)(0.955)} \right]^{0.5}$$

$$= 28{,}957.85 \text{ Mscfd} = \underline{28.96 \text{ MMscfd}}$$

Example 11-2. A portion of a large gas-gathering system consists of a 6.067-in. ID line 9.4 miles long, handling 7.6 MMscfd of gas of average specific gravity equal to 0.64. The pressure at the upstream end of this section is 375 psig, and the average delivery pressure is 300 psig. The average temperature is 73°F. Due to new well completions, it is desired to increase the capacity of this line by 20% by looping with additional 6.067-in. ID pipe. What length is required?

Solution

Note: Weymouth equation parameters are used throughout for the calculations.
Using the approximate relationship of Equation 11-28,

$$x_f \simeq \frac{1 - (1/1.2)^2}{1 - 2^{-2}} = 0.4074$$

Thus, the length of looping pipe required = (0.4074)(9.4) = 3.83 miles

The more accurate method involves using the basic relationships as follows.
Desired flow rate q_{sc} = (1.2)(7.6) = 9.12 MMscfd = 9,120 Mscfd
For γ_g = 0.64, p_{pc} = 671 psia, and T_{pc} = 360°R

For the looped section:

From Equation 11-25, the equivalent length of the looped section is:

$$L_e = (6.067)^{16/3}(9.4x_f)\left[\frac{1}{(6.067)^{8/3} + (6.067)^{8/3}}\right]^{1/0.5}$$

= 2.350 x_f miles = 12,408 x_f ft

Assuming the pressure p_3 at the point 9.4 x_f miles from the inlet (where the loop ends) to be equal to 350 psig (= 364.73 psia), and using Equation 7-32,

$$p_{av} = \left[\frac{(389.73)^3 - (364.73)^3}{(389.73)^2 - (364.73)^2}\right] = 377.34 \text{ psia}$$

So, p_{pr} = 0.57, and T_{pr} = 1.48, and therefore, Z_{av} = 0.940.
Using the Weymouth Equation (Equation 7-34),

$$9,120 = 31.5027 \frac{520}{14.73}\left[\frac{[(389.73)^2 - (p_3)^2](6.067)^{16/3}}{(0.64)(533)(12,408x_f)(0.94)}\right]^{0.5}$$

Gas Gathering and Transport   541

On solving this, we get:

$(p_3)^2 + 17{,}847.084\, x_f = 151{,}866.09\ldots$ Equation A

For the unlooped section:

From Equation 7-32,

$$P_{av} = \left[\frac{(364.73)^3 - (314.73)^3}{(364.73)^2 - (314.73)^2}\right] = 340.31 \text{ psia}$$

$P_{pr} = 0.51$, and $T_{pr} = 1.48$, and therefore, $Z_{av} = 0.945$.
Using the Weymouth Equation (Equation 7-34),

$$9{,}120 = 31.5027\, \frac{520}{14.73} \left[\frac{[(p_3)^2 - (314.73)^2](6.067)^{16/3}}{(0.64)(533)(1 - x_f)(9.4 \times 5{,}280)(0.945)}\right]^{0.5}$$

On solving this, we get:

$(p_3)^2 + 71{,}768.06\, x_f = 170{,}804.15\ldots$ Equation B

On solving Equations A and B simultaneously, the values obtained for the two unknowns are:

$x_f = 0.3512$, and $p_3 = 381.57$ psia.

p_3 is quite close to the assumed value of 364.73 psia: Z factor will not change substantially for such a pressure difference. A second trial is not necessary.
Therefore, the length of the loop required = $(0.3512)(9.4)$ = <u>3.30 miles</u>

Steady-State Flow in Pipeline Networks

Gas transmission systems often form a connected net, flow through which is almost always transient (unsteady). Most design and operation control problems, however, can be solved reasonably well assuming flow to be steady-state. The basic model considers the transmission system to be a pipeline network with two basic elements: nodes and node connecting elements (NCE's). Nodes are defined as the points where a pipe leg ends, or where two or more NCE's join, or where there is an injection or offtake (delivery) of gas. The NCE's include pipe legs, compressor stations, valves, pressure and flow regulators, and underground gas storages.

Before constructing a model of the pipeline network, it is necessary to describe the mathematical models for the individual NCE's. These models are essentially pressure versus rate (throughput) relationships, as described below.

1. *High-pressure pipe leg.* The characteristic equation for a high pressure pipe, according to Equation 7-29, is as follows:

$$p_1^2 - p_2^2 = k_1 q^2 \tag{11-29}$$

or

$$q = \left[\frac{p_1^2 - p_2^2}{k_1}\right]^{0.5} \tag{11-30}$$

where $k_1 = 0.031489 \left(\frac{p_{sc}}{T_{sc}}\right)^2 \frac{\gamma_g(TZ)_{av}fL}{d^5}$ \hfill (11-31)

2. *Low-pressure pipe leg.* For a low pressure pipe leg, with pressure close to atmospheric, $Z_{av} \simeq 1$, and

$$p_1^2 - p_2^2 = (p_1 + p_2)(p_1 - p_2) \simeq 2p_{sc}(p_1 - p_2)$$

Thus, the flow relationship simplifies to

$$p_1 - p_2 = k_2 q^2 \tag{11-32}$$

or

$$q = \left[\frac{p_1 - p_2}{k_2}\right]^{0.5} \tag{11-33}$$

where $k_2 = 0.015744 \dfrac{p_{sc}\gamma_g(TZ)_{av}fL}{T_{sc}^2 d^5}$ \hfill (11-34)

3. *Compressors.* Compressor characteristics vary depending upon the type and the manufacturer. These are usually provided by the manufacturer, and can be approximated as follows:

$$q = \frac{P}{k_3(p_2/p_1)^{k_4} + k_5} \tag{11-35}$$

where P is the compression power and k_3, k_4, and k_5 are compressor constants (see Chapter 9).

4. *Pressure regulators.* Pressure regulators are similar to chokes, and may be described by the flow relationships for chokes. For subcritical flow, Equation 7-87 may be used:

$$q = k_6 p_1 [(p_2/p_1)^{2/\varkappa} - (p_2/p_1)^{(\varkappa+1)/\varkappa}]^{0.5} \tag{11-36}$$

where $k_6 = 974.61\ C_d p_1 d_{ch}^2\ [1/(\gamma_g T_1)]^{0.5}[\varkappa/(\varkappa - 1)]^{0.5}$ \hfill (11-37)

For critical (sonic) flow, the flow relationship given by Equation 7-89 is applicable:

$$q = k_7 p_1 \tag{11-38}$$

where $k_7 = 456.71\ C_d d_{ch}^2/(\gamma_g T_1)^{0.5}$ \hfill (11-39)

5. *Underground gas reservoirs and storages.*

$$q = k_8 (p_1^2 - p_2^2)^n \tag{11-40}$$

where p_1 = average reservoir pressure
p_2 = wellhead pressure
k_8 = productivity index of the reservoir

With these relationships for the components of a gas transmission system, a model can be constructed for the system using the analogy of Kirchhoff's laws for the flow of electricity in electrical networks to gas flow in pipeline networks. According to Kirchhoff's first law, the algebraic sum of gas flows entering and leaving any node is zero:

$$\sum_{i=1}^{m} q_i = 0 \tag{11-41}$$

where m = number of NCE's meeting at the node
q = positive for flow into the node, negative for flow of gas out from the node

By Kirchhoff's second law, the algebraic sum of the pressure drops (taken with consistent signs) around the loop is zero. Thus, if n is the number of NCE's in the loop, then for a high-pressure pipeline:

$$\sum_{i=1}^{n} (p_1^2 - p_2^2)_i = 0 \tag{11-42}$$

and for a low-pressure pipe system:

$$\sum_{i=1}^{n} (p_1 - p_2)_i = 0 \qquad (11\text{-}43)$$

A pipeline distribution system may either be loopless, or contain one or more loops. The application of the relationships developed so far is described below for each of these system types.

Loopless Systems

A loopless pipe system, defined as one where the NCE's joined by nodes form no closed loops, is shown in Figure 11-6. There are n pipe legs, and n + 1 nodes. Gas enters through node 1 and leaves through nodes j, for j = 2, 3, ..., n + 1.

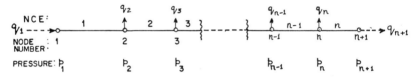

Figure 11-6. Loopless pipeline system.

If one of the terminal pressures, inlet pressure or outlet pressure, is given and the other is to be calculated for a given set of pipe leg parameters and the flow rates into or out of the nodes, then the calculation procedure is quite straightforward. If the inlet pressure, p_1, is known, the pressure at any node j can be computed using Equation 11-29 (for high-pressure pipe legs) summed over the applicable pipe legs in the system:

$$p_j^2 = p_1^2 - \sum_{i=1}^{j-1} k_i q_i^2 \qquad (11\text{-}44)$$

where j = 2, 3, ..., n, n + 1

Similarly, if the outlet pressure, p_{n+1}, is known, Equation 11-45 can be used:

$$p_j^2 = p_{n+1}^2 + \sum_{i=j}^{n} k_i q_i^2 \qquad (11\text{-}45)$$

where j = n, n-1, ..., 2, 1

The problem requires a trial and error type of solution if the maximum throughput through the line at the outlet (node n + 1) is desired for a given set of terminal pressures and flow rates into or out of the intermediate nodes. Hain (1968) describes an efficient procedure for solving this problem:

1. Guesstimate the maximum throughput of pipe leg 1, $q_1^{(1)}$. The superscript (1) indicates that this is a first approximation.
2. Calculate the throughputs for individual pipe legs, $q_i^{(1)}$ using Equation 11-41.
3. Using Equation 11-44, calculate the outlet pressure for the system, $(p_{n+1}^{(1)})$.
4. If $(p_{n+1}^{(1)})^2$ differs from the given outlet pressure p_{n+1}^2 by a value greater than the prescribed tolerance, then correct the throughputs for the individual pipe legs determined in Step 2 using:

$$q_1^{(2)} = q_1^{(1)} + \Delta q \tag{11-46}$$

$$\text{where } \Delta q = \frac{(p_{n+1}^{(1)})^2 - p_{n+1}^2}{2 \sum_{i=1}^{n} k_i q_i^{(1)}} \tag{11-47}$$

5. Repeat Steps 3 and 4 until convergence within a specified tolerance is reached.

In Step 4, the correction Δq becomes more complex for flow systems with a greater variety of NCE's. Hain (1968) gives the following correction for a line containing a compressor station:

$$\Delta q = \frac{[(p_{n+1}^{(1)})^2 - p_{n+1}^2]/2}{[(p_2^2)_c - (p_1^2)_c]/q_c + \sum_{i=1}^{n} k_i q_i^{(1)}} \tag{11-48}$$

where $(p_1)_c$, $(p_2)_c$ = compressor intake and discharge pressures, respectively, psia.

Looped Systems

There are two types of looped pipe systems: single-loop (Figure 11-7a), and multiple-loop (Figure 11-7b). Cross (1936) gave the first solution for low-pressure looped systems, which was later extended to high-pressure systems (Hain, 1968).

546 Gas Production Engineering

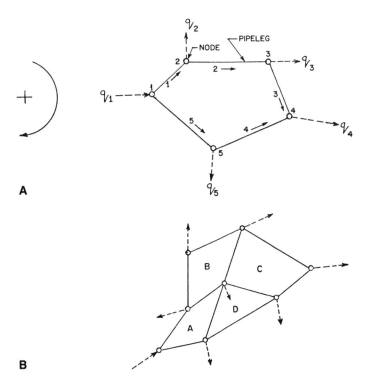

Figure 11-7. Looped systems: (a) single loop, (b) multiple loop.

Consider first the single-loop system shown in Figure 11-7a. In a typical problem, the flow rate q_1 and pressure p_1 for node 1 are known, and it is required to find the flow rates and pressures at all the other nodes. For a looped system, the direction element for flow is important: We have taken clockwise flow to be positive, and counter-clockwise flow to be negative in our analysis here. The arrows on Figure 11-7a indicate the flow directions.

The problem requires a trial and error solution scheme. An initial value for the flow rate in pipe leg 1 is assumed. If this assumed value, $q_1^{(1)}$, differs from the actual throughput by Δq, then by the node law of Equation 11-42 or 11-43 for steady-state flow (Szilas, 1975):

$$\sum_{i=1}^{n} k_i(q_i^{(1)} + \Delta q) \mid q_i^{(1)} + \Delta q \mid = 0 \qquad (11\text{-}49)$$

where n = number of pipe legs in the single-loop system.

Solving Equation 11-49 for Δq, and assuming that $\Delta q << q_i$, we get:

$$\Delta q = \frac{-\sum_{i=1}^{n} k_i \mid q_i^{(1)} \mid q_i^{(1)}}{2 \sum_{i=1}^{n} k_i \mid q_i^{(1)} \mid} \tag{11-50}$$

The gas throughputs for the next iteration, $q_i^{(2)}$, are computed as before (Equation 11-46):

$$q_i^{(2)} = q_i^{(1)} + \Delta q$$

This procedure is repeated until for an iteration k, Δq is less than or equal to a specified tolerance. After this (successful) k-th iteration, the node pressures can be calculated using the relationship (Equation 11-44) for a high-pressure network:

$$p_j^2 = p_1^2 - \sum_{i=1}^{j-1} k_i \mid q_i^{(k)} \mid q_i^{(k)} \tag{11-51}$$

for j = 2, 3, ..., n, n + 1

where k_i for pipe legs are calculated using Equation 11-31 for high-pressure lines. For a low-pressure network, k_i for pipe legs are calculated using Equation 11-34, and the node pressures are computed using Equation 11-52:

$$p_j = p_1 - \sum_{i=1}^{j-1} k_i \mid q_i^{(k)} \mid q_i^{(k)} \tag{11-52}$$

for j = 2, 3, ..., n, n + 1

For a multiple-loop system, an individual Δq is computed for each loop, and the flow corrections are done for the pipe legs loop by loop. A pipe leg which is common to two loops can be handled in two ways:

1. Correct the flow rate using the correction for the first loop, then correct it again using the correction for the second loop (Cross, 1936). Thus, the flow rate is effectively corrected twice at every iteration for any line common to two loops. Note that the flow rate through this common line will be equal in both the loops.
2. Use the sum of the Δq's for the two loops to which the pipe leg is common as the effective flow correction for the pipe leg. This has been

suggested by Renouard and Pernelle (see Szilas, 1975). The Δq values are computed by solving the n linear equations in the n throughput-correction unknowns (Δq_i for i = 1, ..., n) obtained using Equation 11-46 for each of the n pipe legs in the network. These values are then used for updating the pipe leg throughputs. The procedure is repeated until convergence.

Method 1 by Cross (1936) is simple, but converges rather slowly to the final solution and is generally uneconomic for large systems (Szilas, 1975). Method 2, developed to overcome these problems, is approximate and applicable only to relatively non-complex networks for which it may be fairly accurate.

Stoner (1969, 1972) has presented an effective method for handling looped networks with all kinds of NCE's. In this method, the equation of continuity is used to express the flow at each node in the system. The solution to the system of equations is complex, but the method offers the ability to compute any set of unknowns. It thus overcomes the limitation of the Cross method that can only be used to generate throughput or pressure solutions. See Figure 11-8.

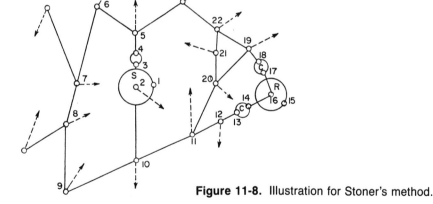

Figure 11-8. Illustration for Stoner's method.

For any node j, the continuity equation (Equation 11-41) expresses the fact that the sum of the inflows and outflows at the node is zero:

$$F_j = \sum_{i=1}^{n} q_{i,j} = 0 \qquad (11\text{-}53)$$

where $q_{i,j}$ is the flow from node i to node j. Flows into the node are considered positive, flows out of the node are negative. F_j thus represents the flow imbalance at the node and will be equal to zero when the system is in balance. For example, consider node 2 that receives gas from underground storage (1,2) and pipe leg (10,2), and delivers gas to compressor intake (3,2), and consumer supply attached directly to node 2. Equation 11-53 for node 2 can now be written as:

$$F_2 = q_{1,2} - q_{3,2} + q_{10,2} - q_2 = 0 \tag{11-54}$$

With the substitution of the appropriate NCE equations (from Equations 11-29 through 11-40), Equation 11-54 becomes:

$$F_2 = (k_8)_{1,2}(p_2^2 - p_1^2)^n S_{1,2} - \frac{P_{3\text{-}4}}{k_3(p_4/p_3)^{k_4} + k_5}$$

$$+ \frac{(p_{10}^2 - p_2^2)^{0.5}}{(k_1)_{10,2}^{0.5}} S_{10,2} - q_2 = 0 \tag{11-55}$$

where $S_{i,j}$ is the sign term that accounts for the flow direction:

$$\begin{aligned} S_{i,j} &= \text{sign } (p_i - p_j) \\ &= +1 \text{ for } p_i \geq p_j \\ &= -1 \text{ for } p_i < p_j \end{aligned} \tag{11-56}$$

Similar equations are written for all the other nodes in the system. Consider a system of n nodes and m NCE's. Steady-state flow through this system of n nodes is mathematically represented by a system of n non-linear equations. There are a total of (2n + m) variables: for each node, there is one pressure and one node flow rate variable, and for each NCE there is a constant k_i defining the flow resistance through it. Thus, we can use this model consisting of n equations in (2n + m) variables to solve for any n variables, given the value of the remaining (n + m) variables.

Each of the node continuity equations, such as Equation 11-55, can be expressed as follows:

$$F_j (x_1, x_2, \ldots, x_n) = 0, \text{ for } j = 1, 2, \ldots, n \tag{11-57}$$

This non-linear system of equations can be solved using various iterative techniques on a computer. Stoner (1969, 1972) used the most popular solution method: Newton-Raphson iteration. The values of the unknowns are computed repeatedly, until the values from any two successive steps converge (i.e., differ by a value less than or equal to a specified tolerance). The values of any unknown at the (k + 1)th iteration is computed as follows:

$$x_i^{(k+1)} = x_i^{(k)} + \Delta x_i^{(k+1)} \tag{11-58}$$

where $\sum_{i=1}^{n} \dfrac{\partial F_j}{\partial x_i} \Delta x_i = -F_j$, for $j = 1, 2, \ldots, n$ (11-59)

where the derivatives $\partial F_j/\partial x_i$ are obtained by differentiating the node continuity equations. The method requires an initial estimate for each of the unknowns, x_i^0. Generally, good initial guesses are required to achieve satisfactory convergence. A standard mathematical technique for improving and accelerating convergence is to introduce an acceleration factor, α_i, in the correction equation (Equation 11-58), as done by Stoner (1969):

$$x_i^{(k+1)} = x_i^{(k)} + \Delta x_i^{(k+1)} \alpha_i \tag{11-60}$$

where α_i is computed using the Δx_i for the current and previous steps. Stoner (1969) proposed the following scheme for obtaining α_i:

Let $A_i = \Delta x_i^{(k+1)}/\Delta x_i^{(k)}$. For the first two iterations, where divergence is most likely to occur, an $\alpha_i = 0.5$ is best to use in order to ensure convergence. In subsequent steps, the value of α_i is determined as below for every other step; for the steps in between, $\alpha_i = 1.0$ is used:

For $A_i \leq -1$, $\quad \alpha_i = 0.5 \,|\, A_i \,|$
For $-1 < A_i < 0$, $\quad \alpha_i = 1.0 - 0.5 \,|\, A_i \,|$
For $0 < A_i < 1$, $\quad \alpha_i = 1.0 + 2.0 \,|\, A_i \,|$
For $A_i \geq 1$, $\quad \alpha_i = 3$ (11-61)

Stoner obtained these specifications for α_i by experimenting with the mathematical model on a computer. Naturally, these are empirical, system-dependent values, and the user may have to do some experimentation to obtain similar or better schemes for the acceleration factor α_i applicable to his or her own system.

In designing and operating transmission systems, it may be useful to ascertain the effect of varying the several different parameters that influence the system, such as pipeline diameters, operating pressures, and so on, and determine an optimum design and operating condition that maximizes the pipeline utility at the lowest cost and satisfies all the requirements. This is known as sensitivity analysis. Sensitivity analysis requires generating a large number of solutions using the computer model for all the possible system configurations, entailing very high computational and associated expenses. Readers interested in this topic are referred to the work by Stoner (1972) who has presented a further development of his steady-state model to enable sensitivity analysis of production and transmission systems.

Gas Gathering and Transport   551

Unsteady-State Flow in Pipelines

Flow is said to be unsteady-state if it is, in addition to flow resistance and the pressure drop, a function of time. Flow through gas transmission systems is usually transient (unsteady-state), primarily because of variations in demand. Unsteady-state flow occurs whenever the rate of withdrawal of gas from a line differs from the rate of supply to it at the inflow end. If the fluid flowing in the line were incompressible (most liquids are almost incompressible), then any imposed change in throughput would be transmitted instantaneously throughout the pipeline, and consequently, the flow would have the same magnitude at any pipeline section, including the head and tail end. For gas, a highly compressible fluid, it takes some length of time for the pressure change at the outlet end to transmit through the line and make itself felt at the head end of the pipeline. The assumption of steady-state flow for such a system, therefore, is valid only for infinitesimally small pipe sections in the flowline.

Fundamental Relationships

Unsteady-state flow of gas can be described using four fundamental equations: the equation of state, equation of continuity, equation of motion, and a relationship accounting for the deviation of the gas from ideal-gas behavior. For flow through long pipelines, it is generally assumed (Streeter and Wylie, 1970) that the flow is substantially isothermal, steady-state friction is valid, the slope for a pipe section is uniform, and that the expansion of pipe walls due to pressure changes is negligible.

The simplest equation of state is commonly used (see Chapter 3):

$$\frac{p}{\rho} = \frac{ZRT}{M} \qquad (11\text{-}62)$$

where the gas compressibility factor Z, a measure of the deviation of the gas from ideality and a function of pressure, temperature, and gas composition, can be expressed in a variety of ways (see Chapter 3). Assuming isothermal flow and a constant composition, this necessary relationship simplifies to:

$$Z = f(p)_{T, y_i} \qquad (11\text{-}63)$$

Several relationships are available for representing this functional dependence of Z on pressure, as given in Chapter 3. In practice, however, an average (constant) value of Z is used, reducing the number of equations to three. Equation 11-62 can now be simplified to (Streeter and Wylie, 1970):

$p/\rho = ZRT/M = B^2/g_c$

or,

$$\rho = pg_c/B^2 \qquad (11\text{-}64)$$

where B = isothermic speed of sound
g_c = conversion factor relating mass and weight (32.17 lbm-ft/lbf-sec^2)

The equation of continuity can be written as:

$$\frac{\partial m}{\partial x} + \frac{\partial(\rho A)}{\partial t} = 0 \qquad (11\text{-}65)$$

where m = mass flow rate of the gas, lbm/sec
x = distance along the flow path, ft
ρ = gas density, lbm/ft^3
A = cross-sectional area of flow, ft^2
t = time, sec

Substituting for ρ from Equation 11-64, the equation of continuity (Equation 11-65) can now be written as:

$$\frac{\partial m}{\partial x} + \frac{\partial(Apg_c/B^2)}{\partial t} = 0$$

or

$$\frac{B^2}{Ag_c}\frac{\partial m}{\partial x} + \frac{\partial p}{\partial t} = 0 \qquad (11\text{-}66)$$

The equation of motion can be expressed as:

$$\frac{\partial p}{\partial x} + \frac{\rho}{g_c}\left[v\frac{\partial v}{\partial x} + \frac{\partial v}{\partial t}\right] + \frac{\rho g}{g_c}\sin\alpha + \frac{fv|v|\rho}{2dg_c} = 0 \qquad (11\text{-}67)$$

where p = pressure, lbf/ft^2; conversion constant of 144 required in Equation 11-67 if psia is used
g = gravitational acceleration, 32.17 ft/sec^2
α = angle of pipe inclination from the horizontal
f = pipe friction factor (Moody)
v = gas flow velocity, ft/sec
d = internal diameter of the pipe, ft

Note that each of the terms in Equation 11-63 has the units of lbf/ft³. Also note that instead of v^2, $v|v|$ is used in order to include the direction of flow in the friction term.

If A is the pipe cross-sectional area, then the flow velocity v is related to the mass flow rate m as follows:

$$v = m/A\rho$$

Substituting for ρ from Equation 11-64 and rearranging, we get

$$v = \frac{mB^2}{Apg_c} \tag{11-68}$$

Substituting Equations 11-64 and 11-68, the equation of motion (Equation 11-67) becomes:

$$\frac{\partial p}{\partial x} + \frac{p}{B^2}\left[\frac{mB^2}{Apg_c}\frac{\partial(mB^2/Apg_c)}{\partial x} + \frac{\partial(mB^2/Apg_c)}{\partial t}\right]$$

$$+ \frac{pg}{B^2}\sin\alpha + \frac{f}{2dg_c}\frac{m|m|B^4}{(Apg_c)^2}\frac{pg_c}{B^2} = 0$$

or,

$$\frac{\partial p}{\partial x} + \frac{p}{B^2}\frac{B^2}{Ag_c}\left[\frac{mB^2}{Ap^2g_c}\frac{\partial m}{\partial x} - \frac{m^2B^2}{Ap^3g_c}\frac{\partial p}{\partial x} + \frac{1}{p}\frac{\partial m}{\partial t} - \frac{m}{p^2}\frac{\partial p}{\partial t}\right]$$

$$+ \frac{pg}{B^2}\sin\alpha + \frac{f}{2dg_c}\frac{m|m|B^2}{A^2pg_c} = 0$$

or,

$$\frac{\partial p}{\partial x} - \frac{p}{Ag_c}\frac{m^2B^2}{Ap^3g_c}\frac{\partial p}{\partial x} - \frac{p}{Ag_c}\frac{m}{p^2}\frac{\partial p}{\partial t} + \frac{p\,mB^2}{Ag_cAp^2g_c}\frac{\partial m}{\partial x}$$

$$+ \frac{p}{Ag_c}\frac{1}{p}\frac{\partial m}{\partial t} + \frac{pg}{B^2}\sin\alpha + \frac{fm|m|B^2}{2dA^2g_c^2p} = 0$$

Substituting for $\partial p/\partial t$ from the continuity equation (Equation 11-66) and rearranging, we get

$$\left[1 - \frac{m^2B^2}{A^2p^2g_c^2}\right]\frac{\partial p}{\partial x} + \frac{2mB^2}{A^2pg_c^2}\frac{\partial m}{\partial x} + \frac{1}{Ag_c}\frac{\partial m}{\partial t}$$

$$+ \frac{pg}{B^2}\sin\alpha + \frac{fm|m|B^2}{2dA^2g_c^2p} = 0 \tag{11-69}$$

554 Gas Production Engineering

In Equation 11-69, the second term in the coefficient of $\partial p/\partial x$ is negligible compared to unity. Also, the $\partial m/\partial x$ term is negligible as compared to the other terms. With these simplifications, the equation of motion for transient gas flow reduces to:

$$\frac{\partial p}{\partial x} + \frac{1}{Ag_c}\frac{\partial m}{\partial t} + \frac{pg}{B^2}\sin\alpha + \frac{fm|m|B^2}{2dA^2g_c^2 p} = 0 \qquad (11\text{-}70)$$

Equations 11-66 and 11-70, describing the transient flow of gas in a pipeline, constitute a system of two non-linear partial differential equations (NL-PDEs). Equation 11-70 is often used in the following form that results on multiplying it throughout by p:

$$\frac{1}{2}\frac{\partial p^2}{\partial x} + \frac{p}{Ag_c}\frac{\partial m}{\partial t} + \frac{p^2 g}{B^2}\sin\alpha + \frac{fm|m|B^2}{2dA^2g_c^2} = 0 \qquad (11\text{-}71)$$

Initial and Boundary Conditions

Initial conditions as well as boundary conditions for the system must be specified in order to obtain the applicable solution to the differential equations just given. The initial conditions of the system are required for resolving initial pressure, velocity, density, compressibility, and other properties as a function of position (x) along the pipeline. Boundary conditions must be specified to obtain a unique solution.

Initial conditions can be specified in two ways:

1. Determine the pressure and flow rates by actual measurement at various points along the pipeline. The initial state of the system will be given by the pressure and flow rate distribution obtained in this manner.
2. Assume flow to be steady-state at the beginning of the unsteady-state flow analysis, i.e., at time t = 0. Use the steady-state flow relationships to calculate the initial pressure distribution (rate is a constant for steady-state flow) in the pipeline.

The first method is difficult to use. Precise measurement at several points is difficult and usually impossible in an installed pipe in the field. Even if this could be done, the pressure and flow-rate profiles obtained may not be at the same instant in time at all points along the pipeline, resulting in an inconsistent specification of the initial state. Therefore, the steady-state initial condition is almost always used.

At least two time-variable boundary conditions must be specified to obtain a unique solution, chosen from among the four variables: input pressure, input flow rate (or velocity), output pressure, and output flow rate (or velocity). Depending upon the choice exercised for specifying the boundary conditions, two cases are possible:

1. Both time-variable boundary conditions are specified at the same end of the pipe, either input or output. In this case, the numerical solution is obtained backwards in time and away from the known boundary. Note that it is not possible to compute the flow parameters at the unknown end of the pipe from their given values at the other end at the same instant in time. We must work backward in time, calculating the pressures and flow rates at various points of the pipeline at time t $-$ Δt on the basis of their known distribution at time t.
2. The more common approach, where one time-variable boundary condition is specified at each end of the pipe. In this case, the numerical computations are not made backwards in time. The two end-point boundary conditions are used to determine the values of pressure and flow rate (or flow velocity) along the length of the pipe. Thus, the pressures and flow rates at various points of the pipeline at time t + Δt are calculated on the basis of their known distribution at time t.

Numerical Solutions

The system of NL-PDE's (Equations 11-66 and 11-70) previously given for transient flow in a gas pipeline cannot be solved analytically. Any analytic solution must incorporate some simplification, or assume some specific set of initial and boundary conditions. Generally, the analytic solutions thus generated reduce computational expense, but are applicable only to the analysis of a subproblem, or a simplified problem. Thus, the equations for transient flow must be solved numerically.

Four types of numerical methods for solving the system of NL-PDE's for transient gas flow have been reported in the literature: (a) explicit finite difference method; (b) implicit finite difference method; (c) method of characteristics; and (d) variational methods. All of these methods proceed in steps, computing the required parameter values (pressure, flow rate) at various points along the pipeline at the instant t + Δt on the basis of the known distribution of these parameters along the pipeline at time t.

Explicit Finite Difference Method

The NL-PDE's are transformed into algebraic equations using finite difference methods such that the values of the unknown parameters (pressure, flow rate) being solved for at time t + Δt depend only upon their values at the preceding time step (at time t). Thus, the equations can be solved individually. The method is faster because it requires lesser computation than the implicit scheme, but is subject to instability and a restricted time step size. For these reasons, and also because of the inaccuracies in computations that generally result from the use of this method, the explicit scheme is rarely used, except in conjunction with the method of characteristics.

The partial differential equations can be discretized along the space (x) and time (t) axes in three ways: *backward difference* scheme, *central difference* scheme, and *forward difference* scheme (see Figure 11-9). Let Δx be the length of each pipe segment; solutions are computed successively at values of time increasing (or decreasing) in time steps of size Δt. Then, using

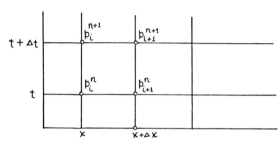

Figure 11-9. Nomenclature for finite difference methods.

the explicit forward finite difference method, the partial differentials in Equations 11-66 and 11-71 can be expanded for any grid point i at time n + 1 as follows:

$$\frac{\partial p}{\partial t} = \frac{p_i^{n+1} - p_i^n}{\Delta t}$$

$$\frac{\partial p^2}{\partial x} = \frac{(p_{i+1}^n)^2 - (p_i^n)^2}{\Delta x}$$

$$\frac{\partial m}{\partial t} = \frac{m_i^{n+1} - m_i^n}{\Delta t}$$

$$\frac{\partial m}{\partial x} = \frac{m_{i+1}^n - m_i^n}{\Delta x} \qquad (11\text{-}72)$$

where the subscript i indicates the grid along the x direction, and superscript n indicates the parameter value at the previous time step.

Thus, p_i^n is the pressure at section i at time step n, p_{i+1}^n is the pressure at grid point i + 1 at time step n, p_i^{n+1} is the pressure at grid point i at time step n + 1 (at which solution is sought), and so on. Δx and Δt indicate the grid-block and time-step size, respectively. The pressure p and mass flow rate m are discretized as follows:

$$p = \frac{p_i^n + p_{i+1}^n}{2}$$

$$m = \frac{m_i^n + m_{i+1}^n}{2} \tag{11-73}$$

The transient flow equations can now be linearized using the relationships of Equations 11-72 and 11-73. The equation of continuity, Equation 11-66, becomes:

$$\frac{B^2}{Ag_c}\left[\frac{m_{i+1}^n - m_i^n}{\Delta x}\right] + \frac{p_i^{n+1} - p_i^n}{\Delta t} = 0$$

Solving for p_i^{n+1}:

$$p_i^{n+1} = p_i^n - \left[\frac{\Delta t}{\Delta x}\right]\frac{B^2}{Ag_c}(m_{i+1}^n - m_i^n) \tag{11-74}$$

Since the values of all the parameters on the right side of Equation 11-74 are known from the solution for the previous time step, p_i^{n+1} can be directly obtained. The equation of motion can also be transformed in a similar manner using Equations 11-72 and 11-73, as below.

$$\frac{(p_{i+1}^n)^2 - (p_i^n)^2}{2\Delta x} + \frac{p_i^n + p_{i+1}^n}{2Ag_c}\left[\frac{m_i^{n+1} - m_i^n}{\Delta t}\right] + \frac{(p_i^n + p_{i+1}^n)^2 g}{4B^2}\sin\alpha$$

$$+ \frac{fB^2}{8dA^2g_c^2}(m_i^n + m_{i+1}^n)|m_i^n + m_{i+1}^n| = 0$$

Solving for the unknown m_i^{n+1} at the new time n + 1, we get

$$m_i^{n+1} = m_i^n - \left[\frac{Ag_c}{p_i^n + p_{i+1}^n}\right](\Delta t/\Delta x)\left[(p_{i+1}^n)^2 - (p_i^n)^2\right] - \frac{Agg_c\Delta t}{2(p_i^n + p_{i+1}^n)B^2}\sin\alpha$$

$$- \frac{fB^2\Delta t}{4dAg_c(p_i^n + p_{i+1}^n)}(m_i^n + m_{i+1}^n)|m_i^n + m_{i+1}^n| \tag{11-75}$$

Additional details may be obtained from the published work of Tuppeck and Kirsche (1962) and Distefano (1970) who have described and used the explicit method.

Implicit Finite Difference Method

In the implicit method, the NL-PDE's are transformed into algebraic equations using finite difference methods such that the values of the unknown parameters (pressure, flow rate) being solved for at time $t + \Delta t$ depend upon their values at the current time $t + \Delta t$ at neighboring points in the pipeline. Thus, the equations must be solved simultaneously. The method is slower because it requires more computation at each time step, but provides stability. The allowable time step size, restricted only by the accuracy desired, is always greater than that for the explicit scheme. Implicit difference procedures have been formulated by Guy (1967), Streeter and Wylie (1970), and Wylie et al. (1971).

Streeter and Wylie (1970) used the central difference scheme for the implicit expansion of the differentials as follows:

$$\frac{\partial p}{\partial t} = \frac{p_i^{n+1} + p_{i+1}^{n+1} - p_i^n - p_{i+1}^n}{2\Delta t}$$

$$\frac{\partial p^2}{\partial x} = \frac{(p_{i+1}^{n+1})^2 + (p_{i+1}^n)^2 - (p_i^{n+1})^2 - (p_i^n)^2}{2\Delta x}$$

$$\frac{\partial m}{\partial t} = \frac{m_i^{n+1} + m_{i+1}^{n+1} - m_i^n - m_{i+1}^n}{2\Delta t}$$

$$\frac{\partial m}{\partial x} = \frac{m_{i+1}^{n+1} + m_{i+1}^n - m_i^{n+1} - m_i^n}{2\Delta x}$$

$$p = \frac{p_i^{n+1} + p_{i+1}^{n+1} + p_i^n + p_{i+1}^n}{4}$$

$$m = \frac{m_i^{n+1} + m_{i+1}^{n+1} + m_i^n + m_{i+1}^n}{4} \qquad (11\text{-}76)$$

Substituting the expressions of Equation 11-76 into the equation of continuity (Equation 11-66) and the equation of motion (Equation 11-71), we obtain:

$$\frac{B^2}{Ag_c\Delta x}(m_{i+1}^{n+1} + m_{i+1}^n - m_i^{n+1} - m_i^n)$$

$$+ \frac{1}{\Delta t}(p_i^{n+1} + p_{i+1}^{n+1} - p_i^n - p_{i+1}^n) = 0 \qquad (11\text{-}77)$$

and

$$\frac{1}{\Delta x}[(p_{i+1}^{n+1})^2 + (p_{i+1}^n)^2 - (p_i^{n+1})^2 - (p_i^n)^2]$$

$$+ \frac{1}{2Ag_c\Delta t}(p_i^{n+1} + p_{i+1}^{n+1} + p_i^n + p_{i+1}^n)(m_i^{n+1} + m_{i+1}^{n+1} - m_i^n - m_{i+1}^n)$$

$$+ \frac{g\sin\alpha}{4B^2}(p_i^{n+1} + p_{i+1}^{n+1} + p_i^n + p_{i+1}^n)^2$$

$$+ \frac{fB^2}{8dA^2g_c^2}(m_i^{n+1} + m_{i+1}^{n+1} + m_i^n + m_{i+1}^n)|m_i^{n+1} + m_{i+1}^{n+1} + m_i^n$$

$$+ m_{i+1}^n| = 0 \quad (11\text{-}78)$$

Equations 11-77 and 11-78 must be solved simultaneously. The parameter values at time-step n (i.e., at time t) are known from the initial conditions or from the previous solution step. There are four unknowns in the two equations for each grid cell i: p_i^{n+1}, p_{i+1}^{n+1}, m_i^{n+1}, and m_{i+1}^{n+1}. For the complete pipeline divided into m grid blocks, the total number of unknowns is 2m + 2, including the two unknowns at the boundary. The total number of equations is also 2m + 2, including the two known boundary conditions. Thus, we have a system of 2m + 2 non-linear algebraic equations to be solved.

These equations can be solved in many different ways, all of which involve iteration (trial and error). Streeter and Wylie (1970) have used Newton-Raphson iteration (described in the section on steady-state flow through looped systems), perhaps the most commonly used method. Additional details for solving such equations can be found in any text on numerical analysis or reservoir simulation, such as Aziz and Settari (1979).

Method of Characteristics

Taylor (1962), Goacher (1969), and Streeter and Wylie (1970) have proposed and used the method of characteristics to solve the transient gas flow problem for pipelines. In this method, paths (called characteristic lines) are defined in the (x, t) plane, along which the system of NL-PDE's is transformed into a system of ordinary differential equations (ODE's). These ODE's can then be solved numerically, using either the explicit or the implicit finite difference method of solution. The procedure used by Streeter and Wylie (1970) is described in the following section.

Equations 11-66 (designated by F_1) and 11-70 (designated by F_2), describing the transient flow of gas in a pipeline, have two independent variables, x and t, and two dependent variables, p and m. These two equations are combined using an unknown multiplier λ, and rearranged as follows:

560 Gas Production Engineering

$$F = \lambda F_1 + F_2$$

$$= \frac{1}{Ag_c}\left[\lambda B^2 \frac{\partial m}{\partial x} + \frac{\partial m}{\partial t}\right] + \lambda\left[(1/\lambda)\frac{\partial p}{\partial x} + \frac{\partial p}{\partial t}\right]$$

$$+ \frac{pg}{B^2}\sin\alpha + \frac{fm|m|B^2}{2dA^2g_c^2p} = 0 \qquad (11\text{-}79)$$

Equation 11-79 can be transformed into two ODE's using a λ given by:

$$\frac{dx}{dt} = \lambda B^2 = \frac{1}{\lambda}$$

implying that

$$\lambda = +\frac{1}{B} \text{ or } -\frac{1}{B} \qquad (11\text{-}80)$$

By putting any of these two values of λ, the quantities in the brackets in Equation 11-79 become total differentials. The controlling equations for use in the method of characteristics now become:

1. The C^+ characteristic line corresponding to $\lambda = +(1/B)$, or $dx/dt = B$, given by

$$\frac{1}{Ag_c}\frac{dm}{dt} + \frac{1}{B}\frac{dp}{dt} + \frac{pg}{B^2}\sin\alpha + \frac{fm|m|B^2}{2dA^2g_c^2p} = 0 \qquad (11\text{-}81)$$

2. The C^- characteristic line corresponding to $\lambda = -(1/B)$, or $dx/dt = -B$, given by

$$\frac{1}{Ag_c}\frac{dm}{dt} - \frac{1}{B}\frac{dp}{dt} + \frac{pg}{B^2}\sin\alpha + \frac{fm|m|B^2}{2dA^2g_c^2p} = 0 \qquad (11\text{-}82)$$

On the x-t plot shown in Figure 11-10, consider points R and S at which m and p are known. By constructing the C^+ characteristic line $dx/dt = B$ through R and the C^- characteristic line $dx/dt = -B$ through S, an intersection P is found at a later time and an intermediate value of x. Since both Equations 11-81 and 11-82 are valid at P, they may be solved simultaneously to yield the new values of m and p at this later time. In this way, the solution can be obtained at a later time, from known values at the previous time step, at all locations in the pipe.

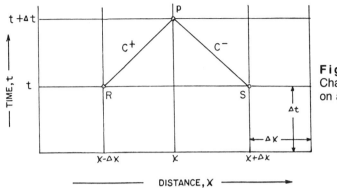

Figure 11-10. Characteristic lines on an x-t diagram.

However, the ordinary differential Equations 11-81 and 11-82 themselves must be solved using some finite difference procedure. Either the explicit or the implicit scheme may be used for this. Streeter and Wylie (1970) have shown solutions using both these schemes. For the explicit method, the C^+ characteristic line (Equation 11-81), on multiplying throughout by B dt, becomes:

$$\frac{B}{Ag_c}(m_P - m_R) + (p_P - p_R) + \left[\frac{pg\sin\alpha}{B^2} + \frac{fm|m|B^2}{2dA^2g_c^2p}\right]Bdt = 0$$

Since dx/dt = B, we can replace B dt by dx to obtain:

$$\frac{B}{Ag_c}(m_P - m_R) + (p_P - p_R) + \left[\frac{pg\sin\alpha}{B^2} + \frac{fm|m|B^2}{2dA^2g_c^2p}\right]dx = 0$$

Discretization along space (x) yields (see Streeter and Wylie, 1970):

$$\frac{B}{Ag_c}(m_P - m_R) + (p_P - p_R) + \frac{fB^2\,\Delta x}{2dA^2g_c^2(p_P + p_R)}$$

$$(m_P|m_P| + m_R|m_R|)\frac{e^s - 1}{s} + \frac{p_P^2}{p_P + p_R}(e^s - 1) = 0 \quad (11\text{-}83)$$

Similarly, on using an explicit expansion scheme, the C^- characteristic line (Equation 11-82) becomes:

$$-\frac{B}{Ag_c}(m_P - m_S) + (p_P - p_S) - \frac{fB^2\,\Delta x}{2dA^2g_c^2(p_P + p_S)}$$

$$(m_P|m_P| + m_S|m_S|)\frac{e^s - 1}{s} - \frac{p_S^2}{p_P + p_S}(e^s - 1) = 0 \quad (11\text{-}84)$$

562 Gas Production Engineering

In Equations 11-83 and 11-84, s is the same as in Chapter 7 for steady-state flow (Equation 7-47), and s = 0 for horizontal flow. For this explicit scheme, Streeter and Wylie (1970) have noted that $\Delta t \leq \Delta x/B$ for stability.
For an implicit finite difference scheme, the equivalent forms for Equations 11-83 and 11-84 can be derived similarly.

Variational Methods

Rachford and Dupont (1974) have presented a variational method that provides a fast and accurate means of simulating transient flow in gas pipelines and networks. Formulated as a Galerkin method, this approach offers significant advantages in terms of computational cost and storage over other methods. Readers are referred to the original paper for a detailed treatment of this special method which is, perhaps, beyond the scope of our present discussion.

Some Approximate Solutions for Transient Flow

There are many situations in pipeline engineering where the flow is unsteady. Approximate relationships have therefore been developed to enable easy estimations for some special cases where it may not be necessary to generate very precise solutions to the unsteady-state flow relationships described earlier. Some such cases are described in the following sections.

Blowdown and Purge

There are instances when it is necessary to blow down and purge a gas line. Izawa (1966) provides some useful solutions for this special case of unsteady-state flow. For subcritical flow ($p_u/p_d < 2$), Izawa (1966) derived the following relationship:

$$V = (\pi/4)d^2 t \, [2g_c Z_{av} RT(1 - p_d/p_u)]^{0.5}$$

where V = gas volume, ft^3
d = internal pipe diameter, ft
t = time, seconds
g_c = constant relating mass and weight, 32.17 lbm-ft^3/lbf-sec^2
Z_{av} = average gas compressibility factor
R = gas constant, 85.56 ft-lbf/lbm-°R
T = temperature, °R
p_u, p_d = upstream and downstream pressures, respectively, psig

Substituting for R and g_c, and changing to customary units of V in Mscf, d in inches and t in minutes, this relationship becomes:

$$V = 0.02428 \, d^2 t \, [Z_{av}T(1 - p_d/p_u)]^{0.5} \qquad (11\text{-}85)$$

For the case of critical flow ($p_u/p_d > 2$), Izawa (1966) has presented the following relationship:

$$V = 0.4524 \, d^2 p_{av} t / T^{0.5} \qquad (11\text{-}86)$$

where, as for the subcritical flow equation (Equation 11-85):

V = gas volume, Mscf
d = internal pipe diameter, in.
p_{av} = average pressure, psig
t = time, min.
T = temperature (°R)

Pressure Surges on Closing a Valve

A pressure surge occurs when a valve on a flowline is closed, and the fluid outflow is stopped. The kinetic energy of the flowing fluid is abruptly converted to internal energy, and a surge or wave travels back, countercurrent to the direction of flow of the fluid. The pressure rise accompanying this flow stoppage is highest at the valve, and decreases gradually towards the upstream end of the pipe where the disturbance is attenuated. The greater the compressibility of the fluid, the greater is the attenuation (or absorption) of the disturbance. Consequently, gas lines do not pose a problem, whereas liquid lines, and sometimes high-pressure gas lines, pose a problem of some concern. In most cases, pressure surges damage pressure-measuring equipment only, but sometimes they may be powerful enough to even rupture the pipe. Stewart (1971) describes this problem and provides a method for computing pressure surges for liquid lines.

Pressure Testing

Pressure testing of a pipeline is commonly done for the detection of leaks. This is a special form of unsteady-state flow where the pressure must be held constant for a sufficient time. Campbell (1984) has given the following relationship for estimating the minimum value of this testing time, $(t_t)_{min}$, necessary:

$$(t_t)_{min} = 3.0 \, d^2 L / p_i \qquad (11\text{-}87)$$

where $(t_t)_{min}$ = minimum required shut-in time for the line, hr
d = internal pipe diameter, in.
L = length of the pipe section that is being tested for leaks, miles
p_i = initial test pressure, psig

A flowline that has been shut-in for at least $(t_t)_{min}$ hours is considered to have no leaks ("tight" line) if the pressure loss is less than the $(\Delta p)_{max}$ psi given below (Campbell, 1984):

$$(\Delta p)_{max} = \frac{tp_i}{949d} \tag{11-88}$$

where t = shut-in time, hr

Handling Variable Consumer Demand—Case of Constant Injection Rate

The two ways of handling daily fluctuations in consumer demand are constant injection rate (but not necessarily pressure) into the pipeline, and variable injection rates (and pressures), to ensure a constant pressure at the delivery (outlet) end of the pipeline.

The advantage of constant injection rate of gas into a pipeline is the ease of control. The pipeline functions as a buffer storage facility, accommodating the difference between the constant input and the fluctuating output. The pressure at the delivery end of the pipeline varies from $(p_2)_{min}$ for the highest delivery rate to $(p_2)_{max}$ for the lowest delivery rate. Potential energy (pressure energy) losses may occur at the delivery end where the pressure will usually exceed $(p_2)_{min}$ during most of the day. These losses can be minimized by using a turbine driving a power generator, instead of pressure reducers such as throttles, for reducing the outlet pressure.

Szilas (1975) presents the following procedure for designing a constant injection rate system for variable consumer demand, a modification of similar developments by Smirnov and Shirkovsky in the USSR. Equation 11-30 is written as follows:

$$q = k \left[\frac{p_1^2 - p_2^2}{Z_{av}} \right]^{0.5} \tag{11-89}$$

where $k = (Z_{av}/k_1)^{0.5}$ is a constant.

Solving Equation 11-89 for p_1:

$$p_1 = \left[p_2^2 + \frac{q^2 Z_{av}}{k^2}\right]^{0.5} \quad (11\text{-}90)$$

Introducing $R_p = p_1/p_2$ in Equation 11-90, we get:

$$p_1 = \frac{qR_p Z_{av}^{0.5}}{k(R_p^2 - 1)^{0.5}} \quad (11\text{-}91)$$

Similarly, we can obtain an expression for p_2 in terms of R_p:

$$p_2 = \left[p_1^2 - \frac{q^2 Z_{av}}{k^2}\right]^{0.5} \quad (11\text{-}92)$$

or

$$p_2 = \frac{qZ_{av}^{0.5}}{k(R_p^2 - 1)^{0.5}} \quad (11\text{-}93)$$

The mean pressure in a gas line is given by (see Equation 7-32):

$$p_{av} = \frac{2}{3}\left[p_1 + \frac{p_2^2}{p_1 + p_2}\right]$$

$$= \frac{2}{3}\left[\frac{p_1^2 + p_1 p_2 + p_2^2}{p_1 + p_2}\right]$$

or,

$$\frac{3}{2}p_{av} = p_2\left[\frac{R_p^2 + R_p + 1}{R_p + 1}\right]$$

Substituting for p_2 from Equation 11-93 and rearranging, we get

$$\frac{3p_{av} k}{2qZ_{av}^{0.5}} = \frac{R_p^2 + R_p + 1}{(R_p + 1)(R_p^2 - 1)^{0.5}} \quad (11\text{-}94)$$

A plot of $3p_{av} k/2qZ_{av}^{0.5}$ versus R_p, as given by Equation 11-79, is shown in Figure 11-11.

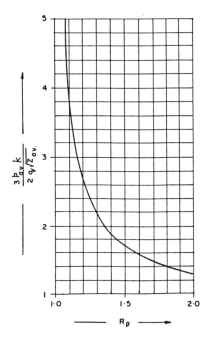

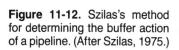

Figure 11-11. Szilas's method for determining the buffer action of a pipeline. (After Szilas, 1975.)

Figure 11-12. Szilas's method for determining the buffer action of a pipeline. (After Szilas, 1975.)

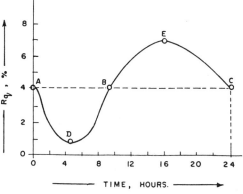

Figure 11-12 shows the daily fluctuation of R_q, the percentage hourly consumption referred to daily consumption. The dotted line parallel to the abscissa indicates the average consumption q_{av}. In segment A-B, consumption is less than q_{av}, and gas accumulates in the pipeline. In segment B-C, consumption is greater than q_{av}, and the gas accumulated in the pipeline at the earlier time is used to cater to this increased demand. Note that at point A, the gas flow (supply or input) into the pipeline is exactly equal to the demand, and gas reserves are zero. Thus, the pressure at output (tail) end of the pipeline at the time corresponding to A must be equal to the minimum output pressure, $(p_2)_{min}$. From Equation 11-90, the pressure at the head end of the pipeline, p_1, is given by:

$$p_1 = \left[(p_2)^2_{min} + \frac{q^2 Z_{av}}{k^2}\right]^{0.5}$$

and the mean pressure in the pipeline, p_{av}, is given by:

$$p_{av} = \frac{2}{3}\left[p_1 + \frac{(p_2)^2_{min}}{p_1 + (p_2)_{min}}\right]$$

The volume of gas stored in the pipeline, V, can be determined as follows:

$$V = \frac{p_{av} T_{sc} AL}{p_{sc} Z_{av} T_{av}}$$

Solving for (p_{av}/Z_{av}):

$$\frac{p_{av}}{Z_{av}} = \frac{p_{sc} T_{av}}{T_{sc} AL} V \qquad (11\text{-}95)$$

where p_{av}, T_{av} = average pressure and temperature in the pipeline, respectively
p_{sc}, T_{sc} = specified standard conditions of pressure and temperature, respectively
Z_{av} = average gas compressibility factor (at p_{av}, T_{av})
A = pipeline cross-sectional area
L = length of the pipeline

From Figure 11-12, we can determine the area embraced by the curve ADB and line AB by planimetering, and convert it to gas volume. This gas volume is the volume of gas stored in the pipeline during the slack-demand period, V_{AB}. This stored volume will be maximum at the time corresponding to point B in Figure 11-12. Therefore, the head and tail end pressures, p_1 and p_2, in the pipeline will also be maximum at B. Using Equation 11-95,

$$\frac{p_{av}}{Z_{av}} = \frac{p_{sc} T_{av}}{T_{sc} AL}[V_A + V_{AB}] \qquad (11\text{-}96)$$

where V_A is the volume of gas stored in the pipeline at point A.

The calculation procedure can now be described from these relationships. First, a plot is made of the fluctuation in demand versus time, as shown in

Figure 11-12. Then, the volume of gas stored in the pipeline during the period of low demand is determined by measuring the appropriate area enclosed by the curve. Then, Equation 11-96 is used to calculate (p_{av}/Z_{av}) from which p_{av} and Z_{av} can be obtained by trial and error. Then, the value of the expression $3p_{av}k/2qZ_{av}^{0.5}$ is calculated. Figure 11-11 can then be used to read off the corresponding value of R_p. Knowing the value of R_p, p_2 can be calculated using Equation 11-93, and then the required upstream pressure p_1 can be obtained $(p_1 = R_p p_2)$.

Let $(p_1)_{max}$ be the maximum feasible pressure at the head end of the pipeline. Now, if $p_1 \leq (p_1)_{max}$, then the quantity of gas V_{AB} stored in the pipeline during the slack-demand period will be sufficient to handle the excess demand in the high-demand period BC. Otherwise, the pipeline must be redesigned.

Handling Variable Consumer Demand—Case of Variable Injection Rate

Batey et al. (1961) described a method for obtaining solutions to the case of variable injection into the pipeline and conducted a sensitivity analysis of the various factors that affect pipeline performance. They represented gas-consumption variations in time by a Fourier function (Figure 11-13), $q_2 = f(t)$. From this supply function and the constant supply pressure p_2, the gas flow rate and pressure versus time functions can be calculated, proceeding step by step for various pipe sections backward along the line from the tail to the head end. Such functions, represented as $q_1 = f^1(t)$ and $p_1 = f^2(t)$, are shown along with the supply function $q_2 = f(t)$ in Figure 11-13. Any nu-

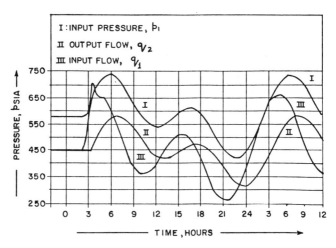

Figure 11-13. Simplified representation of the variable relationships for a line section. (After Batey et al., 1961.)

merical technique may be used for solving the system of equations for obtaining the f^1 and f^2 functions. The pipeline can then be provided optimum control to meet the demand variation at a constant delivery pressure p_2.

Batey et al. (1961) conducted a parametric analysis that enables one to avoid the expensive numeric computations involved in solving the system of equations. To permit Fourier series expansion, the load, i.e., the demand of gas, is expressed in the following functional from:

$$q_o = q_{ss} + q_t \sin(2\pi t/t_{tv}) \tag{11-97}$$

where q_o = flow out of the pipeline section
  q_{ss} = steady or average part of the flow
  q_t = amplitude of the time-variable part of the flow out of the pipeline section
  t = time
  t_{tv} = period of the time-variable output flow

Further, the input flow q_i required for the given output boundary conditions can be expressed as (Batey et al., 1961):

$$q_i = q_{ss} + q_1 \sin(2\pi/t_{tv})(t + \tau_1)$$

$$+ \Sigma_{y=2} \, q_y \sin(2\pi y/t_{tv})(t + \tau_y) \tag{11-98}$$

where q_1 = sine wave amplitude at the input of the pipeline section
  τ_1 = time required for the sine wave to propagate from input to output
  y = y-th harmonic
  q_y = amplitude of the flow corresponding to the y-th harmonic
  τ_y = delay time for the y-th harmonic

Amplitude q_y of the y-th harmonic is generally less than 1% of the amplitude q_1. Thus Equation 11-98 can be simplified to:

$$q_i = q_{ss} + q_1 \sin(2\pi/t_{tv})(t + \tau_1) \tag{11-99}$$

To specify how the sine wave propagates in the pipeline system, values of the amplitude ratio, q_t/q_1, and τ_1 are required.

Batey et al. (1961) studied the effect of the three types of pipeline section parameters that affect flow:

1. Parameters that depend upon line section geometry, such as length, diameter, and relative roughness (ϵ/d).

570   Gas Production Engineering

2. Parameters that depend upon the characteristics of the flowing gas. These include molecular weight, pseudocritical pressure, and the slope of compressibility versus reduced pressure for the gas.
3. Parameters that depend upon the average operating conditions of the pipeline. These are average line pressure, average flow, and average gas temperature.

Figures 11-14 through 11-21 show the results of this parametric analysis by Batey et al. (1961). In Figure 11-14, the amplitude ratio is shown as a function of the period of the time variable output pressure. The amplitude ratio is less than 1, implying that the signal amplitude at the input is greater than the output amplitude. Also, amplitude ratio decreases for high frequencies, indicating that attenuation (damping) increases with frequency. For a given frequency, damping increases as the output pressure decreases, because damping results from energy loss, which is proportional to the square of the velocity (and hence, inversely related to output pressure). Damping also increases with increasing friction factor (Figure 11-16), with decreasing pipe diameter (Figure 11-18).

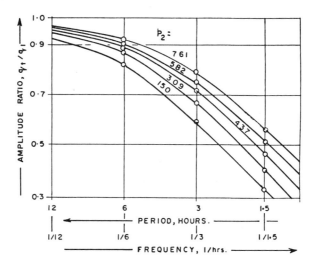

Figure 11-14. Amplitude ratio of input upset to output upset versus period of upset, as a function of the output pressure. (After Batey et al., 1961.)

The corresponding time delays for these variables are shown in Figures 11-15, 11-17, and 11-19. These figures show that higher the frequency (lower the period on the abscissa), lower is the time delay or phase shift of the demand wave (i.e., the phase velocity is higher). For a given frequency,

Gas Gathering and Transport   571

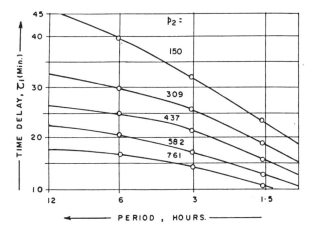

Figure 11-15. Time delay of upset propagation from one end to the other end of a line section versus period, as a function of the output pressure. (After Batey et al., 1961.)

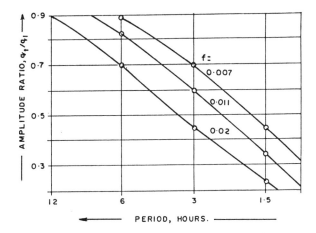

Figure 11-16. Amplitude ratio versus period as a function of the friction factor. (After Batey et al., 1961.)

phase shift decreases with increasing output pressure (Figure 11-15), decreasing friction factor (Figure 11-17), and increasing pipe diameter (Figure 11-19).

Figure 11-20 shows that the amplitude ratio decreases (or, attenuation increases) along the pipeline length. Figure 11-21 shows that the phase shift increases with increasing length along the pipeline. These figures can be

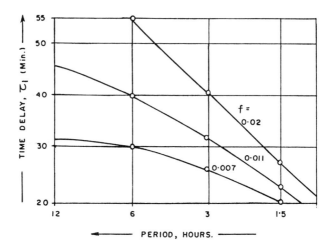

Figure 11-17. Time delay versus period as a function of friction factor. (After Batey et al., 1961.)

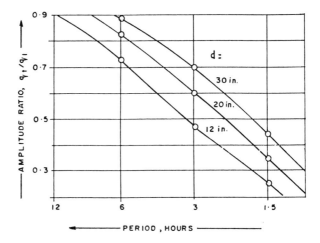

Figure 11-18. Amplitude ratio versus period as a function of pipeline diameter. (After Batey et al., 1961.)

used to determine the operating conditions necessary for meeting consumer demand at a constant, specified contract (output) pressure.

The main advantage of satisfying consumer demand in this way is that no pressure energy needs to be dissipated by throttling at the output end, and the compressor requirements are therefore minimized. However, a sufficiently accurate foreknowledge of the demand wave is required.

Gas Gathering and Transport   573

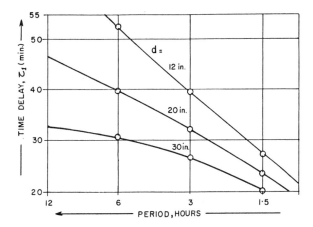

Figure 11-19. Time delay versus period as a function of pipeline diameter. (After Batey et al., 1961.)

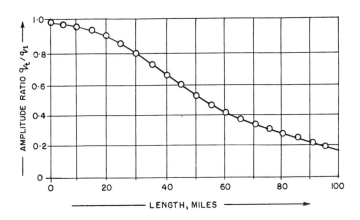

Figure 11-20. Amplitude ratio versus pipeline length. (After Batey et al., 1961.)

Unsteady-State Flow in Pipeline Systems

We have so far described transient flow in a *single* pipeline. As discussed earlier for steady-state flow in pipeline networks, this simple procedure must be modified for application to complex transmission systems where there is injection and offtake of gas at intermediate points along the line, or where

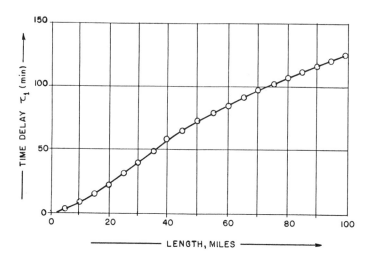

Figure 11-21. Time delay versus pipeline length. (After Batey et al., 1961.)

the flows converge to, or diverge from, nodes. The node law (Equation 11-41) must apply to each of these nodes:

$$m_n + \sum_{i=1}^{n} m_i = 0 \qquad (11\text{-}100)$$

where m_n = mass flow into or out of the node
m_i = mass flow rate in pipe leg i meeting at the node for each of the n pipe legs meeting at the node

The m_i values are computed from the non-linear algebraic equations describing transient flow in a single pipe leg. This makes the description and calculation for a transmission system of even a moderate size very complex. It is essential to incorporate some simplifications into the transient flow equations. Guy (1967) has discussed some of these simplifications that can be usually done without a significant loss in accuracy of the results. The first simplification is to neglect elevation changes that are generally insignificant for gas lines. Thus, the third term in Equation 11-71 can be omitted. Further, the second term, $(p/Ag_c)(\partial m/\partial t)$, describing the change in mass flow rate per unit time, is in most cases smaller than the friction term in Equation 11-71 by an order of magnitude and can also be dropped. The system of equations describing transient flow (Equations 11-66 and 11-71) now simplify to:

$$\frac{\partial p}{\partial t} = -\frac{B^2}{Ag_c}\frac{\partial m}{\partial x} \qquad (11\text{-}101)$$

$$\frac{\partial p^2}{\partial x} = -\frac{fm|m|B^2}{dA^2 g_c^2} \tag{11-102}$$

Every node is assigned half the length of each pipe leg associated with that node. Let V_j be the volume obtained on summing up the volumes of each of the pipe legs that connect to node j. Equation 11-101 can then be written for node j as follows:

$$\frac{Ag_c dx}{B^2} \frac{dp_j}{dt} = \sum_{i=1}^{n} m_{i,j} - m_{o,j}$$

or,

$$\frac{V_j g_c}{B^2} \frac{dp_j}{dt} = \sum_{i=1}^{n} m_{i,j} - m_{o,j} \tag{11-103}$$

where $m_{i,j}$ = flow into node j from pipe leg i
$m_{o,j}$ = flow out of node j

The values of $m_{i,j}$ can be calculated from Equation 11-102 as follows:

$$m_{i,j} = \left[\frac{d_{i,j} A_{i,j}^2 g_c^2}{f_{i,j} B^2} \frac{|p_i^2 - p_j^2|}{L_{i,j}}\right]^{0.5} S_{i,j} \tag{11-104}$$

where $L_{i,j}$ represents the lengths of the individual pipe legs, and $S_{i,j}$ is given by

$$S_{i,j} = \text{sign}(p_i - p_j)$$

Substituting for $m_{i,j}$ from Equation 11-104 into Equation 11-103, and multiplying by $B^2/V_j g_c$:

$$\frac{dp_j}{dt} = \frac{B^2}{V_j g_c} \sum_{i=1}^{n} \frac{g_c}{B} \left[\frac{d_{i,j} A_{i,j}^2}{f_{i,j} L_{i,j}} |p_i^2 - p_j^2|\right]^{0.5} S_{i,j} - \frac{B^2}{V_j g_c} m_{o,j}$$

or

$$\frac{dp_j}{dt} = K_j \sum_{i=1}^{n} [J_{i,j}(|p_i^2 - p_j^2|)^{0.5} S_{i,j}] - \frac{BK_j}{g_c} m_{o,j} \tag{11-105}$$

where $K_j = \dfrac{B}{V_j}$

and

$$J_{i,j} = J_{j,i} = \frac{d_{i,j} A_{i,j}^2}{f_{i,j} L_{i,j}} \qquad (11\text{-}106)$$

Applying finite difference techniques,

$$\frac{dp_j}{dt} = \frac{p_j(t + \Delta t) + p_j(t)}{\Delta t}$$

Using this and rearranging, Equation 11-105 becomes

$$p_j(t + \Delta t) = \Delta t K_j \left\{ \sum_{i=1}^{n} [J_{i,j}(|p_i^2 - p_j^2|)^{0.5} S_{i,j}] - (B/g_c) m_{o,j} \right\} + p_j(t)$$

(11-107)

Equation 11-107 for node j can be expressed as:

$$p_j(t + \Delta t) = C_j + p_j(t) \qquad (11\text{-}108)$$

where $C_j = \Delta t K_j \left\{ \sum_{i=1}^{n} [J_{i,j}(|p_i^2 - p_j^2|)^{0.5} S_{i,j}] - (B/g_c) m_{o,j} \right\}$ (11-109)

Similar equations can be written for all the other nodes in the system. Thus, we will have a system of non-linear algebraic equations representing transient flow in the pipeline network. This system of equations can be solved using finite difference techniques. For the explicit method, Equation 11-108 is expressed as:

$$p_j(t + \Delta t) = C_j(t) + p_j(t)$$

whereas in the implicit method, C_j is evaluated at time $t + \Delta t$, and Equation 11-108 is expressed as:

$$p_j(t + \Delta t) = C_j(t + \Delta t) + p_j(t)$$

Other elements in the system, such as chokes and compressors, can also be incorporated in the previous equations as done for steady-state networks.

Pipeline Economics

Pipeline sizes are generally dictated by the available pressure drop, required flow rate, and the available pipe sizes. If two or more pipe sizes satisfy all the requirements and constraints, then one is chosen based on the economics. An economic pipe diameter minimizes the (discounted or amortized) capital costs and operational (compression) costs. With increasing pipe diameter d, the capital costs increase, whereas the operational costs decrease due to smaller pressure losses. Thus, there exists an optimum d such that the sum of the amortized and operational costs is a minimum.

It is obvious that for such an economic analysis, no general correlation or equation can be easily determined for all possible cases because pipeline costs, amortization (or depreciation) allowances within the corporate tax framework, and operating costs vary widely. A simplified analysis by Peters and Timmerhaus (1980) is presented here for determining the optimum economic pipe diameter for a single pipeline. The concepts developed are subsequently extended to the far more complex problem of the economics of gas transmission systems.

A Simplified Analysis for Optimum Pipe Diameter

Compression Costs

The mechanical energy balance for a flowing system can be reduced to the following form (see Equation 7-11):

$$w_s = \frac{fLv^2(1 + L_{fp})}{2g_c d} + B \qquad (11\text{-}110)$$

where w_s = mechanical work added to the system by the compressor, ft-lbf/lbm
f = dimensionless Moody friction factor
L = length of the pipe, ft
v = flow velocity, ft/sec
L_{fp} = frictional loss due to pipe fittings and bends, expressed as equivalent fractional loss in a straight pipe
g_c = conversion constant relating mass and weight (= 32.17 lbm-ft/lbf-sec^2)
d = inside pipe diameter, ft
B = a constant accounting for all other energy losses in the flowing system

578 Gas Production Engineering

As discussed in Chapter 7 (see Equations 7-14 through 7-21), several relationships can be described for the friction factor, f, in terms of the Reynolds number, N_{Re}. Peters and Timmerhaus (1980) use the following equation for turbulent flow:

$$f = \frac{0.16}{(N_{Re})^{0.16}} \qquad (11\text{-}111)$$

Combining Equations 11-111 and 11-110, and applying the necessary conversion factors, the following equation can be obtained for the annual compression costs:

$$C_{comp} = \frac{0.273 q^{2.84} \rho^{0.84} \mu^{0.16} C_e (1 + L_{fp}) H_y}{d^{4.84} E} + B^* \qquad (11\text{-}112)$$

where C_{comp} = compression cost in dollars/year per foot of pipe length
q = gas flow rate, ft³/sec
ρ = the gas density, lbm/ft³
μ = gas viscosity, cp
C_e = cost of electrical energy, $/kWh
H_y = hours of operation per year
d = pipe diameter, in.
E = compressor efficiency, fraction
B^* = a constant independent of pipe diameter d

Fixed Costs for Piping System

For most types of pipe, the purchase cost per foot of pipe is related to the pipe diameter as follows:

$$C_{pipe} = C_p d^n \qquad (11\text{-}113)$$

where C_{pipe} = purchase cost of new pipe of diameter d inches per foot of pipe length, $/ft
C_p = a constant equal to the purchase price per foot for a 1-in. diameter pipe, $/ft
d = pipe diameter, in.

The annual cost for an installed piping system can be expressed as (Peters and Timmerhaus, 1980):

$$C_{pipe} = (1 + R_p) C_p d^n C_{Fp} \qquad (11\text{-}114)$$

where R_p = ratio of total costs for fittings and installation to the purchase cost for new pipe
C_{Fp} = annual fixed charges, including maintenance, expressed as a fraction of the initial cost for a completely installed pipe

Optimum Economic Pipe Diameter

The total annual cost C_T for the compressor and piping system can be obtained by adding Equations 11-112 and 11-114:

$$C_T = \frac{0.273 q^{2.84} \rho^{0.84} \mu^{0.16} C_e (1 + L_{fp}) H_y}{d^{4.84} E} + B^* + (1 + R_p) C_p d^n C_{Fp} \quad (11\text{-}115)$$

Differentiating C_T expressed by Equation 11-115 with respect to diameter d, setting the resultant expression to zero, and solving for d gives the optimum pipe diameter, d_{opt} in inches, as follows:

$$d_{opt} = \left[\frac{1.32 q^{2.84} \rho^{0.84} \mu^{0.16} C_e (1 + L_{fp}) H_y}{n(1 + R_p) C_p E C_{Fp}} \right]^{1/(4.84 + n)} \quad (11\text{-}116)$$

The value of n for steel pipes is about 1.0 for d < 1 in., and about 1.5 for d ≥ 1 in. Thus, for the commonly used greater than 1-in. diameter pipes, Equation 11-116 becomes (Peters and Timmerhaus, 1980):

$$d_{opt} = q^{0.448} \rho^{0.132} \mu^{0.025} \left[\frac{0.88 C_e (1 + L_{fp}) H_y}{(1 + R_p) C_p E C_{Fp}} \right]^{0.158} \quad (11\text{-}117)$$

An interesting fact to notice about Equation 11-117 is that the optimum pipe diameter is relatively insensitive to most of the parameters involved. Equation 11-117 can be simplified by assuming typical numeric values for the terms involved. Using C_e = \$0.055/kWh, L_{fp} = 0.35, H_y = 8,760 hrs/year, R_p = 1.4, C_p = \$0.45/ft for 1-in. diameter pipe, E = 0.50, C_{Fp} = 0.20, and neglecting the viscosity term which is close to unity for most cases for the small exponent involved, Peters and Timmerhaus (1980) present the following equation for optimum pipe diameter:

$$d = 0.0039 q^{0.45} \rho^{0.13} = 0.0022 m^{0.45} / \rho^{0.31} \quad (11\text{-}118)$$

where m = mass flow rate, lbm/hr
ρ = fluid density, lbm/ft³

This equation is quite good for short pipelengths and where the pressure drop in the pipelength is not large.

Analysis Including Cost of Capital and Corporate Taxes

In the previous analysis, a number of important factors were neglected, such as the time value of money, corporate taxes, costs of compressors, and the cost of capital (or return on investment). Including these factors and using a more accurate expression for the friction loss due to pipe fittings and bends, Peters and Timmerhaus (1980) present the following equation:

$$\frac{d_{opt}^{4.84+n}}{1 + 0.794 L_{fd} d_{opt}} = \frac{1.046 \times 10^{-10} H_y C_e m^{2.84} \mu^{0.16} [Mi^* + (1-t_r)[1 + (i_{dc} + i_{mc})M]]}{n(1+R_p) C_p [i^* + (1-t_r)(i_{dp} + i_{mp})] E \rho^2} \quad (11\text{-}119)$$

where d_{opt} = optimum economic inside pipe diameter, in.
  m = mass flow rate, lbm/hr
  L_{fd} = frictional loss due to fittings and bends expressed as equivalent pipe length in pipe diameters per unit length of pipe ($L_{fd} = L_{fp}/d_{opt}$)
  M = ratio of total compressor installation cost to yearly cost of compression power required
  i^* = rate of return (or cost of capital before taxes) on incremental investment, fraction
  t_r = taxation rate, fraction
  i_{dp} = depreciation rate for the installed piping system, fraction
  i_{dc} = depreciation rate for the compression equipment, fraction
  i_{mp} = fraction of initial cost of installed piping system for annual maintenance
  i_{mc} = fraction of initial cost of compression installation for annual maintenance
  ρ = gas density, lbm/ft^3

All other terms have similar meaning as before.

Thus, for a given throughput, we can compute the optimum pipeline diameter using the relationships given here. But the pipeline chosen may have optimum behavior at some other throughput (that is, it may be able to handle a different throughput more economically). This is an interesting point to note about pipelines: a pipeline's optimum throughput is not the same as the throughput for which it has the most economic diameter.

Gas Transmission Economics

It is relatively easier to estimate the economically optimal dimensions and operating conditions for a single pipeline. A complete transmission system or grid, as described earlier, includes several producing wells, storage wells, and a vast network of pipelines of different dimensions and requirements. Designing such a transmission system to match the requirements is difficult enough; economics introduces even more complexity.

The quantitative analysis described in the previous section assumes that the gas throughput, m or q, is known. In dealing with natural resources such as gas, one soon realizes that throughput is not easy to specify. Some important factors, besides the ones discussed earlier, that play a role in determining the design capacity and size of a gas transmission system are:

1. Available supply of gas: the amount (reserves), current production rate, growth rate in supply due to field development activities and/or new discoveries.
2. Rate of growth in gas demand. The economics of the variation in pipeline capacity factors* has to be evaluated.
3. Availability of capital.
4. Technical problems or feasibility of transmission line construction, such as difficulties in increasing line capacity by installing a second pipeline.

Supply and demand projections are made using complex computer models that account for the vast number of factors that influence such predictions. The supply model involves reservoir modelling for production from known reserves, geologic information for predicting the growth of these reserves, and field development plans. The demand model is even more complex. It deals with global (or local, as the case may be) economic growth, expected energy consumption, energy consumption patterns (share of gas as an energy source), price of gas and its effect on gas production and use, etc. It is thus clear that supply and demand predictions are rarely accurate; some commonsense analysis may often prove more reliable.

Availability of capital is usually not a serious constraint, but may preclude overambitious, ultra-large projects. Technical problems may be too overwhelming in some cases, resulting in alternative feasible designs that may not be optimum. It is easy to imagine how several other factors may also become important in a given situation.

---

* Capacity factor is the ratio of the actual average discharge to the design capacity.

582 Gas Production Engineering

The two major features of a transmission system that is operating in an optimum manner are high load and capacity factors**, and assured, uninterrupted gas supply throughout the range of the demand cycle. A high capacity factor will be obtained if the demand (and supply) is close to design capacity. The load factor can be increased by using underground gas storage, by using some type of gas reserve such as high-pressure gas storage to ensure an excess supply during peak demand, or by using the pipeline as a buffer storage facility. An uninterrupted gas supply to meet varying demand is ensured by considering two factors: *availability*, defined as the fraction of the total time that an uninterrupted supply is achieved, and *reserve factor*, which accounts for gas reserves provided in the form of additional parallel lines, underground storage, pipeline buffer capacity, or other standby supply systems. Higher the availability, lower is the reserve factor required.

Questions and Problems

1. Derive the coefficients indicated in Table 11-1.
2. A 7-in. ID pipeline transports a 0.68 gravity gas from a distance of 22 miles to a liquefaction plant in Saudi Arabia. The first 10 miles of the pipeline is looped with a 4-in. line, the next 5 miles is looped with a 3-in. line, and the last 7 miles is a single unlooped line. The inlet and outlet pressures for the system are 530 psia and 130 psia, respectively. The flowing temperature is 95°F. Calculate the gas flow rate using at least two flow relationships.
3. Two lines AC (5-in. ID, 4.5 miles long) and BC (5-in. ID, 3.8 miles long), emanating from leases A and B, respectively, terminate at the gathering station at C. From C, a single 11.5-mile, 8-in. ID pipeline leads into the regional trunk line at D, where the gas must be delivered at 520 psig. Lease A produces 5 MMscfd of 0.65 gravity gas, while lease B produces 7 MMscfd of 0.68 gravity gas. To what pressure should the gas be compressed at (a) lease A, and (b) lease B, to enable delivery of gas to the trunk line? Assume f = 0.008, ambient and flowing temperature of 85°F, and a perfectly horizontal flow system.
4. A 5-mile long, 8-in. ID pipeline is at an average inclination of 30° from the horizontal, with uphill flow. At a distance of 2 miles from the inlet, it branches out into another 4-mile, 4-in. ID line. Derive a relationship to express the flow rate in the 4-in. line in terms of the flow rate in the 8-in. line.

** Load factor is the ratio of the average rate to the maximum hourly rate of flow.

Solve this expression for the case where: gas gravity = 0.81, gas inlet pressure in the sytem = 950 psig, input gas flow rate = 4 MMscfd, and flowing temperature = 78°F.
5. An 8,000-ft well is equipped with a 3-in. tubing. What minimum size of surface flowline would be required to enable gas production at a rate of 1 MMscfd, with a guaranteed pressure of at least 100 psia at the separator? Use the following data: γ_g = 0.6, bottom-hole pressure = 2,000 psia, bottom-hole temperature = 250°F, surface temperature = 82°F.
Can this surface flowline diameter be considered economic?
6. Write a computer program to solve Equation 11-48 for loopless pipeline systems. Use this program to solve for steady-state flow for a simple pipeline system.

References

Aziz, K. and Settari, A., 1979. *Petroleum Reservoir Simulation*. Elsevier Applied Science Publishers Ltd., England, 476pp.
Batey, E. H., Courts, H. R., and Hannah, K. W., 1961. "Dynamic Approach to Gas-Pipeline Analysis," *Oil & Gas J.*, 59(51, Dec. 18), 65–78.
Campbell, J. M., 1984. *Gas Conditioning and Processing*, Vol. 2. Campbell Petroleum Series, Norman, Oklahoma, 398pp.
Cross, H., 1936. "Analysis of Flow in Networks of Conduits or Conductors," *Bulletin of the Univ. of Illinois*, 286.
Distefano, G. P., 1970. "PIPETRAN, Version IV, A Digital Computer Program for the Simulation of Gas Pipeline Network Dynamics," Cat. No. L20000, March.
Goacher, P. S., 1969. "Steady and Transient Analysis of Gas Flows in Networks," paper GC157 presented at the Research Meet. of Inst. of Gas Engrs., London.
Guy, J. J., 1967. "Computation of Unsteady Gas Flow in Pipe Networks," paper presented at the Inst. of Chem. Engrs., Midlands Branch Conf., U. of Nottingham, April 14.
Hain, H. A., 1968. "How to Determine the Maximum Capability of a Complex Pipeline System," *Pipe Line News*, 9.
Izawa, H. S., 1966. "How to Calculate Gas Volume in Blowdown and Purging," *Oil & Gas J.*, 64(19, May 9), 118–123.
Peters, M. S. and Timmerhaus, K. D., 1980. *Plant Design and Economics for Chemical Engineers*, 3rd ed. McGraw-Hill Book Co., New York, N.Y., 973pp.

Rachford, H. H., Jr. and Dupont, T., 1974. "A Fast, Highly Accurate Means of Modeling Transient Flow in Gas Pipeline Systems by Variational Methods," *Soc. Pet. Eng. J.*, 14(2, April), 165–178.

Stewart, T. L., 1971. "Computer Speeds Surge Calculations," *Oil & Gas J.*, 69(47, Nov 22), 55–60.

Stoner, M. A., 1969. "Steady-State Analysis of Gas Production, Transmission and Distribution Systems," paper SPE 2554 presented at the SPE 44th Ann. Fall Meet., Denver, Colorado, Sept. 28–Oct. 1.

Stoner, M. A., 1972. "Sensitivity Analysis Applied to a Steady-State Model of Natural Gas Transportation Systems," *Soc. Pet. Eng. J.*, 12(2, April), 115–125.

Streeter, V. L. and Wylie, E. B., 1970. "Natural Gas Pipeline Transients," *Soc. Pet. Eng. J.*, 10(4, Dec.), 357–364.

Szilas, A. P., 1975. *Production and Transport of Oil and Gas.* Developments in Petroleum Science, 3, Elsevier Scientific Publ. Co., Amsterdam, 630pp.

Taylor, T. D., Wood, N. E., and Powers, J. E., 1962. "A Computer Simulation of Gas Flow in Long Pipelines," *Soc. Pet. Eng. J.*, 2(4, Dec.), 297–302.

Tuppeck, F. and Kirsche, H., 1962. "Ein Numerisches Verfahren zur Berechnung Instationarei Stromungsvorgange in Ferngasleitungen," *Das Gas Und Wasserfach*, 103(21, June), 523–528.

Wylie, E. B., Stoner, M. A., and Streeter, V. L., 1971. "Network System Transient Calculations by Implicit Methods," *Soc. Pet. Eng. J.*, 11(4, Dec.), 356–362.

APPENDIX A
General Data and Unit Conversion Factors

Appendix A-1
Some Useful Physical Constants

Quantity	Magnitude	In units of
Absolute zero temperature	0.0*	K
	− 273.15*	°C
	0.0*	°R
	− 459.67*	°F
Avogadro's number	6.022 169 E + 23	mole^{-1}
Gas constant (R)	0.082 057 477	(atm liter)/(gmmole K)
	0.084 784	(kg/cm^2 liter)/(gmmole K)
	1.987	(gm cal)/(gmmole K)
	8.314 5	Joules/(gmmole °R)
	1.985 9	Btu/(lbmole °R)
	1545.3	(ft lbf)/(lbmole °R)
	10.732	(psi ft^3)/(lbmole °R)
	0.730 24	(atm ft^3)/(lbmole °R)
Density of dry air	0.001 223 2	gm/cm^3
(at 60°F, 1 atm)	0.076 362	lbm/ft^3
Density of dry air	0.001 293	gm/cm^3
(at 0°C, 1 atm)	0.080 719	lbm/ft^3
Velocity of sound in dry air	33,136	cm/sec
(at 0°C, 1 atm)	1,089	ft/sec
Acceleration due to gravity (g)	9.806 65	m/sec^2
	980.665	cm/sec^2
	32.174 05	ft/sec^2
Dimensional constant (g_c)	1.0*	(kg m)/(N sec^2)
	1.0*	(gm cm)/(dyne sec^2)
	32.174 05	(lbm ft)/(lbf sec^2)
π	22/7*	dimensionless
	3.141 593	dimensionless
e	2.718 282	dimensionless

* Indicates that the number is exact.

Appendix A-2*
General Data for Natural Gas Constituents

Compound	Formula	Molecular weight	Boiling point °F, 14.696 psia	Vapor pressure, 100°F, psia	Freezing point, °F, 14.696 psia	Critical Constants		Volume, cu ft/lb
						Pressure, psia	Temperature, °F	
Methane	CH_4	16.043	−258.69	(5000)	−296.46	667.8	−116.63	0.0991
Ethane	C_2H_6	30.070	−127.48	(800)	−297.89	707.8	90.09	0.0788
Propane	C_3H_8	44.097	−43.67	190.0	−305.84	616.3	206.01	0.0737
n-Butane	C_4H_{10}	58.124	31.10	51.6	−217.05	550.7	305.65	0.0702
Isobutane	C_4H_{10}	58.124	10.90	72.2	−255.29	529.1	274.98	0.0724
n-Pentane	C_5H_{12}	72.151	96.92	15.570	−201.51	488.6	385.7	0.0675
Isopentane	C_5H_{12}	72.151	82.12	20.44	−255.83	490.4	369.10	0.0679
Neopentane	C_5H_{12}	72.151	49.10	35.9	2.17	464.0	321.13	0.0674
n-Hexane	C_6H_{14}	86.178	155.72	4.956	−139.58	436.9	453.7	0.0688
2-Methylpentane	C_6H_{14}	86.178	140.47	6.767	−244.63	436.6	435.83	0.0681
3-Methylpentane	C_6H_{14}	86.178	145.89	6.098	—	453.1	448.3	0.0681
Neohexane	C_6H_{14}	86.178	121.52	9.856	−147.72	446.8	420.13	0.0667
2,3-Dimethylbutane	C_6H_{14}	86.178	136.36	7.404	−199.38	453.5	440.29	0.0665
n-Heptane	C_7H_{16}	100.205	209.17	1.620	−131.05	396.8	512.8	0.0691
2-Methylhexane	C_7H_{16}	100.205	194.09	2.271	−180.89	396.5	495.00	0.0673
3-Methylhexane	C_7H_{16}	100.205	197.32	2.130	—	408.1	503.78	0.0646
3-Ethylpentane	C_7H_{16}	100.205	200.25	2.012	−181.48	419.3	513.48	0.0665
2,2-Dimethylpentane	C_7H_{16}	100.205	174.54	3.492	−190.86	402.2	477.23	0.0665
2,4-Dimethylpentane	C_7H_{16}	100.205	176.89	3.292	−182.63	396.9	475.95	0.0668
3,3-Dimethylpentane	C_7H_{16}	100.205	186.91	2.773	−210.01	427.2	505.85	0.0662
Triptane	C_7H_{16}	100.205	177.58	3.374	−12.82	428.4	496.44	0.0636
n-Octane	C_8H_{18}	114.232	258.22	0.537	−70.18	360.6	564.22	0.0690
Diisobutyl	C_8H_{18}	114.232	228.39	1.101	−132.07	360.6	530.44	0.0676
Isooctane	C_8H_{18}	114.232	210.63	1.708	−161.27	372.4	519.46	0.0656
n-Nonane	C_9H_{20}	128.259	303.47	0.179	−64.28	332.0	610.68	0.0684
n-Decane	$C_{10}H_{22}$	142.286	345.48	0.0597	−21.36	304.0	652.1	0.0679
Cyclopentane	C_5H_{10}	70.135	120.65	9.914	−136.91	653.8	461.5	0.059
Methylcyclopentane	C_6H_{12}	84.162	161.25	4.503	−224.44	548.9	499.35	0.0607
Cyclohexane	C_6H_{12}	84.162	177.29	3.264	43.77	591.0	536.7	0.0586
Methylcyclohexane	C_7H_{14}	98.189	213.68	1.609	−195.87	503.5	570.27	0.0600
Ethylene	C_2H_4	28.054	−154.62	—	−272.45	729.8	48.58	0.0737
Propene	C_3H_6	42.081	−53.90	226.4	−301.45	669.0	196.9	0.0689
1-Butene	C_4H_8	56.108	20.75	63.05	−301.63	583.0	295.6	0.0685

* Reprinted by courtesy of GPSA.

Appendix A 587

Appendix A-2 Continued
General Data for Natural Gas Constituents

Density of liquid, 60°F, 14.696 psia				Temperature Coefficient of Density	Pitzer acentric factor	Compressibility factor of real gas, Z 14.696 psia, 60°F	Gas density, 60°F, 14.696 psia Ideal gas			Specific heat 60°F, 14.696 psia Cp. Btu/lb °F	
Specific gravity 60°F/60°F	lb/gal (Wt in vacuum)	lb/gal (Wt in air)	Gal/lb-mole				Specific gravity Air = 1°	cu ft gas /lb	cu ft gas/gal liquid	Ideal gas	Liquid
0.3	2.5	2.5	6.4	—	0.0104	0.9881	0.5539	23.65	59.	0.5266	—
0.3564	2.971	2.962	10.12	—	0.0986	0.9916	1.0382	12.62	37.5	0.4097	0.9256
0.5077	4.233	4.223	10.42	0.00152	0.1524	0.9820	1.5225	8.606	36.43	0.3881	0.5920
0.5844	4.872	4.865	11.93	0.00117	0.2010	0.9667	2.0068	6.529	31.81	0.3867	0.5636
0.5631	4.695	4.686	12.38	0.00119	0.1848	0.9696	2.0068	6.529	30.65	0.3872	0.5695
0.6310	2.261	5.251	13.71	0.00087	0.2539	0.9549	2.4911	5.260	27.67	0.3883	0.5441
0.6247	5.208	5.199	13.85	0.00090	0.2223	0.9544	2.4911	5.260	27.39	0.3827	0.5353
0.5967	4.975	4.965	14.50	0.00104	0.1969	0.9510	2.4911	5.260	26.17	(0.3866)	0.554
0.6640	5.536	5.526	15.57	0.00075	0.3007	—	2.9753	4.404	24.38	0.3864	0.5332
0.6579	5.485	5.475	15.71	0.00073	0.2825	—	2.9753	4.404	24.15	0.3872	0.5264
0.6689	5.577	5.568	15.45	0.00075	0.2741	—	2.9753	4.404	24.56	0.3815	0.507
0.6540	5.453	5.443	15.81	0.00078	0.2369	—	2.9753	4.404	24.01	0.3809	0.5165
0.6664	5.556	5.546	15.51	0:00075	0.2495	—	2.9753	4.404	24.47	0.378	0.5127
0.6882	5.738	5.728	17.46	0.00069	0.3498	—	3.4596	3.787	21.73	0.3875	0.5283
0.6830	5.694	5.685	17.60	0.00068	0.3336	—	3.4596	3.787	21.57	(0.390)	0.5223
0.6917	5.767	5.757	17.38	0.00069	0.3257	—	3.4596	3.787	21.84	(0.390)	0.511
0.7028	5.859	5.850	17.10	0.00070	0.3095	—	3.4596	3.787	22.19	(0.390)	0.5145
0.6782	5.654	5.645	17.72	0.00072	0.2998	—	3.4596	3.787	21.41	(0.395)	0.5171
0.6773	5.647	5.637	17.75	0.00072	0.3048	—	3.4596	3.787	21.39	0.3906	0.5247
0.6976	5.816	5.807	17.23	0.00065	0.2840	—	3.4596	3.787	22.03	(0.395)	0.502
0.6946	5.791	5.782	17.30	0.00069	0.2568	—	3.4596	3.787	21.93	0.3812	0.4995
0.7068	5.893	5.883	19.39	0.00062	0.4018	—	3.9439	3.322	19.58	(0.3876)	0.5239
0.6979	5.819	5.810	19.63	0.00065	0.3596	—	3.9439	3.322	19.33	(0.373)	0.5114
0.6962	5.804	5.795	19.68	0.00065	0.3041	—	3.9439	3.322	19.28	0.3758	0.4892
0.7217	6.017	6.008	21.32	0.00063	0.4455	—	4.4282	2.959	17.80	0.3840	0.5228
0.7342	6.121	6.112	23.24	0.00055	0.4885	—	4.9125	2.667	16.33	0.3835	0.5208
0.7504	6.256	6.247	11.21	0.00070	0.1955	0.9657	4.2215	5.411	33.85	0.2712	0.4216
0.7536	6.283	6.274	13.40	0.00071	0.2306	—	2.9057	4.509	28.33	0.3010	0.4407
0.7834	6.531	6.522	12.89	0.00068	0.2133	—	2.9057	4.509	29.45	0.2900	0.4322
0.7740	6.453	6.444	15.22	0.00063	0.2567	—	3.3900	3.865	24.94	0.3170	0.4397
—	—	—	—	—	0.0868	0.9938	0.9686	13.53	—	0.3622	—
0.5220	4.352	4.343	9.67	0.00189	0.1405	0.9844	1.4529	49.018	39.25	0.3541	0.585
0.6013	5.013	5.004	11.19	0.00116	0.1906	0.9704	1.9372	6.764	33.91	0.3548	0.535

(table continued)

Gas Production Engineering

Appendix A-2 Continued
General Data for Natural Gas Constituents

Compound	Formula	Molecular weight	Boiling point °F, 14.696 psia	Vapor pressure, 100°F, psia	Freezing point, °F, 14.696 psia	Critical Constants Pressure, psia	Critical Constants Temperature, °F	Critical Constants Volume, cu ft/lb
Cis-2-Butene	C_4H_8	56.108	38.69	45.54	−218.06	610.0	324.37	0.0668
Trans-2-Butene	C_4H_8	56.108	33.58	49.80	−157.96	595.0	311.86	0.0680
Isobutene	C_4H_8	56.108	19.59	63.40	−220.61	580.0	292.55	0.0682
1-Pentene	C_5H_{10}	70.135	85.93	19.115	−265.39	590.0	376.93	0.0697
1,2-Butodiene	C_4H_6	54.092	51.53	(20.)	−213.16	(653.)	(339.)	(0.0649)
1,3-Butodiene	C_4H_6	54.092	24.06	(60.)	−164.02	628.0	306.0	0.0654
Isoprene	C_5H_9	68.119	93.30	16.672	−230.74	(558.4)	(412.)	(0.0650)
Acetylene	C_2H_2	26.038	−119	—	−114.	890.4	95.31	0.0695
Benzene	C_6H_6	78.114	176.17	3.224	41.96	710.4	552.22	0.0531
Toluene	C_7H_8	92.141	231.13	1.032	−138.94	595.9	605.55	0.0549
Ethylbenzene	C_8H_{10}	106.168	227.16	0.371	−138.91	523.5	651.24	0.0564
o-Xylene	C_8H_{10}	106.168	291.97	0.264	−13.30	541.4	675.0	0.0557
m-Xylene	C_8H_{10}	106.168	282.41	0.326	−54.12	513.6	651.02	0.0567
p-Xylene	C_8H_{10}	106.168	281.05	0.347	55.86	509.2	649.6	0.0572
Styrene	C_8H_8	104.152	293.29	(0.24)	−23.10	580.0	706.0	0.0541
Isopropylbenzene	C_9H_{12}	120.195	306.34	0.188	−140.82	465.4	676.4	0.0570
Methyl alcohol	CH_4O	32.042	148.1	4.63	−143.82	1174.2	462.97	0.0589
Ethyl alcohol	C_2H_6O	46.069	172.92	2.3	−173.4	925.3	469.58	0.0580
Carbon monoxide	CO	28.010	−313.6	—	−340.6	507.	−220.	0.0532
Carbon dioxide	CO_2	44.010	−109.3	—	—	1071.	87.9	0.0342
Hydrogen sulfide	H_2S	34.076	−76.6	394.0	−117.2	1306.	212.7	0.0459
Sulfur dioxide	SO_2	64.059	14.0	88.	−103.9	1145.	315.5	0.0306
Ammonia	NH_3	17.031	−28.2	212.	−107.9	1636.	270.3	0.0681
Air	N_2O_2	28.964	−317.6	—	—	547.	−221.3	0.0517
Hydrogen	H_2	2.016	−423.0	—	−434.8	188.1	−399.8	0.5167
Oxygen	O_2	31.999	−297.4	—	−361.8	736.9	−181.1	0.0382
Nitrogen	N_2	28.013	−320.4	—	−346.0	493.0	−232.4	0.0514
Chlorine	Cl_2	70.906	−29.3	158.	−149.8	1118.4	291.	0.0281
Water	H_2O	18.015	212.0	0.9492	32.0	3208.	705.6	0.0500
Helium	He	4.003	—	—	—	—	—	—
Hydrogen chloride	HCl	36.461	−121.	925.	−173.6	1198.	124.5	0.0208

Appendix A   589

Appendix A-2 Continued
General Data for Natural Gas Constituents

Density of liquid, 60°F, 14.696 psia							Gas density, 60°F, 14.696 psia Ideal gas			Specific heat 60°F, 14.696 psia	
Specific gravity 60°F/60°F	lb/gal (Wt in vacuum)	lb/gal (Wt in air)	Gal/lb-mole	Temperature Coefficient of Density	Pitzer acentric factor	Compressibility factor of real gas, Z 14.696 psia, 60°F	Specific gravity Air = 1°	cu ft gas /lb	cu ft gas/gal liquid	Cp. Btu/lb °F Ideal gas	Liquid
0.6271	5.228	5.219	10.73	0.00098	0.1953	0.9661	1.9372	6.764	35.36	0.3269	0.5271
0.6100	5.086	5.076	11.03	0.00107	0.2220	0.9662	1.9372	6.764	34.40	0.3654	0.5351
0.6004	5.006	4.996	11.21	0.00120	0.1951	0.9689	1.9372	6.764	33.86	0.3701	0.549
0.6457	5.383	5.374	13.03	0.00089	0.2925	0.9550	2.4215	5.411	29.13	0.3635	0.5196
0.658	5.486	5.470	9.86	0.00098	0.2485	(0.969)	1.8676	7.016	38.49	0.3458	0.5408
0.6272	5.229	5.220	10.34	0.00113	0.1955	(0.965)	1.8676	7.016	36.69	0.3412	0.5079
0.6861	5.720	5.711	11.91	0.00086	0.2323	(0.962)	2.3519	5.571	31.87	0.357	0.5192
0.615	—	—	—		0.1803	0.9925	0.8990	14.57	—	0.3966	—
0.8844	7.373	7.365	10.59	0.00066	0.2125	0.929	2.6969	4.858	35.82	0.2429	0.4098
0.8718	7.268	7.260	12.68	0.00060	0.2596	0.903	3.1812	4.119	29.94	0.2598	0.4012
0.8718	7.268	7.259	14.61	0.00054	0.3169	—	3.6655	3.574	25.98	0.2795	0.4114
0.8848	7.377	7.367	14.39	0.00055	0.3023	—	3.6655	3.574	26.37	0.2914	0.4418
0.8687	7.243	7.234	14.66	0.00054	0.3278	—	3.6655	3.574	25.89	0.2782	0.4045
0.8657	7.218	7.209	14.71	0.00054	0.3138	—	3.6655	3.574	26.00	0.2769	0.4083
0.9110	7.595	7.586	13.71	0.00057	—	—	3.5959	3.644	27.67	0.2711	0.4122
0.8663	7.223	7.214	16.24	0.00054	0.2862	—	4.1498	3.157	22.80	0.2917	(0.414)
0.796	6.64	6.63	4.83	—	—	—	1.1063	11.84	78.6	0.3231	0.594
0.794	6.62	6.61	6.96	—	—	—	1.5906	8.237	54.5	0.3323	0.562
0.801	6.68	6.67	4.19	—	0.041	0.9995	0.9671	13.55	—	0.2484	—
0.827	6.89	6.88	6.38	—	0.225	0.9943	1.5195	8.623	59.5	0.1991	—
0.79	6.59	6.58	5.17	—	0.100	0.9903	1.1765	11.14	73.3	0.238	—
1.397	11.65	11.64	5.50	—	0.246	—	2.2117	5.924	69.0	0.145	0.325
0.6173	5.15	5.14	3.31	—	0.255	—	0.5880	22.28	114.7	0.5002	1.114
0.856	7.14	7.13	4.06	—	—	0.9996	1.0000	13.10	—	0.2400	—
0.07	—	—	—	—	0.000	1.0006	0.0696	188.2	—	3.408	—
1.14	9.50	9.49	3.37	—	0.0213	—	1.1048	11.86	—	0.2188	—
0.808	6.74	6.73	4.16	—	0.040	0.9997	0.9672	13.55	—	0.2482	—
1.414	11.79	11.78	6.01	—	—	—	2.4481	5.352	63.1	0.119	—
1.000	8.337	8.328	2.16	—	0.348	—	0.6220	21.06	175.6	0.4446	1.0009
—	—	—	—	—	—	—	—	—	—	—	—
0.8558	7.135	7.126	5.11	0.00335	—	—	1.2588	10.41	74.3	0.190	—

(table continued)

Appendix A-2 Continued
General Data for Natural Gas Constituents

Compound	Calorific value, 60°F				Heat of vaporization, 14.696 psia at boiling point, Btu/lb	Refractive index, nD 68°F	Air required for combustion ideal gas cu ft/cu ft	Flammability Limits, Vol % in Air Mixture		ASTM Octane Number		
	Net		Gross					Lower	Higher	Motor method D-357	Research method D-908	
	Btu/cu ft, Ideal gas, 14.696 psia (20)	Btu/cu ft, Ideal gas, 14.696 psia (20)	Btu/lb liquid	(wt in vacuum) Btu/gal liquid								
Methane	909.1	1009.7	—	—	219.22	—	9.54	5.0	15.0	—	—	
Ethane	1617.8	1768.8	—	—	210.41	—	16.70	2.9	13.0	+.05	+1.6	
Propane	2316.1	2517.4	21,513	91,065	183.05	—	23.86	2.1	9.5	97.1	+1.8	
n-Butane	3010.4	3262.1	21,139	102,989	165.65	1.3326	31.02	1.8	8.4	89.6	93.8	
Isobutane	3001.1	3252.7	21,091	99,022	157.53	—	31.02	1.8	8.4	97.6	+.10	
n-Pentane	3707.5	4009.5	20,928	110,102	153.59	1.357,48	38.18	1.4	8.3	62.6	61.7	
Isopentane	3698.3	4000.3	20,889	108,790	147.13	1.353,73	38.18	1.4	(8.3)	90.3	92.3	
Neopentane	3682.6	3984.6	20,824	103,599	135.58	1.342	38.18	1.4	(8.3)	80.2	85.5	
n-Hexane	4403.7	4756.1	20,784	115,060	143.95	1.374,86	45.34	1.2	7.7	26.0	24.8	
2-Methylpentane	4395.8	4748.1	20,757	113,852	138.67	1.371,45	45.34	1.2	(7.7)	73.5	73.4	
3-Methylpentane	4398.7	4751.0	20,768	115,823	140.09	1.376,52	45.34	(1.2)	(7.7)	74.3	74.5	
Neohexane	4382.6	4735.0	20,710	112,932	131.24	1.368,76	45.34	1.2	(7.7)	93.4	91.8	
2,3-Dimethylbutane	4391.7	4744.0	20,742	115,243	136.08	1.374,95	45.34	(1.2)	(7.7)	94.3	+0.3	
n-Heptane	5100.2	5502.9	20,681	118,668	136.01	1.387,64	52.50	1.0	7.0	0.0	0.0	
2-Methylhexane	5092.1	5494.8	20,658	117,627	131.59	1.384,85	52.50	(1.0)	(7.0)	46.4	42.4	
3-Methylhexane	5095.2	5497.8	20,668	119,192	131.11	1.388,64	52.50	(1.0)	(7.0)	55.8	52.0	
3-Ethylpentane	5098.2	5500.9	20,679	121,158	132.83	1.393,39	52.50	(1.0)	(7.0)	69.3	65.0	
2,2-Dimethylpentane	5079.4	5482.1	20,620	116,585	125.13	1.382,15	52.50	(1.0)	(7.0)	95.6	92.8	
2,4-Dimethylpentane	5084.3	5487.0	20,636	116,531	126.58	1.381,45	52.50	(1.0)	(7.0)	83.8	83.1	
3,3-Dimethylpentane	5085.0	5487.8	20,638	120,031	127.21	1.390,92	52.50	(1.0)	(7.0)	86.6	80.8	
Triptane	5081.0	5483.6	20,627	119,451	124.21	1.389,44	52.50	(1.0)	(7.0)	+0.1	+1.8	
n-Octane	5796.7	6249.7	20,604	121,419	129.53	1.397,43	59.65	0.96	—	—	—	
Diisobutyl	5781.3	6234.3	20,564	119,662	122.8	1.392,46	59.65	(0.98)	—	55.7	55.2	
Isooctane	5779.8	6232.8	20,570	119,388	116.71	1.391,45	59.65	1.0	—	100.	100.	
n-Nonane	6493.3	6996.6	20,544	123,613	123.76	1.405,42	66.81	0.87	2.9	—	—	
n-Decane	7188.6	7742.3	20,494	125,444	118.68	1.411,89	73.97	0.78	2.6	—	—	
Cyclopentane	3512.0	3763.7	20,188	126,296	167.34	1.406,45	35.79	(1.4)	—	84.9	+0.1	
Methylcyclopentane	4198.4	4500.4	20,130	126,477	147.83	1.409,70	42.95	(1.2)	8.35	80.0	91.3	
Cyclohexane	4178.8	4480.8	20,035	130,849	153.0	1.426,23	42.95	1.3	7.8	77.2	83.0	
Methylcyclohexane	4862.8	5215.2	20,001	129,066	136.3	1.423,12	50.11	1.2	—	71.1	74.8	
Ethylene	1499.0	1599.7	—	—	207.57	—	14.32	2.7	34.0	75.6	+.03	
Propene	2182.7	2333.7	—	—	188.18	—	21.48	2.0	10.0	84.9	+0.2	
1-Butene	2879.4	3080.7	20,678	103,659	167.94	—	28.63	1.6	9.3	80.8	97.4	
Cis-2-Butene	2871.7	3073.1	20,611	107,754	178.91	—	28.63	(1.6)	—	83.5	100.	
Trans-2-Butene	2866.8	3068.2	20,584	104,690	174.39	—	28.63	(1.6)	—	—	—	
Isobutene	2860.4	2061.8	20,548	102,863	169.48	—	28.63	(1.6)	—	—	—	
1-Pentene	3575.2	3826.9	20,548	110,610	154.46	1.371,48	35.79	1.4	8.7	77.1	90.9	
1,2-Butadiene	2789.0	2940.0	20,447	112,172	(181.)	—	26.25	(2.0)	(12.)	—	—	
1,3-Butadiene	2730.0	2881.0	20,047	104,826	(174.)	—	26.25	2.0	11.5	—	—	
Isoprene	3410.8	3612.1	19,964	114,194	(153.)	1.421,94	33.41	(1.5)	—	81.0	99.1	
Acetylene	1422.4	1472.8	—	—	—	—	11.93	2.5	80.	—	—	
Benzene	3590.7	3741.7	17,992	132,655	169.31	1.501,12	35.79	1.3	7.9	+2.8	—	
Toluene	4273.3	4474.7	18,252	132,656	154.84	1.496,93	42.95	1.2	7.1	+0.3	+5.8	
Ethylbenzene	4970.0	5221.7	18,434	134,414	144.0	1.495,88	50.11	0.99	6.7	97.9	+0.8	
o-Xylene	4958.3	5210.0	18,445	136,069	149.	1.505,45	50.11	1.1	6.4	100.	—	
m-Xylene	4956.8	5208.5	18,441	133,568	147.2	1.497,22	50.11	1.1	6.4	+2.8	+4.0	
p-Xylene	4956.9	5208.5	18,445	133,136	144.52	1.495,82	50.11	1.1	6.6	+1.2	+3.4	
Styrene	4828.7	5030.0	18,150	137,849	(151.)	1.546,82	47.72	1.1	6.1	+0.2	>+3.	
Isopropylbenzene	5661.4	5963.4	18,665	134,817	134.3	1.491,45	57.27	0.88	6.5	99.3	+2.1	

Appendix A   591

Appendix A-2 Continued
General Data for Natural Gas Constituents

Compound	Calorific value, 60°F				Heat of vaporization, 14.696 psia at boiling point, Btu/lb	Refractive index, nD 68°F	Air required for combustion ideal gas cu ft/cu ft	Flammability Limits, Vol % in Air Mixture		ASTM Octane Number	
	Net		Gross					Lower	Higher	Motor method D-357	Research method D-908
	Btu/cu ft, Ideal gas, 14.696 psia (20)	Btu/lb liquid	Btu/cu ft, Ideal gas, 14.696 psia (20)	Btu/gal liquid (wt in vacuum)							
Methyl alcohol	—	—	9,760	64,771	473.	1.3288	7.16	6.72	36.50	—	—
Ethyl alcohol	—	—	12,780	84,600	367.	1.3614	14.32	3.28	18.95	—	—
Carbon monoxide	—	321.	—	—	92.7	—	2.39	12.50	74.20	—	—
Carbon dioxide	—	—	—	—	238.2	—	—	—	—	—	—
Hydrogen sulfide	588.	637.	—	—	235.6	—	7.16	4.30	45.50	—	—
Sulfur dioxide	—	—	—	—	166.7	—	—	—	—	—	—
Ammonia	359.	434.	—	—	587.2	—	3.58	15.50	27.00	—	—
Air	—	—	—	—	92.	—	—	—	—	—	—
Hydrogen	274.	324.	—	—	193.9	—	2.39	4.00	74.20	—	—
Oxygen	—	—	—	—	91.6	—	—	—	—	—	—
Nitrogen	—	—	—	—	87.8	—	—	—	—	—	—
Chlorine	—	—	—	—	123.8	—	—	—	—	—	—
Water	—	—	—	—	970.3	1.3330	—	—	—	—	—
Helium	—	—	—	—	—	—	—	—	—	—	—
Hydrogen Chloride	—	—	—	—	185.5	—	—	—	—	—	—

Appendix A-3
An Overview of the SI System of Units

SI Prefixes

Factor	Prefix	Symbol
10^{18}	exa	E
10^{15}	peta	P
10^{12}	tera	T
10^{9}	giga	G
10^{6}	mega	M
10^{3}	kilo	k
10^{2}	hecto	h
10	deka	da
10^{-1}	deci	d
10^{-2}	centi	c
10^{-3}	milli	m
10^{-6}	micro	μ
10^{-9}	nano	n
10^{-12}	pico	p
10^{-15}	femto	f
10^{-18}	atto	a

(table continued)

Appendix A-3 Continued
An Overview of the SI System of Units

SI Base Quantities and Units

Quantity	SI unit	Symbol
length	meter	m
mass	kilogram	kg
time	second	s
temperature	kelvin	K
amount of substance	mole	mol

The two other base SI units are: *ampere* for electric current, and *candela* for luminous intensity.

Some Common SI Derived Units Used in Gas Engineering

Quantity	Unit	Symbol	Formula
acceleration	meter per second squared	--	m/s^2
area	square meter	--	m^2
density	kilogram per cubic meter	--	kg/m^3
energy, work, quantity of heat	joule	J	N m
entropy	joule per kelvin	--	J/K
force	newton	N	$kg\ m/s^2$
frequency	hertz	Hz	1/s
power	watt	W	J/s
pressure	pascal	Pa	N/m^2
specific heat	joule per kilogram kelvin	--	J/(kg K)
thermal conductivity	watt per meter kelvin	--	W/(m K)
velocity	meter per second	--	m/s
viscosity, dynamic	pascal second	--	Pa s
viscosity, kinematic	square meter per second	--	m^2/s
volume	cubic meter	--	m^3

Appendix A-4
An Overview of the Commonly Used Oil-Field Units

Quantity	Unit	Symbol	Equivalent to
Length	inch	in	--
	foot	ft	12 in
	yard	yd	3 ft
	mile	mi	5280 ft
Area	square (sq.) foot	ft^2	144 in^2
	sq. yard	yd^2	9 ft^2
	acre	acre	43560 ft^2
			4840 yd^2
	sq. mile	mi^2	640 acres
			27,878,400 ft^2
Volume	pint(U.S.)	pt	--
	quart(U.S.)	qt	2 pt
	gallon(U.S.)	gal	4 qt
			0.133 680 6 ft^3
	cubic (cu.) foot	ft^3	1728 in^3
			7.480 520 gal
	barrel(U.S.)	bbl	5.614 584 ft^3
			42 gal
	acre-foot	acre-ft	43560 ft^3
Time	second	s	--
	minute	min	60 s
	hour	hr	3600 s
	day	d	24 hr
			1440 min, or 86,400 s
	year	yr	365.25 d
			(also taken as 365 d)
Velocity	feet per second	ft/s	60 ft/min
			0.681 818 2 mph
	feet per minute	ft/min	60 ft/s
	miles per hour	mph	1.466 667 ft/s
			88 ft/min
Flow rate (volumetric)	cu. feet per second	ft^3/s	3600 ft^3/hr
			26929.872 gph
			15388.495 bpd
	gallons per hour	gal/hr (gph)	3.713 34 E-05 ft^3/s
	barrels per day	bbl/d (bpd)	6.498 36 E-05 ft^3/s

(table continued)

Appendix A-4 Continued
An Overview of the Commonly Used Oil-Field Units

Mass	grain	grain	--
	ounce (av)	oz	437.5 grains.
	ounce (troy)	oz	480 grains.
	pounds (mass)	lbm	16 oz (av)
	ton (short)	ton	2000 lbm
	ton (long)	ton	2240 lbm
Density	pounds per cu. inch	lbm/in^3	--
	pounds per cu. foot	lbm/ft^3	0.133 680 6 lbm/gal
liquids	pounds per gallon	lbm/gal	7.480 520 lbm/ft^3
	API gravity	°API	141.5/(specific gravity) − 131.5
Force	poundal	pdl	0.031 081 lbf
	pound force	lbf	32.174 05 pdls
Pressure	pounds per sq. inch	lbf/in^2	144 lbf/ft^2
			0.068 045 7 atm
	pounds per sq. foot	lbf/ft^2	0.006 944 lbf/in^2
Viscosity	pounds-second per sq. inch	lbf-s/in^2	4633.0632 lbm/(ft-s)
(dynamic)	pounds-second per sq. foot	lbf-s/ft^2	32.174 05 lbm/(ft-s)
	pounds (mass) per ft-s	lbm/ft-s	2.158 39 E−04 lbm/(ft-s)
Viscosity	sq. foot per second	ft^2/s	144 in^2/s)
(kinematic)	sq. inch per second	in^2/s	6.944 444 E−03 in^2/s)
Temperature	absolute, degrees Rankine	°R	°F + 459.67
	degrees Fahrenheit	°F	°R − 459.67
Energy, work	foot-pound (force)	ft-lbf	--
	British thermal unit	Btu	778.169 2 ft-lbf
	horsepower-hour	hp-hr	1.98 E+06 ft-lbf
	therm	therm	1.0 E+05 Btu
	quad	quad	1.0 E+10 therm
Power	horsepower	hp	550 ft-lbf/s
			33,000 ft-lbf/min
			42.407 Btu/min
	Btu per minute	Btu/min	778.169 2 ft-lbf/min

Appendix A-5
"Memory Jogger" for Metric Units

Customary Unit		"BallPark" Metric Values; (Do *Not* Use As Conversion Factors)
acre	4000	square meters
	0.4	hectare
barrel	0.16	cubic meter
British thermal unit	1000	joules
British thermal unit per pound-mass	2300	joules per kilogram
	2.3	kilojoules per kilogram
calorie	4	joules
centipoise	1*	millipascal-second
centistokes	1*	square millimeter per second
darcy	1	square micrometer
degree Fahrenheit (temperature *difference*)	0.5	kelvin
dyne per centimeter	1*	millinewton per meter
foot	30	centimeters
	0.3	meter
cubic foot (cu ft)	0.03	cubic meter
cubic foot per pound-mass (ft³/lbm)	0.06	cubic meter per kilogram
square foot (sq ft)	0.1	square meter
foot per minute	0.3	meter per minute
	5	millimeters per second
foot-pound-force	1.4	joules
foot-pound-force per minute	0.02	watt
foot-pound-force per second	1.4	watts
horsepower	750	watts (¾ kilowatt)
horsepower, boiler	10	kilowatts
inch	2.5	centimeters
kilowatthour	3.6*	megajoules
mile	1.6	kilometers
ounce (avoirdupois)	28	grams
ounce (fluid)	30	cubic centimeters
pound-force	4.5	newtons
pound-force per square inch (pressure, psi)	7	kilopascals
pound-mass	0.5	kilogram
pound-mass per cubic foot	16	kilograms per cubic meter
	260	hectares
section	2.6	million square meters
	2.6	square kilometers
ton, long (2240 pounds-mass)	1000	kilograms
ton, metric (tonne)	1000*	kilograms
ton, short	900	kilograms
*Exact equivalents		

* (From *The SI Metric System of Units and SPE Metric Standard*, 1984; courtesy of *Society of Petroleum Engineers*, Richardson, Texas.)

Gas Production Engineering

Appendix A-6
Alphabetical List of Units (Oil Field to SI Units Conversions)

To Convert From	To	Multiply By**	
abampere	ampere (A)	1.0*	E+01
abcoulomb	coulomb (C)	1.0*	E+01
abfarad	farad (F)	1.0*	E+09
abhenry	henry (H)	1.0*	E−09
abmho	siemens (S)	1.0*	E+09
abohm	ohm (Ω)	1.0*	E−09
abvolt	volt (V)	1.0*	E−08
acre-foot (U.S. survey)[1]	meter^3 (m^3)	1.233 489	E+03
acre (U.S. survey)[1]	meter^2 (m^2)	4.046 873	E+03
ampere hour	coulomb (C)	3.6*	E+03
are	meter^2 (m^2)	1.0*	E+02
angstrom	meter (m)	1.0*	E−10
astronomical unit	meter (m)	1.495 979	E+11
atmosphere (standard)	pascal (Pa)	1.013 250*	E+05
atmosphere (technical = 1 kgf/cm^2)	pascal (Pa)	9.806 650*	E+04
bar	pascal (Pa)	1.0*	E+05
barn	meter^2 (m^2)	1.0*	E−28
barrel (for petroleum, 42 gal)	meter^3 (m^3)	1.589 873	E−01
board foot	meter^3 (m^3)	2.359 737	E−03
British thermal unit (International Table)[2]	joule (J)	1.055 056	E+03
British thermal unit (mean)	joule (J)	1.055 87	E+03
British thermal unit (thermochemical)	joule (J)	1.054 350	E+03
British thermal unit (39°F)	joule (J)	1.059 67	E+03
British thermal unit (59°F)	joule (J)	1.054 80	E+03
British thermal unit (60°F)	joule (J)	1.054 68	E+03
Btu (International Table)-ft/(hr-ft^2-°F) (thermal conductivity)	watt per meter kelvin [W/(m·K)]	1.730 735	E+00
Btu (thermochemical)-ft/(hr-ft^2-°F) (thermal conductivity)	watt per meter kelvin [W/(m·K)]	1.729 577	E+00
Btu (International Table)-in./(hr-ft^2-°F) (thermal conductivity)	watt per meter kelvin [W/(m·K)]	1.442 279	E−01
Btu (thermochemical)-in./(hr-ft^2-°F) (thermal conductivity)	watt per meter kelvin [W/(m·K)]	1.441 314	E−01
Btu (International Table)-in./(s-ft^2-°F) (thermal conductivity)	watt per meter kelvin [W/(m·K)]	5.192 204	E+02
Btu (thermochemical)-in./(s-ft^2-°F) (thermal conductivity)	watt per meter kelvin [W/(m·K)]	5.188 732	E+02
Btu (International Table)/hr	watt (W)	2.930 711	E−01
Btu (thermochemical)/hr	watt (W)	2.928 751	E−01
Btu (thermochemical)/min	watt (W)	1.757 250	E+01
Btu (thermochemical)/s	watt (W)	1.054 350	E+03
Btu (International Table)/ft^2	joule per meter^2 (J/m^2)	1.135 653	E+04
Btu (thermochemical)/ft^2	joule per meter^2 (J/m^2)	1.134 893	E+04
Btu (thermochemical)/(ft^2-hr)	watt per meter^2 (W/m^2)	3.152 481	E+00
Btu (thermochemical)/(ft^2-min)	watt per meter^2 (W/m^2)	1.891 489	E+02
Btu (thermochemical)/(ft^2-s)	watt per meter^2 (W/m^2)	1.134 893	E+04
Btu (thermochemical)/(in.^{2}-s)	watt per meter^2 (W/m^2)	1.634 246	E+06
Btu (International Table)/(hr-ft^2-°F) (thermal conductance)	watt per meter^2 kelvin [W/(m^2·K)]	5.678 263	E+00
Btu (thermochemical)/(hr-ft^2-°F) (thermal conductance)	watt per meter^2 kelvin [W/(m^2·K)]	5.674 466	E+00
Btu (International Table)/(s-ft^2-°F)	watt per meter^2 kelvin [W/(m^2·K)]	2.044 175	E+04
Btu (thermochemical)/(s-ft^2-°F)	watt per meter^2 kelvin [W/(m^2·K)]	2.042 808	E+04

**See footnote on Page 13.

[1]Since 1893 the U.S. basis of length measurement has been derived from metric standards. In 1959 a small refinement was made in the definition of the yard to resolve discrepancies both in this country and abroad, which changed its length from 3600/3937 m to 0.9144 m exactly. This resulted in the new value being shorter by two parts in a million. At the same time it was decided that any data in feet derived from and published as a result of geodetic surveys within the U.S. would remain with the old standard (1 ft = 1200/3937 m) until further decision. This foot is named the U.S. survey foot. As a result, all U.S. land measurements in U.S. customary units will relate to the meter by the old standard. All the conversion factors in these tables for units referenced to this footnote are based on the U.S. survey foot, rather than the international foot. Conversion factors for the land measure given below may be determined from the following relationships:
    1 league = 3 miles (exactly)
    1 rod = 16½ ft (exactly)
    1 chain = 66 ft (exactly)
    1 section = 1 sq mile
    1 township = 36 sq miles

[2]This value was adopted in 1956. Some of the older International Tables use the value 1.055 04 E+03. The exact conversion factor is 1.055 055 852 62* E+03.

* (From *The SI Metric System of Units and SPE Metric Standard*, 1984; courtesy of Society of Petroleum Engineers, Richardson, Texas.)

Appendix A   597

Appendix A-6 Continued
Alphabetical List of Units (Oil Field to SI Units Conversions)

To Convert From	To	Multiply By**
Btu (International Table)/lbm	joule per kilogram (J/kg)	2.326* E+03
Btu (thermochemical)/lbm	joule per kilogram (J/kg)	2.324 444 E+03
Btu (International Table)/(lbm·°F) (heat capacity)	joule per kilogram kelvin [J/(kg·K)]	4.186 8* E+03
Btu (thermochemical)/(lbm·°F) (heat capacity)	joule per kilogram kelvin [J/(kg·K)]	4.184 000 E+03
bushel (U.S.)	meter³ (m³)	3.523 907 E−02
caliber (inch)	meter (m)	2.54* E−02
calorie (International Table)	joule (J)	4.186 8* E+00
calorie (mean)	joule (J)	4.190 02 E+00
calorie (thermochemical)	joule (J)	4.184* E+00
calorie (15°C)	joule (J)	4.185 80 E+00
calorie (20°C)	joule (J)	4.181 90 E+00
calorie (kilogram, International Table)	joule (J)	4.186 8* E+03
calorie (kilogram, mean)	joule (J)	4.190 02 E+03
calorie (kilogram, thermochemical)	joule (J)	4.184* E+03
cal (thermochemical)/cm²	joule per meter² (J/m²)	4.184* E+04
cal (International Table)/g	joule per kilogram (J/kg)	4.186* E+03
cal (thermochemical)/g	joule per kilogram (J/kg)	4.184* E+03
cal (International Table)/(g·°C)	joule per kilogram kelvin [J/(kg·K)]	4.186 8* E+03
cal (thermochemical)/(g·°C)	joule per kilogram kelvin [J/(kg·K)]	4.184* E+03
cal (thermochemical)/min	watt (W)	6.973 333 E−02
cal (thermochemical)/s	watt (W)	4.184* E+00
cal (thermochemical)/(cm²·min)	watt per meter² (W/m²)	6.973 333 E+02
cal (thermochemical)/(cm²·s)	watt per meter² (W/m²)	4.184* E+04
cal (thermochemical)/(cm·s·°C)	watt per meter kelvin [W/(m·K)]	4.184* E+02
capture unit (c.u. = 10^{-3} cm^{-1})	per meter (m^{-1})	1.0* E−01
carat (metric)	kilogram (kg)	2.0* E−04
centimeter of mercury (0°C)	pascal (Pa)	1.333 22 E+03
centimeter of water (4°C)	pascal (Pa)	9.806 38 E+01
centipoise	pascal second (Pa·s)	1.0* E−03
centistokes	meter² per second (m²/s)	1.0* E−06
circular mil	meter² (m²)	5.067 075 E−10
clo	kelvin meter² per watt [(K·m²)/W]	2.003 712 E−01
cup	meter³ (m³)	2.365 882 E−04
curie	becquerel (Bq)	3.7* E+10
cycle per second	hertz (Hz)	1.0* E+00
day (mean solar)	second (s)	8.640 000 E+04
day (sidereal)	second (s)	8.616 409 E+04
degree (angle)	radian (rad)	1.745 329 E−02
degree Celsius	kelvin (K)	$T_K = T_{°C} + 273.15$
degree centigrade (see degree Celsius)		
degree Fahrenheit	degree Celsius	$T_{°C} = (T_{°F} − 32)/1.8$
degree Fahrenheit	kelvin (K)	$T_K = (T_{°F} + 459.67)/1.8$
degree Rankine	kelvin (K)	$T_K = T_{°R}/1.8$
°F·hr·ft²/Btu (International Table) (thermal resistance)	kelvin meter² per watt [(K·m²)/W]	1.781 102 E−01
°F·hr·ft²/Btu (thermochemical) (thermal resistance)	kelvin meter² per watt [(K·m²)/W]	1.762 250 E−01
denier	kilogram per meter (kg/m)	1.111 111 E−07
dyne	newton (N)	1.0* E−05
dyne·cm	newton meter (N·m)	1.0* E−07
dyne/cm²	pascal (Pa)	1.0* E−01
electronvolt	joule (J)	1.602 19 E−19
EMU of capacitance	farad (F)	1.0* E+09
EMU of current	ampere (A)	1.0* E+01
EMU of electric potential	volt (V)	1.0* E−08
EMU of inductance	henry (H)	1.0* E−09
EMU of resistance	ohm (Ω)	1.0* E−09
ESU of capacitance	farad (F)	1.112 650 E−12
ESU of current	ampere (A)	3.335 6 E−10
ESU of electric potential	volt (V)	2.997 9 E+02
ESU of inductance	henry (H)	8.987 554 E+11
ESU of resistance	ohm (Ω)	8.987 554 E+11

(table continued)

Appendix A-6 Continued
Alphabetical List of Units (Oil Field to SI Units Conversions)

To Convert From	To	Multiply By**	
erg	joule (J)	1.0*	E−07
erg/cm²·s	watt per meter² (W/m²)	1.0*	E−03
erg/s	watt (W)	1.0*	E−07
faraday (based on carbon-12)	coulomb (C)	9.648 70	E+04
faraday (chemical)	coulomb (C)	9.649 57	E+04
faraday (physical)	coulomb (C)	9.652 19	E+04
fathom	meter (m)	1.828 8	E+00
fermi (femtometer)	meter (m)	1.0*	E−15
fluid ounce (U.S.)	meter³ (m³)	2.957 353	E−05
foot	meter (m)	3.048*	E−01
foot (U.S. survey)[1]	meter (m)	3.048 006	E−01
foot of water (39.2°F)	pascal (Pa)	2.988 98	E+03
sq ft	meter² (m²)	9.290 304*	E−02
ft²/hr (thermal diffusivity)	meter² per second (m²/s)	2.580 640*	E−05
ft²/s	meter² per second (m²/s)	9.290 304*	E−02
cu ft (volume; section modulus)	meter³ (m³)	2.831 685	E−02
ft³/min	meter³ per second (m³/s)	4.719 474	E−04
ft³/s	meter³ per second (m³/s)	2.831 685	E−02
ft⁴ (moment of section)[3]	meter⁴ (m⁴)	8.630 975	E−03
ft/hr	meter per second (m/s)	8.466 667	E−05
ft/min	meter per second (m/s)	5.080*	E−03
ft/s	meter per second (m/s)	3.048*	E−01
ft/s²	meter per second² (m/s²)	3.048*	E−01
footcandle	lux (lx)	1.076 391	E+01
footlambert	candela per meter² (cd/m²)	3.426 259	E+00
ft-lbf	joule (J)	1.355 818	E+00
ft-lbf/hr	watt (W)	3.766 161	E−04
ft-lbf/min	watt (W)	2.259 697	E−02
ft-lbf/s	watt (W)	1.355 818	E+00
ft-poundal	joule (J)	4.214 011	E−02
free fall, standard (g)	meter per second² (m/s²)	9.806 650*	E+00
cm/s²	meter per second² (m/s²)	1.0*	E−02
gallon (Canadian liquid)	meter³ (m³)	4.546 090	E−03
gallon (U.K. liquid)	meter³ (m³)	4.546 092	E−03
gallon (U.S. dry)	meter³ (m³)	4.404 884	E−03
gallon (U.S. liquid)	meter³ (m³)	3.785 412	E−03
gal (U.S. liquid)/day	meter³ per second (m³/s)	4.381 264	E−08
gal (U.S. liquid)/min	meter³ per second (m³/s)	6.309 020	E−05
gal (U.S. liquid)/hp·hr (SFC, specific fuel consumption)	meter³ per joule (m³/J)	1.410 089	E−09
gamma (magnetic field strength)	ampere per meter (A/m)	7.957 747	E−04
gamma (magnetic flux density)	tesla (T)	1.0*	E−09
gauss	tesla (T)	1.0*	E−04
gilbert	ampere (A)	7.957 747	E−01
gill (U.K.)	meter³ (m³)	1.420 654	E−04
gill (U.S.)	meter³ (m³)	1.182 941	E−04
grad	degree (angular)	9.0*	E−01
grad	radian (rad)	1.570 796	E−02
grain (1/7000 lbm avoirdupois)	kilogram (kg)	6.479 891*	E−05
grain (lbm avoirdupois/7000)/gal (U.S. liquid)	kilogram per meter³ (kg/m³)	1.711 806	E−02
gram	kilogram (kg)	1.0*	E−03
g/cm³	kilogram per meter³ (kg/m³)	1.0*	E+03
gram-force/cm²	pascal (Pa)	9.806 650*	E+01
hectare	meter² (m²)	1.0*	E+04
horsepower (550 ft-lbf/s)	watt (W)	7.456 999	E+02
horsepower (boiler)	watt (W)	9.809 50	E+03
horsepower (electric)	watt (W)	7.460*	E+02
horsepower (metric)	watt (W)	7.354 99	E+02
horsepower (water)	watt (W)	7.460 43	E+02
horsepower (U.K.)	watt (W)	7.457 0	E+02
hour (mean solar)	second (s)	3.600 000	E+03
hour (sidereal)	second (s)	3.590 170	E+03

[3] This sometimes is called the moment of inertia of a plane section about a specified axis.
[4] The exact conversion factor is 1.638 706 4*E−05.

Appendix A 599

Appendix A-6 Continued
Alphabetical List of Units (Oil Field to SI Units Conversions)

To Convert From	To	Multiply By**
hundredweight (long)	kilogram (kg)	5.080 235 E+01
hundredweight (short)	kilogram (kg)	4.535 924 E+01
inch	meter (m)	2.54* E-02
inch of mercury (32°F)	pascal (Pa)	3.386 38 E+03
inch of mercury (60°F)	pascal (Pa)	3.376 85 E+03
inch of water (39.2°F)	pascal (Pa)	2.490 82 E+02
inch of water (60°F)	pascal (Pa)	2.488 4 E+02
sq in.	meter^2 (m^2)	6.451 6* E-04
cu in. (volume; section modulus)$^{(4)}$	meter^3 (m^3)	1.638 706 E-05
in.^{3}/min	meter^3 per second (m^3/s)	2.731 177 E-07
in.^{4} (moment of section)$^{(3)}$	meter^4 (m^4)	4.162 314 E-07
in./s	meter per second (m/s)	2.54* E-02
in./s^2	meter per second^2 (m/s^2)	2.54* E-02
kayser	1 per meter (1/m)	1.0* E+02
kelvin	degree Celsius	$T_{°C} = T_K - 273.15$
kilocalorie (International Table)	joule (J)	4.186 8* E+03
kilocalorie (mean)	joule (J)	4.190 02 E+03
kilocalorie (thermochemical)	joule (J)	4.184* E+03
kilocalorie (thermochemical)/min	watt (W)	6.973 333 E+01
kilocalorie (thermochemical)/s	watt (W)	4.184* E+03
kilogram-force (kgf)	newton (N)	9.806 65* E+00
kgf·m	newton meter (N·m)	9.806 65* E+00
kgf·s^2/m (mass)	kilogram (kg)	9.806 65* E+00
kgf/cm^2	pascal (Pa)	9.806 65* E+04
kgf/m^2	pascal (Pa)	9.806 65* E+00
kgf/mm^2	pascal (Pa)	9.806 65* E+06
km/h	meter per second (m/s)	2.777 778 E-01
kilopond	newton (N)	9.806 65* E+00
kilowatthour (kW-hr)	joule (J)	3.6* E+06
kip (1000 lbf)	newton (N)	4.448 222 E+03
kip/in.^{2} (ksi)	pascal (Pa)	6.894 757 E+06
knot (international)	meter per second (m/s)	5.144 444 E-01
lambert	candela per meter^2 (cd/m^2)	$1/\pi$* E+04
lambert	candela per meter^2 (cd/m^2)	3.183 099 E+04
langley	joule per meter^2 (J/m^2)	4.184* E+04
league	meter (m)	(see Footnote 1)
light year	meter (m)	9.460 55 E+15
liter$^{(5)}$	meter^3 (m^3)	1.0* E-03
maxwell	weber (Wb)	1.0* E-08
mho	siemens (S)	1.0* E+00
microinch	meter (m)	2.54* E-08
microsecond/foot (μs/ft)	microsecond/meter (μs/m)	3.280 840 E+00
micron	meter (m)	1.0* E-06
mil	meter (m)	2.54* E-05
mile (international)	meter (m)	1.609 344* E+03
mile (statute)	meter (m)	1.609 3 E+03
mile (U.S. survey)$^{(1)}$	meter (m)	1.609 347 E+03
mile (international nautical)	meter (m)	1.852* E+03
mile (U.K. nautical)	meter (m)	1.853 184* E+03
mile (U.S. nautical)	meter (m)	1.852* E+03
sq mile (international)	meter^2 (m^2)	2.589 988 E+06
sq mile (U.S. survey)$^{(1)}$	meter^2 (m^2)	2.589 998 E+06
mile/hr (international)	meter per second (m/s)	4.470 4* E-01
mile/hr (international)	kilometer per hour (km/h)	1.609 344* E+00
mile/min (international)	meter per second (m/s)	2.682 24* E+01
mile/s (international)	meter per second (m/s)	1.609 344* E+03
millibar	pascal (Pa)	1.0* E+02
millimeter of mercury (0°C)	pascal (Pa)	1.333 22 E+02
minute (angle)	radian (rad)	2.908 882 E-04

^{1}In 1964 the General Conference on Weights and Measures adopted the name liter as a special name for the cubic decimeter. Prior to this decision the liter differed slightly (previous value, 1.000 028 dm^3) and in expression of precision volume measurement this fact must be kept in mind.
^{2}Not the same as reservoir "perm."

(table continued)

600 Gas Production Engineering

Appendix A-6 Continued
Alphabetical List of Units (Oil Field to SI Units Conversions)

To Convert From	To	Multiply By**
minute (mean solar)	second (s)	6.0* E+01
minute (sidereal)	second (s)	5.983 617 E+01
month (mean calendar)	second (s)	2.628 000 E+06
oersted	ampere per meter (A/m)	7.957 747 E+01
ohm centimeter	ohm meter ($\Omega \cdot m$)	1.0* E-02
ohm circular-mil per ft	ohm millimeter^2 per meter [($\Omega \cdot mm^2$)m]	1.662 426 E-03
ounce (avoirdupois)	kilogram (kg)	2.834 952 E-02
ounce (troy or apothecary)	kilogram (kg)	3.110 348 E-02
ounce (U.K. fluid)	meter^3 (m^3)	2.841 307 E-05
ounce (U.S. fluid)	meter^3 (m^3)	2.957 353 E-05
ounce-force	newton (N)	2.780 139 E-01
ozf·in.	newton meter (N·m)	7.061 552 E-03
oz (avoirdupois)/gal (U.K. liquid)	kilogram per meter^3 (kg/m^3)	6.236 021 E+00
oz (avoirdupois)/gal (U.S. liquid)	kilogram per meter^3 (kg/m^3)	7.489 152 E+00
oz (avoirdupois)/in.^{3}	kilogram per meter^3 (kg/m^3)	1.729 994 E+03
oz (avoirdupois)/ft^2	kilogram per meter^2 (kg/m^2)	3.051 517 E-01
oz (avoirdupois)/yd^2	kilogram per meter^2 (kg/m^2)	3.390 575 E-02
parsec	meter (m)	3.085 678 E+16
peck (U.S.)	meter^3 (m^3)	8.809 768 E-03
pennyweight	kilogram (kg)	1.555 174 E-03
perm (°C)[6]	kilogram per pascal second meter^2 [kg/(Pa·s·m^2)]	5.721 35 E-11
perm (23°C)[6]	kilogram per pascal second meter^2 [kg/(Pa·s·m^2)]	5.745 25 E-11
perm·in. (0°C)[7]	kilogram per pascal second meter [kg/(Pa·s·m)]	1.453 22 E-12
perm·in. (23°C)[7]	kilogram per pascal second meter [km/(Pa·s·m)]	1.459 29 E-12
phot	lumen per meter^2 (lm/m^2)	1.0* E+04
pica (printer's)	meter (m)	4.217 518 E-03
pint (U.S. dry)	meter^3 (m^3)	5.506 105 E-04
pint (U.S. liquid)	meter^3 (m^3)	4.731 765 E-04
point (printer's)	meter (m)	3.514 598* E-04
poise (absolute viscosity)	pascal second (Pa·s)	1.0* E-01
pound (lbm avoirdupois)[8]	kilogram (kg)	4.535 924 E-01
pound (troy or apothecary)	kilogram (kg)	3.732 417 E-01
lbm-ft^2 (moment of inertia)	kilogram meter^2 (kg·m^2)	4.214 011 E-02
lbm-in.^{2} (moment of inertia)	kilogram meter^2 (kg·m^2)	2.926 397 E-04
lbm/ft-hr	pascal second (Pa·s)	4.133 789 E-04
lbm/ft-s	pascal second (Pa·s)	1.488 164 E+00
lbm/ft^2	kilogram per meter^2 (kg/m^2)	4.882 428 E+00
lbm/ft^3	kilogram per meter^3 (kg/m^3)	1.601 846 E+01
lbm/gal (U.K. liquid)	kilogram per meter^3 (kg/m^3)	9.977 633 E+01
lbm/gal (U.S. liquid)	kilogram per meter^3 (kg/m^3)	1.198 264 E+02
lbm/hr	kilogram per second (kg/s)	1.259 979 E-04
lbm/(hp · hr) (SFC, specific fuel consumption)	kilogram per joule (kg/J)	1.689 659 E-07
lbm/in.^{3}	kilogram per meter^3 (kg/m^3)	2.767 990 E+04
lbm/min	kilogram per second (kg/s)	7.559 873 E-03
lbm/s	kilogram per second (kg/s)	4.535 924 E-01
lbm/yd^3	kilogram per meter^3 (kg/m^3)	5.932 764 E-01
poundal	newton (N)	1.382 550 E-01
poundal/ft^2	pascal (Pa)	1.488 164 E+00
poundal-s/ft^2	pascal second (Pa·s)	1.488 164 E+00
pound-force (lbf)[9]	newton (N)	4.448 222 E+00
lbf-ft[10]	newton meter (N·m)	1.355 818 E+00
lbf-ft/in.[11]	newton meter per meter [(N·m)/m]	5.337 866 E+01

[7]Not the same dimensions as "millidarcy-foot."
[8]The exact conversion factor is 4.535 923 7*E-01.
[9]The exact conversion factor is 4.448 221 615 260 5*E+00.
[10]Torque unit; see text discussion of "Torque and Bending Moment."
[11]Torque divided by length; see text discussion of "Torque and Bending Moment."

Appendix A   601

Appendix A-6 Continued
Alphabetical List of Units (Oil Field to SI Units Conversions)

To Convert From	To	Multiply By**
lbf-in.[11]	newton meter (N·m)	1.129 848 E−01
lbf-in./in.[11]	newton meter per meter [(N·m)/m]	4.448 222 E+00
lbf-s/ft^2	pascal second (Pa·s)	4.788 026 E+01
lbf/ft	newton per meter (N/m)	1.459 390 E+01
lbf/ft^2	pascal (Pa)	4.788 026 E+01
lbf/in.	newton per meter (N/m)	1.751 268 E+02
lbf/in.^{2} (psi)	pascal (Pa)	6.894 757 E+03
lbf/lbm (thrust/weight [mass] ratio)	newton per kilogram (N/kg)	9.806 650 E+00
quart (U.S. dry)	meter^3 (m^3)	1.101 221 E−03
quart (U.S. liquid)	meter^3 (m^3)	9.463 529 E−04
rad (radiation dose absorbed)	gray (Gy)	1.0* E−02
rhe	1 per pascal second [1/(Pa·s)]	1.0* E+01
rod	meter (m)	(see Footnote 1)
roentgen	coulomb per kilogram (C/kg)	2.58 E−04
second (angle)	radian (rad)	4.848 137 E−06
second (sidereal)	second (s)	9.972 696 E−01
section	meter^2 (m^2)	(see Footnote 1)
shake	second (s)	1.000 000* E−08
slug	kilogram (kg)	1.459 390 E+01
slug/(ft·s)	pascal second (Pa·s)	4.788 026 E+01
slug/ft^3	kilogram per meter^3 (kg/m^3)	5.153 788 E+02
statampere	ampere (A)	3.335 640 E−10
statcoulomb	coulomb (C)	3.335 640 E−10
statfarad	farad (F)	1.112 650 E−12
stathenry	henry (H)	8.987 554 E+11
statmho	siemens (S)	1.112 650 E−12
statohm	ohm (Ω)	8.987 554 E+11
statvolt	volt (V)	2.997 925 E+02
stere	meter^3 (m^3)	1.0* E+00
stilb	candela per meter^2 (cd/m^2)	1.0* E+04
stokes (kinematic viscosity)	meter^2 per second (m^2/s)	1.0* E−04
tablespoon	meter^3 (m^3)	1.478 676 E−05
teaspoon	meter^3 (m^3)	4.928 922 E−06
tex	kilogram per meter (kg/m)	1.0* E−06
therm	joule (J)	1.055 056 E+08
ton (assay)	kilogram (kg)	2.916 667 E−02
ton (long, 2,240 lbm)	kilogram (kg)	1.016 047 E+03
ton (metric)	kilogram (kg)	1.0* E+03
ton (nuclear equivalent of TNT)	joule (J)	4.184 E+09[12]
ton (refrigeration)	watt (W)	3.516 800 E+03
ton (register)	meter^3 (m^3)	2.831 685 E+00
ton (short, 2000 lbm)	kilogram (kg)	9.071 847 E+02
ton (long)/yd^3	kilogram per meter^3 (kg/m^3)	1.328 939 E+03
ton (short)/hr	kilogram per second (kg/s)	2.519 958 E−01
ton-force (2000 lbf)	newton (N)	8.896 444 E+03
tonne	kilogram (kg)	1.0* E+03
torr (mm Hg, 0°C)	pascal (Pa)	1.333 22 E+02
township	meter^2 (m^2)	(see Footnote 1)
unit pole	weber (Wb)	1.256 637 E−07
watthour (W-hr)	joule (J)	3.60* E+03
W·s	joule (J)	1.0* E+00
W/cm^2	watt per meter^2 (W/m^2)	1.0* E+04
W/in.^{2}	watt per meter^2 (W/m^2)	1.550 003 E+03
yard	meter (m)	9.144* E−01
yd^2	meter^2 (m^2)	8.361 274 E−01
yd^3	meter^3 (m^3)	7.645 549 E−01
yd^3/min	meter^3 per second (m^3/s)	1.274 258 E−02
year (calendar)	second (s)	3.153 600 E+07
year (sidereal)	second (s)	3.155 815 E+07
year (tropical)	second (s)	3.155 693 E+07

[12]Defined (not measured) value.

Appendix A-7
Some Additional Application Standards

Quantity and SI Unit		Customary Unit	Metric Unit SPE Preferred	Metric Unit Other Allowable	Conversion Factor* Multiply Customary Unit by Factor to Get Metric Unit	
Capillary pressure	Pa	ft (fluid)	m (fluid)		3.048*	E−01
Compressibility of reservoir fluid	Pa^{-1}	psi^{-1}	Pa^{-1}	kPa^{-1}	1.450 377 / 1.450 377	E−04 / E−01
Corrosion allowance	m	in.	mm		2.54*	E+01
Corrosion rate	m/s	mil/yr (mpy)	mm/a		2.54*	E−02
Differential orifice pressure	Pa	in. H$_2$O (at 60°F)	kPa	cm H$_2$O	2.488 4 / 2.54*	E−01 / E+00
Gas-oil ratio	m^3/m^3	scf/bbl	"standard" m^3/m^3		1.801 175	E−01$^{(1)**}$
Gas rate	m^3/s	scf/D	"standard" m^3/d		2.863 640	E−02$^{(1)}$
Geologic time	s	yr	Ma			
Head (fluid mechanics)	m	ft	m	cm	3.048* / 3.048*	E−01 / E+01
Heat exchange rate	W	Btu/hr	kW	kJ/h	2.930 711 / 1.055 056	E−04 / E+00
Mobility	m^2/Pa·s	d/cp	μm^2/mPa·s	μm^2/Pa·s	9.869 233 / 9.869 233	E−01 / E+02
Net pay thickness	m	ft	m		3.048*	E−01
Oil rate	m^3/s	bbl/D	m^3/d		1.589 873	E−01
		short ton/yr	Mg/a	t/a	9.071 847	E−01
Particle size	m	micron	μm		1.0*	
Permeability-thickness	m^3	md·ft	md·m	μm^2·m	3.008 142	E−04
Pipe diameter (actual)	m	in.	cm	mm	2.54* / 2.54*	E+00 / E+01
Pressure buildup per cycle	Pa	psi	kPa		6.894 757	E+00$^{(2)}$
Productivity index	m^3/Pa·s	bbl/(psi·D)	m^3/(kPa·d)		2.305 916	E−02$^{(2)}$

Appendix A-7 Continued

Quantity and SI Unit		Customary Unit	Metric Unit		Conversion Factor* Multiply Customary Unit by Factor to Get Metric Unit	
			SPE Preferred	Other Allowable		
Pumping rate	m³/s	U.S. gal/min	m³/h	L/s	2.271 247 6.309 020	E−01 E−02
Revolutions per minute	rad/s	rpm	rad/s		1.047 198 6.283 185	E−01 E+00
Recovery/unit volume (oil)	m³/m³	bbl/(acre-ft)	m³/m³	m³/ha·m	1.288 931 1.288 931	E−04 E+00
Reservoir area	m²	sq mile	km²		2.589 988	E+00
		acre		ha	4.046 856	E−01
Reservoir volume	m³	acre-ft	m³	ha·m	1.233 482 1.233 482	E+03 E−01
Specific productivity index	m³/Pa·s·m	bbl/(D-psi-ft)	m³/(kPa·d·m)		7.565 341	E−02ᵃ⁾
Surface or interfacial tension in reservoir capillaries	N/m	dyne/cm	mN/m		1.0*	E+00
Torque	N·m	lbf-ft	N·m		1.355 818	E+00ᵃ⁾
Velocity (fluid flow)	m/s	ft/s	m/s		3.048*	E−01
Vessel diameter	m					
1-100 cm		in.	cm		2.54*	E+00
above 100 cm		ft	m		3.048*	E−01

*An asterisk indicates the conversion factor is exact using the numbers shown; all subsequent numbers are zeros.

Appendix A   603

* (From *The SI Metric System of Units and SPE Metric Standard*, 1984; courtesy of *Society of Petroleum Engineers*, Richardson, Texas.)

APPENDIX B
Computer Programs (FORTRAN Subroutines)

This section presents a few subroutines in FORTRAN that find use in developing any computer program for computations related to gas properties, production, and flow. Readers are encouraged to use these programs as they are, and also as a part of more complex programming efforts.

Subroutines

CRIPT	Critical pressure and temperature.
GASCOM	Gas compressibility, and compressibility factor.
GASVIS	Gas viscosity. Uses subroutine VISCOR for viscosity correction.
FRICF	Moody friction factor calculation.
HORIZF	Horizontal steady-state single-phase flow using various methods.
CSMITH	Cullender & Smith method for determining bottom-hole pressure.
VERTF	Sukkar & Cornell method for non-horizontal steady-state single-phase flow.
HAGBR	Hagedorn & Brown method for multiphase flow in pipes.
BEGBR	Beggs & Brill method for multiphase flow in pipes.

Auxiliary (Service) Routines

INTP1	Linear interpolation for a single-variable dependent parameter.
INTP2	Lagrange interpolation for a parameter dependent upon two variables.
NUMINT	Romberg numerical integration.

Appendix B 605

General Nomenclature for FORTRAN Code

APR = Absolute pipe roughness, inches.
ANG = Inclination angle of pipe from horizontal, degrees.
APGR = Acceleration pressure gradient, psi/ft.
CR = Reduced gas compressibility, dimensionless.
DENL = Liquid density, lbm/cubic ft.
DELTAP = Pressure difference between inlet and outlet for gas flow, psi.
DND = Dimensionless pipe diameter number.
DNFR = Dimensionless Froude number.
DNGV = Dimensionless gas velocity number.
DNL = Dimensionless liquid viscosity number.
DNLV = Dimensionless liquid velocity number.
DP = Inside pipe diameter, inches.
DPH = Depth of well, or elevation difference, ft.
DR = Reduced gas density, dimensionless.
EPGR = Elevation pressure gradient, psi/ft.
FF = Moody friction factor.
FLH = Liquid holdup, fraction.
FNSLH = No-slip liquid holdup, fraction.
FPGR = Friction pressure gradient, psi/ft.
GGRA = Gas gravity (air = 1 basis), dimensionless.
GDEN = Gas density, lbm/ft^3.
GVIS = Gas viscosity, cp.
IERR = Error indication in subroutine execution.
IFREG = Flow regime indicator.
IOU = I/O code for device on which output is desired (a file, line printer).
ITER = Iteration counter.
MITER = Maximum number of iterations allowed for a trial and error.
MW = Molecular weight, lbm/lbmole.
P = Pressure, psia.
P1 = Pressure at inlet end of pipe, psia.
P2 = Pressure at outlet end of pipe, psia.
PBH = Bottomhole pressure, psia.
PC = Critical pressure, psia.
PL = Length of pipe, ft.
PR = Pseudo-reduced pressure, dimensionless.
PWH = Wellhead pressure, psia.
QGSC = Gas flowrate at standard conditions (14.73 psia, °R), Mscfd.
RNUM = Reynolds number, dimensionless.

606   Gas Production Engineering

SIGMAL = Liquid interfacial tension, dynes/cm.
T = Temperature, °R.
T1 = Temperature at inlet end of pipe, °R.
T2 = Temperature at outlet end of pipe, °R.
TBH = Bottomhole temperature, °R.
TC = Critical temperature, °R.
TOL = Error tolerance between successive iterative solutions.
TPGR = Total pressure gradient, psi/ft (negative) = − (EGR + FGR + AGR).
TR = Pseudo-reduced temperature, dimensionless.
TWH = Wellhead temperature, °R.
VISL = Liquid viscosity, cp.
VM = Fluid (mixture) velocity, ft/sec.
Y = Gas composition (mole fractions in the sequence of Table 3-1) array.
Z = Gas compressibility factor, dimensionless.
ZC = Gas compressibility factor at the critical point (assumed 0.27).

FORTRAN Subroutines

```
C
      SUBROUTINE CRIPT(INDEX,Y,GGRA,PCM,TCM)
C
C-----------------------------------------------------------------
C This subroutine finds the critical pressure and temperature of a gas.
C      from its (i) composition, or (ii) gravity.
C INPUT: Y or GGRA, and
C      INDEX = 1 if gas composition Y is provided (GGRA is computed).
C            = 2 if GGRA is provided.
C OUTPUT:
C Mixture critical pressure, PCM (psia), and temperature, TCM (Rankine).
C DATA USED (for each component): MW, PC, and TC for each component.
C Note: If the gas is sour, supply the mole fractions
C Y(14) of CO2, and Y(15) of H2S, to enable Wichert & Aziz correction.
C-----------------------------------------------------------------
C
      REAL Y(18), MW(18), PC(18), TC(18)
      DATA MW/ 16.043,30.070,44.097,58.124,58.124,72.151,72.151,86.178,
     # 100.205,114.232,128.259,142.286,28.013,44.010,34.076,31.999,
     # 2.016,18.015 /
      DATA PC/ 667.8,707.8,616.3,550.7,529.1,488.6,490.4,436.9,396.8,
     # 360.6,332.0,304.0,493.0,1070.9,1306.0,737.1,188.2,3203.6 /
      DATA TC/ 343.1,549.8,665.7,765.4,734.7,845.4,828.8,913.4,972.5,
     # 1023.9,1070.4,1111.8,227.3,547.6,672.4,278.6,59.9,1165.1 /
```

```
C
      IF(INDEX.EQ.2) GO TO 40
      SUMA = 0.0
      PCM = 0.0
      TCM = 0.0
      DO 20 I=1,18
      SUMA = SUMA + Y(I)*MW(I)
      PCM = PCM + Y(I)*PC(I)
      TCM = TCM + Y(I)*TC(I)
   20 CONTINUE
      GGRA = SUMA/28.97
      GO TO 60
C
   40 PCM = 709.604 - 58.718*GGRA
      TCM = 170.491 + 307.344*GGRA
C
C  Apply Wichert & Aziz (1972) correction for sour gases.
C
   60 A = Y(14) + Y(15)
      B = Y(15)
      EPS = 120.0*(A**0.9 - A**1.6) + 15.0*(B**0.5 - B**4.0)
      TCM = TCM - EPS
      PCM = PCM*TCM/(TCM+EPS + B*EPS*(1.0-B))
C
      RETURN
      END
C
      SUBROUTINE GASCOM(PR,TR,Z,CR)
C
C -----------------------------------------------------------------
C  This subroutine calculates: (i) gas compressibility factor Z
C                         and (ii) reduced gas compressibility CR
C     in the range 0.2 <= PR <= 20.0, 0.7 <= TR <= 3.0.
C
C  METHOD USED: Equation of state methods as follows:
C  For TR <> 1, use DRANCHUK & ABOU-KASSEM's (1975) procedure.
C  For TR = 1, and PR >= 1, use YARBOROUGH & HALL's (1974) method.
C  For TR = 1, and PR < 1, use DRANCHUK ET AL.'s (1974) equation.
C    INPUT : PR, TR.
C    OUTPUT : Z, CR.
C -----------------------------------------------------------------
C
      IF(TR.NE.1.0) GO TO 103
      IF(PR.GE.1.0) GO TO 102
C
C----- DRANCHUK 8-COEFF --------------
C
      CALL ZFACT1(PR,TR,Z,CR)
      RETURN
C
C----- HALL & YARBOROUGH -------------
```

Appendix B   607

```
C
  102 CALL ZFACT2(PR,TR,Z,CR)
      RETURN
C
C----- DRANCHUK 11-COEFF --------------
C
  103 CALL ZFACT3(PR,TR,Z,CR)
      RETURN
      END
C
C ********************************************************************
C
      SUBROUTINE ZFACT1(PR,TR,Z,CR)
C
C ---------------------------------------------------------------------
C Applies Dranchuk et al. (1974) 8-coeff. equation.
C Uses Newton-Raphson method for guessing reduced density DR
C (represented here by the symbol Y), then calculates Comp. factor Z.
C ---------------------------------------------------------------------
C
      REAL Y(105), F(105), DFDY(105)
      DATA A1,A2,A3,A4,A5,A6,A7,A8/ 0.31506237, -1.04670990,-0.57832729,
     #0.53530771, -0.61232032, -0.10488813,0.68157001, 0.68446549/
C
      Y(1)= 0.3723
      TOL = 0.0001
      TR3=TR**3
C
      DO 10 I = 1, 100
      F(I) = Y(I) + (A1+A2/TR+A3/TR3)*(Y(I)**2.0) + (A4+A5/TR)
     # *(Y(I)**3.0) + A5*A6/TR*(Y(I)**6.0) + (A7/TR3)*(Y(I)**3.0)
     # *(1.0+A8*(Y(I)**2.0))*EXP(-A8*(Y(I)**2.0)) - 0.27*PR/TR
C
      DFDY(I) = 1.0 + 2.0*(A1+A2/TR+A3/TR3)*Y(I) + 3.0*(A4+A5/TR)
     # *(Y(I)**2.0) + (6.0*A5*A6/TR)*(Y(I)**5.0) + (A7/TR3)*(Y(I)
     # **2.0)*(3.0+3.0*A8*(Y(I)**2.0)-2.0*(A8**2.0)*(Y(I)**4.0))
     # *EXP(-A8*(Y(I)**2.0))
C
      Y(I+1) = Y(I)- F(I)/DFDY(I)
      ERR = ABS(Y(I+1)- Y(I))
      IF (ERR.LE.TOL) GO TO 20
   10 CONTINUE
C
   20 DR = Y(I+1)
      DR2 = DR**2.0
      Z = 1.0 + (A1+A2/TR+A3/TR3)*DR + (A4+ A5/TR)*DR2
     # + (A5*A6/TR)*(DR**5.0) + (A7/(TR**3.0))*DR2
     # *(1.0+ A8*DR2)*EXP(-A8*DR2)
C
C ----- Computations for CR --------------------------
C
      ZC = 0.270
```

```
      DR4 = DR**4.0
      DZDR = A1 + A2/TR + A3/TR3 + 2.0*(A4+A5/TR)*DR
    # + 5.0*A5*A6*DR4/TR + (2.0*A7*DR/TR3)
    # *(1.0+A8*DR2-(A8**2.0)*DR4)*EXP(-A8*DR2)
      CR = 1.0/PR - (ZC/((Z**2.0)*TR))*(DZDR/(1.0+(DR/Z)*DZDR))
C
      RETURN
      END
C
C ****************************************************************
C
      SUBROUTINE ZFACT2(PR,TR,Z,CR)
C
C ----------------------------------------------------------------
C Follows the procedure from Yarborough and Hall's (1974) paper.
C Nomenclature and methodology similar to subroutine ZFACT1.
C ----------------------------------------------------------------
C
      REAL Y(105), F(105), DFDY(105)
C
      Y(1) = 0.3723
      TOL = 0.0001
      TRI = 1.0/TR
      TRI2 = TRI**2.0
      TRI3 = TRI**3.0
C
      DO 10 I = 1, 100
      F(I) = -0.06125*(PR/TR)*EXP(-1.2*((1.0-TRI)**2.0))
    # + ((Y(I)+Y(I)**2.0+Y(I)**3.0-Y(I)**4.0)/((1.0-Y(I))**3.0))
    # - (14.76/TR-9.76*TRI2+4.58*TRI3)*(Y(I)**2.0)
    # + (90.7/TR-242.2*TRI2+42.4*TRI3)*(Y(I)**(2.18+ 2.82/TR))
C
      DFDY(I) = (1.0+ 4.0*Y(I) + 4.0*(Y(I)**2.0) - 4.0*(Y(I)**3.0)
    # + Y(I)**4.0)/((1.0- Y(I))**4.0) - (29.52/TR-19.52*TRI2
    # +9.16*TRI3)*Y(I) + (2.18+2.82/TR)*(90.7/TR-242.2*TRI2
    # +42.4*TRI3)*(Y(I)**(1.18+ 2.82/TR))
C
      Y(I+1) = Y(I)- F(I)/ DFDY(I)
      ERR = ABS(Y(I+1) - Y(I))
      IF (ERR.LE.TOL) GO TO 20
   10 CONTINUE
C
   20 DR = Y(I+1)
      Z = (0.06125*(PR/TR)*EXP(-1.2*((1.0-TRI)**2.0)))/DR
C
C ----- Computations for CR -------------------------
C
      ZC= 0.2700
      TR2 = TR**2.0
      TR3 = TR**3.0
      DZDY = (4+4*DR-2*(DR**2.0))/((1-DR)**4.0) - (14.76/TR-9.76/TR2
    # +4.58/TR3) + (90.7/TR-242.2/TR2+42.4/TR3)
```

```
      # *(1.18+2.82/TR)*(DR**(0.18+2.82/TR))
      CR = 1.0/PR - (ZC/((Z**2.0)*TR))*(DZDY/(1.0+(DR/Z)*DZDY))
C
      RETURN
      END
C
C ****************************************************************
C
      SUBROUTINE ZFACT3(PR,TR,Z,CR)
C
C -----------------------------------------------------------------
C Dranchuk & Abou-Kassem's (1975) procedure derived from Starling EOS,
C using 11-coeffs.
C -----------------------------------------------------------------
C
      DATA A1,A2,A3,A4,A5,A6,A7,A8,A9,A10,A11 / 0.3265,-1.0700,-0.5339,
     #0.01569,-0.05165,0.5475,-0.7361,0.1844,0.1056,0.6134, 0.7210 /
C
      ITER = 0
      J = 1
      DR = 1.0
      TR2 = TR**2
      TR3 = TR**3
      TR4 = TR**4
      C1 = A7+A8/TR
      C0 = A1*TR +A2 +A3/TR2 +A4/TR3 +A5/TR4
      C2 = A6*TR +C1
      C3 = -C1*A9
      C4 = A10/TR2
C
      IF (PR-30.0) 10,10,200
   10 IF (TR-1.0) 20,20,30
   20 J = 0
      DR = 0.0
      DELDR = 0.1
   30 IF (TR-3.0) 40,40,200
   40 DO 160 ITER = 1,100
      IF (J) 50,50,60
   50 DR1 = DR
      DR = DR+DELDR
   60 DR2 = DR**2
      DR5 = DR**5
      T1 = C0*DR
      T2 = C2*DR2
      T3 = C3*DR5
      T4 = C4*DR2
      T5 = A11*DR2
      T6 = EXP(-T5)
      P = (TR+T1+T2+T3)*DR +T4*DR*(1.0+T5)*T6
      DP = TR+2.0*T1+3.0*T2+6.0*T3+T4*T6*(3.0+3.0*T5-2.0*T5*T5)
      IF (J) 70,70,100
```

```
      70 PRCAL=P/0.27
         IF(ABS(PRCAL-PR)-0.001 ) 170,170,80
      80 IF (PRCAL-PR) 160,170,90
      90 DR = DR1
         DELDR = DELDR/2.0
         GO TO 160
     100 DR1 = DR-(P-0.270*PR)/DP
         IF (DR1) 110,110,120
     110 DR1 = 0.5*DR
     120 IF(DR1-2.2) 140,140,130
     130 DR1 = DR+0.9*(2.2-DR)
     140 IF( ABS(DR-DR1)-0.00001 ) 170,150,150
     150 DR=DR1
     160 CONTINUE
     170 Z = 0.270*PR/(DR*TR)
     200 CONTINUE
C
C ----- Computations for CR -------------------------
C
         ZC = 0.2700
         DR2 = DR**2.0
         DR4 = DR**4.0
         DZDR = A1 + A2/TR + A3/TR3 + A4/TR4 + A5/(TR**5.0)
       # + 2.0*(A6+A7/TR+A8/TR2)*DR - 5.0*A9*(A7/TR+A8/TR2)*DR4
       # + (A10*(DR2/TR3)+A10*A11*(DR4/TR3))*(-2.0*A11*DR)*EXP(-A11*DR2)
       # + (2.0*A10*DR/TR3+4.0*A10*A11*(DR**3.0)/TR3)*EXP(-A11*DR2)
         CR = 1.0/PR - (ZC/((Z**2.0)*TR))* (DZDR/(1.0+(DR/Z)*DZDR))
C
         RETURN
         END
C
         SUBROUTINE GASVIS(PR,TR,T,INDEX,Y,GGRA,GVIS,IERR)
C
C----------------------------------------------------------------------
C  Calculates the viscosity of a gas from its:
C     (i) composition, using Stiel and Thodos (1961) equation.
C     or (ii) gravity using Carr et al. (1954) method.
C     Pressure correction by Carr et al. (1954) is used.
C
C     Valid in the range of 1.05 <= TR <= 3.0, 0.10 <= PR <= 20.0,
C          500 <= T <= 860 (Rankine), and 0.55 <= GGRA <= 1.50.
C  INPUT:
C     PR, TR, T.
C     INDEX = 1 if gas composition is known.
C           = 2 if gas gravity only is known.
C     Component mole fractions, Y,
C  or GGRA alongwith the mole fractions of N2, CO2 and H2S.
C  OUTPUT: Gas viscosity GVIS at PR, TR, and
C     IERR = Error indication - beyond range of method.
C  SUBROUTINES USED: VISCOR, INTP2.
C----------------------------------------------------------------------
```

612   Gas Production Engineering

```
C
      REAL Y(18), MW(18), PC(18), TC(18), VCOMP(18)
      DATA MW/ 16.043,30.070,44.097,58.124,58.124,72.151,72.151,86.178,
     # 100.205,114.232,128.259,142.286,28.013,44.010,34.076,31.999,
     # 2.016,18.015 /
      DATA PC/ 667.8,707.8,616.3,550.7,529.1,488.6,490.4,436.9,396.8,
     # 360.6,332.0,304.0,493.0,1070.9,1306.0,737.1,188.2,3203.6 /
      DATA TC/ 343.1,549.8,665.7,765.4,734.7,845.4,828.8,913.4,972.5,
     # 1023.9,1070.4,1111.8,227.3,547.6,672.4,278.6,59.9,1165.1 /
C
C ---- VISCOSITY AT 1 ATM. & GIVEN TEMP. T ----------------
C
      IF(INDEX.EQ.2) GO TO 100
C
C --Gas viscosity at 1 atm. & given temp. T  using component data ---
C
      DO 20 I=1,18
      SI = TC(I)**(1./6.)/( (MW(I)**(0.5))*(PC(I)**(2./3.)) )
      TRI = T/TC(I)
      IF(TRI.GE.1.5) GO TO 10
      VCOMP(I) = (3.4E-4)*(TRI**0.94)/SI
      GO TO 20
   10 VCOMP(I) = (1.778E-4)*( (4.58*TRI - 1.67)**(5./8.) )/SI
   20 CONTINUE
C
      SUMA = 0.0
      SUMB = 0.0
      DO 30 I=1,18
      SUMA = SUMA + VCOMP(I)*Y(I)*(MW(I)**0.5)
      SUMB = SUMB + Y(I)*(MW(I)**0.5)
   30 CONTINUE
      GVISA = SUMA/SUMB
      GO TO 200
C
C ---- Gas viscosity at 1 atm. & given temp. T using GGRA -----
C
  100 IF( GGRA .LT. 0.55 .OR. GGRA .GT. 1.50) GO TO 150
      IF( T .LT. 500.0 .OR. T .GT. 860.0) GO TO 150
      T = T - 460.0
      GGRA2 = GGRA**2.0
      GGINV = 1.0/(1.0+GGRA)
      VBASIC = 0.0126585 - 0.00611823*GGRA + 0.00164574*GGRA2
     # + 0.0000164574*T - (0.71922E-06)*GGRA*T - (0.609046E-06)*GGRA2*T
      CH2S = 1.0E-6 + ((1.13E-4)*YH2S*GGRA -(3.8E-5)*YH2S +1.0E-6)*GGINV
      CCO2 = (0.000134*YCO2*GGRA - 0.000004*YCO2 + 0.000004*GGRA)*GGINV
     # - 0.000003
      CN2  = (0.000170*YN2*GGRA + 0.000021*YN2 + 0.00001*GGRA)*GGINV
     # - 0.000006
      GVISA = VBASIC + CH2S + CCO2 + CN2
      T = T + 460.0
      GO TO 200
```

Appendix B   613

```
C
  150 IERR = 1
      GVIS = 0.0
      RETURN
C
C ---- VISCOSITY AT GIVEN PRESS. & TEMP. ----------------
C
  200 CALL VISCOR(PR,TR,GVISA,GVIS,IERR)
      RETURN
      END
C
C***************************************************************
C
      SUBROUTINE VISCOR(PR,TR,GVISA,GVIS,IERR)
C
C----------------------------------------------------------------
C Calculates gas viscosity at PR, TR from known viscosity GVISA at
C   1 atmosphere, using Carr et al. (1954) method,
C   valid in the range of 1.05 <= TR <= 3.0, 0.10 <= PR <= 20.0.
C INPUT:
C    PR, TR, GVISA
C OUTPUT:
C    GVIS = Gas viscosity at PR, TR
C    IERR = Error indication - beyond range of table values.
C SUBROUTINES USED: INTP2.
C DATA USED:
C    VISTBL = Table of viscosity ratios as a function of PR, TR.
C    TRTBL = Temperature values in the viscosity table.
C    PRTBL = Pressure values in the viscosity table.
C----------------------------------------------------------------
C
      DIMENSION PRTBL(22), TRTBL(13), VISTBL(22,13)
      INTEGER HP, HT
      DATA PRTBL/ 0.1,0.2,0.3,0.4,0.5,0.6,0.7,0.8,0.9,1.0,1.2,1.4,1.6,
     # 1.8,2.0,3.0,4.0,6.0,8.0,10.0,15.0,20.0 /
      DATA TRTBL/
     # 1.05,1.10,1.15,1.20,1.30,1.40,1.50,1.60,1.75,2.00,2.25,2.50,3.0/
      DATA VISTBL/
     # 1.000,1.012,1.025,1.050,1.075,1.10,1.145,1.195,1.285,1.415,1.76,
     # 2.285,2.865,3.29,3.650,4.760,5.5,6.460,7.150,7.680,8.65,9.370,
     # 1.0,1.011,1.023,1.043,1.065,1.086,1.120,1.150,1.195,1.255,1.435,
     # 1.70,2.070,2.465,2.8,3.85,4.655,5.72,6.50,7.06,8.1,8.88,
     # 1.0,1.01,1.021,1.036,1.055,1.073,1.095,1.12,1.145,1.175,1.28,
     # 1.42,1.59,1.85,2.16,3.225,3.975,5.030,5.82,6.385,7.410,8.18,
     # 1.0,1.009,1.019,1.03,1.045,1.06,1.07,1.085,1.11,1.135,1.195,
     # 1.285,1.425,1.57,1.75,2.6,3.35,4.38,5.125,5.74,6.75,7.5,
     # 1.0,1.008,1.017,1.027,1.04,1.054,1.063,1.075,1.1,1.12,1.155,
     # 1.215,1.285,1.36,1.46,2.02,2.56,3.5,4.185,4.755,5.79,6.5,
     # 1.0,1.007,1.015,1.024,1.035,1.048,1.056,1.067,1.089,1.1,1.135,
     # 1.185,1.235,1.28,1.335,1.69,2.11,2.79,3.38,3.86,4.79,5.41,
     # 1.0,1.006,1.013,1.021,1.03,1.042,1.049, 1.059,1.078,1.1,1.12,
```

614 Gas Production Engineering

```
    # 1.15,1.185,1.22,1.26,1.5,1.785,2.325,2.82,3.23,4.06,4.61,
    # 1.0,1.005,1.011,1.018,1.025,1.036,1.042,1.051,1.067,1.07,1.095,
    # 1.12,1.15,1.18,1.215,1.385,1.595,2.030,2.425,2.77,3.49, 4.025,
    # 1.0,1.004,1.009,1.015,1.021,1.03,1.035,1.043,1.056,1.065,1.09,
    # 1.11,1.125,1.145,1.165,1.28,1.435,1.77,2.095,2.375,2.99,3.5,
    # 1.0,1.003,1.007,1.012,1.017,1.024,1.028,1.035,1.045,1.055,1.06,
    # 1.070,1.08,1.095,1.11,1.205,1.29,1.5,1.725,1.955,2.48,2.925,
    # 1.0,1.002,1.005,1.009,1.013,1.018,1.021,1.027,1.034,1.040,1.045,
    # 1.055,1.065,1.075,1.085,1.145,1.21,1.34,1.485,1.665,2.085,2.46,
    # 1.0,1.001,1.003,1.006,1.009,1.012,1.015,1.019,1.023,1.025,1.03,
    # 1.04,1.05,1.06,1.065,1.105,1.155,1.245,1.36,1.485,1.83,2.15,
    # 1.0,1.0,1.001, 1.003,1.005,1.007,1.009,1.011,1.013,1.015,1.020,
    # 1.025,1.03,1.035,1.04,1.06,1.085,1.14,1.205,1.265,1.495,1.75 /
C
      IERR =0
      IF( TR. LT. 1.05 .OR. TR .GT. 3.00) GO TO 100
      IF( PR. LT. 0.01 .OR. PR .GT. 20.0) GO TO 100
C
C ---- VISCOSITY RATIO ----------------------------------
C
      LP = 2
      HP = 21
      LT = 3
      HT = 11
      DO 10 J = LT, HT
      IF(TRTBL(J).GE.TR) GO TO 20
   10 CONTINUE
      J = HT + 1
   20 IF(PRTBL(LP - 1).LT.PR) GO TO 30
      I = 1
      CALL INTP2(TR,TRTBL(J-2),TRTBL(J-1),TRTBL(J),TRTBL(J+1),
    # VISI,VISTBL(I,J-2),VISTBL(I,J-1),VISTBL(I,J),VISTBL(I,J+1))
      VISGR = 1.00 + (PR*(VISI-1.0))
      GO TO 60
C
   30 IF(PRTBL(HP + 1).GT.PR) GO TO 40
      I = HP + 1
      CALL INTP2(TR,TRTBL(J-2),TRTBL(J-1),TRTBL(J),TRTBL(J+1),
    # VISI,VISTBL(I,J-2),VISTBL(I,J-1),VISTBL(I,J),VISTBL(I,J+1))
      VISGR = VISI
      GO TO 60
C
   40 DO 45 I = LP, HP
      IF(PRTBL(I).GE. PR) GO TO 50
   45 CONTINUE
      I = HP + 1
   50 CALL INTP2(TR,TRTBL(J-2),TRTBL(J-1),TRTBL(J),TRTBL(J+1),
    # VISJ,VISTBL(I,J-2),VISTBL(I,J-1),VISTBL(I,J),VISTBL(I,J+1))
      I = I - 1
      CALL INTP2(TR,TRTBL(J-2),TRTBL(J-1),TRTBL(J),TRTBL(J+1),
    # VISI,VISTBL(I,J-2),VISTBL(I,J-1),VISTBL(I,J),VISTBL(I,J+1))
      VISGR =  VISI + (VISJ-VISI)*(PR -PRTBL(I))/(PRTBL(I+1)- PRTBL(I))
```

```
C
C ---- VISCOSITY AT GIVEN PRESS. & TEMP. ----------------
C
   60 GVIS = VISGR*GVISA
      RETURN
C
  100 IERR = 1
      GVIS = 0.0
      RETURN
      END
C
C*******************************************************************
C
      SUBROUTINE INTP2(X,X1,X2,X3,X4,Y,Y1,Y2,Y3,Y4)
C
C-------------------------------------------------------------------
C  This subroutine finds the value Y for a given X lying in the range
C     of four points: (X1,Y1), (X2,Y2), (X3,Y3), and (X4,Y4),
C     using the generalized Lagrange interpolation equation.
C-------------------------------------------------------------------
C
      A1 = X1- X2
      A2 = X1 -X3
      A3 = X1 -X4
      A4 = X2- X3
      A5 = X2- X4
      A6 = X3-X4
      B1 = X- X1
      B2 = X- X2
      B3 = X- X3
      B4 = X- X4
      Y  = B2/A1*B3/A2*B4/A3*Y1 - B1/A1*B3/A4*B4/A5*Y2
    # + B1/A2*B2/A4*B4/A6*Y3 - B1/A3*B2/A5*B3/A6*Y4
C
      RETURN
      END
C
      SUBROUTINE FRICF(DP,APR,QGSC,GGRA,GVIS,FF)
C
C-------------------------------------------------------------------
C This subroutine finds the Moody friction factor for gas flowing in
C  a pipe, using Colebrooke's equation, With Newton-Raphson iteration.
C INPUT:
C    DP, APR, QGSC, GGRA, GVIS.
C OUTPUT:
C    FF (Moody friction factor).
C-------------------------------------------------------------------
C
      REYN = 20.123*QGSC*GGRA/(GVIS*DP)
      A = APR/(3.7*DP)
      FF = 0.030
      TOL = 0.000005
```

```
C
   10 B = A + 0.628/(REYN*(FF**0.5))
      GG = 1.0/(FF**0.5) + 2.0*ALOG10(B)
      C = 0.314/(REYN*(FF**1.5))
      DGDF = -0.5/(FF**1.5) - (2.0/ALOG(10.0))*(C/B)
      FFN = FF - GG/DGDF
      IF(ABS(FFN-FF).LE.TOL) GO TO 20
      FF = FFN
      GO TO 10
C
   20 FF = FFN
      RETURN
      END
C
      SUBROUTINE HORIZF(DP,PL,APR,INDEX,Y,GGRA,PC,TC,T1,T2,IFIND,
     # IMETH,QGSC,P1,P2,DELTAP)
C
C----------------------------------------------------------------
C    This subroutine finds the pressure drop or flowrate
C for steady-state single-phase gas flow in a horizontal pipe, using:
C    (i) If IMETH = 1, then Weymouth equation.
C    (ii) If IMETH = 2, then Panhandle-A equation.
C    (iii) If IMETH = 3, then Panhandle-B equation.
C    (iv) If IMETH = 4, then Clinedinst equation.
C
C    If IFIND = 1, then compute flowrate (from known P1 and P2).
C             = 2, then compute pressure drop (from known QGSC and P1).
C             = 3, then compute pressure drop (from known QGSC and P2).
C INPUT:
C    DP, PL, APR, INDEX, Y, GGRA, PC, TC, T1, T2, IFIND, IMETH,
C    and one of these three:
C    (i) P1 and P2, or (ii) QGSC and P1, or (ii) QGSC and P2.
C OUTPUT:
C    One of these three, depending upon input:
C    (i) QGSC, or (ii) P1 and DELTAP, or (iii) P2 and DELTAP.
C SUBROUTINES USED:
C    GASCOM, GASVIS, FRICF, NUMINT, INTP1.
C----------------------------------------------------------------
C
      COMMON TRAV
      DIMENSION Y(18), DATARA(100,2)
C
      TOL = 0.05
      MITER = 20
      TAV = T1
      IF(T1.NE.T2) TAV = (T1 - T2)/ALOG(T1/T2)
      TRAV = TAV/TC
C
      IF(IFIND.GT.1) GO TO 100
C
C--- CASE OF FLOW RATE CALCULATION ---------------------
```

Appendix B **617**

```
C
      DELTAP = P1 - P2
      PR1 = P1/PC
      PR2 = P2/PC
      PAV = (2.0/3.0)*(P1**3.0 - P2**3.0)/(P1**2.0 - P2**2.0)
      PRAV = PAV/PC
      CALL GASCOM(PRAV,TRAV,ZAV,CRAV)
      CALL GASVIS(PRAV,TRAV,TAV,INDEX,Y,GGRA,GVISAV,IERR)
      COEF4 = (DP**5.0)/(GGRA*TAV*PL)
      RHS4 = 7.969634*PC*(520.0/14.73)*(COEF4**0.5)
      COEF = (P1**2.0 - P2**2.0)*(COEF4/ZAV)
      RHS = 5.6353821*(520.0/14.73)*(COEF**0.5)
C
C---The unknown rate must be guessed to enable calcn. of FF --
C    except for the case of IMETH = 1.
C
      FF = 0.032/(DP**(1.0/3.0))
      QGSC = RHS/FF**0.5
      IF(IMETH.EQ.1) RETURN
C
C ---Iterate for Flow rate and friction factor ----------
C
      ITER = 0
      IF(IMETH.EQ.4) CALL NUMINT(PR2,PR1,VINT)
  10  ITER = ITER + 1
      IF(IMETH.EQ.4) GO TO 20
      REYN = 20.123*QGSC*GGRA/(GVISAV*DP)
      IF(IMETH.EQ.2) FF = 0.0768/(REYN**0.1461)
      IF(IMETH.EQ.3) FF = 0.00359/(REYN**0.03922)
      QGSCN = RHS/FF**0.5
      GO TO 40
  20  CALL FRICF(DP,APR,QGSC,GGRA,GVISAV,FF)
      QGSCN = (RHS4/FF**0.5)*(VINT**0.5)
C
  40  ERR = ABS(QGSCN - QGSC)/QGSC
      QGSC = QGSCN
      IF(ERR.GT.TOL.AND.ITER.LE.MITER) GO TO 10
      RETURN
C
C--- CASE OF PRESSURE DROP CALCULATION ----------------------
C
  100 COEF = (GGRA*TAV*PL)/(DP**5.0)
      RHS = COEF*( 14.73*QGSC/(5.6353821*520.0) )**2.0
      RHS4 = COEF*( 14.73*QGSC/(7.969634*PC*520.0) )**2.0
C
C--The unknown pressure must be guessed to enable calculation of
C  average Z-factor, and also friction factor (except IMETH=1)
C  which depends upon gas viscosity.
C
      FF = 0.032/(DP**0.333)
      D2P = RHS*FF
```

618 Gas Production Engineering

```
      IF(IFIND.EQ.2) P2 = (P1**2.0 - D2P)**0.5
      IF(IFIND.EQ.3) P1 = (D2P + P2**2.0)**0.5
      ITER = 0
  110 ITER = ITER + 1
      PAV = (2.0/3.0)*(P1**3.0 - P2**3.0)/(P1**2.0 - P2**2.0)
      PRAV = PAV/PC
      CALL GASCOM(PRAV,TRAV,ZAV,CRAV)
      CALL GASVIS(PRAV,TRAV,TAV,INDEX,Y,GGRA,GVISAV,IERR)
      REYN = 20.123*QGSC*GGRA/(GVISAV*DP)
C
      GO TO (140,120,130,150), IMETH
  120 FF = 0.0768/(REYN**0.1461)
      GO TO 140
  130 FF = 0.00359/(REYN**0.03922)
C
  140 D2PN = RHS*ZAV*FF
      IF(IFIND.EQ.2) P2 = (P1**2.0 - D2PN)**0.5
      IF(IFIND.EQ.3) P1 = (D2PN + P2**2.0)**0.5
      ERR = ABS(D2PN - D2P)/D2P
      D2P = D2PN
      IF(ERR.GT.TOL.AND.ITER.LT.MITER) GO TO 110
      DELTAP = P1 - P2
      RETURN
C
C ---- For IMETH = 4 ----------------------
C
  150 CALL FRICF(DP,APR,QGSC,GGRA,GVISAV,FF)
      VINT = RHS4*FF
      PR1 = P1/PC
      PR2 = P2/PC
      IF(IFIND.EQ.3) GO TO 160
      PRG1 = PR1
      PRG2 = 0.5*PR2
      GO TO 170
  160 PRG2 = PR2
      PRG1 = 0.5*PR1
C
  170 DO 190 I = 1,51
      CALL NUMINT(PRG2,PRG1,DATARA(I,2))
      IF(IFIND.EQ.3) GO TO 180
      DATARA(I,1) = PRG2
      PRG2 = PRG2 + 0.02*PR2
      GO TO 190
  180 DATARA(I,1) = PRG1
      PRG1 = PRG1 + 0.02*PR1
  190 CONTINUE
C
      CALL INTP1(2,I,DATARA,X,VINT)
      IF(IFIND.EQ.3) GO TO 200
      PR2 = X
      P2 = PR2*PC
      GO TO 210
```

Appendix B 619

```
  200 PR1 = X
      P1 = PR1*PC
C
  210 DELTAP = P1 - P2
      RETURN
      END
C
C ****************************************************************
C
      SUBROUTINE NUMINT(A,B,VINT)
C
C ---------------------------------------------------------------
C This is a general subroutine for performing numerical integration
C      using the fast and accurate Romberg's method.
C INPUT: Integration limits: lower: A to higher: B.
C OUTPUT: Value of the integrated function, VINT.
C FUNCTION used: FUNCG, which evaluates the integrand for the
C               Clinedinst equation in this case.
C ---------------------------------------------------------------
C
      COMMON TR
      DIMENSION T(11,11)
C
      TOL = 0.0005
      MITER = 10
      C = B - A
      FA = FUNCG(A)
      FB = FUNCG(B)
      ITER = 1
      LVL = 1
      KMAX = 1
      T(1,1) = (C/2.0)*(FA + FB)
      VINTO = T(1,1)
C
   10 ITER = ITER + 1
      LVL = LVL + 1
      KMAX = KMAX + 1
      DX = C/(2.0**(KMAX-1))
      L = IFIX( 2.0**(KMAX-1) - 1 )
      SUMF = 0.0
      DO 20 J = 1,L
      SUMF = SUMF + FUNCG(A + J*DX)
   20 CONTINUE
      T(1,KMAX) = (DX/2.0)*(FA + FB + 2.0*SUMF)
C
      DO 40 L = 2,LVL
      DO 30 K = KMAX-1,1,-1
      EX = 4.0**(L-1)
      FAC = 1.0/(EX - 1)
      T(L,K) = FAC*(EX*T(L-1,K+1) - T(L-1,K))
   30 CONTINUE
   40 CONTINUE
```

620 Gas Production Engineering

```
C
      VINT = T(LVL,1)
      ERR = ABS(VINT-VINTO)/VINT
      IF(ERR.LE.TOL.AND.ITER.LE.MITER) RETURN
      VINTO = VINT
      GO TO 10
      END
C
C *******************************************************************
C
      FUNCTION FUNCG(PR)
      COMMON TR
      IF(PR.LT.0.2) GO TO 20
      CALL GASCOM(PR,TR,Z,CR)
      FUNCG = PR/Z
      RETURN
   20 FUNCG = 0.0
      RETURN
      END
C
C *******************************************************************
C
      SUBROUTINE INTP1(IND,M,DATARA,X,Y)
C
C--------------------------------------------------------------------
C     This is a linear interpolation routine.
C
C     For IND = 1, Y is determined for a given X.
C         IND = 2, X is determined for a given Y.
C INPUT:
C     DATARA = Array of (X,Y) data points.
C     M = number of data points (length of array DATARA).
C--------------------------------------------------------------------
C
      DIMENSION DATARA(100,2)
C
      IF(IND.EQ.2) GO TO 10
      IHV = 1
      Z = X
      GO TO 20
   10 IHV = 2
      Z = Y
C
   20 DO 60 I = 2,M
      A = (Z - DATARA(I-1,IHV))/(DATARA(I,IHV) - DATARA(I-1,IHV))
      IF (A) 30,40,50
   30 Z = DATARA(1,IHV)
      A = 0.0
      GO TO 100
   40 Z = DATARA(I-1,IHV)
      A = 0.0
      GO TO 100
```

```
      50 IF(I.EQ.M.AND.A.GT.1.0) GO TO 80
         IF(A.LT.1.0) GO TO 100
         IF(A.EQ.1.0) GO TO 70
      60 CONTINUE
C
      70 Z = DATARA(I,IHV)
         GO TO 100
      80 Z = DATARA(M,IHV)
         A = 1.0
C
     100 IF(IND.EQ.2) GO TO 110
         Y = DATARA(I-1,2) + A*(DATARA(I,2) - DATARA(I-1,2))
         RETURN
     110 X = DATARA(I-1,1) + A*(DATARA(I,1) - DATARA(I-1,1))
C
         RETURN
         END
C
         SUBROUTINE CSMITH(DP,PL,DPH,APR,INDEX,Y,GGRA,PC,TC,QGSC,PWH,TWH,
       # TBH,ISFCON,PBH)
C
C-----------------------------------------------------------------------
C   This subroutine finds the bottomhole pressure for static or
C   steady-state single-phase flow of gas through a non-horizontal pipe,
C            using Cullender & Smith (1956) method.
C A third-order numerical integration is used instead of Simpson's rule.
C INPUT:
C   DP, PL, DPH, APR, Y, INDEX, GGRA, PC, TC,
C   QGSC, PWH, TWH, TBH (if 0, then it is estimated by this program).
C and ISFCON = 0 for static (no flow) well.
C           = 1 for flowing well.
C OUTPUT: PBH
C
C SUBROUTINES USED: GASCOM, GASVIS, FRICF.
C-----------------------------------------------------------------------
C
         DIMENSION Y(18), A(4)
         DATA A/ 0.125, 0.375, 0.375, 0.125 /
C
         TOL = 0.001
         MITER = 20
C
C--Initial guess of PBH, TBH if unknown ---
C
         IF(PBH.EQ.0.0) PBH = PWH + 0.000025*PWH*DPH
         IF(TBH.EQ.0.0) TBH = TBH + 0.015*DPH
C
         GL = 0.01875*GGRA*DPH
         IF(ISFCON.EQ.1) GL = GL*1000.0
         PRWH = PWH/PC
         TRWH = TWH/TC
         CALL GASCOM(PRWH,TRWH,Z,CRWH)
         PTZ = TWH*Z/PWH
```

622 Gas Production Engineering

```
C
      ITER = 0
      IF(ISFCON.EQ.0) GO TO 20
      PTZ = 1.0/PTZ
      CALL GASVIS(PRWH,TRWH,TWH,INDEX,Y,GGRA,GVISWH,IERR)
      CALL FRICF(DP,APR,QGSC,GGRA,GVISWH,FF)
      FQ2 = 2.6957*(1.0E-6)*(FF/4.0)*(QGSC**2.0)*PL/(DPH*(DP**5.0))
      TKENER = 1.111*1.875*(1.0E-6)*GGRA*(QGSC**2.0)/(PWH*(DP**4.0))
      PTZ = (PTZ + TKENER)/(FQ2 + 0.001*(PTZ**2.0))
   20 VINTG2 = A(1)*PTZ
      PRES = PWH
      TEMP = TWH
      VINTG1 = VINTG2
C
C-----Integration loop ---------------
C
   30 ITER = ITER + 1
      PINCR = (PBH - PWH)/3.0
      TINCR = (TBH - TWH)/3.0
C
      DO 50 I = 2,4
      PRES = PRES + PINCR
      TEMP = TEMP + TINCR
      PRED = PRES/PC
      TRED = TEMP/TC
      CALL GASCOM(PRED,TRED,Z,CRED)
      PTZ = TEMP*Z/PRES
      IF(ISFCON.EQ.0) GO TO 40
      PTZ = 1.0/PTZ
      CALL GASVIS(PRED,TRED,TEMP,INDEX,Y,GGRA,GVISCO,IERR)
      CALL FRICF(DP,APR,QGSC,GGRA,GVISCO,FF)
      FQ2 = 2.6957*(1.0E-6)*(FF/4.0)*(QGSC**2.0)*PL/(DPH*(DP**5.0))
      TKENER = 1.111*1.875*(1.0E-6)*GGRA*(QGSC**2.0)/(PRES*(DP**4.0))
      PTZ = (PTZ + TKENER)/(FQ2 + 0.001*(PTZ**2.0))
   40 VINTG1 = VINTG1 + A(I)*PTZ
   50 CONTINUE
C
C-----Newton-Raphson iteration-----
C
      FUNC = GL - (PBH - PWH)*VINTG1
      PNEW = PBH + FUNC/PTZ
      ERR = ABS(PBH - PNEW)/PBH
      PBH = PNEW
      IF(ERR.LT.TOL.OR.ITER.GT.MITER) GO TO 100
      PRES = PWH
      TEMP = TWH
      VINTG1 = VINTG2
      GO TO 30
C
  100 CONTINUE
      RETURN
      END
```

Appendix B 623

```
C
      SUBROUTINE VERTF(DP,PL,DPH,APR,INDEX,Y,GGRA,PC,TC,T1,T2,
     # IFIND,QGSC,P1,P2,DELTAP)
C
C------------------------------------------------------------------------
C This subroutine finds the pressure drop or flowrate for steady-state
C single-phase gas flow in a non-horizontal pipe using Sukkar-Cornell
C (1955) method, with the integrand evaluated by numerical integration.
C
C If IFIND = 1, then compute flowrate (from known P1 and P2).
C          = 2, then compute pressure drop (from known QGSC and P1).
C          = 3, then compute pressure drop (from known QGSC and P2).
C NOTE:
C (1) P1,T1 is upstream, P2,T2 is downstream. DPH is positive
C     for upwards flow from 1 to 2, negative otherwise.
C     So, for a flowing well, 1 is bottomhole, 2 is wellhead.
C (2) Subroutine NUMINT should be suitably modified to accept
C     Function FUNCSC (instead of FUNCG). Also, the COMMON
C     declaration in NUMINT should include variable BB.
C
C INPUT:
C    DP, PL, DPH, APR, Y, INDEX, GGRA, PC, TC, T1, T2, IFIND, IMETH,
C    and one of these three:
C    (i) P1 and P2, or (ii) QGSC and P1, or (ii) QGSC and P2.
C OUTPUT:
C    One of these three, depending upon input:
C    (i) QGSC, or (ii) P1 and DELTAP, or (iii) P2 and DELTAP.
C SUBROUTINES USED:
C    GASCOM, GASVIS, FRICF, NUMINT, INTP1.
C------------------------------------------------------------------------
C
      COMMON TRAV, BB
      DIMENSION Y(18), DAT(100,2)
C
      TOL = 0.001
      MITER = 20
C
      TAV = T1
      IF(T1.NE.T2) TAV = (T1 - T2)/ALOG(T1/T2)
      TRAV = TAV/TC
      RHS = 0.01875*GGRA*DPH/TAV
      ERHS = EXP(2.0*RHS)
C
      IF(IFIND.EQ.1) GO TO 100
C
C--- CASE OF PRESSURE DROP CALCULATION ---------------------
C
C Assume a value for unknown pressure to enable av. viscosity calcn.
C
      FF = 0.032/(DP**(1.0/3.0))
      XX = (2.5272E-05)*FF*GGRA*TAV*PL*(ERHS-1.0)*(QGSC**2.0)
      XX = XX/(2.0*RHS*(DP**5.0))
```

624 Gas Production Engineering

```
      IF(IFIND.EQ.2) P2 = ( (P1**2.0 - XX)/ERHS )**0.5
      IF(IFIND.EQ.3) P1 = ( (P2**2.0)*ERHS + XX )**0.5
C
C ---Iteration procedure for Pressure and friction factor -----
C
      ITER = 0
   10 ITER = ITER + 1
      PAV = (2.0/3.0)*(P1**3.0 - P2**3.0)/(P1**2.0 - P2**2.0)
      PRAV = PAV/PC
      PR1 = P1/PC
      PR2 = P2/PC
      CALL GASVIS(PRAV,TRAV,TAV,INDEX,Y,GGRA,GVISAV,IERR)
      CALL FRICF(DP,APR,QGSC,GGRA,GVISAV,FF)
      BB = (6.6663E-04)*FF*PL*(QGSC**2.0)*(TAV**2.0)
      BB = BB/(DPH*(DP**5.0)*(PC**2.0) )
C
      IF(IFIND.EQ.2) GO TO 20
      PRG1 = 0.5*PR1
      DO 15 I = 1,51
      CALL NUMINT(PR2,PRG1,DAT(I,2))
      DAT(I,1) = PRG1
      PRG1 = PRG1 + 0.02*PR1
   15 CONTINUE
      CALL INTP1(2,I,DAT,X,RHS)
      PR1 = X
      P1N = PR1*PC
      ERR = ABS(P1N - P1)/P1N
      P1 = P1N
      GO TO 50
C
   20 PRG2 = 0.5*PR2
      DO 25 I = 1,51
      CALL NUMINT(PRG2,PR1,DAT(I,2))
      DAT(I,1) = PRG2
      PRG2 = PRG2 + 0.02*PR2
   25 CONTINUE
      CALL INTP1(2,I,DAT,X,RHS)
      PR2 = X
      P2N = PR2*PC
      ERR = ABS(P2N - P2)/P2N
      P2 = P2N
C
   50 IF(ERR.GT.TOL.AND.ITER.LT.MITER) GO TO 10
      DELTAP = P1 - P2
      RETURN
C
C--- CASE OF FLOW RATE CALCULATION ---------------------
C
  100 PAV = (2.0/3.0)*(P1**3.0 - P2**3.0)/(P1**2.0 - P2**2.0)
      PRAV = PAV/PC
      PR1 = P1/PC
      PR2 = P2/PC
      CALL GASVIS(PRAV,TRAV,TAV,INDEX,Y,GGRA,GVISAV,IERR)
```

Appendix B   625

```
C
C  Assume a value for unknown rate to enable calcn. of FF and B
C
       FF = 0.032/(DP**(1.0/3.0))
       XX = 2.0*RHS*(DP**5.0)*(P1**2.0 - (P2**2.0)*ERHS )
       YY = (2.5272E-05)*FF*GGRA*TAV*PL*(ERHS - 1.0)
       QGSC = (XX/YY)**0.5
C
C ---Iteration procedure for Flow rate and friction factor -----
C
       QGSCG = 0.5*QGSC
       DO 110 I = 1,51
       CALL FRICF(DP,APR,QGSCG,GGRA,GVISAV,FF)
       BB = (6.6663E-04)*FF*PL*(QGSCG**2.0)*(TAV**2.0)
       BB = BB/(DPH*(DP**5.0)*(PC**2.0) )
       CALL NUMINT(PR2,PR1,DAT(I,2))
       DAT(I,1) = QGSCG
       QGSCG = QGSCG + 0.02*QGSC
  110 CONTINUE
       CALL INTP1(2,I,DAT,X,RHS)
       QGSC = X
       DELTAP = P1 - P2
C
       RETURN
       END
C
C************************************************************************
C
       FUNCTION FUNCSC(PR)
       COMMON TR, BB
       IF(PR.LT.0.2) GO TO 20
       CALL GASCOM(PR,TR,Z,CR)
       FUNCSC = Z/(PR + BB*(Z**2.0)/PR)
       RETURN
   20 FUNCSC = 0.0
       RETURN
       END
C
       SUBROUTINE HAGBR(DP,ANG,APR,GDEN,GVIS,DENL,VISL,SIGMAL,VM,FNSLH,P,
      # IFREG,FLH,EPGR,FPGR,APGR,TPGR)
C
C-----------------------------------------------------------------------
C  This subroutine calculates the liquid holdup and pressure gradient
C  for multiphase flow using the Hagedorn and Brown (1965) correlation.
C  The Duns and Ros (1961) equation is used for computing the
C  acceleration pressure gradient.
C  INPUT:
C     DP, ANG, APR, GDEN, GVIS, DENL, VISL, SIGMAL, VM, FNSLH,
C       and Pressure P at which the gradient is desired to be computed.
C  OUTPUT:
C     IFREG = 1 for single-phase liquid
C           = 2 for single-phase gas
C           = 3 for two-phase bubble flow
```

626 Gas Production Engineering

```
C          = 4 for other type of two-phase flow.
C    and FLH, EPGR, FPGR, APGR, and TPGR.
C
C    SUBROUTINES USED: INTP1
C-----------------------------------------------------------------------
C
      DIMENSION DATHL(12,2), DATCNL(10,2), DATSI(12,2),
     # DATHL1(100,2), DATCN1(100,2), DATSI1(100,2)
C
C    Data arrays for liquid holdup correlation
C
      DATA DATHL/
     # 0.2,0.5,1.0,2.0,5.0,10.0,20.0,50.0,100.0,200.0,300.0,1000.0,
     # 0.04,0.09,0.15,0.18,0.25,0.34,0.44,0.65,0.82,0.92,0.96,1.0 /
      DATA DATCNL/
     # 0.002,0.005,0.01,0.02,0.03,0.06,0.1,0.15,0.2,0.4,
     # 0.0019,0.0022,0.0024,0.0028,0.0033,0.0047,0.0064,0.008,0.009,
     # 0.0115 /
      DATA DATSI/
     # 0.01,0.02,0.025,0.03,0.035,0.04,0.045,0.05,0.06,0.07,0.08,0.09,
     # 1.0,1.1,1.23,1.4,1.53,1.6,1.65,1.68,1.74,1.78,1.8,1.83 /
C
C Calculate liquid (VSL) and gas (VSG) superficial velocities.
C
      VSL = VM*FNSLH
      VSG = VM - VSL
C
C Calculate the required dimensionless numbers.
C
      C1 = (DENL/SIGMAL)**0.5
      C2 = (DENL/SIGMAL)**0.25
      C3 = ( DENL*(SIGMAL**3.0) )**0.25
      DND = 10.0727*DP*C1
      DNGV = 1.938*VSG*C2
      DNLV = 1.938*VSL*C2
      DNL = 0.15726*VISL/C3
C
C ANGR is the inclination angle in radians.
C
      ANGR = ANG*22.0/(7.0*180.0)
C
C-- Determine flow regime type --------------------
C
      IF(FNSLH.LT.1.0) GO TO 10
      FLH = 1.0
      IFREG = 1
      DENMIX = DENL
      GO TO 60
C
   10 IF(FNSLH.GT.0.0) GO TO 20
      FLH = 0.0
      IFREG = 2
```

Appendix B   627

```
      DENMIX = GDEN
      GO TO 60
C
   20 A = 1.071 - 0.2218*12.0*(VM**2.0)/DP
      IF(A.LT.0.13) A = 0.13
      B = 1.0 - FNSLH
      IF(B.GE.A) GO TO 30
C
C-- Griffith (1962) correlation for bubble flow -----
C
      IFREG = 3
      VS = 0.80
      FGH = 0.5*( 1.0 + VM/VS - ((1.0+VM/VS)**2.0 - 4.0*VSG/VS)**0.5 )
      FLH = 1.0 - FGH
      IF(FLH.LT.FNSLH) FLH = FNSLH
      DENMIX = FLH*DENL + FGH*GDEN
      RNUM = 1488.0*DENL*(VSL/FLH)*(DP/12.0)/VISL
      FF = 1.14 - 2.0*ALOG10(APR/DP + 21.25/RNUM**0.9)
      FF = 1.0/FF**2.0
      EPGR = DENMIX*SIN(ANGR)/144.0
      FPGR = FF*DENL*(VSL/FLH)**2.0/(2.0*32.2*12.0*DP)
      CFF = 0.0
      GO TO 70
C
C-- Hagedorn & Brown correlation -----
C
   30 IFREG = 4
      DO 40 I = 1,10
      DATCN1(I,1) = ALOG(DATCNL(I,1))
      DATCN1(I,2) = ALOG(DATCNL(I,2))
   40 CONTINUE
      DO 50 I = 1,12
      DATHL1(I,1) = ALOG(1.0E-05*DATHL(I,1))
      DATHL1(I,2) = DATHL(I,2)
      DATSI1(I,1) = DATSI(I,1)
      DATSI1(I,2) = DATSI(I,2)
   50 CONTINUE
C
C-- Liquid holdup calculations -----
C
      X = ALOG(DNL)
      CALL INTP1(1,10,DATCN1,X,Y)
      X = ALOG( DNLV*EXP(Y)*((P/14.73)**0.1)/(DND*(DNGV**0.575)) )
      CALL INTP1(1,12,DATHL1,X,FLH)
      X = DNGV*(DNL**0.38)/(DND**2.14)
      CALL INTP1(1,12,DATSI1,X,SI)
      IF(SI.LT.1.0) SI = 1.0
      FLH = FLH*SI
      IF(FLH.LT.0.0) FLH = 0.0
      IF(FLH.GT.1.0) FLH = 1.0
      IF(FLH.GT.FNSLH) GO TO 60
      FLH = FNSLH
```

628 Gas Production Engineering

```
      C
         60 DMIXNS = FNSLH*DENL + (1.0 - FNSLH)*GDEN
            DENMIX = FLH*DENL + (1.0 - FLH)*GDEN
            VISMIX = (VISL**FLH)*(GVIS**(1.0 - FLH))
            RNUM = 1488.0*DMIXNS*VM*(DP/12.0)/VISMIX
            FF = 1.14 - 2.0*ALOG10(APR/DP + 21.25/RNUM**0.9)
            FF = 1.0/FF**2.0
            EPGR = DENMIX*SIN(ANGR)/144.0
            FPGR = FF*(DMIXNS**2.0)*(VM**2.0)/(2.0*32.2*12.0*DP*DENMIX)
            VSG = (1.0 - FNSLH)*VM
            CFF = DENMIX*VM*VSG/(32.2*144.0*P)
            IF(CFF.GT.0.95) GO TO 80
         70 TPGR = - (EPGR + FPGR)/(1.0 - CFF)
            APGR = - TPGR*CFF
            RETURN
      C
         80 WRITE(IOU,200)
        200 FORMAT(/2X,'CRITICAL FLOW REGION IS BEING APPROACHED....'/7X,
           # '...ABORTING SUBROUTINE HAGBR.'/)
            RETURN
            END
      C
            SUBROUTINE BEGBR(DP,ANG,APR,GDEN,GVIS,DENL,VISL,SIGMAL,VM,FNSLH,P,
           # IFREG,FLH,EPGR,FPGR,APGR,TPGR)
      C
      C-----------------------------------------------------------------
      C  This subroutine calculates the pressure gradient for multiphase flow
      C        using Beggs and Brill (1973) correlation.
      C  INPUT:
      C    DP, ANG, APR, GDEN, GVIS, DENL, VISL, SIGMAL, VM, FNSLH,
      C    and Pressure P at which the gradient is desired to be computed.
      C  OUTPUT:
      C    IFREG = 1 for single-phase liquid
      C          = 2 for single-phase gas
      C          = 3 for distributed (multiphase) flow
      C          = 4 for intermittent (multiphase) flow
      C          = 5 for segregated (multiphase) flow
      C          = 6 for transition flow (Treated as intermittent).
      C  and FLH, EPGR, FPGR, APGR, and TPGR.
      C
      C-----------------------------------------------------------------
      C
      C  Calculate liquid (VSL) and gas (VSG) superficial velocities.
      C
            VSL = VM*FNSLH
            VSG = VM - VSL
      C
      C  Calculate the required dimensionless numbers.
      C
            DNLV = 1.938*VSL*((DENL/SIGMAL)**0.25)
            DNFR = 12.0*(VM**2.0)/(32.2*DP)
      C
      C  ANGR is the inclination angle in radians.
```

Appendix B 629

```
C
      ANGR = ANG*22.0/(7.0*180.0)
C
C-- Determine flow regime type --------------------
C
      IFREG = 5
      IF(FNSLH.GT.0.99999) IFREG = 1
      IF(FNSLH.LT.1.0E-05) IFREG = 2
      IF(IFREG.GT.2) GO TO 10
      FLH = FNSLH
      GO TO 100
C
   10 AL1 = 316.0*(FNSLH**0.302)
      AL2 = 0.0009252/(FNSLH**2.46842)
      AL3 = 0.10/(FNSLH**1.45155)
      AL4 = 0.50/(FNSLH**6.7380)
C
      INDIC = 0
      AINDIC = AL1
      IF(FNSLH.LT.0.01) GO TO 20
      IF(FNSLH.GT.0.4) AINDIC = AL4
      IF(DNFR.GE.AL2.AND.DNFR.LT.AL3) INDIC = 1
      IF(DNFR.GE.AL3.AND.DNFR.LT.AINDIC) IFREG = 4
      IF(DNFR.GE.AINDIC) IFREG = 3
      GO TO 30
   20 IF(DNFR.GE.AL1) IFREG = 3
   30 IF(INDIC.EQ.1.AND.IFREG.EQ.5) IFREG = 6
C
      II = IFREG - 2
      GO TO (40,50,60,50), II
C
C-- For distributed flow -----
   40 A = 1.065
      B = 0.5824
      C = 0.0609
      D = 1.0
      E = 0.0
      F = 0.0
      G = 0.0
      GO TO 70
C-- For intermittent flow -----
   50 A = 0.845
      B = 0.5351
      C = 0.0173
      D = 2.96
      E = 0.305
      F = -0.4473
      G = 0.0978
      GO TO 70
C-- For segregated flow -----
   60 A = 0.98
      B = 0.4846
      C = 0.0868
```

630 Gas Production Engineering

```
      D = 0.011
      E = -3.768
      F = 3.539
      G = -1.614
C
   70 FLH0 = A*(FNSLH**B)/DNFR**C
      IF(FLH0.LT.FNSLH) FLH0 = FNSLH
      IF(ANGR.GE.0.0) GO TO 80
C--For all flow types for downhill flow--
      D = 4.70
      E = -0.3692
      F = 0.1244
      G = -0.5056
      GO TO 90
C
   80 IF(ANGR.NE.0.0) GO TO 90
      FLH = FLH0
      GO TO 100
C
C--Flow is non-horizontal ------
C
   90 X = D*(FNSLH**E)*(DNLV**F)*(DNFR**G)
      CFAC = (1.0 - FNSLH)*ALOG(X)
      IF(CFAC.LT.0.0) CFAC = 0.0
      X = SIN(1.8*ANGR)
      SI = 1.0 + CFAC*(X - 0.333*X**3.0)
      IF(SI.LT.0.0) SI = 0.0
      FLH = FLH0*SI
      IF(FLH.GT.1.0) FLH = 1.0
C
C Mixture properties and friction factor
C
  100 DMIXNS = FNSLH*DENL + (1.0 - FNSLH)*GDEN
      DENMIX = FLH*DENL + (1.0 - FLH)*GDEN
      VMIXNS = FNSLH*VISL + (1.0 - FNSLH)*GVIS
      RNUM = 1488.0*DMIXNS*VM*(DP/12.0)/VMIXNS
      FF = 1.14 - 2.0*ALOG10(APR/DP + 21.25/RNUM**0.9)
      FF = 1.0/FF**2.0
      IF(IFREG.LE.2) GO TO 130
C
      Y = FNSLH/(FLH**2.0)
      IF(Y.GT.1.0.AND.Y.LT.1.2) GO TO 110
      YLN = ALOG(Y)
      S = YLN/(-0.0523 + 3.182*YLN - 0.8725*(YLN**2.0)
      #       + 0.01853*(YLN**4.0))
      GO TO 120
  110 S = ALOG(2.2*Y - 1.2)
  120 FF = FF*EXP(S)
C
C Calculation of pressure gradients
C
  130 EPGR = DENMIX*SIN(ANGR)/144.0
```

```
      FPGR = FF*DMIXNS*(VM**2.0)/(2.0*32.2*12.0*DP)
      CFF = DENMIX*VM*VSG/(32.2*144.0*P)
      IF(CFF.GT.0.95) GO TO 140
      TPGR = - (EPGR + FPGR)/(1.0 - CFF)
      APGR = - TPGR*CFF
      RETURN
C
  140 WRITE(IOU,200)
  200 FORMAT(/2X,'CRITICAL FLOW REGION IS BEING APPROACHED....'/7X,
     # '...ABORTING SUBROUTINE BEGBR.'/)
      RETURN
      END
```

Index

A

Abandonment pressure, 32
Absorber, 212–213, 218–226, 261–262, 265–267
Absorption, 260–262, 263–271
Absorption dehydration. See Dehydration.
Acentric factor, 44–45, 50, 64
Acid gases, 7, 255
 removal of. See Desulfurization processes.
Additive injection, 208–210
 requirements, 209–210
 techniques, 208–209
 types, 208
Adiabatic exponent, 358, 407, 409–412
ADIP process, 264
Adsorption dehydration. See Dehydration.
Aerosols, 128
Alkanol-amine process, 263–267
Alkazid processes, 268
Allowable pressure in pipes, 284
Allowable velocity in pipes, 284–285
Alumina, 236
Amines, 263–267
Annular flow, 347–348

API gravity, 366
Aquasorption, 261
Average pressure in gas pipe, 288–290

B

Beam pumping, 389
Binary interaction coefficient, 44
Blowdown of pipe, 562–563
Boiling point
 cubic-average, 35
 molar-average, 35
 normal, 34, 35
Bottom-hole pressure
 flowing. See Flowing bottom-hole pressure.
 static. See Static bottom-hole pressure.
Brake horsepower, 423–424, 440, 448
Brown's chart, 48
Bubble point curve, 22, 31, 36
 prediction of, 36

C

Campbell method, 171, 174–178
Centrifugal compressors. See Compressors.
Centrifuge, 96–98

Characteristics. *See* Method of characteristics.
Chemical absorption, 263–271
Chemical potential, 26–27
Chokes, 101, 126, 250
  flow through, 354–359
  purpose, 354–355
  subsurface, 354–355
  surface, 354–355
Cleaning. *See* Gas cleaning.
Clearance, 426–428
Clinedinst equation, 292
Commercial fuel costs, 4
Composition-composition diagrams, 23
Compressibility, 68–71
Compressibility factor. *See* Z-factor.
Compression
  adiabatic, 406–408
  efficiency, 411, 423–424, 426–428
  isentropic, 406–408
  isothermal, 406
  optimum ratio, 425
  polytropic, 408
  processes, 406–412
    characteristics, 408–409
  stages, 413, 424–425, 440
Compressors
  aftercooler, 422
  axial-flow, 401
  centrifugal, 400–401, 403–406, 438–447
    design, 438–447
  continuous-flow, 394–395, 400–403
  design
    analytic method, 414–417, 438–439
    centrifugal, 438–447
    charts, 423, 428–435, 441–445
  Mollier charts, 417–423, 439
    reciprocating, 424–438
    rotary, 448–449
  dynamic, 400–402
  intercooler, 413, 421
  efficiency, 423–424
  ejector, 402–403
  mixed-flow, 402
  multistage, 413, 424–425, 440
  positive-displacement, 394–400
  reciprocating, 396, 403–406, 424–438
    clearance, 426–428
    horsepower required, 426, 428–435
    speed/stroke length, 435
    volumetric efficiency, 426–428
  rotary, 396–400
    design, 448–449
    helical-lobe, 400
    liquid-piston, 399
    sliding-vane, 398
    spiral-lobe, 400
    two-impeller, 399
  selection, 403–406
  thermal, 402
  types, 394–403
Computer programs, 604–631
  compressibility and Z-factor, 607–611
  critical pressure and temperature, 606–607
  friction factor, 615–616
  interpolation
    Lagrange, 615
    linear, 620–621
  multiphase flow
    Beggs-Brill, 628–631
    Hagedorn-Brown, 625–628
  nomenclature for, 605–606
  numerical integration, Romberg, 619–620

Index   635

steady-state flow
  horizontal, 616–619
  vertical: Cullender-Smith,
    621–622
  vertical: Sukkar-Cornell,
    623–625
  viscosity, 611–615
Concentric orifice, 462–463
Condensates. See Gas condensates.
Conservation of mass, law of, 29
Constant pressure process, 20
Constants, physical, 585
Contactor, 212–213, 218–226,
  261–262, 265–267
Contaminants, 4
Convergence pressure, 135–136,
  164
Conversion factors, 593–603
  oil-field to SI, 596–603
Corresponding states, law of, 44,
  50
Cricondenbar, 22, 34–35
Cricondentherm, 22, 34–35
Critical
  compressibility factor, 45, 50
  flow, 357–358, 456, 543, 563.
    See also Flow, regimes.
  locus, 23, 136
  point, 20–23, 35
  pressure, 20, 45, 46–49
  saturation, 32
  temperature, 20, 45, 46–49
Critical-flow prover, 456–457
Cullender-Smith method, 338–340,
  345–347
Cunningham correction, 129
Cyclone, 98

D

Daltons' law, 29
DEA. See Amines.
DEG. See Glycol.
Dehydration, 207–208, 208–251,
  521
  absorption, 210–235
    flow scheme, 211–213
    plant design, 215–232
    plant problems, 213–214
    additive injection, 208–210
  adsorption, 235–250
    adsorbent (desiccant) capacity,
      242, 244–245
    adsorbent (desiccant)
      properties, 242
    adsorbent types, 236
    analysis of process, 240–241
    bed design, 243–250
    bed regeneration cycle, 238,
      239–240
    breakthrough time, 245–246
    chemical, 235
    cycle time, 242
    design variables, 241–242
    flow scheme, 237–239
    minimum bed length, 246
    physical, 235
    regeneration calculations,
      247–250
  expansion refrigeration, 251
  solvent properties, 210
Densitometer, 524
Density, 66
Desiccants. See Dehydration,
  adsorption.
Desulfurization processes
  alkanol-amine process, 263–267
  alkazid, 268
  Benfield, 267
  carbonate, 267–271
  Catacarb, 267
  chemical absorption, 263–271
  G-V, 267–268
  Holmes-Stretford, 268–271

hot carbonate, 267-268
iron-sponge, 258-259
molecular sieves, 259-260
physical absorption, 260-262
selection, 257-258
selectivity, 257
selexol, 261-263
solid-bed (adsorption), 258-260
types, 256
water wash (aquasorption), 261
Dew point, 22, 36, 170
curve, 22, 36
depression, 170
prediction of, 36
Differential liberation, 116-117
Differential pressure, 454-457
Diluents, 4
DIPA. See Amines.
Directional flow. See Multiphase flow.
Distribution coefficient. See Equilibrium ratio.
Drag coefficient, 97
Dry ice, 20

E

Eccentric orifice, 462-463
Economics
gas transmission, 581-582
optimum pipe diameter, 577-580
EG. See Glycol.
Ejectors, 402-403
Elbow meter, 459
Electric precipitators, 134
Embrittlement, of steel, 128
EMR. See Eykman molecular refraction.
EMR index, 57
Energy balance, 275-277

Energy consumption, worldwide, 2-3
Enhanced oil recovery, 33
EOS. See Equation of state.
Equation of continuity, 552
Equation of motion, 552
Equation of state, 39-46
Benedict-Webb-Rubin, 42-43, 56
general form, 43-45
hydrate prediction, 206
Peng-Robinson, 43
Redlich-Kwong, 43
solution methods, 45-46
Starling-Carnahan, 55
Starling, 56
Van der Waals, 41-42
water content, 178-181
Equilibrium ratio, 27-29, 36, 45, 135, 189-193, 194-197
charts, 137-163
thermodynamic criteria of, 25-27
vapor-solid, 189-193, 194-197
Equilibrium vaporization ratio. See Equilibrium ratio.
Equivalent diameter of pipe, 532, 538
Equivalent length of pipe, 532, 537
Erosional velocity, 284-285
Expansion factor, 67
Expansion refrigeration, 125-128, 250-251
Expansion, temperature drop due to, 127
Eykman molecular refraction, 50, 57-62

F

Fanning friction factor, 282-283
Fillers, 4

Filters, 133
  diatomaceous earth, 267
Finite difference methods,
    555-562, 576
  backward difference, 556
  central difference, 556
  explicit, 555-559
  forward difference, 556
  implicit, 558-559
Flange taps, 463
Flash calculations, 29-31,
    134-136, 164-166
Flash liberation, 117
Flash tank, 212, 261-263
Flow
  annulus, 347-348
  liquid present in small amount,
    366-367
  modeling of, 532-550, 551-576
  multiphase. See Multiphase flow.
  nozzles, 455
  pipe
    steady-state. See Steady-state
      flow.
    unsteady-state. See
      Unsteady-state flow.
  pipeline systems. See Pipeline
    systems.
  pressure profile, 289
  regimes, 277-281, 282-283, 368
  restriction/chokes, 354-359
    critical flow, 357-358
    subcritical flow, 357-358
    steady-state. See Steady-state
      flow.
    temperature profile, 359-362
    unsteady-state. See
      Unsteady-state flow.
Flowing bottom-hole pressure,
    341-347
  average approximate, 341-342

Cullender-Smith method,
    345-347
  Sukkar-Cornell method, 344-345
Flowing temperature in pipe. See
    Temperature profile.
Flow measurement
  mass, 523-524
  methods, 454-461
  natural gas liquids, 524-525
  problems, 521-522
  two-phase, 525-526
Flowmeters
  attributes, 452-453
  accuracy, 452-453
  calibration curve, 453
  linearity, 453
  precision, 452-453
  rangeability, 453
  repeatability, 453
  critical-flow prover, 456-457
  differential pressure, 454-457
  elbow meter, 459
  flow nozzles, 455
  mass, 523-524
  inferential, 523-524
  true, 523
  natural gas liquids, 524-525
  orifice meter, 454. See also
    Orifice meter.
  orifice well tester, 456
  pitot tube, 455-456
  positive-displacement, 457-458
  reciprocating-piston, 457-458
  rotameter, 459-460
  rotary meter, 457-458
  selection, 454
  turbine meter, 458
  two-phase, 525-526
  ultrasonic meter, 461-462
  venturi meter, 455
  vortex-shedding, 460-461

Formation volume factor, 67
Fortran programs. See Computer programs.
Fracturing, 13, 14
  proppant, 13
Friction, 276
Friction factor, 282–283
  chart, 278
Fugacity, 27–29, 45, 46, 179–181
Fugacity coefficient, 28, 29, 45, 46, 179–181
  chart, 181

G

Gas cleaning, 128–134
  methods, 130
  steps in, 128
Gas compressibility, 68–71
Gas compressibility factor. See Z-factor.
Gas condensates, 11, 32
  reservoir, 32
Gas constant, 26–27, 40, 79
Gas density, 66
Gas deviation factor. See Z-factor.
Gas drive reservoir
  dissolved, 31
  internal, 31
  solution, 31
Gas formation volume factor, 67
Gas gravity of total wellstream, 366
Gas horsepower, 423–424, 439–440
Gas hydrates. See Hydrates.
Gas lift, 390
Gas-liquid flow. See Multiphase flow.
Gas-oil ratio, 32–33
  producing, 32

Gas production
  system optimization, 33
  worldwide, 5–6
Gas relief valves, 91
Gas reservoir, 32
Gas specific heat. See Specific heat.
Gas transmission economics, 581–582
Gas, type of
  condensates, 11
  dissolved or associated, 11
  free gas, 9
  in brine, 14
  in solution, 9
  non-associated, 10
Gas viscosity, 72–79
  Carr et al. method, 72–75
  dynamic, 72
  kinematic, 72
  Lee et al. method, 78–79
  single-component, 75–76
Gathering systems, 529–531
  axial, 529–530
  common-line, 530–531
  radial, 529–530
  trunk-line, 530–531
  well-center, 530–531
Gels, 236–237
Giammarco-Vetrocoke process, 267–268
Gibbs free energy, 25
Gibbs phase rule, 18–19
Glycol, 208, 210–232
  contactor, 218–226
  corrosion, 214
  decomposition, 214
  flash separator, 231–232
  injection pump, 209
  losses, 213
  plant design, 215–232

plant problems, 213–214
pump, 214, 230–231
reboiler, 226–228
reconcentration, 214
stripping still, 229–230
vapor flooding, 214
GOR. See Gas-oil ratio.
Gravity settling, 98–100

H

Hammerschmidt's equation, 209
Holmes-Stretford process, 268–271
 effluent stream, 270–271
 Stretford solution, 269
Horizontal flow, 285–305, 368–371
Horizontal separator. See
 Separators.
Horsepower, 416, 423–424,
 438–440
Hydrates, 183–208, 521
 conditions promoting, 185–186
 curve, 184–185
 formulas, 184
 from sudden expansion, 186
 phase behavior, 184–185
 prediction, 186–207
  approximate, 186–187
  equations of state, 206
  Katz et al., 189–193
  McLeod-Campbell, 204–206
  Trekell-Campbell, 193,
   197–207
 prevention. See Dehydration.
 without sudden expansion, 186
Hydraulic radius, 277–279

I

Ideal gas, 40
 deviation from ideality, 40–41

Ideal gas law, 40
Ideal horsepower, 416, 423–424,
 438–440
Impingement, 100, 130–133
Inclined flow, 305–348, 381–383
Inhibitors, 7, 208–210, 521
Intensive properties, 20
Iron-sponge process, 258–259
Isentropic exponent, 358, 407,
 409–412
Isobaric process, 20
Isobaric specific heat. See Specific
 heat.

J

Joule-Thomson effect, 101, 126,
 250, 360–361
 temperature drop upon
  expansion, 127

K

Katz et al. method, 189–193
Kay's rule, 46, 51–52
Kirchhoff's laws, 543–544, 574
K-values. See Equilibrium ratio.

L

Laminar flow. See Flow, regimes.
Latent heat
 fusion, 20
 vaporization, 20
Law of conservation of mass, 29
Law of corresponding states, 44, 50
Lever's rule, 24
Liquid
 discharge valves, 91
 loading. See Liquid loading.
 recovery maximization, 116

settling volume, 103–104, 114, 117, 119, 121–122
slugs, 91
surging, 90
Liquid loading, 385–391
  minimum rate for prevention, 386–388
  prevention, 386–388
  unloading, 388–391
    beam pumping, 389
    gas lift, 390
    plunger lift, 389–390
    small tubing, 391
    surfactant/soap injection, 391
Lockhart-McHenry method, 165–166
Log-mean temperature, 286
Looped pipes. *See* Pipeline systems.
Low temperature separation, 125–128, 250–251
LTX units, 125–128, 250–251

M

McCabe-Thiele diagram, 220, 226–227
McLeod-Campbell method, 204–206
McKetta-Wehe correlation, 171, 172
MDEA. *See* Amines.
MEA. *See* Amines.
Measurement. *See* Flow measurement.
Measurement devices. *See* Flowmeters.
Method of characteristics, 559–562
MFH. *See* Fracturing.
Miscible injection, 33
Mist extractor. *See* Separation.

Molecular sieves, 237, 259–260
  acid-resistant, 237
  poisoning, 260
Mollier charts, 417–423, 439
Moody friction factor, 278, 282–283
Multiphase flow, 365–391
  directional wells, 381
  hilly terrain/inclined, 382–383
  horizontal, 368–371
  liquid loading. *See* Liquid loading.
  pressure traverse curves, 369–379
  small liquid presence, 366–367
  vertical, 372–380
Multipole correction factor, 64

N

Natural gas
  constituents, 4, 7
    physical data, 586–591
  consumption, 2–3
  contaminants, 4
  costs, 4
  development, 1–4
  diluents, 4
  field processing, 89
  fillers, 4
  history, 1
  hydrates. *See* Hydrates.
  occurrence, 9–14
    coal, 14
    geologic structures, 8–10
    geopressured aquifers, 14
    reservoirs, 10
    tight sands, 13
    tight shales, 13–14
  origin, 8
  production, 5–6

production system, 14–15
water content, 169–183
Newton's law, 129
Non-ideal gas, 40
Nozzles, 455

O

Oil-field units, 593–594
  conversion to SI, 596–603
Optimum pipe diameter, 577–580
Orifice meter, 454, 462–522
  accuracy, 520–521
  calculations, 467–519
  installation, 465–466
  factors, 470–519
    basic, 470–475, 486–489
    expansion, 479–485, 494–497
    flowing temperature, 498
    gauge location, 517–518
    manometer, 501, 517
    orifice thermal expansion, 518–519
    pressure-base, 489
    Reynolds number, 475–479, 490–493
    specific gravity, 498
    supercompressibility, 498–516
    temperature-base, 498
  pressure measurement/recording, 466–467
    direct-reading chart, 466
    square-root chart, 466
  pressure tap locations, 463–464
    corner taps, 464
    flange taps 463
    pipe taps, 463
    vena contracta, 463–464
  rounded-edged, 457
  selection, 518–519
  sharp-edged, 456–457, 462
  square-edged, 462
  straightening vanes, 464–465
  types, 462–463
    concentric, 462–463
    eccentric, 462–463
    segmental, 462–463
Orifice well tester, 456

P

Panhandle A equation, 291, 537
Panhandle B equation, 291, 537
Paraffinic, 35
Parallel pipes. See Pipeline systems.
Partial pressure, 28, 170
Particulate matter, 128
Peng-Robinson equation, 43
Phase behavior
  applications, 31–33
  multicomponent systems, 21–25
  single-component systems, 19–21
Phase diagrams, 18–25
  pressure-temperature diagram, 19, 21–22
  pressure-composition diagram, 24–25
  ternary diagram, 24
Phase rule, 18–19
Physical absorption, 260–262
Physical constants, 585
Physical data for gas constituents, 586–591
Pipeline
  advantages of, 275
  aging/corrosion, 531
  allowable pressure, 284
  allowable velocity, 284–285
  average pressure, 288–290
  blowdown/purge, 562–563
  capacity factor, 581
  closing valve, 563

economics, 577–582
efficiency, 292, 303
flow modeling, 532–550,
  551–576
flow through
  steady-state. See Steady-state
    flow.
  unsteady-state. See
    Unsteady-state flow.
load factor, 582
pressure deration, 531
pressure profile, 289
pressure testing, 563–564
reserve factor, 582
roughness, 281
storage/buffer capacity, 567, 582
transmission factor, 303
trash, 7, 128
variable demand. See Variable
  consumer demand.
Pipeline systems, flow through
steady-state, 532–550. See also
  Steady-state flow.
  looped, 535–537
  networks, 541–550
    looped, 545–550
    loopless, 544–545
  parallel, 533–535
  series, 532–533
unsteady-state, 551–576. See also
  Unsteady-state flow.
  approximate solutions,
    562–572
  networks, 573–576
  pipes, 551–562
Pipe taps, 463
Pitot tube, 455–456
Plunger lift, 389–390
Polytropic efficiency, 411
Polytropic exponent, 409–412

Polytropic head, 438–439
Positive-displacement, 394–396,
  457–458
Pressure traverse curves. See
  Multiphase flow.
Processing of natural gas
  dehydration. See Dehydration.
  desulfurization. See
    Desulfurization processes.
Pseudocritical, 34
  pressure, 46–49
  temperature, 46–49
Pulsating flow, 521–522

R

Real gas behavior, 40
Reciprocating compressors. See
  Compressors.
Reciprocating-piston meter,
  457–458
Redlich-Kwong equation, 43
Reduced density, 56
Reduced pressure, 50
Reduced temperature, 50
Refrigeration. See Expansion
  refrigeration.
Regeneration, 239–240, 258–261
Relative paraffinicity, 35
Reserves
  classification, 11
  possible, 13
  potential, 11, 13
  probable, 11
  proved, 11–12
  worldwide, 12
Reservoir
  abandonment pressure, 32
  condensate, 32
  engineering, 33

gas, 32
oil, 31, 32
undersaturated oil, 31
Residence time, 96, 104
Restrictions, gas flow through, 354–359
critical, 357–358
subcritical, 357–358
Retrograde
condensation, 32
region, 22
Revaporization, 32
Reynolds number, 129, 277–281
Robinson et al. correlation, 171, 173–175
Rotameter, 459–460
Rotary meter, 457–458
Roughness, of pipe, 281

S

Scrubbers, 133–134, 211–212, 215–218
cartridge, 134
dry, 134
oil-bath, 134
specifications 216–218
Segmental orifice, 462–463
Selexol process, 261–263
Separation
baffle
angle, 92
centrifugal, 91
conical, 91
plate, 100
centrifuge, 96–98
coefficient, 99, 102
differential, 116–117
economics, 101
equipment, 90–91

factors affecting, 100–101
flash, 117
gravity settling, 98–100
impingement, 100, 130–133
low temperature, 125–128
mist extractor, 90, 91, 95, 131–133
fiber, 133
sizing, 102–103
vane-type, 131–133
wire mesh, 100, 131–133
optimum, 119, 123
primary, 90–91, 92
principles, 95–100
stage, 114–125
Separators
allowable velocity, 101
design, 101–114
design considerations, 104–105
functions, 90
gas capacity, 100, 102–103, 107–112
high-pressure, 105
horizontal, 92–95
advantages of, 94
disadvantages of, 95
double-tube, 93–94
single-tube, 92–93
specifications, 118–121
liquid accumulation section, 90–91, 93, 95
liquid capacity, 103–104, 112–113
liquid discharge (dump) valve, 105
liquid settling volume, 103–104, 114, 117, 119, 121–122
low-pressure, 105
manufacturers' charts, 107–113
minimum diameter, 104

minimum height (length), 104
operating conditions control, 101
performance charts, 107–113
primary separation section,
    90–91, 92
retention time, 104
settling section, 90–91, 93, 95
spherical, 95
    advantages of, 95
    disadvantages of, 95
    specifications, 122
types, 91–95
vertical, 91–92
    advantages of, 91
    disadvantages of, 92
    specifications, 113–117
Series pipes. See Pipeline systems.
Sharma-Campbell method,
    179–181
Silica gel, 236–237
SI units, 591–592
    conversion from oil-field units
    to, 596–603
Slugging flow, 522
Slugs, 90, 91
Solution loading, 265
Souders-Brown equation, 99, 102
Sour gas, 51, 255, 522
Sour gas correction factor, 52
Specific heat, 20, 79–85, 407
    constant pressure, 79, 80–83
    constant volume, 79, 83–84
    heat capacity difference, 79
    molal single-component, 82
Spherical separator. See Separators.
Square-root chart, 466
Stage separation, 114–125
Standard conditions, 67, 451–452
Standard pressure, 451–452
Standard temperature, 451–452

Standing-Katz Z-factor chart. See
    Z-factor.
Static bottom-hole pressure,
    307–341
    Cullender-Smith method,
        338–340
    Sukkar-Cornell integral, 310–335
    Sukkar-Cornell method,
        308–309, 336
Steady-state flow
    hilly terrain, 348–354
        flow correction, 350–351
        general method, 351
        static correction, 349
    horizontal, 285–305
        Clinedinst equation, 292
        effect of assumptions, 288
        Panhandle A equation, 291
        Panhandle B equation, 291
        Weymouth equation, 290
    vertical/inclined, 305–348
        flowing pressure. See Flowing
            bottom-hole pressure.
        limitations due to
            assumptions, 348
        static pressure. See Static
            bottom-hole pressure.
Stewart-Burkhardt-Voo mixing
    rule, 62
Stokes-Cunningham equation, 129
Stokes law, 129
Stoner's method, 548–550
Straightening vanes, 464–465
Stress corrosion, 265
Stripping gas, 231
Subroutines. See Computer
    programs.
Sukkar-Cornell integral, 310–335
Sukkar-Cornell method, 308–309,
    336, 344–345

Sulfinol process, 264
Sulfur removal. See Desulfurization processes.
Supercompressibility factor, 67
Superficial velocity, 382-383
Surfactant/soap injection, 391
Surging, 90
Sweetening of gas. See Desulfurization processes.
Sweet gas, 50, 255

T

TEA. See Amines.
TEG. See Glycol.
T_4EG. See Glycol.
Temperature drop due to expansion, 127
Temperature profile, 359-362
  horizontal pipe, 360-362
  vertical pipe/wells, 362
Terminal velocity, 129
Thermodynamic equilibrium, 25-27
Thomas et al. equation, 47
Throttling effect. See Joule-Thomson effect.
Tie-lines, 24
Toxicity of H_2S, 255
Transition flow. See Flow, regimes.
Transmission economics, 581-582
Transmission factor, 303
Traverse curves. See Multiphase flow.
Tray efficiency, 226
Trekell-Campbell method, 193, 197-207
Triple point, 20
Turbine meter, 458
Turbulent flow. See Flow, regimes.

Two-phase flow. See Multiphase flow.

U

Ultrasonic meter, 461-462
Units
  conversions, oil-field to SI, 596-603
  oil-field, 593-594
  SI, 591-592
Universal gas constant, 26-27, 40, 79
Unsteady-state flow, 551-576
  approximate solutions, 562-572
  blowdown/purge, 562-563
  closing valve, 563
  pressure testing, 563-564
  variable demand. See Variable consumer demand.
  in networks, 573-576
  in pipes, 551-562
  initial and boundary conditions, 554-555
  numerical solutions, 555-562

V

Van der Waals
  criteria for critical point, 42, 44
  equation, 41-42
  forces, 185
Vapor-solid equilibrium ratio, 189-193, 194-197
Variable consumer demand
  constant injection rate case, 564-568
  variable injection rate case, 568-573
Variational methods, 562

Vena contracta, 463–464
Venturi meter, 455
Vertical flow, 305–348, 372–380
Vertical separator. See Separators.
Viscosity. See Gas viscosity.
Volumetric efficiency, 426–428
Vortex detection, 461
Vortex-shedding meter, 460–461

W

Water content of gases, 169–183
   estimation
      Campbell, 171, 174–178
      empirical plots, 171
      equations of state, 178–181
      McKetta-Wehe, 171, 172
      partial pressure, 170–171
      Robinson et al., 171, 173–175
      Sharma-Campbell, 179–181
   factors, 169–170
Water wash, 261
Watson characterization factor, 35
Weber number, 387
Well inhibitors, 7
Weymouth equation, 290, 531–532, 537
World
   energy consumption, 2–3
   gas production, 5–6
   gas reserves, 12

Z

Z-factor, 41, 46, 50–66
   Buxton-Campbell method, 64–66
   curve fits for Standing-Katz chart, 54
   EMR method, 57–62
   from equations of state, 55–66
   Standing-Katz chart, 51
   Stewart-Katz chart, 51
   Stewart-Burkhardt-Voo method, 62–63
   Wichert-Aziz correction, 52
Z-factor related properties, 66–68
Z-values. See Z-factor.